MW01344675

THE TRIBOLOGY HANDBOOK

THE TRIBOLOGY HANDBOOK

Second edition

Edited by

M. J. NEALE
OBE, BSc(Eng), DIC, FCGI, WhSch, FEng, FIMechE

Butterworth-Heinemann
Linacre House, Jordan Hill, Oxford OX2 8DP
225 Wildwood Avenue, Woburn, MA 01801-2041
A division of Reed Educational and Professional Publishing Ltd

A member of the Reed Elsevier plc group

OXFORD AUCKLAND BOSTON
MELBOURNE JOHANNESBURG NEW DELHI

First published 1973
Second edition 1995
Reprinted 1997, 1999

British Library Cataloguing in Publication Data
A catalogue record for this book is available from the
British Library

Library on Congress Cataloguing in Publication Data
A catalogue record for this book is available from the
Library of Congress

ISBN 0 7506 1198 7

Printed and bound in Great Britain

Contents

Contents

Editor's Preface

This second revised edition of the *Tribology Handbook* follows the pattern of the original, first published over twenty years ago. It aims to provide instant access to essential information on the performance of tribological components, and is aimed particularly at designers and engineers in industry.

Tribological components are those which carry all the relative movements in machines. Their performance, therefore, makes a critical contribution to the reliability and efficiency of all machines. Also because they are the local areas of machines, where high forces and rapid movements are transmitted simultaneously, they are also the components most likely to fail, because of the concentration of energy that they carry. If anything is wrong with a machine or its method of use, these components are the mechanical fuses, which will indicate the existence of a problem. If this happens, guidance on the performance that these components would be expected to provide, can be invaluable.

Designers of machines should also find the contents helpful, because they provide an atlas of component performance, aimed at providing the guidance needed when planning the feasibility of various possible layouts for a machine design.

In a book of this size it is not possible to cover the whole of the technology of tribological components. More focused design procedures, standards and text books will do this, and hopefully guide engineers in how to get their designs close to the optimum. In a sense the objective of this handbook is to make sure that they do not get it wrong.

The format of the book is original and has possibly set an example on the presentation of technical information in the form of an atlas. Like an atlas it is intended to provide guidance on where you are or should be, more or less at a glance, rather than to be read like a novel from cover to cover. The presentation of information in this form has been quite a challenge to the contributors who have responded well and the editor would like to record his appreciation of their work and of all the people who have helped him in the preparation of the book.

The editor, who has spent over forty years solving problems with machinery around the world, has found the information in this book of tremendous value. He hopes that it will be equally helpful to its readers with both design and problem solving. For those engineers in countries, who are now moving towards industrialisation, it is hoped, also, that it will provide a useful summary of the experience of those who have been doing it for a little longer.

Michael Neale
Neale Consulting Engineers Ltd
Farnham, Surrey UK

Contributors

Section	Author
Selection of bearing type and form	M. J. Neale OBE, BSc(Eng), DIC, FCGI, WhSch, FEng, FIMechE
Selection of journal bearings	M. J. Neale OBE, BSc(Eng), DIC, FCGI, WhSch, FEng, FIMechE
Selection of thrust bearings	P. B. Neal BEng, PhD, CEng, MIMechE
Plain bearing materials	P. T. Holligan BSc(Tech), FIM, J. M. Conway Jones BSc, PhD, DIC, ACGI
Dry rubbng bearings	J. K. Lancaster PhD, DSc, FInstP
Porous metal bearings	V. T. Morgan AIM, MIMechE
Grease, wick and drip fed journal bearings	W. H. Wilson BSc(Eng), CEng, MIMechE
Ring and disc fed journal bearings	F. A. Martin CEng, FIMechE
Steady load pressure fed journal bearings	F. A. Martin CEng, FIMechE
High speed bearings and rotor dynamics	M. J. Neale OBE, BSc(Eng), DIC, FCGI, WhSch, FEng, FIMechE
Crankshaft bearings	D. de Geurin CEng, FIMechE
Plain bearing form and installation	J. M. Conway Jones BSc, PhD, DIC, ACGI
Oscillatory journal bearings	K. Jakobsen LicTechn
Spherical bearings	D. Bastow BSc(Eng), CEng, FIMechE, MConsE, MSAE, MSIA(France)
Plain thrust bearings	P. B. Neal BEng, PhD, CEng, MIMechE
Profiled pad thrust bearings	P. B. Neal, BEng, PhD, CEng, MIMechE
Tilting pad thrust bearings	A. Hill CEng, FIMechE, FIMarE
Hydrostatic bearings	W. B. Rowe BSc, PhD, DSc, CEng, FIMechE, FIEE
Gas bearings	A. J. Munday BSc(Tech), CEng, MIMechE
Selection of rolling bearings	D. G. Hjertzen CEng, MIMechE
Rolling bearing materials	D. B. Jones CEng, MIMechE, P. L. Hurricks BSc, MSc
Rolling bearing installation	C. W. Foot CEng, MIMechE
Slide bearings	F. M. Stansfield BSc(Tech), CEng, MIMechE, A. E. Young BEng, CEng, MIMechE, AMCT
Instrument jewels	G. F. Tagg BSc, PhD, CEng, FInstP, FIEE, FIEEE
Flexures and knife edges	A. B. Crease MSc, ACGI, CEng, MIMechE
Electromagnetic bearings	G. Fletcher BSc, CEng, MIMechE
Bearing surface treatments and coatings	M. J. Neale OBE, BSc(Eng), DIC, FCGI, WhSch, FEng, FIMechE
Belt drives	T. H. C. Childs BA, MA, PhD, CEng, FIMechE, MInstP
Roller chain drives	M. C. Christmas BSc, CEng, MIMechE, MIMgt
Gears	A. Stokes
Flexible couplings	M. J. Neale OBE, BSc(Eng), DIC, FCGI, WhSch FEng, FIMechE
Self-synchronising clutches	J. Neeves BA(Eng)
One-way clutches	T. A. Polak MA, CEng, MIMechE
Friction clutches	T. P. Newcomb DSc, CEng, FIMechE, FInstP, CPhys R. T. Spurr DSc, PhD, DIC, FInstP, CPhys H. C. Town CEng, FIMechE, FIProdE
Brakes	T. P. Newcomb DSc, CEng, FIMechE, FInstP, CPhys R. T. Spurr DSc, PhD, DIC, FInstP, CPhys
Screws	M. J. Neale OBE, BSc(Eng), DIC, FCGI, WhSch, FEng, FIMechE

Contributors

Section	Author
Cams and followers	T. A. Polak MA, CEng, MIMechE, C. A. Beard CEng, FIMechE, AFRAeS
Wheels rails and tyres	W. H. Wilson BSc(Eng), CEng, MIMechE
Capstans and drums	C. M. Taylor BSc(Eng) MSc. PhD, DEng, CEng, FIMechE
Wire ropes	D. M. Sharp
Control cables	G. Hawtree C. Derry
Damping devices	J. L. Koffman DiPlIng, CEng, FIMechE
Pistons	B. L. Ruddy BSc, PhD, CEng, MIMechE G. Longfoot CEng, MIMechE
Piston rings	R. Munro BSc, PhD, CEng, MIMechE B. L. Ruddy, BSc, PhD, CEng, MIMechE D. C. Austin
Cylinders and liners	E. J. Murray BSc(Eng), CEng, MIMechE N. Tommis AIM, MIEI, AIMF
Selection of seals	B. S. Nau BSc, PhD, ARCS, CEng, FIMechE, MemASME
Sealing against dirt and dust	W. H. Barnard BSc(Lond), CEng, MIMechE
Oil flinger rings and drain grooves	A. B. Duncan BSc, CEng, FIMechE
Labyrinths, brush seals and throttling bushes	B. S. Nau BSc, PhD, ARCS , CEng, FIMechE, MemASME
Lip seals	E. T. Jagger BSc(Eng), PhD, CEng, FIMechE
Mechanical seals	A. Lymer BSc(Eng), CEng, FIMechE, W. H. Wilson BSc(Eng), CEng, MIMechE
Packed glands	R. Eason CEng, MIMechE
Mechanical piston rod packings	J. D. Summers-Smith BSc, PhD, CEng, FIMechE R. S. Wilson MA
Soft piston seals	R. T. Lawrence MIED
Selection of lubricant type	A. R. Lansdown MSc, PhD, FRIC, FInstPet
Mineral oils	T. I. Fowle BSc (Hons), ACGI, CEng, FIMechE
Synthetic oils	A. R. Lansdown BSc, PhD, FRIC, FInstPet
Greases	N. Robinson & A. R. Lansdown BSc, PhD, FRIC, FInstPet
Solid lubricants and coatings	J. K. Lancaster PhD, DSc, FInstP
Other liquids	D. T. Jamieson FRIC
Plain bearing lubrication	J. C. Bell BSc, PhD
Rolling bearing lubrication	E. L. Padmore CEng, MIMechE
Gear and roller chain lubrication	J. Bathgate BSc, CEng, MIMechE
Slide lubrication	M. J. Neale OBE, BSc(Eng), DIC, FCGI, WhSch, FEng, FIMechE
Lubrication of flexible couplings	J. D. Summers-Smith BSc, PhD, CEng, FIMechE
Wire rope lubrication	D. M. Sharp
Selection of lubrication systems	W. J. J. Crump BSc, ACGI, FInstP
Total loss grease systems	P. L. Langborne BA, CEng, MIMechE
Total loss oil and fluid grease systems	P. G. F. Seldon CEng, MIMechE
Dip splash systems	J. Bathgate BSc, CEng, MIMechE
Mist systems	R. E. Knight BSc, FCGI
Circulation systems	D. R. Parkinson FInstPet
Commissioning lubrication systems	N. R. W. Morris
Design of storage tanks	A. G. R. Thomson BSc(Eng), CEng, AFRAeS

Contributors

Section	Author
Selection of oil pumps	A. J. Twidale
Selection of filters and centrifuges	R. H. Lowres CEng, MIMechE, MIProdE, MIMarE, MSAE, MBIM
Selection of heaters and coolers	J. H. Gilbertson CEng, MIMechE, AMIMarE
A guide to piping design	P. D. Swales BSc, PhD, CEng, MIMechE
Selection of warning and protection devices	A. J. Twidale
Running in procedures	W. C. Pike BSc, MSc, ACGI, CEng, MIMechE
Lubricant change periods and tests	J. D. Summers-Smith BSc, PhD, CEng, FIMechE
Biological deterioration of lubricants	E. C. Hill MSc., FInstPet
Lubricant hazards; fire, explosion and health	J. D. Summers-Smith BSc, PhD, CEng, FIMechE
Lubrication maintenance planning	R. S. Burton
High pressure and vacuum	A. R. Lansdown MSc, PhD, FRIC, FInstPet & J. D. Summers-Smith BSc., PhD, CEng, FIMechE
High and low temperatures	M. J. Todd MA.
World ambient climatic data	A. G. R. Thomson BSc(Eng), CEng, AFRAeS
Industrial plant environmental data	R. L. G. Keith BSc
Chemical effects	H. H. Anderson BSc(Hons), CEng, FIMechE
Storage	C. E. Carpenter FRIC
Failure patterns and failure analysis	J. D. Summers-Smith BSc, PhD, CEng, FIMechE M. J. Neale OBE, BSc(Eng), DIC, FCGI, WhSch, FEng, FIMechE
Plain bearing failures	P. T. Holingan BSc(Tech), FIM
Rolling bearing failures	W. J. J. Crump BSc, ACGI, FInstP
Gear failures	T. I. Fowle BSc(Hons), ACGI, CEng, FIMechE H. J. Watson BSc(Eng), CEng, MIMechE
Piston and ring failures	M. J. Neale OBE, BSc(Eng), DIC, FCGI, WhSch, FEng, FIMechE
Seal failures	B. S. Nau BSc, PhD, ARCS, CEng, FIMechE, MemASME
Wire rope failures	S. Maw MA, CEng, MIMechE
Brake and clutch failures	T. P. Newcombe DSc, CEng, FIMechE, FInstP R. T. Spurr BSc, PhD
Fretting problems	R. B. Waterhouse MA, PhD, FIM
Maintenance methods	M. J. Neale OBE, BSc(Eng), DIC, FCGI, WhSch, FEng, FIMechE
Condition monitoring	M. J. Neale OBE, BSc(Eng), DIC, FCGI, WhSch, FEng, FIMechE
Operating temperature limits	J. D. Summers-Smith BSc, PhD, CEng, FIMechE
Vibration analysis	M. J. Neale OBE, BSc(Eng), DIC, FCGI, WhSch, FEng, FIMechE
Wear debris analsis	M. H. Jones BSc(Hons), CEng, MIMechE, MInstNDT M. J. Neale OBE, BSc(Eng), DIC, FCGI, WhSch, FEng, FIMechE
Performance analysis	M. J. Neale OBE, BSc(Eng), DIC, FCGI, WhSch, FEng, FIMechE
Allowable wear limits	H. H. Heath FIMechE
Repair of worn surfaces	G. R. Bell BSc, ARSM, CEng, FIM, FWeldI, FRIC

Contributors

Section	Author
Wear resistant materials	H. Hocke CEng, MIMechE, FIPlantE, MIMH, FIL M. Bartle CEng, MIM, DipIM, MIIM, AMWeldI
Repair of plain bearings	P. T. Holligan BSc(Tech), FIM
Repair of friction surfaces	T. P. Newcomb DSc, CEng, FIMechE, FInstP R. T. Spurr BSc, PhD
Industrial flooring materials	A. H. Snow FCIS, MSAAT
The nature of surfaces and contact	J. A. Greenwood BA, PhD
Surface topography	R. E. Reason DSc, ARCS, FRS
Hardness	M. J. Neale OBE, BSc(Eng), DIC, FCGI, WhSch, FEng, FIMechE
Friction mechanisms, effect of lubricants	D. Tabor PhD, ScD, FInstP, FRS
Frictional properties of materials	D. Tabor PhD, ScD, FInstP, FRS
Viscosity of lubricants	H. Naylor BSc, PhD, CEng, FIMechE
Methods of fluid film formation	D. Dowson CBE, BSc, PhD, DSc, FEng, FIMechE, FRS
Mechanisms of wear	K. H. R. Wright PhD, FInstP
Heat dissipation from bearing assemblies	A. B. Crease MSc, ACGI, CEng, MIMechE
Shaft deflections and slopes	M. F. Madigan BSc
Shape tolerances of typical components	J. J. Crabtree BSc(Tech)Hons.
S.I. units and conversion factors	M. J. Neale OBE, BSc(Eng), DIC, FCGI, WhSch, FEng, FIMechE

Selection of bearing type and form A1

Bearings allow relative movement between the components of machines, while providing some type of location between them.

The form of bearing which can be used is determined by the nature of the relative movement required and the type of constraints which have to be applied to it.

Relative movement between machine components and the constraints applied

Constraint applied to the movement	*Continuous movement*	*Oscillating movement*
About a point	The movement will be a rotation, and the arrangement can therefore make repeated use of accurate surfaces	If only an oscillatory movement is required, some additional arrangements can be used in which the geometric layout prevents continuous rotation
About a line	The movement will be a rotation, and the arrangement can therefore make repeated use of accurate surfaces	If only an oscillatory movement is required, some additional arrangements can be used in which the geometric layout prevents continuous rotation
Along a line	The movement will be a translation. Therefore one surface must be long and continuous, and to be economically attractive must be fairly cheap. The shorter, moving component must usually be supported on a fluid film or rolling contact for an acceptable wear rate	If the translational movement is a reciprocation, the arrangement can make repeated use of accurate surfaces and more mechanisms become economically attractive
In a plane	If the movement is a rotation, the arrangement can make repeated use of accurate surfaces	If the movement is rotational and oscillatory, some additional arrangements can be used in which the geometric layout prevents continuous rotation
	If the movement is a translation one surface must be large and continuous and to be economically attractive must be fairly cheap. The smaller moving component must usually be supported on a fluid film or rolling contact for an acceptable wear rate	If the movement is translational and oscillatory, the arrangement can make repeated use of accurate surfaces and more mechanisms become economically attractive

For both continuous and oscillating movement, there will be forms of bearing which allow movement only within a required constraint, and also forms of bearing which allow this movement among others.

The following tables give examples of both these forms of bearing, and in the case of those allowing additional movement, describe the effect which this can have on a machine design.

Selection of bearing type and form

Examples of forms of bearing suitable for continuous movement

Constraint applied to the movement	*Examples of arrangements which allow movement only within this constraint*	*Examples of arrangements which allow this movement but also have other degrees of freedom*	*Effect of the other degrees of freedom*
About a point	Gimbals 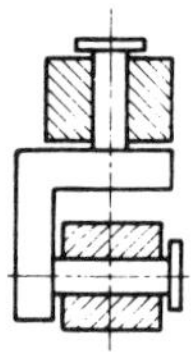	Ball on a recessed plate 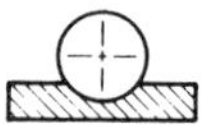	Ball must be forced into contact with the plate
About a line	Journal bearing with double thrust location 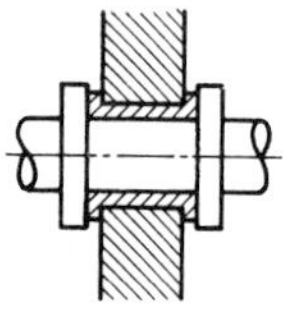	Journal bearing 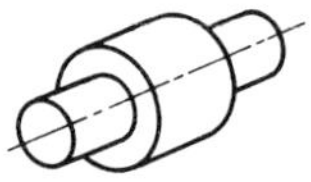	Simple journal bearing allows free axial movement as well
	Double conical bearing 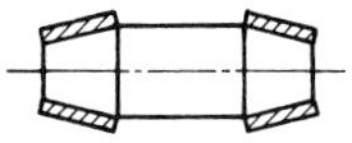	Screw and nut 	Gives some related axial movement as well
		Ball joint or spherical roller bearing 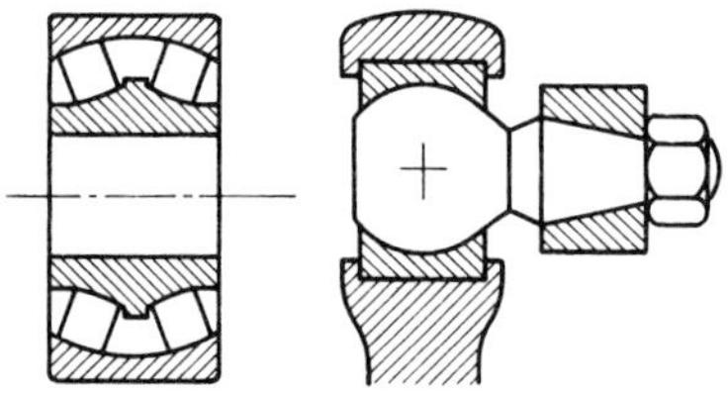	Allows some angular freedom to the line of rotation
Along a line	Crane wheel restrained between two rails 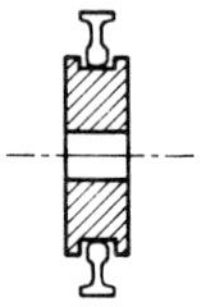	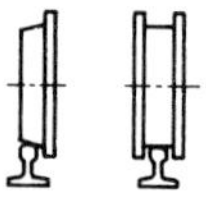 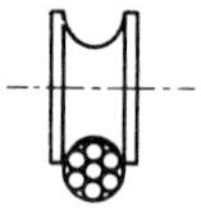  Railway or crane wheel on a track	These arrangements need to be loaded into contact. This is usually done by gravity. Wheels on a single rail or cable need restraint to prevent rotation about the track member
		Pulley wheel on a cable	
		Hovercraft or hoverpad on a track	
In a plane (rotation)	Double thrust bearing 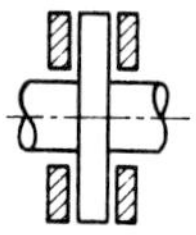	Single thrust bearing 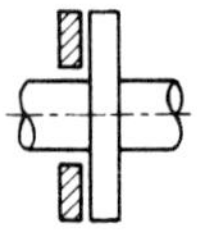	Single thrust bearing must be loaded into contact
In a plane (translation)		Hovercraft or hoverpad 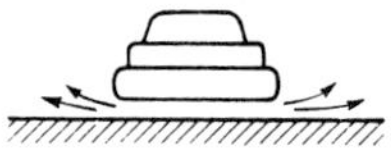	Needs to be loaded into contact usually by gravity

Examples of forms of bearing suitable for oscillatory movement only

Constraint applied to the movement	*Examples of arrangements which allow movement only within this constraint*	*Examples of arrangements which allow this movement but also have other degrees of freedom*	*Effect of the other degrees of freedom*
About a point	Hookes joint 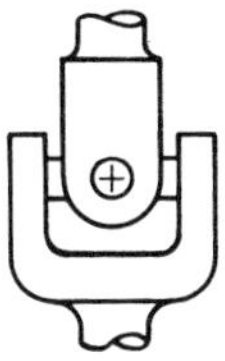	Cable connection between components 	Cable needs to be kept in tension
About a line	Crossed strip flexure pivot 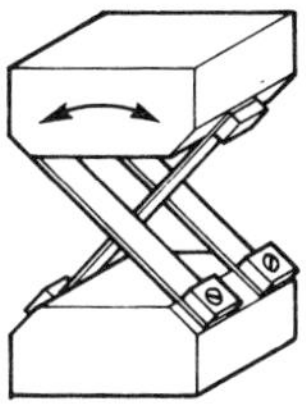	Torsion suspension 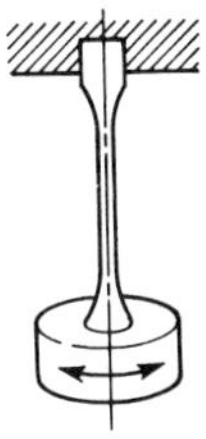	A single torsion suspension gives no lateral location
		Knife-edge pivot 	Must be loaded into contact
		Rubber bush 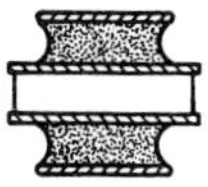	Gives some axial and lateral flexibility as well
		Rocker pad 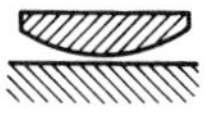	Gives some related translation as well. Must be loaded into contact
Along a line	Crosshead and guide bars 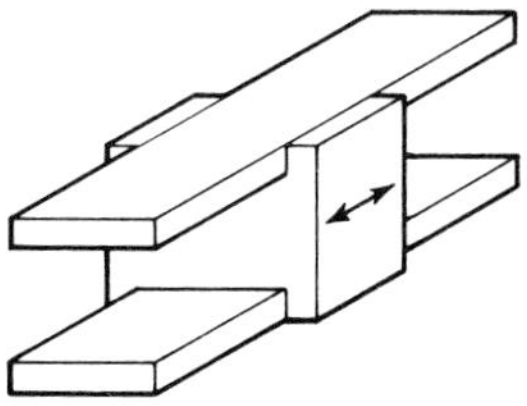	Piston and cylinder 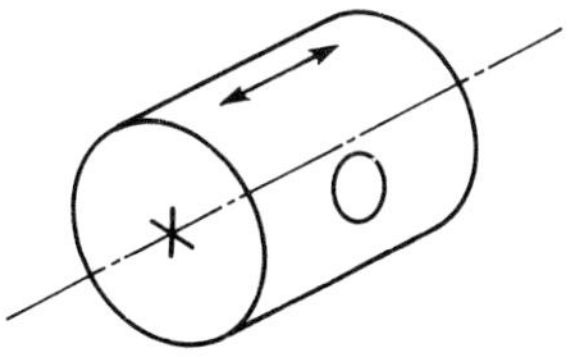	Piston can rotate as well unless it is located by connecting rod
In a plane (rotation)		Rubber ring or disc 	Gives some axial and lateral flexibility as well
In a plane (translation)	Plate between upper and lower guide blocks 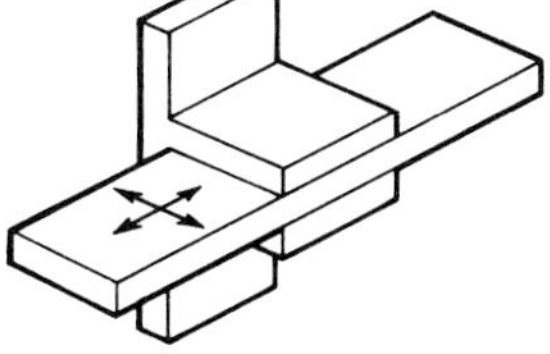	Block sliding on a plate 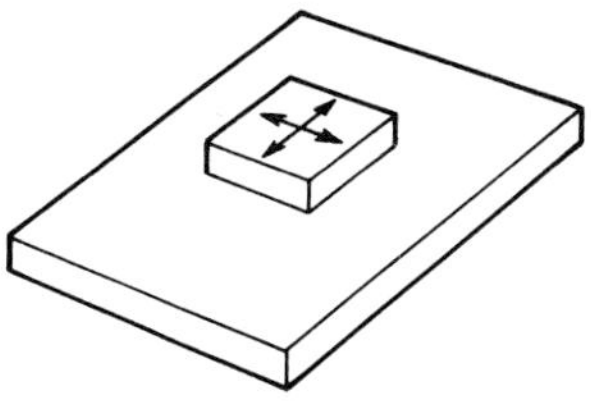	Must be loaded into contact

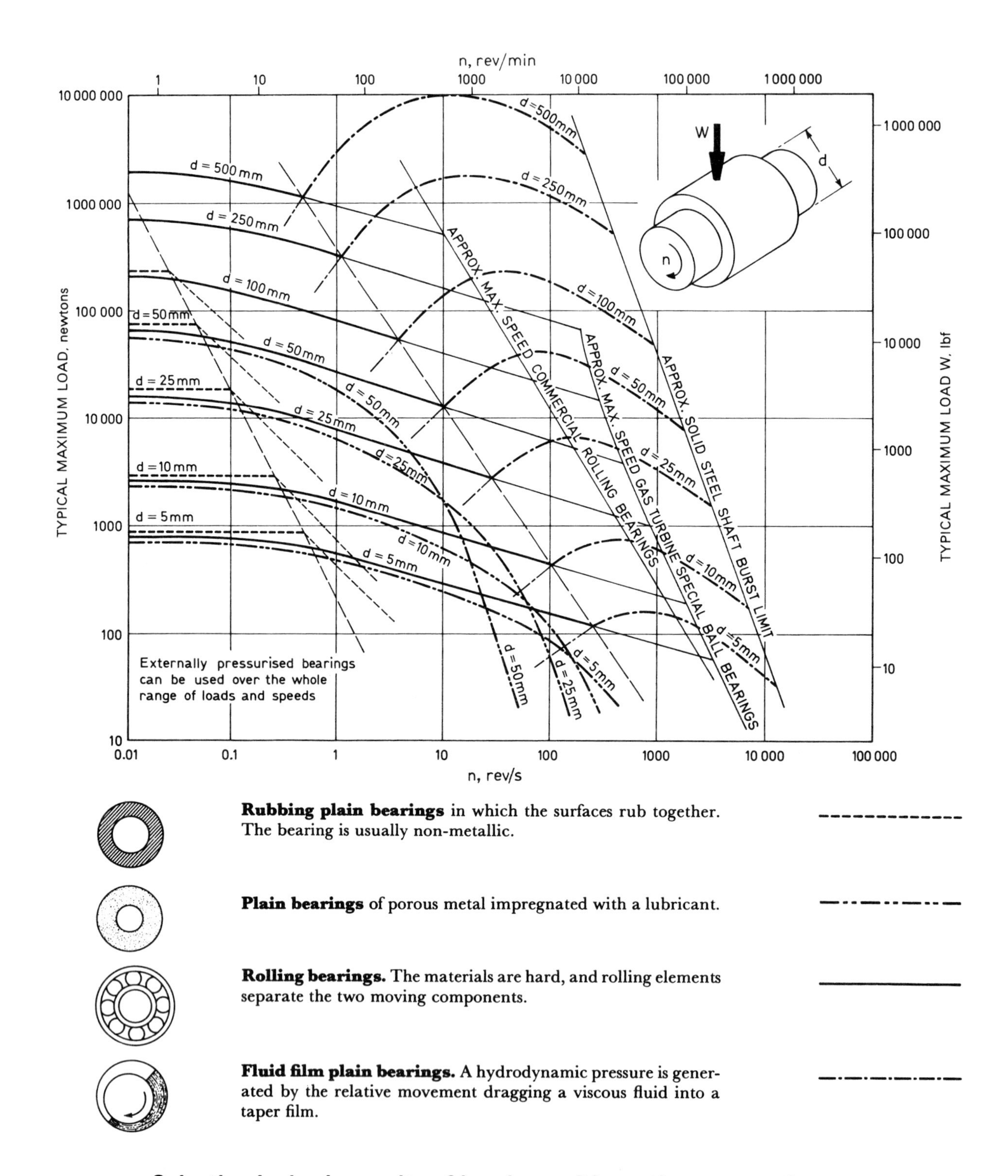

Selection by load capacity of bearings with continuous rotation

This figure gives guidance on the type of bearing which has the maximum load capacity at a given speed and shaft size. It is based on a life of 10 000 h for rubbing, rolling and porous metal bearings. Longer lives may be obtained at reduced loads and speeds. For the various plain bearings, the width is assumed to be equal to the diameter, and the lubricant is assumed to be a medium viscosity mineral oil.

In many cases the operating environment or various special performance requirements, other than load capacity, may be of overriding importance in the selection of an appropriate type of bearing. The tables give guidance for these cases.

Selection of journal bearings with continuous rotation for special environmental conditions

Type of bearing	High temp.	Low temp.	Vacuum	Wet and humid	Dirt and dust	External Vibration	Type of bearing
Rubbing plain bearings (non-metallic)	Good up to the temperature limit of material	Good	Excellent	Good but shaft must be incorrodible	Good but sealing helps	Good	
Porous metal plain bearings oil impregnated	Poor since lubricant oxidises	Fair; may have high starting torque	Possible with special lubricant	Good	Sealing essential	Good	
Rolling bearings	Consult makers above 150°C	Good	Fair with special lubricant	Fair with seals	Sealing essential	Fair; consult makers	
Fluid film plain bearings	Good to temperature limit of lubricant	Good; may have high starting torque	Possible with special lubricant	Good	Good with seals and filtration	Good	
Externally pressurised plain bearings	Excellent with gas lubrication	Good	No; lubricant feed affects vacuum	Good	Good; excellent when gas lubricated	Excellent	
General comments	Watch effect of thermal expansion on fits			Watch corrosion		Watch fretting	

Selection of journal bearings with continuous rotation for special performance requirements

Type of bearing	Accurate radial location	Axial load capacity as well	Low starting torque	Silent running	Standard parts available	Simple lubrication	Type of bearing
Rubbing plain bearings (non-metallic)	Poor	Some in most cases	Poor	Fair	Some	Excellent	
Porous metal plain bearings oil impregnated	Good	Some	Good	Excellent	Yes	Excellent	
Rolling bearings	Good	Yes in most cases	Very good	Usually satisfactory	Yes	Good when grease lubricated	
Fluid film plain bearings	Fair	No; separate thrust bearing needed	Good	Excellent	Some	Usually requires a circulation system	
Externally pressurised plain bearings	Excellent	No; separate thrust bearing needed	Excellent	Excellent	No	Poor; special system needed	

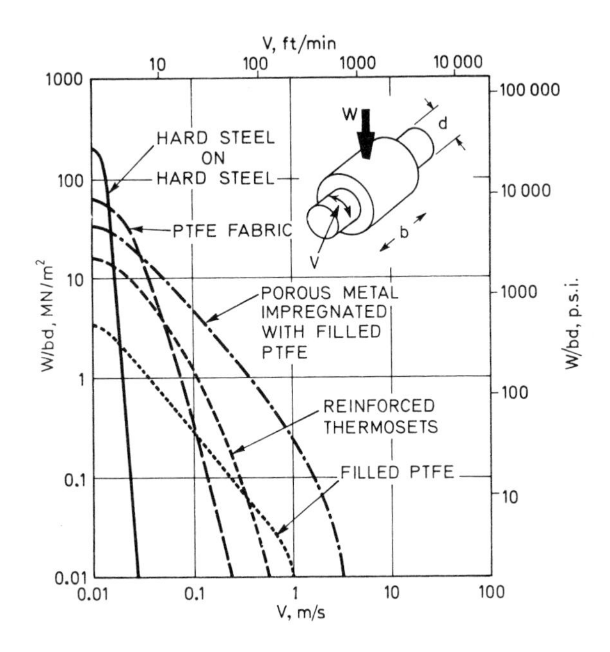

Selection of rubbing plain bearing materials for bushes with oscillatory movement, *by maximum pressure and maximum value of average sliding speed. Rolling bearings in an equivalent arrangement usually can carry about 10 MN/m²*

W/bD, MN/m²
W/bD, p.s.i.
RUBBER
BUSHES
CROSSED
STRIP
FLEXURES
MAX. DEFLECTION EACH WAY, DEGREES

Selection of flexure bearings by external load pressure and the required deflection. *If the centre of rotation does not have to be held constant, single strips or cables can be used*

Selection of journal bearings with oscillating movement for special environments or performance

Type of bearing	*Low friction*	*High temp.*	*Low temp.*	*Dirt and dust*	*External Vibration*	*Wet and humid*	*Type of bearing*
Rubbing plain bearings	Good with PTFE	Good to the temp. limit of material	Very good	Good but sealing helps	Very good	Good but shaft must be incor-rodible	
Porous metal plain bearings oil impregnated	Good	Poor since lubricant oxidises	Fair; friction can be high	Sealing is essential	Good	Good	
Rolling bearings	Very good	Consult makers above 150°C	Good	Sealing is essential	Poor	Good with seals	
Rubber bushes	Elastically stiff	Poor	Poor	Excellent	Excellent	Excellent	
Strip flexures	Excellent	Good	Very good	Excellent	Excellent	Good; watch corrosion	
Knife edge pivots	Very good	Good	Good	Good	Poor	Good; watch corrosion	

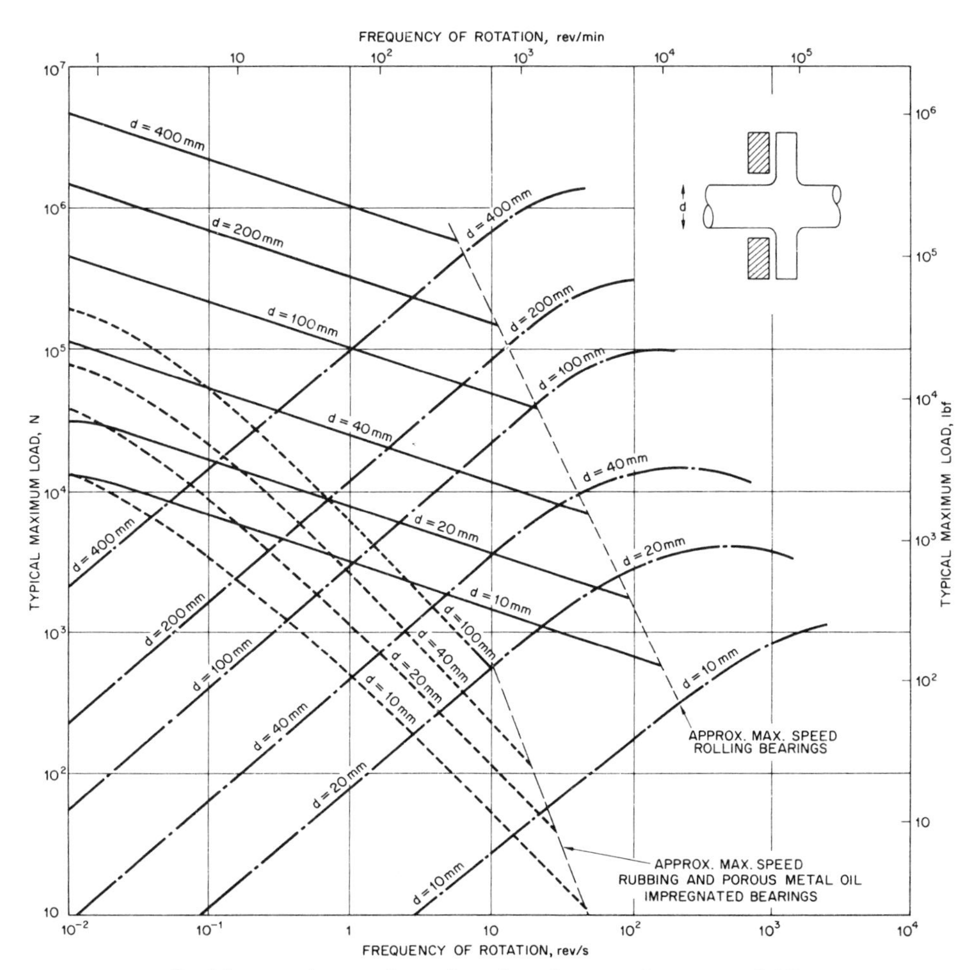

Guide to thrust bearing load-carrying capability

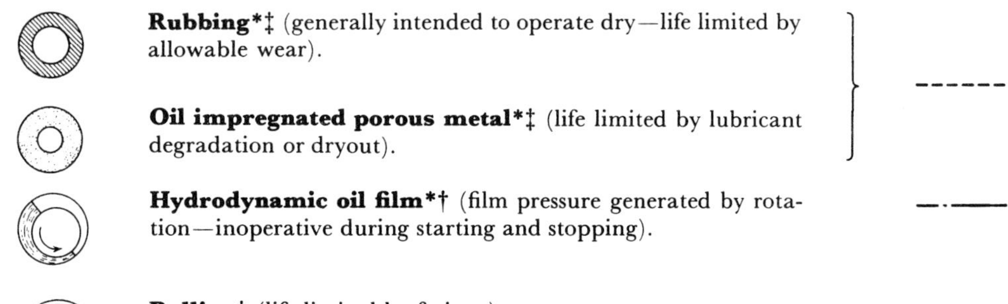

Rubbing*‡ (generally intended to operate dry—life limited by allowable wear).

Oil impregnated porous metal*‡ (life limited by lubricant degradation or dryout).

(Rubbing and oil impregnated porous metal: ------------)

Hydrodynamic oil film*† (film pressure generated by rotation—inoperative during starting and stopping). (—·—·—·—)

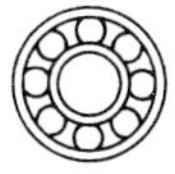

Rolling‡ (life limited by fatigue). (———)

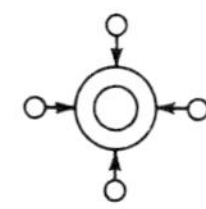

Hydrostatic (applicable over whole range of load and speed—necessary supply pressure 3–5 times mean bearing pressure).

* Performance relates to thrust face diameter ratio of 2.
† Performance relates to mineral oil having viscosity grade in range 32–100 ISO 3448
‡ Performance relates to nominal life of 10 000 h.

This figure gives guidance on the maximum load capacity for different types of bearing for given speed and shaft size.

In many cases the operating environment or various special performance requirements, other than load capacity, may be of overriding importance in the selection of an appropriate type of bearing. The tables give guidance for these cases.

Thrust bearing selection for special environmental conditions

Type of bearing	*High temperature*	*Low temperature*	*Vacuum*	*Wet and humid*	*Dirt and dust*	*External vibration*
Rubbing bearing (non-metallic)	Good—to temperature limit of material	Good	Excellent	Good with suitable shaft and runner material	Good—sealing helps	Good
Oil-impregnated porous metal bearing	Poor—lubricant oxidation	Fair—starting torque may be high	Possible with special lubricant	Good	Sealing necessary	Good
Rolling bearing	Above 100°C reduced load capacity Above 150°C consult makers	Good	Requires special lubricant	Fair with seals	Sealing necessary	Consult makers
Hydrodynamic film bearing	Good—to temperature limit of lubricant	Good—starting torque may be high	Possible with special lubricant	Good	Sealing necessary	Good
Hydrostatic film bearing	Good—to temperature limit of lubricant	Good	Not normally applicable	Good	Good—filtration necessary	Good
General comments	Consider thermal expansion and fits			Consider corrosion		Consider fretting

Thrust bearing selection for special performance requirements

Type of bearing	*Accuracy of axial location*	*Low starting torque*	*Low running torque*	*Silent running*	*Suitability for oscillatory or intermittent movement*	*Availability of standard parts*	*Simplicity of lubrication system*
Rubbing bearing (non-metallic)	Limited by wear	Poor	Poor	Fair	Yes	Some	Excellent
Oil-impregnated porous metal bearing	Good	Fair	Good	Good—until bearing dry-out	Yes	Good	Excellent
Rolling bearing	Good	Good	Good	Usually satisfactory	Yes	Excellent	Good—when grease lubricated
Hydrodynamic film bearing	Good	Fair	Good	Good	No	Some	Usually requires circulation system
Hydrostatic film bearing	Excellent	Excellent	Good	Good	Yes	No	Special system necessary

Requirements and characteristics of lubricated plain bearing materials

Physical property	Significance of property in service	Characteristics of widely used materials: White metals	Copper-base alloys	Aluminium-base alloys
Fatigue strength	To sustain imposed dynamic loadings at operating temperature	Adequate for many applications, but falls rapidly with rise of temperature	Wide range of strength available by selection of composition	Similar to copper-base alloys by appropriate selection of composition
Compressive strength	To support uni-directional loading without extrusion or dimensional change	As above	As above	As above
Embedd-ability	To tolerate and embed foreign matter in lubricant, so minimising journal wear	Excellent—unequalled by any other bearing materials	Inferior to white metals. Softer weaker alloys with low melting point constituent, e.g. lead; superior to harder stronger alloys in this category. These properties can be enhanced by provision of overlay, e.g. lead–tin or lead–indium, on bearing surface where appropriate	Inferior to white metals. Alloys with high content of low melting point constituent, e.g. tin or cadmium; superior in these properties to copper-base alloys of equivalent strength. Overlays may be provided in appropriate cases to enhance these properties
Conform-ability	To tolerate some mis-alignment or journal deflection under load			
Compati-bility	To tolerate momentary boundary lubrication or metal-to-metal contact without seizure			
Corrosion resistance	To resist attack by acidic oil oxidation products or water or coolant in lubricant	Tin-base white metals excellent in absence of sea-water. Lead-base white metals attacked by acidic products	Lead constituent, if present, susceptible to attack. Resistance enhanced by lead–tin or lead–tin–copper overlay	Good. No evidence of attack of aluminium-rich matrix even by alkaline high-additive oils

Physical properties, forms available, and applications of some white metal bearing alloys

Type of bearing	Physical properties: Melting range, °C	Hardness H_v at 20°C	Coefficient of expansion $\times 10^{-6}$/°C	Forms available	Applications
Tin-base white metal ISO 4381 Sn Sb8 Cu4 tin 89% antimony 7.5% copper 3.5%	239–312	~23–25	~23	Lining of thin-walled steel-backed half-bearings, split bushes and thrust washers; lining of bronze-backed components, unsplit bushes	Crankshaft bearings of ic engines and reciprocating compressors within fatigue range; FHP motor bushes; gas turbine bearings (cool end); camshaft bushes; general lubricated applications
Tin-base white metal ISO 4381 Sn Sb8 Cu4 Cd tin 87% antimony 8% copper 4% cadmium 1%	239–340	~27–32	~23	Lining of medium and thick-walled half-bearings and bushes; lining of direct-lined housings and connecting rods	Crankshaft and cross-head bearings of medium and large diesel engines within fatigue range; marine gearbox bearings; large plant and machinery bearings; turbine bearings
Lead-base white metal ISO 4381 Pb Sb10 Sn6 tin 6% antimony 10% copper 1% lead 83%	245–260	~26	~28	'Solid' die-castings; lining of steel, cast iron, and bronze components	General plant and machinery bearings operating at lower loads and temperatures

Note: for the sake of brevity the above table lists only two tin-base and one lead-base white metal. For other white metals and applications refer to ISO 4381 and bearing suppliers.

Physical properties, forms available, and applications of some copper-base alloy bearing materials

Type of bearing	*Physical properties*			*Forms available*	*Applications*
	Melting range, °C	*Hardness* H_v *at* 20°C	*Coefficient of expansion* $\times 10^{-6}$/°C		
Lead bronze ISO 4382/1 Cu Pb20 Sn5 copper 75% tin 5% lead 20%	Matrix ~900 Lead constituent ~327	45–70	~18	Machined cast components; as lining of steel- backed components	Machined bushes, thrust washers, slides, etc., for high-duty applications; as lining of thin and medium-walled heavily loaded IC engine crankshaft bearings, small-end and camshaft bushes, gearbox bushes; gas-turbine bearings, etc.
Lead bronze ISO 4382/1 Cu Pb10 Sn10 copper 80% tin 10% lead 10%	Matrix ~820 Lead constituent ~327	65–90	~18	As above. Hard, strong bronze	Machined bush and thrust washer applications; as lining of thin-walled split bushes for small-ends, camshafts, gearboxes, linkages, etc.
Lead bronze ISO 4382/1 Cu Pb9 Sn5 copper 85% tin 5% lead 9%	Matrix ~920 Lead constituent ~327	45–70	~18	Machined cast components, bars, tubes	Bushes, thrust washers, slides for wide range of applications
Phosphor bronze ISO 4382/1 Cu Sn10 P copper; remainder tin 10% min. phosphorus 0.5% min.	~800	70–150	~18	Machined cast components; bushes, bars, tubes, thrust washers, slides, etc.	Heavy load, high-temperature bush and slide applications, e.g. crankpress bushes, rolling mill bearings, gudgeon-pin bushes, etc.
Copper lead ISO 4383 Cu Pb30 copper 70% lead 30%	Matrix ~1050 Lead constituent ~327	35–45	Lining ~16	As lining of thin-, medium- or thick-walled half-bearings, bushes, and thrust washers	Crankshaft bearings for high- and medium-speed petrol and diesel engines; gas-turbine and turbo-charger bearings; compressor bearings; camshaft and rocker bushes. May be used with or without overlay
Lead bronze ISO 4383 Cu Pb24 Sn4 copper 74% lead 24% tin 4%	Matrix ~900 Lead constituent ~327	40–55	Lining ~18	As above	As above, for more heavily loaded applications, usually overlay plated for crank-shaft bearing applications

Note: for details of other lead bronzes, gunmetals, leaded gunmetals, and aluminium bronzes, consult ISO 4382/1, ISO 4383, EN 133 and bearing supplier.

Physical properties, forms available, and applications of some aluminium-base alloy bearing materials

Type of bearing	*Physical properties*			*Forms available*	*Applications*
	Melting range, °C	*Hardness H_v at 20°C*	*Coefficient of expansion $\times 10^{-6}$/°C*		
Duralumin-type material, no free low melting-point constituents	~550–650	~80–150 depending on composition and heat treatment	~22–24	Cast or wrought machined components	Bushes, slides, etc., for slow speed heavily loaded applications, e.g. small ends of medium and large diesel engines; general machinery bushes, etc.
Low tin aluminium alloy ISO 4383 Al Sn6 Cu tin 6% copper 1% nickel 1% aluminium remainder	Matrix ~650 Tin eutectic ~230	~45–60	~24	Cast or rolled machined components. As lining of steel-backed components. Usually overlay plated for crankshaft bearing applications	Unsplit bushes for small-ends, rockers, linkages, gearboxes. Crankshaft half-bearings for diesel engines and linings of thin- and medium-walled steel-backed crankshaft bearings for heavily loaded diesels and compressors; also as linings of steel-backed split bushes.
Aluminium silicon-cadmium alloy ISO 4383 Al Si4 Cd silicon 4% cadmium 1% aluminium remainder	Matrix ~650 Cadmium ~320	~55	~22	As lining of thin-walled half-bearings, split bushes and thrust washers. Usually overlay plated for crankshaft applications	Heavily loaded diesel engine crankshaft bearings; small-end, gearbox, rocker bushes, etc.
Aluminium tin silicon alloy ISO 4383 tin 10 or 12% silicon 4% copper 1 or 2% aluminium remainder	Matrix ~650 Tin eutectic ~230	55–65	~24	As lining of thin-walled half-bearings, split bushes and thrust washers. Used without an overlay	Heavily loaded crankshaft bearings for high-speed petrol and diesel engines. Used without an overlay
High tin aluminium alloy ISO 4383 Al Si20 Cu tin 20% copper 1% aluminium remainder	Matrix ~650 Tin eutectic ~230	~40	~24	As lining of thin- and medium-walled half-bearings, split bushes and thrust washers. Usually used without an overlay	Moderately loaded crankshaft bearings for high-speed petrol and diesel engines. Camshaft gearbox and linkage bushes; thrust washers

Overlay plating

Functions of an overlay

1. To provide bearing surface with good frictional properties, i.e. compatibility
2. To confer some degree of embeddability
3. To improve load distribution
4. To protect lead in lead-containing interlayer materials (e.g. copper–lead, lead bronze) from corrosion

Thickness of overlay

0.017 mm (0.0007 in) to 0.040 mm (0.0015 in) depending upon bearing loading and type and size of bearing

Typical overlay compositions

1. 10–12% tin, remainder lead
2. 10% tin, 2% copper, remainder lead
   Tin and copper may be higher for increased corrosion resistance.
   Where the maximum corrosion resistance is required with lead-tin-copper overlays or copper-lead a nickel interlayer 0.001 mm thick is used beneath the overlay
3. 5–8% indium, remainder lead
4. 20–40% tin, remainder aluminium applied by vapour deposition (sputter) on aluminium alloy substrates

Relative load-carrying capacities of bearing materials, and recommended journal hardness

Material	Maximum dynamic loading		Recommended journal hardness
	MN/m^2	lbf/in^2	H_v min at 20°C
Tin and lead-base white metal linings ~0.5 mm (0.020 in) thick	10.3–13.7	1500–2000	Soft journal (~140) satisfactory
As above ~0.1 mm (0.004 in) thick	>17.2	>2500	As above
70/30 copper–lead on steel	24–27.5	3500–4000	~250
70/30 copper–lead on steel, overlay plated	27.5–31	4000–4500	~230
Lead bronze, 20–25% lead, 3–5% tin, on steel	35–42	5000–6000	~500
As above, overlay plated	42–52	6000–7500	~230
Low lead (~10%) lead bronzes, steel-backed	>48	>7000	~500
Aluminium–6% tin, 'solid' or steel-backed	~42	~6000	~500
As above, overlay plated	45–52	6500–7500	~280
Aluminium–20% tin (reticular structure) on steel	~42	~6000	~230
Aluminium–tin–silicon on steel	~52	~7500	~250
Phosphor–bronze, chill or continuously cast	~62	~9000	~500

Note: the above figures must be interpreted with caution, as they apply only to specific testing conditions. They should not be used for design purposes without first consulting the appropriate bearing supplier.

Fatigue strength and relative compatibility of some bearing alloys (courtesy: Glacier Metal Company Limited)

Material	Fatigue rating(1)		Seizure load(2)	
	MN/m^2	lbf/in^2	MN/m^2	lbf/in^2
Tin-base white metal	35	5000	14	2000 not seized
Lead-base white metal	35	5000	14	2000 not seized
70/30 copper–lead, unplated	95	13 500	11–14	1600–2000
70/30 copper–lead, overlay plated	70 119	10 000 overlay 17 000 copper–lead	—(3) —(3)	—(3) —(3)
Lead bronze (22% lead, 4% tin), unplated	—	—	5.5–11	800–1600
Lead bronze (22% lead, 4% tin), overlay plated	70 125	10 000 overlay 18 000 lead bronze	—(3)	—(3)
Lead bronze (10% lead, 10% tin), unplated	—	—	3–8.5	400–1200
6% tin–aluminium, unplated	105	15 000	5.5–14	800–2000
6% tin–aluminium, overlay plated	76 114	11 000 overlay 16 500 tin–aluminium	—(3) —(3)	—(3) —(3)
Aluminium–tin–silicon	118	13000	10–14	1500–2000
Aluminium–20% tin	90	13000	14	2000 usually

Notes: (1) Fatigue ratings determined on single-cylinder test rig. Not to be used for engine design purposes.
(2) Seizure load determined by stop–start tests on bushes. Maximum load on rig 14 MN/m^2 (2000 lbf/in^2).
(3) Overlay does not seize, but wears away. Seizure then occurs between interlayer and journal at load depending upon thickness of overlay, i.e. rate of wear. The overlay thickness on aluminium–tin is usually less than that on copper–lead and lead bronze, hence the slightly higher fatigue rating.

Characteristics of rubbing bearing materials

Material	Maximum P loading		PV value		Maximum temperature °C	Coefficient of friction	Coefficient of expansion $\times 10^{-6}$/°C	ments	Application
	lbf/in^2	MN/m^2	lbf/in^2 × ft/min	MN/m^2 × m/s					
Carbon/graphite	200–300	1.4–2	≯3000 for continuous operation; 5000 for short period llfe	0.11; 0.18	350–500	0.10–0.25 dry	2.5–5.0	For continuous dry operation P≯ 200 lbf/in^2 (1.4 MN/m^2), V≯ 250 ft/min (1.25 m/s)	Food and textile machinery where contamination by lubricant inadmissible; furnaces, conveyors, etc. where temperature too high for conventional lubricants; where bearings are immersed in liquids, e.g. water, acid or alkaline solutions, solvents, etc.
Carbon/graphite with metal	450–600	3–4	4000 for continuous operation; 6000 for short period	0.145; 0.22	130–350	0.10–0.35 dry	4.2–5.0	Permissible peak load and temperature depend upon metal impregnant	
Graphite impregnated metal	10 000	70	8000–10 000	0.28–0.35	350–600	0.10–0.15 dry; 0.020–0.025 grease lubricated	12–13 with iron matrix; 16–20 with bronze matrix	Operates satisfactorily dry within stated limits; benefits considerably if small quantity of lubricant present, i.e. higher PV values	Bearings working in dusty atmospheres, e.g. coal-mining, foundry plant, steel plant, etc.
Graphite/thermo-setting resin	300	2	~10 000	~0.35	250	0.13–0.5 dry	3.5–5.0	Particularly suitable for operation in sea-water or corrosive fluids	
Reinforced thermo-setting plastics	5000	35	~10 000	~0.35	200	0.1–0.4 dry; 0.006 claimed with water lubrication	25–80 depending on plane of reinforcement	Values depend upon type of reinforcement, e.g. cloth, asbestos, etc. Higher PV values when lubricated	Water-lubricated roll-neck bearings (esp. hot rolling mills), marine sterntube and rudder bearings; bearings subject to atomic radiation
Thermo-plastic material without filler	1500	10	~1000	~0.035	100	0.1–0.45 dry	~100	Higher PV values acceptable if higher wear rates tolerated. With initial lubrication only, PV values up to 20 000 can be imposed	Bushes and thrust washers in automotive, textile and food machinery—linkages where lubrication difficult

Characteristics of rubbing bearing materials (continued)

Material	*Maximum* P *loading*		PV *value*		*Maximum temperature* °C	*Coefficient of friction*	*Coefficient of expansion* $\times 10^{-6}$/°C	*Comments*	*Applications*
	lbf/in^2	MN/m^2	lbf/in^2 × ft/min	MN/m^2 × m/s					
Thermo-plastic with filler or metal-backed	1500–2000	10–14	1000–3000	0.035–0.11	100	0.15–0.40 dry	80–100	Higher loadings and *PV* values sustained by metal-backed components, especially if lubricated	As above, and for more heavily loaded applications
Thermo-plastic with filler bonded to metal back	20 000	140	10 000	0.35	105	0.20–0.35 dry	27	With initial lubrication only. *PV* values up to 40 000 acceptable with re-lubrication at 500–1000 h intervals	For conditions of intermittent operation or boundary lubrication, or where lubrication limited to assembly or servicing periods, e.g. ball-joints, suspension and steering linkages, king-pin bushes, gearbox bushes, etc.
Filled PTFE	1000	7	Up to 10 000	Up to 0.35	250	0.05–0.35 dry	60–80	Many different types of filler used, e.g. glass, mica, bronze, graphite. Permissible *PV* and unit load and wear rate depend upon filler material, temperature, mating surface material and finish	For dry operation where low friction and low wear rate required, e.g. bushes, thrust washers, slideways, etc., may also be used lubricated
PTFE with filler, bonded to steel backing	20 000	140	Up to 50 000 continuous rating	Up to 1.75	280	0.05–0.30 dry	20 (lining)	Sintered bronze, bonded to steel backing, and impregnated with PTFE/lead	Aircraft controls, linkages; automotive gearboxes, clutch, steering suspension, bushes, conveyors, bridge and building expansion bearings
Woven PTFE reinforced and bonded to metal backing	60 000	420	Up to 45 000 continuous rating	Up to 1.60	250	0.03–0.30 dry	—	The reinforcement may be interwoven glass fibre or rayon	Aircraft and engine controls, linkages, automotive suspensions, engine mountings, bridge and building expansion bearings

Notes: (1) Rates of wear for a given material are influenced by load, speed, temperature, material and finish of mating surface. The *PV* values quoted in the above table are based upon a wear rate of 0.001 in (0.025×10^{-3} m) per 100 h, where such data are available. For specific applications higher or lower wear rates may be acceptable—consult the bearing supplier.
(2) Where lubrication is provided, either by conventional lubricants or by process fluids, considerably higher *PV* values can usually be tolerated than for dry operation.

MATERIALS

Usually composites based on polymers, carbons, and metals.

The properties of typical dry rubbing bearing materials

Type	*Examples*	*Max. static load*		*Max. service temp.*	*Coeff. exp.*	*Heat conductivity*		*Special features*
		MN/m^2	10^3 lbf/in^2	°C	10^6/°C	W/m°C	Btu/ft h °F	
Thermoplastics	Nylon, acetal, UHMWPE	10	1.5	100	100	0.24	0.14	Inexpensive
Thermoplastics +fillers	Above+MoS_2, PTFE, glass, graphite, etc.	15–20	2–3	150	60–100	0.24	0.14	Solid lubricants reduce friction
PTFE+fillers	Glass, bronze, mica, carbon, metals	2–7	0.3–1	250	60–100	0.25–0.5	0.15–0.3	Very low friction
High temperature polymers (+fillers)	Polyimides polyamide-imide PEEK	30–80	4.5–12	250	20–50	0.3–0.7	0.2–0.4	Relatively expensive
Thermosets +fillers	Phenolics, epoxies +asbestos, textiles, PTFE	30–50	4.5–7.5	175	10–80	0.4	0.25	Reinforcing fibres improve strength
Carbon–graphite	Varying graphite content; may contain resin	1–3	0.15–0.45	500	1.5–4	10–50	6–30	Chemically inert
Carbon–metal	With Cu, Ag, Sb, Sn, Pb	3–5	0.45–0.75	350	4–5	15–30	9–18	Strength increased
Metal–solid lubricant	Bronze–graphite -MoS_2; Ag–PTFE	30–70	4.5–10	250–500	10–20	50–100	30–60	High temperature capability
Special non-machinable products	Porous bronze/ PTFE/Pb	350	50	275	20	42	24	Need to be considered at the design stage
	PTFE/glass weave+resin	700	100	250	12	0.24	0.14	
	Thermoset+ PTFE surface	50	7.5	150	10	0.3	0.2	
	Metal+filled PTFE liner	7	1	275	100	0.3	0.2	

Notes:
All values are approximate; properties of many materials are anisotropic.
Most materials are available in various forms: rod, sheet, tube, etc.
For more detailed information, consult the supplier, or ESDU Data Item 87007.

EFFECT OF ENVIRONMENT

Type of material	*Temp. above 200°C*	*Temp. below −50°C*	*Radiation*	*Vacuum*	*Water*	*Oils*	*Abrasives*	*Acids and alkalis*
Thermoplastics + fillers	Few suitable	Usually good	Usually poor	Most materials suitable; avoid graphite as fillers	Often poor; watch finish of mating surface	Usually good	Poor to fair; rubbery materials best	Fair to good
PTFE + fillers	Fair	Very good	Very poor					Excellent
Thermosets + fillers	Some suitable	Good	Some fair					Some good
Carbon–graphite	Very good; watch resins and metals	Very good	Very good; avoid resins	Useless	Fair to good	Good	Poor	Good, except strong acids

PERFORMANCE

Best criterion of performance is a curve of P against V for a specified wear rate. The use of $P \times V$ factors can be misleading.

Curves relate to journal bearings with a wear rate of 25 μm (1 thou.)/100 h—unidirectional load; 12.5 μm (0.5 thou.)/100 h—rotating load

Counterface finish 0.2–0.4 μm cla (8–16 μin).

A Thermoplastics
B PTFE
C PTFE + fillers
D Porous bronze + PTFE + Pb
E PTFE–glass weave + thermoset
F Reinforced thermoset + MoS_2
G Thermoset/carbon-graphite + PTFE

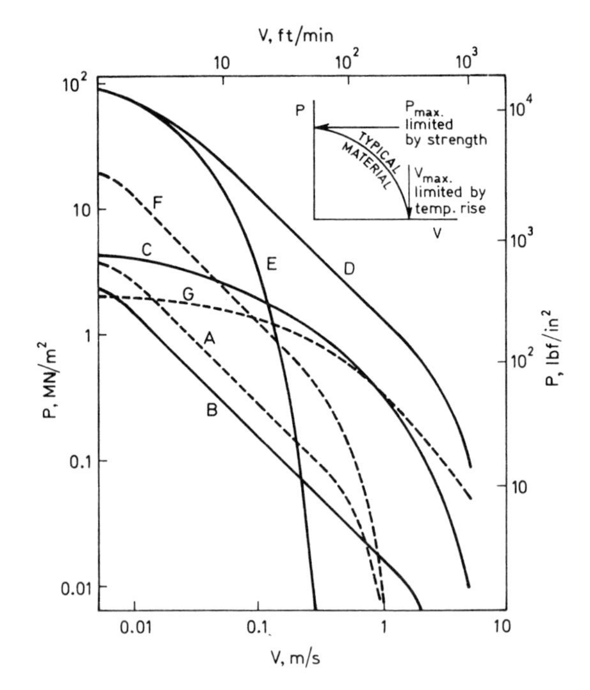

WEAR

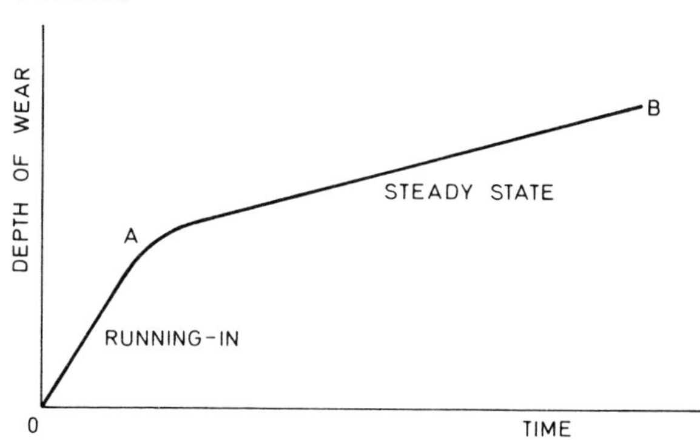

'Running-in' wear *O–A* is very dependent upon counterface roughness. Approximately, wear rate α (cla roughness)

'Steady-state' wear *A–B* depends on (*i*) mechanical properties of the material and its ability to (*ii*) smooth the counterface surface and/or (*iii*) transfer a thin film of debris.

In general, the steady-state wear rate, depth/unit time = KPV (*a.b.c.d.e*). *K* is a material constant incorporating (*i*), (*ii*), and (*iii*) above. Wear-rate correction factors *a,b,c,d,e*, depend on the operating conditions as shown below.

Approximate values of wear-rate correction factors

Factor	Condition		Value
a, Geometrical factor	continuous motion	rotating load	0.5
		unidirectional load	1
	oscillatory motion		2

Factor	Condition	Value
b, Heat dissipation factor	metal housing, thin shell, intermittent operation	0.5
	metal housing, continuous operation	1
	non-metallic housing, continuous operation	2

Factor	Temperature	PTFE–base	carbon–graphite, thermosets
c, Temperature factor	20°C	1	1
	100°C	2	3
	200°C	5	6

Factor	Condition	Value
d, Counterface factor	stainless steels, chrome plate	0.5
	steels	1
	soft, non-ferrous metals (Cu alloys, Al alloys)	2–5

Factor	Condition	Value
e, Surface finish factor	0.1–0.2 μm cla (4–8 μin)	1
	0.2–0.4 μm cla (8–16 μin)	2–3
	0.4–0.8 μm cla (16–32 μin)	4–10

Note: Factors do not apply to metal–solid lubricant composites.

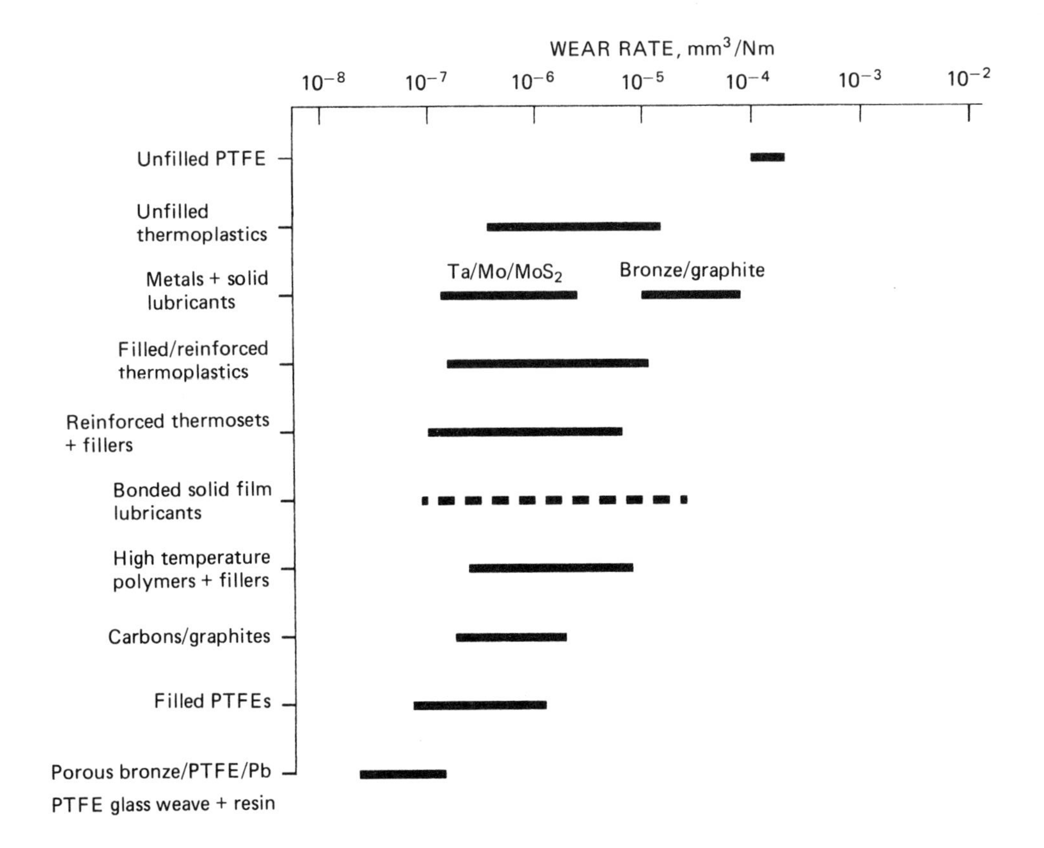

Order-of-magnitude wear rates of dry bearing material groups. *At light loads and low speeds (frictional heating negligible) against smooth (0.15 μm Ra) mild steel*

POINTS TO NOTE IN DESIGN

Choose length/diameter ratio between $\frac{1}{2}$ and $1\frac{1}{2}$.
Minimise wall thickness to aid heat dissipation.

Possibility of dimensional changes after machining
- moisture absorption
- high expansion coefficients
- stress relaxation

Machining tolerances may be poor: 25–50 μm (1–2 thou.) for plastics; better for carbons.

Suitable housing location methods are
- plastics—mechanical interlock or adhesives
- metal-backed plastics—interference fit
- carbon–graphite—press or shrink fit

Avoid soft shafts if abrasive fillers present, e.g. glass.
Minimise shaft roughness: 0.1–0.2 μm cla (4–8 μin) preferred.

Allow generous running clearances
- plastics, 5 μm/mm (5 thou./in). min, 0.1 mm (4 thou.)
- carbon–graphite, 2 μm/mm (2 thou./in). min, 0.075 mm (3 thou.)

Contamination by fluids, or lubrication, usually lowers friction but:

increases wear of filled PTFE's and other plastics containing PTFE, graphite or MoS_2;
decreases wear of thermoplastics and thermosets without solid lubricant fillers.

DESIGN AND MATERIAL SELECTION

Having determined that a self-lubricating porous metal bearing may be suitable for the application, use Fig. 6.1 to assess whether the proposed design is likely to be critical for either load capacity or oil replenishment. With flanged bearings add together the duty of the cylindrical and thrust bearing surfaces.

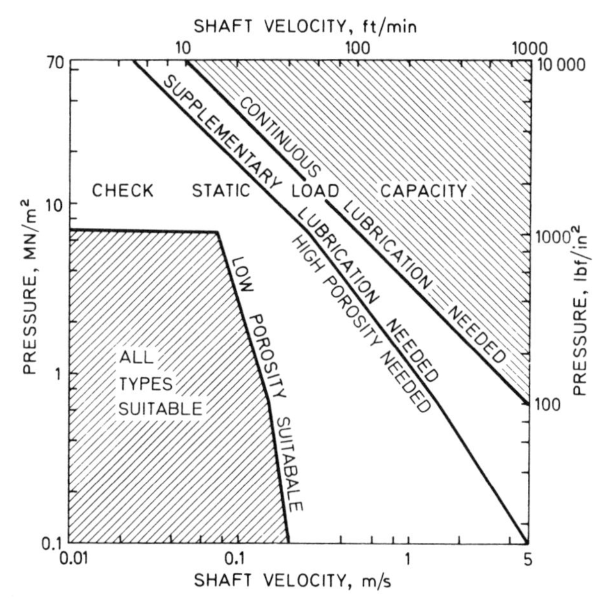

Fig. 6.1. A general guide to the severity of the duty. *At high pressures and particularly high velocities the running temperature increases, which requires provision for additional lubrication to give a satisfactory life. Attention to the heat conductivity of the assembly can reduce the problem of high running temperatures. High porosity bearings contain more oil but have lower strength and conductivity. The data are based on a length to diameter ratio of about 1, and optimisation of the other design variables*

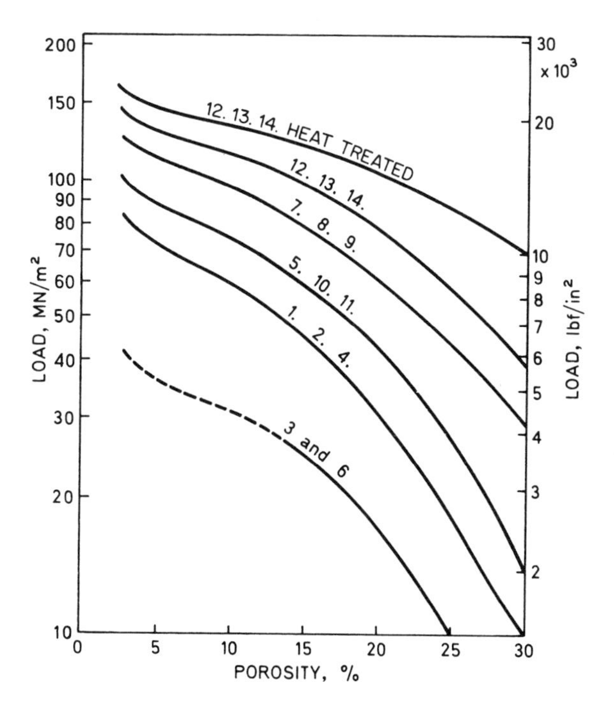

Fig. 6.2. A general guide to the maximum static load capacity (including impact loads) of a wide range of compositions and porosities. *The curves are based on a length to diameter of about 1, and assume a rigid housing. Note that all compositions are not available in all porosities and sizes*

Bearing strength

Figure 6.2 give the relationship between the maximum static load capacity and porosity for the fourteen different standard compositions listed in Table 6.1. Wherever possible select one of these preferred standards for which the design data in Fig. 6.3 and 6.4 apply. Having made the choice, check with the manufacturers that at the wall thickness and length-to-diameter ratio, the static load capacity is acceptable.

Wall thickness, L/d ratio, tolerances

The length, diameter and composition determine the minimum wall thickness which can be achieved, and avoid a very large porosity gradient in the axial direction. Porosity values are quoted as average porosity, and the porosity at the ends of the bearing is less than in the centre. As most properties are a function of the porosity, the effect of the porosity gradient on the performance has to be separately considered. The dimensional tolerances are also a function of the porosity gradient, wall thickness, length-to-diameter ratio, composition, etc.

Figure 6.3(a) gives the general case, and manufacturers publish, in tabular form, their limiting cases. A summary of these data is given in Fig. 6.4 for cylindrical and flanged bearings in the preferred standard composition and porosities indicated in Table 6.1. Clearly the problem is a continuous one, hence, when dealing with a critical design, aim for *L/d* about unity and avoid the corners of the stepped relationship in Fig. 6.4.

The corresponding limiting geometries and tolerances for thrust bearings and self-aligning bearings are given in Figs 6.3(b) and 6.3(c). In all cases avoid the areas outside the enclosed area.

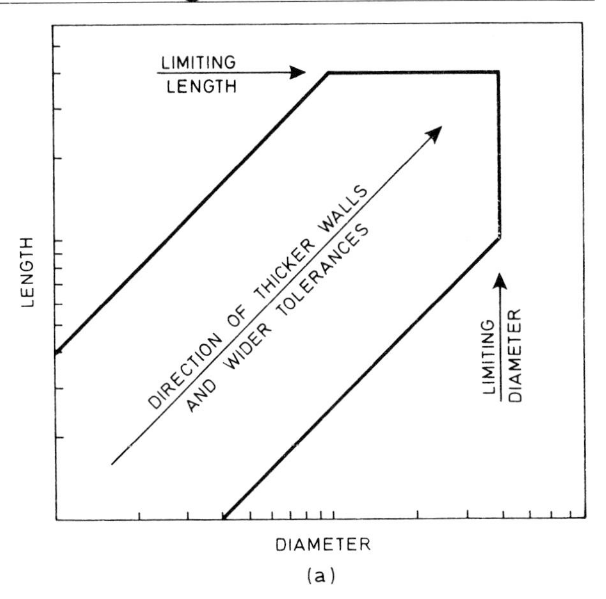

***Fig. 6.3a. General effect of length and diameter on the minimum wall thickness and dimensional tolerance.** The stepped relationships present in Fig. 6.4 arise from a tabular interpretation of the continuous effect shown in Fig. 6.3a*

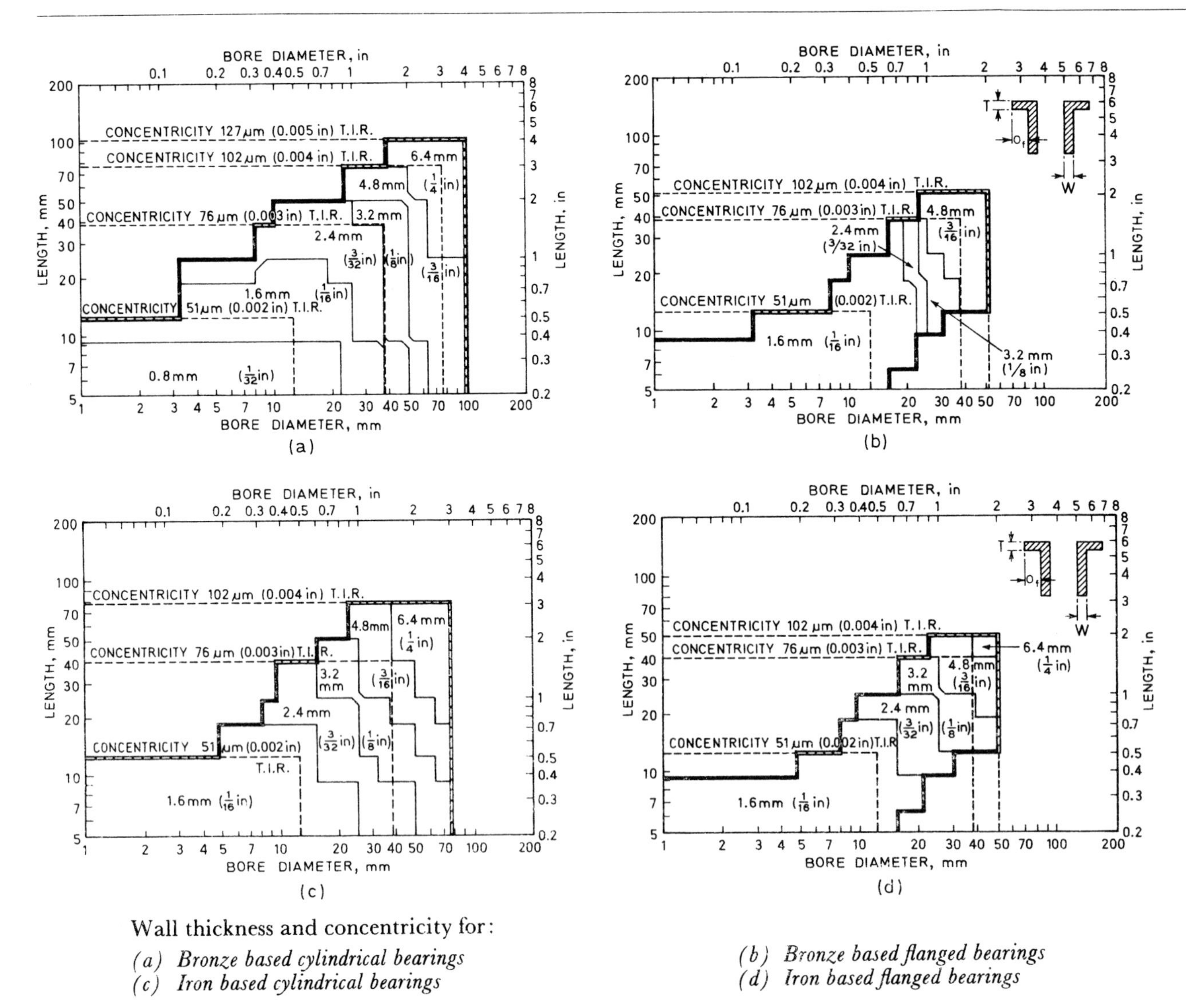

Wall thickness and concentricity for:

(a) Bronze based cylindrical bearings
(b) Bronze based flanged bearings
(c) Iron based cylindrical bearings
(d) Iron based flanged bearings

Fig. 6.4. Recommended minimum wall thicknesses and standard tolerances of diameters, length and concentricity for bearings to the preferred standards in Table 6.1.

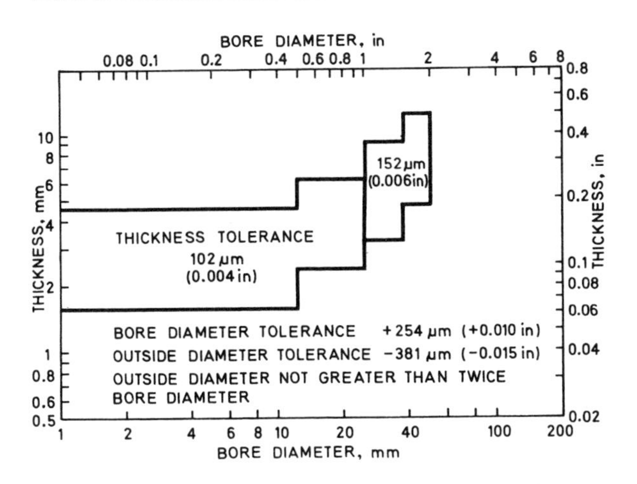

Fig. 6.3b. The thickness of thrust bearings or washers is a function of their diameter, as shown by the area bounded above

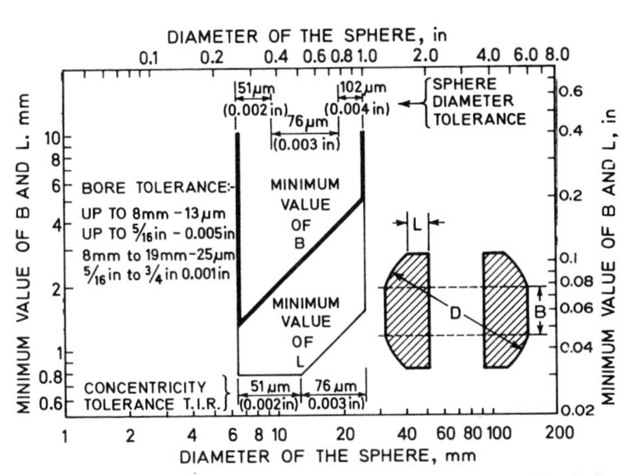

Fig. 6.3c. Self aligning bearings are standard between $\frac{1}{4}$" and 1" sphere diameter. *Within this range the minimum length of the flat on the end faces (L) and the sphere diameter (B) is given above, together with concentricity, bore, and sphere diameter tolerances*

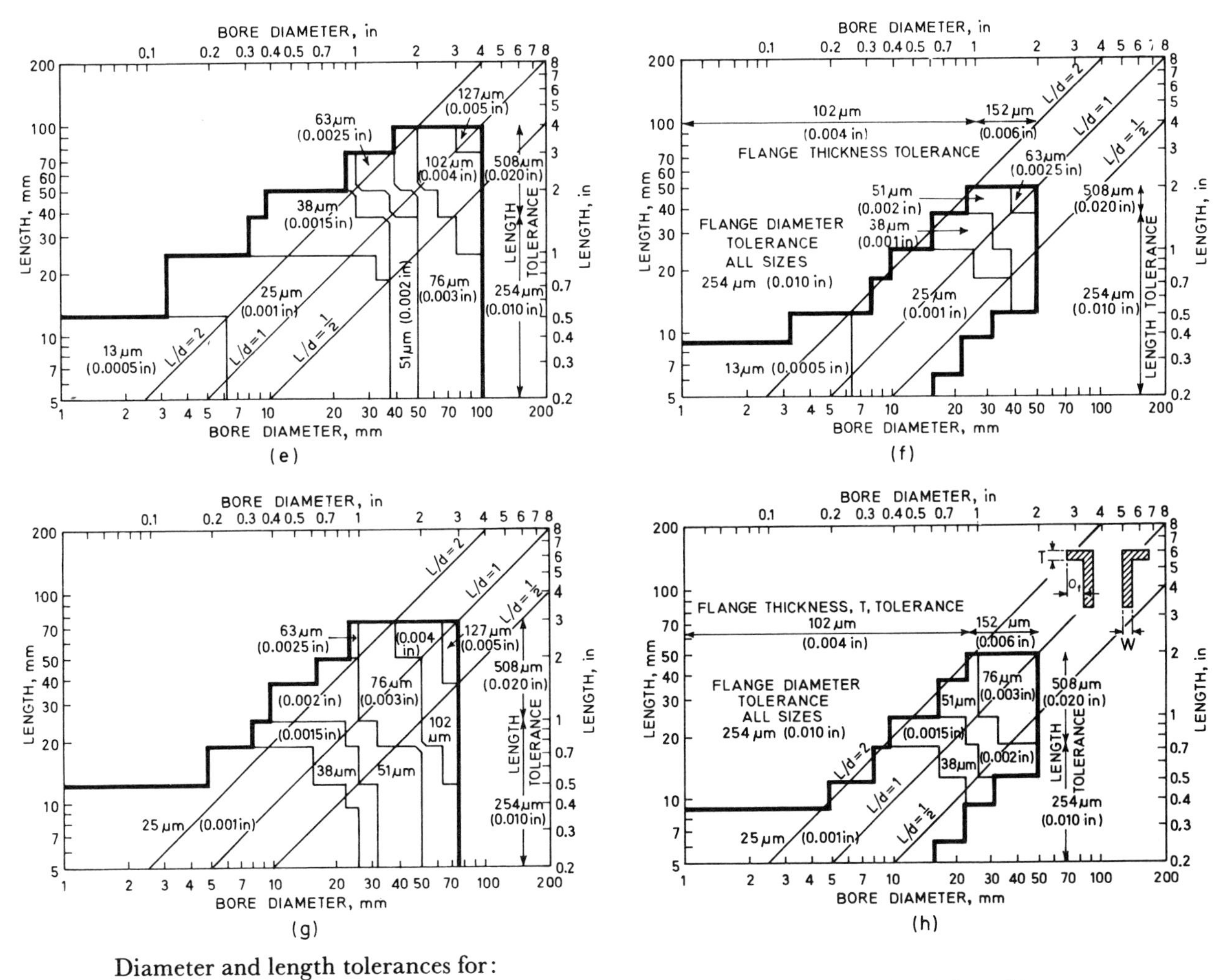

Diameter and length tolerances for:

(e) Bronze based cylindrical bearings
(g) Iron based cylindrical bearings
(f) Bronze based flanged bearings
(h) Iron based flanged bearings

Note: (1) For flanged bearings these data apply only where $T \simeq W$ and where $T \leqslant 0_f \leqslant 3T$.
(2) Smaller tolerance levels can usually be supplied to special order.
(3) Data for other compositions or porosities are available from the manufacturers.

Composition and porosity

The graphited tin bronze (No.1 in Table 6.1) is the general purpose alloy and gives a good balance between strength, wear resistance, conformability and ease of manufacture. Softer versions have lead (No. 4) or reduced tin (No. 2). Graphite increases the safety factor if oil replenishment is forgotten, and the high graphite version (No. 3) gives some dry lubrication properties at the expense of strength.

Where rusting is not a problem, the cheaper and stronger iron-based alloys can be used. Soft iron (No. 5) has a low safety factor against oil starvation, especially with soft steel shafts. Graphite (Nos. 6 and 10) improves this, but reduces the strength unless the iron is carburised during sintering (No. 11). Copper (Nos. 7, 8 and 9) increases the strength and safety factor. If combined with carbon (Nos. 12, 13 and 14) it gives the greatest strength especially after heat treatment.

Table 6.1 Typical specifications for porous metal bearing materials

No. ref. Fig. 6.2	*Composition*	*Notes on composition*
1	**89/10/1 Cu/Sn/graphite**	General purpose bronze (normally supplied unless otherwise specified). Reasonably tolerant to unhardened shafts
2	91/8/1 Cu/Sn/graphite	Lower tin bronze. Reduced cost. Softer
3	85/10/5 Cu/Sn/graphite	High graphite bronze. Low loads. Increased tolerance towards oil starvation
4	86/10/3/1 Cu/Sn/Pb/graphite	Leaded bronze. Softer. Increased tolerance towards misalignment
5	>99% iron (soft)	Soft iron. Cheaper than bronze. Unsuitable for corrosive conditions. Hardened shafts preferred
6	$97\frac{1}{2}/2\frac{1}{2}$ Fe/graphite	Graphite improves marginal lubrication and increases tolerance towards unhardened shafts
7	98/2 Fe/Cu	Increasing copper content increases strength and cost. This series forms the most popular range of porous iron bearings. Hardened shafts preferred
8	**2% to 25% Cu in Fe**	
9	75/25 Fe/Cu	
10	89/10/2 Fe/Cu/graphite	High graphite improves marginal lubrication and increases tolerance towards unhardened shafts
11	99/0.4 Fe/C	Copper free, hardened steel material
12	97/2/0.7 Fe/Cu/C	Hardened high strength porous steels. Increasing copper content gives increasing strength and cost
13	2% to 10% Cu in 0.7 C/Fe	
14	89/10/0.7 Fe/Cu/C	

Note: These typical specifications are examples of materials listed in various relevant standards such as: ISO 5755/1, BS 5600/5/1, DIN 30 910/3, MPIF/35, ASTM B 438, ASTM 439. Most manufacturers offer a wide choice of compositions and porosities.

LUBRICATION

As a general recommendation, the oil in the pores should be replenished every 1000 hours of use or every year, whichever is the sooner. However, the data in Fig. 6.5 should be used to modify this general recommendation. Low porosity bearings should be replenished more frequently. Bearings running submerged or receiving oil-splash will not require replenishment. See the notes in Table 6.1 about compositions which are more tolerant to oil starvation. Figure 6.6 gives details of some typical assemblies with provision for supplementary lubrication.

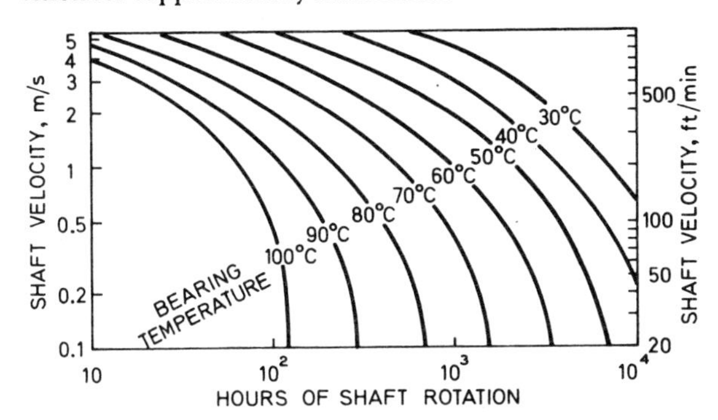

Fig. 6.5. The need to replenish the oil in the pores arises because of oil loss (which increases with shaft velocity) and oil deterioration (which increases with running temperature). *The above curves relate to the preferred standard bearing materials in Table 6.1*

Selection of lubricant

1. Figure 6.7 gives general guidance on the choice of oil viscosity according to load and temperature.
2. Lubricants must have high oxidation resistance.
3. Unless otherwise specified, most standard porous metal bearings are impregnated with a highly refined and oxidation-inhibited oil with an SAE 20/30 viscosity.
4. Do not select oils which are not miscible with common mineral oils unless replenishment by the user with the wrong oil can be safeguarded.
5. Do not use grease, except to fill a blind cavity of a sealed assembly (see Fig. 6.6).
6. Avoid suspensions of solid lubricants unless experience in special applications indicates otherwise.
7. For methods of re-impregnation—consult the manufacturers.

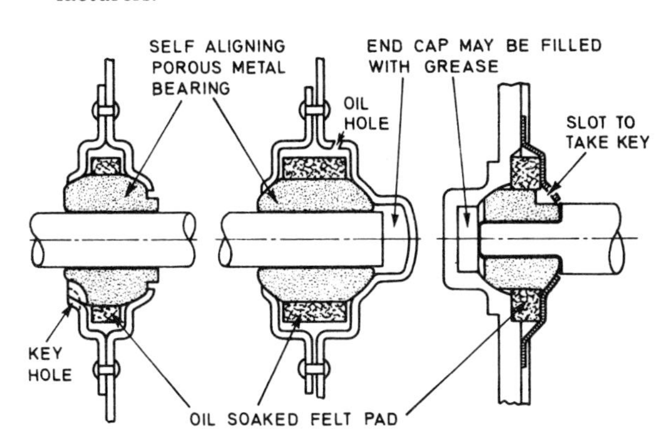

ASSEMBLIES OF SELF-ALIGNING POROUS METAL BEARINGS WITH PROVISION FOR ADDITIONAL LUBRICATION

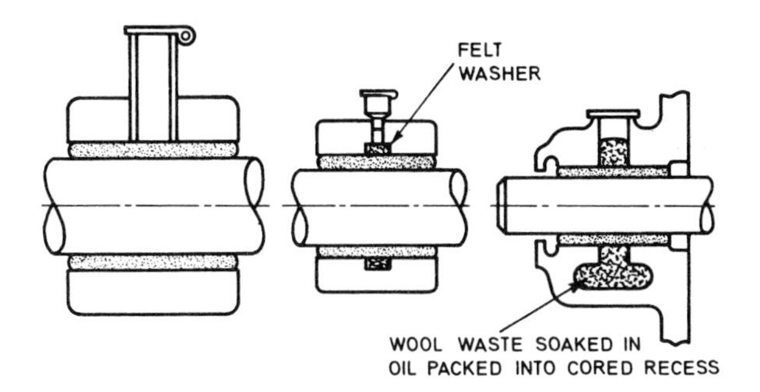

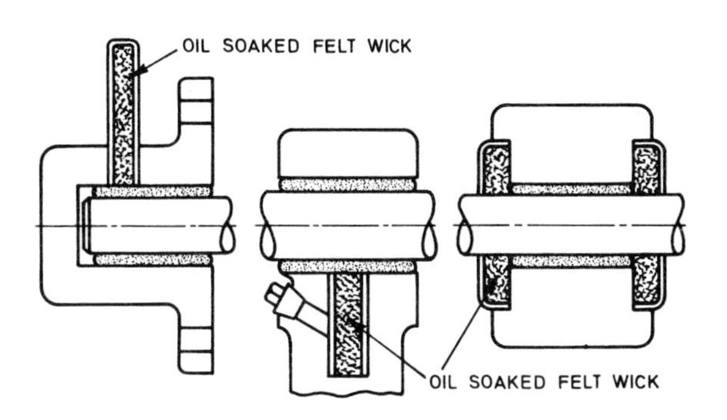

TYPICAL METHODS OF SUPPLEMENTING AND REPLENISHING THE OIL IN THE PORES OF A FORCE FITTED BEARING

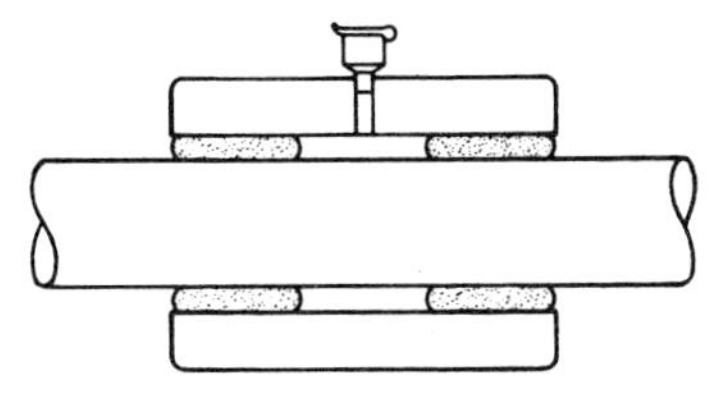

SIMPLE LUBRICATION ARRANGEMENT WHICH CAN BE EMPLOYED WITH A PAIR OF FORCE FITTED POROUS METAL BEARINGS

Fig. 6.6. Some typical assemblies showing alternative means of providing supplementary lubrication facilities

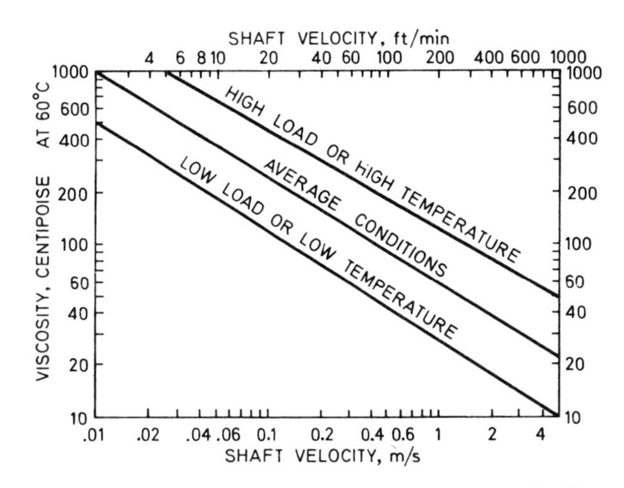

Fig. 6.7. General guide to the selection of oil viscosity expressed in centipoises at 60°C

INSTALLATION

General instructions

1 Ensure that the bearings are free of grit, and wash in oil if not held in dust-free storage. Re-impregnate if held in stock for more than one year or if stored in contact with an oil absorbent material.

2 With a self-aligning assembly (see example in Fig. 6.6):
 (*a*) ensure that the sphere is able to turn freely under the action of the misalignment force;
 (*b*) check that the static load capacity of the housing assembly is adequate;
 (*c*) note that the heat dissipation will be less than a force-fitted assembly and hence the temperature rise will be higher.

3 With a force-fitted assembly (see examples in Fig. 6.6):
 (*a*) select a mean diametral interference of $0.025+0.0075\sqrt{D}$ mm ($0.001+0.0015\sqrt{D}$ inches);
 (*b*) check that the stacking of tolerances of housing and bearing (see Fig. 6.4) keeps the interference between about half and twice the mean interference;
 (*c*) allow adequate chamfer on the housing (see Table 6.2 for details);
 (*d*) estimate the bore closure on fitting using the F factor from Fig. 6.8 and the extremes of interference from (*b*) above. Select a fitted bore size which is not smaller than 'the unfitted bore size minus the bore closure'. Check at the extremes of the tolerances of interference and bore diameter (see Fig. 6.4);
 (*e*) estimate the diameter of the fitting mandrel shown in Fig. 6.9, by adding to the desired bore size, a spring allowance which varies with the rigidity of the porous metal (Fig. 6.2) and the housing, as given in Table 6.3;
 (*f*) check that the differential thermal expansion between the housing and bearing over the expected temperature range does not cause a loss of interference in service (use the expansion coefficient of a non-porous metal of the same composition for all porosities);
 (*g*) for non-rigid housings, non-standard bearings or where the above guidance does not give a viable design, consult the manufacturers.

4 Never use hammer blows, as the impact force will generally exceed the limiting load capacity given in Fig. 6.2. A steady squeezing action is recommended.

5 Select a mean running clearance from Fig. 6.10, according to shaft diameter and speed. Check that the stacking of tolerances and the differential expansion give an acceptable clearance at the extremes of the design. Note that excessive clearance may give noisy running with an out-of-balance load, and that insufficient clearance gives high torque and temperature.

6 Specify a shaft-surface roughness of about 0.8 μm (32 micro-inches) cla, remembering that larger diameters can tolerate a greater roughness, and that a smaller roughness gives better performance and less running-in debris. In critical applications (Fig. 6.1), iron based bearings using steel shafts need a smoother shaft finish than bronze based bearings.

Table 6.2 Minimum housing chamfers at 45°

Housing diameter, D	*Length of chamfer*
Up to 13 mm ($\frac{1}{2}$ in)	0.8 mm ($\frac{1}{32}$ in)
13 mm to 25 mm ($\frac{1}{2}$ in to 1 in)	1.2 mm ($\frac{3}{64}$ in)
25 mm to 51 mm (1 in to 2 in)	1.6 mm ($\frac{1}{16}$ in)
51 mm to 102 mm (2 in to 4 in)	2.4 mm ($\frac{3}{32}$ in)
Over 102 mm (4 in)	3.2 mm ($\frac{1}{8}$ in)

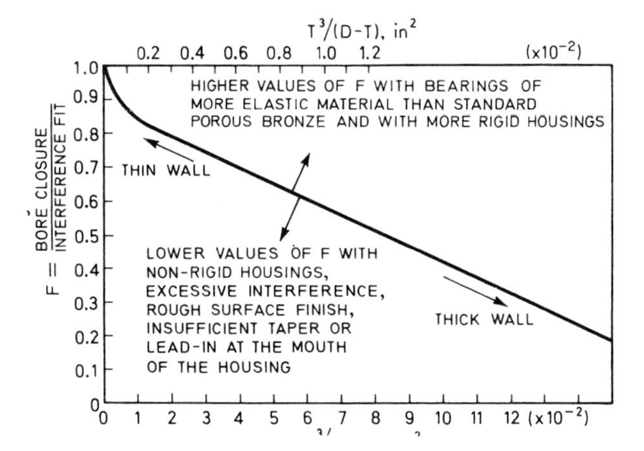

Fig. 6.8. Ratio of interference to bore closure, F, as a function of the wall thickness, W, and outside diameter, D, of the porous metal bearing

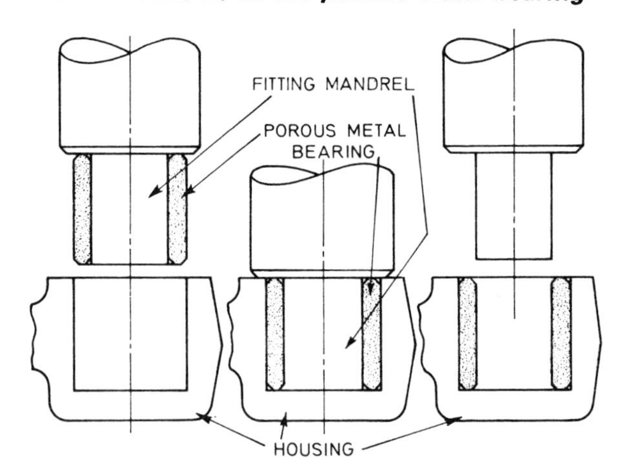

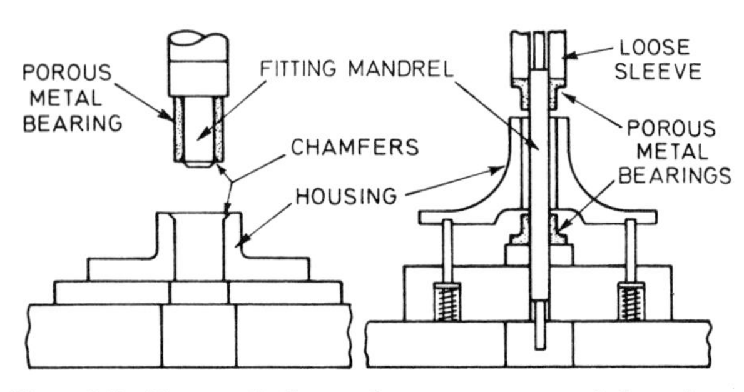

Fig. 6.9. Force fitting of porous metal bearings using a fitting mandrel to control the fitted bore diameter and to achieve alignment of a pair of bearings

Table 6.3 Spring allowance on force fitting mandrel

Static load capacity (Fig. 6.2)		*Spring allowance*
Up to 20 MN/m^2	(3 000 p.s.i.)	0.01%
20 to 40 MN/m^2	(6 000 p.s.i.)	0.02%
40 to 80 MN/m^2	(12 000 p.s.i.)	0.04%
Over 80 MN/m^2	(12 000 p.s.i.)	0.06%

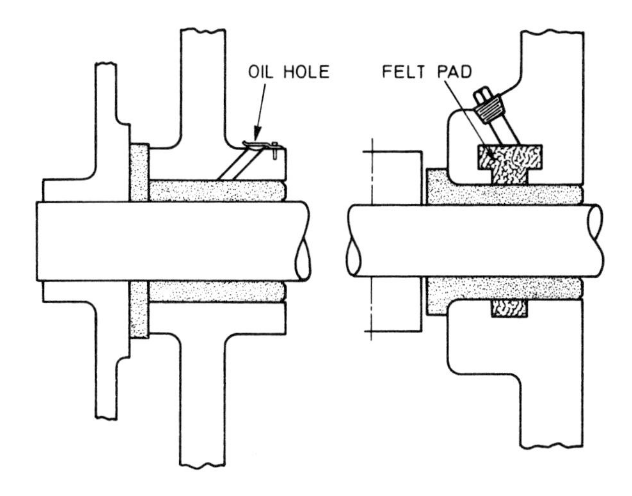

THRUST LOADS ARE CARRIED BY EITHER A SEPARATE THRUST WASHER OR THE USE OF A FLANGED CYLINDRICAL BEARING

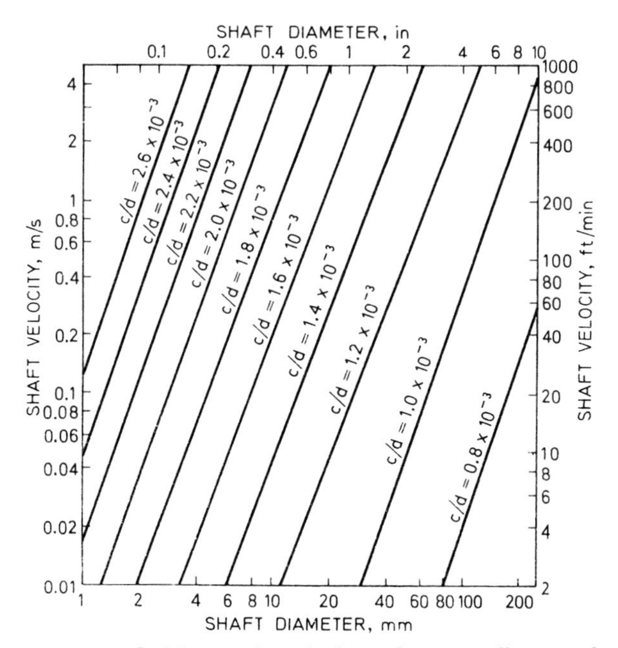

Fig. 6.10. Guide to the choice of mean diametral clearance expressed as the clearance ratio c/d

Bore correction

If after fitting, it is necessary to increase the bore diameter, this should not be done by a cutting tool, which will smear the surface pores and reduce the free flow of oil from the porosity to the working surface. Suitable burnishing tools for increasing the bore diameter, which do not close the pores, are given in Fig. 6.11.

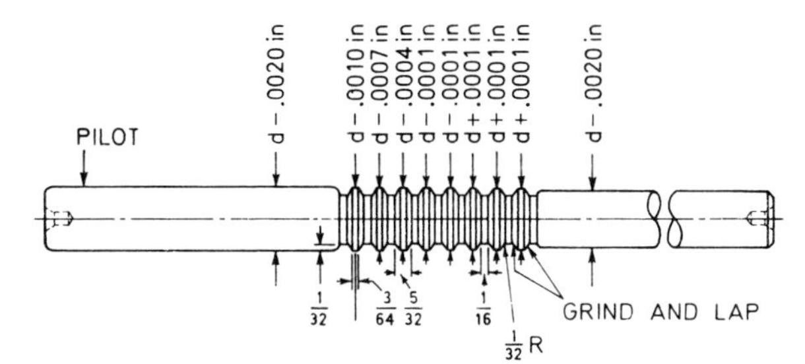

d = FINISHED DIAMETER OF BEARING

BUTTON TYPE DRIFT

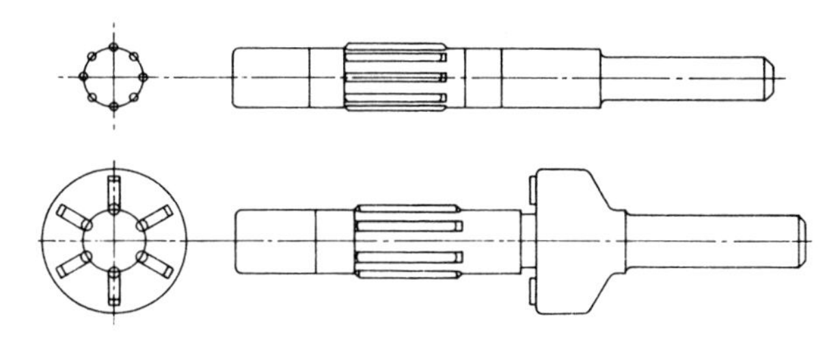

ROLLER TYPE BURNISHING TOOL

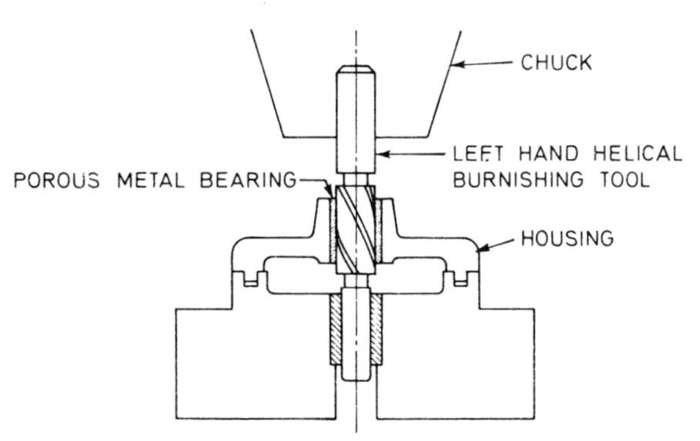

Fig. 6.11. Tools for increasing the bore diameter and aligning a fitted assembly

GENERAL NOTE

The previous sections on design, materials and lubrication give general guidance applicable to normal operating conditions with standard materials, and therefore cover more than half of the porous metal bearings in service. There are, however, many exceptions to these general rules, and for this reason the manufacturers should be consulted before finalising an important design.

A7 Grease, wick and drip fed journal bearings

Journal bearings lubricated with grease, or supplied with oil by a wick or drip feed, do not receive sufficient lubricant to produce a full load carrying film. They therefore operate with a starved film as shown in the diagram:

As a result of this film starvation, these bearings operate at low film thicknesses.

To make an estimate of their performance it is, therefore, necessary to take particular account of the bearing materials and the shaft and bearing surface finishes as well as the feed rate from the lubricant feed system.

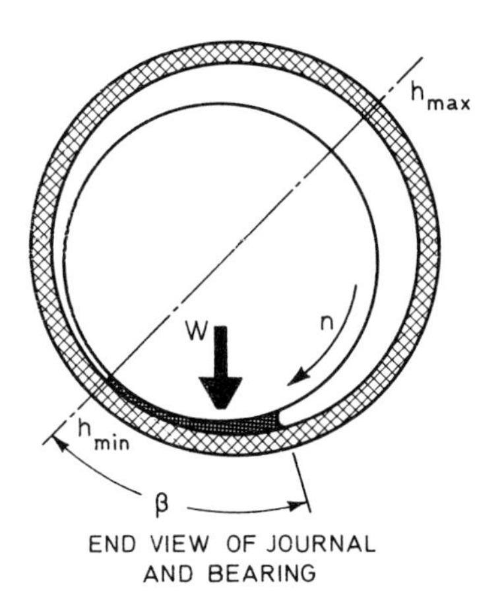

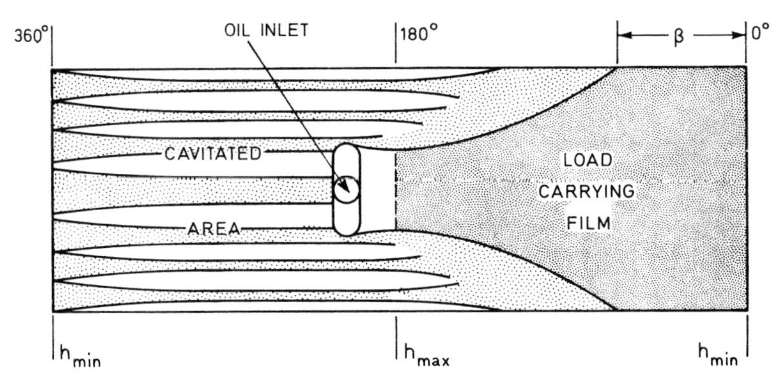

STARVED FILM

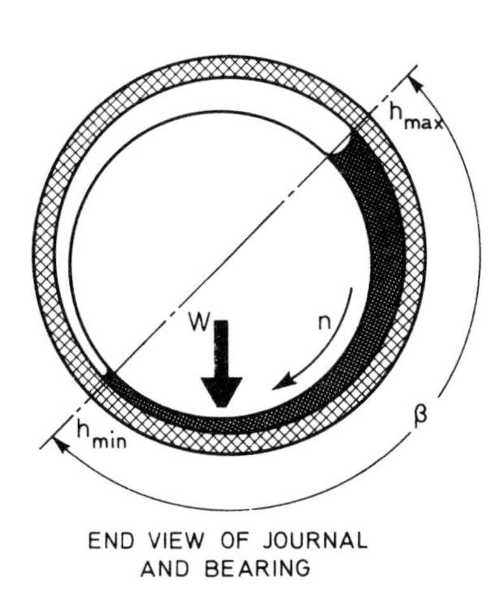

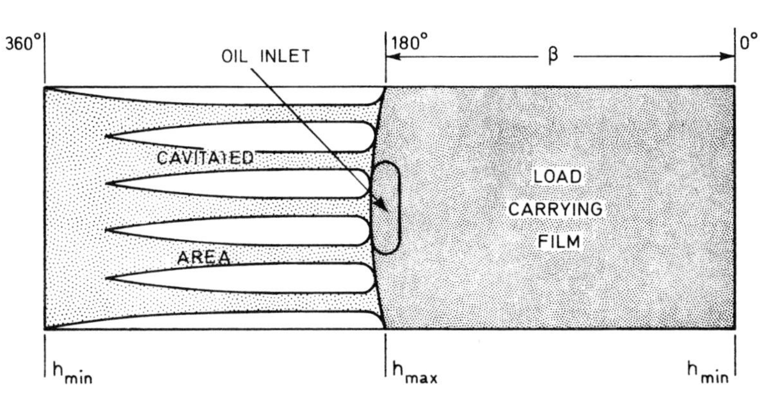

FULL FILM
(AS OBTAINED WITH A PRESSURE OIL FEED)

AN APPROXIMATE METHOD FOR THE DESIGN OF STARVED FILM BEARINGS

Step 1

Check the suitability of a starved film bearing for the application using Fig. 7.1.

Note: in the shaded areas attention should be paid to surface finish, careful running-in, good alignment and the correct choice of materials for bearing and journal.

Bearing width to diameter ratio, b/d, should be between 0.7 and 1.3.

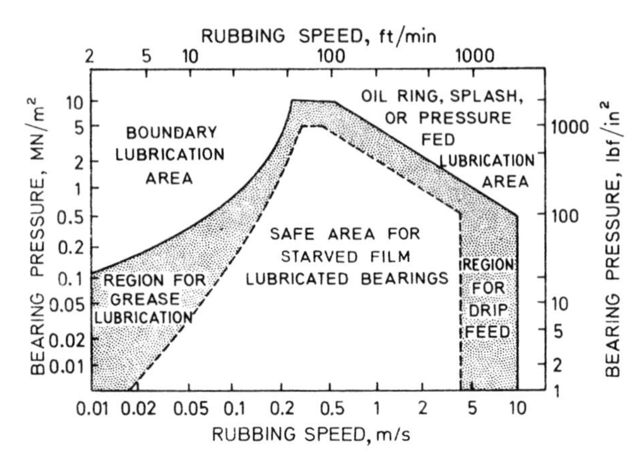

Fig. 7.1. A guide to the suitability of a 'starved' bearing

Step 2

Select a suitable clearance C_d, knowing the shaft diameter (Fig. 7.2) and the manufacturing accuracy.

Note: the lowest line in Fig. 7.2 gives clearance suitable only for bearings with excellent alignment and manufacturing precision.
For less accurate bearings, the diametral clearance should be increased to a value in the area above the lowest line by an amount, $= Mb +$ the sum of out-of-roundness and taper on the bearing and journal.

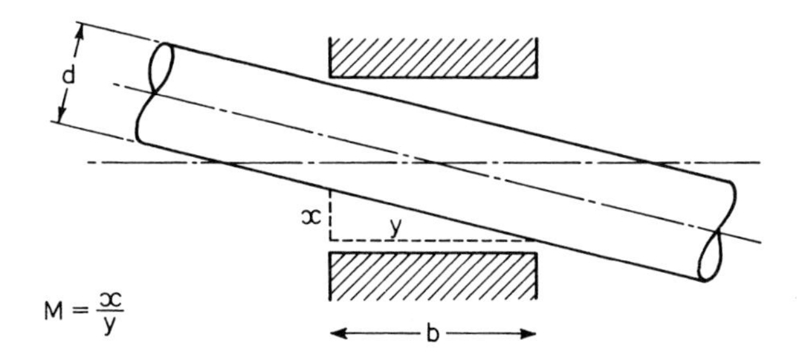

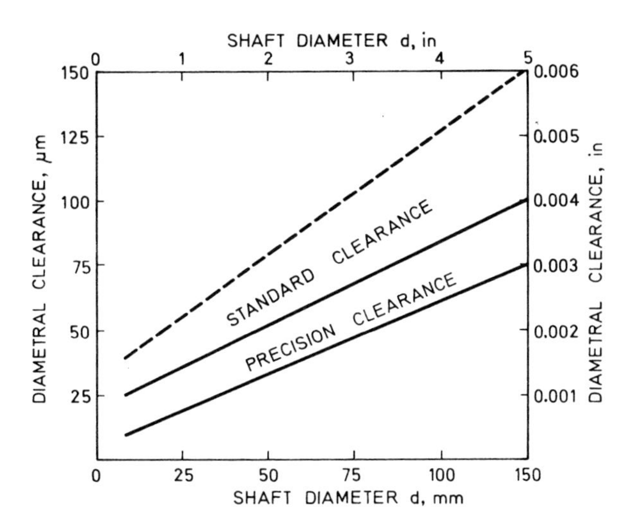

Fig. 7.2. Guidance on choice of clearance

Step 3

Choose the minimum permissible oil film thickness h_{min} corresponding to the materials, the surface roughnesses and amount of misalignment of the bearing and journal.
Minimum oil film thickness

$$h_{min} = k_m\,(R_p\,\text{journal} + R_p\,\text{bearing}) + \frac{Mb}{2}$$

Table 7.1 Material factor, k_m

Bearing lining material	k_m
Phosphor bronze	1
Leaded bronze	0.8
Tin aluminium	0.8
White metal (Babbitt)	0.5
Thermoplastic (bearing grade)	0.6
Thermosetting plastic	0.7

Note: journal material hardness should be five times bearing hardness.

Table 7.2 Surface finish, predominant peak height, R_p

Surface type	*Micro-inch* cla	μm RMS	*Class*	R_p μm	R_p μin
Turned or rough ground	100	2.8	6	12	480
Ground or fine bored	20	0.6	8	3	120
Fine ground	7	0.19	10	0.8	32
Lapped or polished	1.5	0.04	12	0.2	8

Step 4

Assume a lubricant running-temperature of about 50 to 60°C above ambient and choose a type and grade of lubricant with references to Tables 7.3 and 7.4. Note the viscosity corresponding to this temperature from Fig. 7.3.

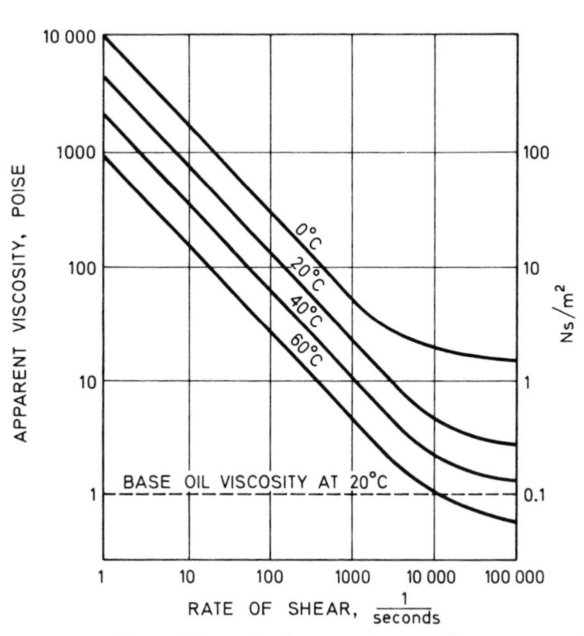

Fig. 7.3. The effect of shear rate on the apparent viscosity of a typical No. 2 NLGI consistency grease

Table 7.3 Guidance on the choice of lubricant grade

Lubricant running temperature	Grease		Oil	
	Type	*Grade (NLGI No.)*	*Types*	*Viscosity grade* ISO 3448
Up to 60°C	Calcium based 'cup grease'		Mineral oil with fatty additives	
< 0.5 m/s		1 or 2		68
> 0.5 m/s		0		32
60°C to 130°C	Lithium hydroxystearate based grease with high V.I. mineral oil and antioxidant additives		Good quality high V.I. crankcase or hydraulic oil with antioxidant additives (fatty oils for drip-fed bearings)	
< 0.5 m/s		3		150
> 0.5 m/s		3		68
Above 130°C	Clay based grease with silicone oil	3	Best quality fully inhibited mineral oil, synthetic oil designed for high temperatures, halogenated silicone oil	150

Notes: for short term use and total loss systems a lower category of lubricant may be adequate.
A lubricant should be chosen which contains fatty additives, i.e. with good 'oiliness' or 'lubricity'.
The use of solid lubricant additives such as molybdenum disulphide and graphite can help (but not where lubrication by wick is used).

Table 7.4 Factors to consider in the choice of grease as a lubricant

Feature	*Advantage*	*Disadvantage*	*Practical effect*
h_{min} Minimum film thickness	Fluid film lubrication maintained at lower W' values		Grease lubrication is better for high load, low-speed applications
C_d/d Clearance diameter ratio	Larger clearances are permissible	Overheating and feeding difficulties arise with small clearances	Ratios 2 to 3 times larger than those for oil lubricated bearings are common
Lubricant supply	Much smaller flow needed to maintain a lubricant film. Rheodynamic flow characteristics lead to small end-loss and good recirculation of lubricant	Little cooling effect of lubricant, even at high flow rates	Flow requirement 10 to 100 times less than with oil. Long period without lubricant flow possible with suitable design
μ Friction coefficient			
(*a*) at start-up	(*a*) Lubricant film persists under load with no rotation		(*a*) Lower start-up torque
(*b*) running		(*b*) Higher effective viscosity leads to higher torque	(*b*) Higher running temperatures
W' Bearing load capacity number		Calculated on the basis of an 'effective viscosity' value dependent on the shear rate and amount of working. Gives an approx. guide to performance only	Prediction of design performance parameters poor

Step 5

With reference to the formulae on Fig. 7.4 calculate W' from the dimensions and operating conditions of the bearing, using the viscosity just obtained. Obtain the appropriate misalignment factor M_w from Table 7.5. Calculate W' (misaligned) by multiplying by M_w. Use this value in further calculations involving W'.

Notes: M_w is the available percentage of the load capacity W' of a correctly aligned bearing.
Misalignment may occur on assembly or may result from shaft deflection under load.

Table 7.5 Values of misalignment factor M_w at two ratios of minimum oil film thickness/diametral clearance

$M \times b/c_d$	$h_{min}/C_d = 0.1$	$h_{min}/C_d = 0.01$
0	100	100
0.05	65	33
0.25	25	7
0.50	12	3
0.75	8	1

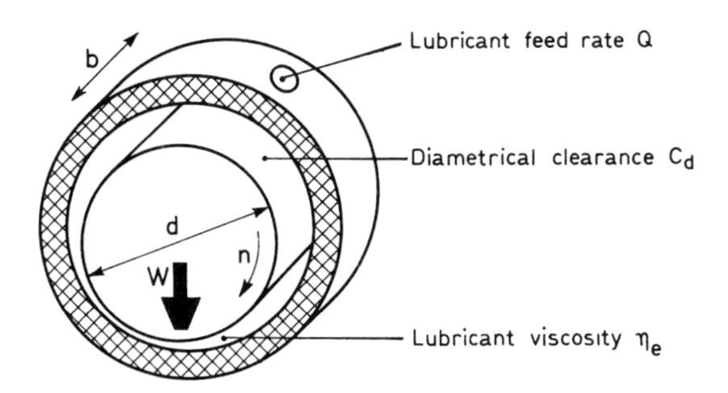

$Q' =$	$\dfrac{176.6Q}{bd\,nC_d}$	$\dfrac{2Q}{\pi\,bd\,nC_d}$
$W' =$	$\dfrac{4.137 \times 10^8 W}{\eta_e\,nbd}\left(\dfrac{Cd}{d}\right)^2$	$\dfrac{W}{\eta_e\,nbd}\left(\dfrac{Cd}{d}\right)^2$
UNITS	gall/min inch lbf rev/min cP	m^3/s m N rev/s Ns/m^2

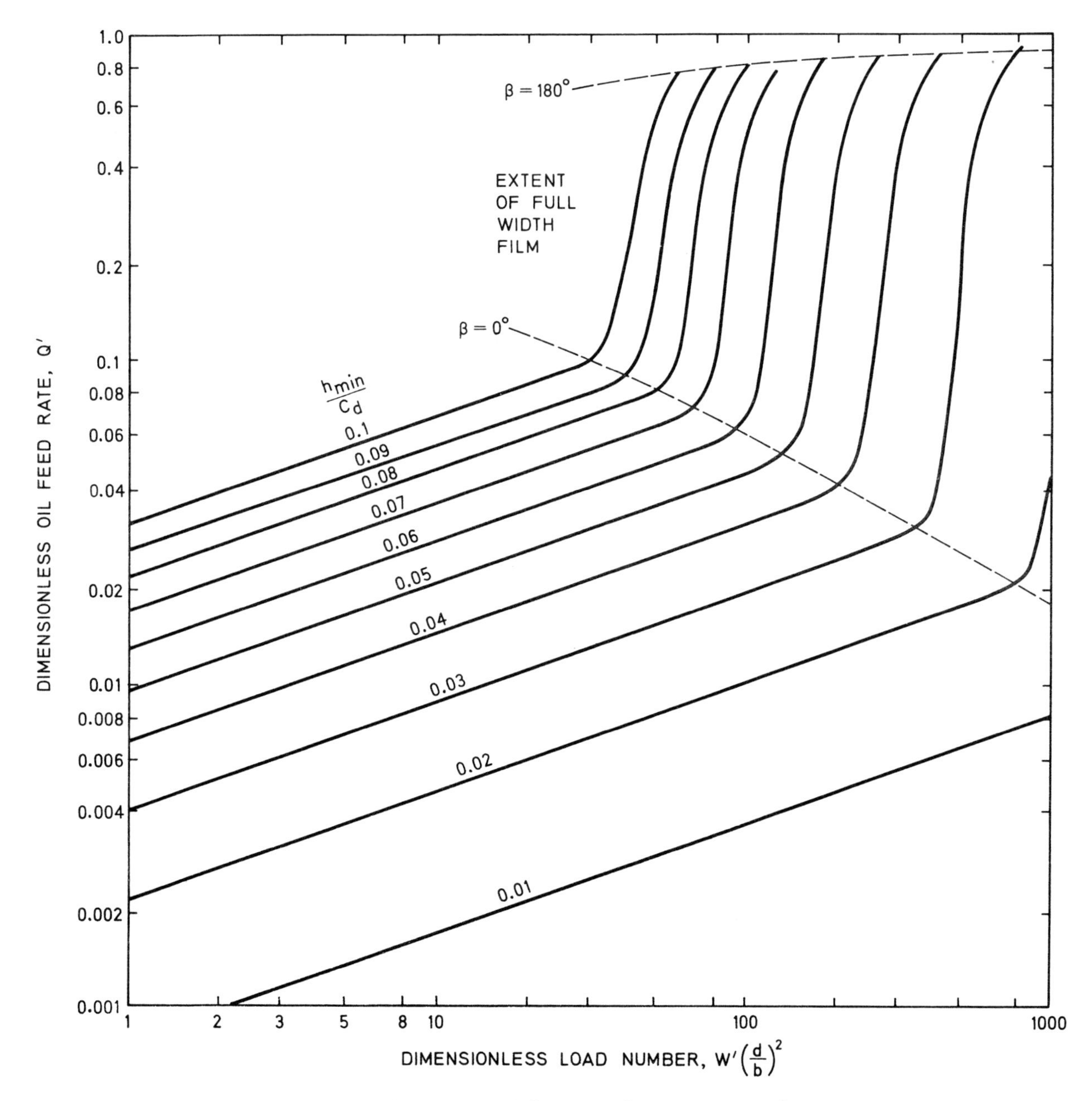

Fig. 7.4. Minimum oil flow requirements to maintain fluid film conditions, with continuous rotation, and load steady in magnitude and direction (*courtesy:* Glacier Metal Co Ltd)

Step 6

From Fig. 7.5 read the value of F' corresponding to this W'. Calculate the coefficient of fiction $\mu = F' C_d/d$.

Calculate the power loss H in watts

$H =$	$1.9 \times 10^{-3}\, \pi \mu W d n$	$\pi \mu W d n$
Units	in lbf rev/min	m N rev/s

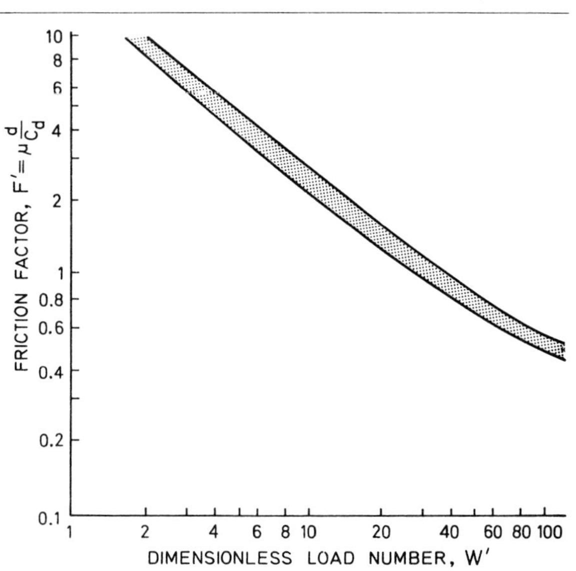

Fig. 7.5. Relationship between friction-factor and dimensionless load number (W' is defined on Fig. 7.4)

Step 7

It is assumed that all the power loss heat is dissipated from the housing surface. From Fig. 7.6 find the value of housing surface temperature above ambient which corresponds to the oil film temperature assumed in step 4. Read off the corresponding heat dissipation and hence derive the housing surface area using the power loss found in step 6. If this area is too large, a higher oil film temperature must be assumed and steps 4–7 repeated. It may be necessary to choose a different grade of lubricant to limit the oil film temperature.

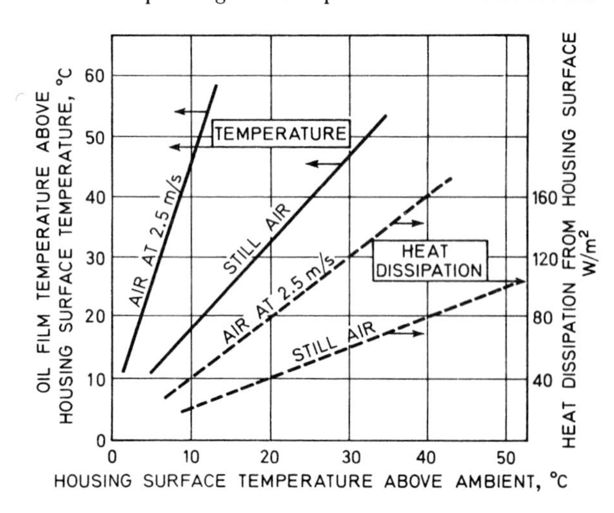

Fig. 7.6. A guide to the heat balance of the bearing housing

Step 8

Using Fig. 7.4 read off Q' and calculate Q the minimum oil flow through the film corresponding to the dimensionless load number, $W' \times \left(\frac{d}{b}\right)^2$ and the value for h_{min}/C_d.

A large proportion of this flow is recirculated around the bearing and in each meniscus at the ends of the bearing.

An estimate of the required additional oil feed rate from the feed arrangement is given by $Q/10$ and this value may be used in step 9.

For grease lubrication calculate the grease supply rate per hour required Q_g from

$$Q_g = k_g \times C_d \times \pi \times d \times b$$

Table 7.6 Values of k_g for grease lubrication at various rotational speeds

Journal speed rev/min	k_g
Up to 100	0.1
250	0.2
500	0.4
1000	1.0

Under severe operating conditions such as caused by running at elevated temperatures, where there is vibration, where loads fluctuate or where the grease has to act as a seal against the ingress of dirt from the environment, supply rates of up to ten times the derived Q_g value are used.

Step 9

Select a type of lubricant supply to give the required rated lubricant feed using Tables 7.7 and 7.8 and Figs. 7.7, 7.8 and 7.9.

Where the rate of lubricant supply to the bearing is known, Fig. 7.4 will give the load number corresponding to a particular h_{min}/C_d ratio. The suggested design procedure stages should then be worked through, as appropriate.

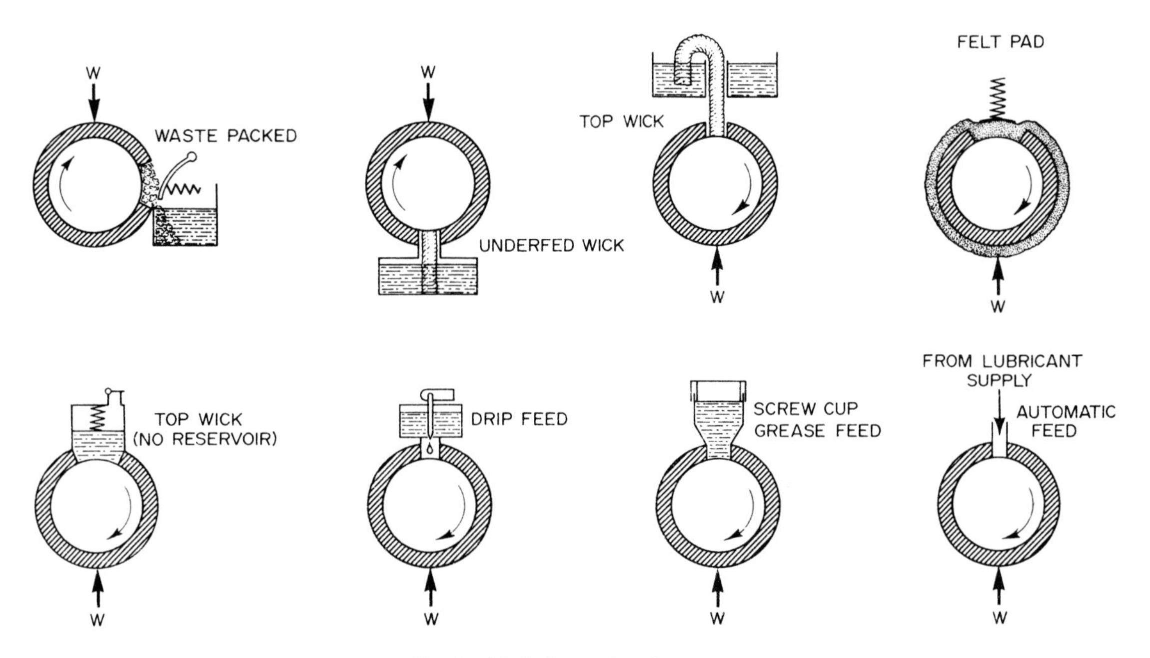

Fig. 7.7 Typical lubricant feed arrangements

Table 7.7 Guidance on the choice of lubricant feed system

Type	*Lubricant supply method*	*Cost*	*Toleration of dirty environment*	*Maintenance needs rating*	*Lubricant flow characteristics*
Wool waste lubricated	Capillary	Expensive housing design	Fair. Waste acts as an oil filter	Good. Infrequent refilling of reservoir	Very limited rate controlled by height of oil in reservoir. Recirculation possible. Varies automatically with shaft rubbing speed. Stops when rotation ceases
Wick lubricated (with reservoir)	Capillary and siphonic	Moderate	Fair. Wick acts as an oil filter	Good. Infrequent refilling of oil reservoir	Limited rate and control (ref. Fig. 7.8). Recirculation possible with underfed wick type. Varies slightly with shaft rubbing speed. Underfed type stops when rotation ceases (not siphonic)
Wick or pad lubricated (no reservoir)	Capillary	Cheap	Fair. Wick act as an oil filter	Fair. Reimpregnation needed occasionally	Very limited rate, decreasing with use. Varies slightly with shaft rubbing speed. Stops when rotation ceases. Recirculation possible
Grease lubricated	Hand-operated grease gun or screw cup	Very cheap	Good. Grease acts as a seal	Poor. Regular regreasing needed	Negligible flow, slumping only. Rheodynamic, i.e. no flow at low shear stress hence little end flow loss from bearing
Drip-feed lubricated	Gravity, through a controlled orifice	Cheap for simple installations	Poor	Poor. Regular refilling of reservoir needed	Variable supply rate. Constant flow at any setting. Total loss, i.e. no recirculation. Flow independent of rotation
Automatic feed (oil or grease)	Pump-applied pressure	Expensive ancillary equipment needed	Fair	Good. Supply system needs occasional attention	Wide range of flow rate. Can vary Automatically. Total loss. Can stop or start independently of rotation

Table 7.8 The comparative performance of various wick and packing materials

Type	*Felt, high density (sg 3.4)*	*Felt, low density (sg 1.8 to 2.8)*	*Gilled thread*	*Wool waste*	*Cotton lamp wick*
Height of oil lift (dependent on wetting and size of capillary channels)	Very good	Fair	Good	Poor	Fair
Rate of flow	Very good	Fair	Good	Poor	Fair
Oil capacity	Low	High	Low	Moderate (3 times weight of waste)	Fair
Suitability for use as packing	Poor (tendency to glaze)	Poor (tendency to glaze)	Poor	Good (superior elasticity)	Poor

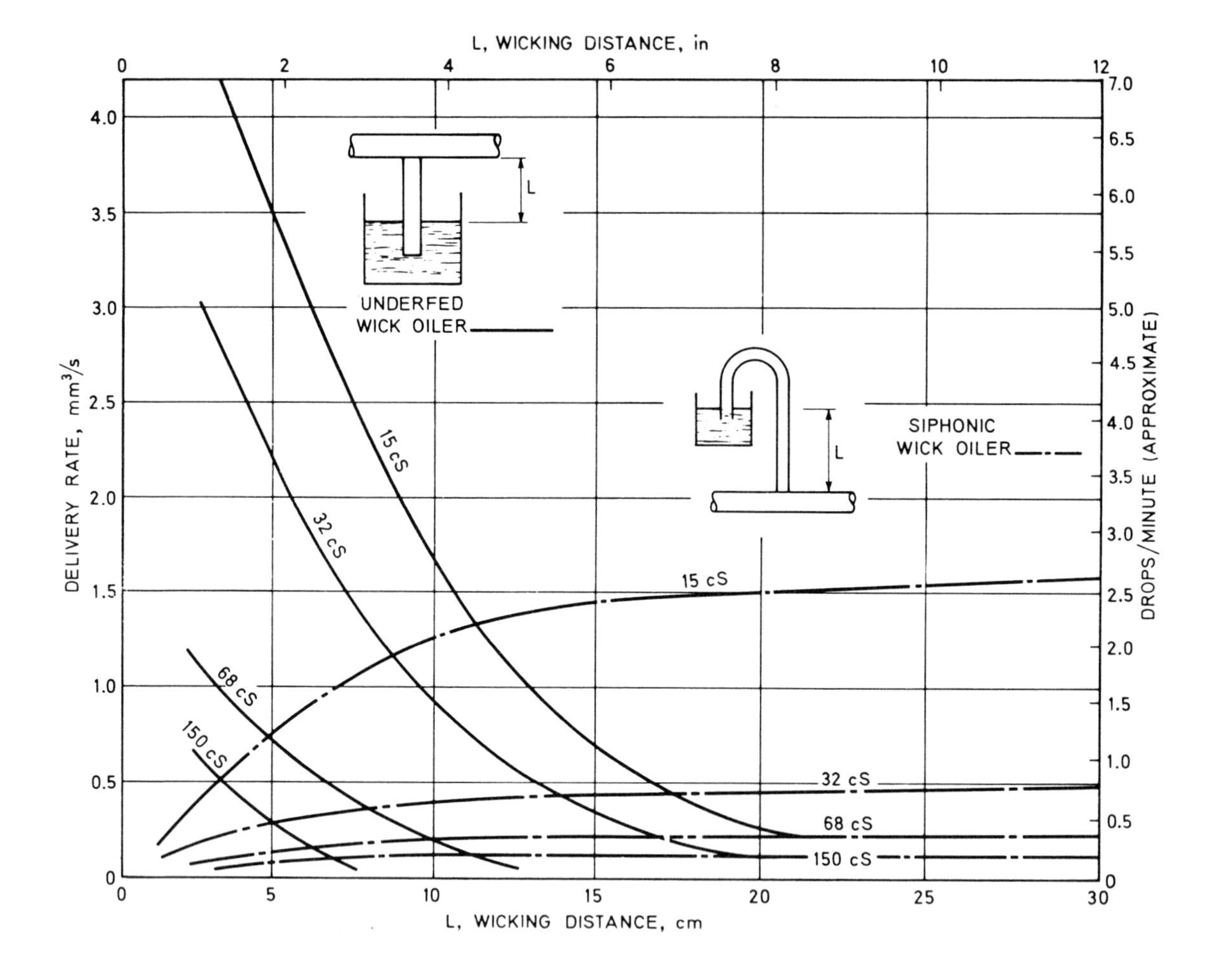

Fig. 7.8. Oil delivery rates for SAE F1 felt wicks, density 3.4g/cc, cross-sectional area 0.65cm² (0.1 in²), temperature 21°C, viscosity at 40°C (ISO 3448) (Data from the American Felt Co.)

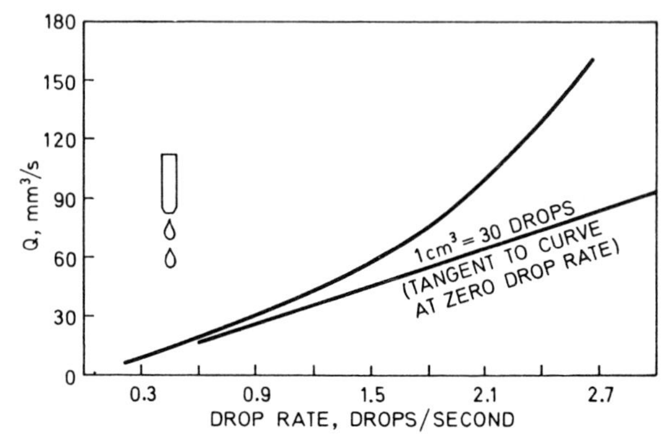

Fig. 7.9. Effect of drop rate on oil drop size, temperature 27°C. Oil viscosity and lubricator tip shape have little effect on drop size over the normal working ranges

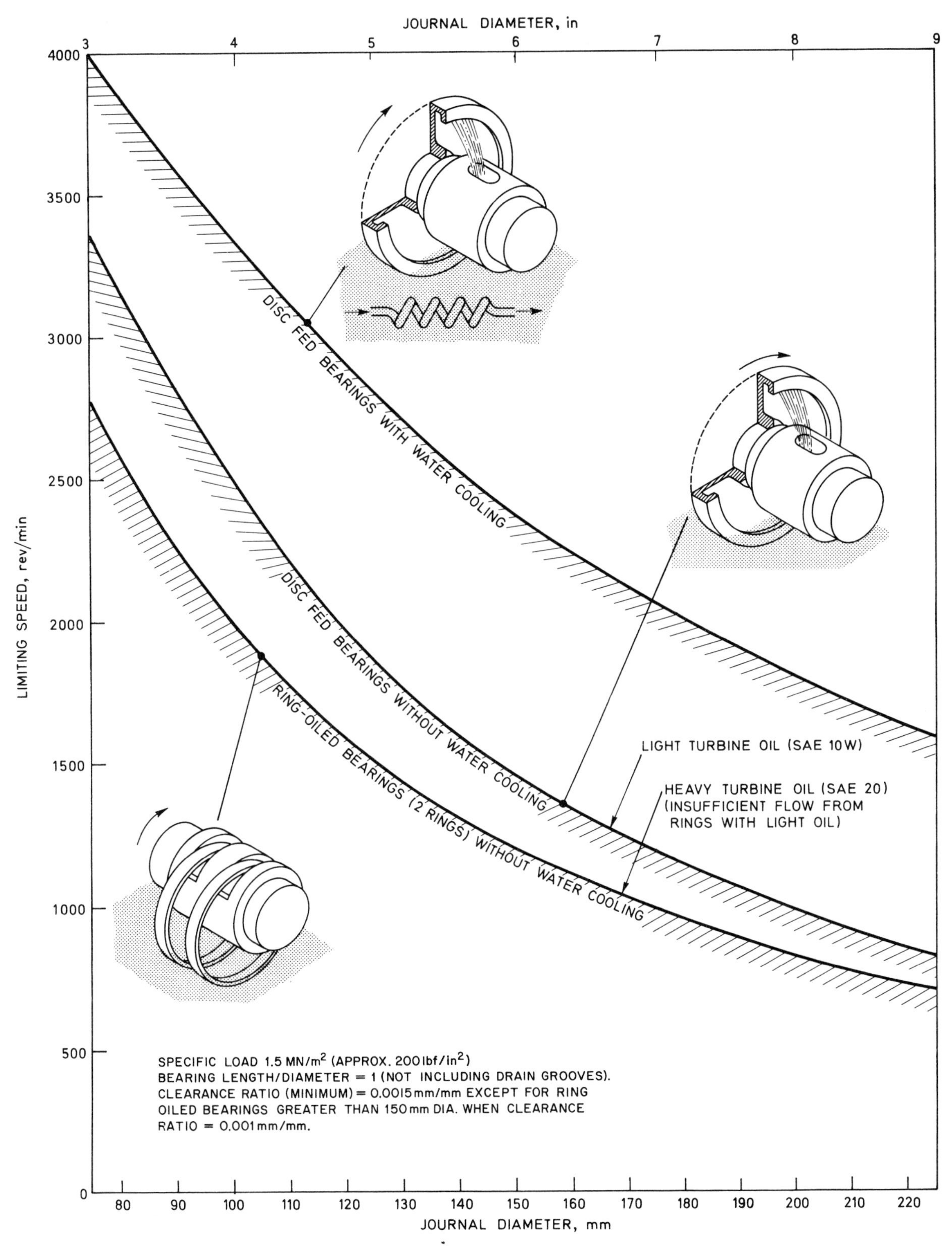

Fig. 8.1. General guide to limiting speed for ring and disc lubricated bearings

Disc fed—water cooled:
The above curves give some idea of what can be achieved, assuming there is sufficient oil to meet bearing requirement. It is advisable to work well below these limits. Typical maximum operating speeds used in practice are 75% of the above figures.

Ring and disc fed—without water cooling:
For more detailed information see Fig. 8.2. The limiting speed will be reduced for assemblies incorporating thrust location — see Fig. 8.5.

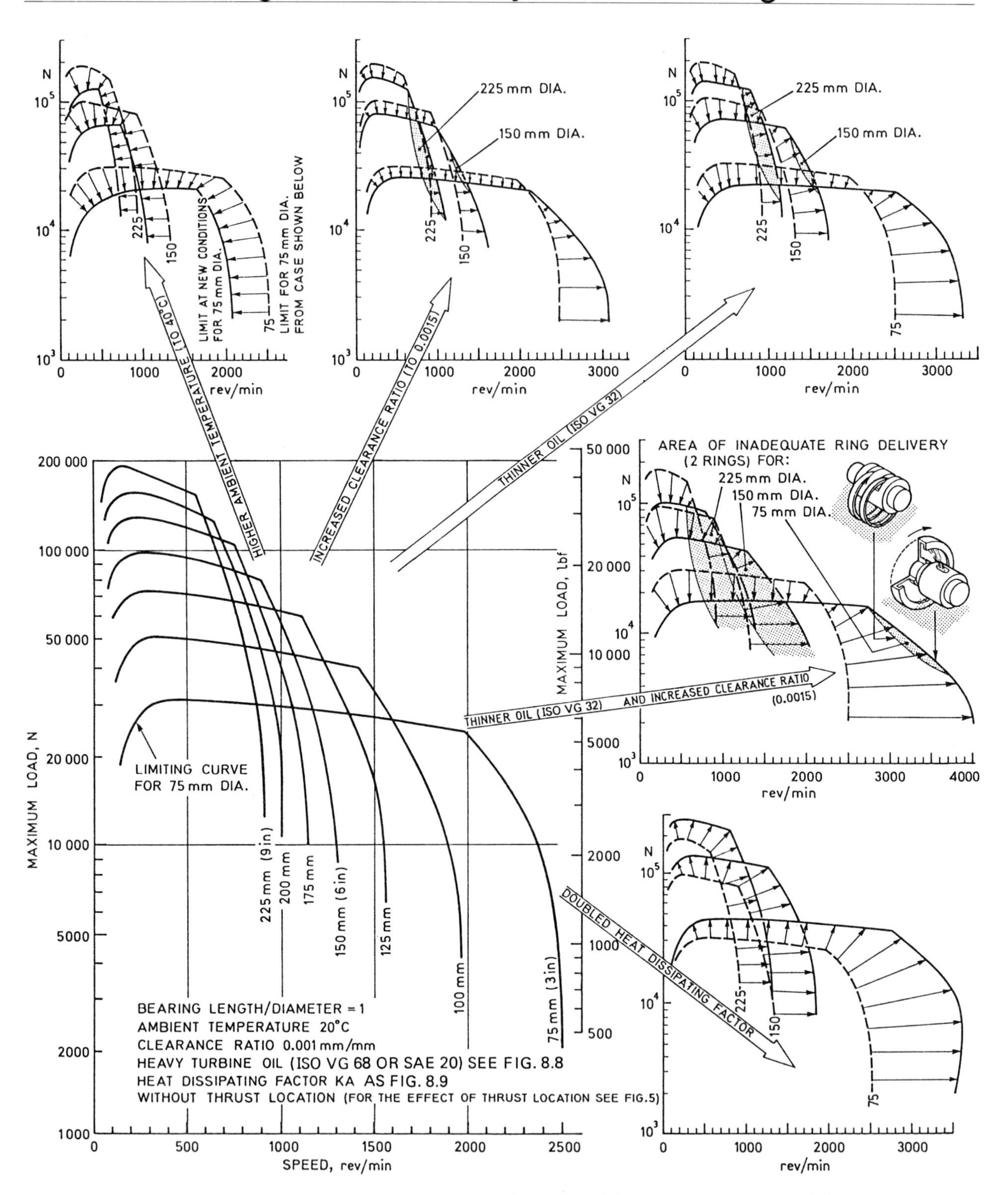

Fig. 8.2. Load capacity guidance for self-contained journal bearing assemblies

Disc fed:
For any diameter work below appropriate limiting curve* (oil film thickness and temperature limits).

Ring oiled (2 rings):
For any diameter work below appropriate limiting curve* and avoid shaded areas (inadequate supply of lubricant from rings).

In these areas disc fed bearings should be used instead.

* These limits assume that the bearing is well aligned and adequately sealed against the ingress of dirt. Unless good alignment is achieved the load capacity will be severely reduced. In practice, the load is often restricted to 1.5 to 2 MN/m^2 (approx. 200 to 300 lbf/in^2) to allow for unintentional misalignment, starting and stopping under load and other adverse conditions.

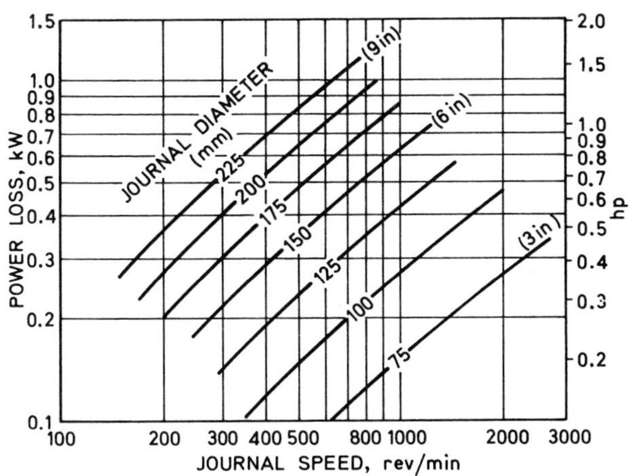

Fig. 8.3. Guide for power loss in self-contained bearings

Specific Load = 1.5 MN/m²
Bearing length/diameter = 1
Ambient temperature = 20°C (for ambient temperature 40°C take 80% of losses shown)
Clearance ratio = 0.001 mm/mm (for clearance ratio of 0.0015 mm/mm take 95% of losses shown)
Heavy turbine oil (ISO VG68 or SAE 20) (for light turbine oil take 85% of losses shown)
Heat dissipating factor as Fig. 8.9 (for effect of heat dissipating factor see Fig. 8.4)
The power loss will be higher for assemblies incorporating thrust location — see Fig. 8.6

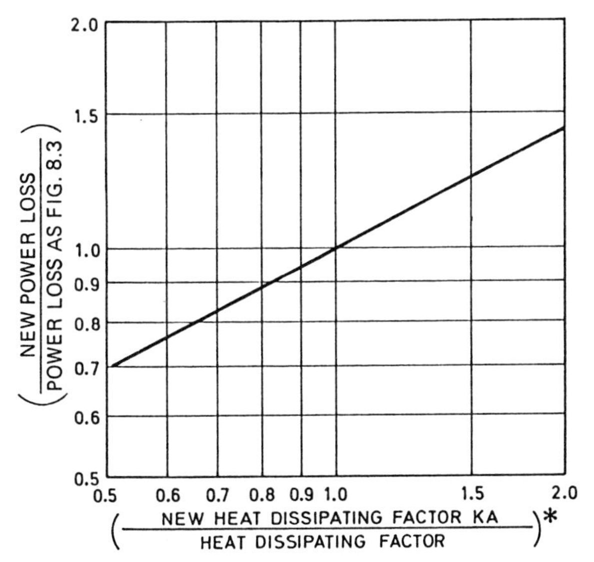

Fig. 8.4. Showing how power loss in self-contained bearings (without thrust) is affected by heat dissipating factor KA

The heat dissipating casing area A and/or the heat transfer coefficient K may both differ from the values used to derive the load capacity and power loss design charts.

Figure 8.4 shows how change in KA affects power loss.

*The ratio

$$\frac{\text{New heat dissipating factor } KA}{\text{Heat dissipating factor}} \text{ in Fig. 8.4}$$

is given by

$$\frac{K \text{ for actual air velocity (Fig. 8.9}(b))}{18 \text{ (for still air)}} \times \frac{\text{actual casing area}}{\text{casing area (Fig. 8.9}(a))}$$

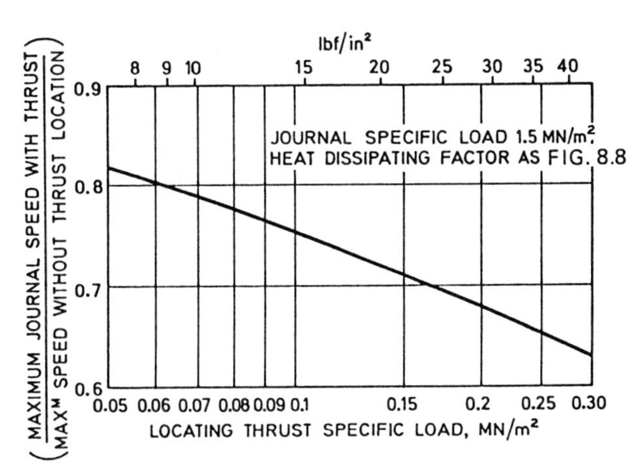

Fig. 8.5. Reduced limiting speed where assembly includes thrust location

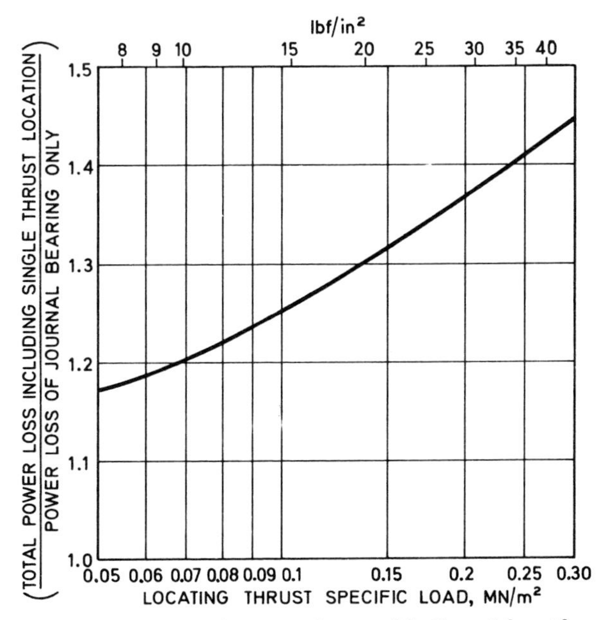

Fig. 8.6. Increased power loss with thrust location (single thrust plain washer — for typical dimensions see Fig. 8.7)

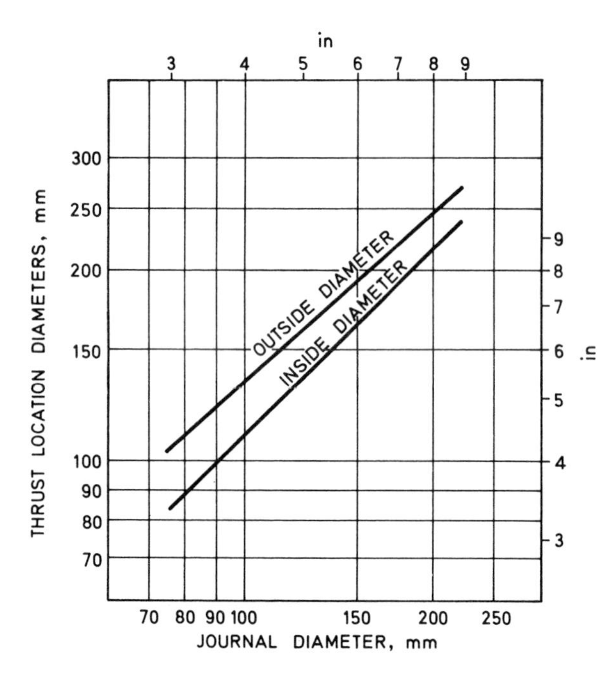

Fig. 8.7. Typical dimensions of plain thrust annulus as used in Figs. 8.5 and 8.6

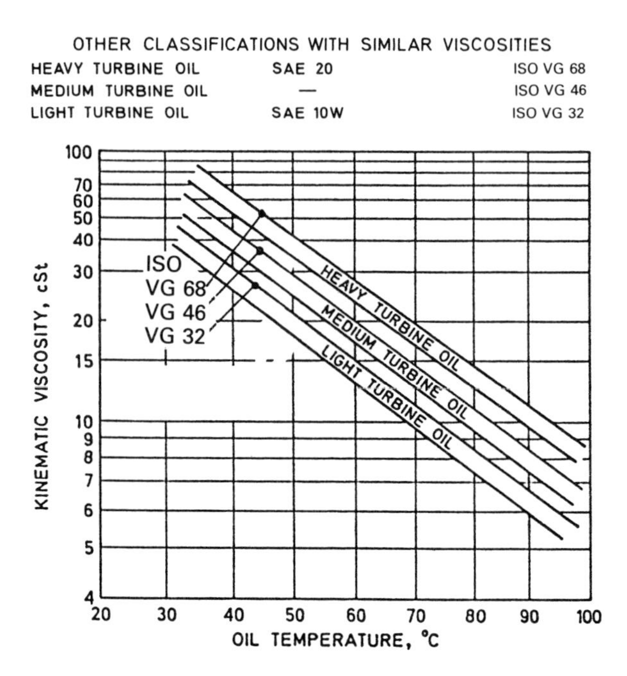

Fig 8.8. Turbine and other oil viscosity classifications

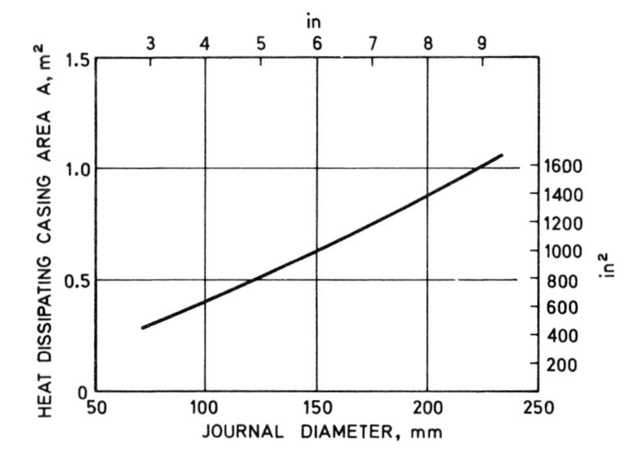

Fig. 8.9(a). Typical heat dissipating area of casing as used in the design guidance charts

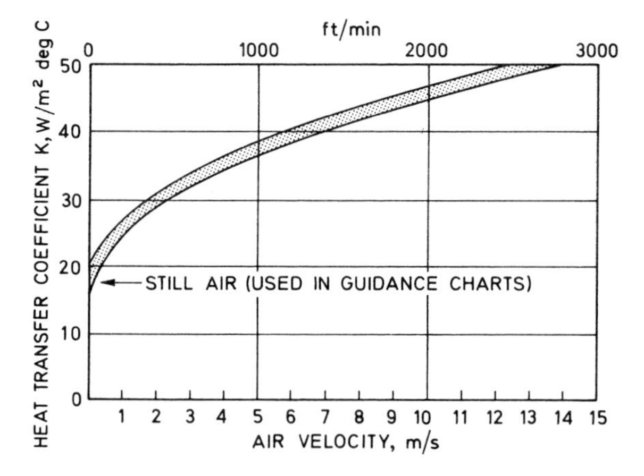

Fig. 8.9(b). Guidance on heat transfer coefficient K, depending on air velocity

The heat dissipating factor KA used in the design guidance charts was based on the area diameter relationship in Fig. 8.9(a) and a heat transfer coefficient for still air of 18 W/m^2 degC as shown in Fig. 8.9(b). The effect of different dissipating areas or air velocity over the casing may be judged:

for load capacity Fig. 8.2 (doubled heat dissipating factor KA)

for power loss Fig. 8.4.

HYDRODYNAMIC BEARINGS

Principle of operation

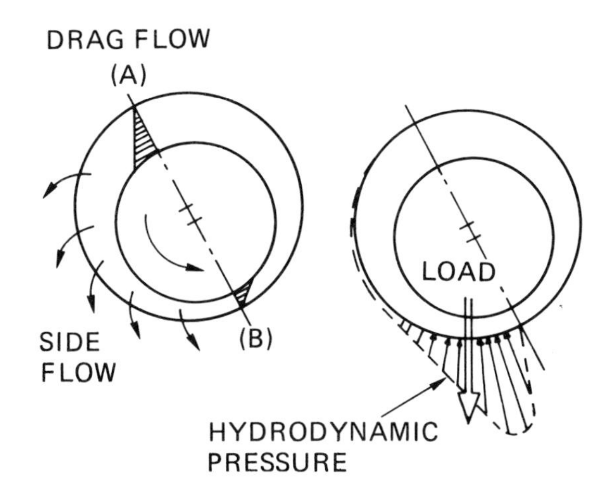

(1) On 'start up' the journal centre moves forming a converging oil film in the loaded region
(2) Oil is dragged into the converging film by the motion of the journal (see velocity triangle at 'A'). Similarly, a smaller amount passes through the minimum film (position 'B'*). As the oil is incompressible a hydrodynamic pressure is created causing the side flow
(3) The journal centre will find an equilibrium position such that this pressure supports the load

*The drag flow at these positions is modified to some extent by the hydrodynamic pressure.

Fig. 9.1 Working of hydrodynamic bearings – explained simply

The load capacity, for a given minimum film thickness increases with the drag flow and therefore increases with journal speed, bearing diameter and bearing length. It also increases with any resistance to side flow so will increase with operating viscosity. The bearing clearance may influence the load capacity either way. If the minimum film thickness is small and the bearing long then increasing the clearance could result in a decrease in load capacity, whereas an increase in clearance for a short bearing with a thick film could result in an increase in load capacity.

GUIDE TO PRELIMINARY DESIGN AND PERFORMANCE

The following guidance is intended to give a quick estimate of the bearing proportions and performance and of the required lubricant.

GUIDE TO GROOVING AND OIL FEED ARRANGEMENTS

An axial groove across the major portion of the bearing width in the unloaded sector of the bearing is a good supply method. A 2-axial groove arrangement, Fig. 9.2, with the grooves perpendicular to the loading direction is an arrangement commonly used in practice. The main design charts in this section relate to such a feed arrangement. A circumferential grooved bearing is used when the load direction varies considerably or rotates, but has a lower load capacity. However, with a 2-axial grooved bearing under small oil film thickness conditions, the load angle may be up to ±30° from the centre without significantly deviating the bearing. The lubricant is pumped into the feed grooves at pressures from 0.07 to 0.35 MN/m^2.

0.1 MN/m^2 is used in the following design charts together with a feed temperature of 50°C.

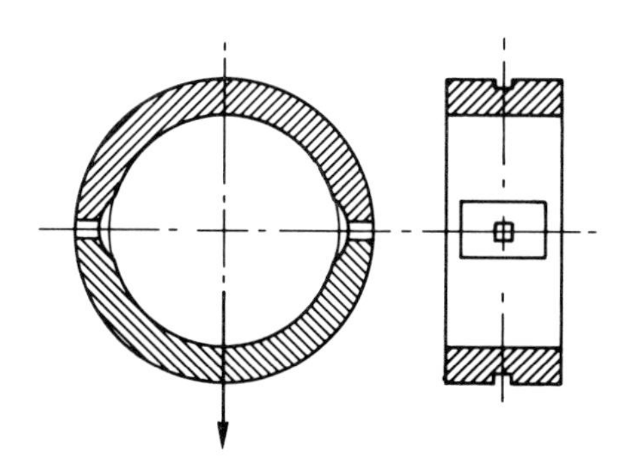

Fig. 9.2. Example of 2-axial groove bearing with load mid-way between grooves

BEARING DESIGN LIMITS

Figure 9.3 shows the concept of a *safe operating region* and Fig. 9.6 gives practical general guidance (also shows how the recommended operating region changes with different variables).

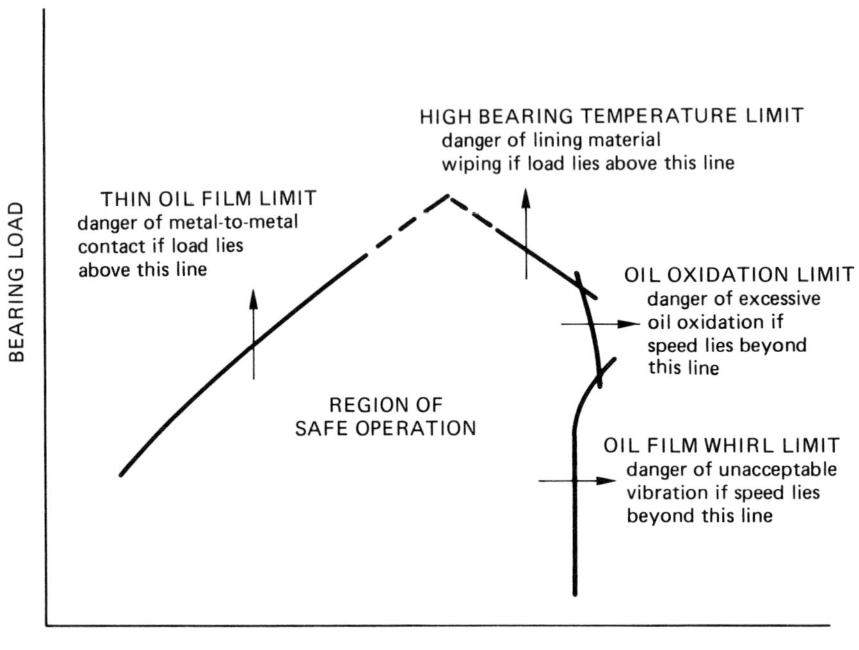

Fig. 9.3. Limits of safe operation for hydrodynamic journal bearings

Thin film limit – danger of metal to metal contact of the surfaces resulting in wear.
Background Safe limit taken as three times the peak-to-valley (R_{max}) value of surface finish on the journal. The factor of three, allowing for small unintentional misalignment and contamination of the oil is used in the general guide, Fig. 9.6. A factor of two may be satisfactory for very high standards of build and cleanliness. R_{max} depends on the trend in R_a values for different journal diameters as shown in Fig. 9.4 together with the associated machining process.

High bearing temperature limit – danger of bearing wiping at high speed conditions resulting in 'creep' or plastic flow of the material when subjected to hydrodynamic pressure. Narrow bearings operating at high speed are particularly prone to this limit.

Background The safe limit is well below the melting point of the bearing lining material. In the general guide, Fig. 9.6, whitemetal bearings are considered, with the bearing maximum temperature limited to 120°C. For higher temperatures other materials can be used: aluminium-tin (40% tin) up to 150°C and copper-lead up to 200°C. The former has the ability to withstand seizure conditions and dirt, and the latter is less tolerant so a thin soft overlay plate is recommended, togetjer with a hardened shaft and good filtration.

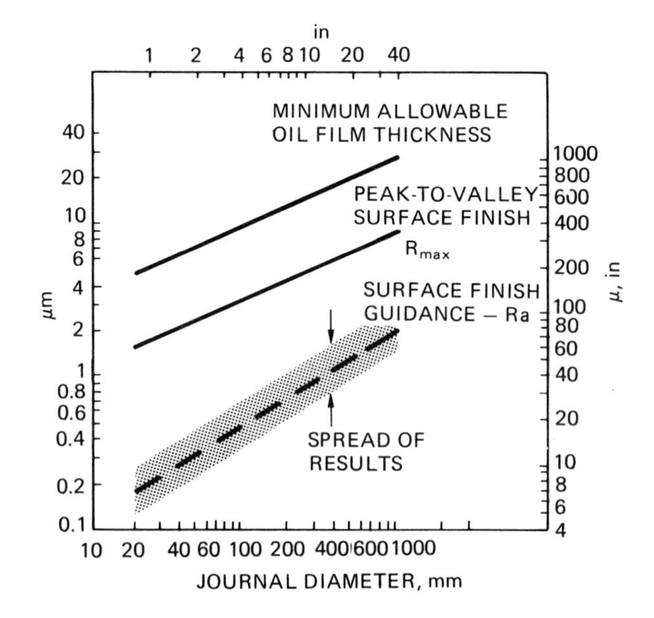

Fig. 9.4. Guidance on allowable oil film thickness dependent on surface finish

High temperature – oil oxidation limit – danger of excessive oil oxidation.
Background Industrial mineral oils can rapidly oxidize in an atmosphere containing oxygen (air). There is no precise limit; degradation is a function of temperature and operating period. Bulk drain temperature limit in the general guide. Fig. 9.6, is restricted to 75–80°C (assuming that the bulk temperatures of oil in tanks and reservoirs is of the same order).

Oil film whirl limit – danger of oil film instability.
Background Possible problem with lightly loaded bearings/rotors at high speeds.

RECOMMENDED MINIMUM DIAMETRAL CLEARANCE

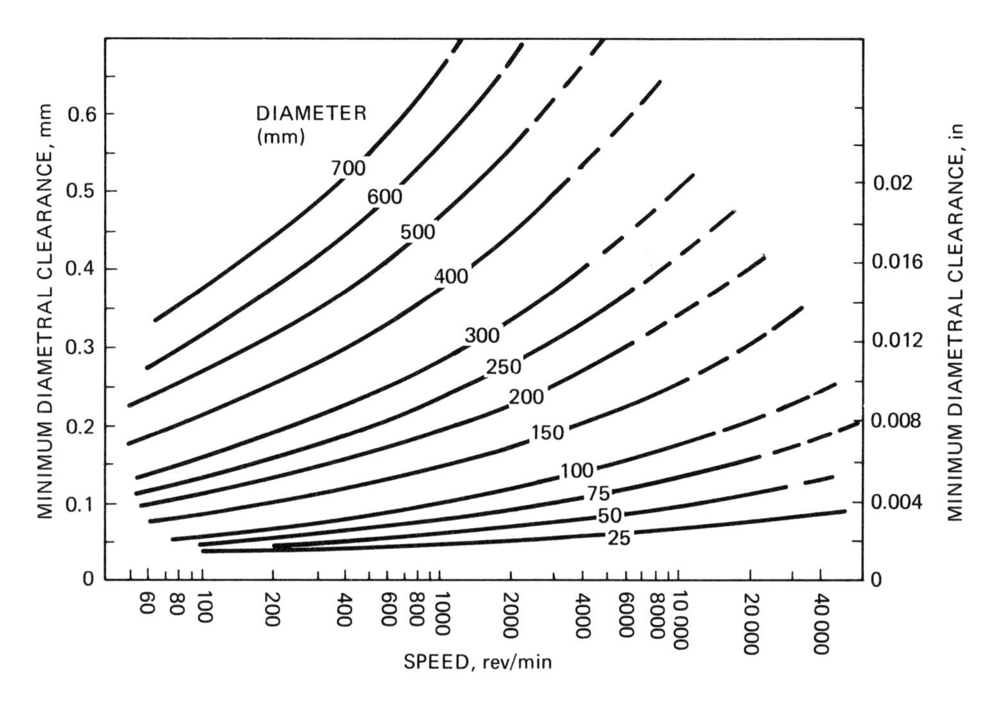

Fig. 9.5. Recommended minimum clearance for steadily loaded bearings
(dashed line region – possibility of non-laminar operation)

GUIDE FOR ESTIMATING MAXIMUM CLEARANCE

Trends in bearing clearance tolerances – for bearing performance studies

Bearing type	A bearing where the housing bore tolerance has little effect on bearing clearance (thickwalled or bored on assembly). Small tolerance range		A bearing which conforms to the housing bore. It has a larger tolerance as the wall thickness and housing bore must also be considered
Typical tolerance (mm) on diametral clearance	$\left(\frac{(\text{Bearing diameter, mm})^{1/3}}{80}\right)$	to	$\left(\frac{(\text{Bearing diameter, mm})^{1/3}}{60}\right)$

Maximum diametral clearance = Minimum recommended clearance (Fig. 9.5) + Tolerance (see trends above)

PRACTICAL GUIDE TO REGION OF SAFE OPERATION (INDICATING ACCEPTABLE GEOMETRY AND OIL GRADE)

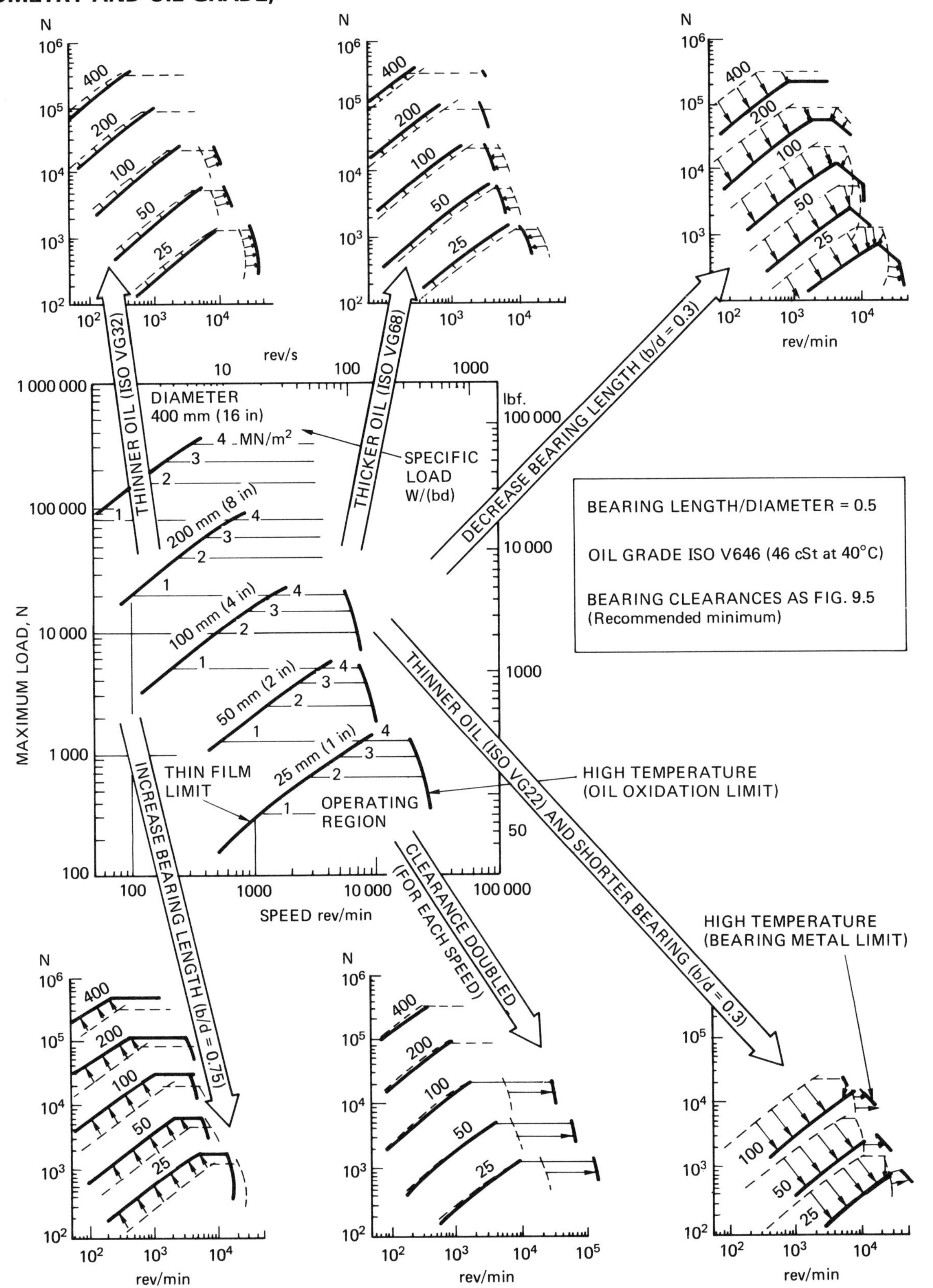

Fig. 9.6. Guide to region of safe operation (showing the effect of design changes) *Work within the limiting curves*

2 axial groove bearing – Groove length 0.8 of bearing length and groove width 0.25 of bearing diameter
Oil feed conditions at bearing – Oil feed pressure 0.1 MN/m^2 and oil feed temperature 50°C

BEARING LOAD CAPACITY

Operating load

The bearing load capacity is often quoted in terms of specific load (load divided by projected area of the bearing, W/bd) and it is common practice to keep the specific load below 4 MN/m^2. This is consistent with the practical guide shown in Fig. 9.6 which also shows that loads may have to be much lower than this in order to work within an appropriate speed range.

Guide to start-up load limit

For whitemetal bearings the start-up load should be limited to the following values:

Specific load limit at start-up*		*MN/m^2*
Frequent stops/starts	Several a day	1.4
Infrequent stops/starts	One a day or less	2.5

* Other limits at operating speeds must still be allowed for as shown in Fig. 9.6

BEARING PERFORMANCE

Figures 9.7 to 9.9 give the predicted minimum film thickness, power loss and oil flow requirements for a 2-axial grooved bearing with the groove geometry and feed conditions shown in Fig. 9.7. Any diametral clearance ratio C_d/d can be considered; however, the maximum should be used when estimating flow requirements. In some cases it may be necessary to judge the influence of different load line positions (at thick film conditions) or misalignment; both are considered in Figs. 9.10 to 9.12. A design guide 'check list' is given below.

DESIGN GUIDE CHECK LIST

(i) Using minimum clearance

	Information
See recommended minimum clearance	Fig. 9.5
Check that the bearing is within a safe region of operation adjust (or choose) geometry and/or oil as found necessary	Fig. 9.6
Predict oil film thickness ratio (minimum film thickness/diametral clearance)	Fig. 9.7
Predict power loss	Fig. 9.8
Allow for non-symmetric load angle (relative to grooves), if necessary	Figs 9.10 and 9.11
Allow for the influence of misalignment on film thickness, if necessary	Fig. 9.12
Check that modified minimum film thickness is acceptable	Fig. 9.4

(ii) Using maximum clearance

	Information
Calculate maximum clearance	Fig. 9.5 and tolerance equation
Predict film thickness and check that it is acceptable	Figs 9.7 and 9.4
Predict oil flow requirements	Fig. 9.9
Allow for non-symmetric load line and/or misalignment if necessary	Figs 9.10, 9.11 and 9.12
Check that modified minimum film thickness is acceptable	Fig. 9.4

GUIDE TO OPERATING MINIMUM FILM THICKNESS

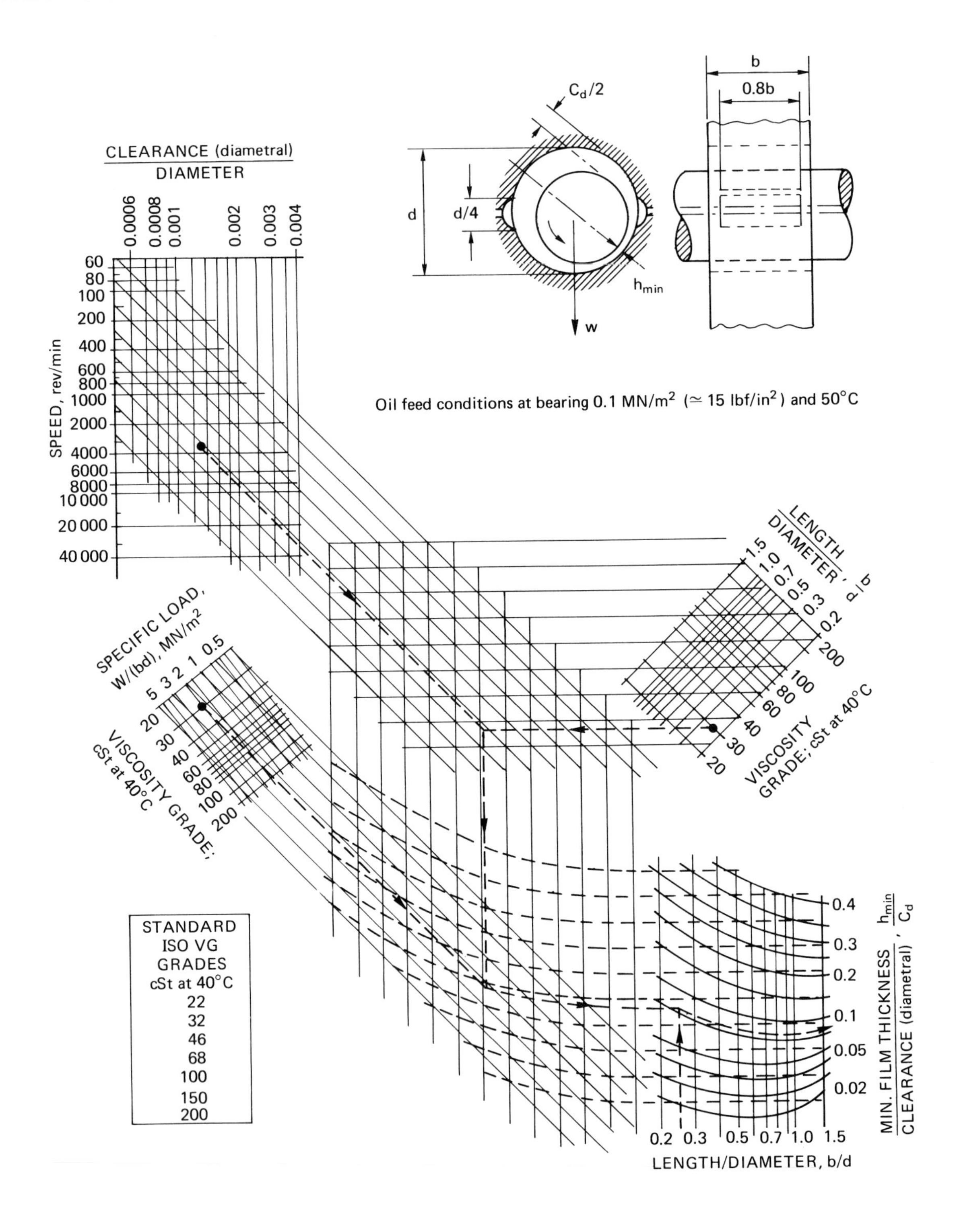

Fig. 9.7. Prediction of minimum oil film thickness for a centrally loaded bearing (mid-way between feed grooves) and for an aligned journal (laminar conditions)

GUIDE TO POWER LOSS

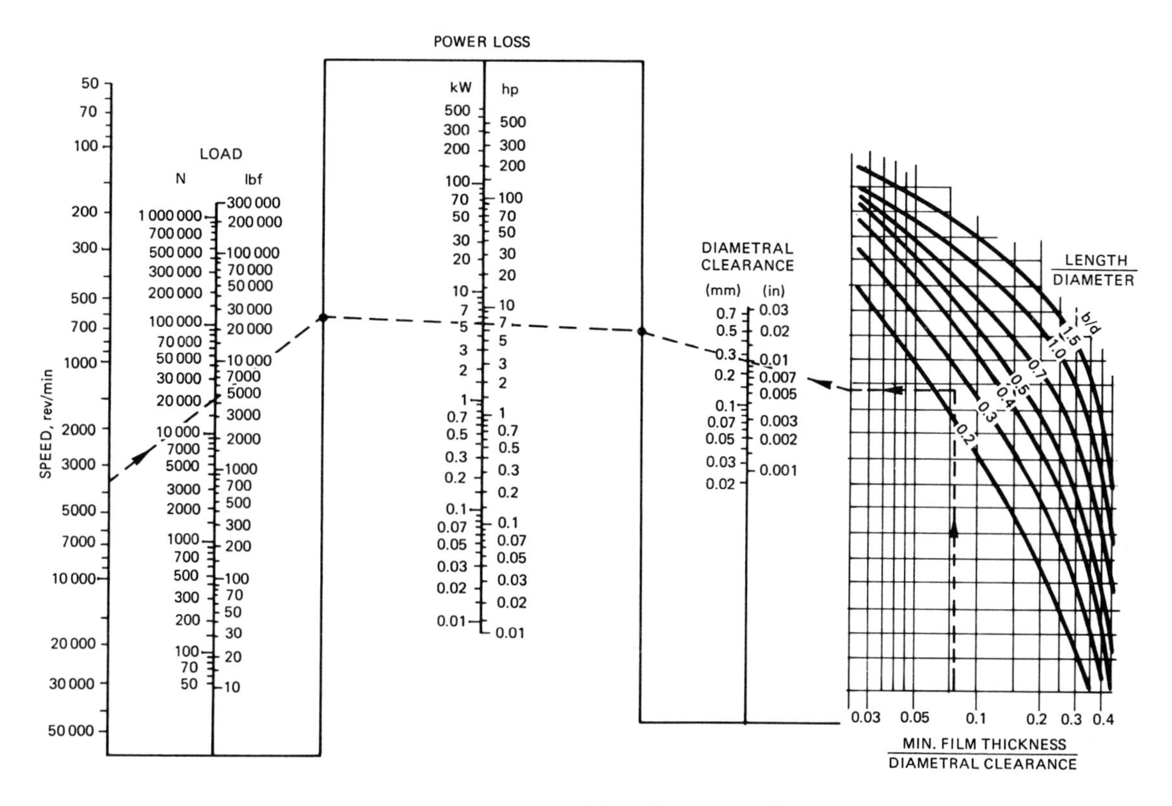

Fig. 9.8. Prediction of bearing power loss

GUIDE TO OIL FLOW REQUIREMENT

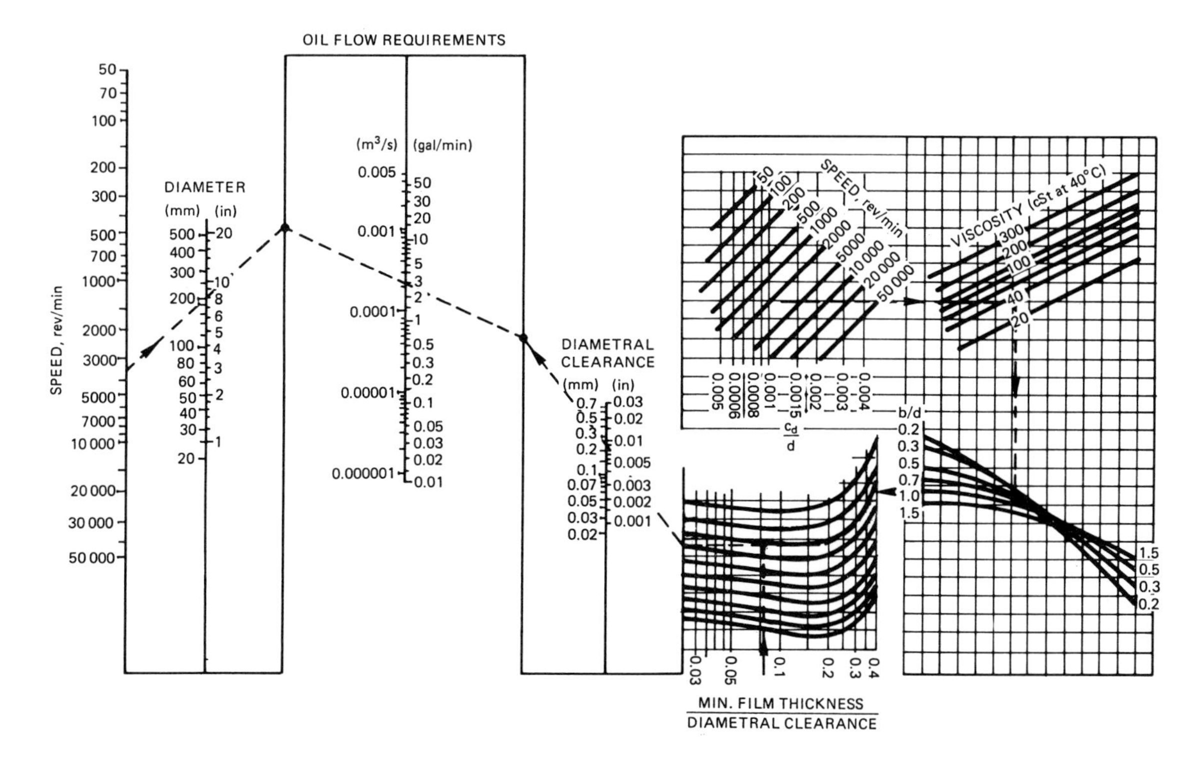

Fig. 9.9. Prediction of bearing oil flow requirement

EFFECT OF LOAD ANGLE ON FILM THICKNESS

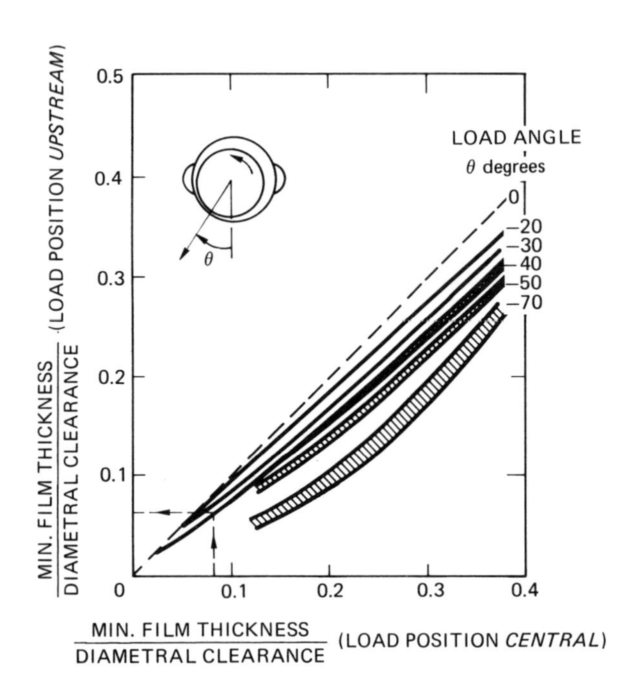

Fig. 9.10. Influence of load position on minimum film thickness (with load position upstream of centre)

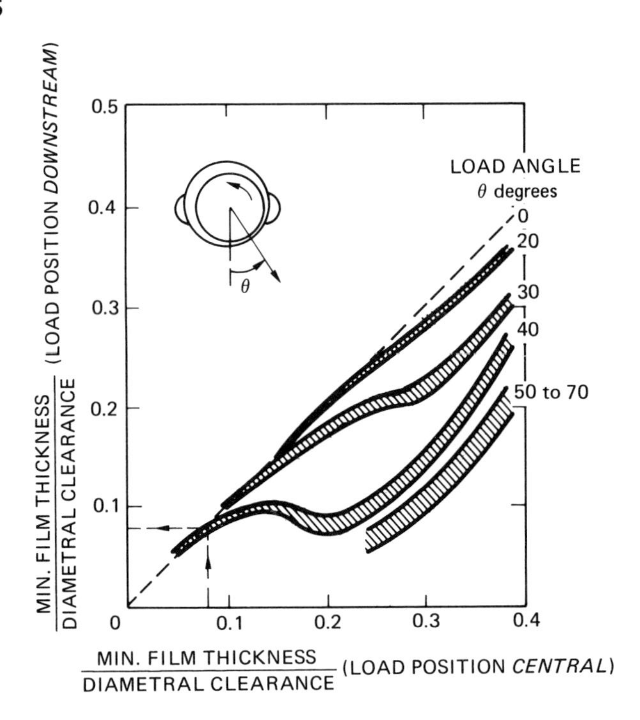

Fig. 9.11. Influence of load position on minimum film thickness (with load position downstream of centre)

EFFECT OF MISALIGNMENT ON FILM THICKNESS

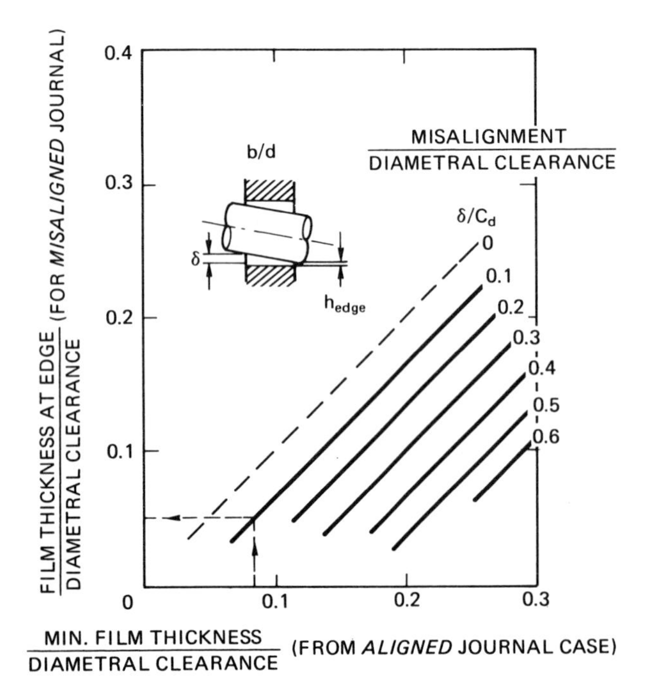

Fig. 9.12. Influence of misalignment on minimum film thickness

Bearings in high speed machines tend to have high power losses and oil film temperatures.
Machines with high speed rotating parts tend to be prone to vibration and this can be reduced by the use of bearings of an appropriate design.

Avoiding problems which can arise from high power losses and temperatures

Possible problem	*Conditions under which it may occur*	*Possible solutions*
Loss of operating clearance when starting the machine from cold	Designs in which the shaft may heat up and expand more rapidly than the bearing and its housing. Tubular shafts are prone to this problem	1. Avoid designs with features that can cause the problem 2. Design with diametral clearances towards the upper limit
	Housings of substantial wall thickness e.g. a housing outside diameter > 3 times the shaft diameter	3. Use a lubricant of lower viscosity if possible 4. Control the acceleration rate under cold starting conditions
	Housings with a substantial external flange member in line with the bearing	5. Preheat the oil system and machine prior to starting
Loss of operating clearance caused by the build up of corrosive deposits on a bearing or seal The deposit usually builds up preferentially at the highest temperature area, such as the position of minimum oil film thickness	Corrosion and deposition rates increase at higher operating temperatures A corrosive material to which the bearing material is sensitive needs to be present in the lubricating oil	1. Determine the chemical nature of the corrosion and eliminate the cause, which may be:- (a) an external contaminant mixing with the lubricant (b) an oil additive 2. Change the bearing material to one that is less affected by the particular corrosion mechanism 3. Attempt to reduce the operating temperature
Loss of operating clearance from the build up of deposits from microbiological contaminants	The presence of condensation water in the lubricating oil and its build up in static pockets in the system	1. Modify the oil system to eliminate any static pockets, particularly in the oil tank
The deposit usually builds up down-stream of the minimum film thickness where any water present in the lubricant tends to evaporate	Temperatures in low pressure regions of the oil films which exceed the boiling point of water	2. Occasional treatment of the lubrication system with biocides 3. Raise the oil system temperature if this is permissible
Increased operating temperatures arising from turbulence in the oil film	High surface speeds and clearances combined with low viscosity lubricants (see Fig. 10.1)	1. Check whether reduced bearing diameter or clearance may be acceptable 2. Accept the turbulence but check that the temperature rises are satisfactory
Increased operating temperatures arising from churning losses in the bearing housing	Thrust bearings are particularly prone to this problem because they are usually of a larger diameter than associated journal bearings	1. Keep the bearing housing fully drained of oil 2. In bearings with separate pads such as tilting pad thrust bearings, feed the pads by individual sprays

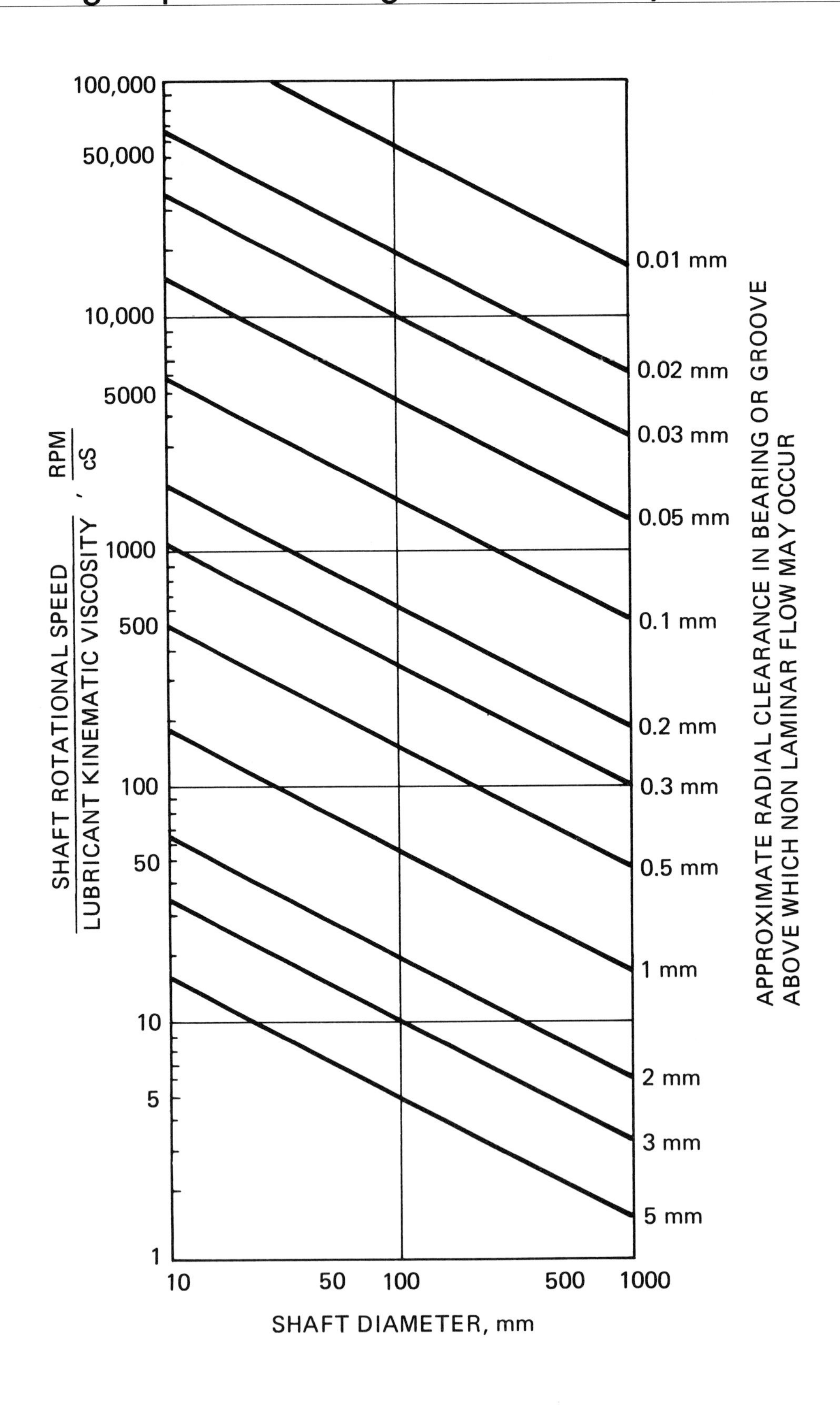

Fig. 10.1. Guidance on the occurrence of non laminar flow in journal bearings

Shaft lateral vibrations which may occur on machines with high speed rotors

Type of vibration	*Cause of the vibration*	*Remarks*
A vibration at the same frequency as the shaft rotation which tends to increase with speed	Unbalance of the rotor	Can be reduced by improving the dynamic balance of the rotor
A vibration at the same frequency as the shaft rotation which increases in amplitude around a particular speed	The rotor, as supported in the machine, is laterally flexible and has a natural lateral resonance or critical speed at which the vibration amplitude is a maximum	The response of the rotor in terms of vibration amplitude will depend on a balance between the damping in the system and the degree of rotor unbalance
A vibration with a frequency of just less than half the shaft rotational speed which occurs over a range of speeds	The rotor is supported in lightly loaded plain journal bearings which can generate half speed vibration (see Fig. 10.2). The actual frequency is generally just less than half shaft speed due to damping	An increase in the specific bearing loading by a reduction in bearing width can help. Alternatively bearings with special bore profiles can be used (see Fig. 10.3)
A vibration with a frequency of about half the shaft rotational speed, which shows a major increase in amplitude above a particular speed	The rotor is supported in lightly loaded plain journal bearings and is reaching a rotational speed, equal to twice its critical speed, when the major vibration increase occurs	This severe vibration arises from an interaction between the bearings and the rotor. The critical speed of the rotor resonates with half speed vibration of the bearings Machines with plain journal bearings generally have a maximum safe operating speed of twice their first critical speed

The diagram shows the mechanism of operation of a plain journal bearing when supporting a steady load from the shaft.

The shaft rotational movement draws the viscous lubricant into the converging clearance and generates a film pressure to support the load

OIL FEED IN THIS ZONE

W

N

REGION OF MINIMUM OIL FILM THICKNESS

RESULTANT PRESSURE GENERATION WHICH SUPPORTS THE LOAD, W

CONVERGING OIL FILM

If a load is applied which rotates at half the shaft speed, the working of the bearing is not easy to visualise

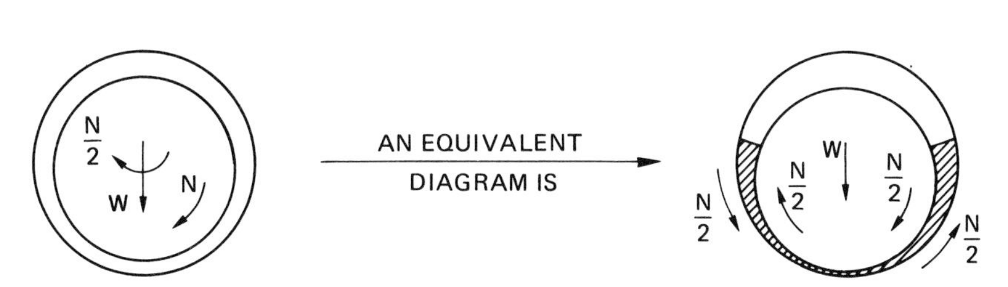

Plain journal bearings cannot support loads which rotate at half the shaft speed.

Half speed loads arise if a bearing carries simultaneously a steady load and a load rotating at shaft speed, which are of equal magnitude

In this arrangement there is no resultant dragging of viscous lubricant into the loaded region, and no load carrying film is generated

If a bearing is lightly loaded the shaft tends to sit near the centre of the clearance space when operating. Any tendency then, to precess or vibrate at half the shaft rotational frequency, builds up in amplitude, because the bearing cannot provide a restoring force for loads/movements at this frequency.

Lightly loaded journal bearings tend therefore to generate shaft vibrations with a frequency of about half the shaft rotational speed

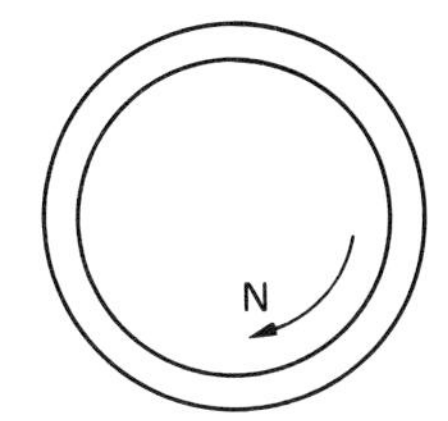

Fig. 10.2. The mechanism by which lightly loaded plain journal bearings tend to vibrate the shaft at about half its rotational speed

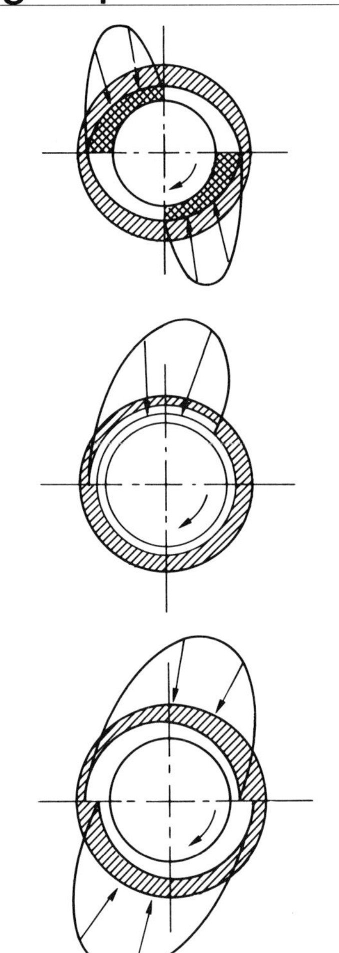

Lemon bore bearing
If the bearing is machined with shims between the joint faces, which are then removed for installation, the resultant bore is elliptical. When the shaft rotates, hydrodynamic clamping pressures are generated which restrain vibration

Dammed groove bearing
A wide and shallow central part circumferential groove in the upper half of the bearing terminates suddenly, and generates a hydrodynamic pressure which clamps downwards onto the shaft

Offset halves bearing
This can be made by machining the bore of two half bearing shells with a lateral offset and then rotating one shell about a vertical axis. This produces two strong converging oil films with high clamping pressures. This bearing, however, demands more oil flow than most other types. There are advantages in rotating the shells in the housing so that the pressure pattern that is generated aligns with the external load

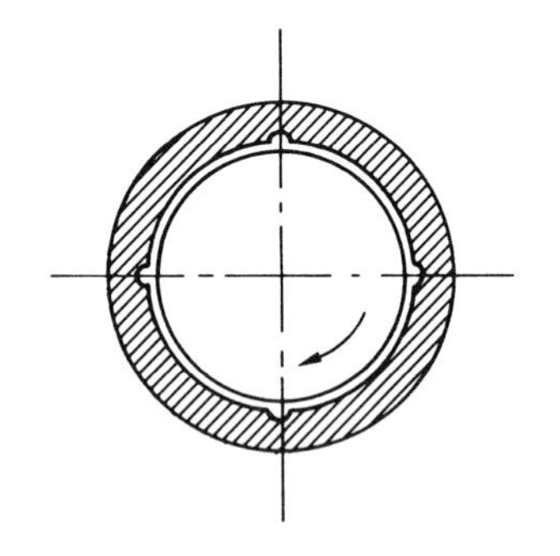

Multi-lobed bearing
A number of pads with a surface radius that is greater than that of the shaft are machined onto the bearing bore. This requires a broach or special boring machine. Each pad produces a converging hydrodynamic film with a clamping pressure which stabilises the shaft

Preset of pads
For increased effectiveness the pads of multi lobe and tilting pad journal bearings need to be preset towards the shaft.

The typical presets commonly used are in the range of 0.6 to 0.8

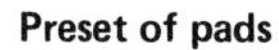

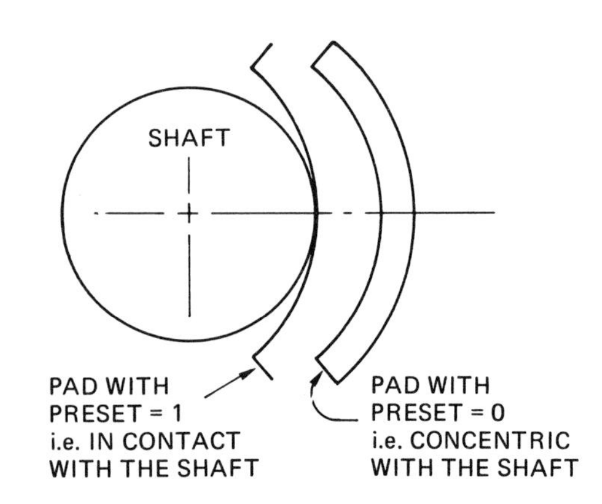

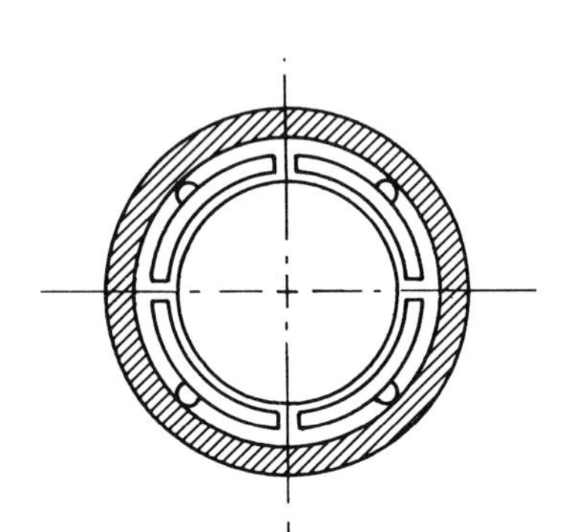

Tilting pad bearing
The shaft is supported by a number of separate pads able to pivot relative to an outer support housing. Each can generate stabilising hydrodynamic pressures

Fig. 10.3. Bearings with special bore profiles to give improved shaft stability

ROTOR CRITICAL SPEEDS

The speed of a rotor at which a resonant lateral vibration occurs corresponds to the natural resonant frequency in bending of the rotor in its supports. This frequency corresponds closely to the ringing tone frequency which can be excited by hitting the rotor radially with a hammer, while it is sitting in its bearing supports. If the supports have different flexibilities in, for example, vertical and horizontal directions, such as may occur with floor mounted bearing pedestals, there will be two critical speeds.

The critical speed of a rotor can be reduced substantially by adding overhung masses such as drive flanges or flexible couplings at the ends of the rotor shaft. Figure 10.4 gives guidance on these effects.

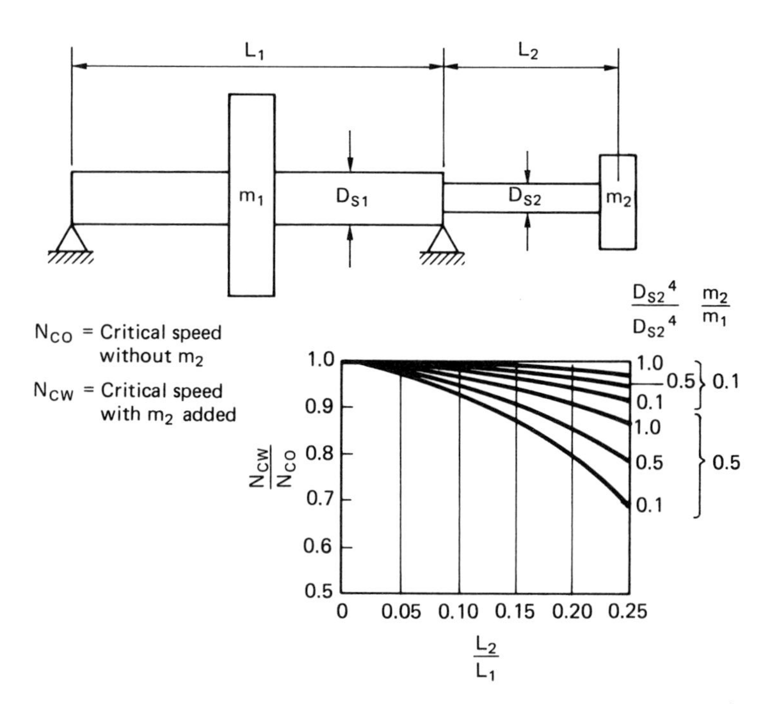

To determine the effect of additional mass at the shaft end, calculate $\frac{L_2}{L_1}$ and $\frac{D_{S2}^4}{D_{S1}^4}$ for the rotor, and also the mass ratio $\frac{m_2}{m_1}$, and read off an approximate value for the critical speed reduction from the graph.

Fig. 10.4. The effect of an overhung mass such as a flexible coupling on the critical speed of a shaft

ROTORDYNAMIC EFFECTS

A full rotordynamic analysis of a machine tends to be complex largely because plain journal bearings give cross coupling effects. That is, a force applied by the shaft to the oil film, produces motion not only in line with the force, but also at right angles to it. This arises from the nature of the action of the hydrodynamic films in which the resultant pressure forces are not in line with the eccentric position of the shaft, within the bearing clearance.

Basic points of guidance for design can, however, be stated:

1 The most important performance aspect is the rotor response, in terms of its vibration amplitude.
2 The response is very dependent on the design of the journal bearings and the amount of damping that they can provide. Bearings with full oil films provide the most damping.
3 The likely mode shapes of the shaft need to be considered and the bearings should be positioned away from the expected position of any nodes. This is because at these positions the shaft has negligible radial movement when vibrating and bearings positioned at these nodes can therefore provide very little damping.

Crankshaft bearings

SELECTION OF PLAIN OR ROLLING BEARINGS

Characteristic	*Rolling*	*Plain*
Relative cost	High	Low
Weight	Heavier	Lighter
Space requirements: Length Diameter	 Smaller Larger	 Larger Smaller
Shaft hardness	Unimportant	Important with harder bearings
Housing requirements	Usually not critical	Rigidity and clamping most important
Radial clearance	Smaller	Larger
Toleration of shaft deflections	Poor	Moderate
Toleration of dirt particles	Poor with hard particles	Moderate, depending on bearing hardness
Noise in operation	Tend to be noisy	Generally quiet
Running friction: Low speeds High speeds	 Very low May be high	 Generally higher Moderate at usual crank speeds
Lubrication requirements	Easy to arrange. Flow small except at high speeds	Critically important pressure feed and large flow
Assembly on crankshaft	Virtually impossible except with very short or built-up crankshafts	Bearings usually split, and assembly no problem

At the present time the choice is almost invariably in favour of plain bearings, except in special cases such as very high-speed small engines, and particularly petroil two-strokes.

SELECTION OF TYPE OF PLAIN BEARING

Journal bearings

Direct lined	*Insert liners*
Accuracy depends on facilities and skill available	Precision components
Consistency of quality doubtful	Consistent quality
First cost may be lower	First cost may be higher
Repair difficult and costly	Repair by replacement easy
Liable to be weak in fatigue	Will generally sustain higher peak loads
Material generally limited to white metal	Range of available materials extensive

Thrust bearings

Flanged journal bearings	*Separate thrust washer*
Costly to manufacture	Much lower first cost
Replacement involves whole journal/thrust component	Easily replaced without moving journal bearing
Material of thrust face limited in larger sizes	Range of materials extensive
Aids assembly on a production line	Aligns itself with the housing

SELECTION OF PLAIN BEARING MATERIALS

Properties of typical steel-backed materials

Lining materials	*Nominal composition* %	*Lining or overlay thickness* mm	in	*Relative fatigue strength*	*Guidance peak loading limits* MN/m²	lbf/in²	*Recommended Journal hardness* V.P.N.
Tin-based white metal	Sn 87 Sb 9 Cu 4 Pb 0.35 max.	Over 0.1 Up to 0.1	Over 0.004 Up to 0.004	1.0 1.3	12–14 14–17	1800–2000 2000–2500	160 160
Tin-based white metal with cadmium	Sn 89 Sb 7.5 Cu 3 Cd 1		No overlay	1.1	12–15	1800–2200	160
Sintered copper–lead, overlay plated with lead–tin	Cu 70 Pb 30	0.05 0.025	0.002 0.001	1.8 2.4	21–23 28–31	3000–3500(1) 4000–4500(1)	230 280
Cast copper-lead, overlay plated with lead–tin or lead–indium	Cu 76 Pb 24	0.025	0.001	2.4	31	4500(1)	300
Sintered lead–bronze, overlay plated with lead–tin	Cu 74 Pb 22 Sn 4	0.025	0.001	2.4	28–31	4000–4500(1)	400
Aluminium–tin	Al 60 Sn 40	No overlay		1.8	21–23	3000–3500	230
Aluminium–tin	Al 80 Sn 20	No overlay		3	42	6000	230
Aluminium–tin (1) plated with lead–tin	Al 92 Sn 6 Cu 1 Ni 1	No overlay		3.5	48	7000	400
Aluminium–tin–silicon (2)	Al 82 Sn 12 Si 4 Cu 2	No overlay		3.7	52	7500	250

(1) Limit set by overlay fatigue in the case of medium/large diesel engines.
(2) Particularly suitable for use with nodular iron crankshafts.

Suggested limits are for big-end applications in medium/large diesel engines and are not to be applied to crossheads or to compressors. Maximum design loadings for main bearings will generally be 20% lower.

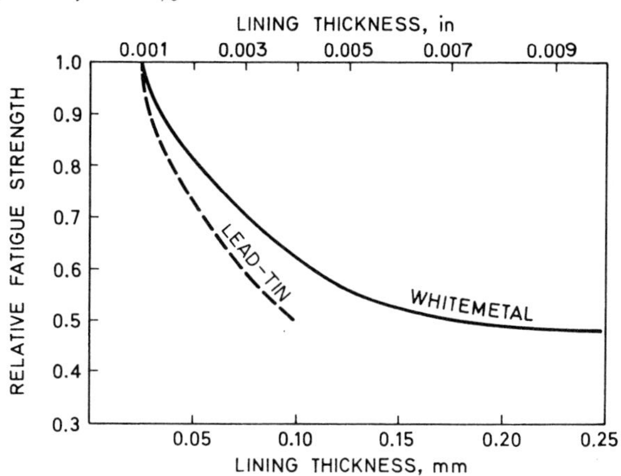

The fatigue strength of bearing metals varies with their thickness. As indicated in the previous table the fatigue strength of the overlay on a bearing material often provides a limit to the maximum load.

The graph shows the relative fatigue strength of two types of overlay material.

Automotive diesel and petrol engines

In smaller engines, thinner overlay plating can be used and the maximum loading is then no longer limited by fatigue of the overlay. With the exception of white metals the specific loading limits for smaller automotive diesel and petrol engines may by considerably higher than the values quoted in the table for medium/large diesel engines. Lead bronzes and tin–aluminium materials are available suitable for loads up to 7500 p.s.i. in these applications.

EFFECT OF FILM THICKNESS ON PERFORMANCE

Many well-designed bearings of modern engines tend to reach their limit of performance because of thin oil film conditions rather than by fatigue breakdown.

A vital factor determining the film thickness is the precise shape of the polar load diagram compared with the magnitude of the load vector. With experience it is possible to assess, from the diagram, whether conditions are likely to be critical and thus to determine whether computation of oil film thickness, peak pressures and power losses is advisable. The existence of load vectors rotating at half shaft speed is undesirable since they tend to reduce the oil film thickness.

Factors promoting thin oil films

Polar load diagrams, with features, are shown below.

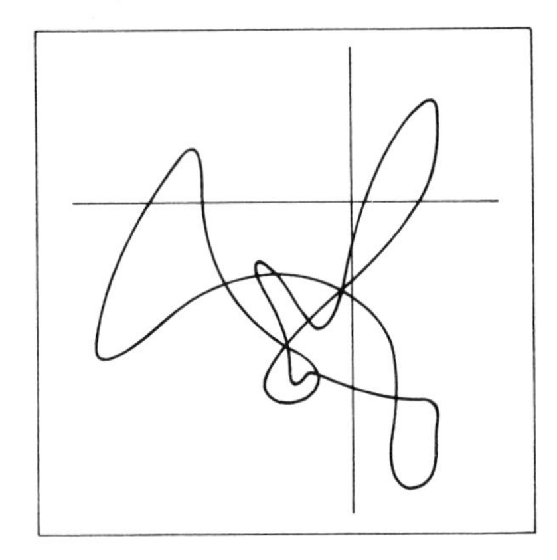

Two firing peaks combined with arc where load vector is travelling approximately at half shaft speed. (Diagram is typical of V-engine main bearing).

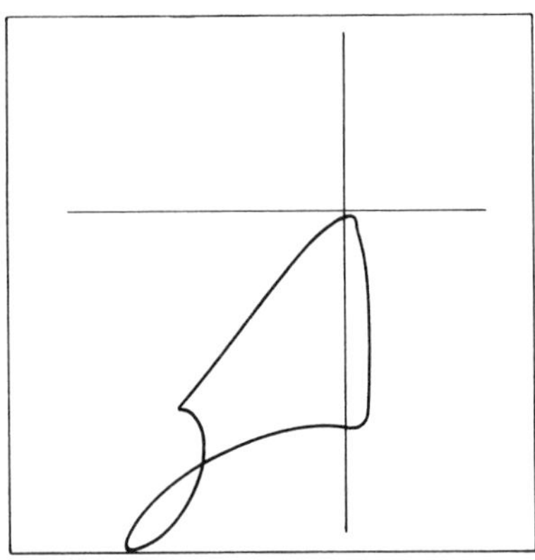

No half-speed vector but forces of large magnitude mostly directed in a limited quadrant. Journal dwells in one position in bearing during cycle, giving no chance for squeeze-film effects to assist.

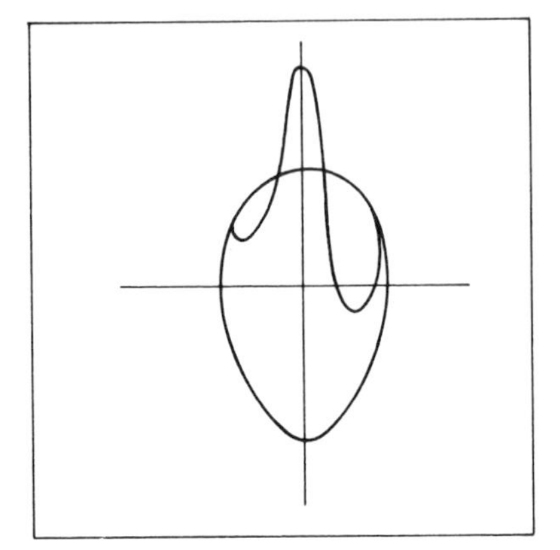

Big-end diagram with large inertia loop, promoting heavy loading on one area of crankpin throughout cycle. The resulting local overheating reduces the oil viscosity and gives thin oil films.

Computation of film thickness

If the film thickness conditions are likely to be critical, computation of the theoretical film thickness is desirable and several computer programmes are available.

Typical computer programme assumptions	*Interpretation of results*
Bearings are circular Journals are circular Perfect alignment Uniform viscosity around the bearing Effect of pressure on viscosity is ignored Surfaces are rigid Crankcase and crankshaft do not distort	The results cannot be used absolutely to determine whether any given hearing will or will not withstand the conditions of operation However, they are useful as a means of comparison with similar engines, or to indicate the effect of design changes. They provide a guided estimate of the probability of success

Depending on which programme is used, most bearings which are known to operate satisfactorily have computed minimum film thicknesses that are no lower than the following:

Mains 0.0025–0.0042 mm (0.0001–0.000 17 in)
Big ends 0.002–0.004 mm (0.000 08–0.000 15 in)

Means of improving oil film thickness

Modify crankshaft balance to reduce magnitude of rotating component of force.

Modify firing order, e.g. to eliminate successive firing of two or more cylinders adjacent to one bearing.

Increase bearing area to reduce specific loading, increase of land width being more effective than increase of diameter.

For big-end bearing, reduce reciprocative and/or rotating mass of con-rod.

Aim at acceptable compromise between factors including:

- Firing loads
- Inertia loads
- Crankshaft stiffness
- Torsional characteristics
- Engine balance
- Stiffness of big-end eye
- Overall length of assembly

Each factor must be considered both on its own merits and in relation to others on which it has an effect.

BEARING FEATURES

General rules for grooving in crankshaft bearings

Never use a groove unless for a valid reason, and under no circumstances use a groove reaching in an axial direction.

For many engine applications a plain, central circumferential groove is used, e.g. to permit oil to be fed to a main bearing and thence, without interruption, to a big-end bearing via a drilled connecting rod to a small-end bearing.

Recommended groove configurations

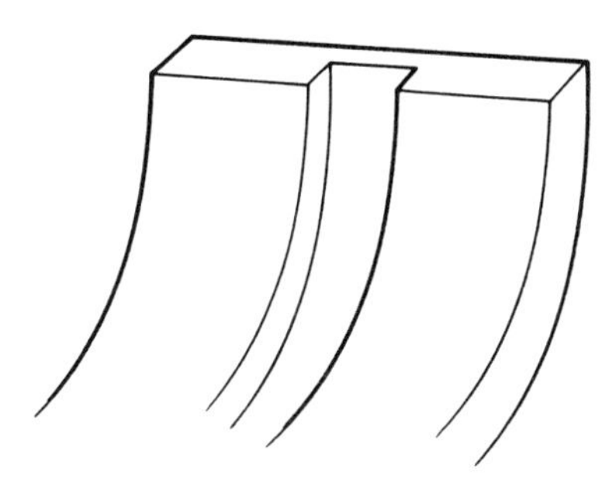

The groove may be formed within the bearing wall provided the wall thickness is adequate.

Undesirable grooving features

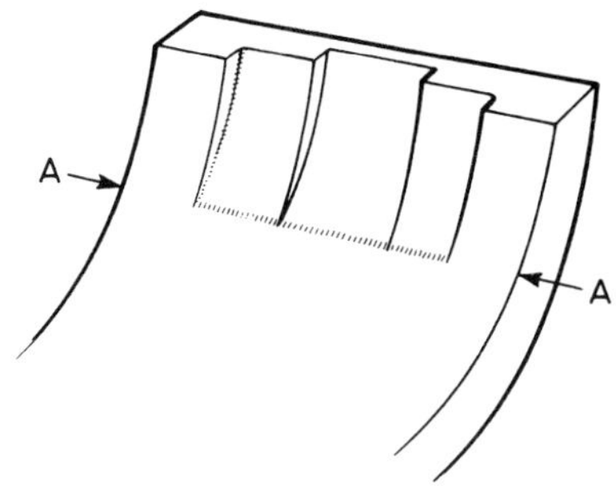

Gutterways or oil pockets at the horns. These do not trap dirt and are not needed for axial distribution of oil in a typical modern bearing. They can cause local overloading along line *A–A*, and fatigue breakdown.

Recommended groove configurations

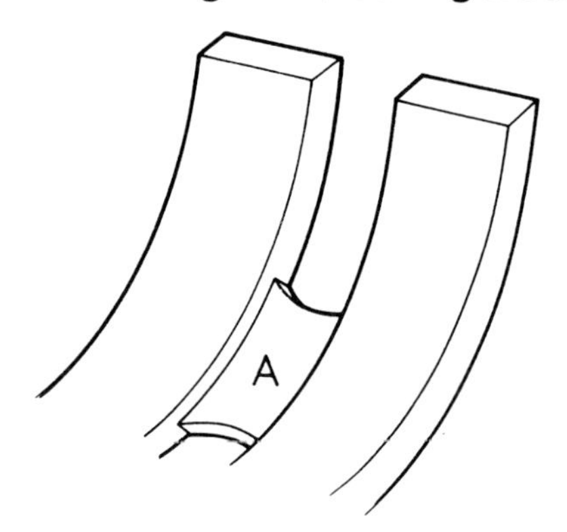

With thin shells the groove can take the form of a slot, communicating if necessary with a corresponding groove round the housing. The two halves of the shell are connected by bridge pieces as at *A*.

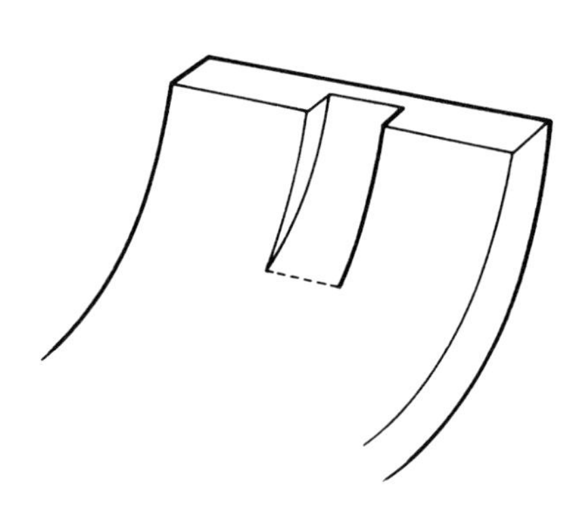

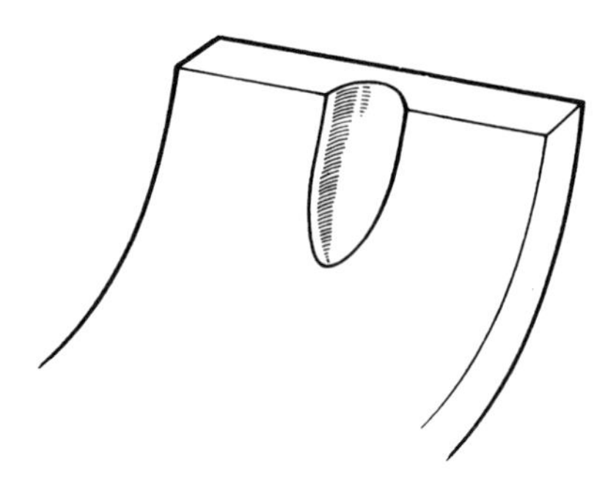

In order to leave a heavily loaded area of a bearing uninterrupted, a circumferential groove may be partial rather than continuous. The ends of such a groove must be blended gently into the plain surface to avoid sudden starting or stopping of the oil flow into or out of the groove and consequent erosion of the bearing.

Undesirable grooving features

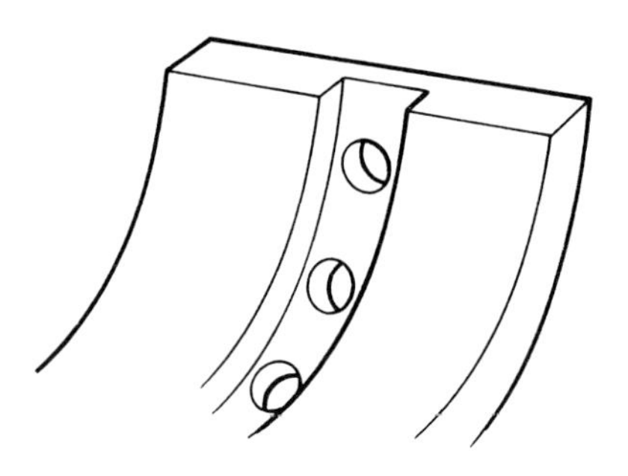

Rows of holes in oil groove, communicating with corresponding groove in bearing housing, can promote local recirculation of oil, leading to excessive temperatures.

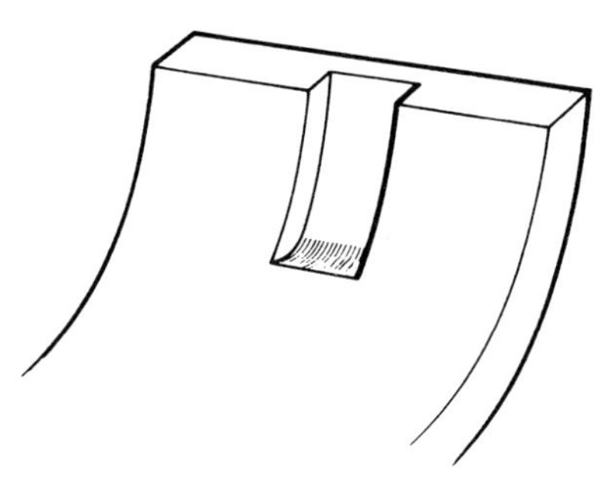

Abrupt endings to partial oil grooves can cause erosion of the surface downstream of the groove end.

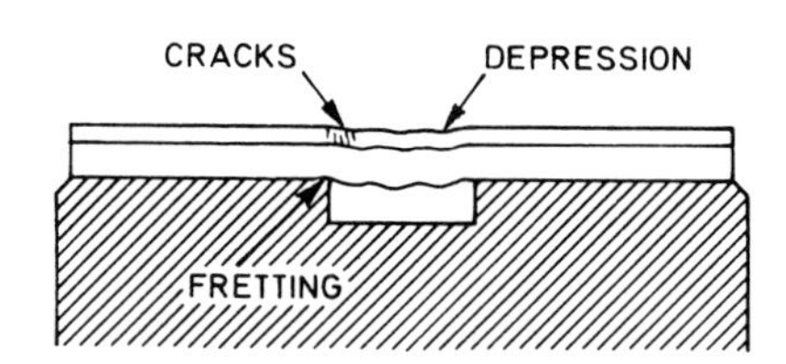

Any hole, slot or groove in the housing behind a bearing is dangerous because it promotes undesirable deflection, fretting and, in extreme cases, cracking of the bearing.

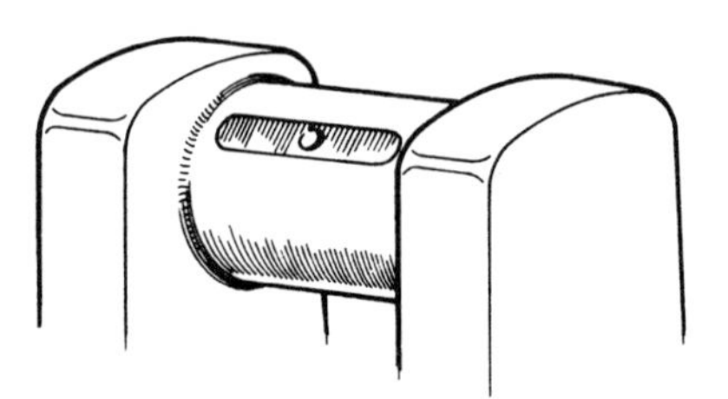

Any flat or depression in a journal surface, other than a blended hole in line with the bearing oil groove, is liable to cause severe local overheating.

Locating devices

Recommended

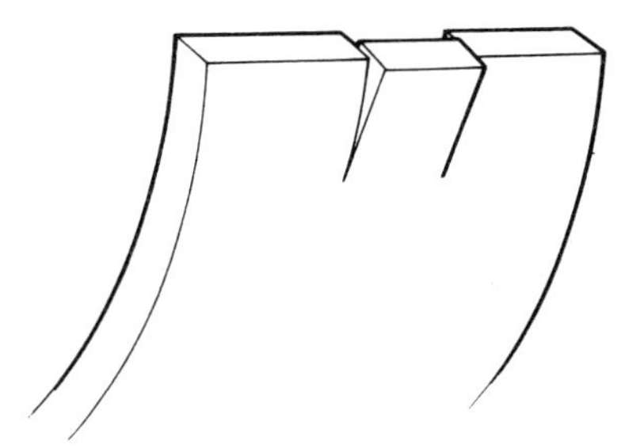

Nick, tang or lug locates shell in axial direction only. Lugs should be on one joint face of each shell with both lugs positioned at the same housing joint face.

Back face of nick must not foul slot or notch in housing, and nick should not be in heavily loaded area of bearing.

Bearing must be arranged with joint face in line with housing joint face.

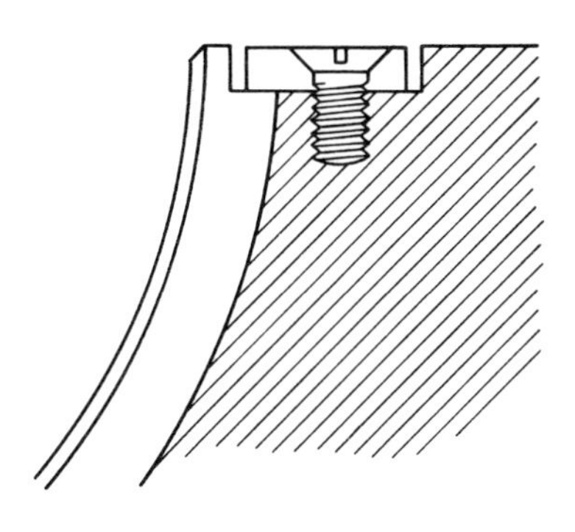

Button stops—useful for large and heavy bearing shells—to retain upper halves in housings against falling out by own weight. Button may be any shape but is usually circular, and must be securely screwed to housing with small clearance all round and at upper face.

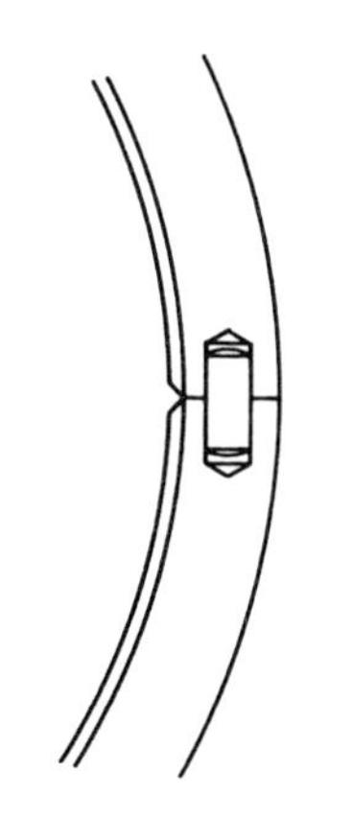

Dowel, tight fitting in one joint face, clearance in the other, locates shells to each other. More susceptible to damage during assembly, does not locate shells in housing and cannot be used on thin shells.

Not recommended

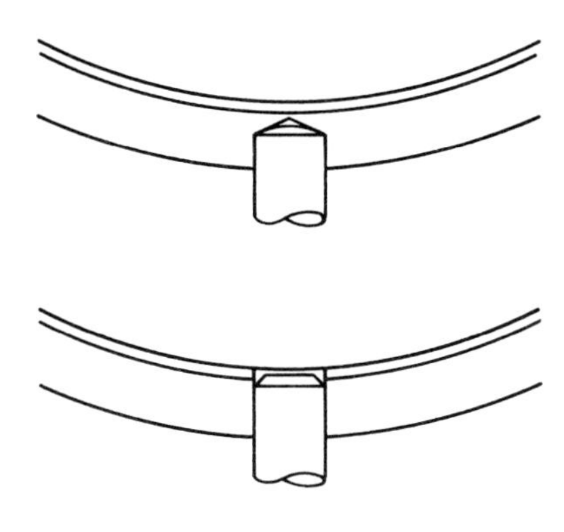

Dowel in crown—weakens shell, may induce fatigue or, if hole pierces shell, may cause erosion and oil film breakdown. Will not prevent a seized bearing from rotating, and when this happens damage is aggravated.

Redundant locating devices result in 'fighting' between them and possible damage to bearings.

The correct conditions of installation are particularly important with dynamically loaded bearings.

LUBRICANT FEED SYSTEM

Except for small low-cost machines, oil feed is by a pressurised system consisting of a sump or reservoir, a mechanical pump with pressure-regulating valve and by-pass, one or more filters and an arrangement of pipes or ducts. The capacity of the system must be adequate to feed all bearings and other components even after maximum permissible wear has developed.

A guide to the flow through a conventional central circumferentially grooved bearing is given by

$$Q = \frac{kpC_d^3}{\eta} \cdot \frac{d}{b} (1.5\,\varepsilon^2 + 1)$$

where Q = flow rate, m^3/sec (gal/min)
p = oil feed pressure, N/m^2 (lbf/in^2)
C_d = diametral clearance, m (in)
η = dynamic viscosity, Ns/m^2 (cP)
d = bearing bore, m (in)
b = land width, m (in)
ε = eccentricity ratio
k = constant
= 0.0327 for SI units
(4.86×10^4 for Imperial units)

For most purposes it is sufficient to calculate the flow for a fully eccentric shaft, i.e. where $\varepsilon = 1$.

As a guide to modern practice, in medium/large diesel engines, oil-flow requirements at 3.5×10^5 N/m^2 (50 lbf/in^2) pressure are as follows:

Bedplate gallery to mains (with piston cooling), 0.4 l/min/h.p. (5 gal/h/h.p.)
Mains to big end (with piston cooling), 0.27 l/min/h.p. (3.5 gal/h/h.p.)
Big ends to pistons (with oil cooling), 0.15 l/min/h.p. (2 gal/h/h.p.)
With uncooled pistons, total flow, 0.25 l/min/h.p. (3 gal/h/h.p.)

Velocity in ducts

On suction side of pump, 1.2 m/s (4ft/s).
On delivery side of pump, 1.8–3.0 m/s (6–10 ft/s).

Pressure

Modern high-duty engines will generally use delivery pressures in the range 2.8×10^5 to 4.2×10^5 N/m^2 (40–60 lbf/in^2) but may be as high as 5.6×10^5 N/m^2 (80 lbf/in^2).

Filtration

With the tendency to operate at very thin minimum oil films, filtration is specially important.

Acceptable criteria are that full-flow filters should remove:

100% of particles over 15 μm
95% of particles over 10 μm
90% of particles over 5 μm

Continuous bypass filtration of approximately 10% of the total flow may be used in addition.

INFLUENCE OF ENGINE COMPONENT DESIGN ON BEARING DESIGN AND PERFORMANCE

Housing tolerances

Geometric accuracy (circularity, parallelism, ovality) should be to H6 tolerances.

Lobing or waviness of the surface not to exceed 0.0001 of diameter.

Run-out of thrust faces not to exceed 0.0003 of diameter, total indicator reading.

Surface finish:

Journals 0.2–0.25 μm Ra (8–10 μin cla)
Gudgeon pins 0.1–0.16 μm Ra (4–6 μin cla)
Housing bores 0.75–1.6 μm Ra (30–60 μin cla)

Alignment of adjacent housing should be within 1 in 10 000 to 1 in 12 000.

Bearing housing bolts

These must be stressed to take with safety the sum of the loads due to compressing the bearing in its housing, and those due to the dynamic forces acting on the journal.

Housing stiffness

Local deflections under load have a disastrous effect on the fatigue strength of a bearing, causing fretting on the back, increased operating temperature and lower failure load. Housings, particularly connecting-rod eyes, should be as nearly as possible of uniform stiffness all round and the bolts must be positioned so as to minimise distortion of the bore. The effect of tightening bolts positioned too far from the bore is shown, exaggerated, below.

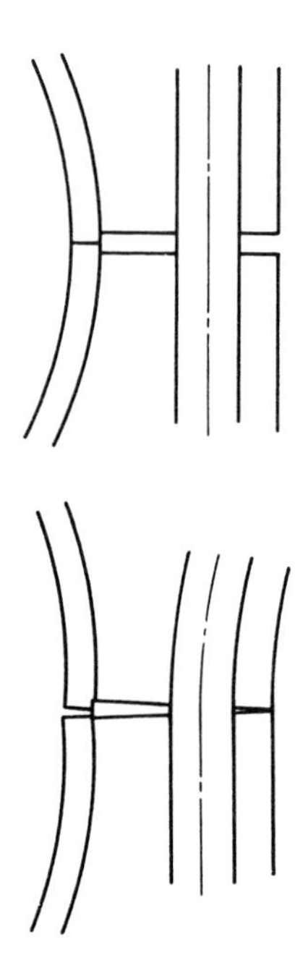

Crankshaft details

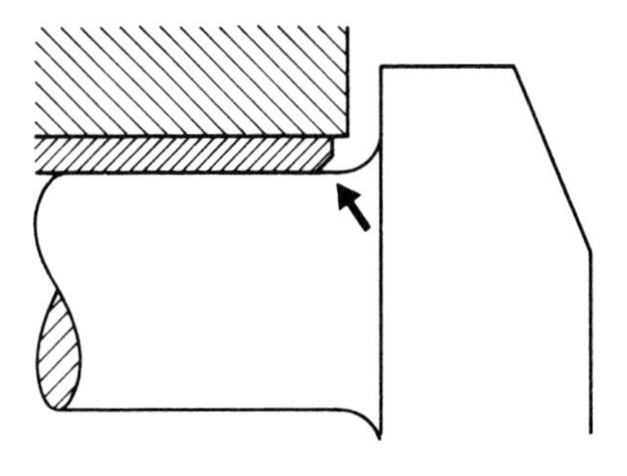

Normal fillet must clear chamfer on bearing under all conditions.

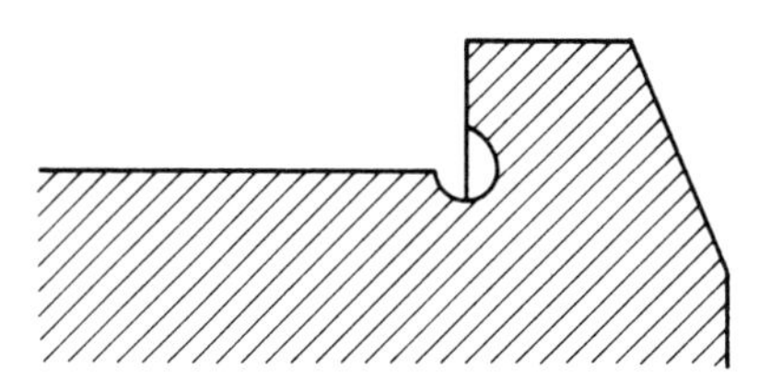

Undercut fillet, useful to help obtain maximum possible bearing width, and facilitates re-grinding.

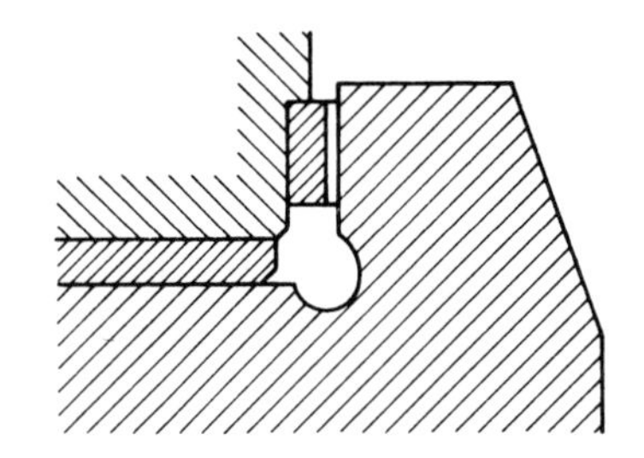

Thrust faces on crank cheeks must always be wider, radially, than soft surface of thrust ring or flange, to avoid throttling oil outlet when wear develops.

Grinding

Direction of grinding, especially on nodular iron shafts, is important and should be in the same direction as the bearing will move relative to the shaft.

Complete removal of friable or 'white' layer from nitrided journals is essential. This implies fine grinding or honing, approximately 0.025–0.05 mm (0.001–0.002 in) from surface.

A12 Plain bearing form and installation

SELECTION

Types of replaceable plain bearing

	Bush			*Half bearing*		
	SOLID	SPLIT	CLENCHED	THINWALL	MEDIUM WALL	THICK WALL
Normal housing	Solid	Solid	Solid	Split	Split	Split
Location feature	None	None	None	Nick	Nick or button stop	Dowel or button stop
Bearing material	Monometal*	Bimetal	Bimetal	Bimetal or trimetal	Monometal, bimetal or trimetal	Bimetal or trimetal
Manufacture	Cast or machine from solid	Wrap from flat strip	Wrap from flat strip	Press from flat strip	Cast lining on preformed blanks	Cast lining on preformed blanks
Forming of outside diameter	Machined	As pressed	Ground	As pressed	Fine-turned or ground	Fine turned
Other points to note when selecting a suitable type	Mainly below 50 mm dia.	Cheaper than solid in quantity	Higher precision	Cheapest for mass production	Moderate quantities 10–100 off	Made in pairs for 1–10 off

* Rarely large solid bimetal bushes are made for end assembly in solid housings. The wall thickness would approximate to that of medium-wall half bearings.

Double split bushes, with halves directly bolted together have been used in the past for turbine bearings, and some gearbox and ring oiled bearings. These are not now common in modern designs.

Double-flanged half bearings (thin and thick wall) and single flanged (solid) bushes are also used.

An oil distribution groove may be cut circumferentially in the outside diameter of thick-wall bearings only.

Choice of wall thickness

In the past, the shaft was chosen as the basis. The emphasis is now changing to standardising on the housing size.

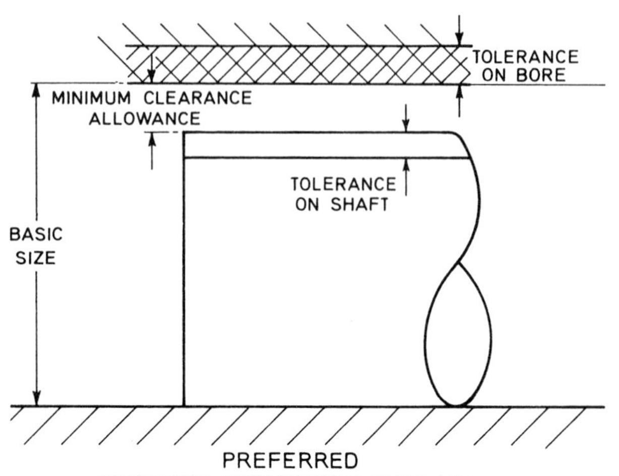

PREFERRED
FACILITATES INTERCHANGEABILITY AND STANDARDISATION OF BEARINGS

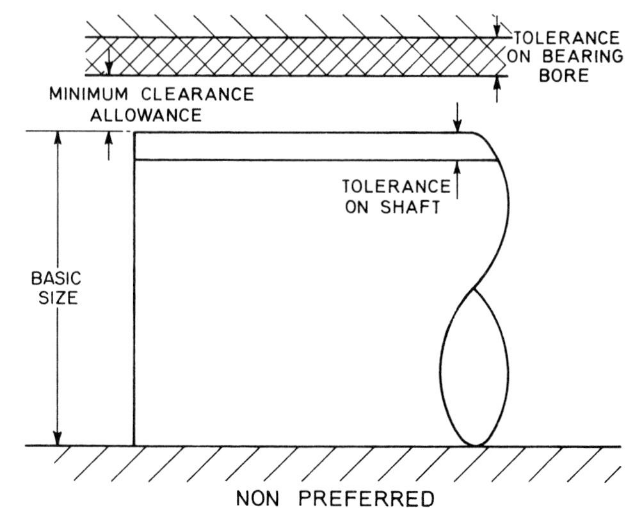

NON PREFERRED
PRIMARY USE WHEN PREGROUND BAR STOCK CAN BE USED FOR JOURNAL

Solid bushes and thick-wall bearings

Bore and outside diameter in metric sizes are chosen from preferred number series, the wall thickness varying accordingly but generally, within range shown in figures.

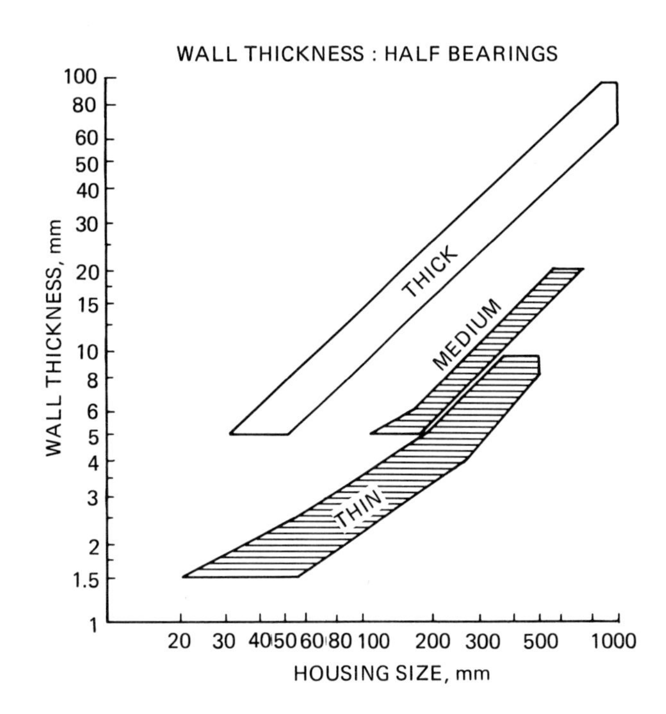

Wrapped bushes, thin- and medium-wall half bearings

The housing for metric sizes is based on preferred number series (ISO Recommendation R497 Series R40′) and wall thickness chosen from preferred thicknesses in table, but generally within range shown in figures.

Preferred wall thickness (mm) and typical tolerances

Wrapped bushes		*Half bearings*			
		Thin wall		*Medium wall*	
Thickness	*Tolerance*	*Thickness*	*Tolerance*	*Thickness*	*Tolerance*
0.75*	0.050	1.5	0.008	4.0	0.020
1.0	0.050	1.75	0.008	5.0	0.020
1.5	0.050	2.0	0.008	6.0	0.020
2.0	0.080	2.5	0.010	8.0	0.025
2.5	0.100	3.0	0.015	10.0	0.025
3.0	0.100	3.5	0.015	12.5	0.030
3.5	0.100	4.0	0.020	15.0	0.030
		5.0	0.020	20.0	0.040
		6.0	0.025	25.0	0.040
		8.0	0.025		
		10.0	0.030		
		12.0	0.030		

* Non-preferred

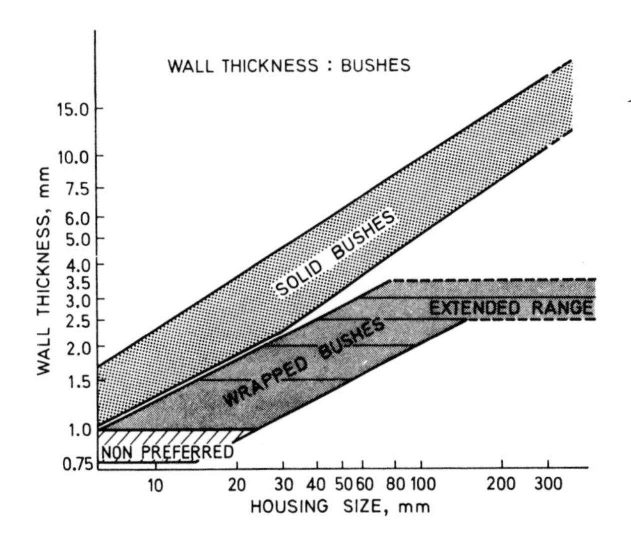

Wall thickness tolerance

Bushes are generally supplied with an allowance for finishing the bore *in situ* by boring, broaching, reeming, or ball burnishing, although prefinished bushes may be available to special orders.

Half bearings are generally machined to close tolerances on wall thickness and may be installed without further machining.

METHODS OF MEASUREMENT AND CHOICE OF HOUSING FITS

Ferrous-backed bearings in ferrous housings

Solid bushes

Outside diameter is measured.

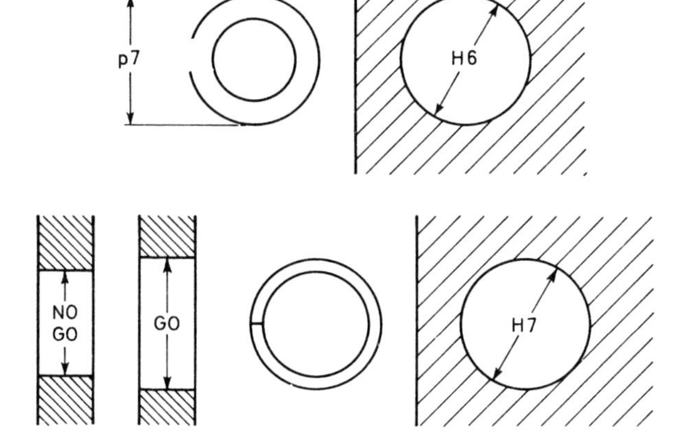

Split bushes

Outside diameter is measured in GO and NO GO ring gauges.

Typical ring gauge sizes (mm) are:

NO GO ring gauge
Below 30 ϕ Maximum housing $+0.035$
Above 30 ϕ Maximum housing $+0.008\sqrt{D}$

GO ring gauge
Below 30 ϕ Maximum housing $+0.060$
Above 30 ϕ Maximum housing $+0.015\sqrt{D}$

Where D = housing diameter (mm).
Split bushes may also be checked under load in a manner similar in principle to that used for thin-wall half bearings.

Thick-wall half bearings

Outside diameter is measured.
Split line not truly on centre line.
Bearings must be kept and stamped as pairs.

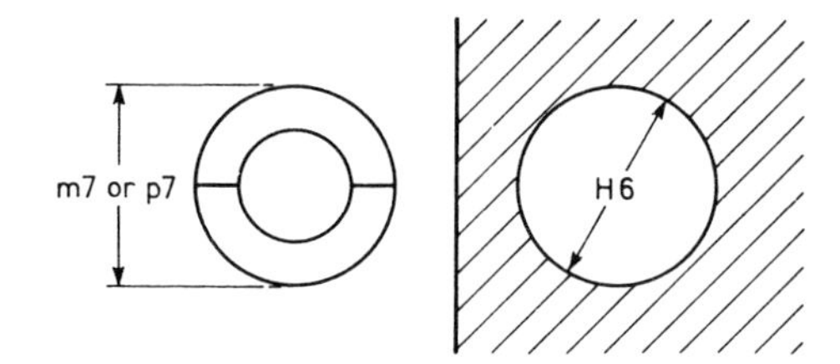

Thin-wall half bearing

Peripheral length of each half is measured under load W_c

$$W_c = 6000\,\frac{b \times s}{D}$$

W_c = checking load (N)
b = width (mm)
s = wall thickness (mm)
D = housing diameter (mm)

'Nip' or 'Crush' is the amount by which the total peripheral length of both halves under no load exceeds the peripheral length of the housing. A typical minimum total nip n (mm) is given by: $n = 4.4 \times 10^{-5}\,\frac{D^2}{s}$ or 0.12 mm whichever is larger.

For bearings checked under load W_c

$$A = \frac{n}{2} - 0.050 \text{ mm}$$

The tolerance for A is grade 7 on peripheral length.

The nip value n gives a minimum radial contact pressure between bearing and housing of approximately 5 MN/m^2 for $D > 70$ mm increasing to 10 MN/m^2 at $D = 50$.

For white-metal lined bearings or similar lightly loaded bearings, minimum contact pressure and minimum nip values may be halved.

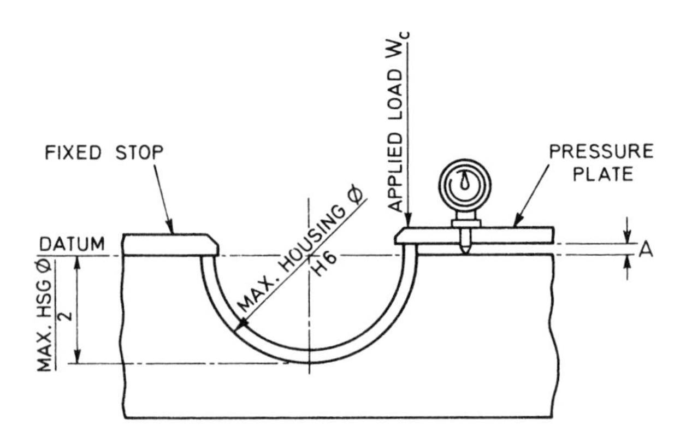

Non-ferrous backed bearings and housings

Allowance must be made in each individual case for the effects of differential thermal expansion. Higher interference fits may be necessary, and possible distortions will require checking by fitting tests.

Free spread

Thin- and medium-wall bearings are given a small amount of extra spread across the joint face to ensure that both halves assemble correctly and do not foul the shaft in the region of the joint when bolted up.

Free spread may be lost when bearing heats up, particularly for copper and aluminium based lining alloys. The loss depends upon method of forming bearing, and thickness of lining. Initial minimum free spread should exceed the likely loss if bearing is required to be reassembled.

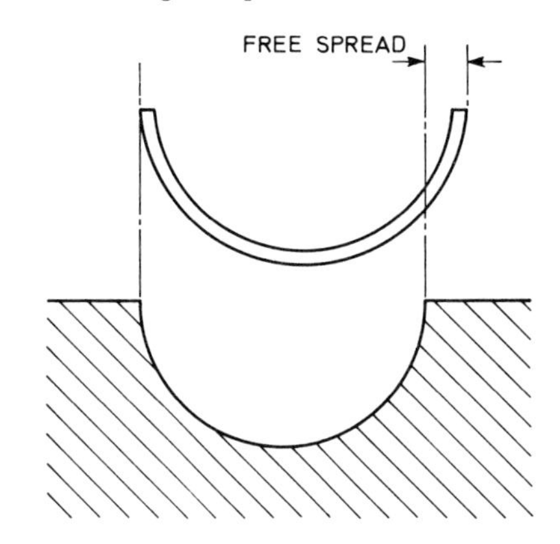

Typical minimum free spread (mm)

Outside diameter	*Thin-wall aluminium or copper based lining*	*Medium-wall white-metal lining*
50 ϕ	0.2–0.3	—
100 ϕ	1.0–2.0	0.2–0.5
300 ϕ	4.0–8.0	1.0–2.0

Note: these figures are for guidance only and specific advice should be obtained from the bearing manufacturers.

THE DESIGN OF THE HOUSING AND THE METHOD OF CLAMPING HALF BEARINGS

Locating the housing halves

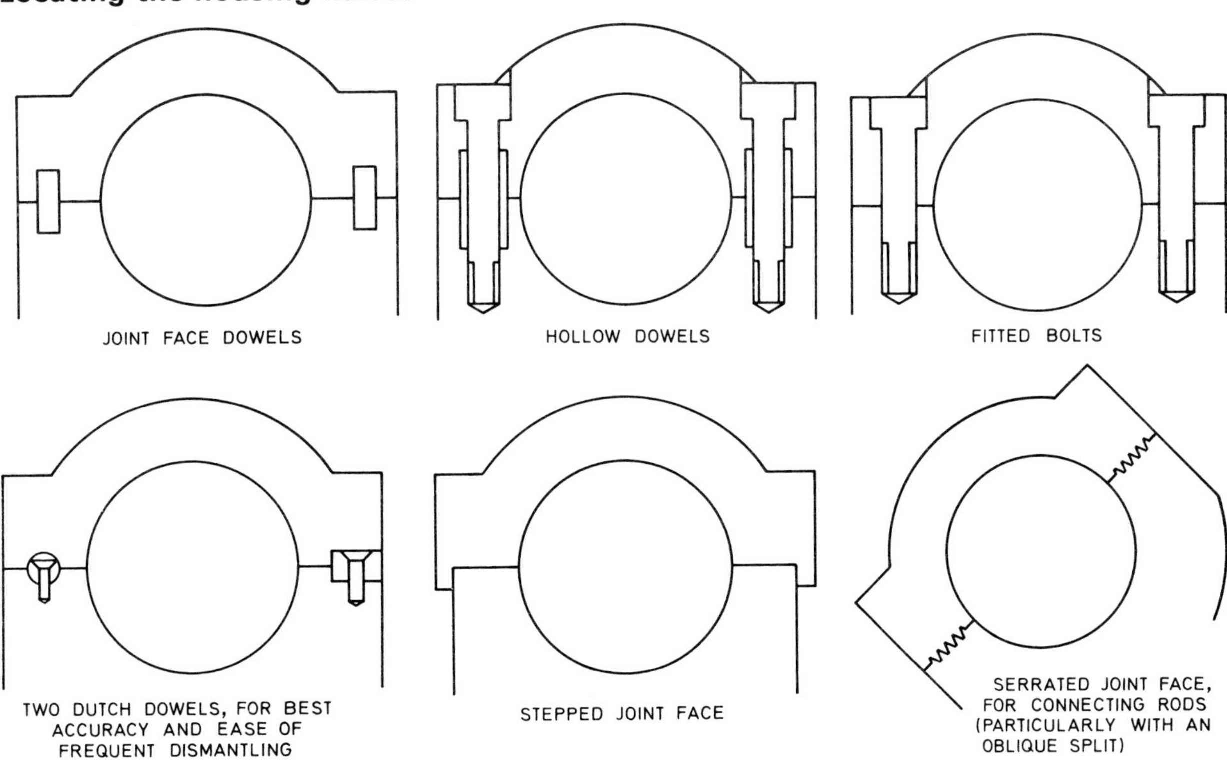

Positioning the bolts

These should be close to the back of the bearing.

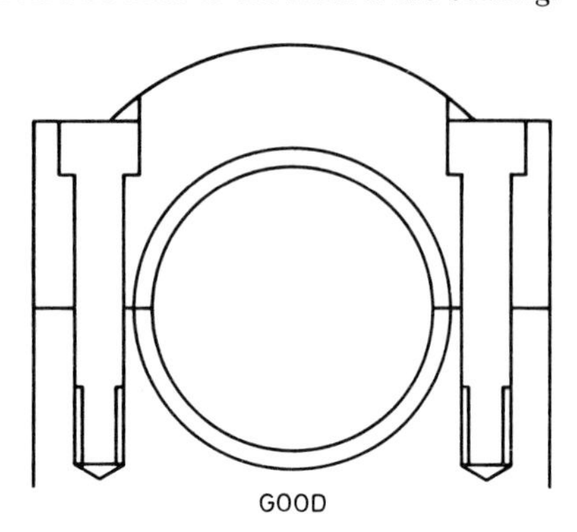

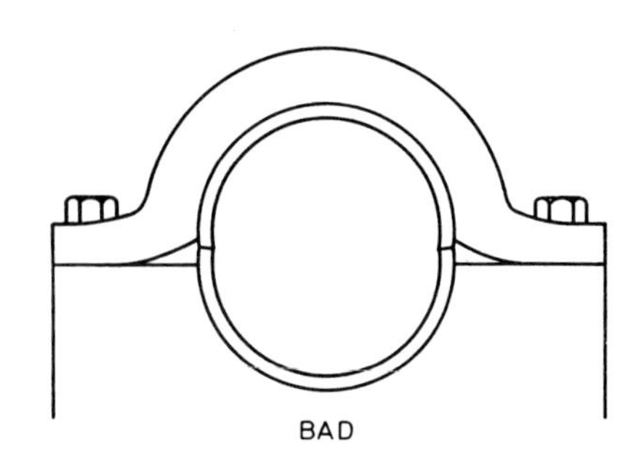

Choice of bolting loads

Load to compress nip

$$W = \frac{E\,b\,t\,m}{\pi\,(D-t)\times 10^6}$$

or

$$W = b\,t\,\phi \times 10^{-6}$$

whichever is smaller.

Extremely rigid housing (very rare)

Bolt load per side

$$W_b = 1.3\,W$$

(to allow for friction between bearing and housing).

Normal housing with bolts close to back of bearing

Bolt load per side

$$W_b \simeq 2\,W$$

(to allow for friction, and relative moments of bearing shell and line of bolt about outer edge of housing).

Note: If journal loads react into housing cap

$$W_b = 2\,W + \tfrac{1}{2}\,W_j$$

Housing material

For non-ferrous housings, or for non-ferrous backed bearings in ferrous housings, allowance must be made for the effects of thermal expansion causing loss of interference fit. Specific advice should be obtained from bearing manufacturers.

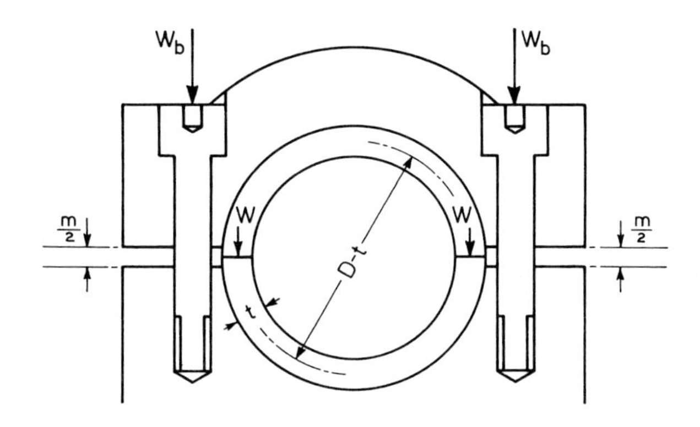

D = housing diameter (mm)

t = steel thickness $+\frac{1}{2}$ lining thickness (mm)

W = compression load on each bearing joint face (N)

m = sum of maximum circumferential nip on both halves (mm) [m = $n+\pi$(housing diametral tolerance) + tolerance on nip]

E = elastic modulus of backing (N/m^2) (For steel $E = 0.21\times 10^{12}$ N/m^2)

b = bearing axial width (mm)

W_b = bolt load required on each side of bearing to compress nip (N)

W_j = Maximum journal load to be carried by cap (N)

ϕ = yield stress of steel backing (N/m^2)

ϕ varies with manufacturing process and lining, e.g.

white-metal lined bearing
$\phi = 350\times 10^6$ N/m^2

copper based lined bearing
$\phi = 300$–400×10^6 N/m^2

aluminium based lined bearing
$\phi = 600\times 10^6$ N/m^2

Dynamic loads

Dynamic loads may result in flexing of the housing if this is not designed to have sufficient rigidity. Housing flexure can result in fretting between bearing back and housing, and in severe cases, in fatigue of the housing. Increase of interference fit with a corresponding increase in bolting load, may give some alleviation of fretting but stiffening the housing is generally more effective. Poor bore contour and other machining errors of the housing can lead to fretting or loss of clearance. Surface finish of housing bore should be 1 to 1.5 μm cla. Oil distribution grooves in the bore of the housing should be avoided for all except thick-wall half bearings.

Rotating loads

High rotating loads (in excess of 4–6 MN/m^2) such as occur in the planet wheels of epicyclic gearboxes may result in bushes very slowly creeping round in their housings and working out of the end. Location features such as dowels are unable to restrain the bush and it is normal practice in these circumstances to cast the lining material on the pin.

PRECAUTIONS WHEN FITTING PLAIN BEARINGS

Half bearings

(1) *Ensure cleanliness:* between bearing and housing, between bearing bore and joint face of bearing and housing, and in oil-ways and oil system.
(2) *Check free spread:* bearing should be in contact with housing in region of joint face.
(3) *Check nip:* tighten bolts, slacken one side to hand tight, use feeler gauge on housing joint. Tighten bolts and repeat on other side.
(4) *Tighten bolts:* to specified stretch or torque.
(5) *Check clearance:* between journal and assembled bore with leads or similar, particularly checking for distortions (local loss of clearance) near oil grooves or bearing joint face. (For new design of housing conduct fitting test: measure bore diameter with shaft removed.)
(6) *Oil surface:* before turning shaft.

Bushes

(1) *For insertion:* use oil or other suitable lubricant to facilitate pressing-in and prevent scoring of bush outside diameter. Use a 15° lead in chamfer in housing, and press in square.
(2) *Check assembled bore size:* with comparator or plug gauge.
(3) *Remove burrs and sharp edges from shaft:* to avoid damage to bush bore during assembly.
(4) *Check cleanliness:* in bush bore, at bush ends and in lubricant supply drillings.

SPHERICAL BEARINGS

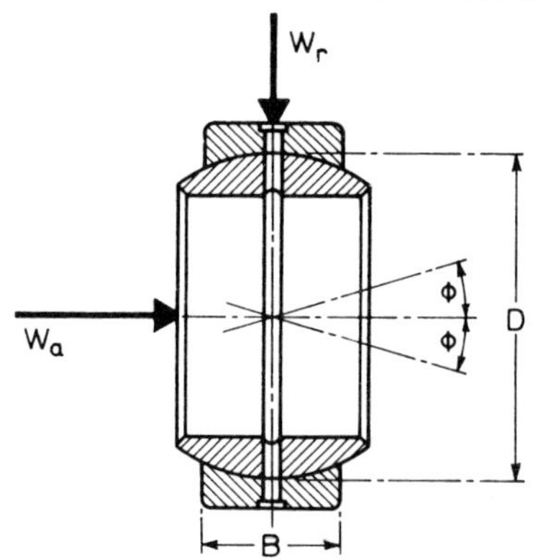

EQUIVALENT BEARING PRESSURE

$$p = \frac{W_r^2 + 6W_a^2}{W_r BD}$$

PROVIDED THAT

$W_a < W_r$

AS A RULE: $\Phi < 8°$

Bearing life calculations:

$$L = f \cdot \left(\frac{p_o}{p}\right)^3 \times 10^5$$

L = bearing life, i.e. average number of oscillations to failure assuming unidirectional loading

f = life-increasing factor depending on periodical relubrication (ref. table below)

p_o = maximum bearing pressure if an average bearing-life of 10^5 number of oscillations is to be expected (ref. table below)

p = equivalent bearing pressure (ref. formula at left)

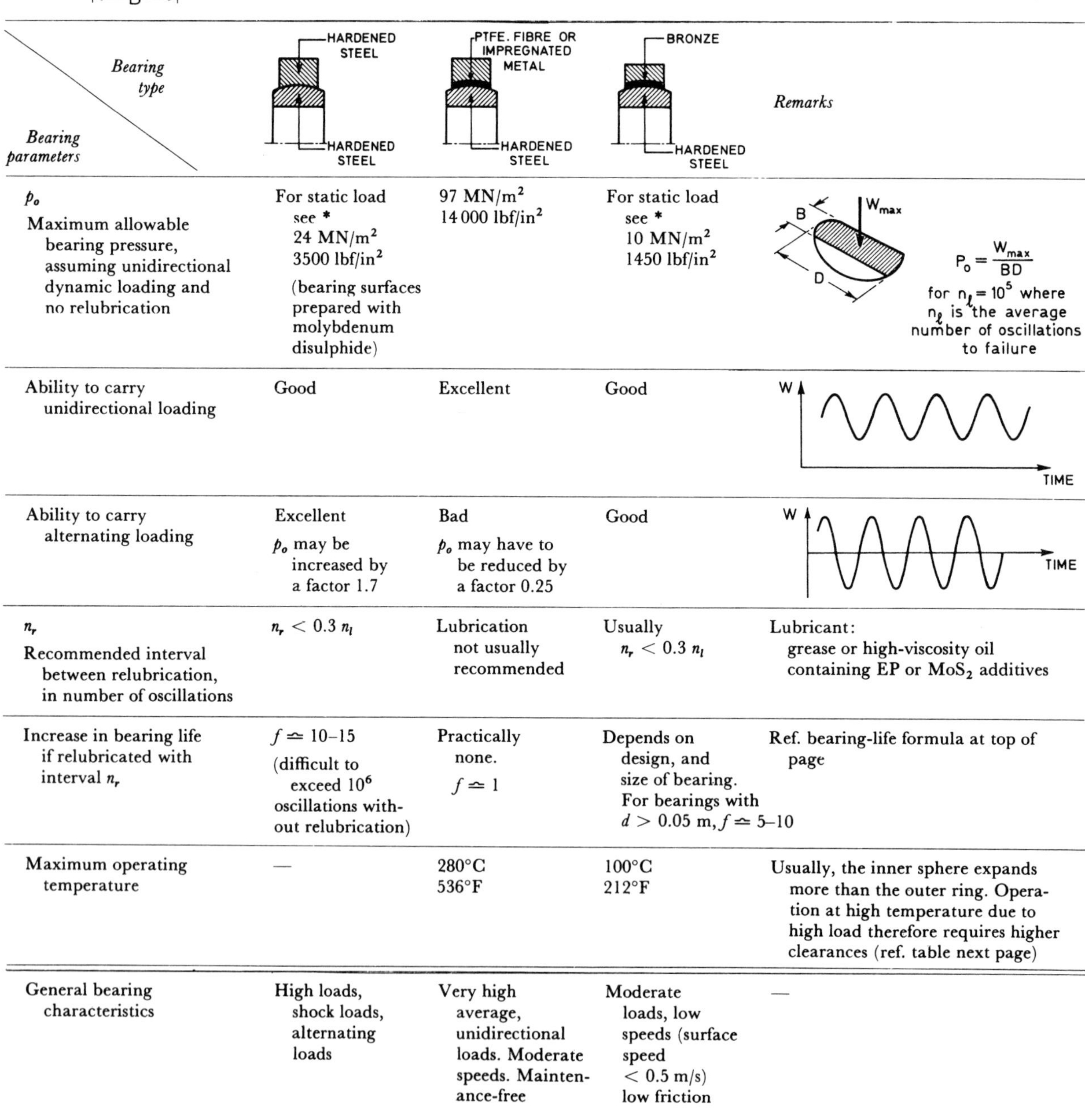

Bearing parameters / *Bearing type*	Hardened steel / hardened steel	PTFE. fibre or impregnated metal / hardened steel	Bronze / hardened steel	*Remarks*
p_o Maximum allowable bearing pressure, assuming unidirectional dynamic loading and no relubrication	For static load see * 24 MN/m^2 3500 lbf/in^2 (bearing surfaces prepared with molybdenum disulphide)	97 MN/m^2 14 000 lbf/in^2	For static load see * 10 MN/m^2 1450 lbf/in^2	$P_o = \frac{W_{max}}{BD}$ for $n_l = 10^5$ where n_l is the average number of oscillations to failure
Ability to carry unidirectional loading	Good	Excellent	Good	
Ability to carry alternating loading	Excellent p_o may be increased by a factor 1.7	Bad p_o may have to be reduced by a factor 0.25	Good	
n_r Recommended interval between relubrication, in number of oscillations	$n_r < 0.3\ n_l$	Lubrication not usually recommended	Usually $n_r < 0.3\ n_l$	Lubricant: grease or high-viscosity oil containing EP or MoS_2 additives
Increase in bearing life if relubricated with interval n_r	$f \simeq 10$–15 (difficult to exceed 10^6 oscillations without relubrication)	Practically none. $f \simeq 1$	Depends on design, and size of bearing. For bearings with $d > 0.05$ m, $f \simeq 5$–10	Ref. bearing-life formula at top of page
Maximum operating temperature	—	280°C 536°F	100°C 212°F	Usually, the inner sphere expands more than the outer ring. Operation at high temperature due to high load therefore requires higher clearances (ref. table next page)
General bearing characteristics	High loads, shock loads, alternating loads	Very high average, unidirectional loads. Moderate speeds. Maintenance-free	Moderate loads, low speeds (surface speed < 0.5 m/s) low friction	—

* The table is based on dynamic load conditions. For static load conditions, where the load-carrying capacity of the bearing is based on bearing-surface permanent deformation, not fatigue, the load capacity of steel bearings may reach $10 \times p_o$ and of aluminium bronze $5 \times p_o$.

Spherical bearings are commonly used where the oscillatory motion is a result of a misalignment, which may be intentional or not. A lateral oscillatory motion is then often combined with a rotatary motion about the bearing axis. If the frequency of rotation is high, it may require a separate bearing fitted inside the spherical assembly.

Clearance

Bore		*Radial clearance C_r*					
D		*C2*		*Normal*		*C3*	
From	Up to	Min.	Max.	Min.	Max.	Min.	Max.
mm		μm		μm		μm	
—	12	8	32	32	68	68	104
12	20	10	40	40	82	82	124
20	35	12	50	50	100	100	150
35	60	15	60	60	120	120	180
69	90	18	72	72	142	142	212
90	140	18	85	85	165	165	245
140	240	18	100	100	192	192	284
240	300	18	110	110	214	214	318

Clearance definitions :-

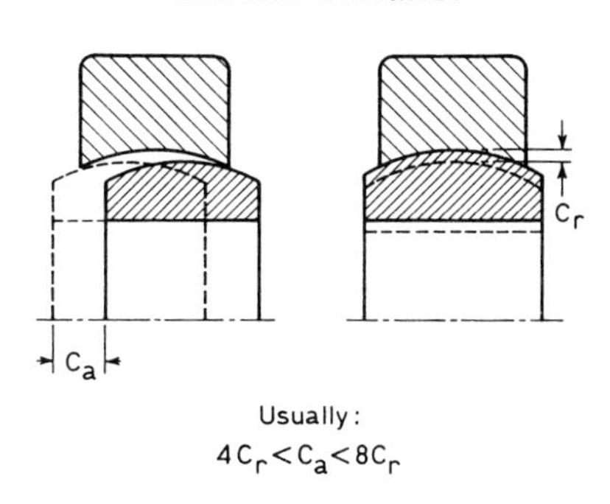

C2 CLEARANCE may somtimes be chosen if the loading is alternating

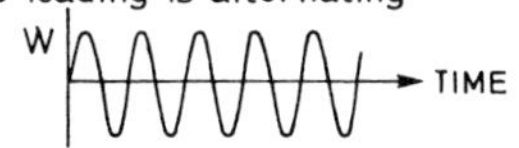

C3 CLEARANCE is sometimes required if the temperature difference between inner sphere and outer ring is large, or the bearing works in dirty conditions

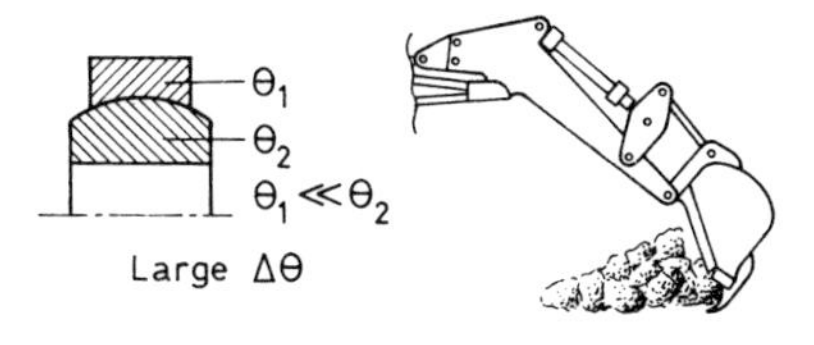

Fit

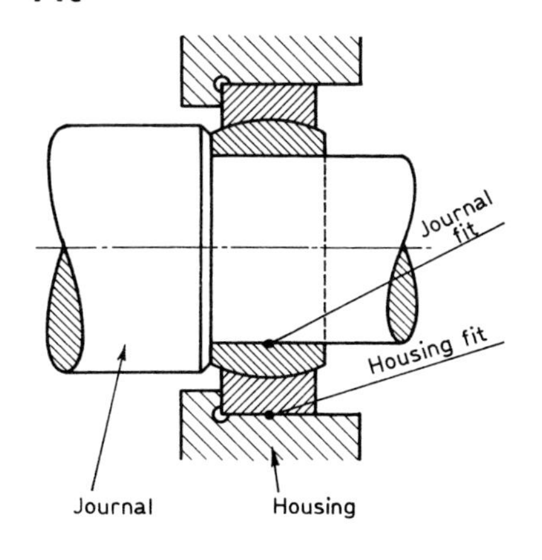

Contact point	*Most common fit*			*Housing made of light alloy*	*Push fit in housing*	*Push fit on journal*
	C2	*Normal*	*C3*			
Recommended journal fit	*h6*	*m6*	*m6*	—	*m6*	*h6* Journal should be hardened
Recommended housing fit	*J7*	*K7* or *M7*	*M7*	*N7**	*H7*	*M7*

* In general, if the housing is made of light-metal alloy or other soft material, or the housing is thin-walled, a tighter fit is required. Recommended: one degree tighter fit than otherwise would have been chosen, i.e. *M* is replaced by *N*, *J* by *K*, etc.

ROLLING BEARINGS

Rolling bearings (i.e. ball, roller and needle bearing) may also be used for oscillatory motion, but preferably where the load is unidirectional, or at least varies with moderate gradient.

Static (or near static) loading

Bearings are selected on the basis of their load-carrying capacity.

Examples:

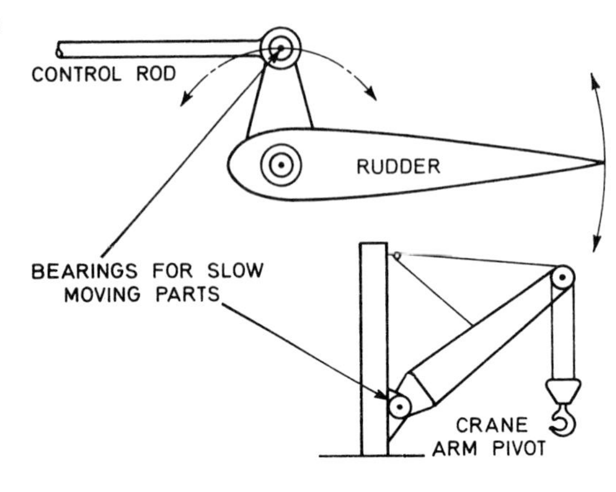

Rule:

Use manufacturers figures for the static load coefficient C_o multiplied by a factor f such that:

- $f = 0.5$ for sensitive equipment (weights, recorders, etc.)
- $f = 1.0$ for crane arms, etc.
- $f = 5.0$ for emergency cases on control rods (e.g. for aircraft controls).

Dynamic loading

Bearings are selected on the basis of their required life before failure.

Examples:
small-end bearings in engines and compressors
plunger-pin bearings in crank-operated presses
connecting rods in textile machinery
connecting rods in wood reciprocating saws

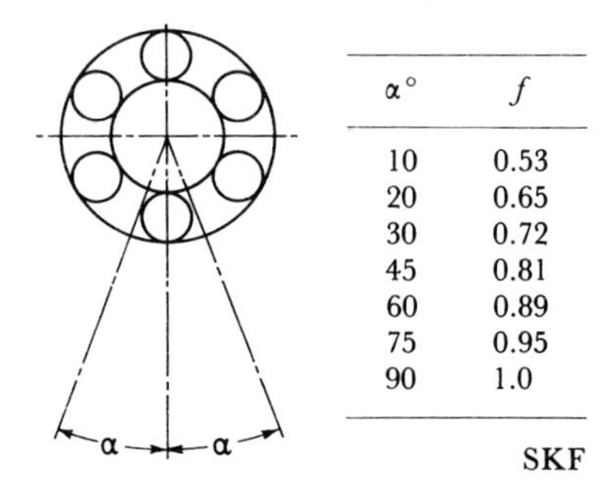

$\alpha°$	f
10	0.53
20	0.65
30	0.72
45	0.81
60	0.89
75	0.95
90	1.0

SKF

Rule:

(1) If $\alpha \geqslant 90°$, each oscillation is considered as a complete revolution, and the bearing life is determined as if the bearing was rotating.

(2) If $\alpha < 90°$, the equivalent load on the bearing is reduced by a factor f, taken from the table above. The calculations are then carried out as under (1).

ENGINE SMALL-END BEARINGS

Type of engine	*2-Stroke*		*4-Stroke*	
	Small and medium engines	*Large engines*		
Type of bearing and oil grooving	NEEDLE BEARINGS: single	A series of axial oil grooves interconnected by a circumferential groove	PLAIN BEARING	Single central circumferential oil groove. For small bearings, sometimes only oil hole(s)
Type of gudgeon pin	double	d, D, B	$d \approx B$ $d \approx D/3$	Surface finish better than 0.05 µm CLA Material: Surface hardened steel
*Bearing material and allowable pressure**	21–35 MN/m² 3000–5000 lbf/in². The load capacity varies, refer to manufacturer's specifications	Phosphor bronze 25–30 MN/m² 3500–4300 lbf/in²	Lead bronze up to 25 MN/m² 3500 lbf/in²	Phosphor bronze up to 50 MN/m² 7100 lbf/in²
Type of friction	Rolling friction	Mixed to boundary friction	Mostly mixed lubrication, but may be fully hydrodynamic under favourable condition	
Diametral clearance	Clearances not applicable. Fits: Pin/Piston *J6*; Ring/conrod *P7*	≃ 1 µm/mm of *d*	1–1.5 µm/mm of *d*	≃ 1 µm/mm of *d*
Remarks	Needle bearings are best suited where the loading is uni-directional	This type of bush is often made floating in a fixed steel bush	Liable to corrosion in plain mineral oil. An overlay of lead-base white metal will reduce scoring risk	

* Bearing pressure is based on projected bearing area, i.e. $B \times d$ (ref. sketch above).

ENGINE CROSSHEAD BEARINGS

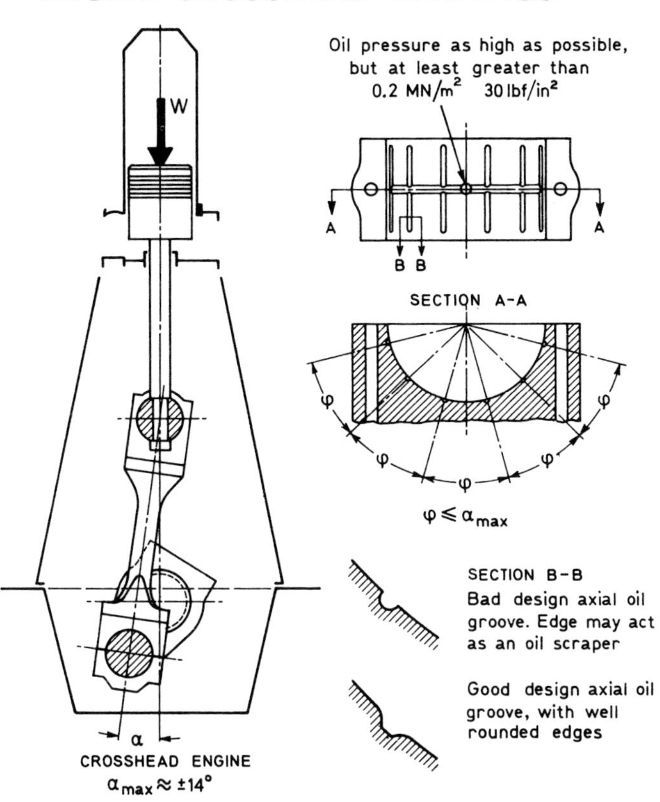

Bearing materials*	Maximum allowable peak pressure	Diametral clearance (in μm/mm of pin dia.)	Remarks
White metal (tin base)	7 MN/m^2 1000 lbf/in^2	Δ_1 0.5–0.7	Excellent resistance against scoring. Corrosion resistant. Low fatigue strength
Copper–lead	14 MN/m^2 2000 lbf/in^2	$2\Delta_2$ ≃ 1	High-strength bearing metal sensitive to local high pressure. Liable to corrosion by acidic oil unless an overlay of lead–tin or lead–indium is used (≃ 25 μm)
Tri-metal e.g. steel copper–lead white metal (lead)	14 MN/m^2 2000 lbf/in^2	$2\Delta_2$ ≃ 1	Same as above, but better resistance to corrosion, wear and scoring. Installed as bearing shells, precision machined

* Tin–aluminium is also a possible alloy for crosshead bearings, and spherical roller bearings have been used experimentally.

EXAMPLE OF OIL GROOVE DESIGN IN A CROSSHEAD BEARING OF CA 300mm DIAMETER. ALL MEASUREMENTS ARE IN MILLIMETRES (SECTION B–B)

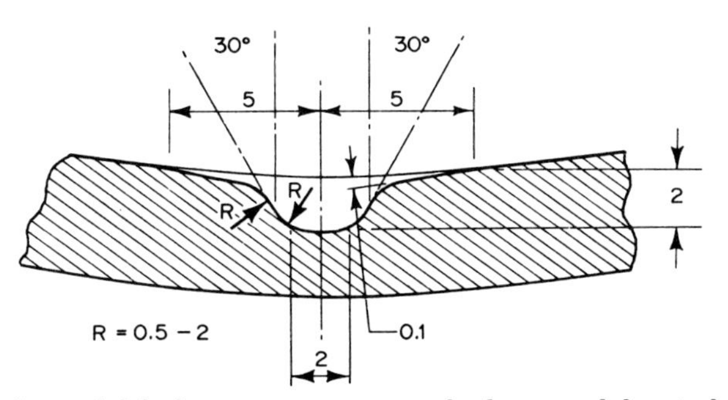

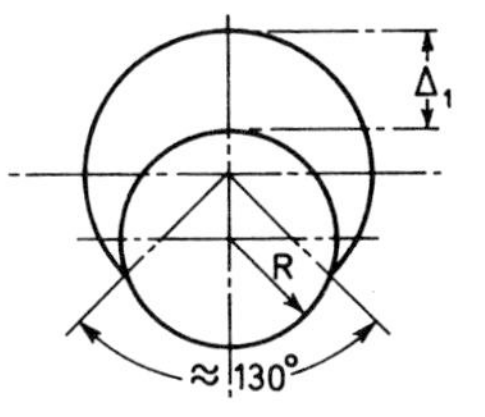

Old (and still common) practice:

Bearing metal scraped to conformity with wristpin over an arc of ≃ 120–150°. Mostly used for large, two-stroke marine diesel engines. Works mainly with boundary friction.

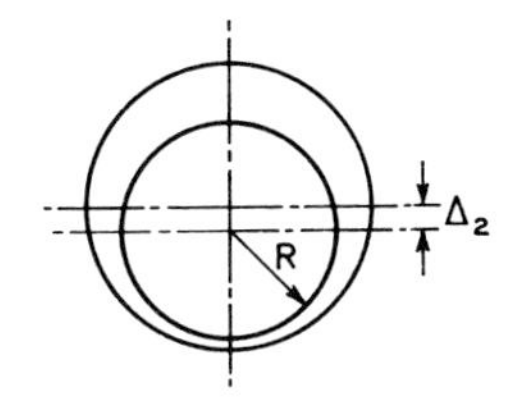

New practice:

Bearing precision machined to an exact cylinder with radius slightly greater than wristpin radius. Mainly hydrodynamic lubrication. In common use for 4-stroke engines, and becoming common on 2-stroke engines.

Local high pressures and thermal instability

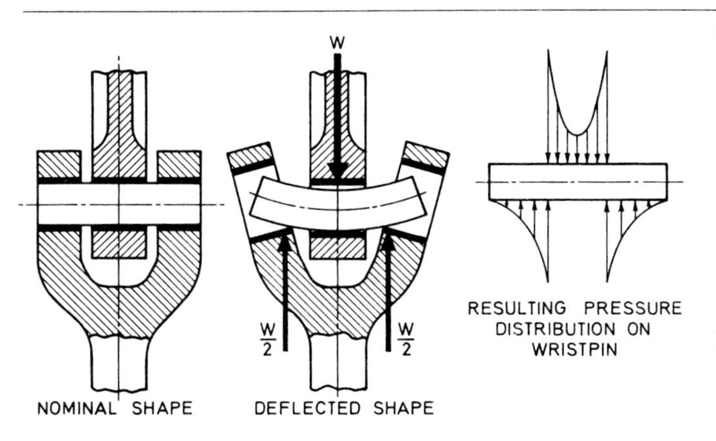

Oscillating bearings in general, and crosshead bearings in particular, have a tendency to become thermally unstable at a certain load level. It is therefore of great importance to avoid local high pressures due to wear, misalignments or deflections such as shown in the figure at left.

In critical machine components, such as crosshead bearings, temperature warning equipment should be installed

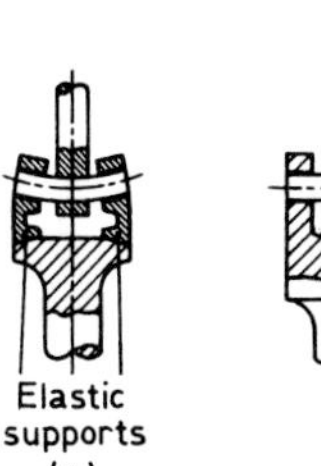

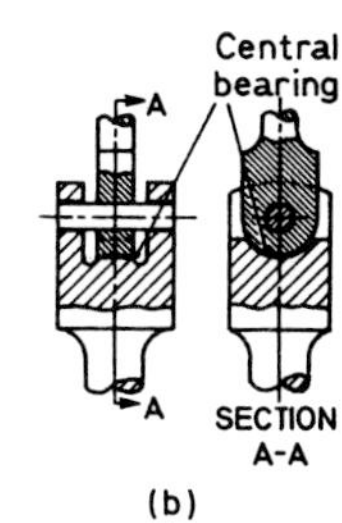

Two possible solutions used in crosshead bearings:
(a) Elastic bearing supports
(b) Upper end of connecting rod acts as a partial bearing (central loading)

Squeeze action

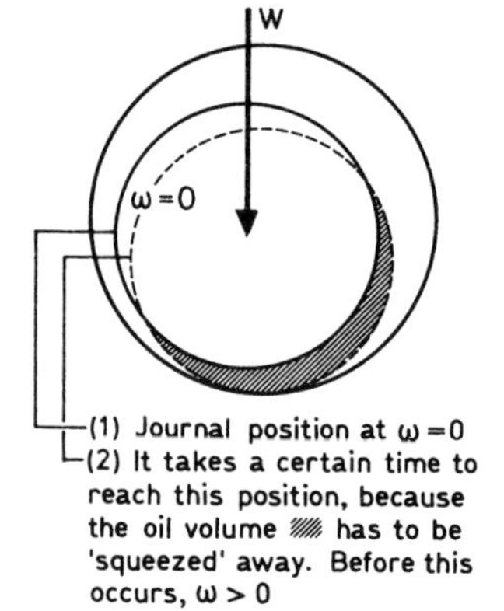

In spite of the fact that the angular velocity of the journal is zero twice per cycle in oscillating bearings, such bearings may still work hydrodynamically. This is due to the squeeze action, shown in the diagram. This squeeze action plays an important role in oscillating bearings, by preventing excessive metallic contact

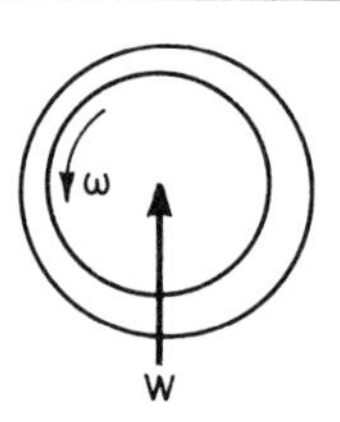

THE LOAD REVERSES ITS DIRECTION DURING THE CYCLE

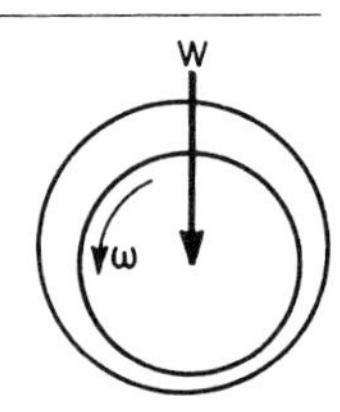

On some bearings, the load reverses direction during the cycle. This will help to create a thicker oil film at velocity reversal, and thus the squeeze action will be more effective. Load reversal takes place in piston pin and crosshead bearings in 4-stroke engines, but not in 2-stroke engines. The latter engines are therefore more liable to crosshead bearing failure than the former

OSCILLATORY BEARINGS WITH SMALL RUBBING VELOCITY

In oscillatory bearings with small rubbing velocities, it is necessary to have axial oil grooves in the loaded zone, particularly if the load is unidirectional.

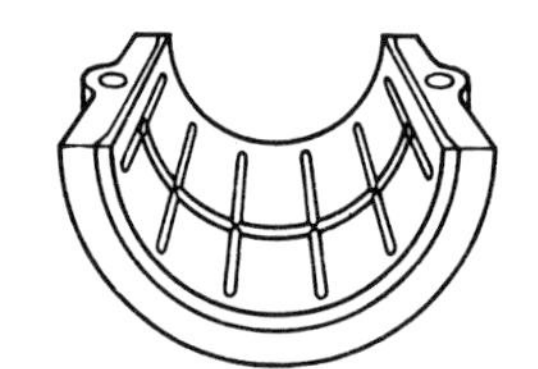

EXAMPLES OF GROOVE PATTERN

Oil lubricated bearing

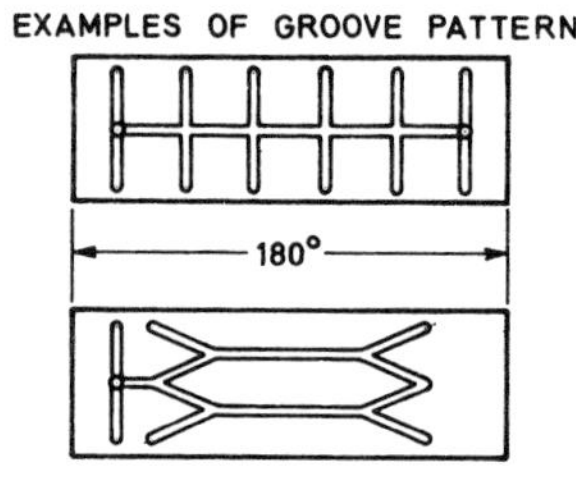

Grease lubricated bearing

Grease lubricated bearing

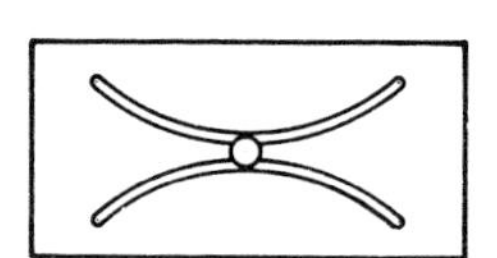

Bronze is a common material in oscillatory journal bearings with small rubbing velocity and large, unidirectional loading.

Bearings are often made in the form of precision machined bushings, which may be floating.

W up to 60 MN/m^2; 8500 lbf/in^2.

$\Delta \simeq 1.2\ \mu m/mm$ of D.

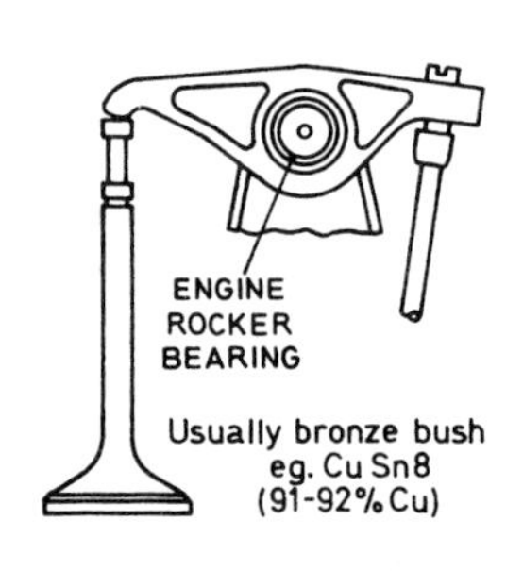

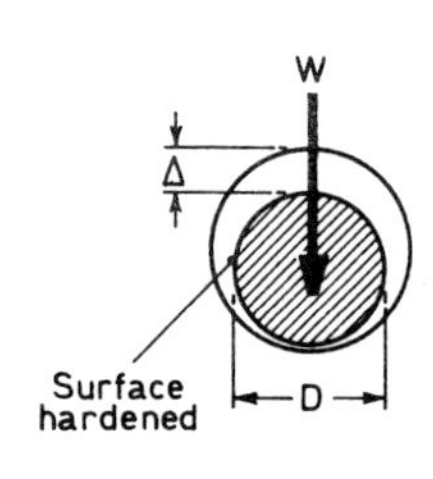

Example of floating bush

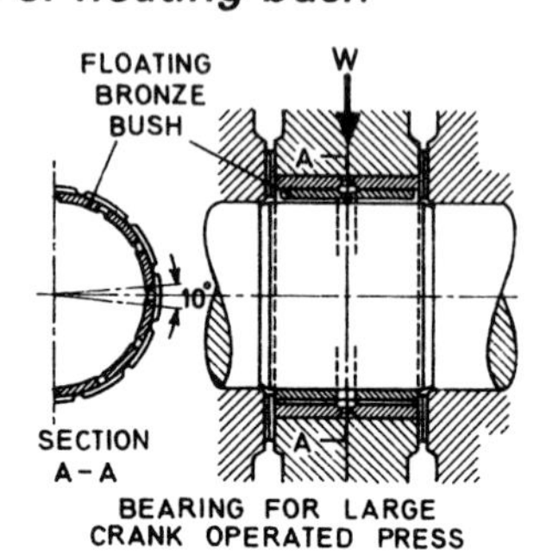

In the example left the projected bearing area is 0.018m^2 (278 in^2) and the bearing carries a load of 2 MN ($\simeq$ 450.000 lbf).

$$W = 2\ MN\ (\simeq 450\,000\ lbf)$$

The bush has axial grooves on inside *and* outside. On the outside there is a circumferential groove which interconnects the outer axial grooves and is connected to the inner axial grooves by radial drillings.

SPHERICAL BEARINGS FOR OSCILLATORY MOVEMENTS (BALL JOINTS)

Types of ball joints

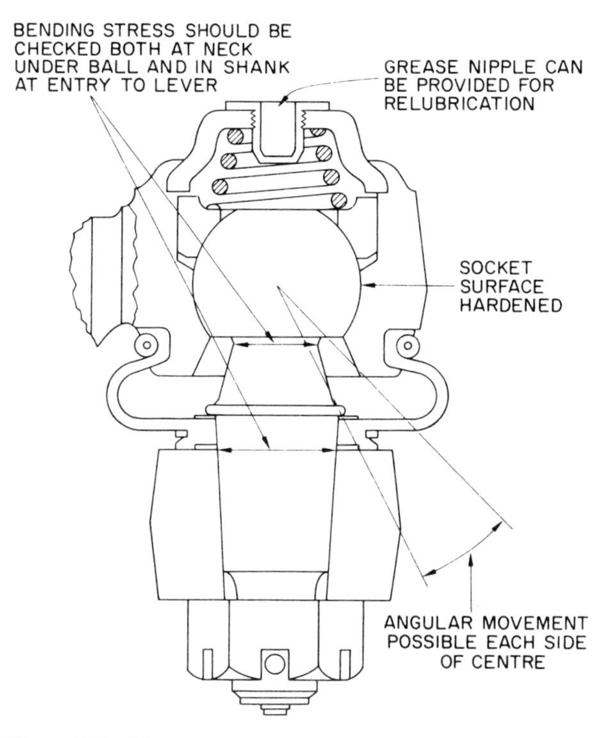

Fig. 14.1. Transverse type ball joint with metal surfaces (*courtesy:* Automotive Products Co. Ltd)

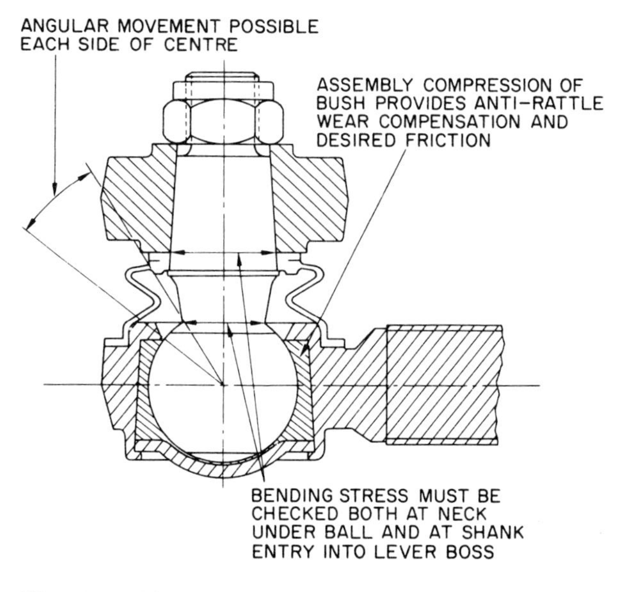

Fig. 14.2. Transverse steering ball joint (*courtesy:* Cam Gears Ltd)

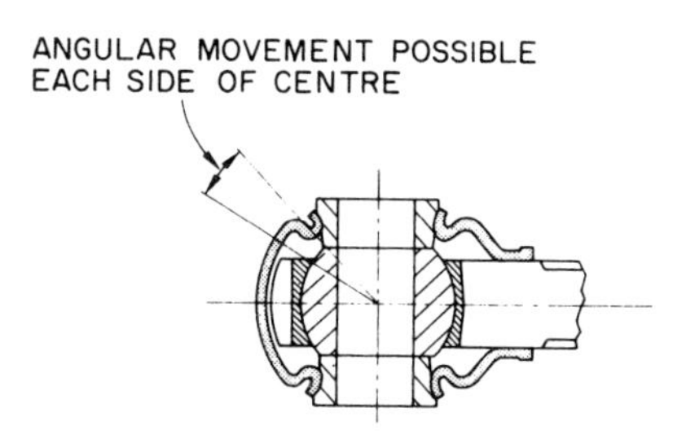

Fig. 14.3. Straddle type joint shown with gaiters and associated distance pieces (*courtesy:* Rose Bearings Ltd)

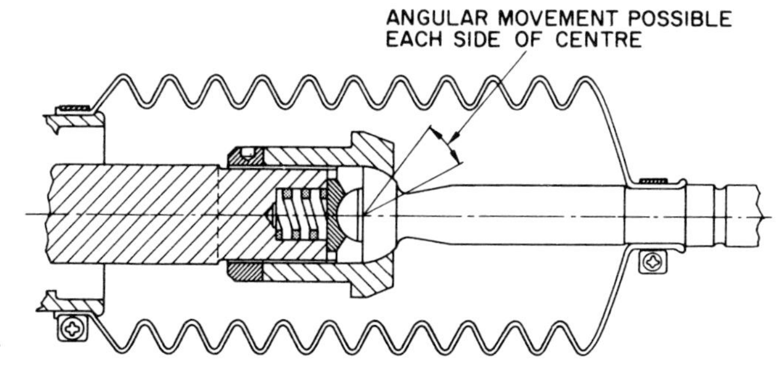

Fig. 14.4. Axial ball joint (*courtesy:* Cam Gears Ltd)

Selection of ball joints

The many different forms of ball joints developed for a variety of purposes can be divided into two main types, *straddle mounted* [rod ends], and *overhung*. They may be loaded perpendicularly to, or in line with the securing axis. Working loads on ball joints depend upon the application, the working pressures appropriate to the application, the materials of the contacting surfaces and their lubrication, the area factor of the joint and its size. The area factor, which is the projected area of the tropical belt of width L divided by the area of the circle of diameter D, depends upon the ratio L/D. The relationship is shown in the graph (Fig. 14.6).

Transverse types are seldom symmetrical and probably have a near equatorial gap (Fig. 14.1) but their area factors can be arrived at from Fig. 14.6 by addition and subtraction or by calculation. For straddle and transverse type joints, either the area factor or an actual or equivalent L/D ratio could be used to arrive at permissible loadings, but when axially loaded joints are involved it is more convenient to use the area factor throughout and Fig. 14.6 also shows the area factor–L/D ratio relationship for axially loaded joints.

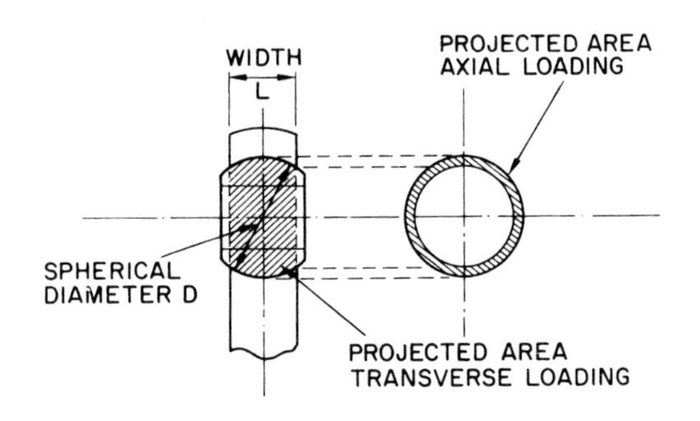

Fig. 14.5. Ball joint parameters

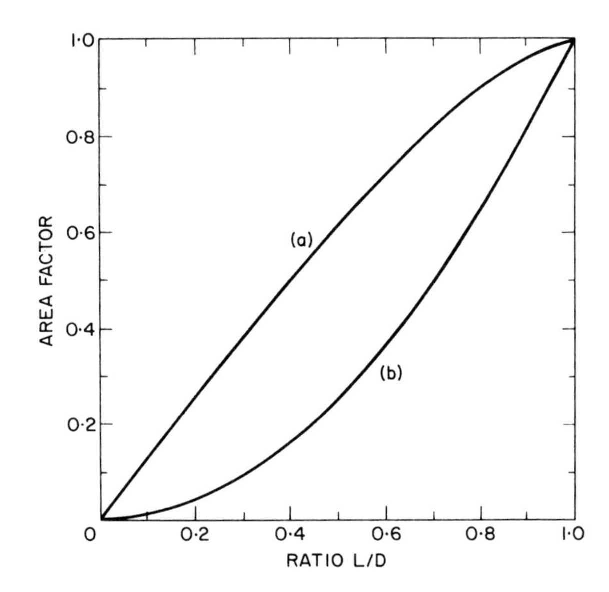

Fig. 14.6. Area factors (a) transverse and straddle type ball joints (b) axial type ball joints

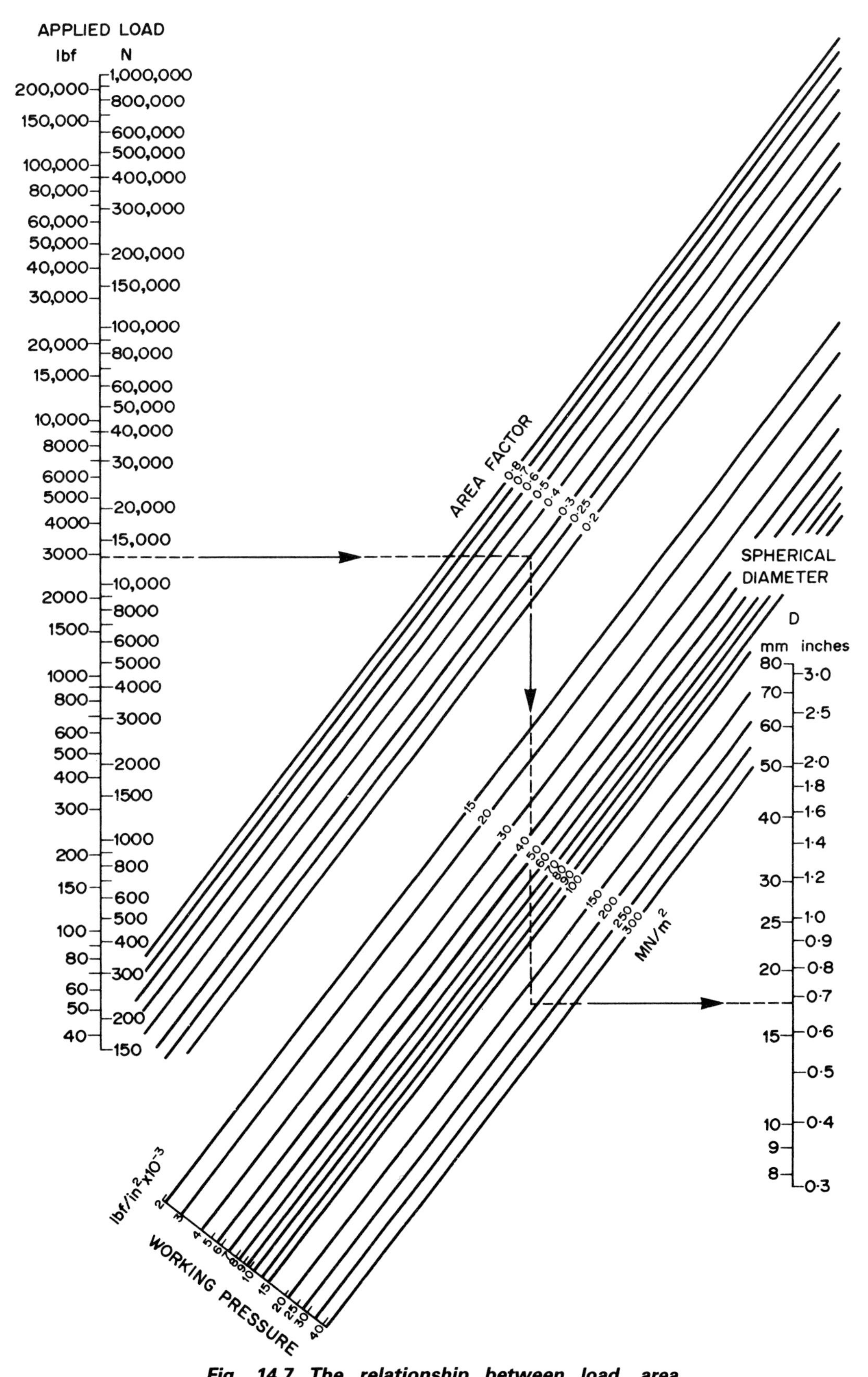

Fig. 14.7. The relationship between load, area factor, working pressure and spherical diameter of ball joints

A guide to the selection and performance of ball joints

Type	*Straddle or rod end*	*Axial*	*Transverse*
Angle	±10° to ±15° with minimum shoulder on central pin, ±30° to 40° with no shoulder on central pin	±25° to ±30°	±10° to ±15° low angle ±25° to ±30° high angle
Main use	Linkages and mechanisms	Steering rack end connections	Steering linkage connections, suspension and steering articulations
Lubrication	Grease	Grease	Lithium base grease on assembly. Largest sizes may have provision for relubrication
Enclosure and protection	Often exposed and resistant to liquids and gases. Rubber gaiters available (Fig. 14.3)	Rubber or plastic bellows, or boot	Rubber or plastic seals, or bellows
Materials	*Inner.* Case or through hardened steel, hardened stainless steel, hardened sintered iron; possibly chromium plated *Outer bearing surfaces.* Aluminium bronze, naval bronze, hardened steel, stainless steel, sintered bronze, reinforced PTFE	*Ball.* Case-hardened steel *Bushes.* Case or surface hardened steel, bronze, plastic or woven impregnated	*Ball.* Case-hardened steel *Bushes.* Case or surface hardened steel, bronze, plastic or woven impregnated
Working pressures	Limiting static from 140 MN/m^2 to 280 MN/m^2 on projected area depending on materials. Wear limited on basis of 50×10^3 cycles of ±25 at 10 cycles/min from 80MN/m^2 to 180 MN/m^2 depending on materials	35 to 50 MN/m^2 on maximum or measured forces	Approximately 15 MN/m^2 on plastic, 20 MN/m^2 on metal surface projected areas. Bending stress in the neck or shank which averages 15 times the bearing pressure limits working load. Fatigue life must also be considered
Area factors	0.42 to 0.64 with radial loads and 0.12 to 0.28 with axial loads	0.25	0.55 large angles 0.7 small angles
Remarks	No provision to take up wear which probably determines useful life. Use Fig. 14.7 for selection or consult manufacturer	Spring loaded to minimise rattle and play	Steering and suspension joints spring loaded to minimise rattle and play and to provide friction torque. Some plastic bush joints rely on compression assembly for anti-rattle and wear compensation

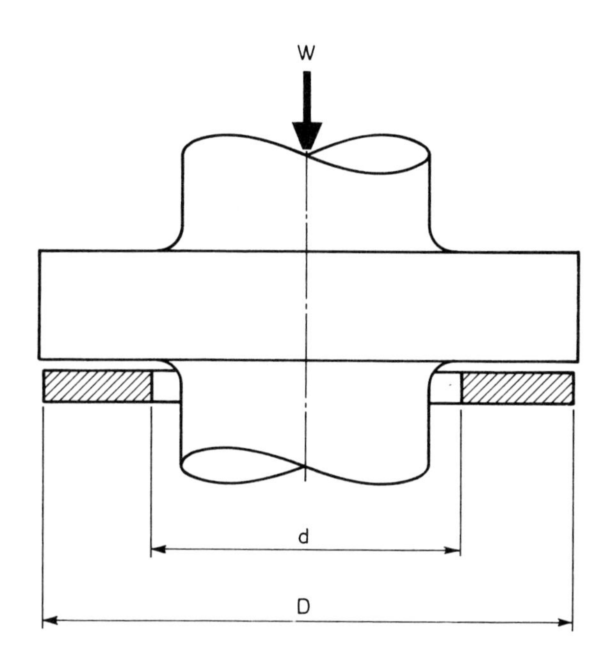

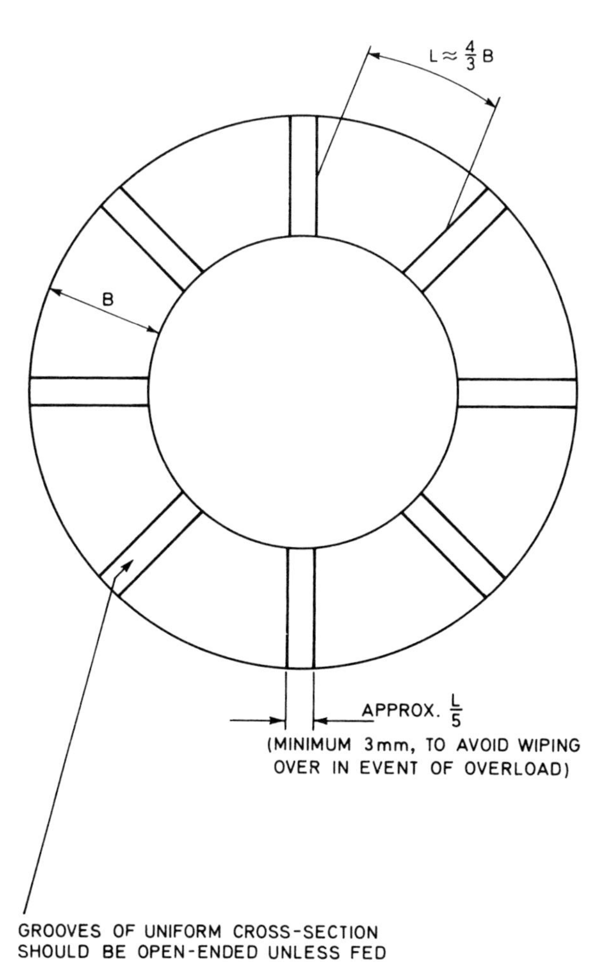

Plain thrust washers are simple and occupy little axial space. Their performance cannot be predicted with accuracy because their operation depends upon small-scale surface undulations and small dimensional changes arising from thermal expansion whilst running.

Thrust washers with radial grooves (to encourage hydrodynamic action) are suitable for light loads up to 0.5 MN/m^2 (75 lbf/in^2), provided the mean runner speed is not less than the minimum recommended below according to lubricant viscosity.

Minimum sliding speeds to achieve quoted load capacity

Viscosity grade ISO 3448	*Minimum sliding speed* $= \pi n d_m$	
	m/s	in/s
100	2.5	100
68	4	160
46	6	240
32	8	320

Suitable materials

0.5 mm (approx) white metal on a steel backing, overall thickness 2–5 mm, with a Mild Steel Collar.

or

Lead bronze washer with a hardened steel collar.

Recommended surface finish for both combinations
Bearing 0.2–0.8 μm Ra
(8–32 μin cla)
Collar 0.1–0.4 μm Ra
(4–16 μin cla)

Estimation of approximate performance

Recommended maximum load:

$$W = K_1(D^2 - d^2)$$

Approximate power loss in bearing:

$$H = K_2\, n\, d_m\, W$$

Lubricant flow rate to limit lubricant (oil) temperature rise to 20°C:

$$Q = K_3\, H$$

Symbol and meaning	*SI units*	*Imperial units*
W load	N	lbf
H power loss	W	h.p.
Q flow rate	m^3/s	g.p.m.
n rotational speed	rev/s	r.p.s.
d, D	mm	in
d_m $(D+d)/2$	mm	in
K_1	0.3	48
K_2	70×10^{-6}	11×10^{-6}
K_3	0.03×10^{-6}	0.3

Lubricant feeding

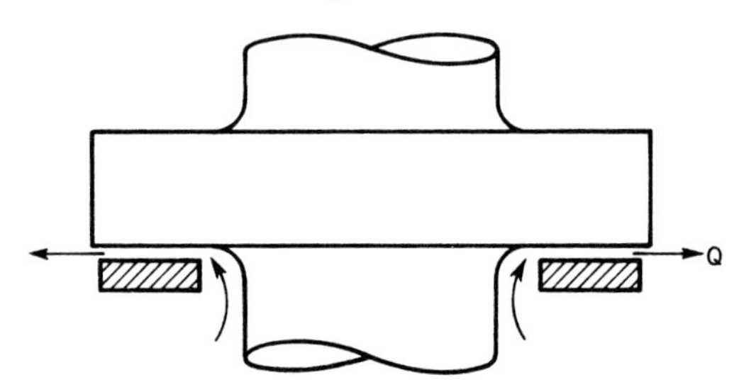

Lubricant should be fed to, or given access to, inner diameter of the bearing so that flow is outward along the grooves.

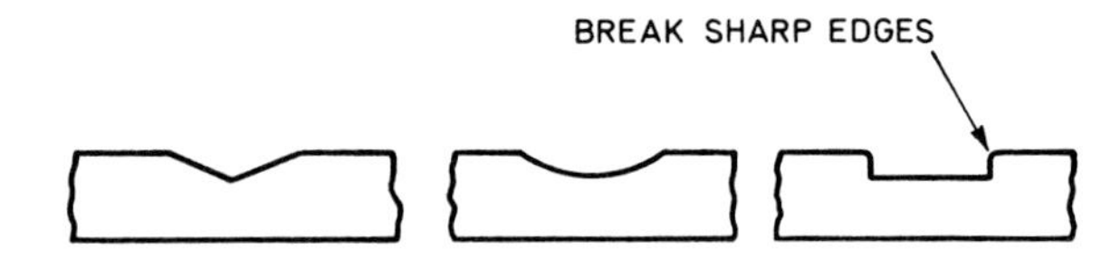

Suitable groove profiles

For horizontal-shaft bearings the grooves may have to be shallow (0.1 mm) to prevent excess drainage through the lower grooves, which would result in starvation of the upper pads. For bearings operating within a flooded housing a groove depth of 1 mm is suitable.

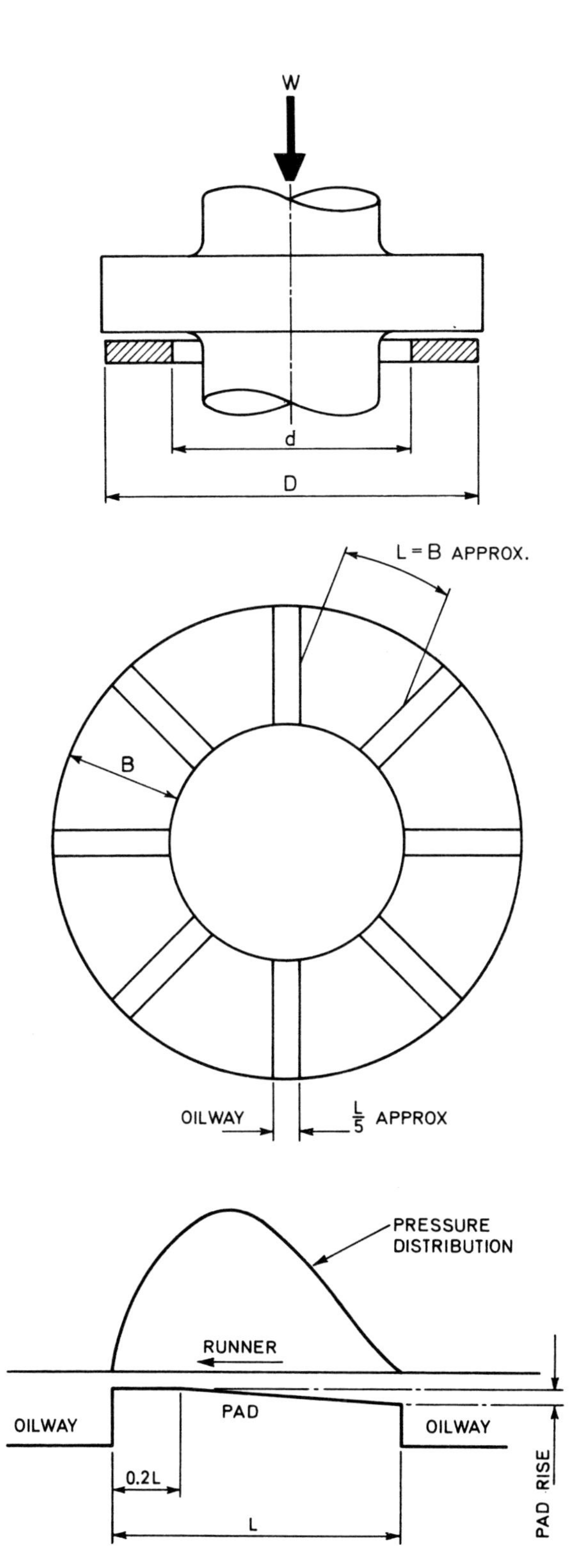

Fig. 16.1 Bearing and pad geometry

BEARING TYPE AND DESCRIPTION

The bearing comprises a ring of sector-shaped pads. Each pad is profiled so as to provide a convergent lubricant film which is necessary for the hydrodynamic generation of pressure within the film. Lubricant access to feed the pads is provided by oil-ways which separate the individual pads. Rotation of the thrust runner in the direction of decreasing film thickness establishes the load-carrying film. For bi-directional operation a convergent–divergent profile must be used (see later). The geometrical arrangement is shown in Fig. 16.1

FILM THICKNESS AND PAD PROFILE

In order to achieve useful load capacity the film thickness has to be small and is usually in the range 0.005 mm (0.0002 in) for small bearings to 0.05 mm (0.002 in) for large bearings. For optimum operation the pad rise should be of the same order of magnitude. Guidance on suitable values of pad rise is given in Table 16.1.

The exact form of the pad surface profile is not especially important. However, a flat land at the end of the tapered section is necessary to avoid excessive local contact stress under start-up conditions. The land should extend across the entire radial width of the pad and should occupy about 15–20% of pad circumferential length.

Table 16.1 Guidance on suitable values of pad rise

Bearing inner diameter d		*Pad rise*	
mm	inch	mm	inch
25	1	0.015–0.025	0.0006–0.001
50	2	0.025–0.04	0.001 –0.0016
75	3	0.038–0.06	0.0015–0.0025
100	4	0.05 –0.08	0.002 –0.0032
150	6	0.075–0.12	0.003 –0.0048
200	8	0.10 –0.16	0.004 –0.0064
250	10	0.12 –0.20	0.005 –0.008

It is important that the lands of all pads should lie in the same plane to within close tolerances; departure by more than 10% of pad rise will significantly affect performance (high pads will overheat, low pads will carry little load). Good alignment of bearing and runner to the axis of runner rotation (to within 1 in 10^4) is necessary. Poorly aligned bearings are prone to failure by overheating of individual pads.

GUIDE TO BEARING DESIGN

Bearing inner diameter should be chosen to provide adequate clearance at the shaft for oil feeding and to be clear of any fillet radius at junction of shaft and runner.

Bearing outer diameter will be determined, according to the load to be supported, as subsequently described. Bearing power loss is very sensitive to outer diameter, and conservative design with an unnecessarily large outside diameter should therefore be avoided.

Oil-ways should occupy about 15–20% of bearing circumference. The remaining bearing area should be divided up by the oil-ways to form pads which are approximately 'square'. The resulting number of pads depends upon the outer/inner diameter ratio—guidance on number of pads is given in Fig. 16.4

Safe working load capacity

Bearing load capacity is limited at low speed by allowable film thickness and at high speed by permissible operating temperature.

Guidance on safe working load capacity is given for the following typical operating conditions:

Lubricant (oil) feed temperature, 50°C
Lubricant temperature rise through bearing housing, 20°C

in terms of a basic load capacity W_b for an arbitrary diameter ratio and lubricant viscosity (Fig. 16.2), a viscosity factor (Fig. 16.3) and a diameter ratio factor (Fig. 16.4). That is:

$$\text{Safe working load capacity} = W_b \times (\text{Viscosity factor}) \times (\text{Diameter ratio factor})$$

Example

To find the necessary outer diameter D for a bearing of inner diameter $d = 100$ mm to provide load capacity of 10^4 N when running at 40 rev/s with oil of viscosity grade 46 (ISO 3448):

From Fig. 16.2, $W_b = 3.8 \times 10^4$ N.
From Fig. 16.3, viscosity factor 0.75.

Necessary diameter ratio factor

$$= \frac{10^4}{3.8 \times 10^4 \times 0.75} = 0.35$$

From Fig. 16.4, D/d required is 1.57.
Therefore, outer diameter D required is 157 mm.

Power loss

For a bearing designed in accordance with the recommendations given, the power loss may be estimated by:

$$H = K_1\, K_2\, K_3\, d^3\, n^{1.66}$$

where the symbols have the meanings given in the following table.

Symbol	*Meaning*	*SI units*	*Imperial units*
d	Inner diameter	mm	in
n	Speed	rev/s	r.p.s.
H	Power loss	W	h.p.
K_1		5.5×10^{-6}	0.12×10^{-3}
K_2	Table 16.2		
K_3	Table 16.3		

Table 16.2 Viscosity grade factor for power loss

Viscosity grade ISO 3448	K_2
32	0.64
46	0.78
68	1.0
100	1.24

Table 16.3 Diameter ratio factor for power loss

D/d	K_3
1.5	0.42
1.6	0.54
1.7	0.66
1.8	0.80
1.9	0.95
2.0	1.1
2.2	1.5
2.4	1.9
2.6	2.4

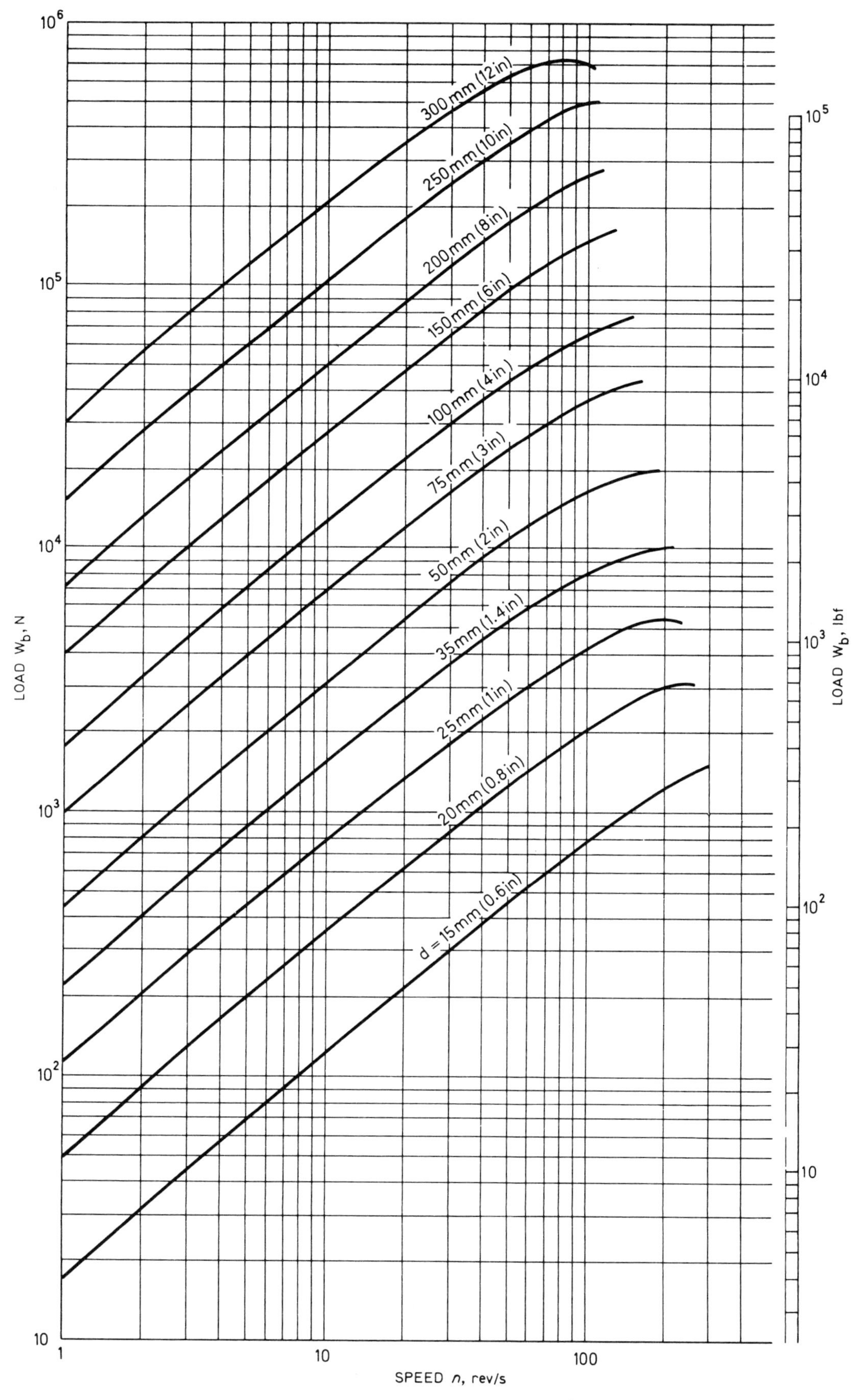

Fig. 16.2. Basic load capacity W_b

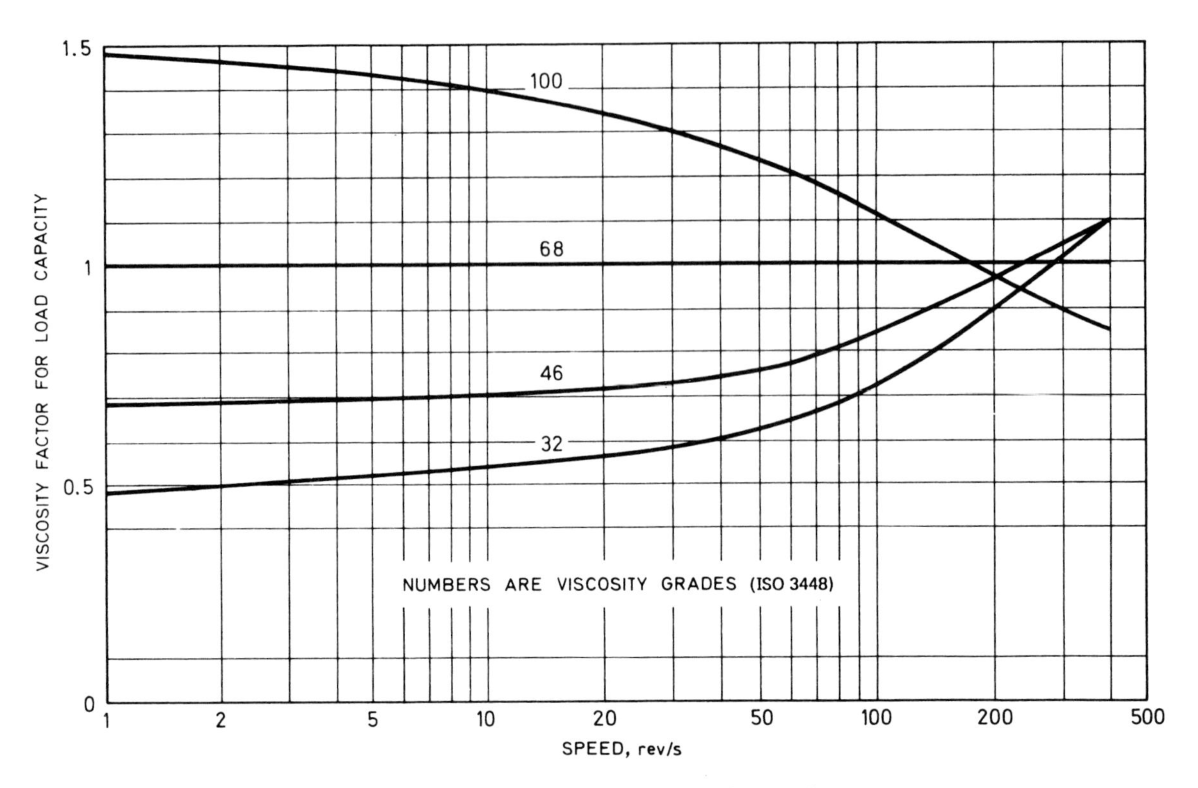

Fig. 16.3. Viscosity factor for load capacity

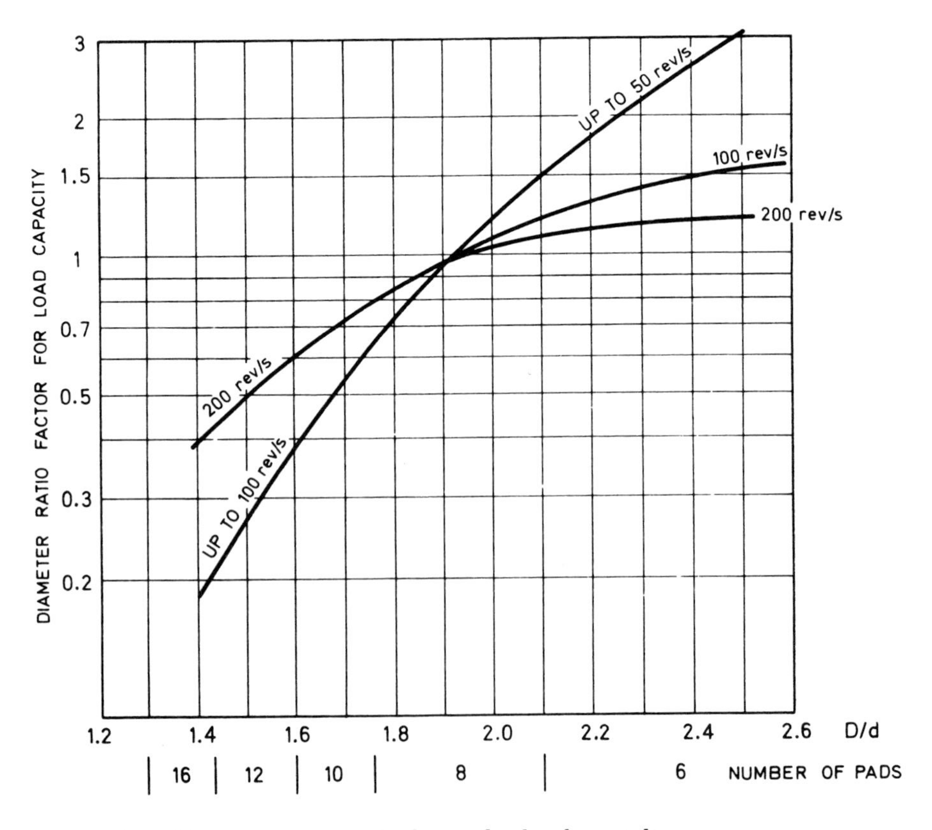

Fig. 16.4. Diameter ratio factor for load capacity

LUBRICANT FEEDING

Lubricant should be directed to the inner diameter of the bearing so that it flows radially outward along the oil-ways. The outlet from the bearing housing should be arranged to prevent oil starvation at the pads.

At high speed, churning power loss can be very significant and can be minimised by sealing at the shaft and runner periphery to reduce the area of rotating parts in contact with lubricant.

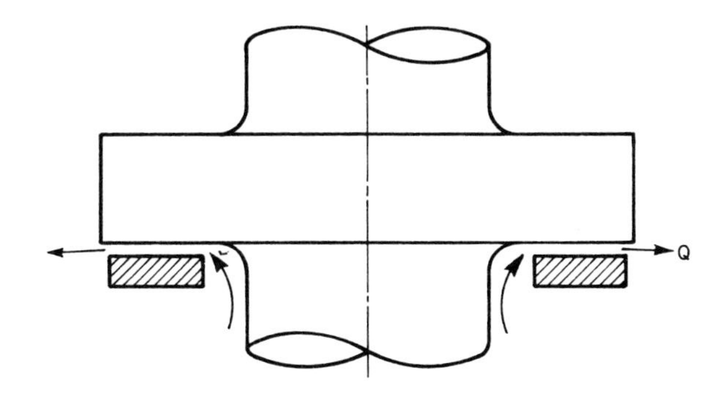

Lubricant feed rate

A lubricant temperature rise of 20°C in passing through the bearing housing is typical. For a feed temperature of 50°C the housing outlet temperature will then be 70°C, which is satisfactory for general use with hydrocarbon lubricants. The flow rate necessary for 20°C temperature rise may be estimated by

$$Q = KH,$$

where $K = 0.3 \times 10^{-7}$ for Q in m^3/s, H in W

or $K = 0.3$ for Q in gal/min, H in h.p.

BEARINGS FOR BI-DIRECTIONAL OPERATION

For bi-directional operation a tapered region at both ends of each pad is necessary. In consequence each pad should be circumferentially longer than the corresponding uni-directional pad. The ratio (mean circumferential length/radial width) should be about 1.7 with central land 20% of length. This results in a reduction of the number of pads in the ring, i.e. about $\frac{2}{3}$ the number of pads of the corresponding uni-directional bearing. The resulting load capacity will be about 65%, and the power loss about 80%, of the corresponding uni-directional bearing.

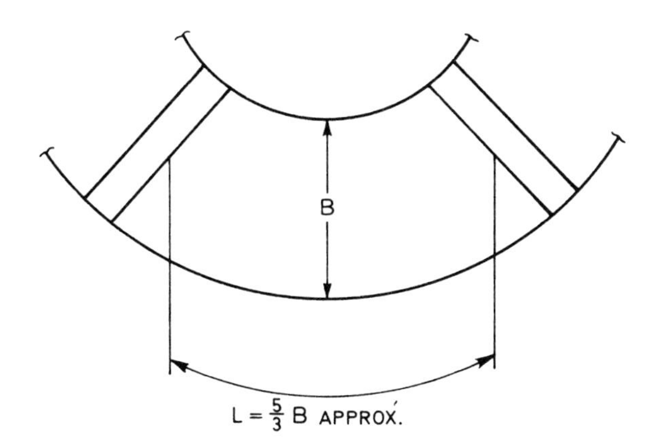

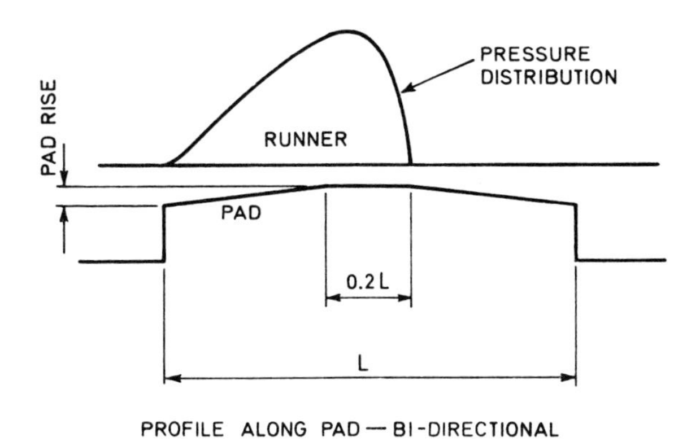

PROFILE ALONG PAD — BI-DIRECTIONAL

A17 Tilting pad thrust bearings

The tilting pad bearing is able to accommodate a large range of speed, load and viscosity conditions because the pads are pivotally supported and able to assume a small angle relative to the moving collar surface. This enables a full hydrodynamic fluid film to be maintained between the surfaces of pad and collar. The general proportions and the method of operation of a typical bearing are shown in Fig. 17.1. The pads are shown centrally pivoted, and this type is suitable for rotation in either direction.

Each pad must receive an adequate supply of oil at its entry edge to provide a continuous film and this is usually achieved by immersing the bearing in a flooded chamber. The oil is supplied at a pressure of 0.35 to 1.5 bar (5–22 lbf/in^2) and the outlet is restricted to control the flow. Sealing rings are fitted at the shaft entry to maintain the chamber full of oil. A plain journal bearing may act also as a seal. The most commonly used arrangements are shown in Fig. 17.2.

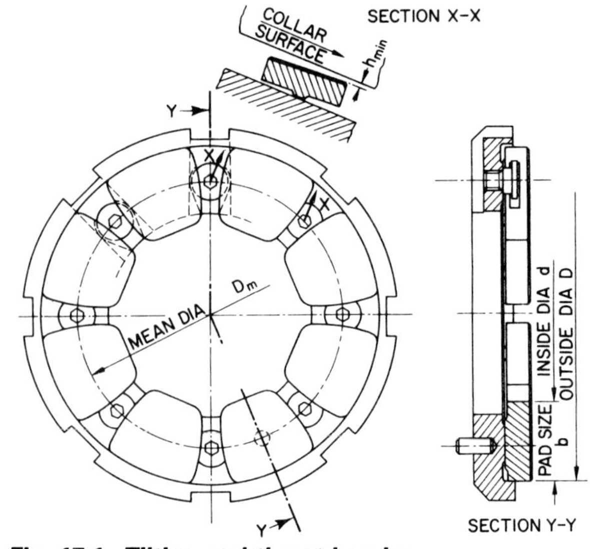

Fig. 17.1. Tilting pad thrust bearing

Fig. 17.2. Typical mounting and lubrication arrangements

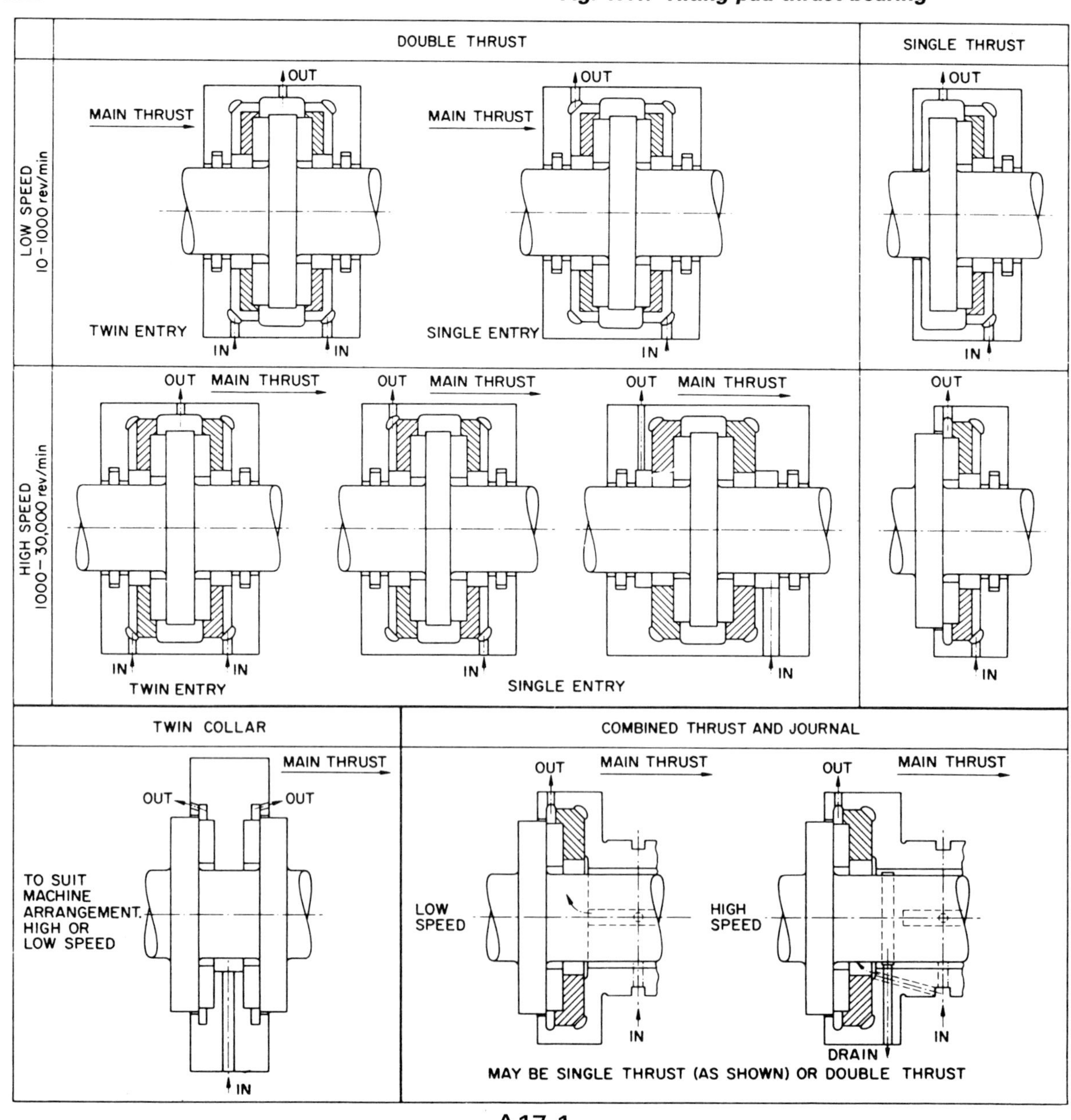

SELECTION OF THRUST BEARING SIZE

The load carrying capacity depends upon the pad size, the number of pads, sliding speed and oil viscosity. Using Figs. 17.3–17.7 a bearing may be selected and its load capacity checked. If this capacity is inadequate then a reiterative process will lead to a suitable bearing.

(1) Use Fig. 17.3(a) for first approximate selection.
(2) From Fig. 17.3(b) select diameters D and d.
(3) From Fig 17.4 find thrust ring mean diameter D_m.
(4) From Fig. 17.5 find sliding speed.
(5) For the bearing selected calculate:

$$\text{Specific load, } p\ (\text{MN/m}^2) = \frac{\text{Thrust load (N)}}{\text{Thrust surface (mm}^2)}$$

Check that this specific load is below the limits set by Figs. 17.6(a) and 17.6(b) for safe operation.

Note: these curves are based on an average turbine oil having a viscosity of 25 cSt at 60°C, with an inlet to the bearing at 50°C.

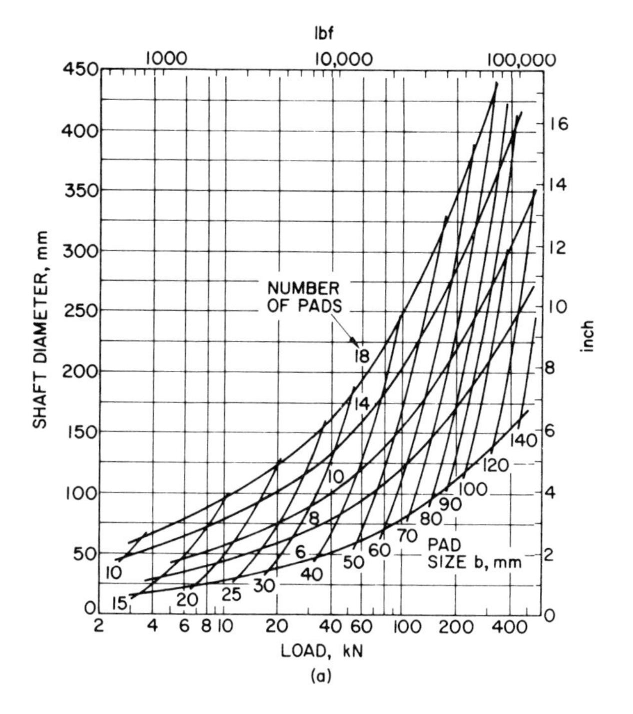

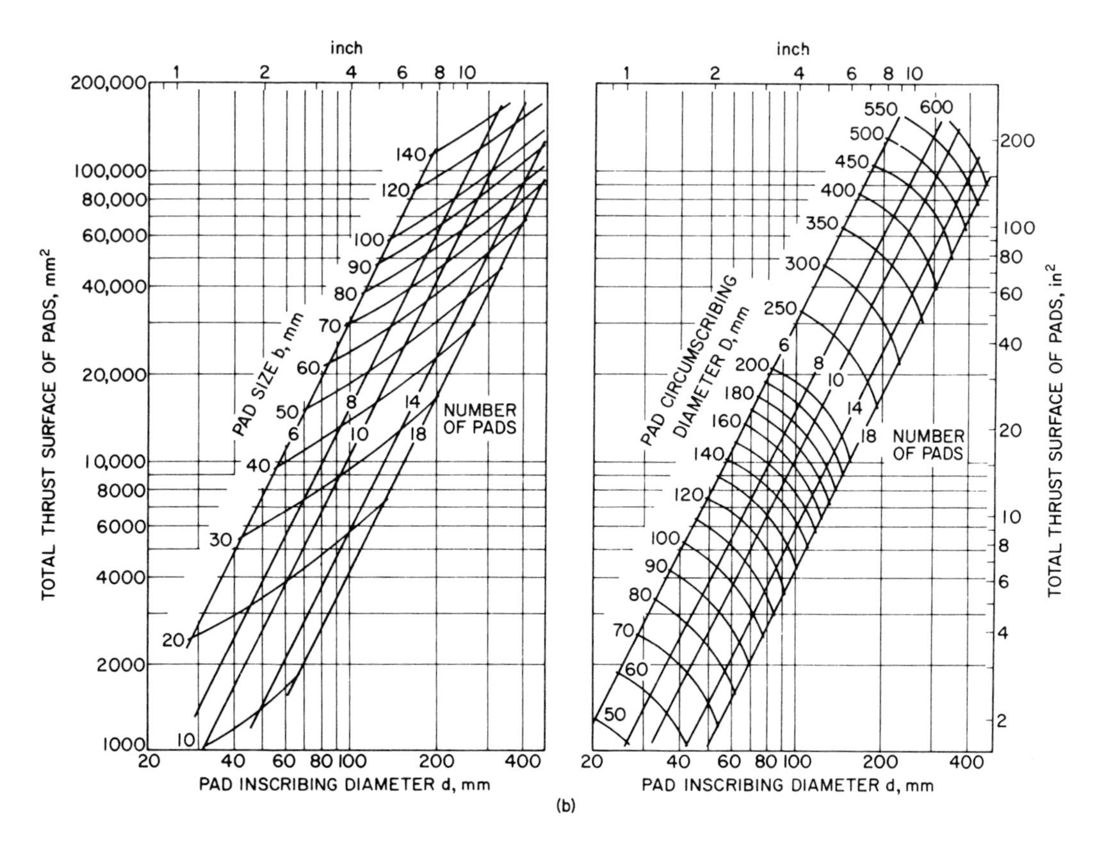

Fig. 17.3. First selection of thrust bearing

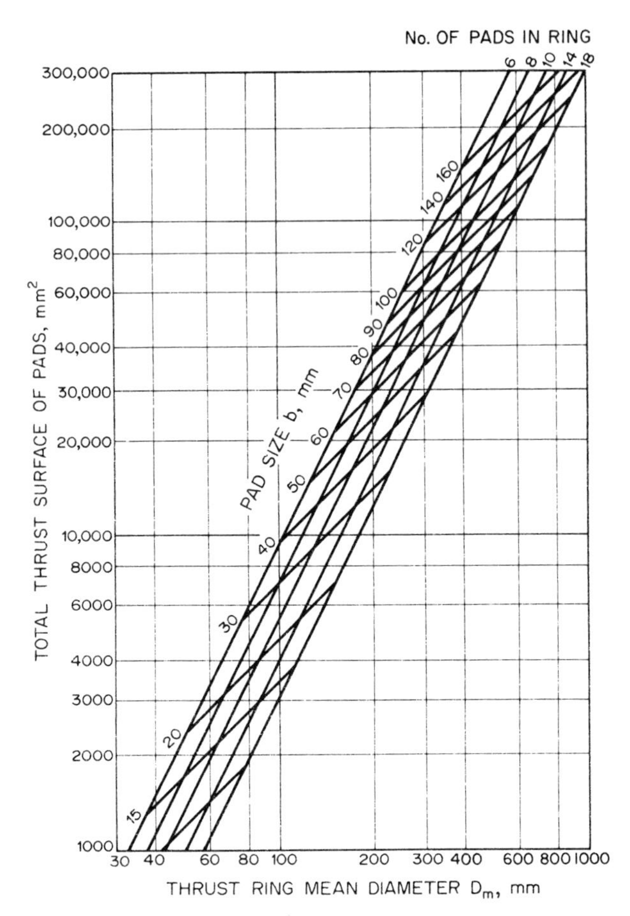

Fig. 17.4. Thrust ring mean diameter

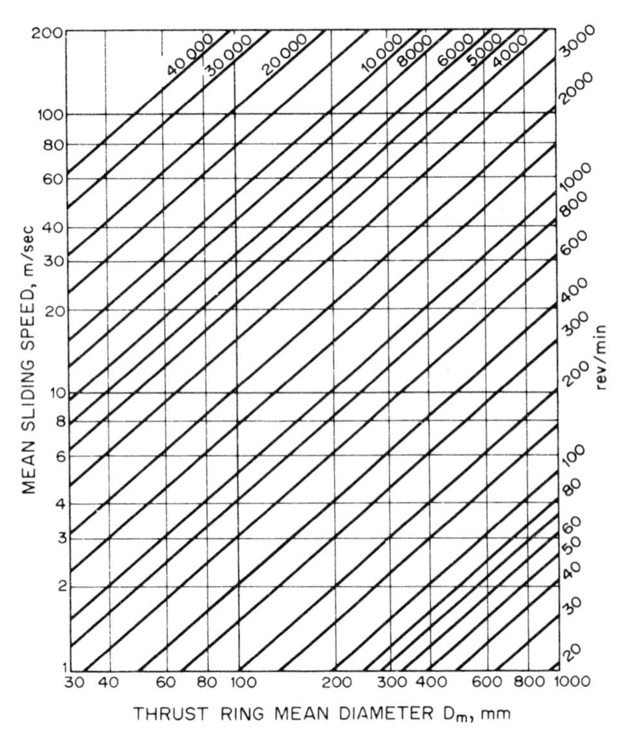

Fig. 17.5. Sliding speed

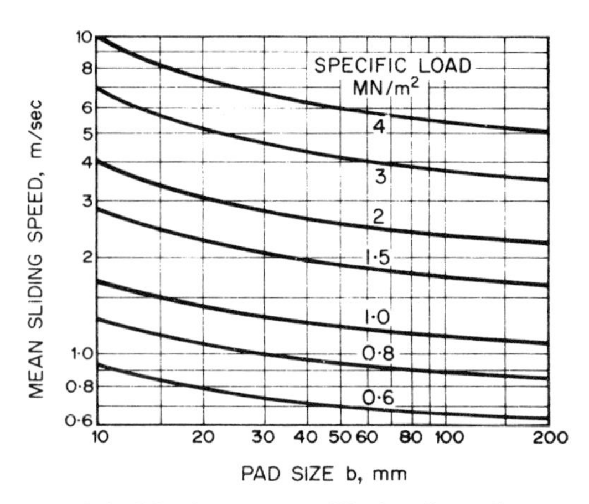

Fig. 17.6(a). Maximum specific load at slow speed to allow an adequate oil film thickness

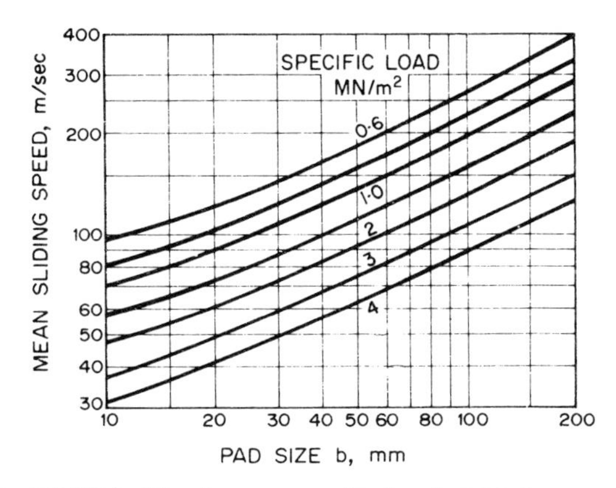

Fig. 17.6(b). Maximum specific load at high speed to avoid overheating in oil film

The load carrying capacity varies with viscosity, and for different oils may be found by applying the correction factors in Fig. 17.7:

Maximum specific load = Specific load (Fig. 17.6) × factor f (Fig. 17.7).

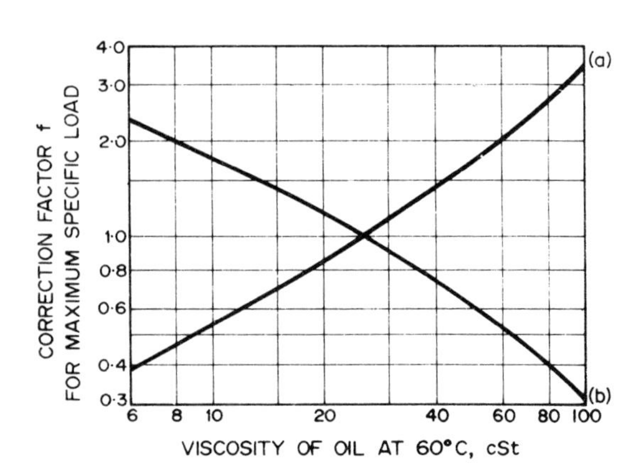

Fig. 17.7. Maximum safe specific load:
slow speed = f (curve (a)) × spec. load Fig. 17.6(a)
high speed = f (curve (b)) × spec. load Fig. 17.6(b)

CALCULATION OF POWER ABSORBED

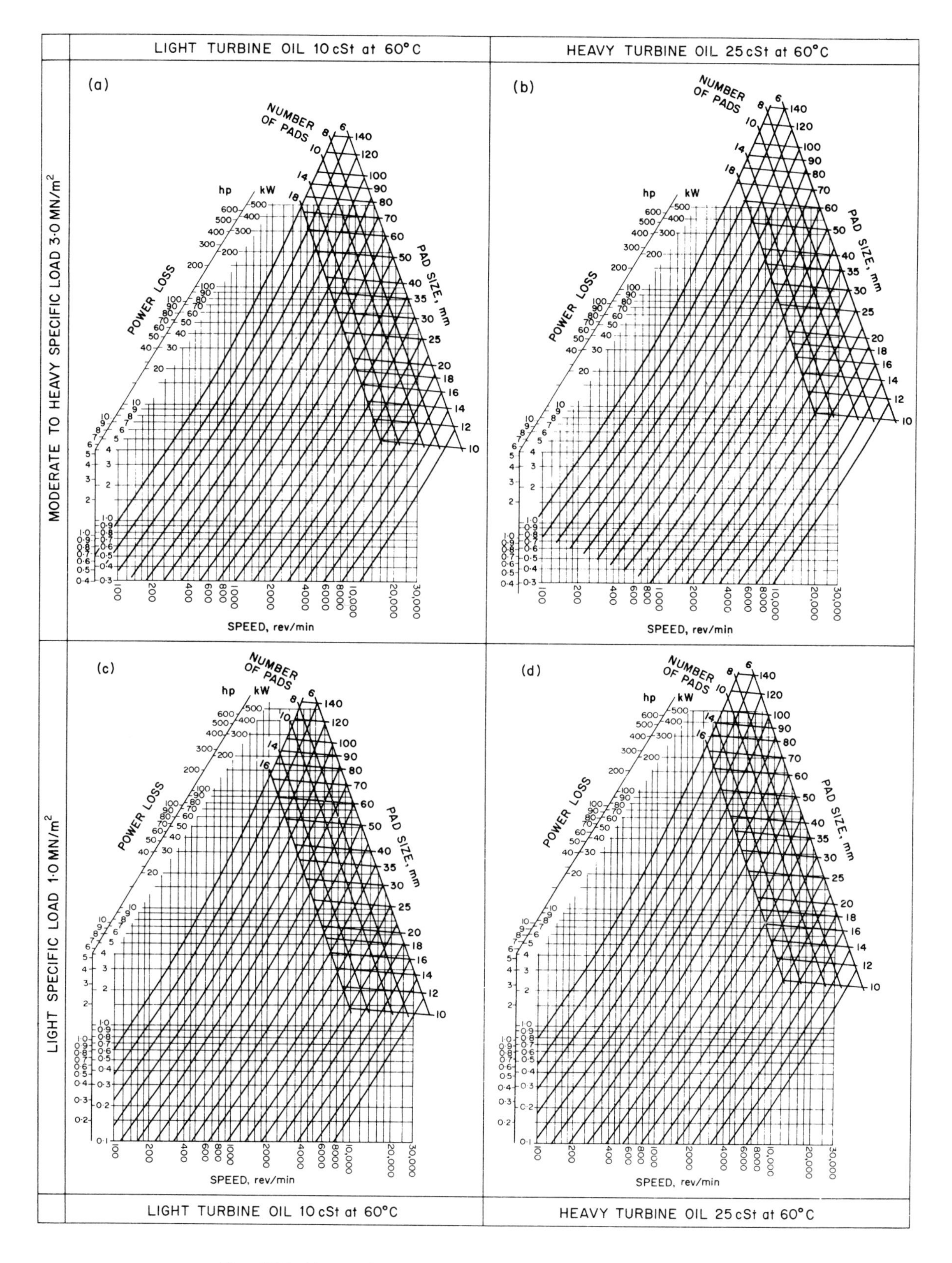

Fig. 17.8. Power loss guidance charts, double thrust bearings

The total power absorbed in a thrust bearing has two components:

(1) Resistance to viscous shear in the oil film.
(2) Fluid drag on exposed moving surfaces—often referred to as 'churning losses'.

The calculations are too complex to be included here and data should be sought from manufacturers. Figure 17.8 shows power loss for typical double thrust bearings.

Figure 17.9 shows the components of power loss and their variation with speed.

Note: always check that operating loads are in the safe region.

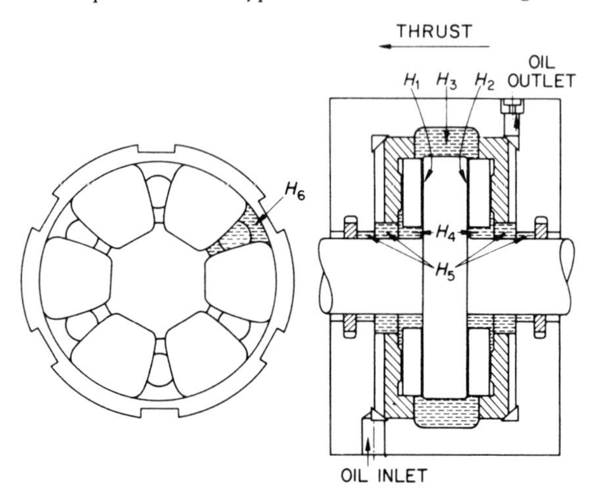

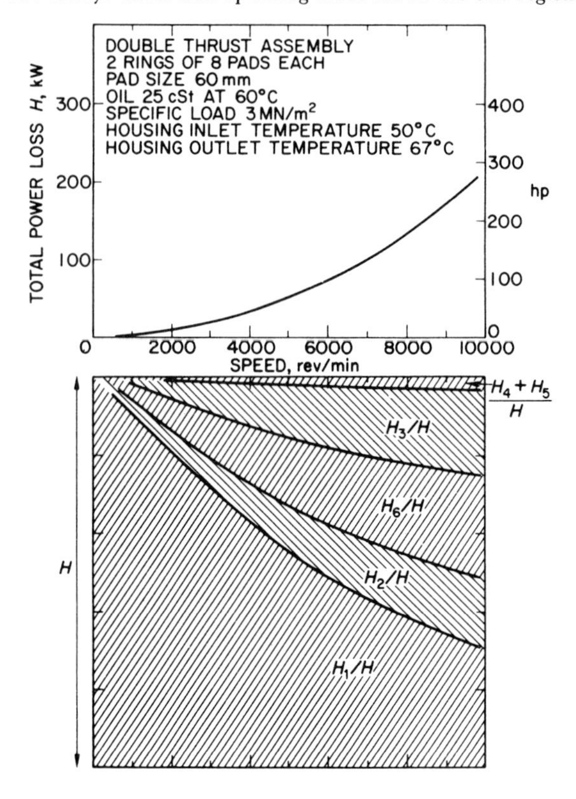

H = *total power loss*
H_1 = *film shear at main face*
H_2 = *film shear at surge face*
H_3 = *drag loss at rim of collar*
H_4 = *drag loss at inside of pads*
H_5 = *drag loss along shaft*
H_6 = *drag loss between pads*

Fig. 17.9 Components of power loss in a typical double thrust bearing

High speed bearings

In low speed bearings component 2, the churning loss, is a negligible proportion of the total power loss but at high speeds it becomes the major component. It can be reduced by adopting the arrangement shown in Fig. 17.10. Instead of the bearing being flooded with oil, the oil is injected directly on to the collar face to form the film. Ample drain capacity must be provided to allow the oil to escape freely.

Figure 17.11 compares test results for a bearing having 8 pads, $b = 28$ mm, at an oil flow of 45 litres/min.

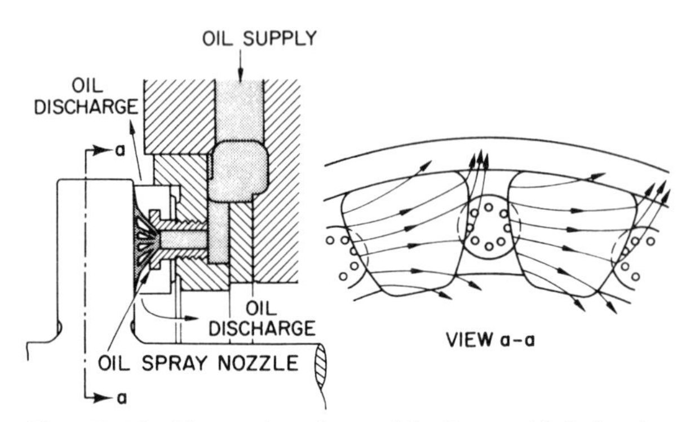

Fig. 17.10. Thrust bearing with directed lubrication

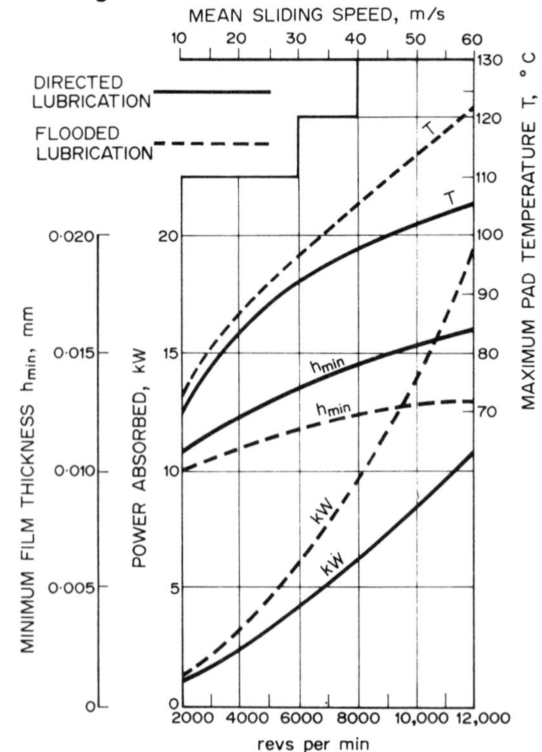

Fig. 17.11. Comparison between flooded and directed lubrication

Oil flow

Oil is circulated through the bearing to provide lubrication and to remove the heat resulting from the power loss.
It is usual to supply oil at about 50°C and to allow for a temperature rise through the bearing of about 17°C.

There is some latitude in the choice of oil flow and temperature rise, but large deviations from these figures will affect the performance of the bearing.

The required oil flow may be calculated from the power loss as follows:-

$$\text{Oil flow (litres/min)} = 35.8 \times \frac{\text{Power Loss (kW)}}{\text{Temperature Rise (°C)}}$$

$$\text{Oil Flow (US gals/min)} = 12.7 \times \frac{\text{Power Loss (hp)}}{\text{Temperature Rise (°F)}}$$

EQUALISED PAD BEARINGS

Where the bearing may be subject to misalignment, either due to initial assembly, or to deflection of the supporting structure under load an alternative construction can be adopted, although with the disadvantages of increased size and expense.

The equalised pad bearing is shown in Fig. 17.12. The pads are supported on a system of interlinked levers so that each pad carries an equal share of the load.

Misalignment of the order of up to 0.1° (0.002 slope) can be accepted. Above this the equalising effect will diminish.

In practice the ability to equalise is restricted by the friction between the levers, which tend to lock when under load. Thus the bearing is better able to accept initial misalignment than deflection changes under load.

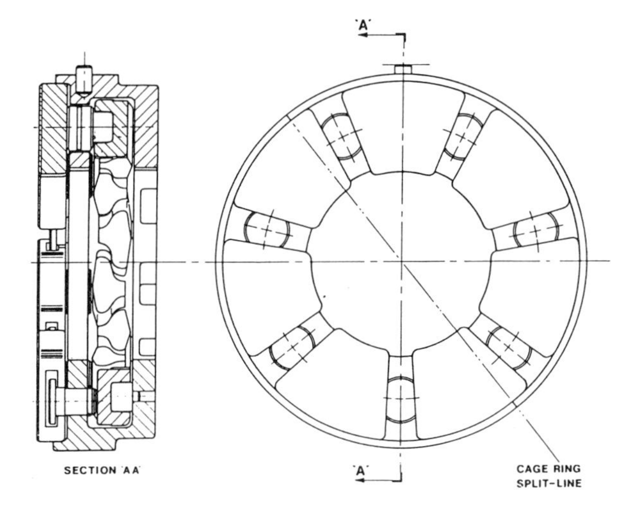

Fig. 17.12. An equalised pad thrust bearing

STARTING UNDER LOAD

In certain applications, notably vertical axis machines the bearing must start up under load. The coefficient of friction at break-away is about 0.15 and starting torque can be calculated on the Mean Diameter.

The specific load at start should not exceed 70% of the maximum allowable where acceleration is rapid and 50% where starting is slow.

Where load or torque at start are higher than acceptable, or for large machines where starting may be quite slow a jacking oil system can be fitted. This eliminates friction and wear.

BEARINGS FOR VERY HIGH SPEEDS AND LOADS

Traditionally the thrust pads are faced with whitemetal and this is still the most commonly used material. But, with increasingly higher specific loads and speeds the pad surface temperature will exceed the permissible limit for whitemetal – usually a design temperature of 130°C.

Two alternative approaches are available:-

1 The pad temperature may be reduced by

   (a) Directed lubrication – see Figs. 17.10 and 17.11.

   (b) Adopting offset pivots; accepting their disadvantages.

   (c) Changing the material of the pad body to high conductivity. Copper – Chromium alloy.

2 Alternatively the pads can be faced with materials able to withstand higher temperatures but at increased cost.

   (a) **40% Tin–Aluminium** will operate 25°C higher than whitemetal. Has comparable boundary lubrication tolerance and embeddability with better corrosion resistance.

   (b) **Copper–Lead** will operate 40°C higher than whitemetal. Poorer tolerance to boundary lubrication and embeddability. Requires the collar face to be hardened.

   (c) **Polymer based upon PEEK** can be used at temperatures up to 200°C and above. Comparable embeddability and better tolerance to boundary lubrication. Suitable for lubrication by water and mainly low viscosity process fluids.

   (d) **Ceramic Pads and Collar Face**, made from silicon – carbide, these can be used up to 380°C and specific loads up to about 8 MPa (1200 p.s.i.). They are chemically inert and suitable for lubrication by low viscosity fluids such as water, most process fluids and liquified gases.

LOAD MEASUREMENT

The bearings can be adapted to measure thrust loads using either electronic or hydraulic load cells. The latter can provide very effective load equalisation under misalignment and may be used to change the axial stiffness at will to avoid resonant vibration in the system.

In a hydrostatic bearing the surfaces are separated by a film of lubricant supplied under pressure to one or more recesses in the bearing surface. If the two bearing surfaces are made to approach each other under the influence of an applied load the flow is forced through a smaller gap. This causes an increase in the recess pressure. The sum of the recess pressure and the pressures across the lands surrounding the recess build up to balance the applied load. The ability of a bearing film to resist variations in gap with load depends on the type of flow controller.

LOAD CAPACITY

Figures 18.1 and 18.2 give an approximate guide to the load capacities of single plane pads and journal bearings at various lubricant supply pressures. Approximate rules are:

(1) The maximum mean pressure of a plane pad equals one-third the supply pressure.

(2) The maximum mean pressure on the projected area of journal bearings and opposed pad bearings equals one-quarter the supply pressure.

An approximate guide to average stiffness λ may be obtained by dividing the approximate load by the design film thickness $\lambda = W/h_o$.

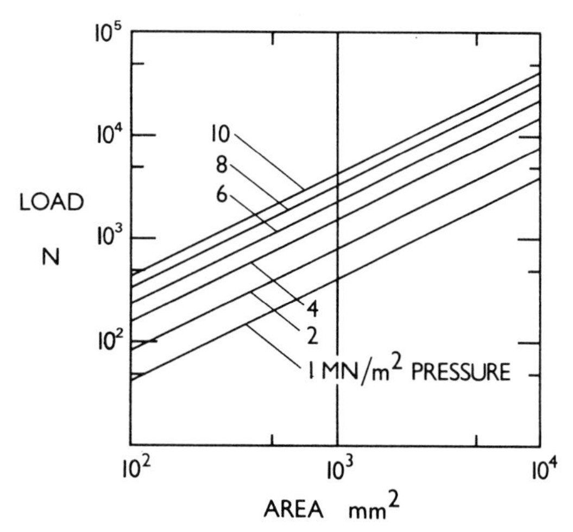

Fig. 18.1. Plane pad bearing load capacity

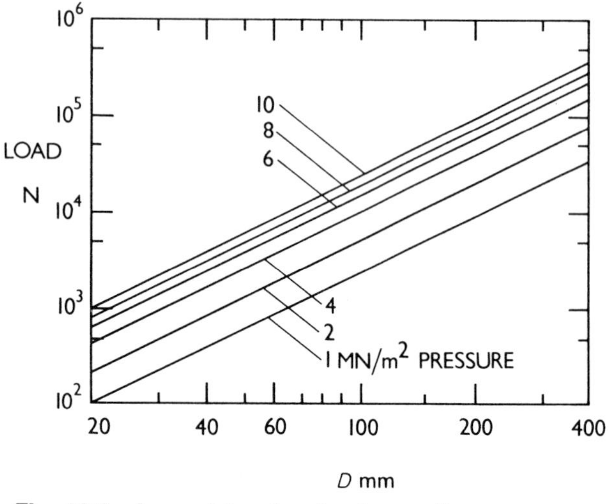

Fig. 18.2. Journal bearing load capacity

CONTROL CIRCUITS

Each recess in a bearing must have its own flow controller, as illustrated in Figs. 18.3 and 18.4, so that each recess may carry a load independently of the others.

Flow controllers in order of increasing bearing stiffness are as follows:

(1) Laminar restrictors (capillary tubing).

(2) Orifices (length to diameter ratio $\ll 1$).

(3) Constant flow (fixed displacement pumps or constant flow valves).

(4) Pressure sensing valves.

Figure 18.4(a) illustrates a typical circuit for capillary or orifice control. The elements include a filter (FLT), a motor (M), a fixed displacement pump (PF), an inlet strainer (STR), and a flow relief valve set to maintain the operating supply pressure at a fixed maximum p_f. Figure 18.4(b) shows a circuit involving a constant flow control valve with pressure compensation (PC).

The control circuit must be designed to provide the necessary value of recess pressure p_o at the design bearing clearance h_o. It is first necessary to calculate the flow from the bearing recess at the design condition Q_o. The values of Q_o and p_o are then employed in the calculation of the restrictor dimensions and in selection of other elements in the circuit.

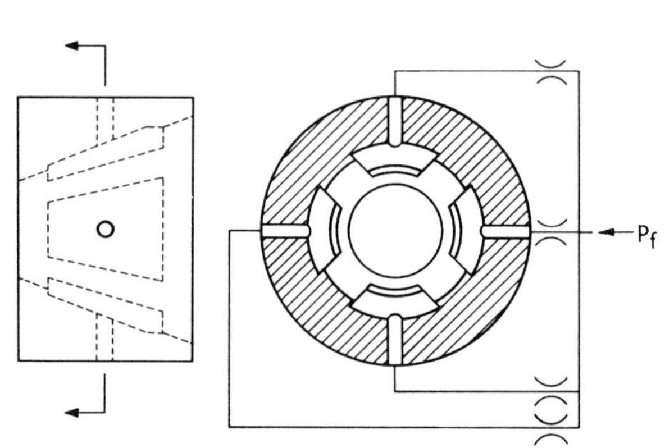

Fig. 18.3. A conical hydrostatic journal bearing showing recesses and restrictions

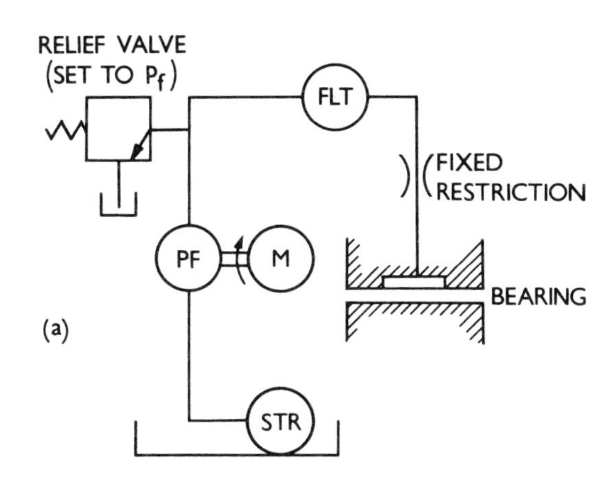

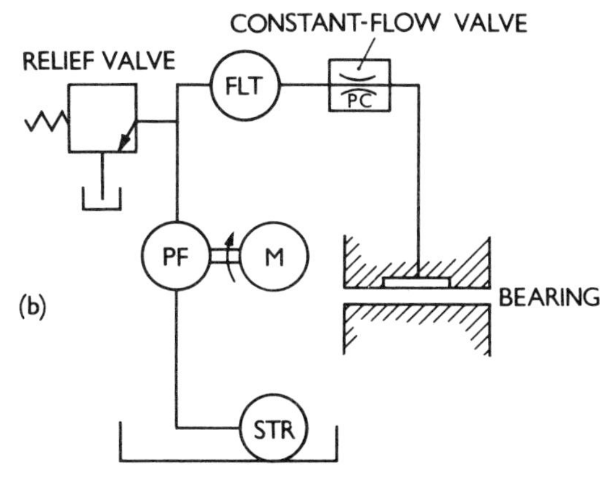

Fig. 18.4. Layout of typical control circuits

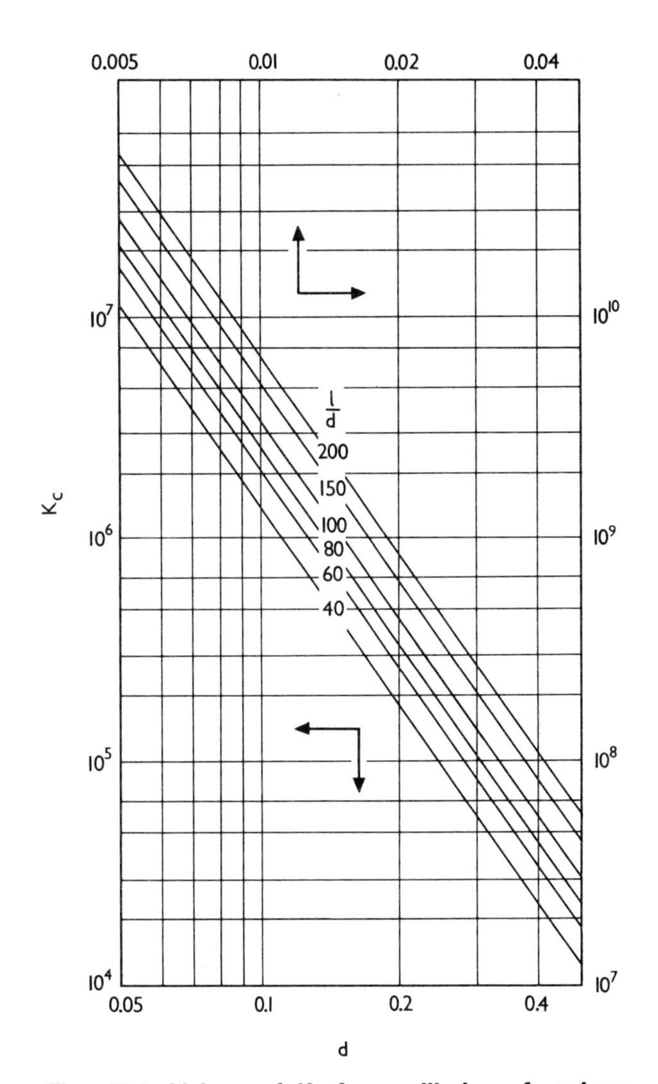

Fig. 18.5. Values of K_c for capillaries of various dimensions

Design of control restrictors

Dimensions of capillary and orifice restrictors may be calculated from the following equations

Capillary flow $$Q_o = \frac{p_f - p_o}{K_c \eta}$$

where $$K_c = \frac{128\,l}{\pi d^4}$$

Suitable values for length and diameter are presented in Fig. 18.5. The l/d ratio should preferably be greater than 100 for accuracy and the Reynolds Number should be less than

1000 where $R_e = \dfrac{\rho v d}{\eta}$, ρ = density, v = average velocity, d = bore diameter and η = dynamic viscosity.

Orifice flow $$Q_o = \frac{\rho}{2(C_f A)^2}\left(p_f - p_o\right)^{\frac{1}{2}}$$

A = cross-sectional area of the orifice.

Values of the flow coefficient C_f are presented in Fig. 18.6 according to the values of R_e.

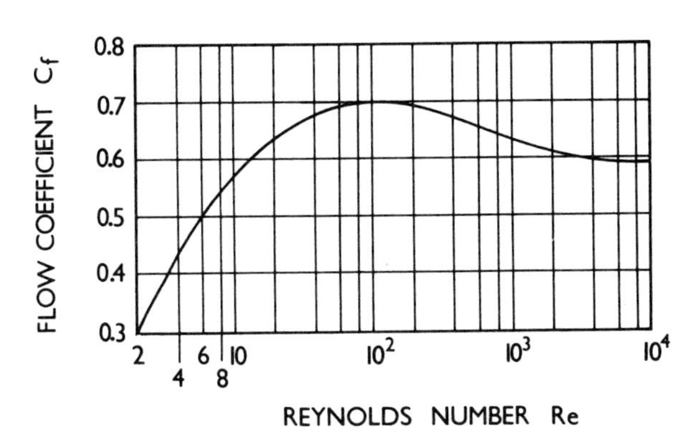

Fig. 18.6. Values of the flow coefficient C_f for orifices

Calculation of bearing stiffness

Bearing characteristics are dependent not only on the type of control device. The characteristics are also dependent on the design pressure ratio $\beta = \frac{p_o}{p_f}$. Figures 18.8 and 18.9 show the variation of dimensionless stiffness parameter $\bar{\lambda}$ with the film thickness h and design pressure ratio β. Taking all considerations into account including manufacturing tolerances it is recommended to aim for $\beta = 0.5$. The relationship between stiffness λ and the dimensionless value $\bar{\lambda}$ is

$$\lambda = \frac{P_f A_e}{h_o} . \bar{\lambda}$$

where A_e is the effective area of the pad over which p_f may be assumed to act. For a plane pad in which the recess occupies one-quarter of the total bearing pad area A the effective area is approximately $A/2$ which may be deduced by assuming that p_f extends out to the mid-land boundary. For more accuracy the effective area may be expressed as $A_e = A\bar{A}$ which defines a dimensionless area factor $\bar{A}$. Some computed values of $\bar{A}$ are presented for plane pads in Figs. 18.7 and 18.10.

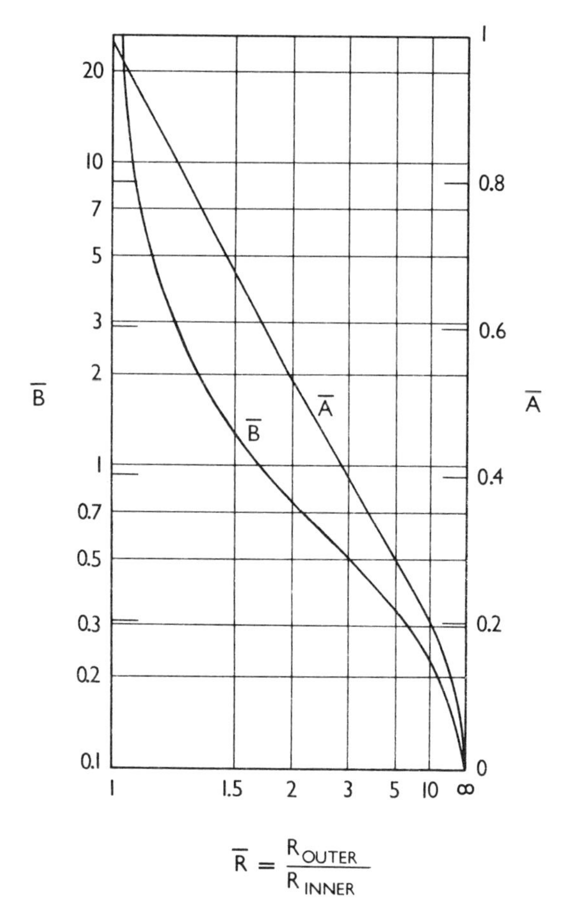

Fig. 18.7. Pad coefficient for a circular pad

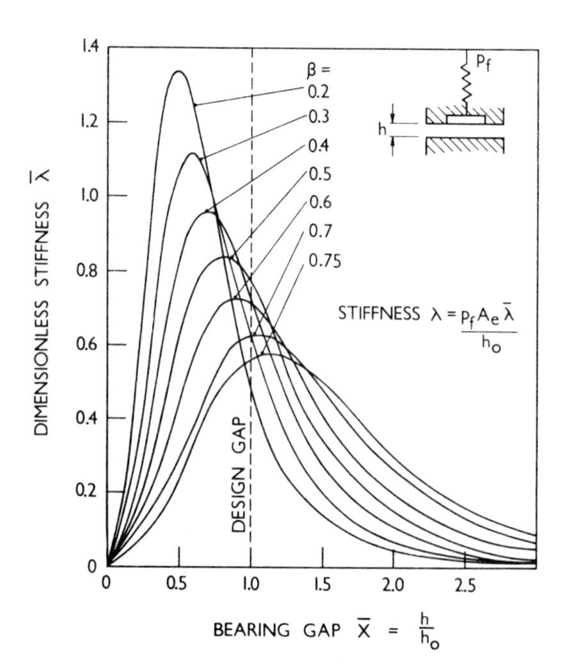

Fig. 18.8. Stiffness parameters for capillary-compensated single pad bearings

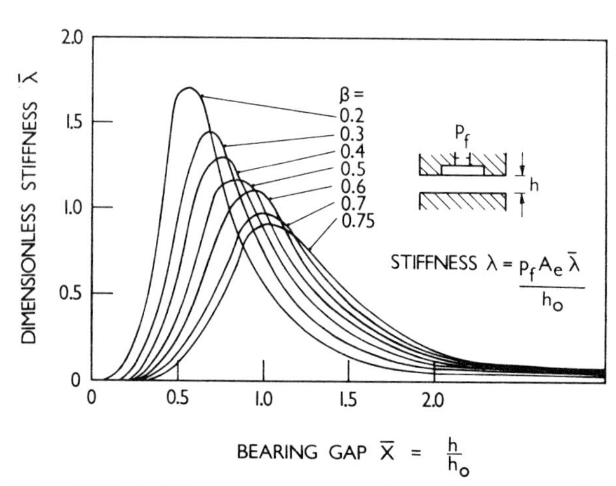

Fig. 18.9. Stiffness parameters for orifice-controlled single pad bearings

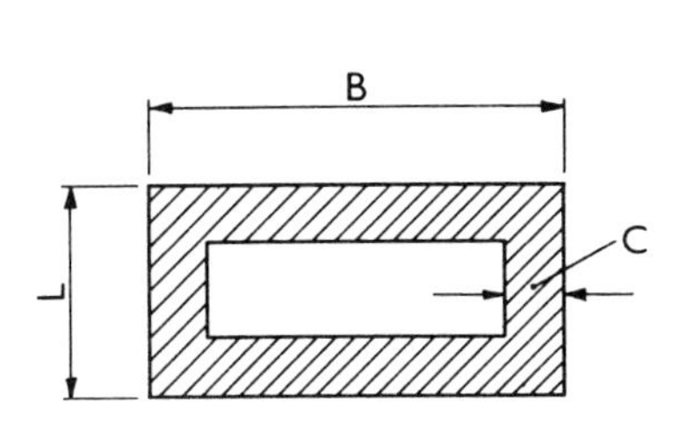

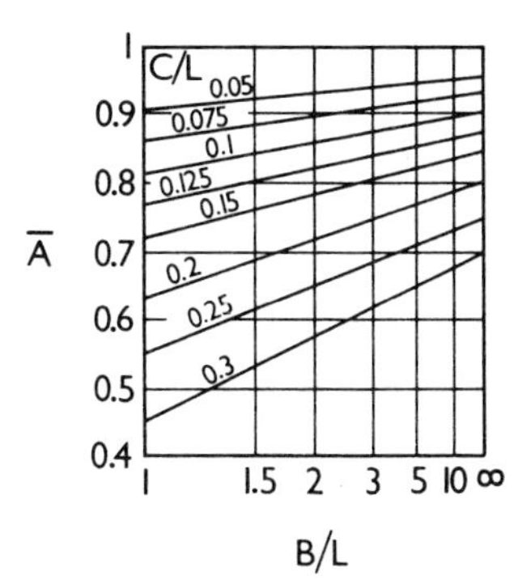

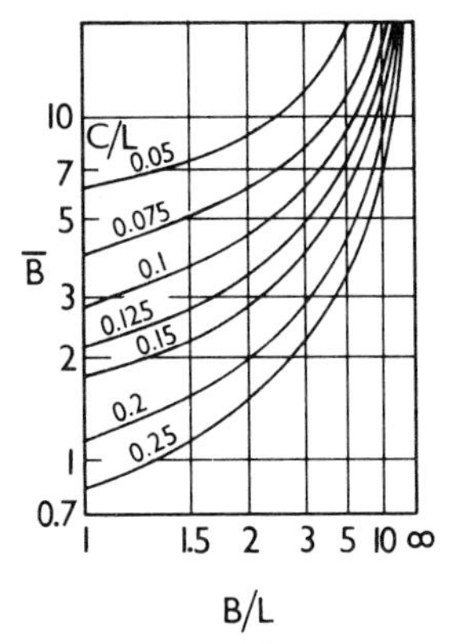

Fig. 18.10. Pad coefficient for a rectangular pad. For a rectangular pad with uniform land width it is recommended that C / L < 0.25

PLANE HYDROSTATIC PAD DESIGN

The performance of plane pad bearings may be calculated from the following formulae:

Load: $$W = p_f\, A\,.\,\bar{A}\,.\,\bar{P}$$

Flow: $$Q = \frac{p_f\, h_o^3}{\eta}\,.\,\bar{P}\,.\,\bar{B}$$

where $\bar{A}$ is a factor for effective area ($A_e = A\bar{A}$)
$\bar{B}$ is a factor for flow
$\bar{P} = \dfrac{p}{p_f}$ and varies with film thickness
h_o = design film thickness

Figures 18.7 and 18.10 give values of $\bar{A}$ and $\bar{B}$ for circular and rectangular pads of varying land widths.

The relationship between $\bar{P}$ and h depends on β and curves are presented in Figs. 18.11 and 18.12 for capillary and orifice control.

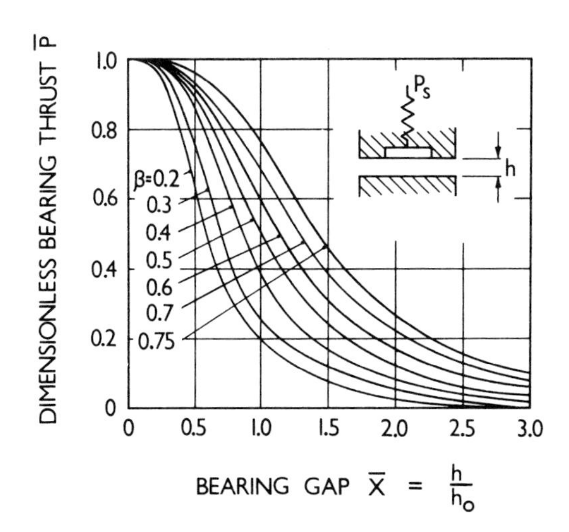

Fig. 18.11. Variation of thrust pad load capacity with film thickness using capillary control

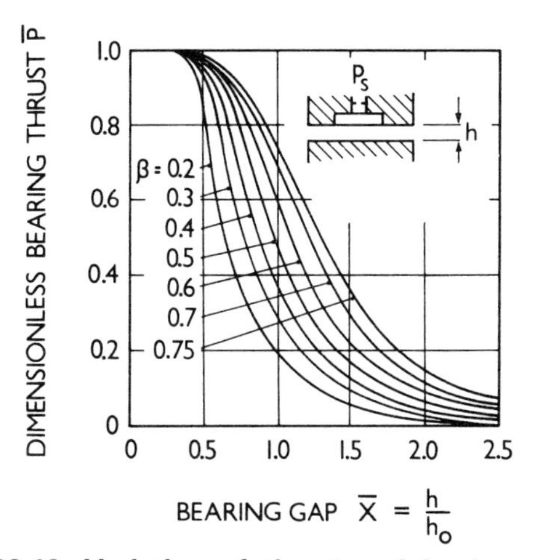

Fig. 18.12. Variation of thrust pad load capacity with film thickness using orifice control

For bearings which operate at speed it is important to optimise the design to minimise power dissipation and to prevent cavitation and instability problems. The optimisation required may be achieved by selecting values of viscosity η and film thickness h_o to satisfy the following equation:

$$\frac{\eta\, v}{p_f\, h_o^2} = \left(\frac{\beta\, \bar{B}}{A_f}\right)^{\frac{1}{2}}$$

where A_f = (total area) $-\frac{3}{4}$(recess area)
= effective friction area

Recess depth = 20 × bearing clearance h_o
v = linear velocity of bearing

The above relationship minimises total power which is the sum of friction power and pumping power. A further advantage of optimisation is that it ensures that temperature rise does not become excessive. For optimised bearings the maximum temperature rise as the lubricant passes through the bearing may be calculated from

$$\Delta T = \frac{2\, p_f}{J\, \rho\, C_v}$$

where J = mechanical equivalent of heat
C_v = specific heat

DESIGN OF HYDROSTATIC JOURNAL BEARINGS

The geometry and nomenclature of a cylindrical journal bearing with n pads are illustrated in Fig. 18.13. For journal bearings the optimum value of design pressure ratio is $\beta = 0.5$ as for other hydrostatic bearings. Other values of β will reduce the minimum film thickness and may reduce the maximum load. The following equations form a basis for safe design of journal bearings with any number of recesses and the three principal forms of flow control (refer to Fig. 18.13 and Table 18.1).

Load: $W = p_f A_e . \bar{W}$

where $\bar{W}$ is a load factor which normally varies from 0.30 to 0.6 a better guide is $\bar{W} = \dfrac{\bar{\lambda}'}{2}$

$\bar{\lambda}$ = dimensionless stiffness parameter from Table 18.1

$\bar{\lambda}'$ = value of $\bar{\lambda}$ for capillary control and $\beta = 0.5$,

$A_e = D\,(L-a)$.

Concentric stiffness: $\lambda = \dfrac{p_f A_e}{C} . \bar{\lambda}$

where $C = h_o$ = radial clearance.

Flow-rate: $Q = \dfrac{p_f C^3}{\eta} . n \beta \bar{B}$

where $\bar{B} = \dfrac{\pi D}{6 a n}$

is the flow factor for one of the n recesses.

$$\gamma = \frac{n a (L-a)}{\pi D b}$$

is a circumferential flow factor. If the dimension 'b' is too small the value γ will be large and the bearing will be unstable.

The recommended geometry for a journal bearing (see Fig. 18.13).

$$a = \frac{L}{4}, \quad \frac{L}{D} = 1, \quad b = \frac{\pi D}{4 n}$$

Journal bearings which operate at speed should be optimised for minimum power dissipation and low temperature rise for the same reasons as given under the previous paragraph headed 'Plane Hydrostatic Pad Design'. Values of viscosity and clearance should be selected so that:

$$\frac{\eta N'}{p_f} \left(\frac{D}{C_D}\right)^2 = \frac{1}{4 \pi} \left(\frac{n \beta \bar{B}}{\bar{A}_f}\right)^{\frac{1}{2}}$$

where N' = rotational speed in rev/sec

$\bar{A}_f = [(\text{total area}) - \frac{3}{4}(\text{recess area})]/D^2$

Recess depth = 20 × radial clearance. Maximum temperature rise may be calculated as for plane pads.

Table 18.1 Dimensionless stiffness $\bar{\lambda}$ (for a journal bearing with n pads)

n	*Capillary*	*Orifice*	*Constant flow*
4	$\dfrac{3.82 \beta (1-\beta)}{1+\gamma (1-\beta)}$	$\dfrac{7.65 \beta (1-\beta)}{2-\beta+2 \gamma (1-\beta)}$	$\dfrac{3.82 \beta}{1+\gamma}$
5	$\dfrac{4.12 \beta (1-\beta)}{1+0.69 \gamma (1-\beta)}$	$\dfrac{8.25 \beta (1-\beta)}{2-\beta+1.38 \gamma (1-\beta)}$	$\dfrac{4.25 \beta}{1+0.69 \gamma}$
6	$\dfrac{4.30 \beta (1-\beta)}{1+0.5 \gamma (1-\beta)}$	$\dfrac{8.60 \beta (1-\beta)}{2-\beta+\gamma (1-\beta)}$	$\dfrac{4.30 \beta}{1+0.5 \gamma}$

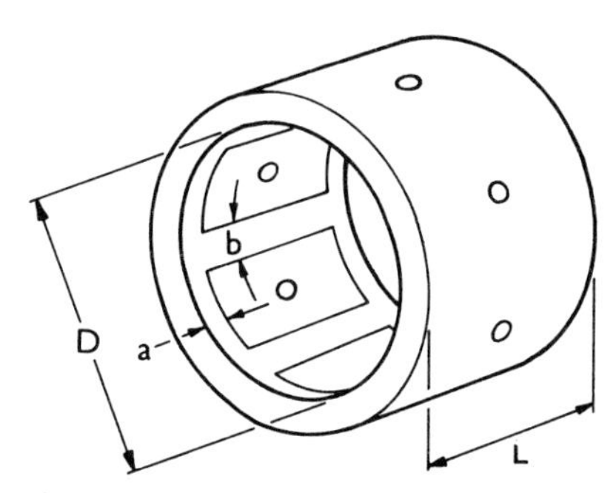

Fig. 18.13. Typical hydrostatic journal bearing

EXTERNALLY PRESSURISED GAS BEARINGS

Externally pressurised gas bearings have the same principle of operation as hydrostatic liquid lubricated bearings. There are three forms of external flow restrictor in use.

Restrictor type		*Pocketed orifice (simple orifice)*	*Unpocketed orifice (annular orifice)*	*Slot*
Flow restriction		$\alpha \frac{1}{\pi d^2/4}$	$\alpha \frac{1}{\pi dh}$	$\alpha \frac{-y}{az^3}$
Geometrical restrictions		$\frac{d^2}{4} < d\delta$ i.e. $\delta > \frac{d}{4}$ $\frac{\pi d^2}{4} < \pi d_p h$	When $\frac{\pi d^2}{4} < \pi dh$ i.e. $d < 4h$ orifice reverts to simple type giving increased bearing stiffness	y is generally large for practical values of z resulting in large overall bearing size
Steady state load capacity	Journal and back-to-back thrust	Generally the highest	33% less than pocketed	Highest for very low b/D journals
	Single thrust	Can approach that of slot type by spreading pressure with surface grooves	Same as pocketed	Generally the highest
Steady state stiffness K	Journal and back-to-back thrust	Generally the highest	33% less than pocketed	Highest for very low b/D journals
	Single thrust	Highest	33% less than pocketed	
Damping		Both pockets and surface grooves will reduce damping		
Stability		Both pockets and surface grooves can result in instabilities		
Dynamics		Overall damping decreases with rotation in journal bearings. All bearings are subject to resonance at the natural frequency $\omega_n = \sqrt{(K/m)}$: symmetrical rigidly supported journal systems have 'whirl' at a speed $2 \times \omega_n$		
Lubricant		Any clean gas: design charts presented are for air.		

Externally pressurised gas thrust bearings

There are two main types of externally pressurised gas thrust bearings. These are the central recess and annular bearing as shown on the right. The central recess bearing is fed from a central orifice, while the annular bearing is fed from a ring of orifices. The stiffness of these bearings varies with clearance as shown below.

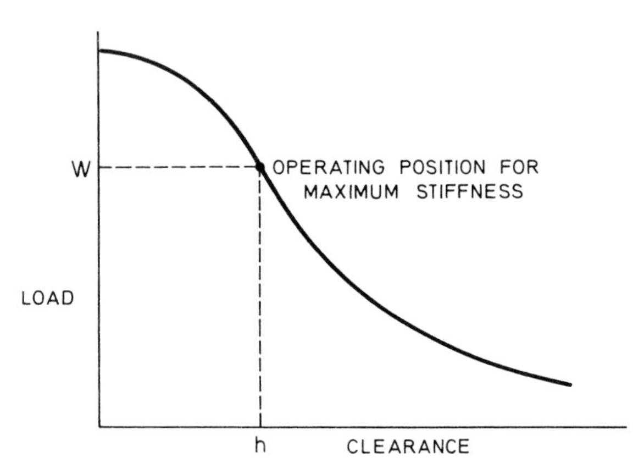

GENERAL LOAD CLEARANCE CHARACTERISTIC

All design data are given for operation at the position of maximum stiffness and are approximate.

Load carried $W = C_L' . \pi r_0^2 P_f$

where P_f = supply pressure (gauge)

Stiffness $K = 1.42\ W/h$ for pocketed orifices

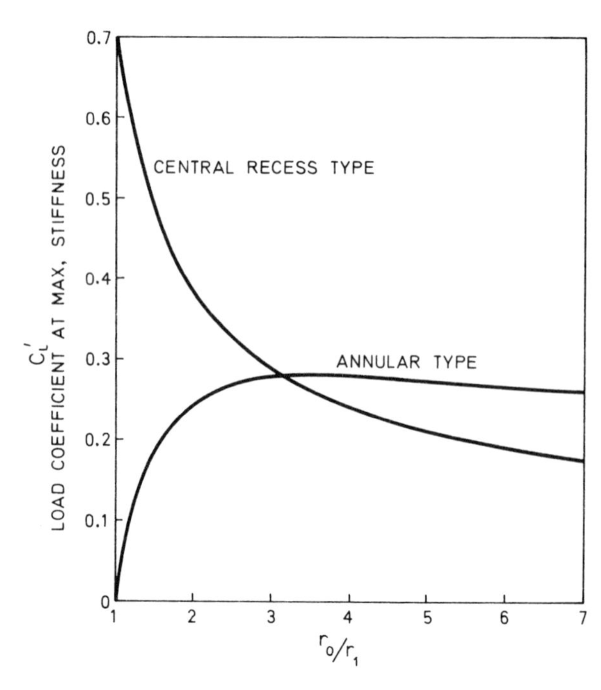

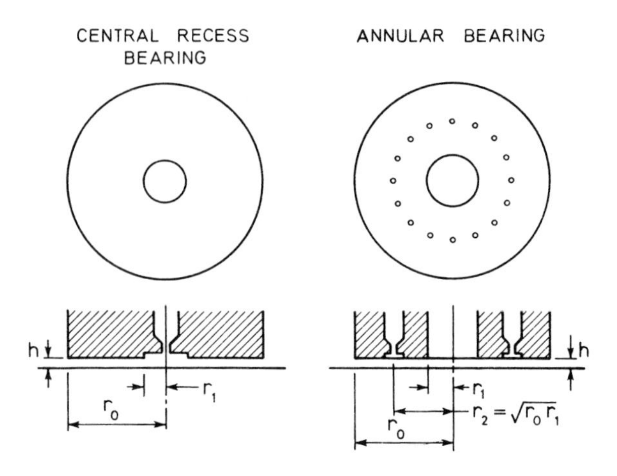

Orifice dia. d^* for a central recess bearing with clearance $h_0 = 25\ \mu m$ (0.001 in) for *air*, 15°C, ambient pressure 1 bar, is given below.

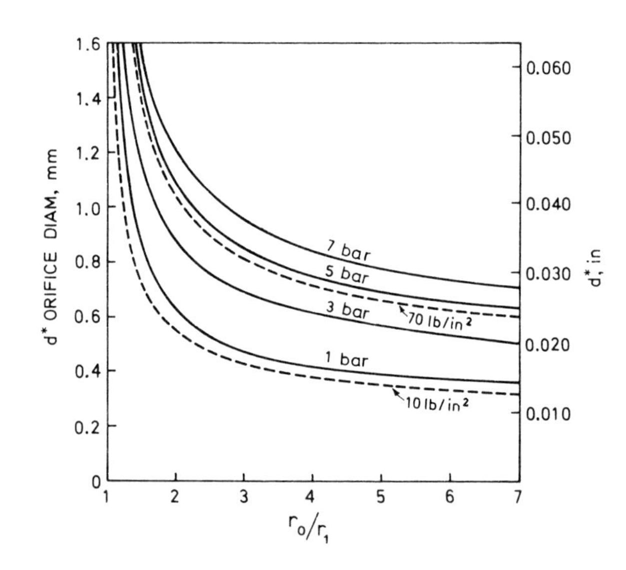

For other values of clearance h new orifice dia.;

$$d = d^* \left(\frac{h}{h_0}\right)^{\frac{3}{2}}$$

For annular bearing with N pocketed orifices;

$$d = d^* . \left(\frac{h}{h_0}\right)^{\frac{3}{2}} . \frac{2}{\sqrt{N}}$$

For annular bearing with N unpocketed orifices;

$$d = \frac{d^{*2} . h^2}{N h_0^3}$$

and stiffness $K = 0.95 \dfrac{W}{h}$

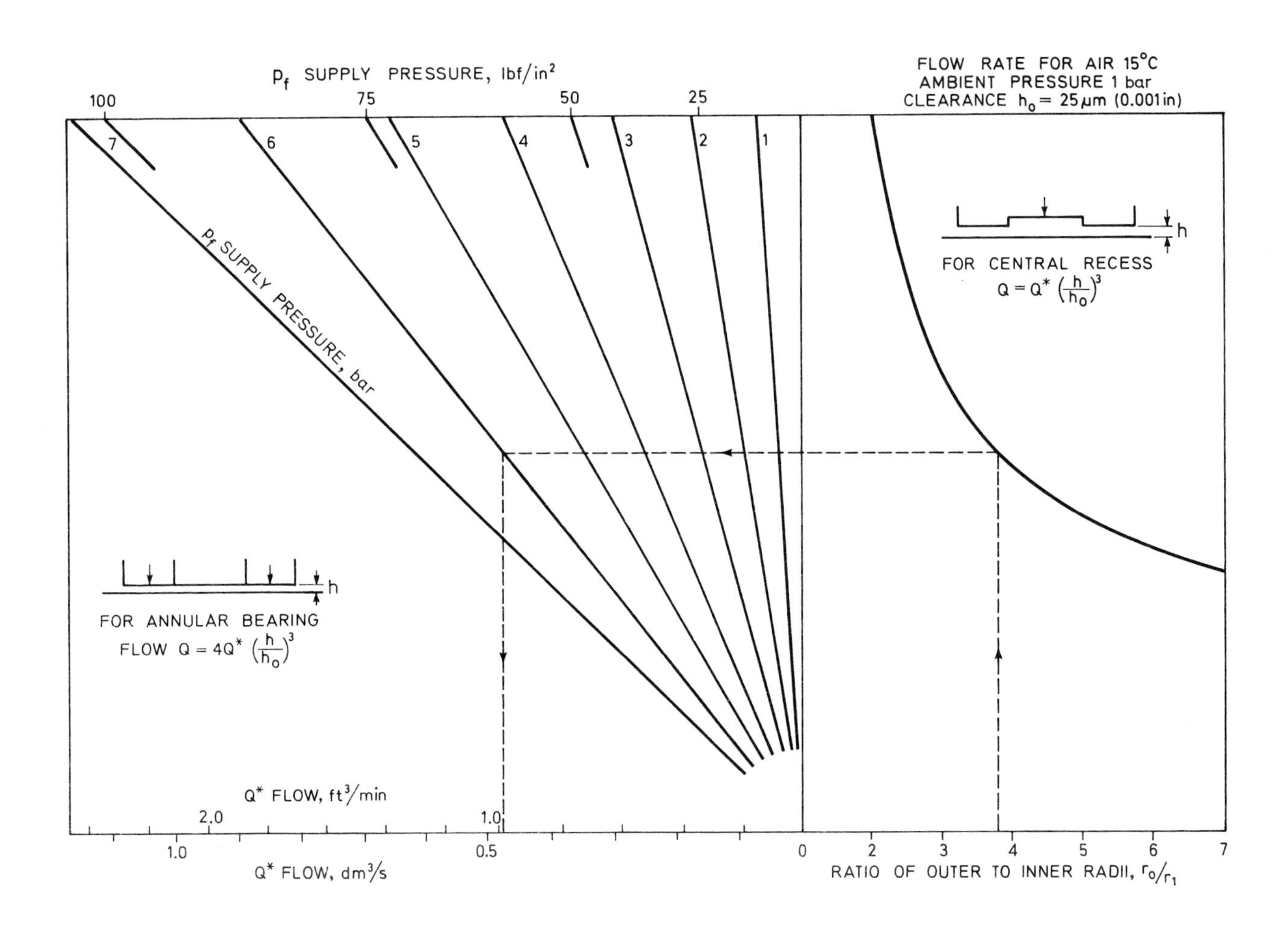

Example of thrust bearing design

Load to be carried weighs 1000 N. Bearing must have central hole 50 mm dia.: 5 bar supply pressure available. Design for maximum stiffness.

Take maximum load coefficient available as starting point. $C_L' = 0.28$. Then since $W = 1000$ N; $P_f = 5$ bar;

$$r_0^2 = \frac{1000}{0.28\pi \,.\, 5.10^5} = 2.10^{-3}\ \text{m}^2,$$

i.e. $r_0 = 0.045$ m and $r_0/r_1 = 0.045/0.025 = 1.8$. But C_L' was taken at $r_0/r_1 = 3$: bearing cannot operate at maximum C_L' with 5 bar supply pressure. Design can therefore be off maximum. C_L' or at lower supply pressure.

(1) Take $r_0/r_1 = 2$, then $C_L' = 0.25$, $r_0 = 50$ mm: $W = 0.25 \quad \pi \quad (0.050)^2, \quad 5.10^5 = 1000$ N. To operate with 20 μm clearance; then from d^* graph, $d^* = 1.08$ mm. Thus for this annular bearing with say, 16 orifices; orifice diameter

$$d = 1.08\left(\frac{20}{25}\right)^{\frac{3}{2}} \times \frac{2}{(16)^{\frac{1}{2}}} = 0.4\ \text{mm}:$$

$$\text{Stiffness} = \frac{1.42 \times 1000}{0.020} = 70\ \text{kN/mm}$$

or; for greater stability operate with unpocketed orifices. Then for $N = 40$ say,

$$d = \frac{1.08^2}{40}\ \frac{0.020^2}{0.025^3} = 0.75\ \text{mm [check } d > 4h\text{]}:$$

$$\text{Stiffness} = \frac{0.95 \times 1000}{0.020} = 46\ \text{kN/mm}.$$

Flow rate is not dependent upon orifice type, thus flow rate for either type

$$Q = 4Q^*\left(\frac{20}{25}\right)^2,$$

now $Q^* = 0.65$ dm³/s, therefore

$$Q = 4 \times 0.65 \times 0.51 = 1.3\ \text{dm}^3/\text{s}.$$

(2) For $r_0/r_1 = 3$, i.e. $r_0 = 75$ mm, and $C_L' = 0.28$,

$$P_f = \frac{1000}{0.28 \,.\, \pi\ (0.075)^2} = 2\ \text{bar};$$

for operation at 20 μm clearance stiffness is unchanged from (1). Flow $Q^* = 0.12$ dm³/s, therefore actual flow rate

$$Q = 4 \times 0.12\left(\frac{20}{25}\right)^3 = 0.25\ \text{dm}^3/\text{s},$$

cf. 1.3 dm³/s for arrangement (1).

Externally pressurised—gas journal bearings

This section gives an approximate guide to load carrying capacity and flow requirements for a design which is optimised for load capacity and stiffness.

Values are given for a bearing with two rows of 8 orifices as shown on the right, and with $l/b = \frac{1}{4}$, $P_f^* = 6.9$ bar (100 p.s.i.).

The load W^* at an eccentricity $\varepsilon = \dfrac{2e}{C_d} = 0.5$ is given below.

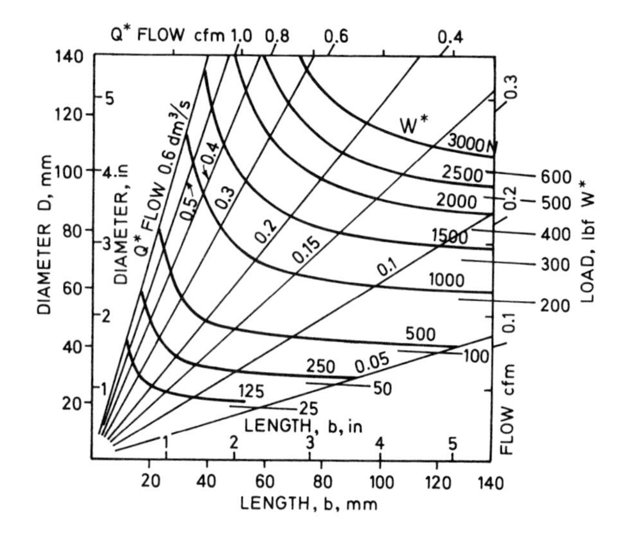

Load for other supply pressures for pocketed orifices

$$W = W^* \frac{P_f}{P_f^*}$$

Load at $\varepsilon = 0.5$ for unpocketed orifices

$$W = \tfrac{2}{3} W^* \frac{P_f}{P_f^*}$$

for higher load capacity:
(load at $\varepsilon = 0.9$) = 1.28 (load at $\varepsilon = 0.5$).

$$\text{Stiffness} = \frac{(\text{load at } \varepsilon = 0.5)}{C_d/4}$$

Flow Q^* is given for $C_d^* = 25\ \mu\text{m}$ (0.001 in), air, 15°C, $p_a = 1$ bar.

At other clearances, flow $Q = Q^* \left(\dfrac{C_d}{C_d^*}\right)^3$.

For other numbers of orifices per row use graph below.

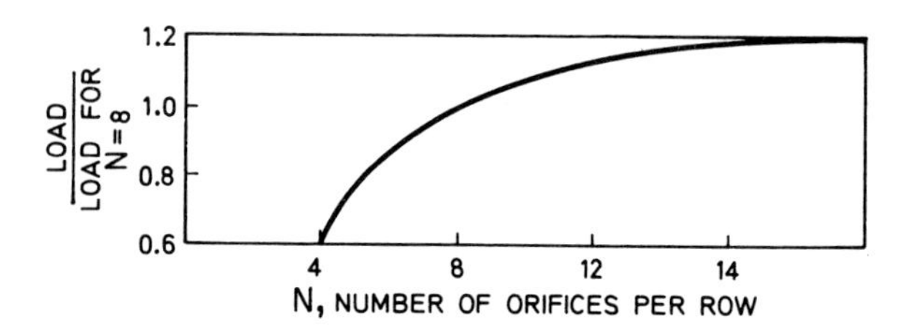

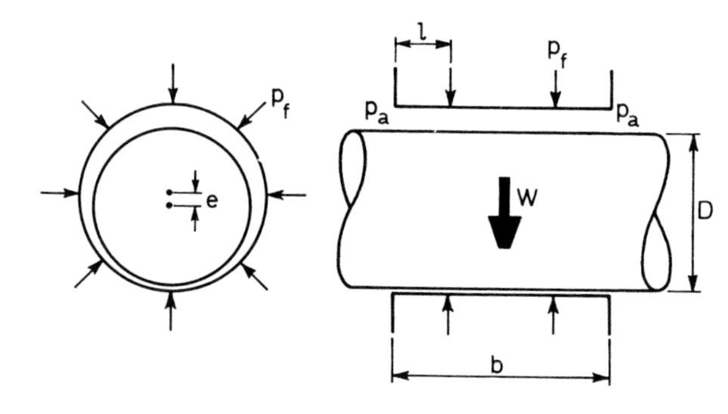

Orifice size d^* is given for:

$b/D = 1$, 8 orifices per row
air, 15°C, $p_a = 1$ bar
$C_d^* = 25\ \mu\text{m}$ (0.001 in)

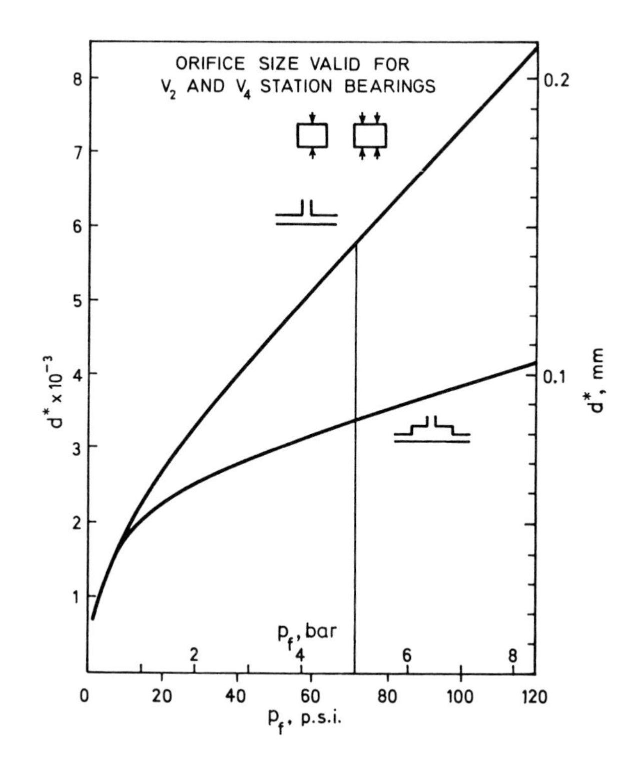

Under other operating conditions, for pocketed orifices:

$$d = d^* \cdot \left(\frac{C_d}{C_d^*}\right)^{\frac{3}{2}} \cdot \left(\frac{8}{N}\frac{D}{b}\right)^{\frac{1}{2}}$$

for unpocketed orifices

$$d = d^* \cdot \frac{C_d}{C_d^*} \cdot \frac{8D}{Nb}$$

In general, orifices are positioned at either $l/b = \frac{1}{4}$ (quarter station configuration).
or a single row at $l/b = \frac{1}{2}$ (half station)

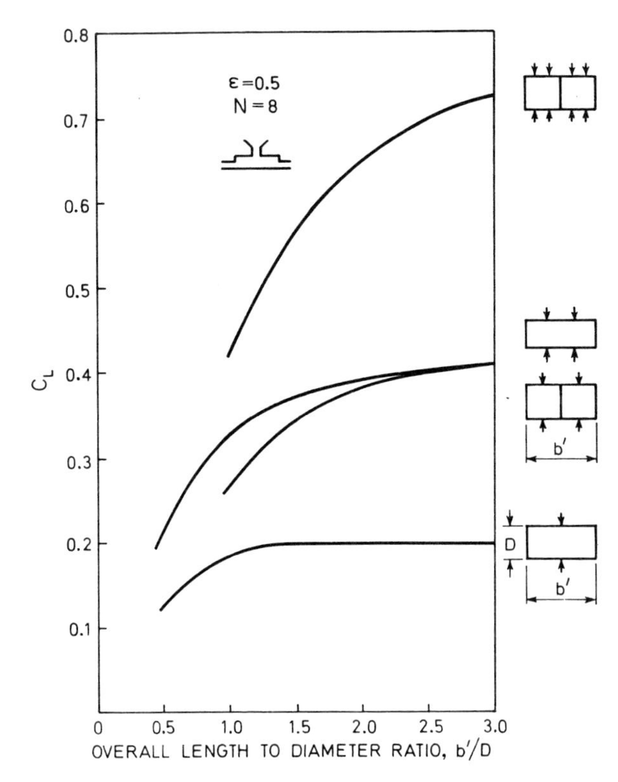

The load coefficient C_L for various combinations of half and quarter stations is used to assess the best arrangement for the highest load capacity bearing with a fixed shaft diameter. Load carrying capacity at $\varepsilon = 0.5$ and 8 orifices per row (pocketed) $W = C_L . P_f . D^2$.

For low pumping power use half stations. For greater load capacity and stiffness from smallest area bD use quarter stations.

Use of two short bearings side by side can give higher load capacity and stiffness for given length but with greatly increased air flow.

Air flow rate is given for bearing with $b/D = 1$, $l/b = \frac{1}{4}$ and $C_d^* = 25\ \mu m$ (0.001 in).

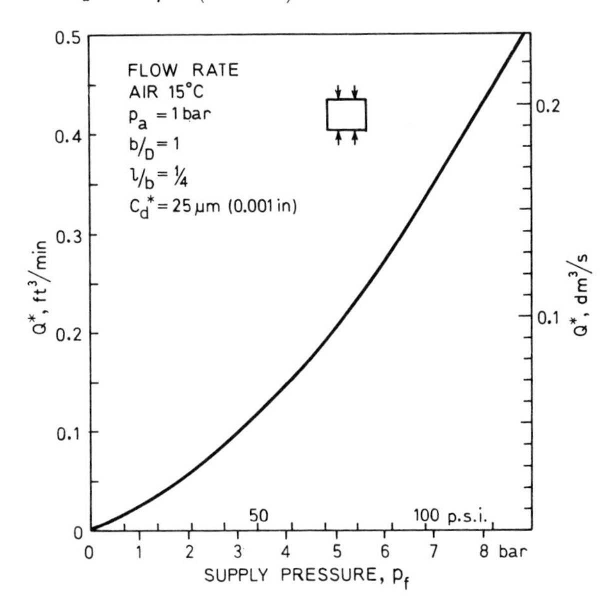

Flow rate for other operating conditions is given by

$$Q = Q^* \cdot \left(\frac{C_d}{C_d^*}\right)^3 \cdot \frac{D}{b}$$

For $\frac{1}{2}$ station bearings

$$Q = Q^* \left(\frac{C_d}{C_d^*}\right)^3 \frac{D}{2b}$$

Flow rate is not dependent upon orifice type.

Example of journal bearing design

Design wanted for spindle bearing. Radial stiffness must be greater than 200 kN/mm to achieve required resonant frequency; 5 bar supply pressure is available. Minimum shaft dia. set by shaft stiffness is 100 mm and minimum practicable clearance $C_d = 30\ \mu m$. What is the shortest length of bearing which can be used?

(*i*) $$\text{Stiffness} = \frac{W}{C_d/4},$$

therefore bearing can be defined by load which can be carried at $\varepsilon = 0.5$,

i.e. $$W = (200.10^6)\frac{(30.10^{-6})}{4} = 1500 \text{ N}.$$

Using minimum shaft dia., $D = 100$ mm and if $P_f = 5$ bar

$$C_L = \frac{W}{P_f D^2} = \frac{1500}{(5.10^5)(0.1)^2} = 0.3$$

Therefore the shortest bearing system will be, a single $\frac{1}{4}$ station bearing with $b/D = 0.8$, i.e. $b = 80$ mm from the graph of C_L. Orifice size for this bearing is obtained via the graph of d^*, $d^* = 0.085$ mm for $b/D = 1$ and $C_d^* = 25\ \mu m$ and 8 orifices per row.

With our values for b/D and C_d;

$$d = 0.085\left(\frac{30}{25}\right)^{\frac{3}{4}}\left(\frac{1}{0.8}\right)^{\frac{1}{4}} = 0.125 \text{ mm}.$$

This is an acceptable value. We could increase stiffness by increasing the number above 8 per row but may introduce manufacturing problems.

(*ii*) If unpocketed orifices were required for improved stability, then:

$$\text{Stiffness} = \frac{2}{3}\frac{W}{C_d/4} \quad \therefore\ W = \tfrac{3}{2}.1500 = 2250 \text{ N}$$

and $$C_L = \tfrac{3}{2}\,0.3 = 0.45.$$

We must use two $\frac{1}{4}$ station bearings side by side; overall $b'/D = 1.2$, $b' = 120$ mm. Each bearing has $b/D = 0.6$. $d^* = 0.145$. Therefore actual orifice dia. required

$$d = 0.145\ \frac{30}{25}\ \frac{1}{0.6} = 0.29 \text{ mm}$$

at 8 orifices/row.

Alternatively the shaft dia. could be increased to allow the use of a single $\frac{1}{4}$ station bearing, saving on air flow rate; since a short bearing is required take $C_L = 0.32$ at $b/D = 1$. Then $W = 0.32 \times 5 \times 10^5\ D^2$ and since $W = 2250$ N,

$$D^2 = \frac{2250}{0.32 \times 5 \times 10^5} = 1.41\ 10^{-2}\ \text{m}^2,$$

$D = 119$ mm, $b = 119$ mm.

Points to note in designing externally pressurised gas bearings

Designing for	*Points to note*
High speed	Power absorbed and temperature rise in bearing due to viscous shear. Resonance and whirl in cylindrical and conical modes. Variation in clearance due to mechanical and thermal changes
Low friction	Friction torque $= \dfrac{\pi^2 \eta D^3 bn}{C_d}$ Note that flow rate αC_d^3
No inherent torque	Feedholes must be radial. Journal circular and thrust plates without swash. Clearance and supply pressure low to reduce gas velocity in clearance
High load capacity	Spread available pressure over whole of available surface
High stiffness	Need small clearance and pocketed feedholes. In journals and back-to-back thrust bearings use large areas. In single thrust bearings maximum stiffness is proportional to $\dfrac{\text{load}}{\text{clearance}}$ at maximum stiffness condition. The performance of back-to-back thrust bearing arrangements can be deduced from the data given for single sided thrust bearings
Low compressor power	Low clearance, large length/diameter ratio journals and long flow path thrust bearings
High accuracy of rotation	Stator circularity is less important than that of rotor. About 10:1 ratio between mzc value of rotor surface and locus of rotor

SELF-ACTING GAS BEARINGS

1. The mechanism of operation is the same as that of liquid lubricated hydrodynamic bearings.
2. Self-acting gas bearings can be made in a considerable variety of geometric forms which are designed to suit the application.

   The simplest type of journal bearing is a cylindrical shaft running in a mating bearing sleeve. Design data for this type is well documented and reliable. Tilting pad journal and thrust bearings form another well-known group, having higher speed capabilities than the plain journal bearing.

   Another type of self-acting gas bearing uses one form or another of spiral grooves machined into the bearing surfaces. The function of the grooves is to raise the pressure within the bearing clearance due to the relative motion of one surface to the other. Bearings of this type have been applied to cylindrical journals using complementary flat thrust faces. Other shapes that have been frequently used are spherical and conical in form; in which case the bearing surfaces carry both radial and axial load.
3. Gases lack the cooling and boundary lubrication capabilities of most liquid lubricants, so the operation of self-acting gas bearings is restricted by start/stop friction and wear.

   If start/stop is performed under load then the design is limited to about 48 kN/m^2 (7 lbf/in^2) of projected bearing area depending upon the choice of materials. In general the materials used are those of dry rubbing bearings; either hard/hard combination such as ceramics with or without a molecular layer of boundary lubricant or a hard/soft combination using a plastics surface.
4. It is vital that the steady state and dynamic loads being applied to a self-acting gas bearing are known precisely since if contact between the bearing surfaces results from the response to external forces or whirl instabilities then failure of the bearing may occur. Any design should therefore include a dynamic analysis of the system of bearings and rotor.
5. For full design guidance it is advisable to consult specialists on gas bearings or tribology.

DETERMINATION OF BASIC DYNAMIC LOADING RATING

The following nomogram can be used for determining the ratio

$$\frac{C}{P} = \frac{\text{basic dynamic load rating of the required bearing}}{\text{equivalent load to be carried by the bearing}}$$

from the rotational speed and the required rating life. The International Organisation for Standardisation (ISO) defines the rating life of a group of apparently identical rolling bearings as that completed or exceeded by 90% of that group before first evidence of fatigue develops. The median life is approximately 5 times the rating life.

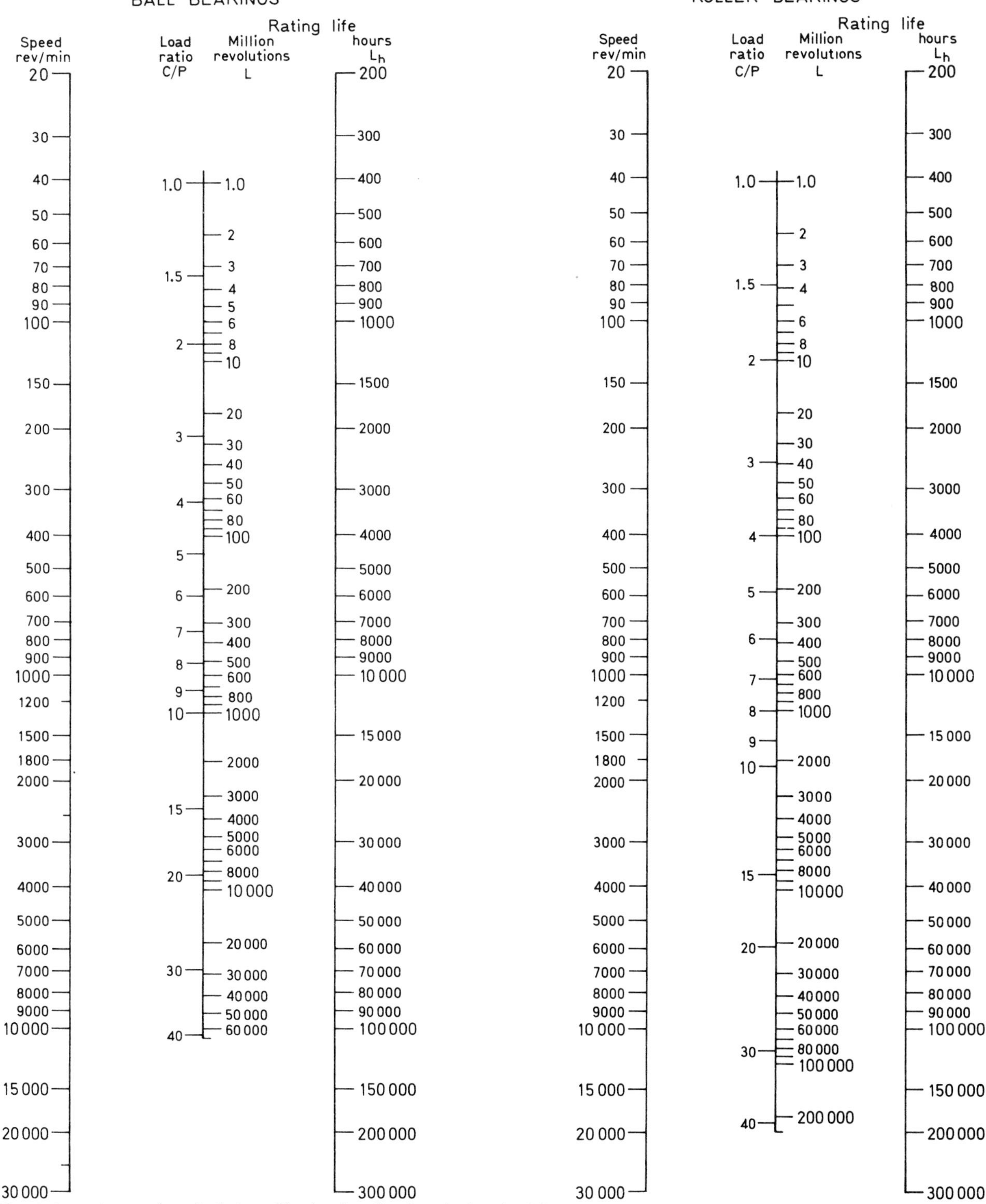

Note: information on the calculation of load rating C and equivalent load P can be obtained from ISO 281:1990. Values of C for various types of bearing can be obtained from the bearing manufacturers.

SELECTION OF TYPE OF BEARING REQUIRED

Table 20.1 Guide for general application

Design of bearing	*Bearing bore*	*Sealed* (Se) *or shielded* (S)	*Load capability*		*Allowable misalignment (degrees)*[4]	*Coefficient of friction*[5]	*Bearing section*
			Radial	*Axial*			
Single row deep groove ball	Cylindrical	1 or 2 Se 1 or 2 S	Light and medium	Light and medium	0.01 to 0.05	0.0015	
Self-aligning double row ball	Cylindrical or tapered	2 S	Light and medium	Light	2 to 3	0.0010	
Angular contact single row ball	Cylindrical	—	Medium[1]	Medium and heavy	—	0.0020	
Angular contact double row ball	Cylindrical	—	Medium	Medium	—	0.0024	
Duplex	Cylindrical	—	Light[2]	Medium	—	0.0022	
Cylindrical roller single row	Cylindrical	—	Heavy	—[3]	0.03 to 0.10	0.0011	
Cylindrical roller, double row	Cylindrical	—	Heavy	—	—	0.0011	
Needle roller single row	Cylindrical	—	Heavy	—	—	0.0025	
Tapered roller single row	Cylindrical	—	Heavy[1]	Medium and heavy	—	0.0018	
Spherical roller double row	Cylindrical or tapered	—	Very heavy	Light and medium	1.5 to 3.0	0.0018	
Thrust ball single row	Cylindrical	—	—	Light and medium	—	0.0013	
Angular contact thrust ball double row	Cylindrical	—	—	Medium	—	0.0013	
Spherical roller thrust	Cylindrical	—	Not exceed 55% of simultaneously acting axial load	Heavy	1.5 to 3.0	Refer to manufacturer	

Information given in this chart is for general guidance only.

(1) Must have simultaneously acting axial load or be mounted against opposed bearing.
(2) Must carry predominant axial load.
(3) Cylindrical roller bearings with flanges on inner and outer rings can carry axial loads providing the lubrication is adequate.
(4) The degree of misalignment permitted is dependent on internal design and manufacturers should be consulted.
(5) The friction coefficients given in this table are approximate and will enable estimates to be made of friction torque in different types of bearings.

$$\text{Friction torque} = \mu P \frac{d}{2} 10^{-3}\ \text{Nm}$$

Where P = bearing load, N
d = bearing bore diameter, mm
μ = friction coefficient.

Speed limits for radial bearings

Relatively high speeds in general applications can be achieved without resorting to special measures. The following information can be used as a guide to determining the speed limits for various types of bearings and assumes adequate care is taken with control of internal clearance, lubrication, etc. Figure 20.3 gives an approximate guide to limiting speeds and is based on loading giving a rating life of 100 000 hours. The curves are related to the expression

$$nd_m = f_1 f_2 A$$

where:

n = rotational speed, rev/min
d_m = mean diameter of bearing, 0.5 $(D+d)$ mm
D = bearing outside diameter, mm
d = bearing bore diameter, mm
A = factor depending on bearing design (Tables 20.2 and 20.3)
f_1 = correction factor for bearing size (Fig. 20.1); this has been taken into account in the preparation of the curves (Fig. 20.3).
f_2 = correction factor for load (Fig. 20.2).
The rating life L_h is obtained from the nomogram.

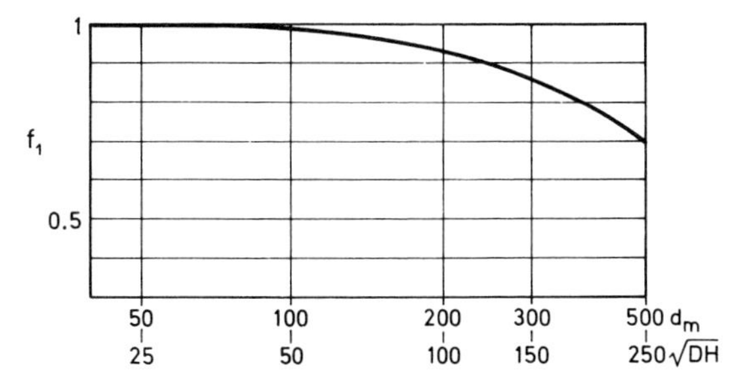

Fig. 20.1.

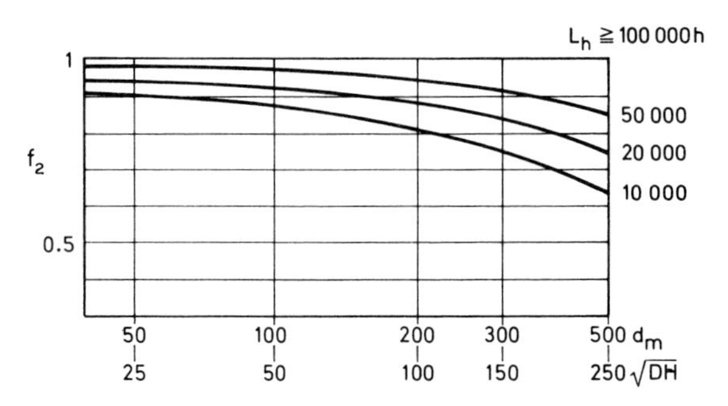

Fig. 20.2.

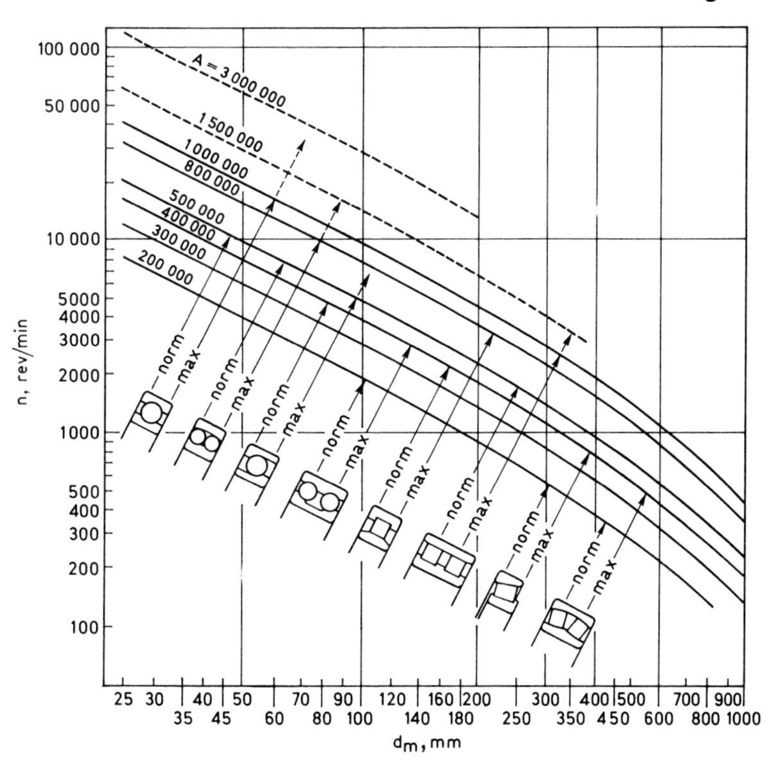

Fig. 20.3.

The limit curves, which are shown by a line of dashes, namely $A = 1\,500\,000$ and $A = 3\,000\,000$, indicate extreme values which have been achieved in a few specialist applications. The design, mounting and operation of such applications require considerable care and experience.

For spherical and tapered roller bearings the approximate speed limits apply to predominantly radial loads. Lower limits apply when the loads are predominantly axial.

When it is necessary to mount bearings such as angular contact single row ball bearings in pairs, the speed limit indicated by the curves is reduced by approximately 20%.

Table 20.2

Design of bearing	*Factor A*		*Remarks*
Single row deep groove ball bearings	Normal	500 000	Pressed steel cage
	Maximum	1 000 000	Solid cage
	Maximum	1 500 000	Solid cage, spray lubrication
Self-aligning double row ball bearings	Normal	500 000	Pressed steel cage
	Maximum	800 000	Solid cage
Angular contact single row ball bearings, $\alpha = 40$	Normal	400 000	Pressed steel cage
	Maximum	650 000	Solid cage, carefully controlled oil lubrication
$\alpha = 15$	Maximum	1 000 000	Solid cage
Angular contact double row ball bearings,	Normal	200 000	
	Maximum	400 000	C3 clearance (greater than normal)
Cylindrical roller single row bearings	Normal	400 000	Pressed steel cage
	Maximum	800 000	Solid cage
Cylindrical roller double row bearings	Normal	500 000	
	Maximum	1 000 000	Oil lubrication
	Maximum	1 300 000	Spray lubrication, maximum precision
Spherical roller bearings	Normal	200 000	For a predominantly axial load, limits of 150 000 (grease lubrication) and 250 000 (oil lubrication) are used
	Maximum	400 000	
Tapered roller single row bearings	Normal	200 000	For a predominantly axial load 20–40% lower values should be assumed, depending on difficulty of working conditions
	Maximum	400 000	

Table 20.3

Type of bearing	*Factor A*		*Remarks*
Angular contact thrust ball bearings	Normal	250 000	Grease lubrication
	Normal	300 000	Oil lubrication
	Maximum	400 000	Oil lubrication, cooling
Thrust ball bearings	Normal	100 000	
	Maximum	200 000	Solid cage
Spherical roller thrust bearings	Normal	200 000	Oil lubrication Good natural cooling usually sufficient
	Maximum	400 000	Effective cooling required

Speed limits for thrust bearings

An approximate guide to limiting speeds can be obtained from the curves in Fig. 20.4. These are based on the expression

$$n\sqrt{DH} = f_1 f_2 A,$$

where H equals the height of bearing in mm (for double row angular contact thrust ball bearings $H/2$ replaces H). All other factors are as for radial bearings.

Ball and roller thrust bearings for high-speed applications must always be preloaded or loaded with a minimum axial force and the bearing manufacturer should be consulted.

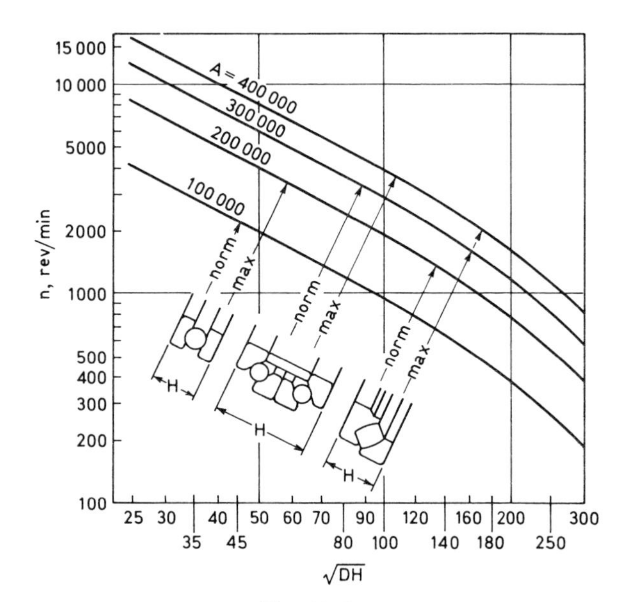

Fig. 20.4.

LUBRICATION

Grease lubrication

Grease lubrication is generally used when rolling bearings operate at normal speeds, loads and temperature conditions. For normal application the bearings and housings should be filled with grease up to 30–50% of the free space. Overpacking with grease will cause overheating. When selecting a grease, the consistency, rust-inhibiting properties and temperature range must be carefully considered. The relubrication period for a grease-lubricated bearing is related to the service life of the grease and can be estimated from the expression:

$$t_f = k\left(\frac{14 \times 10^6}{n\sqrt{d}} - 4\,d\right)$$

where:

t_f = service life of grease or relubrication interval, hours
k = factor dependent on the type of bearing (Table 20.4)
n = speed, rev/min
d = bearing bore diameter, mm

or from the curves given in Fig. 20.5.

Table 20.4

Bearing type	*Factor k for calculation of relubrication interval*
Spherical roller bearings, tapered roller bearings	1
Cylindrical roller bearings, needle roller bearings	5
Radial ball bearings	10

The amount of grease required for relubrication is obtained from:

$$G = 0.005\,DB$$

where:

G = weight of grease, g
D = bearing outside diameter, mm
B = bearing width, mm

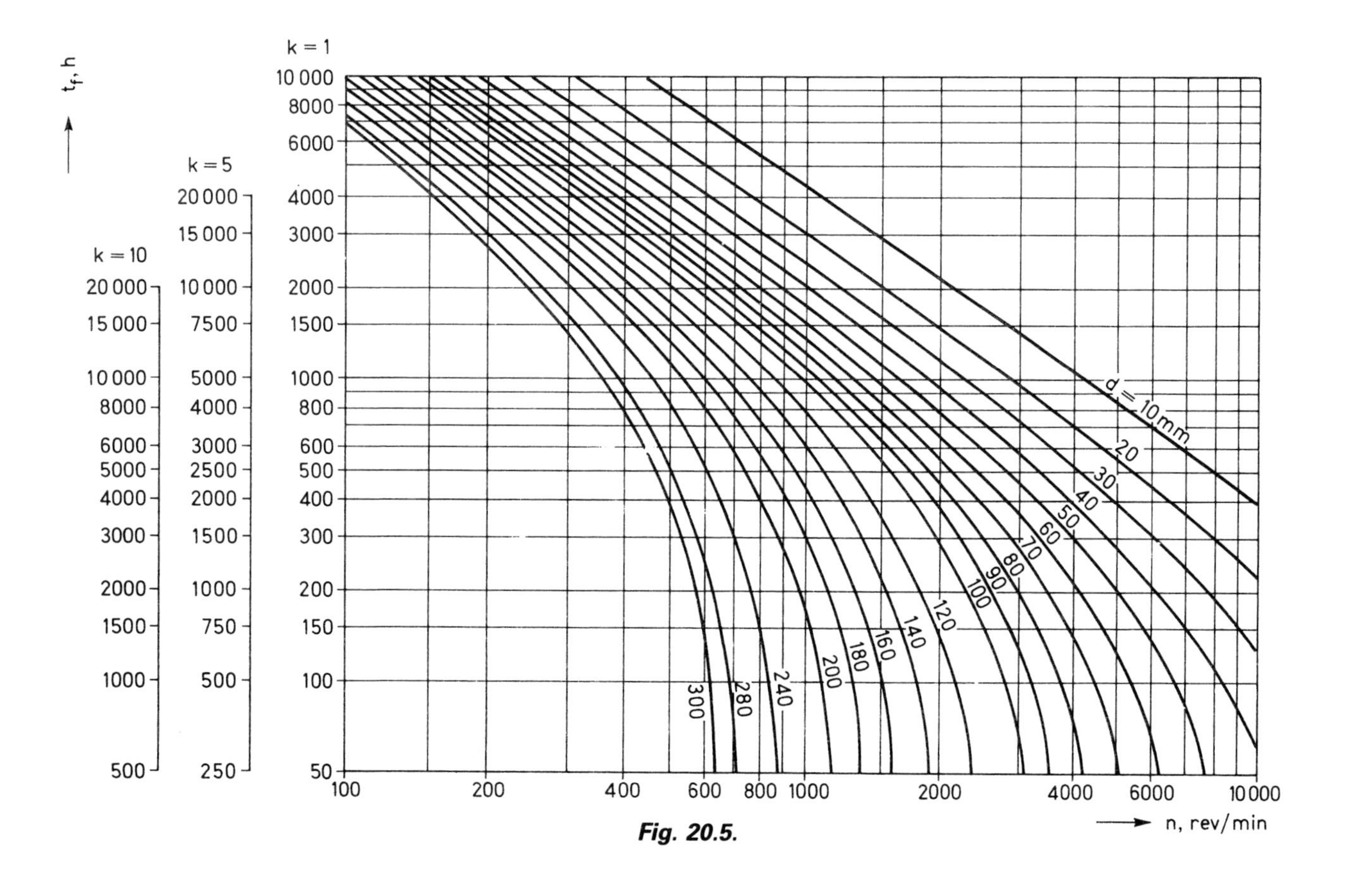

Fig. 20.5.

Oil lubrication

Oil lubrication is used when operating conditions such as speed or temperature preclude the use of grease. Fig. 20.6 gives a guide to suitable oil viscosities for rolling bearings taking into account the bearing size and operating temperature.

In the figure d = bearing bore diameter mm
n = rotational speed rev/min

An example is given below and shown on the graph by means of the lines of dashes.

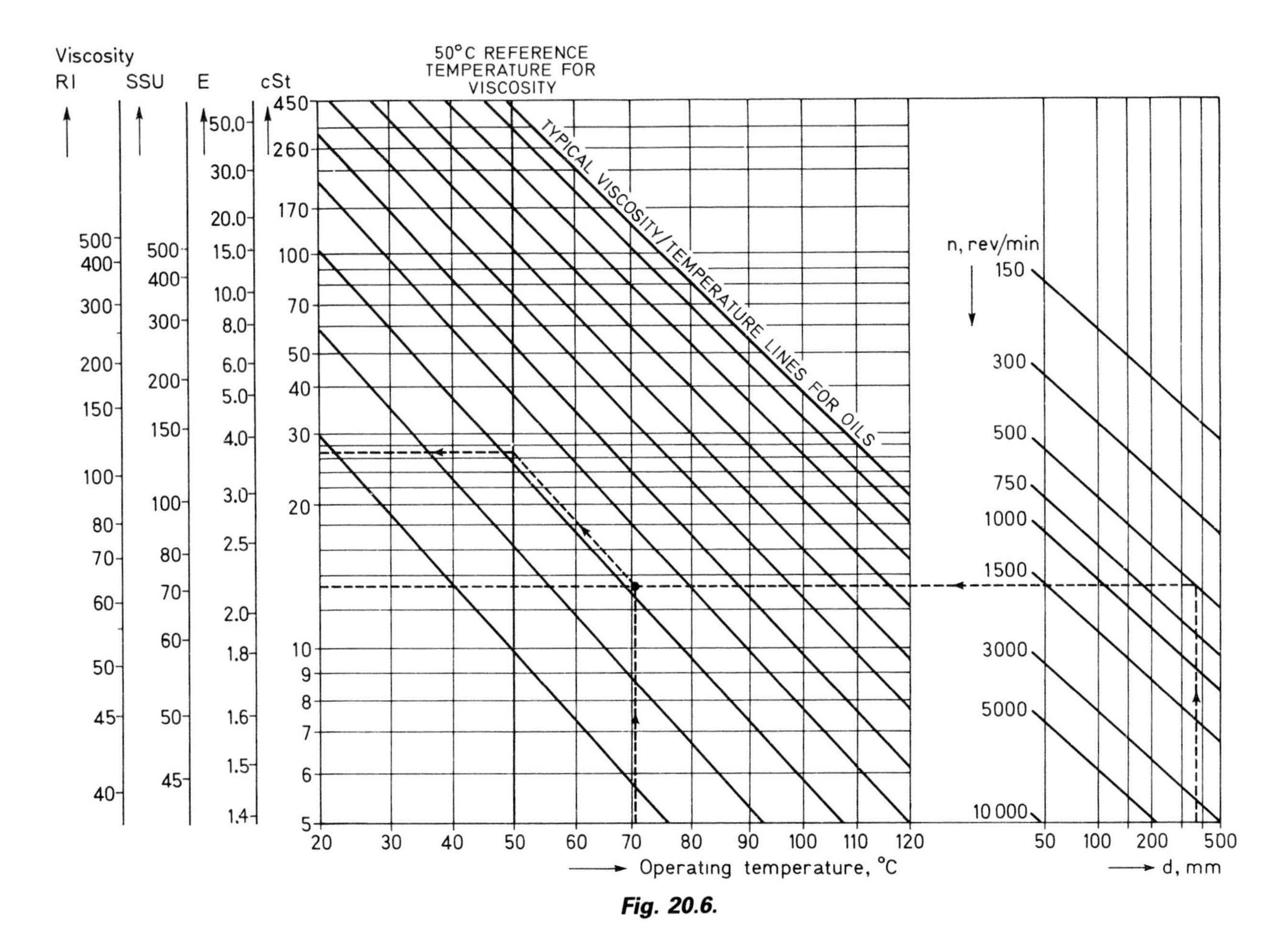

Fig. 20.6.

Generally the oil viscosity for medium and large size bearings should not be less than 12 centistokes at the operating temperature. For small high-speed bearings less viscous oils are used in order to keep friction to a minimum.

Example: A bearing having a bore diameter d = 340mm and operating at a speed n = 500 rev/min requires an oil having a viscosity of 13·2 centistokes at the operating temperature. If this operating temperature is assumed to be 70°C an oil having a viscosity of 26 centistokes at 50°C should be selected.

Acknowledgement is made to SKF (U.K.) Limited for permitting the use of graphical and tabular material.

COMPOSTION AND PROPERTIES

The stress levels in rolling bearings limit the choice of materials to those with a high yield and high creep strength. In addition, there are a number of other important considerations in the selection of suitable materials, including the following:

High impact strength
Good wear resistance
Corrosion resistance
Dimensional stability
High endurance under fatigue loading
Uniformity of structure

Steels have gained the widest acceptance as rolling contact materials as they represent the best compromise amongst the requirements and also because of economic considerations. The through hardening steels are listed in Table 21.1. The 535 grade is most popular (formerly EN31), but is not satisfactory at elevated temperatures due to loss of hardness and fatigue resistance. High-speed tool steels containing principally tungsten and molybdenum are superior to other materials at elevated temperatures, assuming adequate lubrication. Heat treatment is very important to give a satisfactory carbide structure of optimum hardness in order to achieve the maximum rolling contact fatigue life. Steelmaking practice can also influence results, e.g. refinements which improve mechanical properties and increase resistance to failure are vacuum melting and modification.

Table 21.1 Through hardening rolling-element bearing steels

Specification	*Maximum operating temperature* °C	*Composition %* C	Si	Mn	Cr	W	V	Mo	Ni	*Other elements*	*Notes*
SAE 50100*	120	0.95/ 1.10	0.20/ 0.35	0.25/ 0.45	0.40/ 0.60						Used for small thin parts, needle rollers etc.
SAE 51100	140	0.95/ 1.10	0.20/ 0.35	0.25/ 0.45	0.90/ 1.25						
534A99†	160	0.95/ 1.10	0.10/ 0.35	0.25/ 0.40	1.20/ 1.60						Most popular steels, used for majority of ball and roller bearings. Equivalent grades, previously SAE 52100 and EN31 respectively
535A99	160	0.95/ 1.10	0.10/ 0.35	0.40/ 0.70	1.20/ 1.60						
A‡	260	0.95/ 1.10	0.25/ 0.45	0.25/ 0.55	1.30/ 1.60					Al 0.75/ 1.25	Modified 535 A99 grades, generally by increase of Mn or Mo in heavy sections to give increase of hardenability
B	180	0.90/ 1.05	0.50/ 0.70	0.95/ 1.25	0.90/ 1.15						
C	180	0.85/ 1.00	0.60/ 0.80	1.40/ 1.70	1.40/ 1.70						
D	180	0.64/ 0.75	0.20/ 0.35	0.25/ 0.45	0.15/ 0.30			0.08/ 0.15	0.07/ 1.00		An alternative to 535A99, little used
5Cr5Mo	310	0.65	1.20	0.27	4.60		0.55	5.2			Intermediate high temperature steels. M50 most widely used
M50*	310	0.80	0.25	0.30	4.0		1.0	4.25			
BT1§	430	0.70	0.25	0.30	4.0	18.0	1.0				High hardenability steel used for large sections
M10	430	0.85	0.30	0.25	4.0		2.0	8.0			
BM1	450	0.80	0.30	0.30	4.0	1.50	1.0	8.0			
BM2	450	0.83	0.30	0.30	3.85	6.15	1.90	5.0			
4Cr7W2V4Mo5Co	540	1.07	0.02	0.30	4.40	6.80	2.0	3.90		Co 5.2	Superior hot hardness to the Mo base tool steels (660 VPN after 500 h at 540°C)

Increasing hardenability ↓

* AISI—SAE Steel Classification.
† BS 970 Pt II.
‡ Proprietary Steels, no specifications.
§ BS 4659.

Table 21.2 Physical characteristics of 535 grade (formerly EN31)

Property	Temperature	Value
Young's Modulus	20°C	204 GN/m^2 (13 190 tsi)
	100°C	199 GN/m^2 (12 850 tsi)
Thermal expansion	20 to 150°C	$13.25 \times 10^{-6}/°C$
Specific heat	20 to 200°C	0.107 cal/gm °C
Thermal conductivity	0.09 cgs	
Electrical resistivity	20°C	35.9 $\mu\Omega/cm^3$
Density	20°C	7.7 gm/cc

Carburising is an effective method of case hardening, and is used for convenience in the manufacture of large rolling elements and the larger sizes of race. Conventional bearing steels pose through hardening and heat treatment difficulties in the larger sizes and sections. A deep case with the correct structure, supported on a satisfactory transition zone, can be as effective as a through hardened structure. Carburising steels are listed in Table 3, many of these being alternative to the through hardening steels used for normal bearing sizes. Other surface modifying techniques such as nitriding and carbide deposition are not effective.

Table 21.3 Carburising rolling-element bearing steels suitable for a maximum operating temperature of 180°C

(Arrow down the table: Increasing hardenability; Increasing shock resistance; Increasing core hardness)

Specification	*Composition %* C	Si	% Mn	Cr	Ni	Mo	*Notes*
SAE 1015	0.13/0.18		0.30/0.60				Little used. Mainly for small thin parts such as needle rollers etc.
SAE 1019	0.15/0.20		0.70/1.00				
SAE 1020	0.18/0.23		0.30/0.60				
SAE 1024	0.19/0.25		1.35/1.65				
SAE 1118	0.14/0.20		1.30/1.60				
SAE 4023	0.20/0.25	0.20/0.35	0.70/0.90			0.20/0.30	
SAE 4027	0.25/0.30	0.20/0.35	0.70/0.90			0.20/0.30	
SAE 4422	0.20/0.25	0.20/0.35	0.70/0.90			0.35/0.45	
SAE 5120	0.17/0.22	0.20/0.35	0.70/0.90	0.70/0.90			
SAE 4118	0.18/0.23	0.20/0.35	0.70/0.90	0.40/0.60		0.08/0.15	
805 M17*	0.14/0.20	0.10/0.35	0.60/0.95	0.35/0.65	0.35/0.75	0.15/0.25	
805 M20	0.17/0.23	0.10/0.35	0.60/0.95	0.35/0.65	0.35/0.75	0.15/0.25	
805 M22	0.19/0.2	0.10/0.35	0.60/0.95	0.35/0.65	0.35/0.75	0.15/0.25	
665 M17	0.14/0.20	0.10/0.35	0.35/0.75		1.5/2.0	0.20/0.30	Equivalent grades. Most used
665 M20	0.17/0.23	0.10/0.35	0.35/0.75		1.5/2.0	0.20/0.30	
665 M23	0.20/0.26	0.10/0.35	0.35/0.75		1.5/2.0	0.20/0.30	
SAE 4720	0.17/0.22	0.20/0.35	0.50/0.70	0.35/0.55	0.9/1.20	0.15/0.25	Modified 665 grades. Increased Cr and/or Ni contents increase hardenability on heavier sections
SAE 4820	0.18/0.23	0.20/0.35	0.50/0.70		3.25/3.75	0.20/0.30	
SAE 4320	0.17/0.22	0.20/0.35	0.45/0.65	0.40/0.60	1.65/2.0	0.20/0.38	
SAE 9310	0.08/0.13	0.20/0.35	0.45/0.65	1.0/1.4	3.0/3.5	0.08/0.15	For very high shock resistance and core hardness, can increase nickel content to 4%
SAE 3310	0.08/0.13		0.45/0.60	1.4/1.75	3.25/3.75		

* BS 970 Pt III

For use in a corrosive environment, martensitic stainless steels are used, Table 21.4. These generally have a poorer fatigue resistance than the through hardening steels at ambient temperatures, but are preferred for elevated temperature work.

In special applications, where the temperature or environment rules out the use of steels, refractory alloys, cermets and ceramics can be used. However, there are no standards laid down for the choice of very high temperature bearing materials and it is left to the potential user to evaluate his own material. Some refractory alloys which have been successfully used are listed in Table 21.5 and some ceramic materials in Table 21.6. Generally their use has not been widespread due to their limited application, high cost and processing difficulties, and with some their brittle nature.

Table 21.4 Stainless rolling-element bearing steels (martensitic)

Material	Maximum operating temperature °C	Composition %								Notes
		C	Si	Mn	Cr	W	V	Mo	Other	
440C*	180	1.03	0.41	0.48	17.3		0.14	0.50		Conventional material low hot hardness. Satisfactory for cryogenic applications
14Cr4Mo	430	0.95/ 1.20	<1.00	<1.00	13.0/ 16.0		<0.15	3.75/ 4.25		Modified 440C can be used as a high temperature bearing steel or corrosion resistant tool steel. Same workability as 440C
(*a*) 12Cr5Mo	430	0.70	1.00	0.30	12.0		—	5.25		Modified 14/4 grades. These have greater hardenability and increased hot hardness a→d as well as corrosion resistance
(*b*) 15Cr4Mo1V	430	1.15	0.30	0.50	14.5		1.10	4.0		
(*c*) 15Cr4Mo2V	430	1.20	0.30	0.50	14.5		2.0	4.0		
(*d*) 15Cr4Mo5Co	500	1.10/ 1.15	<0.15	<0.15	14.0/ 16.0	2.0/ 2.5	2.5/ 3.0	3.75/ 4.25	Co 5.0/ 5.5	

* AISI Steel Classification.

Table 21.5 Refractory alloy rolling-bearing materials

Material composition*	Maximum operating temperature °C	Composition %									Notes
		C	Cr	W	Ni	Fe	Co	Mo	Ti	Al	
Haynes 6B	540	1.1	30	4.5	3.0	3.0	Δ				Cobalt base, general proprietary compositions. Some with high carbon contents cannot be forged or machined so have to be cast and ground
Haynes 6K	to 310	1.6	31	—	—	—	Δ				
Stellite 100	,,	2	34	19	—	—	43				
Haynes 25 (Forging Stock)	,,	0.1	20	15	10	3	Δ				
Stellite 3	,,	2.4	30	13	—	—	52				
Rene 41	,,	0.09	19	—	Δ	—	11	10	3.1	1.5	Nickel base
TZM	Not known	99% Mo, 0.5% Ti, 0.08% Zr									Not for use in oxidising atmosphere at high temperature. Works well in liquid sodium or potassium

* Proprietary materials, no standard specifications. Δ means remaining % to 100%.

Table 21.6 Cermet and ceramic rolling-element bearing materials

Material	Maximum operating temperature °C	Composition	Notes
Alumina	above 800°C	99% Al_2O_3, hot pressed	
Titanium carbide cermet	,,	44% Ti; 4.5% Nb; 0.3% Ta; 11.0% C; 33% Ni; 7% Mo	Most successful in high temperature field
Tungsten carbide	,,	90/94% WC; 6.0/9.0% C	Brittle, oxidised at temperature. Works best with small carbide size and minimum of matrix material
Zirconium oxide	,,	Zr_2O_3	Little experience so far
Silicon nitride	,,	Si_3N_4 reaction bonded	Light loads, excellent resistance to thermal shock

MATERIAL SELECTION AND PERFORMANCE

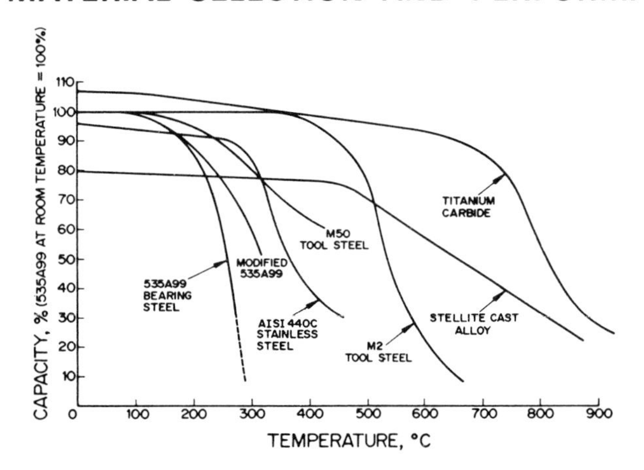

Fig. 21.1. Load capacity of rolling-element bearing materials

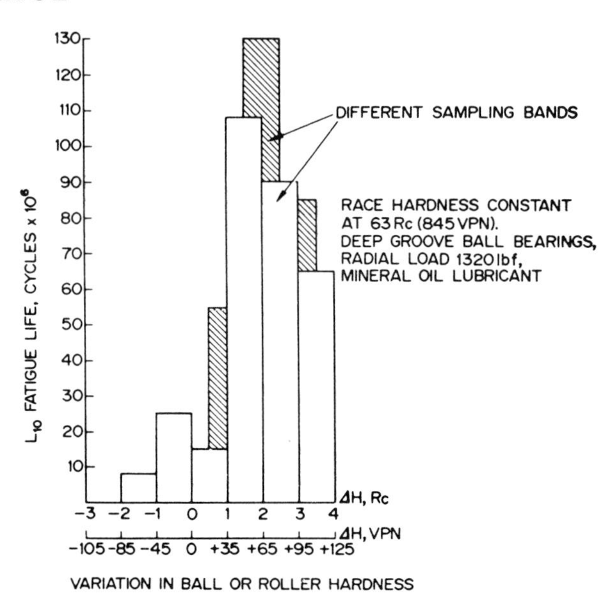

Fig. 21.2. Effect of hardness differences in 534A99 rolling-element bearing material

Rolling-element bearings are used where low friction, moderate loads and accurate positioning of a shaft are required. Because of the high contact stresses between the rolling elements and the races, precision can only be maintained at temperature with a high hardness. The effect of temperature on the load capacity of rolling contact bearings is shown in Fig. 21.1 where a significant drop in load limit can be seen above a temperature level characteristic for each material.

Through hardening 535A99 steel (EN31) is the most commonly used steel for conventional bearings. To ensure maximum life, the hardnesses of both races and rolling elements should be within an optimum hardness range, the balls or rollers being up to 10% harder than the races, Fig. 21.2. This corresponds to two points on the Rockwell C scale or 70–80 on the Vickers hardness scale when properly matched. However, current heat treatment procedures result in bearing components with significant hardness variations and bearings with a controlled hardness difference need to be specially ordered.

The combination of materials used in rolling contact seems to be important, the elastic and plastic material properties having a significant effect on the incidence of failure. For instance, a ceramic on ceramic combination will generally give a very short life because of the very high contact stresses produced, whereas ceramic rolling elements used in combination with steel races will give a considerably increased life. Information on some material combinations is given in Table 21.7.

Table 21.7 Roller / ball and race material combinations

Maximum operating temperature °C	*Race material*	*Rolling element material*	*Notes*
150	Carburised	535A99 or BTI	Used for the smaller size of rolling element
—	M10	535A99	Not recommended. Reduced rolling contact fatigue lives below that of the tool steel—tool steel combination
—	BTI	535A99	
—	M50	535A99	
480	BM2	TiC	Performance equivalent to 535A99 on 535A99 at room temperature
Up to 480	14/4	TiC	
650	Chromium carbide	Stellite 100	Limited life
820	TiC Cermet	Al_2O_3	Limited life

SELECTION OF CAGE MATERIALS

The most commonly used materials are steel, brass, bronze and plastics. Steel retainers are generally manufactured from riveted strips, while bronze and plastic cages are usually machined.

There is a number of important considerations in the selection of materials, including the following:

Resistance to wear	To withstand the effect of sliding against hardened steel
Strength	To enable thin sections to be used
Resistance to environment	To avoid corrosion, etc.
Suitability for production	Usually machined or fabricated

Table 21.8 Typical materials and their limitations

Material	*Temperature* °C	*Wear resistance*	*Oxidation resistance*	*Remarks*
Low carbon steel	260	Fair	Poor	Standard material for low-speed or non-critical applications
Iron silicon bronze	320	Good 150°C Excellent 260°C	Excellent	Jet engine applications as well as other medium-speed, medium-temperature bearings
S Monel	535	Fair	Excellent	Excellent high temperature strength
AISI 430 stainless steel	535	Poor	Excellent	Standard material for 440C stainless steel bearings—low speed
17–4–Ph stainless steel	535	Poor in air	Excellent	Good high temperature performance. Good wear resistance
Non-metallic retainers, fabric base phenolic laminates	135	Excellent	—	High-speed bearing applications
Silver plate	Possibly 180	Excellent to 150°C	—	Has been used in applications where marginal lubrication was encountered during part of the operating cycle

SHAFT AND HOUSING DESIGN

Rigidity

1 Check the shaft slope at the bearing positions due to load deflection, unless aligning-type bearings are to be used.
2 Check that the housing gives adequate support to the bearing outer ring, and that housing distortion under load will not cause distortion of the bearing outer ring.
3 Design the housing so that the resultant bearing slope is subtractive — see Figs. 22.1(a) and 22.1(b).

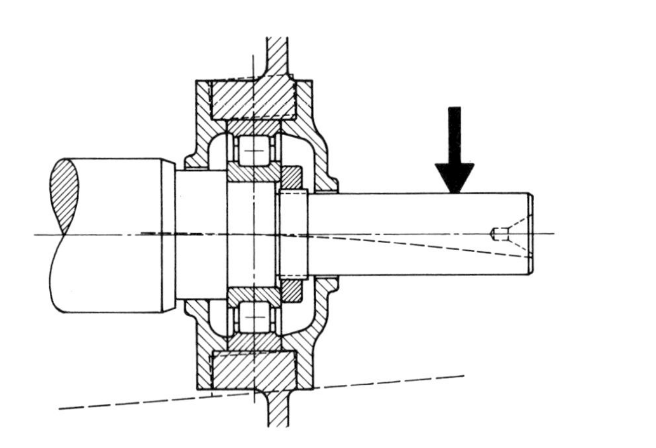

Fig. 22.1(a). Incorrect — slopes adding

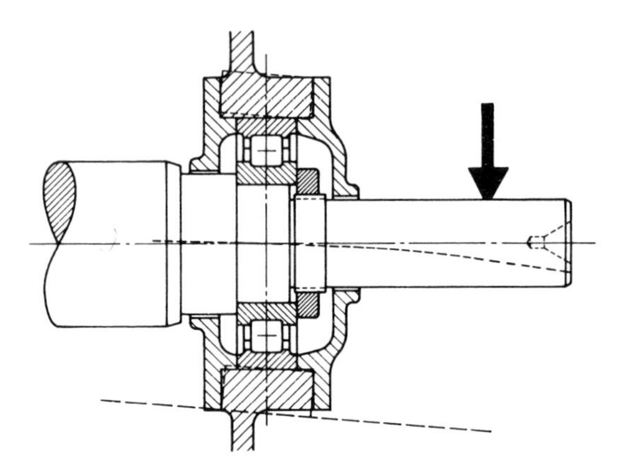

Fig. 22.1(b). Correct — slopes subtracting

Alignment

1 For rigid-type bearings, calculate the shaft and housing slopes due to load deflection.
2 Determine the errors of housing misalignment due to tolerance build-up.
3 Ensure that the total misalignment does not exceed the values given in Table 22.1.

Table 22.1 Approximate maximum misalignments for rigid bearings

Rigid bearing type	*Permitted misalignment*
Radial ball bearings	1.0 mrad
Angular contact ball bearings	0.4 mrad
Radial roller bearings	0.4 mrad
Needle roller bearings	0.1 mrad

Seatings

1 The fits indicated in Table 22.2 should be used to avoid load-induced creep of the bearing rings on their seatings.

Table 22.2 Selection of seating fit

Rotating member	*Radial load*	*Shaft seating*	*Housing seating*
Shaft	Constant direction	Interference fit	Sliding or transition fit
Shaft	Rotating	Clearance fit	Interference fit
Shaft *or* housing	Combined constant direction and rotating	Interference fit	Interference fit
Housing	Constant direction	Clearance fit	Interference fit
Housing	Rotating	Interference fit	Sliding or transition fit

2 Bearings taking purely axial loads may be made a sliding fit on both rings as there is no applied creep-inducing load.
3 Select the shaft and housing seating limits from Tables 22.3 and 22.4, respectively, these having been established to suit the external dimensions, and internal clearances, of standard metric series bearings.
4 Where a free sliding fit is required to allow for differential expansion of the shaft and housing use H7.

Table 22.3 Shaft seating limits for metric bearings (values in micro-metres)

Shaft mm	over	—	6	10	18	30	50	80	120	150	180	250	315
	incl.	6	10	18	30	50	80	120	150	180	250	315	400
Int. fit	grade	j5	j5	j5	j5	j5	k5	k5	k5	m5	m5	n6	n6
	limits	+3 −2	+4 −2	+5 −3	+5 −4	+6 −5	+15 +2	+18 +3	+21 +3	+33 +15	+37 +17	+66 +34	+73 +37
Sliding fit	grade	g6	g6	g6	g6	g6	g6	g6	g6	g6	g6	g6	g6
	limits	−4 −12	−5 −14	−6 −17	−7 −20	−9 −25	−10 −29	−12 −34	−14 −39	−14 −39	−15 −44	−17 −49	−18 −54

EXAMPLE
Interference fit shaft 35 mm dia. tolerance from table = +6/−5μm. Therefore, shaft limit = 35.006/34.995 mm.

Table 22.4 Housing seating limits for metric bearings (values in micro-metres)

Hsg mm	over	—	6	10	18	30	50	80	120	180	250	315	400	500	630
	incl.	6	10	18	30	50	80	120	180	250	315	400	500	630	800
Int. fit	grade	M6	M6	M6	M6	M6	M6	M6	M6	M6	M6	M6	M6	M6	M6
	limits	−1 −9	−3 −12	−4 −15	−4 −17	−4 −20	−5 −24	−6 −28	−8 −33	−8 −37	−9 −41	−10 −46	−10 −50	−26 −70	−30 −80
Transition fit	grade	J6	J6	J6	J6	J6	J6	J6	J6	J6	J6	J6	J6	H6	H6
	limits	+5 −3	+5 −4	+6 −5	+8 −5	+10 −6	+13 −6	+16 −6	+18 −7	+22 −7	+25 −7	+29 −7	+33 −7	+44 −0	+50 −0

EXAMPLE
Transition fit housing 72 mm dia. tolerance from table = +13/−6μm. Therefore, housing limit = 72.013/71.994 mm.

Seatings (continued)

4 Control the tolerances for out-of-round and conicity errors for the bearing seatings. These errors in total should not exceed the seating dimensional tolerances selected from Tables 22.3 and 22.4.
5 Adjust the seating limits if necessary, to allow for thermal expansion differences, if special materials other than steel or cast iron are involved. Allow for the normal fit at the operating temperature, but check that the bearing is neither excessively tight nor too slack at both extremes of temperature. Steel liners, or liners having an intermediate coefficient of thermal expansion, will ease this problem. They should be of at least equivalent section to that of the bearing outer ring.
6 Avoid split housings where possible. Split housings must be accurately dowelled before machining the bearing seatings, and the dowels arranged to avoid the two halves being fitted more than one way round.

Abutments

1 Ensure that these are sufficiently deep to provide adequate axial support to the bearing faces, particularly where axial loads are involved.

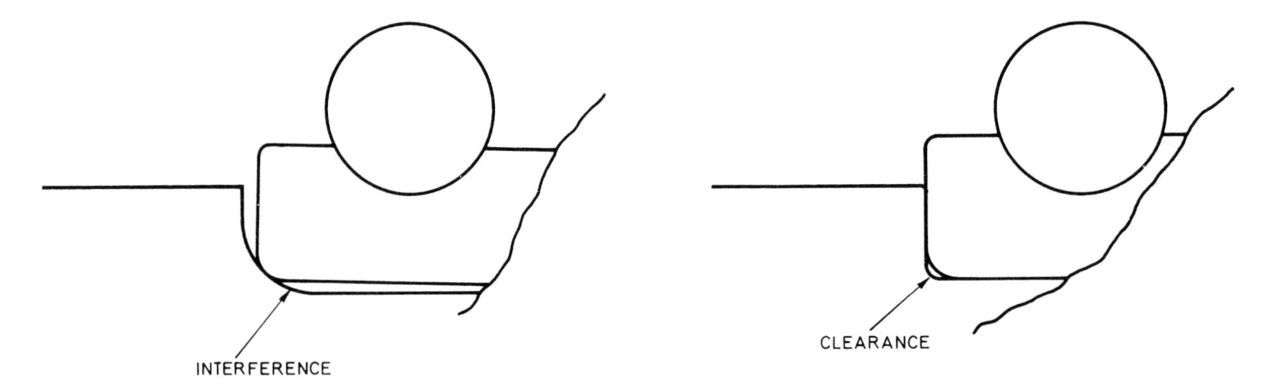

Fig. 22.2(a). Incorrect

Fig. 22.2(b). Correct

2 Check that the seating fillet radius is small enough to clear the bearing radius — see Figs. 22.2(a) and 22.2(b). Values for maximum fillet radii are given in the bearing manufacturers' catalogues and in ISO 582 (1979).
3 Design suitable grooves into the abutments if bearing extraction is likely to be a problem.

BEARING MOUNTINGS

Horizontal shaft

1 The basic methods of mounting illustrated in Figs. 22.3(a) and 22.3(b) are designed to suit a variety of load and rotation conditions. Use the principles outlined and adapt these mountings to suit your particular requirements.
2 The type of mounting may be governed more by end-float or thermal-expansion requirements than considerations of loading and rotation.

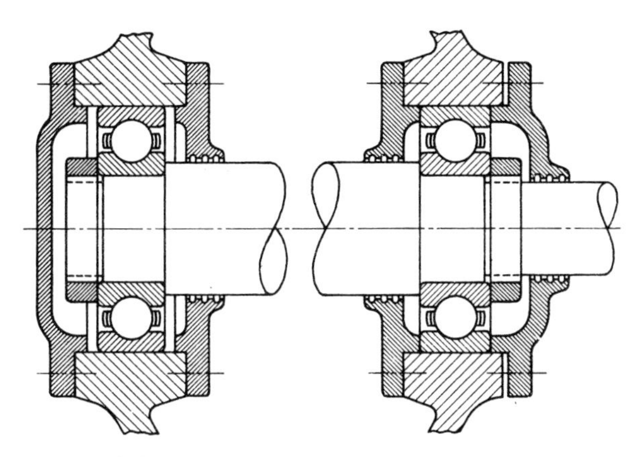

Fig. 22.3(a). Two deep groove radial ball bearings

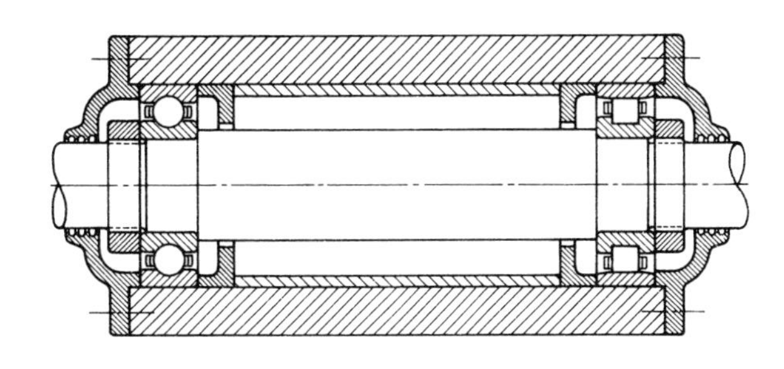

Fig. 22.3(b). One ball bearing with one cylindrical roller bearing

Condition	*Suitability*
Rotating shaft	Yes
Rotating housing	No
Constant direction load	Yes
Rotating load	No
Radial loads	Moderate capacity
Axial loads	Moderate capacity
End-float control	Moderate
Relative thermal expansion	Moderate

Condition	*Suitability*
Rotating shaft	Yes
Rotating housing	Yes
Constant direction load	Yes
Rotating load	Yes
Radial loads	
Location bearing	Moderate capacity
Non-location bearing	Good capacity
Axial loads	Moderate capacity
End-float control	Moderate
Relative thermal expansion	Yes

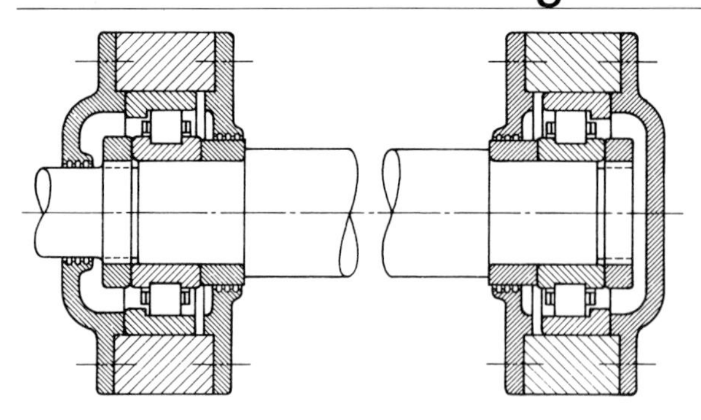

Fig. 22.3(c). Two lip-locating roller bearings

Condition	Suitability
Rotating shaft	Yes
Rotating housing	No
Constant direction load	Yes
Rotating load	No
Radial loads	Good capacity
Axial loads	Low capacity
End-float control Relative thermal expansion	Sufficient end float required to allow for tolerances and temperature

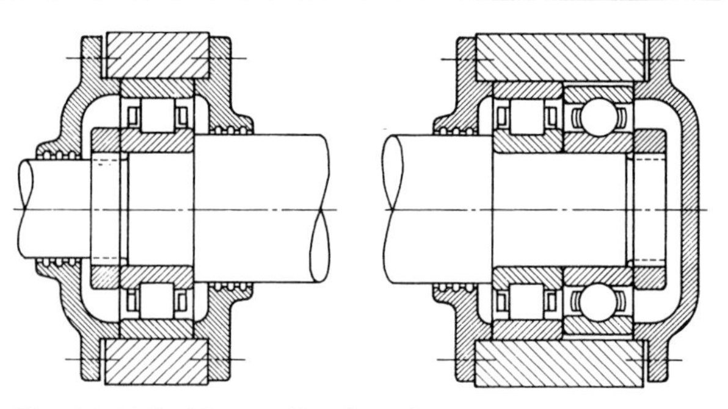

Fig. 23.3(d). Two roller bearings with 'loc' location pattern ball bearing which has reduced o.d. so that it does not take radial loads

Condition	Suitability
Rotating shaft	Yes
Rotating housing	Yes
Constant direction load	Yes
Rotating load	Yes
Radial loads	Good capacity
Axial loads	Moderate capacity
End-float control	Moderate
Relative thermal expansion	Yes

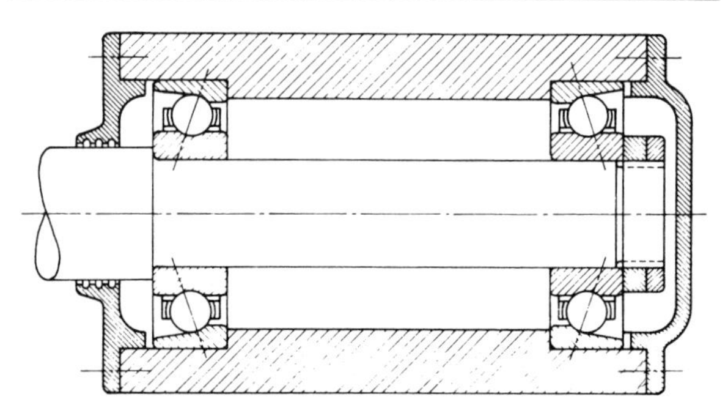

Fig. 22.3(e). Two angular contact ball bearings

Condition	Suitability
Rotating shaft	No
Rotating housing	Yes
Constant direction load	Yes
Rotating load	No
Radial loads	Moderate capacity
Axial loads	Good capacity
End-float control	Good
Relative thermal expansion	Allow for this in the initial adjustment

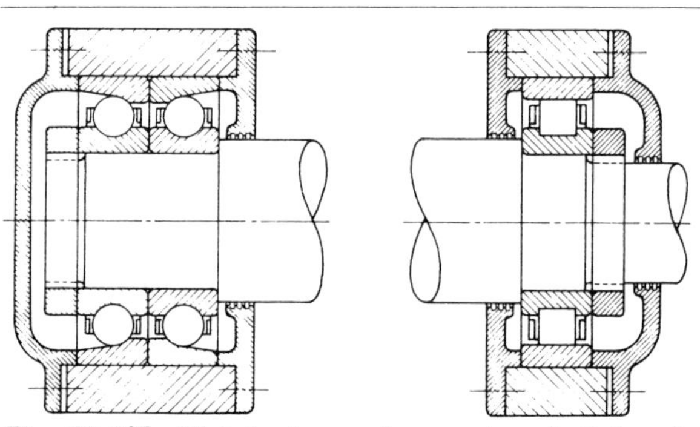

Fig. 22.3(f). Matched angular contact ball bearing unit with roller bearing

Condition	Suitability
Rotating shaft	Yes
Rotating housing	Yes
Constant direction load	Yes
Rotating load	Yes
Radial loads	Good capacity
Axial loads	Good capacity
End-float control	Good
Relative thermal expansion	Yes

Vertical shaft

1 Use the same principles of mounting as indicated for horizontal shafts.
2 Where possible, locate the shaft at the upper bearing position because greater stability is obtained by supporting a rotating mass at a point above its centre of gravity.
3 Take care to ensure correct lubrication and provide adequate means for lubricant retention. Use a No. 3 consistency grease and minimise the space above the bearings to avoid slumping.
4 Figure 22.4 shows a typical vertical mounting for heavily loaded conditions using thrower-type closures to prevent escape of grease from the housings.

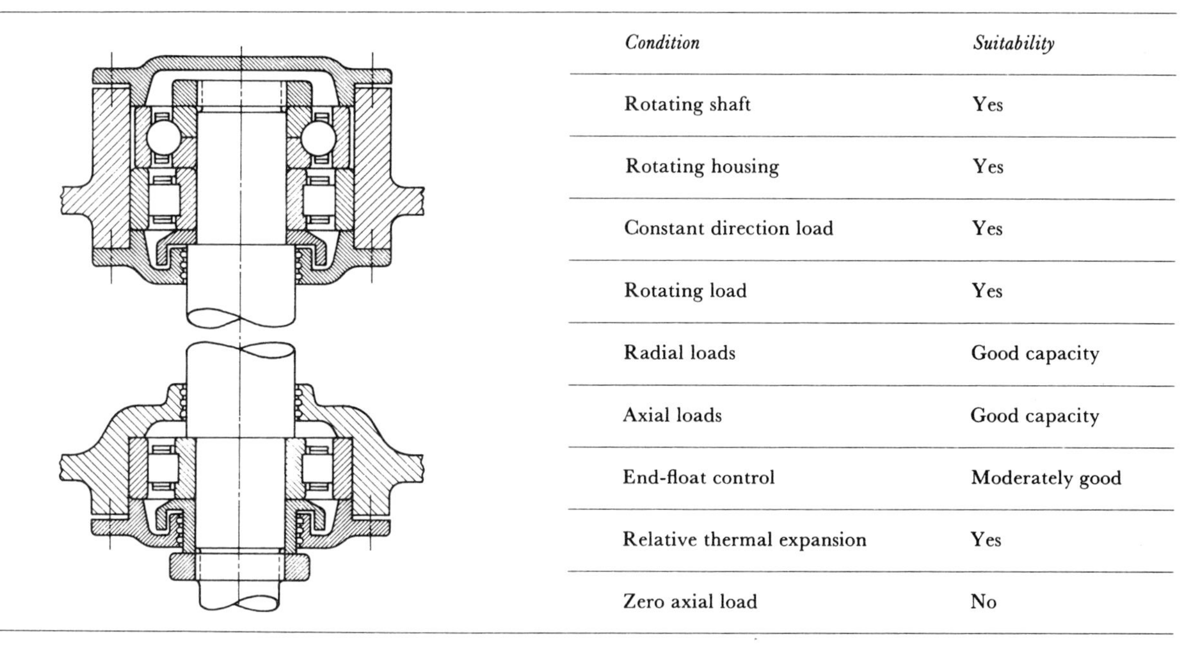

Condition	*Suitability*
Rotating shaft	Yes
Rotating housing	Yes
Constant direction load	Yes
Rotating load	Yes
Radial loads	Good capacity
Axial loads	Good capacity
End-float control	Moderately good
Relative thermal expansion	Yes
Zero axial load	No

Fig. 22.4. Vertical mounting for two roller bearings and one duplex location pattern bearing, which has reduced o.d. so that it does not take radial loads

5 For high speeds use a stationary baffle where two bearings are used close together. This will minimise the danger of all the grease slumping into the lower bearing (Fig. 22.5).

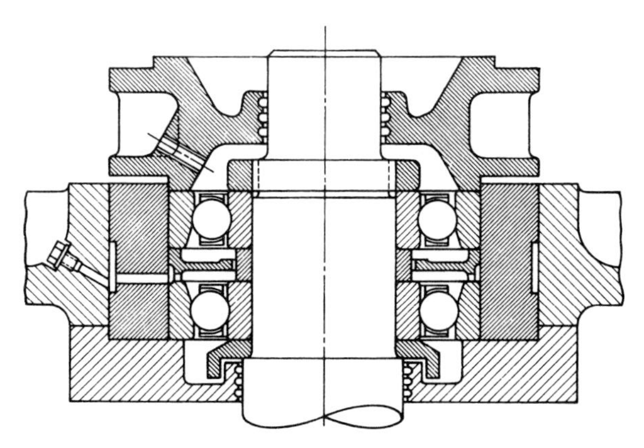

Fig. 22.5. Matched angular contact unit with baffle spacer

Fixing methods

Type of fixing	*Description*
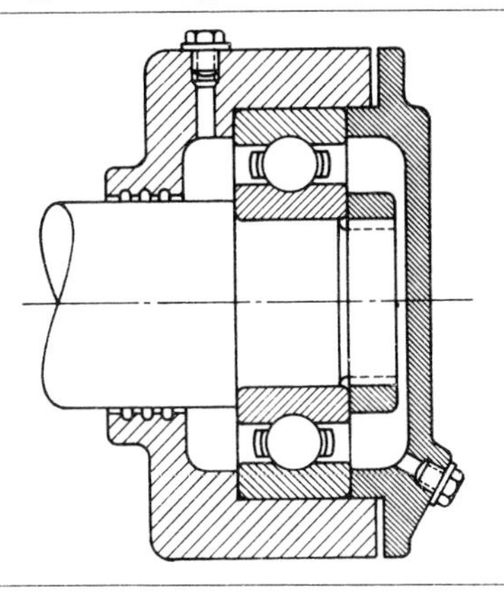	Shaft—screwed nut provides positive clamping for the bearing inner ring Housing—the end cover should be spigoted in the housing bore, *not* on the bearing o.d., and bolted up uniformly to positively clamp the bearing outer ring squarely
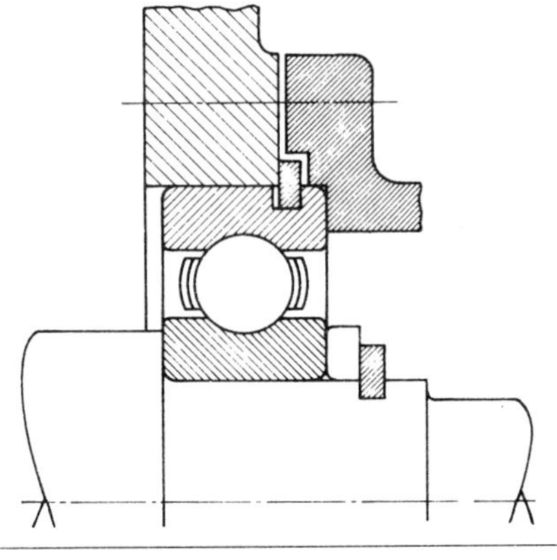	Circlip location can reduce cost and assembly time. Shaft—use a spacer if necessary to provide a suitable abutment. Circlips should not be used if heavy axial loads are to be taken or if positive clamping is required (e.g. paired angular contact unit). Housing shows mounting for snap ring type of bearing
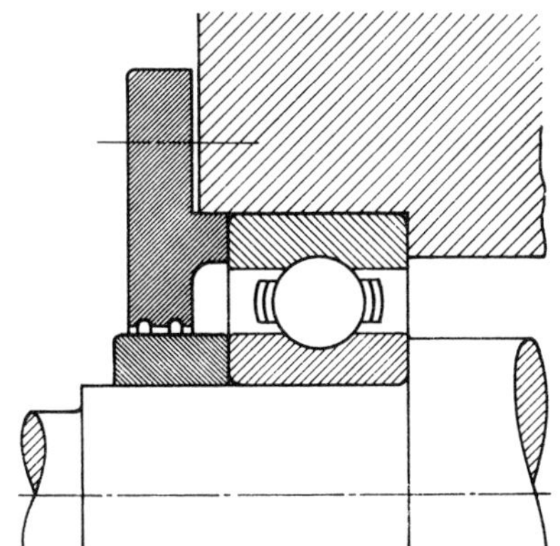	Interference fit rings are sometimes used as a cheap and effective method of locating a bearing ring axially. The degree of interference must be sufficient to avoid movement under the axial loads that apply. Where cross-location is employed, the bearing seating interference may give sufficient axial location
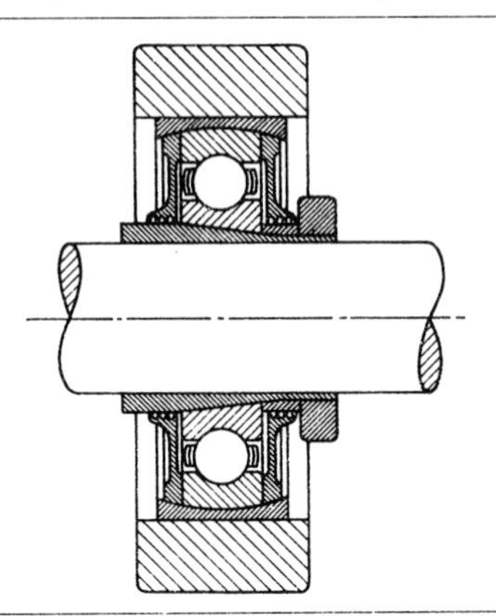	Bearing with tapered clamping sleeve. This provides a means for locking a bearing to a parallel shaft. The split tapered sleeve contracts on to the shaft when it is drawn through the mating taper in the bearing bore by rotation of the screwed locking nut

Fig. 22.6. Methods of fixing bearing rings

Sealing arrangements

1 Ensure that lubricant is adequately retained and that the bearings are suitably protected from the ingress of dirt, dust, moisture and any other harmful substances. Figure 22.7 gives typical sealing methods to suit a variety of conditions.

Sealing arrangement | ***Description***

(a) (b) (c)

(a) Shielded bearing—metal shields have running clearance on bearing inner ring. Shields non-detachable; bearing 'sealed for life'
(b) Sealed bearing—synthetic rubber seals give rubbing contact on bearing inner ring, and therefore improved sealing against the ingress of foreign matter. Sealed for life
(c) Felt sealed bearing—gives good protection in extremely dirty conditions

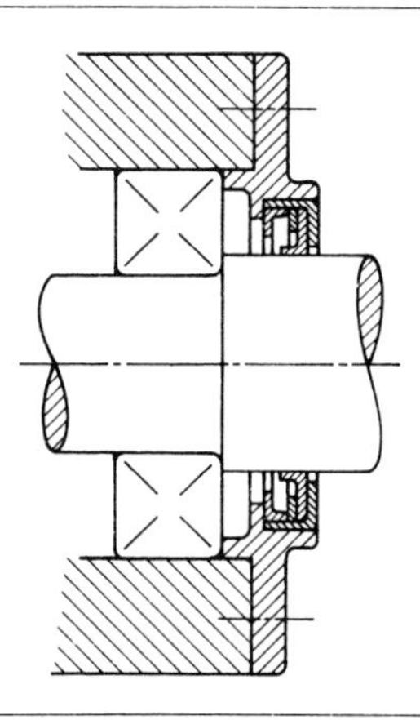

Proprietary brand rubbing seals are commonly used where oil is required to be retained, or where liquids have to be prevented from entering the bearing housing. Attention must be given to lubrication of the seal, and the surface finish of the rubbing surface

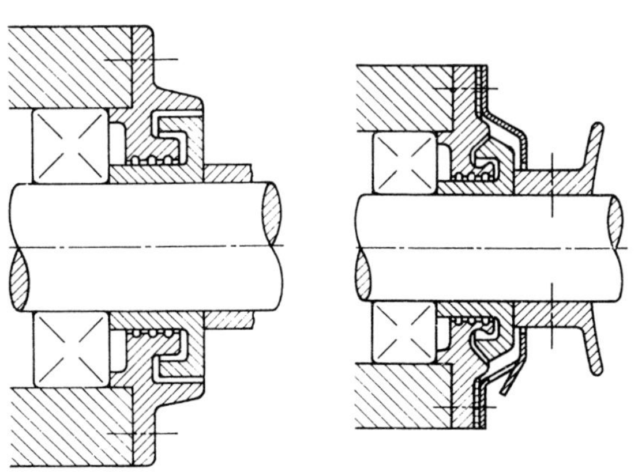

Labyrinth closures of varying degrees of complexity can be designed to exclude dirt and dust, and splashing water. The diagram shown on the left is suitable for dusty atmospheres, the one on the right has a splash guard and thrower to prevent water ingress. The running clearances should be in the region of 0.2 mm and the gap filled with a stiff grease to improve the seal effectiveness

Fig. 22.7. Methods of sealing bearing housing

BEARING FITTING

1 Ensure cleanliness of all components and working areas in order to avoid contamination of the bearings and damage to the highly finished tracks and rolling elements.
2 Check that the bearing seatings are to the design specification, and that the correct bearings and grades of clearance are used.
3 Never impose axial load through the rolling elements when pressing a bearing on to its seating—apply pressure through the race that is being fitted — see Figs. 22.8(a) and 22.8(b). The same principles apply when extracting a bearing from its seatings.

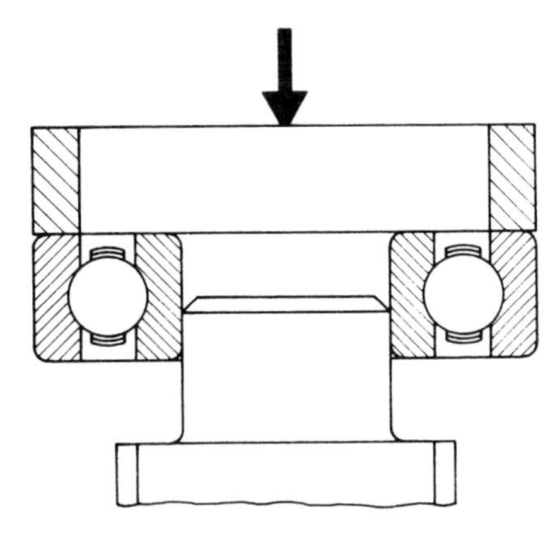

Fig. 22.8(a). Incorrect — load applied through outer ring when fitting inner ring

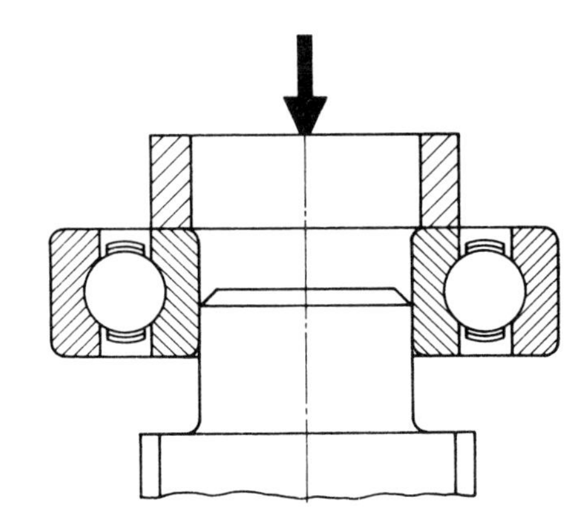

Fig. 22.8(b). Correct — load applied through ring being fitted

4 When shrink-fitting bearings on their seatings, never heat the bearings above 120°C and always ensure the bearing is firmly against its abutment when it has cooled down.
5 Where bearing adjustment has to be carried out, ensure that the bearings are not excessively preloaded against each other. Ideally, angular contact bearings should have just a small amount of preload in the operating conditions, so it is sometimes necessary to start off with a degree of end float to allow for relative thermal expansion.
6 Ensure that the bearings are correctly lubricated. Too much lubricant causes churning, overheating and rapid oxidation and loss of lubricant effectiveness. Too little lubricant in the bearing will cause premature failure due to dryness.

Table 23.1 The selection of the type of slideway by comparative performance

	Plain		Rolling element		Hydrostatic	
	Metal/metal	*Plastic/metal*	*Non recirculating*	*Recirculating*	*Liquid*	*Gas*
Stroke	Any	Any	Short	Any	Any[1]	Any[1]
Lubricant	Oil, grease, use transverse oil grooves	Oil, grease, dry	Oil (oil mist), grease	Oil (oil mist), grease	Any (non corrosive)	Air (clean and dry)
Load capacity	Medium[2]	Medium, high at low speed[2]	Medium (consult maker)	Medium, high (consult maker)	Can be very high[3]	Medium low[4]
Speed	Medium, high[6]	Medium	Any (consult maker)	Any (consult maker)	Low, medium[5]	Any
Typical friction	See note 6					
Stiffness	High	High[7]	High	High	High (keep h small)[8]	Low, medium (keep h very small)[8]
Transverse damping	Good	Good	Low, medium varies with preload	Low, medium varies with preload	Good	Low, medium
Accuracy of linear motion	Good if ways ground or scraped	Good, beware variation in thickness of adhesive, etc.	Virtually that of guideway	Virtually that of guideway	Excellent, averages local geometrical errors.[9] May run warm	Excellent, averages local geometrical errors.[9] Runs cool
Materials	Any good bearing combination	Metal is usually CI or steel with finish better than 0.25μm cla[9]	Hardened steel (R_c60) guideways (proprietary insert strips available)	Hardened steel (R_c60) guideways (proprietary insert strips available)	Any[10]	Any[10]

Table 23.1—continued

	Plain		Rolling element		Hydrostatic	
	Metal/metal	*Plastic/metal*	*Non recirculating*	*Recirculating*	*Liquid*	*Gas*
Wear rate	Low/medium[11]	Low/medium[11]	Low[11]	Low[11]	Virtually none	Virtually none
Installation	Easy	Moderate	Moderate	Moderate	Requires pump, etc.	Requires air supply, etc.
Preload used (on opposed faces)	Negligible: it increases frictional resistance	Negligible: it increases frictional resistance	Needed to eliminate backlash, excess reduces life	Needed to eliminate backlash, excess reduces life	Inherent high, can distort a weak structure	Inherent
Protection required	Wipers, covers	Wipers,[12] covers	Wipers, covers	Wipers, covers	Covers, filter fluid for re-use	Wipers[12]
Initial cost	Low	Low, medium	Medium	Medium, high[13]	Medium, high[14]	Medium, high

(1) Fluid, at relatively high pressure is supplied to the *shorter* of the sliding members.
(2) Typically 50 to 500 kN m^{-2} for machine tools, otherwise use PV value for the material pair for boundary lubrication of a collar-type thrust bearing.
(3) Ultimate, typically 0.5 × supply pressure × area; working ≏ 0.25 to 0.5 × ultimate.
(4) Limited, often, by air line pressure and area available.
(5) Prevent air entrainment by flooding the leading edge if the slide velocity exceeds the fluid velocity in the direction of sliding.
(6) Liable to stick-slip at velocities below 1 mm s^{-1}, use slideway oil with polar additive, stiffen the drive so that

$$[\text{drive stiffness } (\text{N m}^{-1})/\text{driven mass } (\text{kg})]^{1/2} > 300$$

(7) Provided plastic facing or insert is in full contact with backing.
(8) $h > 3 \times$ geometrical error of bearing surfaces.
(9) Some sintered and PTFE impregnated materials must not be scraped or ground. Some resins may be cast, with high accuracy, against an opposing member (or against a master) and need no further finishing.
(10) Use a good bearing combination in case of fluid supply failure or overload. Consider a cast resin – see note 9.
(11) May be excessive if abrasive or swarf is present.
(12) Wiper may have to operate dry.
(13) Cost rises rapidly with size.
(14) Cost rises rapidly with size but more slowly than for rolling element bearings; may share hydraulic supplies.

Combined bearings

Hydrostatic (liquid) bearings are usually controlled by a restrictor (as illustrated) or by using constant flow pumps, one dedicated to each pocket.

Hydrostatic bearing style pockets supplied at constant pressure can be combined with a plain bearing to give a 'pressure assist', i.e., 'load relief', feature whilst still retaining the high stiffness characteristic of a plain bearing; a combination useful for cases of heavy dead-weight loading.

Hydrostatic bearing style lands, supplied via small pockets at a usually low constant pressure can be fitted around, or adjacent to, rolling element bearings to give improved damping in the transverse direction, used rarely and only when vibration mode shape is suitable.

Table 23.2 Notes on the layout of slideways

(generally applicable to all types shown in Table 23.1)

Geometry	*Surfaces*	*Notes*
SINGLE SIDED 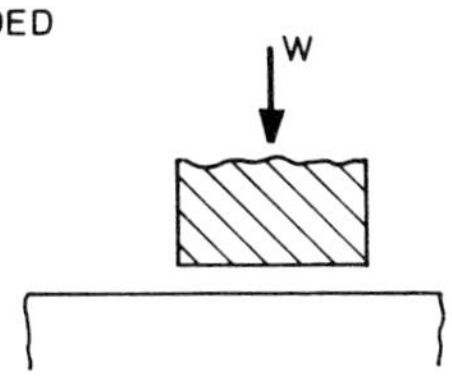		This basic single-sided slide relies largely on mass of sliding member to resist lifting forces, plain slides tend to rise at speed, hydrostatics soft under light load. Never used alone but as part of a more complex arrangement
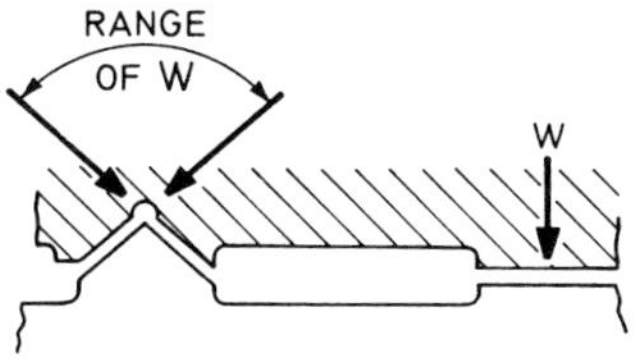	3	Direction of net load limited, needs accurate *V* angle, usually plain slide
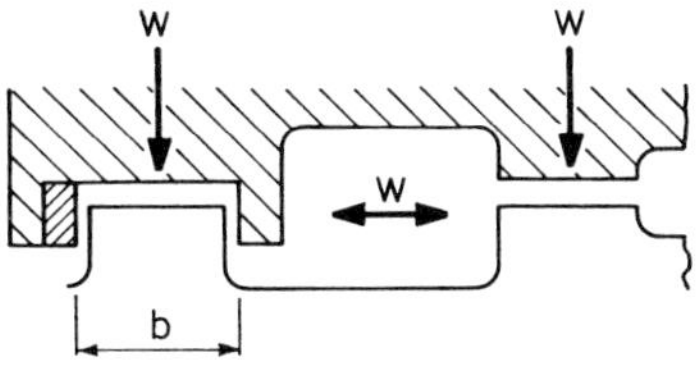	4	Easy to machine; the double-sided guide slide needs adjustment, e.g. taper gib, better if *b* is small in relation to length
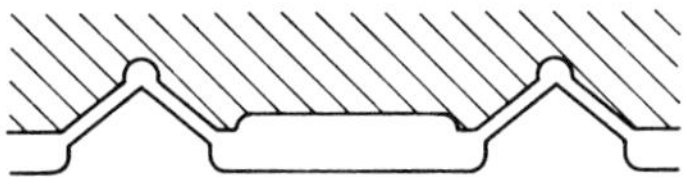	4	Used for intermittent movement, often clamped when stationary, usually plain slide
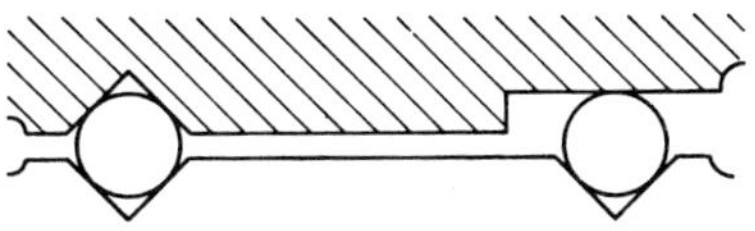	3 or 4	Accurate location, 3 ball support for instruments (2 balls in the double vee, 1 ball in the vee-flat)
DOUBLE SIDED 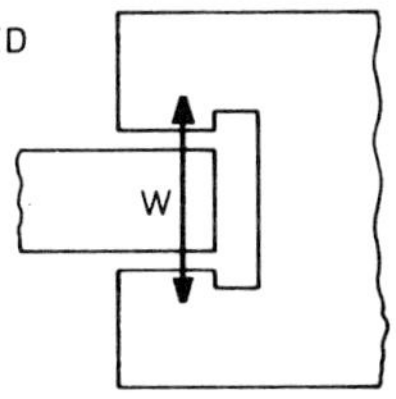		Resists loads in both directions
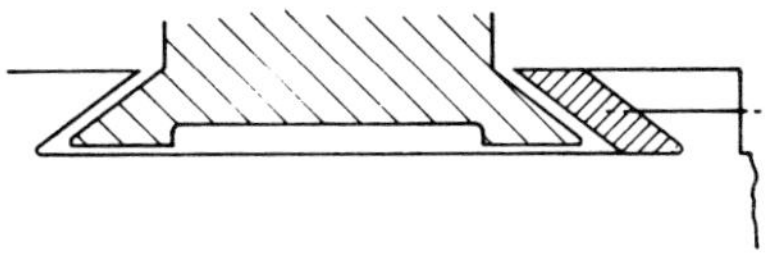	4	Generally plain, adjusted by parallel gib and set screws, very compact
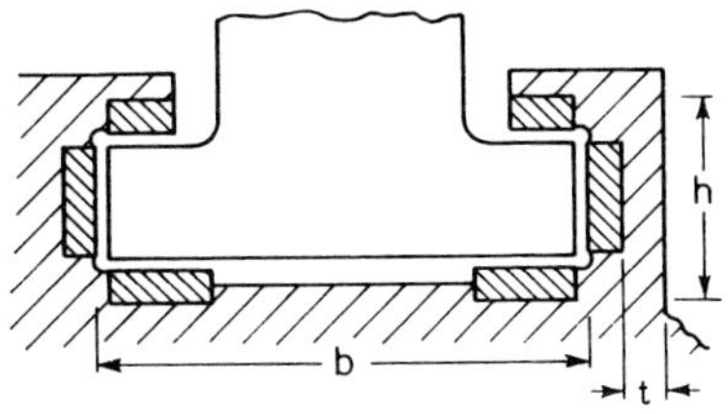	6	All types, *h* large if separate thick pads used, make *t* sufficiently large so as to prevent the structure deforming, watch for relative thermal expansion across *b* if *b* is large relative to the clearance

***Table 23.2** — continued*

Geometry	*Surfaces*	*Notes*
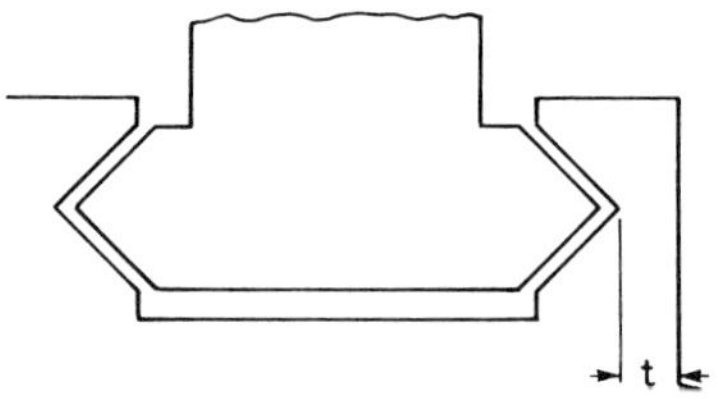	4	Usually plain or hydrostatic; watch thickness *t*. If hydrostatic an offset vertical load causes horizontal deflection also
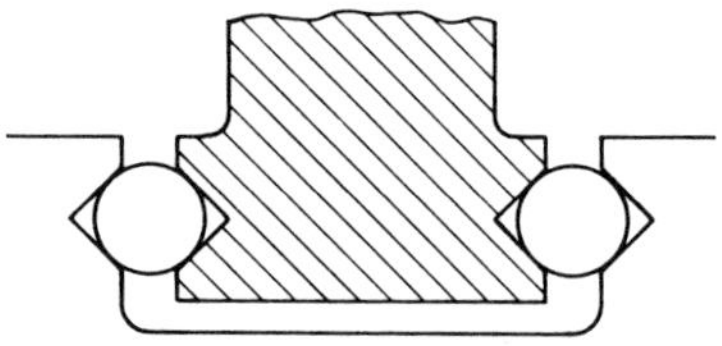	4	Ball bearings usually but not always, non recirculating, crossed-axis rollers are also used instead of balls Proprietary ball and roller, recirculating and non-recirculating units of many types involving both 2- and 4-track assemblies, complete with rails, are also available
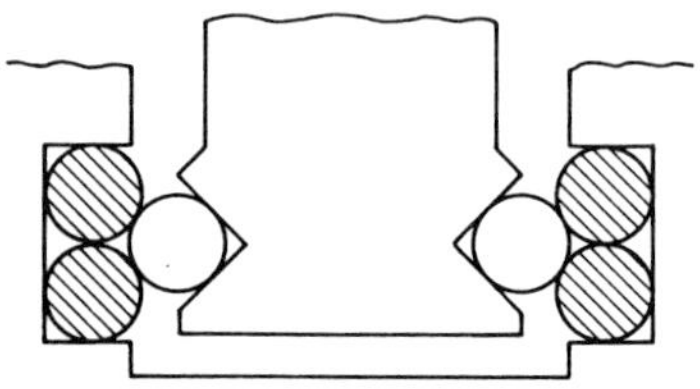		Rod guides must be in full contact with main body: (rod guides are shown cross hatched)
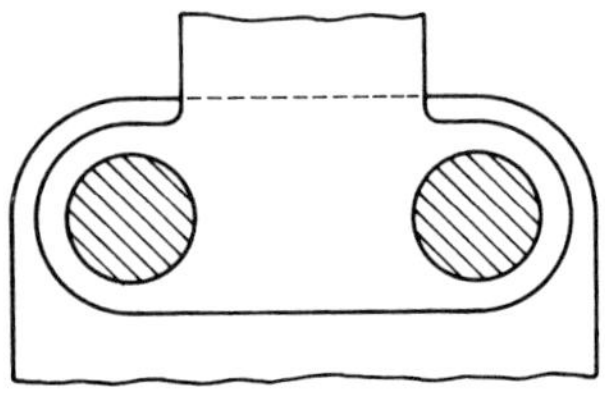		Most types (including plain, hydrostatic, ball bushes or hour-glass-shaped rollers) bars liable to bend, bar centres critical, gaps or preload adjustment not easy 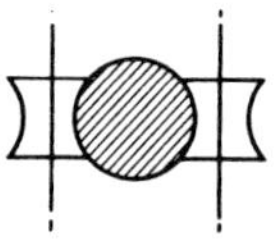
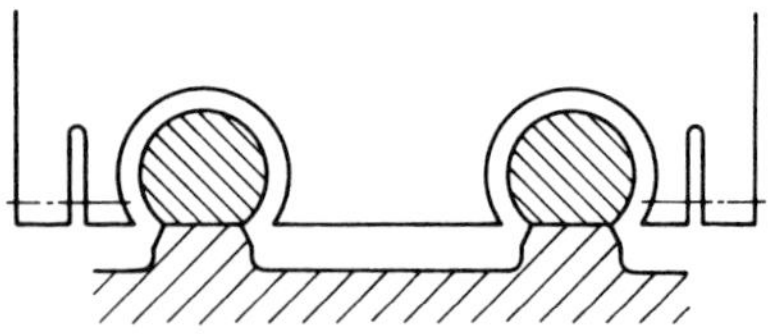		Not usually hydrostatic, bars supported but might rock, bearings weaker, clearance adjustment is easy by 'springing' the slotted housing
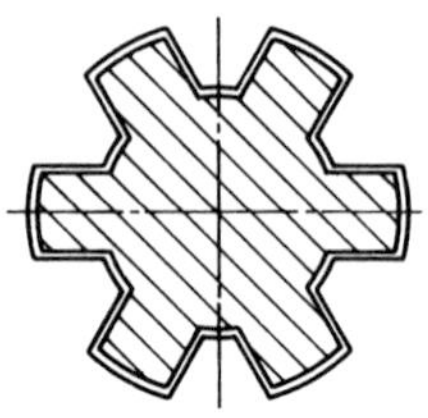		Plain or roller, torsional stiffness determined by bar usually
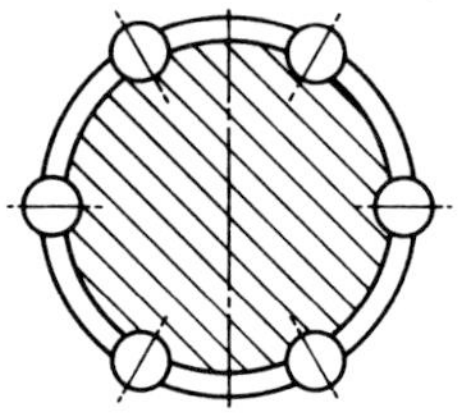		Recirculating ball usually medium torque loads permitted except when plain bars are used

MATERIALS

Usual combinations	*Special precautions*
The most usual combination is that of a steel pivot and a synthetic sapphire jewel. The steel must be of high quality, hardened and tempered, with the tip highly polished. The jewel also must be highly polished. Diamond jewels are sometimes used for very heavy moving systems. A slight trace of a good quality lubricating oil such as clock oil or one of the special oils made for this purpose, improves the performance considerably	The sapphire crystal has natural cleavage planes, and the optic axis, i.e. the line along which a ray of light can pass without diffraction, is at right angles to these planes. The angle between this optic axis and the line of application of the load is called the optic axial angle α, and experiment has conclusively demonstrated that, for the best results as regards friction, wear etc., this angle should be 90 degrees, and any departure from this produces a deterioration in performance 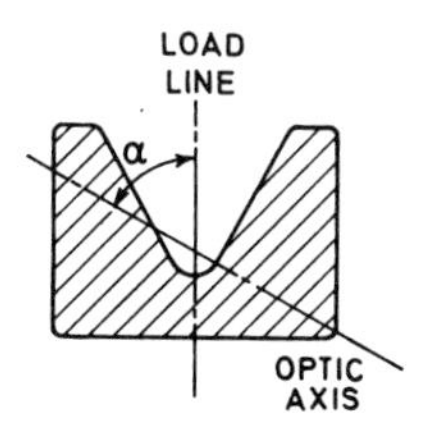

OPERATING CONDITIONS

Arrangement		*Remarks*
Vertical shaft	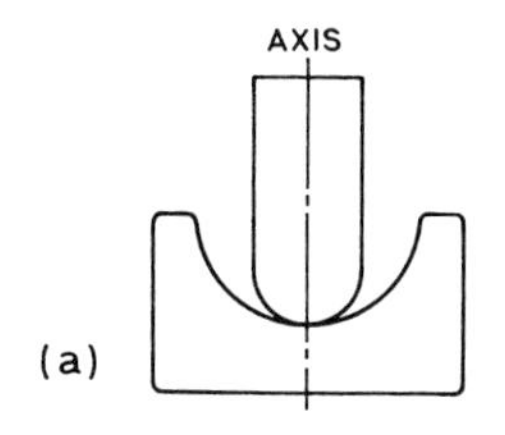	The pivot is cylindrical with a spherical end. The jewel is a spherical cup. This is used, for example, in compasses and electrical integrating metres. The optical axial angle can be controlled in this case
	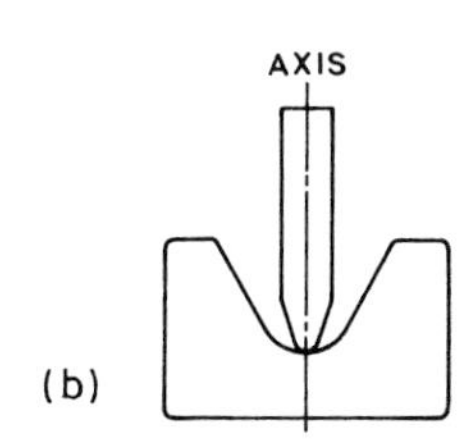	The pivot is cylindrical ending in a cone with a hemispherical tip. The jewel recess is also conical with a hemispherical cup at the bottom of the recess. This is used in many forms of indicating instrument and again the optic axial angle can be controlled
Horizontal shaft	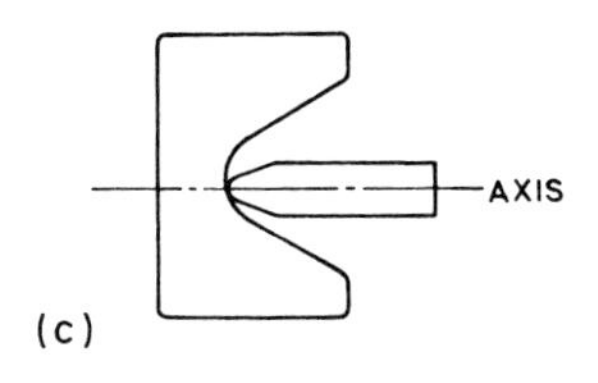	The pivot and jewel are the same as for the vertical shaft. In this case the optic axial angle cannot be controlled since the jewel is usually rotated for adjustment, so that the load on the jewels must be reduced in this case

PERFORMANCE CHARACTERISTICS

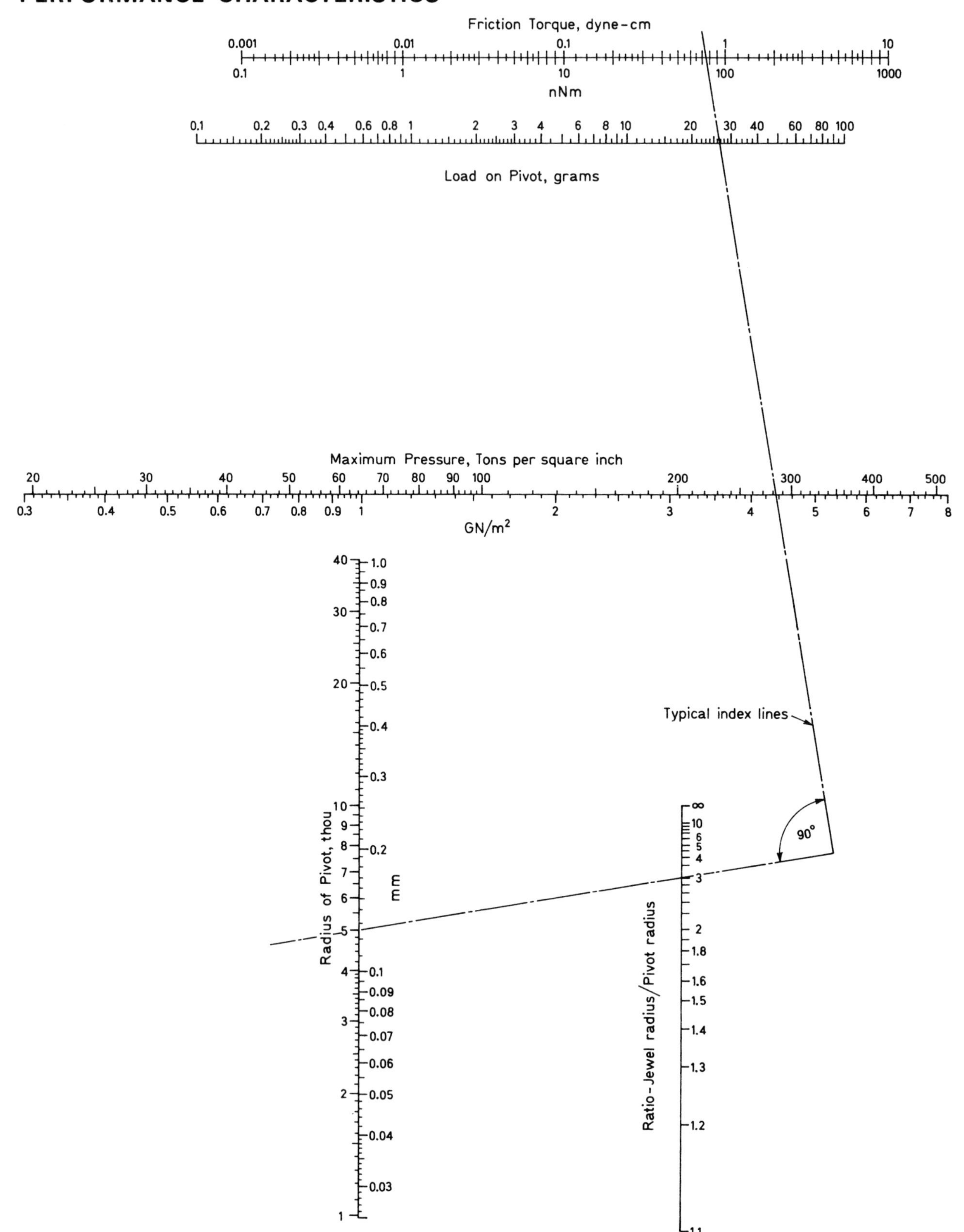

* The friction torque in the nomogram is calculated for a coefficient of friction of 0.1. For a dry jewel and pivot multiply friction torque by 1.6 and for an oiled combination by 1.4.

DESIGN

There are two important quantities which must be considered in designing a jewel/pivot system and in assessing its performance. These are the maximum pressure exerted between the surfaces of the jewel and pivot, and the friction torque between them. These depend on the dimensions and the elastic constants of the two components and can be determined by the use of the nomogram. This is of the set-square index type, one index line passes through the values of the pivot radius and the ratio jewel radius/pivot radius.

The second index line, at right angles to the first, passes through the value of the load on the jewel, and will then also pass through the values of maximum pressure and friction torque. The example shown is where the pivot radius is 5 thou. of an inch (0.127 mm), the ratio jewel radius/pivot radius is 3, and the load on the pivot is 27 grams. The resulting pressure is 282 tons per square inch (4.32 GN/m^2), and the friction torque 0.77 dyne-cm (77 nNm).

Loading	*Remarks*
Static	It is generally considered that the crushing strength of steel is about 500 tons per square inch, and experiment has shown that the sapphire surface cannot sustain pressures much above this without damage. If a safety factor of 2 is introduced then the maximum pressure should not exceed 250 tons per square inch. Unfortunately, an alteration in jewel and pivot design aimed at reducing the pressure, results in an increase in friction torque and vice versa, so that a compromise is usually necessary
Impact	All calculations have been based on static load on the jewel. Impact due to setting an instrument down on the bench, transport etc., can increase the pressure between jewel and pivot very considerably, and in many cases the jewel is mounted with a spring loading, so as to reduce the maximum force exerted on it. In general, the force required to move the jewel against the spring should not be more than twice the static load of the moving system. This spring force must then be taken as the load on the pivot 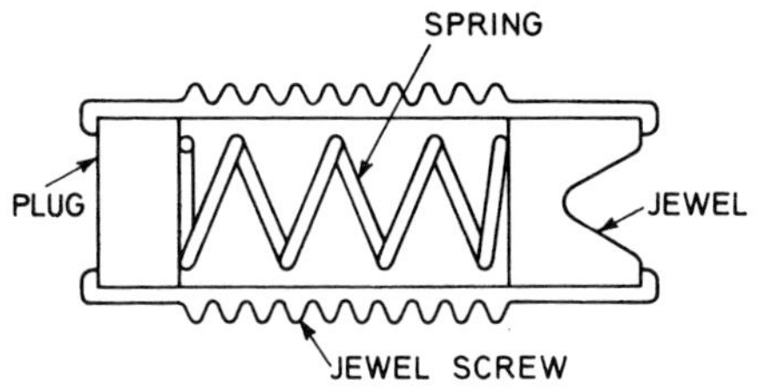

MATERIALS FOR FLEXURE HINGES AND TORSION SUSPENSIONS

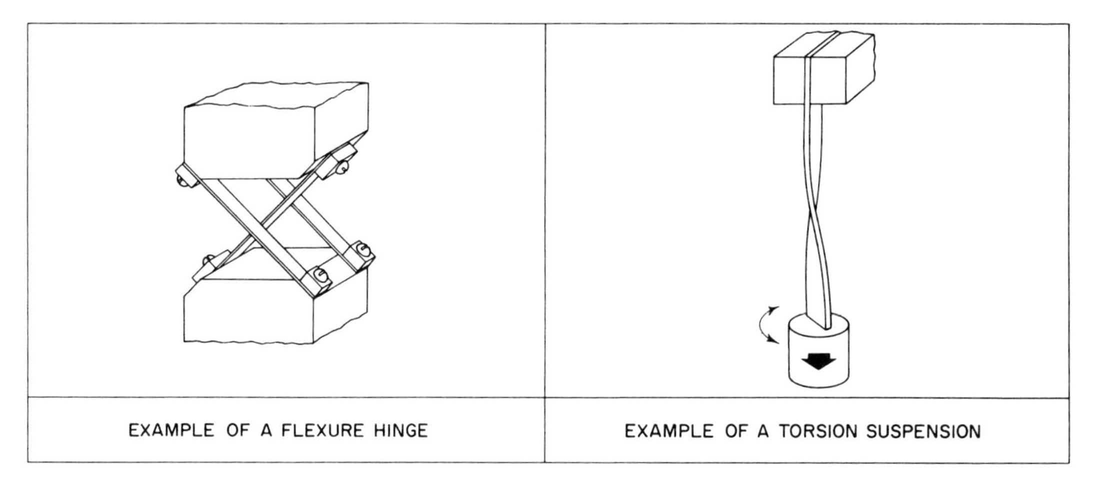

EXAMPLE OF A FLEXURE HINGE — EXAMPLE OF A TORSION SUSPENSION

Flexure hinges and torsion suspensions are devices which connect or transmit load between two components while allowing limited relative movement between them by deflecting elastically.

Selection of the most suitable material from which to make the elastic member will depend on the various requirements of the application and their relative importance. Common application requirements and the corresponding desired properties of the elastic member are listed in Table 25.1.

Table 25.1 Important material properties for various applications of flexure hinges and torsion suspensions

Application requirement	*Desired material property*
1. Small size	High maximum permissible stress, f_{max} = yield strength, f_Y *unless* the application involves a sufficiently large number of stress cycles for fatigue to be the critical condition, in which case: f_{max} = fatigue strength, f_F
2. Flexure hinge with maximum movement for a given size	High f_{max}/E: E = Young's Modulus
3. Flexure hinge with the maximum load capacity for a given size and movement	High f^3_{max}/E^2
4. Flexure hinge with minimum stiffness (for a given pivot geometry)	High 1/E: note that stiffness can be made zero or negative by suitable pivot geometry design
5. Torsion suspension with minimum stiffness for a given suspended load	High $f^2_{max} \times U/G$: G = shear modulus, U = aspect ratio (width/thickness) of suspension cross-section. U is not a material property but emphasises the value of being able to manufacture the suspension material as thin flat strip
6. Elastic component has to carry an electric current	High electrical conductivity, k_e
7. Elastic component has to provide a heat path	High thermal conductivity, k_t
8. Elastic component has to provide the main reactive force in a sensitive measurement or control system	Negligible hysteresis and elastic after-effect; non magnetic
9. As 8 and may be subject to temperature fluctuations	Low temperature coefficient of thermal expansion and elastic modulus (E or G)
10. As 6 but current has to be measured accurately by system of which elastic component is a part	Low thermoelectric e.m.f. against copper (or other circuit conductor) and low temperature coefficient of electrical conductivity
11. Elastic component has to operate at high or low temperature	As for 1–10 above, but properties, for example strength, must be those at the operating temperature
12. Elastic component has to operate in a potentially corrosive environment (includes 'normal' atmospheres)	Appropriate, good, corrosion resistance, especially if requirements 8 or 10 have to be met

Table 25.2 Relevant properties of some flexure materials

Material	*Yield strength*[1] f_y		*Fatigue strength*[2] f_f		*Young's Modulus* E (For G see note 7)		*Thermal conductivity* k_t		*Electrical conductivity* k_e	*Atmospheric corrosion resistance*[4]	*Approximate maximum continuous operating temperature in air*	
	$N/m^2 \times 10^7$	$lbf/in^2 \times 10^3$	$N/m^2 \times 10^7$	$lbf/in^2 \times 10^3$	$N/m^2 \times 10^{10}$	$lbf/in^2 \times 10^6$	W/m °C	Btu/h ft °F	%IACS[3]		°C	°F
Spring steels 0.6–1.0C 0.3–0.9Mn	80–210	120–300	40–70	60–100	21	30	45	26	9.5	P	230	450
Carbon chromium stainless steel (BS 420 S45)	150	200	60	85	21	30	24	14	2.8	M	540	1000
High strength alloy steels: nickel maraging steel	210	300	66	96	19	27	17	10	4	P	480	900
DTD 5192 (NCMV)	210	300	80	115	21	30	35	20	6	P	400	750
Inconel X	165	240	65	95	21	31	12	7	1.7	E	650	1200
High strength titanium alloy	95	140	65	95	11	16	9	5	1.1	G	480	900
High strength aluminium alloy	50	73	15	22	7.2	10.4	120	70	30	P	200	400
Beryllium copper	90	135	38	55	12.5	18	100	60	25	G	230	450
Low beryllium copper	65	95	24	35	11.5	16.5	170	100	45	G	200	400
Phosphor bronze (8% Sn; hard)	60	90	20	29	11	16	55	32	12	G	180	350
Glass fibre reinforced nylon (40% G.F.)	20	30	NA	NA	1.2	1.8	0.35	0.20	negligible	E	110	230
Polypropylene	3.7	5.4[5]	NA	NA	0.14	0.2[5]	0.17	0.1	negligible	E[6]	50	120

Notes:
1. Very dependent on heat treatment and degree of working. Figures given are typical of fully heat treated and processed strip material of about 0.1 in thickness at room temperature. Thinner strip and wire products can have higher yield strengths.
2. Fatigue strengths are typical for reversed bending of smooth finished specimens subjected to 10^7 cycles. Fatigue strengths are reduced by poor surface finish and corrosion, and may continue to fall with increased cycles above 10^7.
3. Percentage of the conductivity of annealed high-purity copper at 20°C.
4. Order of resistance on following scale: P—poor, M—moderate, G—good, E—excellent. Note, however, that protection from corrosion can often be given to materials which are poor in this respect by grease or surface treatments.
5. At high strain rate. Substantial creep occurs at much reduced stress levels, probably restricting applications to where the steady load is zero or very small, and the deflections are of short duration.
6. But the material deteriorates rapidly in direct sunlight.
7. Modulus of Rigidity, $G = E/2\,(1 + \text{Poisson's ratio}, \upsilon)$. For most materials $\upsilon \simeq 0.3$, for which $G \simeq E/2.6$.

NA Data not available.

The main properties of interest for selection of materials for flexures and torsion suspensions are given in Table 25.2 and Table 25.3. Values given are intended to provide a comparison of different materials, but they are only typical and should not be used to specify minimum properties.

Table 25.3 Some materials used to meet special requirements in accurate instruments or control systems

Requirement	*Material(s)*
Minimum hysteresis and elastic after-effect	91.5% platinum, 8.5% nickel alloy. Platinum/silver alloy, 85/15 to 80/20. Quartz
Zero thermoelectric e.m.f. against copper	Copper
Maximum corrosion resistance (instrument torsion suspensions)	Gold or platinum alloys, quartz
Maximum electrical conductivity	Silver
Torsionless suspensions (e.g. for magnetometers)	Stranded silk and other textile fibres)

MATERIALS FOR KNIFE EDGES AND PIVOTS

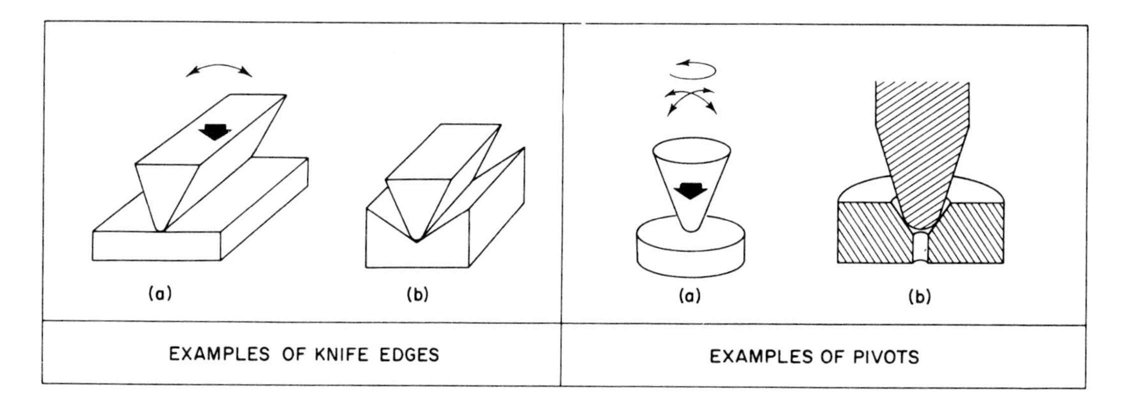

Knife edges and pivots are bearings in which two members are loaded together in nominal line or point contact respectively, and can tilt relative to one another through a limited angle by rotation about the contact; a pivot can also rotate freely about the load axis.

The main requirement of the materials for this type of bearing is high hardness, so that high load capacity can be provided, while keeping the width of the contact area small for low friction torque and high positional accuracy of the load axis.

Table 25.4 Important material properties for various applications of knife edges and pivots

Application requirement	*Desired material property*
1. High load capacity for a given bearing geometry	High $\frac{hardness^2}{modulus\ of\ elasticity}$
2. Ability to tolerate overload, impact or rough treatment generally	A measure of ductility in compression, so that overload can be accommodated by plastic deformation rather than chipping or fracture
3. Requirements 1 and 2 together (for example for weigh-bridges, strength-testing machines, etc.)	High hardness together with some ductility. In practice various metallic materials with hardnesses greater than 60 Rc (690 Knoop) are usually specified
4. Very low friction with useful load capacity where freedom from impact and overloading can be expected (for example in sensitive force balances and other delicate equipment)	Very high $\frac{hardness^2}{modulus\ of\ elasticity}$ using various brittle materials having exceptionally high hardness
5. High wear resistance	High hardness is generally beneficial
6. Little indentation of block by knife edge or pivot	Hardness of block > hardness of knife edge or pivot. (This is nearly always desirable; the differential should be at least 5%)
7. The two members of the bearing have to slide relative to one another at the contact (see examples (b)) and must be metallic to withstand impact, etc.	Low tendency to adhesion to avoid high sliding friction and wear; in practice it is often sufficient to avoid using identical materials
8. Bearing to be used in a sensitive force balance	Non magnetic; should not absorb moisture or be subject to any other weight variation. (Agate, for example, is unsatisfactory in the latter respect since it is hygroscopic)
9. Bearing to be used in a potentially corrosive environment (includes 'normal' atmospheres)	Good corrosion resistance, especially if requirement 8 has to be met

Table 25.5 Relevant properties of some knife edge and pivot materials

Material	*Hardness* H, *Knoop*	*Modulus of elasticity*, E		*Load capacity factor*, H^2/E *(arbitrary units)*	*Ductility*	*Approximate maximum continuous operating temperature in air*		*Corrosion resistance*[1]
		($N/m^2 \times 10^{10}$)	($lbf/in^2 \times 10^6$)			°C	°F	
High carbon steel	to 690	21	30	2.3	Some	250	500	Poor
Tool steels	to 850	21	30	3.4	Some	650	1200	Poor–Good
Stainless steel (440C)	660	21	30	2.1	Some	430	800	Moderate
Agate	730	7.2	10.4	7.4	None	575[2]	1070[2]	Excellent
Synthetic corundum (Al_2O_3)	2100	38	55	11.6	None	1500	2700	Excellent
Boron carbide	2800	45	65	17.4	None	540	1000	Excellent
Silicon carbide	2600	41	60	16.4	None	800	1470	Excellent
Hot pressed silicon nitride	2000	31	45	13	None	1300	2400	Excellent

Notes: 1. Materials with poor corrosion resistance can often be protected by grease, oil bath or surface treatments (such as chromising of steels).
2. Phase change temperature.

A26 Electromagnetic bearings

Electromagnetic bearings use powerful electromagnets to control the position of a steel shaft. Sensors are used to detect the shaft position and their output is used to control the currents in the electromagnets in order to hold the shaft in a fixed position. Steady and variable loads can be supported, and since no liquid lubricant is involved, new machine design arrangements become possible.

RADIAL BEARING CONFIGURATION

Four electromagnets are arranged around the shaft to form the bearing. Each electromagnet is driven by an amplifier. In horizontal shaft applications, the magnet centrelines are orientated at 45° to the perpendicular such that forces due to gravity are acted on by the upper two adjoining magnets. This adds to the load capability and increases the stability of the system.

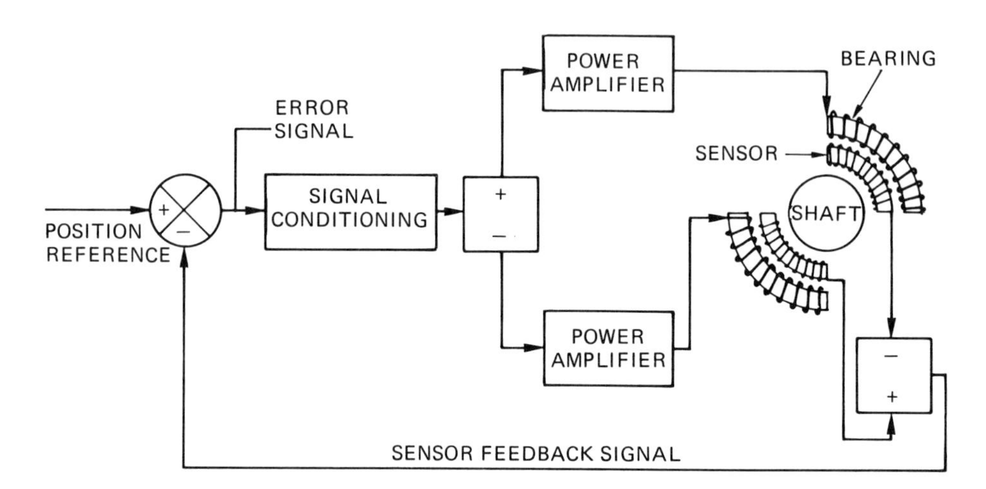

Fig. 26.1. Two of the four magnets of a radial bearing with their associated control system

Opposite electromagnets are adjusted to pull against one another in the absence of any externally applied force (the bias force). When an externally applied force causes a change in position of the shaft it is sensed by position transducers which, via the electronic control system, cause an increase in one current and a decrease in the other current flowing through the respective electromagnets. This produces a differential force to return the shaft to its original position. The signals from the position transducers continuously update this differential force to produce a stable system.

Typical radial bearing applications

A main field of application is on high speed rotating machines such as compressors, turbo-expanders, pumps and gas turbines.

Bearing bore sizes	40 – 1500 mm
Radial clearance gaps	0.1 – 5.0 mm
Speeds	400 – 120,000 RPM
Temperatures	185° C to 480° C
Load	up to 80 kN

AXIAL BEARING CONFIGURATION

A flat, solid ferromagnetic disc, secured to the shaft is used as the collar for the axial thrust bearing. Solid disc electromagnets are situated either side of the collar and operate in a similar manner to those in a radial bearing but in one dimension only.

POSITION TRANSDUCERS

Two dimensions are controlled at each radial bearing location and one dimension is controlled at the axial bearing. One transducer could be used for each dimension if it were totally linear and free from drift due to ageing or temperature effects. Two transducers per dimension are, however, used in practice because they require only that a balance or difference be maintained, thus cancelling unwanted offsets. A passive bridge system such as this greatly increases accuracy and reliability without undue increase in cost or complication.

PERFORMANCE RELATIVE TO HYDRODYNAMIC BEARINGS

Requirement	*Magnetic bearings*	*Hydrodynamic bearings*
1. High loads	Load capacity low, but bearing area could be higher than with conventional bearings (see 3 below)	High load capacity (except at low speeds)
2. High speeds	Limited mainly by bursting speed of shaft; system response to disturbance must be considered carefully	Shear losses can become significant
3. Sealing	No lubricant to seal, and the bearing can usually operate in the process fluid	Seals may need to be provided
4. Unbalance response	Shaft can be made to rotate about its inertial centre, so no dynamic load transmitted to the frame	Synchronous vibration results from unbalance
5. Dynamic loads	Damping can be tuned, but adequate response at high frequency may not be possible	Damping due to squeeze effects is high, and virtually instantaneous in its effect
6. Losses	Very low rotational losses at shaft, and low power consumption in magnets/electronics	Hydrodynamic and pumping losses can be significant, particularly at high speeds
7. Condition monitoring	Rotor position and bearing loads may be obtained from the control system	Vibration and temperature instrumentation can be added
8. Reliability and maintainability	Magnets and transducers do not contact the shaft so operating damage is unlikely; electronics may be sited in any convenient position	Very reliable with low maintenance requirements

Particular features of electromagnetic bearings

No mechanical contact.
No oil contamination of process fluid.
Shaft position does not change with speed.
Wide speed range including high speeds.
Can accept wide range of temperature.
Can provide a machine diagnostic output.
Requires a very reliable power supply and/or emergency support bearings.
Can produce electromagnetic radiation interference.
Requires space for its control system.

A27 Bearing surface treatments and coatings

Table 27.1 The need for surface treatments and coatings

Type of application	*Example*	*Function of coating or treatment*
General use on many components		Allows components to be designed with a more optimum balance between the bulk and surface properties of the material
Lubricated plain bearing systems using high strength harder bearing materials	Crankshafts in heavy duty engines	Allows the shaft surface hardness to be increased to about five times the hardness of the bearing material, which is required for good compatibility
Lubricated components with small areas of contact with hard surfaces on both components in order to carry the contact pressures	Spur gears, cams and followers	Allows operation with low elastohydrodynamic film thickness with a reduced risk of scuffing and wear
Components operating at high loads and low speed or oscillating motion, and with only occasional lubrication	Bearings in mechanical linkages and roller chains, etc.	Aids oil retention on the surfaces and reduces the risk of seizure and wear
Surfaces which have intermittent rubbing contact with a reciprocating component	Cylinder liners in I.C. engines and actuators	Aids oil retention on the surfaces and reduces the risk of scuffing
Surfaces of components that are subject to fretting movements in contact with others	Connecting rod big end housing bores in high speed engines	Provides a surface layer that can allow small rubbing movements without the build up of surface damage
Components in contact with moving fluids containing abrasive material	Rotors and casings of pumps and fans	Gives abrasive wear resistance to selected areas of the surfaces of larger components
Components handling abrasive solid materials	Earth-moving machines. Coal and ore mills	Provides a hard abrasion resistant surface on a tough base material
Cutting tools	Drills and milling cutters	Provides a hard surface layer resistant to adhesion and abrasion and by providing a hard outer skin helps to retain sharpness
High temperature components with relative movements	Furnace conveyors. Boiler or reactor internals	Provides a surface resistant to adhesion and wear in the absence of conventional lubricants

Table 27.2 Types of surface treatments and coatings

General type	*Materials which can be coated or treated*	*Examples*	
Modifying the material at the surface without altering its chemical constituents	Ferrous materials, steels and cast irons	Induction hardening Flame hardening Laser hardening Shot peening	See Table 27.3
Adding new material to the surface, to change its chemical composition and properties	Ferrous materials	Carburising Nitriding Boronising Chromising	See Table 27.4
	Aluminium alloys	Anodising	
Placing a layer of new material on the surface	Ferrous materials Non–ferrous materials Plastics	Electroplating Physical vapour deposition Chemical vapour deposition Plasma spraying Flame spraying Vacuum deposition	See Table 27.5

Table 27.3 Surface treatment of ferrous materials

Process	*Mechanism*	*Hardness and design depth*	*Aspects*
Induction hardening of medium carbon alloy steels	Rapid heating followed by quenching produces martensite in the surface	Up to 600 Hv at depths up to 5 mm	Provides a hard layer deep enough to contain the high sub-surface shear stresses which occur in concentrated load contacts. Very large components can be treated
Flame hardening of medium carbon alloy steels	Surface heating followed by a quenching process produces martensite in the surface	Up to 500 Hv at depths up to 3 mm	Provides a hard wear-resistant surface layer. Particularly suitable for components with rotational symmetry that can be spun
Laser hardening of medium carbon steels and cast irons	A scanned laser beam focused on the surface. The very local heated area is self quenched by its surroundings	Up to 800 Hv at depths up to 0.75 mm	Suitable for local hardening of small areas of special components
Shot peening of ferrous materials	Work hardens the surface and leaves it in compression	Up to 0.5 mm approx	Gives increased resistance to fatigue and stress corrosion

Table 27.4 Diffusion of materials into surfaces

Process	*Mechanism*	*Surface effects*	*Design aspects*
Carburising of steels	Carbon is diffused into the surface at temperatures around 900°C followed by quenching and tempering	Hard surfaces at depths up to several mm depending on the diffusion time. Dimensional changes of ± 0.1% may occur	Suitable, for example, for gears to increase surface fatigue strength or pump parts to increase abrasive wear resistance
Carbonitriding of steels	Similar to carburising but with addition of nitrogen as well as carbon	Slightly harder surfaces are obtainable	Can be oil quenched with some reduction in distortion of the component
Boronising of steels	Diffusion of boron into the surface to form iron boride	Surface hardnesses up to 1200 Hv	The core material needs to be relatively hard to support the surface layer
Vanadium, niobium or chromium diffusion into steels	Salt bath treatment at 1020°C to produce thin surface layers of the metallic carbide	Layer hardness up to 3500 Hv at up to 15 μm thickness	Used to reduce wear of the surfaces of metal forming tools
Chromising, aluminising and siliconising of steels	High temperature pack processes for diffusing these materials into the surface	Increased surface hardness	Improves high temperature corrosion resistance and reduces fretting damage
Nitriding of steels which contain chromium or aluminium	Treatment at about 550°C in cracked ammonia gas or cyanide salt bath to produce chromium or aluminium nitrides in the surface layers	Hardness up to 850 Hv with surface layers up to 0.3 mm deep	Forms a brittle white surface layer which needs to be removed before use in tribological applications
Nitrocarburising of steels including plain carbon steels (tufftriding)	Salt bath, gas or plasma treatments are available	Hard surface layer about 20 μm thick with hardness of about 700 Hv	Reduces fretting damage. The surface layer can be oxidised and impregnated with lubricant to produce a low friction corrosion resistant surface
Ion implantation of steel	Surface bombardment by a high energy nitrogen ion beam at 150°C produces up to 30% implanted material at the surface	Surface effects to a depth of 0.1 μm	Can improve resistance to abrasion and fatigue
Sulfinuz treatment of steel	Salt bath treatment at about 600°C to add carbon, nitrogen and sulphur	Surface layer contains sulphides which act as a solid lubricant in the surface	Can give improved scuffing resistance with little risk of component distortion during treatment
Phosphating of steel	Phosphate layer produced by chemical or electrochemical action	Surface layer which is porous and helps to retain lubricant	Assists running in and reduces risk of scuffing
Electrolytic deposition and diffusion of material into non-ferrous metals	Copper, aluminium and titanium can be treated at temperatures in the range 200–400°C	Creates surface layers at selected positions	Gives improved scuffing and wear resistance
Hard anodising of aluminium	The component is made the anode for electrolytic treatment in sulphuric acid. The oxygen generated at the surface produces a hard porous oxide layer	The hard layer is usually 25–75 μm thick	The coating is porous and can be overlaid with PTFE to to give a low friction surface

Table 27.5 The coating of surfaces

Process	*Mechanism*	*Surface effects*	*Design aspects*
Electroplating with hard chromium	Plating is carried out at less than 100°C so there are no distortion problems. Current density can be adjusted to produce tensile stresses which, with diamond honing produce regular distributed cracks for good oil retention	Surface hardness of up to 900 Hv and thicknesses of up to 1 mm. Also used in thin, as deposited layers about 50 µm thick	A good surface for cylinder liners but needs to be cleaned up by Si C honing. A good piston ring coating for use with cast iron liners
Electroplating with tin, lead or silver	Can be plated on to any metals	Soft surface layers with thicknesses typically up to 50 µm	Gives good resistance to fretting and improves the bedding in and running properties of harder bearing materials
Electroless nickel plating	A low temperature process with good throwing power to follow component shapes. Can also be used with hard wear-resistant particles or PTFE particles dispersed in the coating	The hardness is about 550 Hv as deposited but can be increased to 1000 Hv by heat treatment	Good for cylinder liners and all component surfaces requiring improved wear resistance. Inclusion of PTFE gives lower friction
Physical vapour deposition of coatings such as titanium nitride	The coating is transferred to the component via a glow discharge in a vacuum chamber. Deposition temperature is 250–450°C	Ti.N coatings have a hardness of about 3000 Hv and are usually 2–4 µm thick	Ideal for cutting tools and for components subject to adhesive wear or low stress abrasion such as hard particles sliding over the surface. It needs an undercoat of electroless nickel if full corrosion resistance is required
Chemical vapour deposition of metals and ceramics	The coating is produced by decomposition of a reactive gas at the component surface. The operation occurs within a reaction chamber at 800°C minimum temperature	Coatings of Ti.N TiC. Al_2O_3. CrN WC and CrC in thicknesses of the order of 5–10 µm	Very hard wear resistant coatings can be created by this process
Spray coating using a gas flame or electric arc	The coating material is supplied to the process as a powder or as a wire or wires. The powder coating may then be subsequently fused to the surface	Coatings can be of hardnesses up to 900 Hv in thicknesses in the range of 0.05–1.0 mm	Suitable for covering large areas of component surfaces
Plasma spray coating	Uses an ionised inert gas to produce very high temperatures which enable oxides and other ceramics to be coated onto metals. Component heating is minimal	Typical coating thickness 0.1 mm	Particularly suitable for improving the abrasive wear resistance of components
Plasma arc spraying	Carried out in a partially evacuated chamber or under an inert gas shield. The component gets hot and good bonding is achieved. Higher temperatures can be achieved by having an electric current flow between the arc and the surface. The performance of the materials can be further improved by hot isostatic pressing at temperatures of the order of 1000°C and above	Up to 10 mm thicknesses of very hard ceramic materials can be deposited	Suitable for high temperature gas turbine components to give improved oxidation and fretting resistance. Also abrasion resistant components for mining and agricultural machinery

Table 27.5 The coating of surfaces (continued)

Process	*Mechanism*	*Surface effects*	*Design aspects*
High velocity explosive coating	The coating spray is driven by successive explosive hydrocarbon combustion cycles. This gives coatings with very low porosity and good adhesion	Coatings of CrC, WC and cobalt-based cermets about 0.1 μm thick	Produces very tough abrasion resistant coatings
Thermochemically formed coatings	Ceramic coatings sprayed on and then heat treated at temperatures of about 500°C	Coatings such as chromium oxide of 1100Hv about 0.1 mm thick	Produces a corrosion and abrasion resistant coating
Laser cladding and alloying	The application of coatings derived from powders in small controlled areas	About 0.1 mm thick as a maximum	Good local corrosion and abrasion resistance
Weld coatings	Layers of hard material applied from welding rods and wires by gas or electric welding	Thick layers up to 30 mm thick can be applied selectively	A relatively adaptable manual process suitable for the repair and improved wear resistance of heavy duty components subject to abrasive wear
Cladding	Hard surface materials attached to components by welding, adhesives, or roll bonding	Thick wear-resistant materials can be attached to surfaces	Components, chutes, etc., requiring abrasion resistance

BELT DRIVE DESIGN

(a) Basic 2-shaft drive

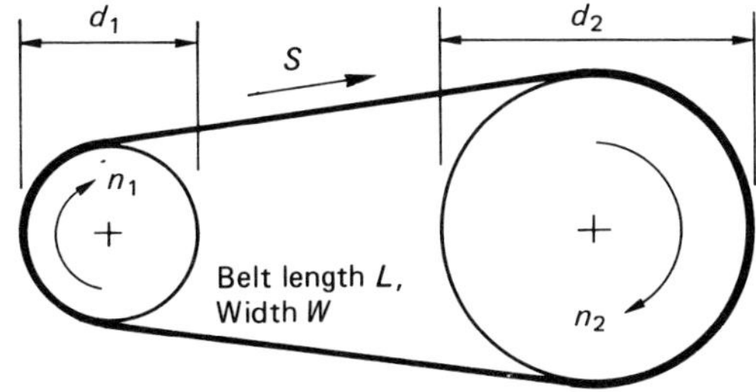

d in mm, n in rev/min
Belt speed $S = nd/19100$ m/s

(c) Serpentine layout to drive more than one shaft or reverse direction of rotation

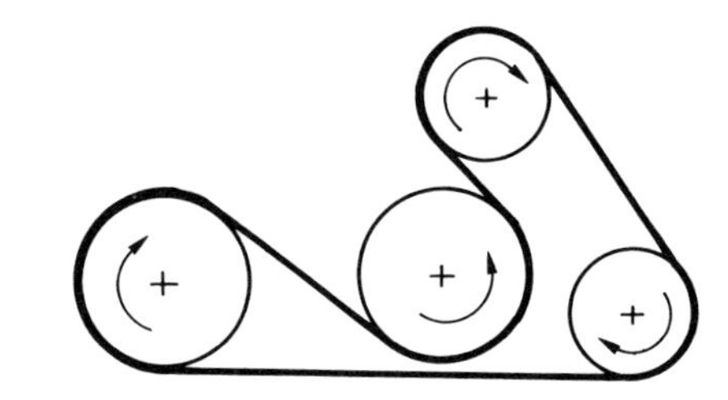

(b) Idler added to increase arc of contact or control belt vibration

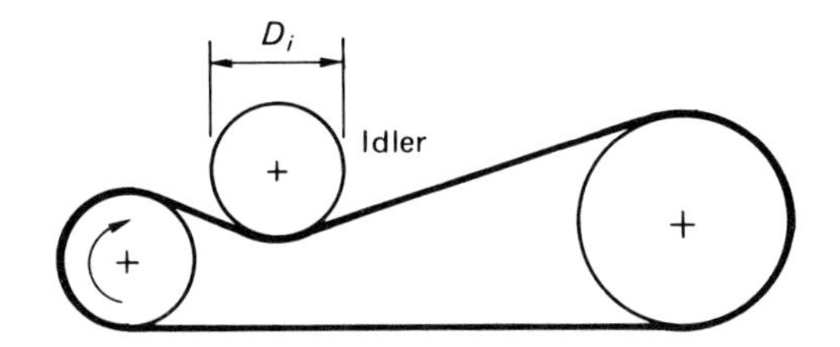

(d) Out-of-plane layout

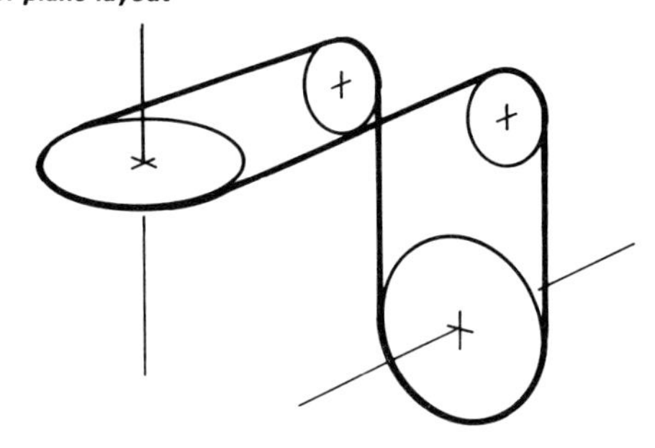

Figure 1.1 Some belt drive layouts

Belt selection, pulley size and shaft load calculation procedures for drives with two shafts

Specify rated power and pulley rev/min requirements

A typical two shaft belt arrangement is as in Figure 1.1(a).

Calculate design power from rated power × service factors

Some guides recommend selection on basis of rated power. This can result in smaller diameter pulleys and lighter section belts but a wider drive.

Select belt section from design power rating charts, on basis of design power and rev/min of fastest pulley. Choose pulley diameters greater than minimum recommended.

Pulley diameter and belt section selection is a compromise between smaller diameter pulleys for compactness, but limited by overloading of the tension members in the belts, and larger pulleys giving lower bearing loads but requiring more space. Larger pulley diameters may always be chosen, subject only to allowable belt speeds, and the designer is advised to check the effect of varying diameter by trial calculations.

Select belt width (or number of belts, for Vee belts) from belt capacity tables

Calculate belt tensions and shaft loading to limit slip (or to ensure good meshing of synchronous belts)

Belt width and tension selection are closely related and depend on arc of contact as well as power and rev/min.

BELT TYPES AND MATERIALS SELECTION

Standard Vee, Vee-ribbed and synchronous (timing) belts have a high modulus wound-cord tension member (glass fibre, polymeric or steel members are used according to application), a rubber carcass and a woven fabric backing cover. The drive surface of synchronous belts is strengthened by a woven fabric cover, that of Vee-belts may be covered or uncovered (raw), while Vee-ribbed belts are generally raw. The flat belts considered here have polyamide strip tension members and either rubber or leather drive surfaces. This handbook does not consider all variations of belt section and materials that exist. Users should consult catalogues for the full range of constructions and materials.

The operating temperature range of belts is typically −20°C to +70°C, but materials may be formulated for both lower (−40°C) or higher (+120°C) temperatures. The static conductivity of belts is regulated by standards (e.g. ISO 1813, ISO 284, ISO 9563). Belts should be shielded from oil but most types are resistant to small amounts of contamination.

Power transmission efficiencies of 95% to 98% in steady operating conditions are achievable with all well maintained belt drives.

Table 1.1 Selection of the type of belt

WEDGE AND CLASSICAL VEE BELTS

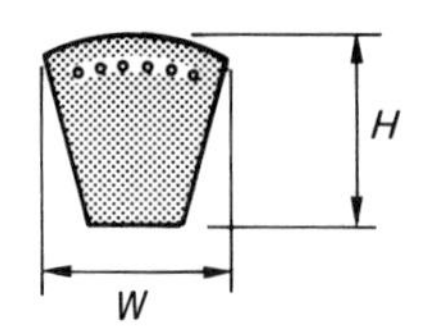

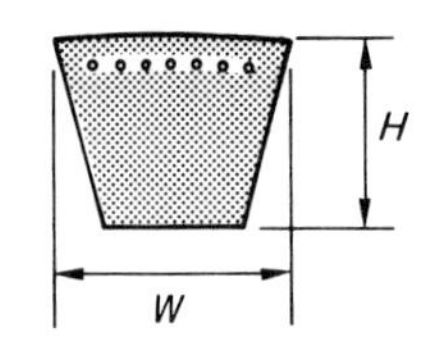

These are the standard choice for large power transmissions where slip in the event of shock loading is needed as overload protection. Wedge belts with a larger ratio of H to W than classical Vee belts give more compact drives but cannot be used in layouts requiring reverse bending of the belt. Stock pulley sizes allow speed ratios up to 8:1. Recommended maximum belt speeds are 30 to 40 m/s.

Further information
BS 3790, ISO 1081, DIN 2211, RMAIP 20
ISO 4184, DIN 7753, RMAIP 23
and manufacturer's catalogues

VEE-RIBBED BELTS

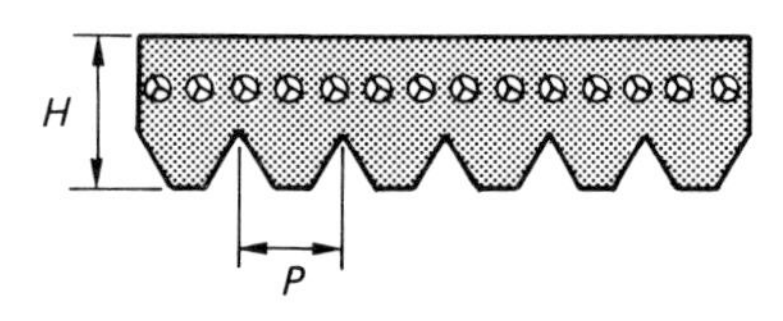

These have been developed to combine the grip of Vee belts and almost the flexibility of flat belts and find application where space is confined (smaller diameter pulleys) or where some serpentine layout capability is needed. Stock pulley sizes allow higher speed ratios than for Vee belts, up to 25:1 depending on belt section. Recommended maximum belt speeds are 35 to 45 m/s.

Further information
ISO 8372, DIN 7876, RMAIP 26
ISO 9982 and manufacturer's catalogues

FLAT BELTS

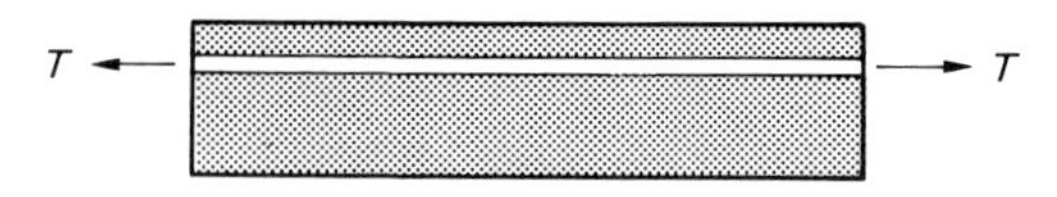

The ease of joining polyamide strip tension member belts enables them to be made of virtually any length; their flexibility makes them suitable for highly serpentine or out of plane layouts. At low belt speeds, drives are less compact and bearing loads are higher than for other belt types, but as speed increases above 20 m/s.,flat belts come into their own. Speeds up to 70 m/s are possible.

The cross-sections of flat belts are not controlled by Standards. Flat belts are rated by $k_{1\%}$, the load per unit width to stretch the belt 1%.
Further information from manufacturer's catalogues

SYNCHRONOUS BELTS

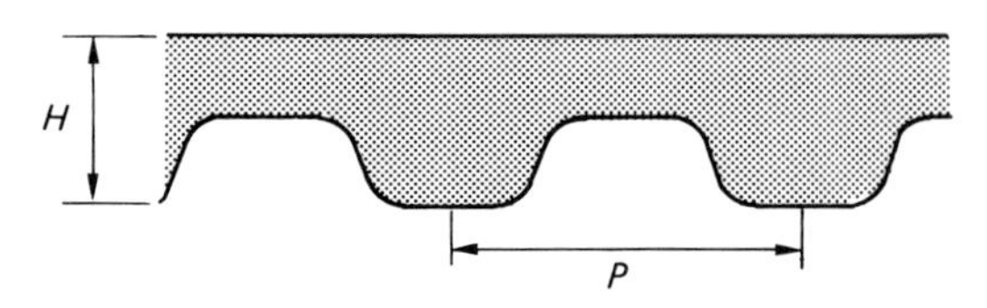

Synchronous belts are now developed to a similar power capacity as Vee belts. They are clearly essential if synchronous motion is needed but can suffer from tooth failure in conditions of extreme shock loading. The earlier developed trapezoidal toothed belt (illustrated) has now been displaced by curvilinear toothed belts in new drive designs. Typical maximum speed ratios are 10:1 and belt speeds are up to 60 m/s.

Further information
BS 4548, ISO 5296, DIN 7721
and manufacturer's catalogues

SERVICE FACTORS (SF) FOR VARIOUS APPLICATIONS

Design power = SF × Rated power.

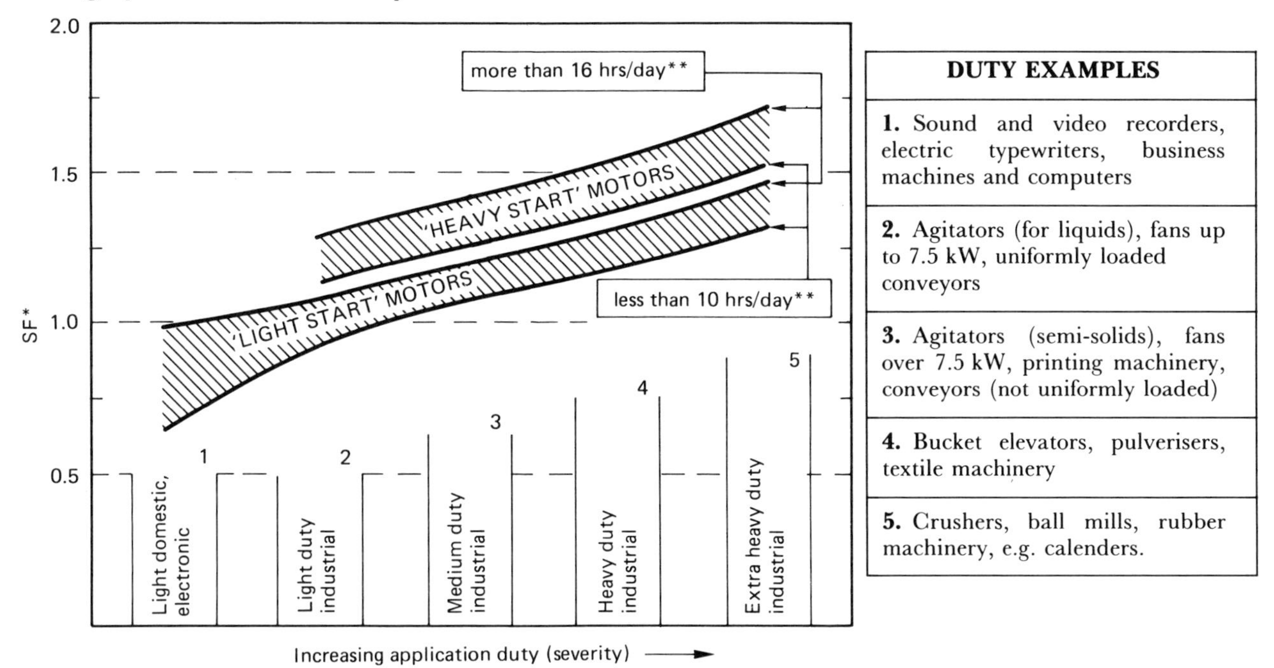

DUTY EXAMPLES
1. Sound and video recorders, electric typewriters, business machines and computers
2. Agitators (for liquids), fans up to 7.5 kW, uniformly loaded conveyors
3. Agitators (semi-solids), fans over 7.5 kW, printing machinery, conveyors (not uniformly loaded)
4. Bucket elevators, pulverisers, textile machinery
5. Crushers, ball mills, rubber machinery, e.g. calenders.

* Some design procedures include arc of contact factors in SF. This is not followed here. SF may be increased by up to 0.6 for speed-up drives and if the belt layout requires reverse bending. SF also depends slightly on belt length and other factors: this detail is not considered here.
** It is not customary to include operating time/day in SF for flat belt drives. A mid-range value may be chosen for these.

Figure 1.2 Service factors for various duties

BELT SELECTION FROM DESIGN POWER RATING CHARTS

The belt section is selected on basis of the region which contains the intersection of design kW with fastest pulley rev/min. Regions can overlap – more than one belt section may be a possibility.

Drives may be designed with any pulley diameters > d_{min} for that section. It is usually desirable to choose the largest diameter pulleys for which space is available, subject to the maximum recommended belt speed not being exceeded.

The labels $d_1 \times w_1$, $d_2 \times w_2$ give indications of likely combinations of smallest pulley diameter and belt width for 180° arc of contact drives.

The design power rating charts on the following pages are all consistent with the service factors SF above. They differ from some catalogues' charts which use differently based SF values.

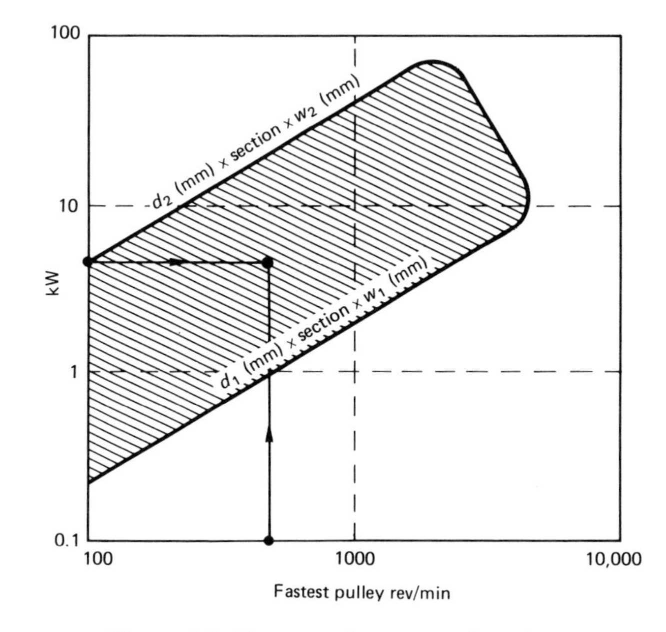

Figure 1.3 The use of power rating charts

DESIGN POWER RATINGS FOR VARIOUS TYPES OF BELT

Wedge belts

Section	*SPZ*	*SPA*	*SPB*	*SPC*	*8V*****
W mm	9.5	13	16	22	26
H mm	8	10	14	18	23
β°	40	40	40	40	40
d_{min} mm*	67	100	160	224	335
(d_{min} mm	56	80	112)		
D_i min	reverse bending is not allowed				
L_{min} mm**	525	750	1260	2000	2520
L_{max} mm**	3560	4500	9010	12500	11410
W_{max} mm***	72	90	152	306	343

* The d values here and in the power rating chart are for covered belts. For raw-edge moulded-cog wedge belts smaller sizes are appropriate, as shown by (bracketed values).

** Typical stock values (also covered by Standards); other lengths to special order.

*** Max. number of belts per pulley × width of one pulley groove l, (Table 1.2).

**** Rubber manufacturers of America (RMA) standard.

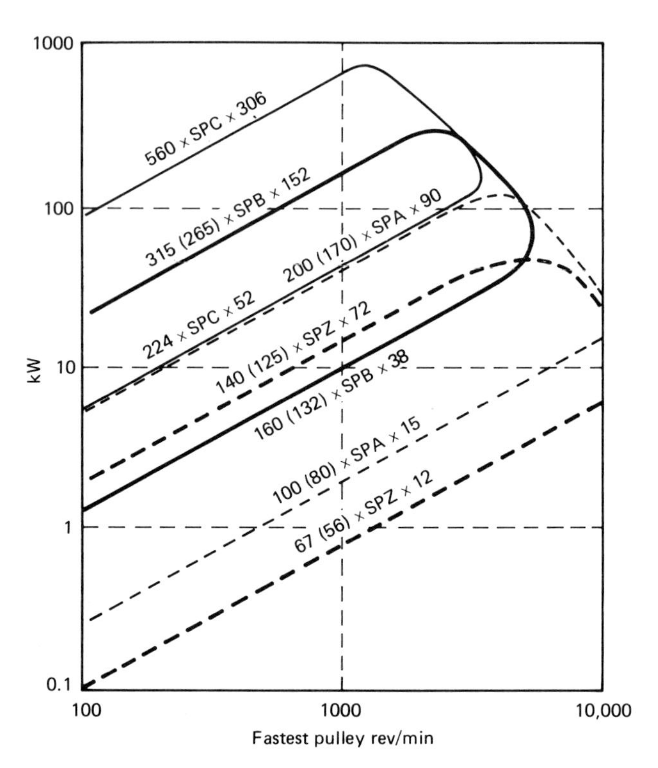

Figure 1.4 Power rating of wedge belts

Classical vee belts

Section	*Z*	*A*	*B*	*C*	*D*
W mm	10	13	17	22	32
H mm	6	8	11	14	19
β°	40	40	40	40	40
d_{min} mm	50	75	125	200	355
D_i min	> Smallest loaded pulley diameter				
L_{min} mm**	270	415	613	920	2570
L_{max} mm**	1540	5510	12000	10700	15200
W_{max} mm***	72	90	152	306	444

, * As above.

β is the total included angle of the belt profile.

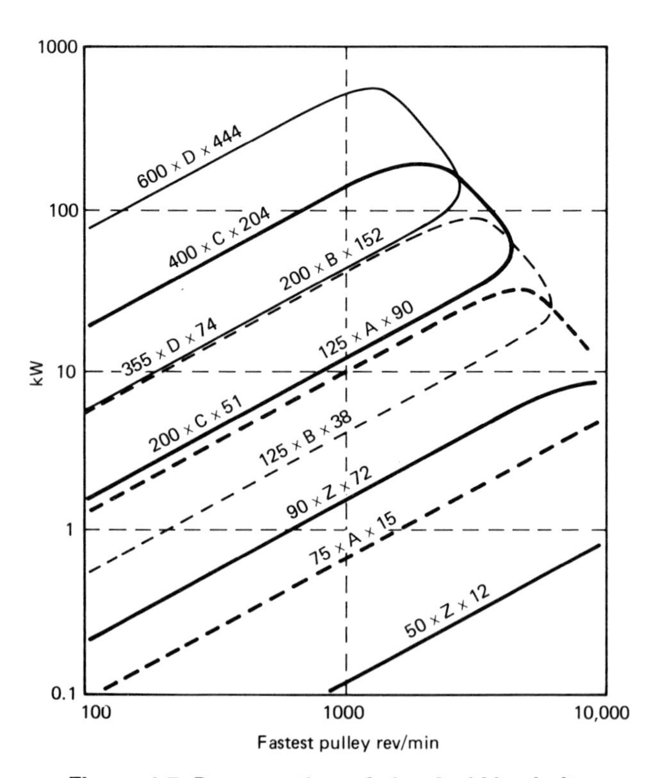

Figure 1.5 Power rating of classical Vee belts

Vee-ribbed belts

Section	*J*	*K*	*L*	*M*
P mm	2.3	3.6	4.7	9.4
H mm	Varies with manufacturer.			
β°	40	40	40	40
d_{min} mm*	20	38	76	180
$D_{i,min}$ mm*	32	75	115	267
L_{min} mm*	450	–**	1250	2250
L_{max} mm*	2450	–**	5385	12217
W_{max} mm***	46	72	94	188

* Varies with manufacturer.
** *K* is designed for automotive use but can be used for general machinery; some manufacturers also have an H-section, with $P = 1.6$ mm, and $d_{min} = 12.7$ mm.
*** W_{max} is 20 ribs for stock belts.

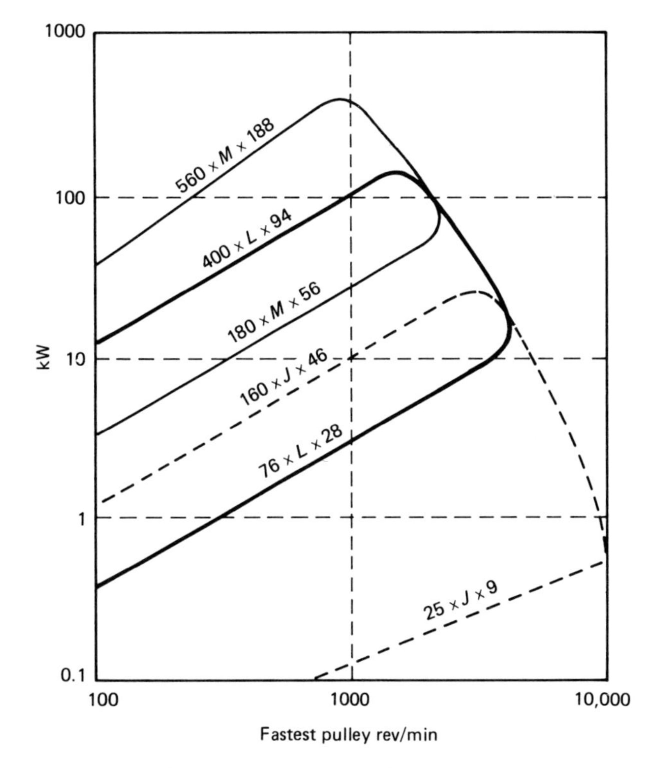

Figure 1.6 Power rating of Vee-ribbed belts

Polyamide core flat belts

*Section**	*a*	*b*	*c*	*d*
$k_{1\%}$ N/mm**	3	7	15	25
d_{min} mm***	45	90	180	360
$D_{i,\,min}$ mm***	45	90	180	360
L_{min} mm, L_{max} mm	there is virtually no limit on belt length			
W_{max} mm	< smallest pulley diameter			

* *a*, *b*, *c*, *d* have no Standards significance. The letters refer to Figure 1.7 and Figure 1.12.
** See Table 1.1. The $k_{1\%}$ values given here are a selection from those available. Values of 5, 10, 20, 33 and 40 N/mm may be obtained from a range of manufacturers.
*** Minimum pulley diameters vary with belt thickness, *H*. *H* can take a wide range of values for any $k_{1\%}$, as the ratio of friction material to tension member thickness varies with application. d_{min} and $D_{i,\,min}$ values given here are smallest values.

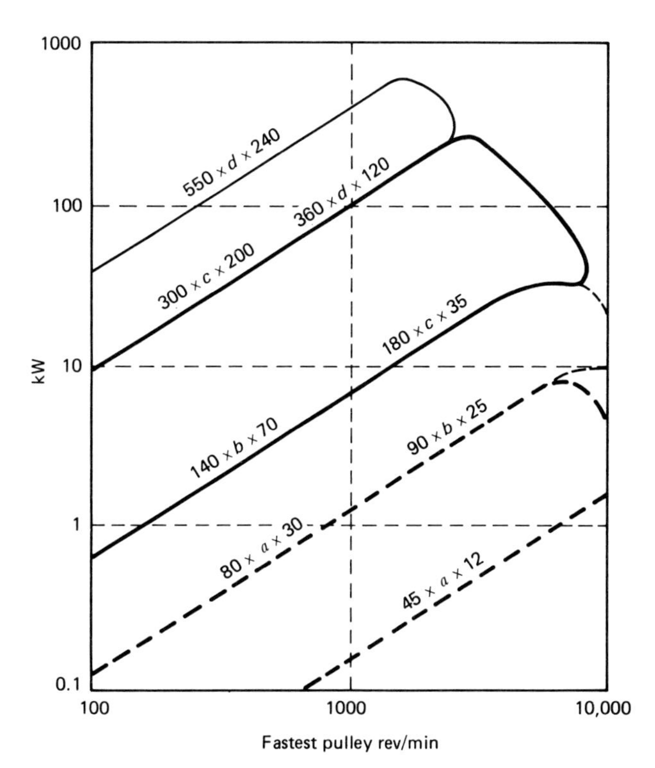

Figure 1.7 Power rating of flat belts with polyamide cores

Curvilinear tooth belts

The pitches below are de facto but not formal standards

P mm	3	5	8	14	20
H mm*	2.4	3.8	6	10	13.2
d_{min} mm	9.5	22	56	125	216
D_i mm	> smallest loaded pulley diameter				
L_{min} mm**	129	245	320	966	
L_{max} mm**	1863	2525	4400	6860	
W_{max} mm	< smallest pulley diameter				

* Varies slightly with manufacturer.
** Stock lengths, varies with manufacturer, other lengths to special order.

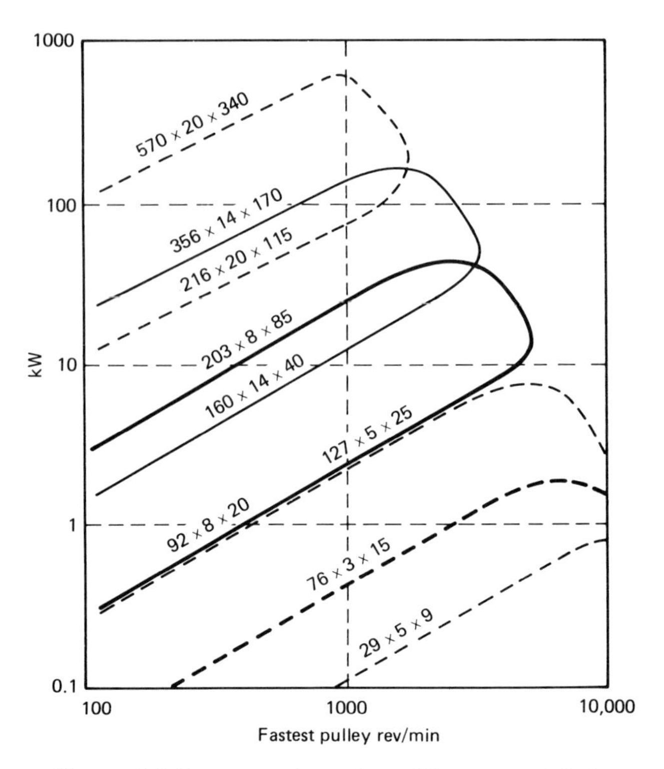

Figure 1.8 Power rating of curvilinear tooth belts

Trapezoidal tooth belts

Section	*XL*	*L*	*H*	*XH*	*XXH*
P mm	5.08	9.52	12.7	22.22	31.75
H mm	2.3	3.6	4.3	11.2	15.7
d_{min} mm	16	36	65	156	222
D_i, mm	> smallest loaded pulley diameter				
L_{min} mm**	152	314	609	1289	1778
L_{max} mm**	685	1524	4318	44445	4572
W_{max} mm	< smallest pulley diameter				

** As above.
The d values in the Figures 1.8 and 1.9 are nominal. Actual values give an integral number of teeth per pulley.

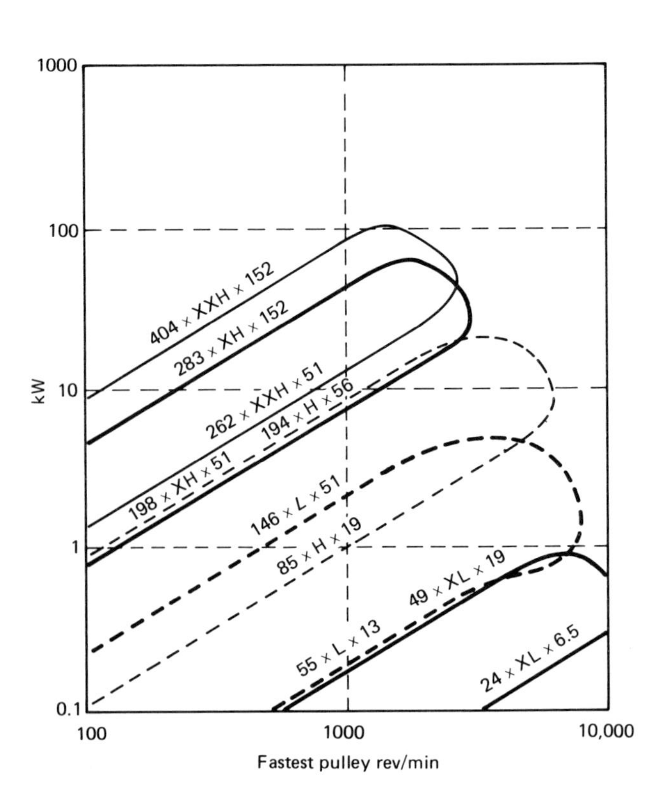

Figure 1.9 Power rating of trapezoidal tooth belts

BELT TENSIONS

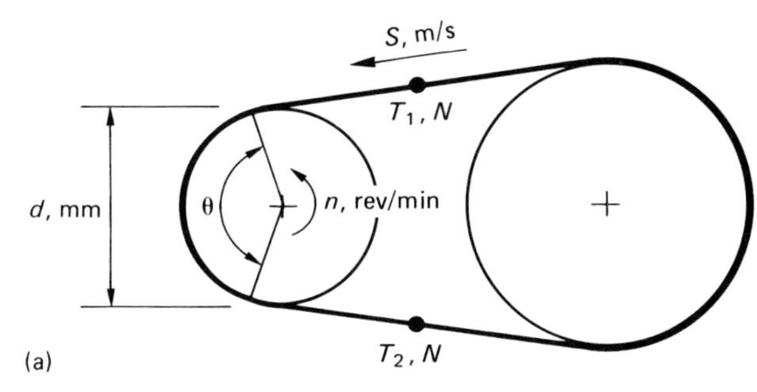

(a)

Tension sum $T_1 + T_2$ must be large enough to limit slip or, for synchronous belts, poor meshing.

$$T_1 + T_2 \geqslant (T_1 - T_2)/\lambda \quad (2)$$

where traction coefficient λ varies with belt type and arc of contact.

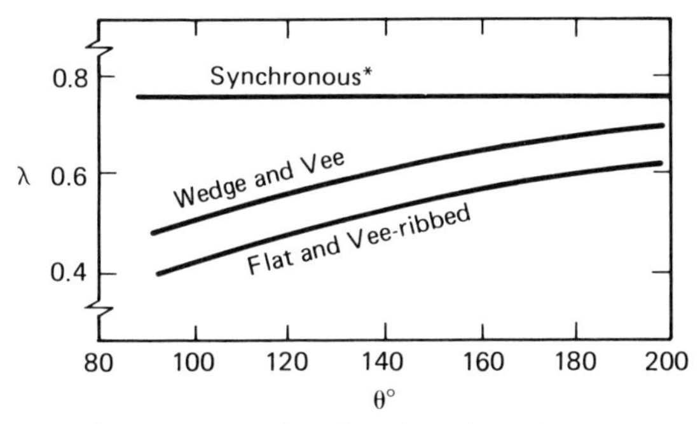

(b)

Figure 1.10 Traction coefficient at various arcs of contact

Tension difference $T_1 - T_2$ arises from torque or power transmission

$$(T_1 - T_2).d = 19.1 \times 10^6 \left(\frac{\text{kW}}{n}\right) \quad (1)$$

$$\text{or } (T_1 - T_2) = 10^3 \left(\frac{\text{kW}}{S}\right)$$

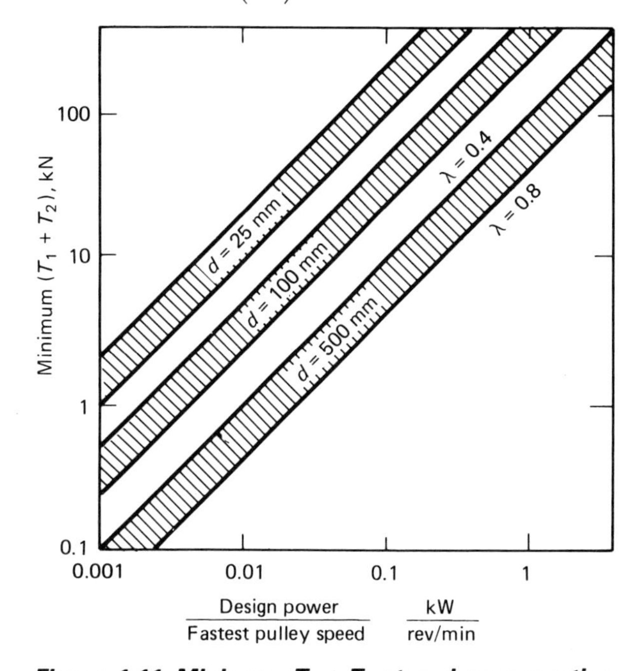

Figure 1.11 Minimum $T_1 + T_2$ at various operating conditions

BELT WIDTH

Belt width must be large enough to support tension. From (1) and (2) above, tension increases as kW/($nd\lambda$). Design guides tabulate allowable kW/(nd) per belt or mm belt width, for $\theta = 180°$ arc of contact and varying n and d. Figure 1.12 gives values of F^*, from which such tables can be created.

To use the chart below, use θ for the smallest pulley to estimate λ_θ. Calculate

$$F = 19.1 \times 10^6 \left(\frac{\text{kW}}{nd}\right)\left(\frac{\lambda_{180}}{\lambda_\theta}\right)$$

Belt width = F/F^*, mm.

Bending a belt round a pulley increases tension member strain. Thus F^* reduces with reducing pulley diameter. Values below are mean values. F^* also reduces with increasing belt speed (see Figure 1.13)

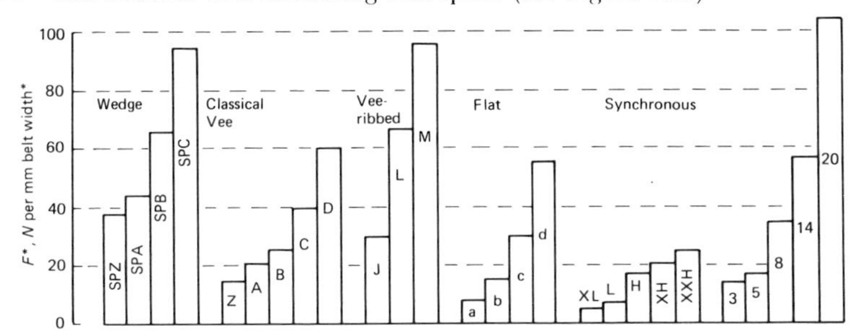

For wedge, Vee and Vee-ribbed belts, it is more usual to record F^* as N per belt or per rib. This can be derived from the above by multiplying by belt or rib width (mm). The data for wedge belts are for covered types: for raw-edge moulded-cog wedge belts of SPZ, SPA and SPB Section, F^* should be increased by 25–30%.

Figure 1.12 F* the allowable belt tension difference per unit width for various types of belt

SHAFT LOADING

Static or installed shaft load

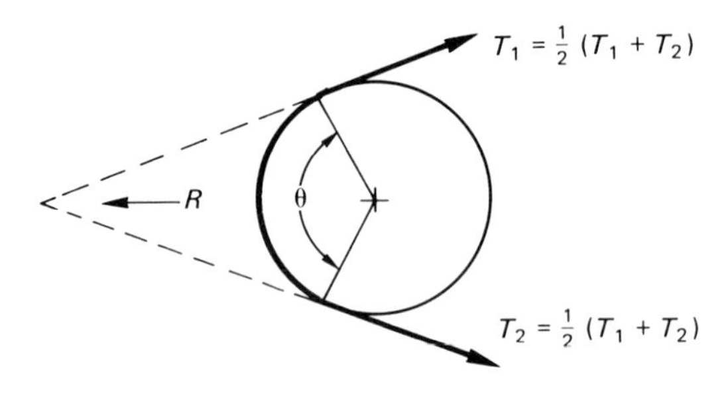

$R = (T_1 + T_2)\sin(\theta/2)$

Except for synchronous belts, $(T_1 + T_2)$ is obtained from Figure 1.11. For synchronous belts, $(T_1 + T_2)$ is 50%–90% of value from Figure 1.11 increasing with severity of shock loading: $(T_1 + T_2)$ rises to 100% during power transmission.

Dynamic or running shaft load

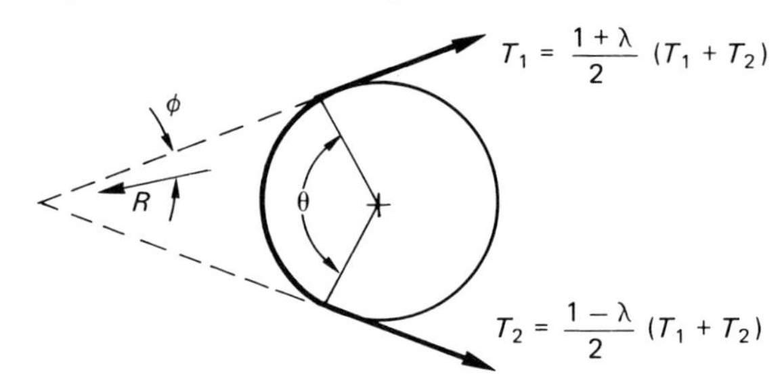

$$R = (T_1 + T_2)\left\{\frac{(1 + \lambda^2)}{2} - (1 - \lambda^2)\frac{\cos\theta}{2}\right\}^{1/2}$$

$\sin\phi = \sin\theta\,(R/T_2)$

SPEED EFFECTS

Allowable kW/(nd) per belt or belt width reduces as S increases above 1 m/s, for Vee, Vee-ribbed and synchronous belts for two reasons. Up to $\simeq$ 20 m/s a faster speed simply means the belt is used more in a given time. Derating maintains its absolute life time. Over 20 m/s centrifugal loading becomes more significant. For both reasons belt width and shaft loadings are increased relative to the values obtainable from Figure 1.11 and Figure 1.12 by a speed dependent factor f.

Flat belts are, in practice, only de-rated for centrifugal loading.

ISO 5292 provides a model for the power rating of Vee-belts with respect to both speed and pulley diameter. It could be applied to all belt types.

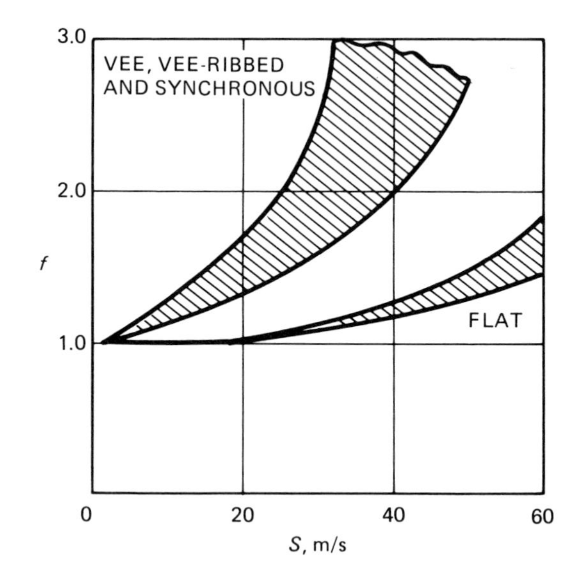

Figure 1.13 The derating in belt performance at higher speeds

Pulley materials for high belt speeds

The maximum safe surface speed of cast iron pulleys is 40 m/s. Steel of 430 MPa tensile strength may be used up to 50 m/s and aluminium alloys of 180 MPa tensile strength up to 60 m/s. Aluminium alloys are not recommended for uncovered rubber drive faces because of wear/abrasion problems with aluminium oxide. For operation up to 70 m/s special designs using high strength steel or aluminium alloys are required. Plastics are commonly used for low-speed, low power applications.

PULLEY DESIGN

Table 1.2 Wedge, Vee and Vee-ribbed pulley groove dimensions (mm)

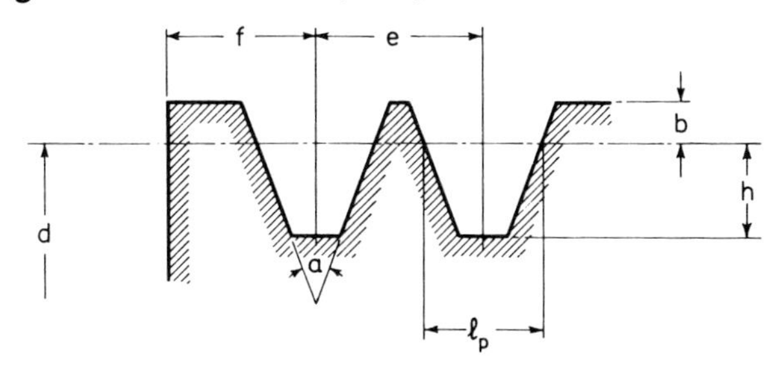

	l_p	b_{min}	h_{min}	e	f
Z	8.5	2.08	7.1	12	8
A	11	3.3	8.7	15	12.5
B	14	4.2	10.8	19	17
C	19	5.7	14.3	25.5	24
D	27	8.1	19.9	37	29
SPZ	8.5	2	9	12	8
SPA	11	2.8	11	15	10
SPB	14	3.5	14	19	12.5
SPC	19	4.8	19	25.5	17
J		−0.38*	2.18	2.34	1.8
L		−0.74	5.39	4.70	3.3
M		−1.47	11.05	9.40	6.4

* For Vee-ribbed belts the pitch line lies outside the pulley, and the groove tips and roots are rounded.

Table 1.3 Vee and wedge groove angles α related to pitch diameter (mm)

	32°	*34°*	*36°*	*38°*	*40°*
Z SPZ		<80		>80	
A SPA		<118		>118	
B SPB		<190		>190	
C SPC		>315		>315	
D			<500	>500	
J					>20
L					>76
M					>180

Users should not attempt to manufacture their own synchronous pulleys.

Flat belt pulley crowning requirements

Both pulleys should be crowned. If one pulley is flat the crown of the other should be increased by 50%. A similar increase should also be applied when the shafts are vertical or when the centre distance is short.

Further information ISO 100, DIN 111

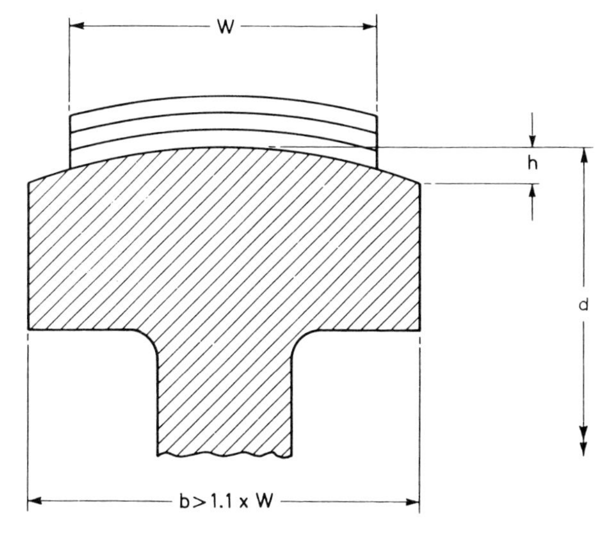

Avoid flanges if possible. If used, the corner must be undercut and b ⩾ 1.2*W*.

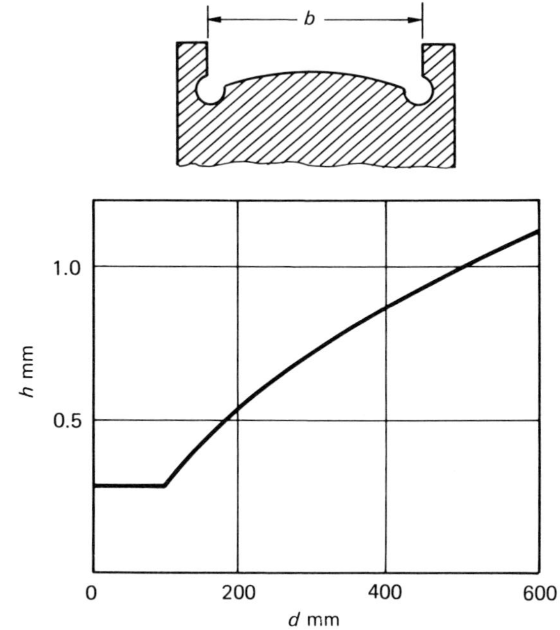

Figure 1.14 The amount of crowning required for flat belt pulleys

Surface finish

Standard recommendations for pulley drive face surface finish vary from 1.6 to 6.3 µm R_a (R_z from 6.3 to 25 µm), with 1.6 to 4 µm R_a being most common.

Further information
ISO 254, ISO R468, DIN 111, DIN 4768
and manufacturers' catalogues

FITTING AND TAKE-UP ALLOWANCES

Fitting allowance (mm)

Section	Fitting allowance
A	19-25
B	25-40
C	38-50
D	50-75
SPZ	12-20
SPA	20-25
SPB	25-30
SPC	40-50
J	8-20
L	22-30
M	38-80

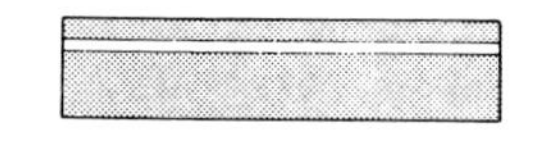

2-3% *L*

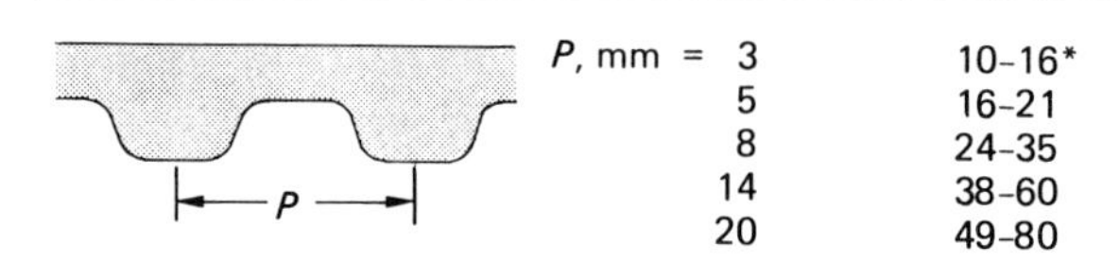

P, mm =	
3	10-16*
5	16-21
8	24-35
14	38-60
20	49-80

*For flanged pulleys. For unflanged pulleys, 2-5 mm is sufficient for all sections

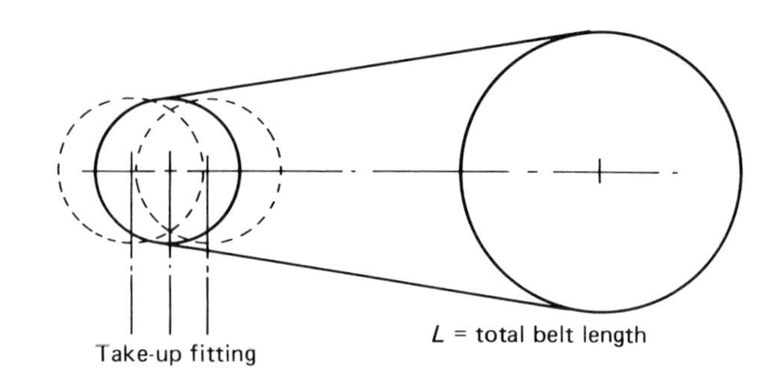

Take-up allowance

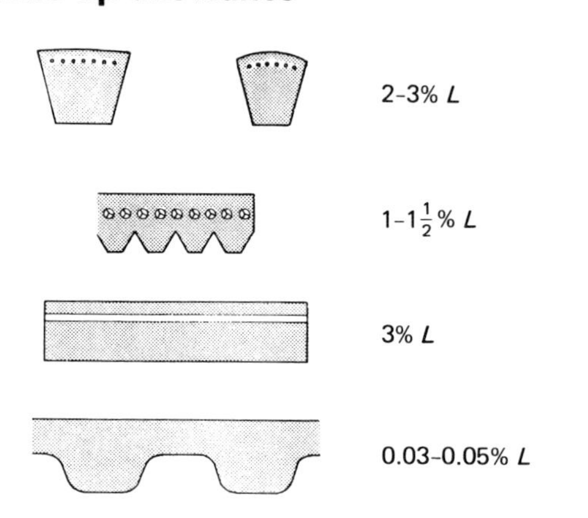

DESIGNING MULTI-DRIVE SYSTEMS

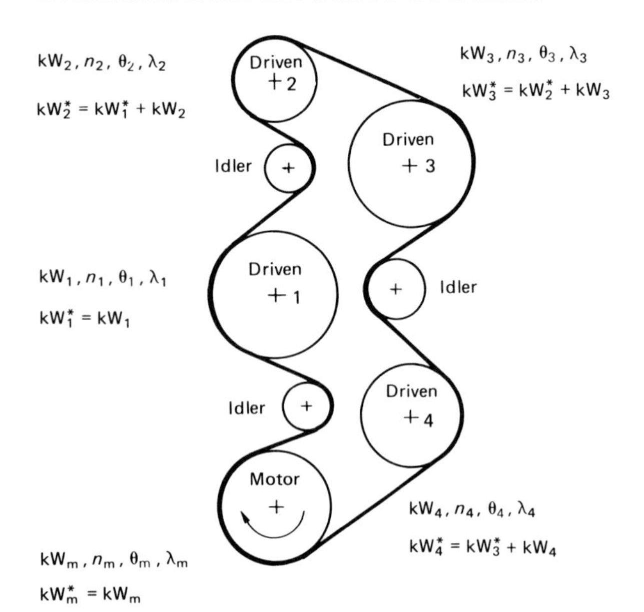

For each driven shaft and the motor shaft, write down its transmitted design kW, its rev/min n, its approximate arc of contact θ, its recommended λ (from Figure 1.10) and kW*, the accumulated kW, as illustrated.

Select belt section from the power rating charts for the most severe combination at any shaft of kW* and n. For speed reduction drives, the motor shaft will always be the most severe.

Estimate the number of belts (or belt width) and belt tension as for a 2-shaft drive based on kW, n, d, λ for the motor pulley.

The resulting tension may not prevent slip on one or more of the driven shafts. Check this by calculating, for each driven shaft, i, and the motor shaft m, the quantities f_i and f_m below. Select the largest ratio of f_i/f_m. If this is greater than 1.0, increase the number of belts (or belt width) and belt tension by this factor.

It may be advantageous to drive the shafts by separate belts. The complexity may be such that a specialist should be consulted, particularly if any $\theta < 90°$ or if, for a synchronous belt, less than 6 teeth are in mesh on any pulley.

$$f_i = \mathrm{kW}_i \frac{1 + \lambda_i}{\lambda_i} - 2\,\mathrm{kW}_i^* \qquad f_m = \mathrm{kW}_m \frac{1 - \lambda_m}{\lambda_m}$$

SELECTION OF CHAIN DRIVES

The following selection procedure gives guidance for Industrial applications for the selection of chain drives comprised of roller chains and chain wheels conforming to ISO 606.

Chain selected using this method will have a life expectancy, with proper installation and lubrication, of 15 000 hours.

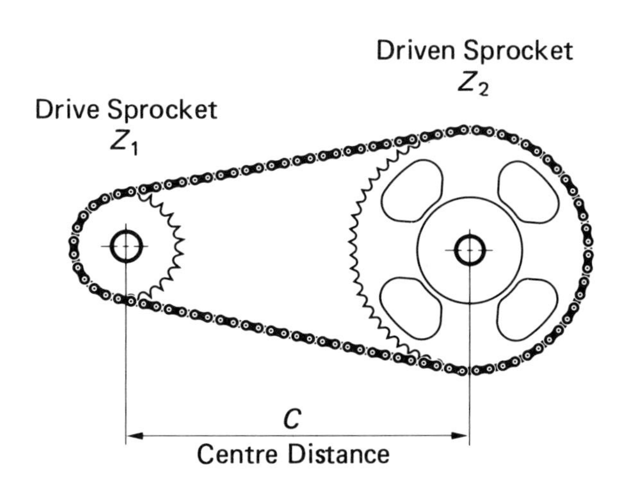

Basic information required

In order to select a chain drive the following essential information must be known:-

(a) The power, in kilowatts, to be transmitted.
(b) The speed of the driving and driven shafts in rev/min.
(c) The characteristics of the drive.
(d) Centre Distance.

From this basic information, the driver sprocket speed and selection power to be applied to the ratings charts, are derived.

Z_1 – No of teeth on drive sprocket
Z_2 – No of teeth on driven sprocket
C – Centre Distance mm
P – Chain Pitch mm
i = Drive Ratio
L = Chain Length Pitches

Table 2.1 Drive ratios i, relative to one, using preferred sprockets

No. of teeth on driven sprocket Z_2	*No. of teeth on drive sprocket Z_1*					
	15	*17*	*19*	*21*	*23*	*25*
25	–	–	–	–	–	1,00
38	2,53	2,23	2,00	1,80	1,65	1,52
57	3,80	3,35	3,00	2,71	2,48	2,28
76	5,07	4,47	4,00	3,62	3,30	3,04
95	6,33	5,59	5,00	4,52	4,13	3,80
114	7,60	6,70	6,00	5,43	4,96	4,56

Step 1 Select drive ratio and sprockets

Table 2.1 may be used to choose a ratio based on the standard wheel sizes available.

It is best to use an odd number of teeth combined with an even number of chain pitches.

Ideally, chain wheels with a minimum of 17 teeth should be chosen. If the chain drive operates at high speed or is subjected to impulsive loads, the smaller wheels should have at least 25 teeth and should be hardened.

It is recommended that chain wheels should have a maximum of 114 teeth.

For large ratio drives, check that the angle of lap on Z is not less than 120°.

Drive ratios can otherwise be calculated using the formula

$$i = \frac{Z_2}{Z_1}$$

Step 2 Calculate selection power

SELECTION POWER = Power to be transmitted $\times f_1 \times f_2$ (kW)

Where f_1 and f_2 are given in Tables 2.2 and 2.3

This selection power can then be used in Figure 2.1 and Figure 2.2 to select a suitable drive arrangement

Step 3 Select drive

From the rating charts, Figure 2.1 and Figure 2.2, select the smallest pitch of simple chain to transmit the SELECTION POWER at the speed of the driving sprocket Z_1.

This normally results in the most economical drive selection.

Should the SELECTION POWER be greater than that shown, then consider a multiplex chain selected from the ratings charts.

Chain manufacturers should be consulted if any of the following apply.

(a) More than 1 driven sprocket
(b) Power or speeds above or below chart
(c) When volume production is envisaged
(d) Ambient conditions other than normal

This section has been compiled with the assistance of Renold Chains

Table 2.2 Application factor f_1

		Characteristics of driver		
		Smooth running	*Slight shocks*	*Moderate shocks*
	Driven machine characteristics	Electric motors. Steam and gas turbines. Internal combustion engines with hydraulic coupling.	Internal combustion engines with 6 cyls. or more with mechanical coupling. Electric motors with frequent starts. (2+ per day)	Internal combustion engines with less than 6 cyls. with mechanical coupling.
Smooth running	Centrifugal pumps and compressors. Printing machines. Paper calenders. Uniformly loaded conveyors. Escalators. Liquid agitators and mixers. Rotary driers. Fans.	1.0	1.1	1.3
Moderate shocks	Pumps and compressors (3+ Cyls). Concrete mixing machines. Non uniformly loaded conveyers. Solid agitators and mixers.	1.4	1.5	1.7
Heavy shocks	Planers. Excavators. Roll and ball mills. Rubber processing machines. Presses and shears. 1 & 2 cyl pumps and compressors. Oil drilling rigs.	1.8	1.9	2.1

Factor f_1 takes account of any dynamic overloads depending on the chain operating conditions. The value of factor f_1 can be chosen directly or by analogy using Table 2.2.

Table 2.3 Tooth factor f_2 for standard wheel sizes

Z_1	f_2
15	1.27
17	1.12
19	1.0
21	0.91
23	0.83
25	0.76

The tooth factor f_2 allows for the choice of a smaller diameter wheel which will reduce the maximum power capable of being transmitted, since the load in the chain will be higher.

Tooth factor f_2 is calculated using the formula

$$f_2 = \frac{19}{Z_1}$$

(Note that this formula arises due to the fact that selection rating curves shown in Figure 2.1 and Figure 2.2 are those for a 19 tooth wheel).

Recommended centre distances for drives

Pitch	(in) (mm)	$\frac{3}{8}$ 9,525	$\frac{1}{2}$ 12,70	$\frac{5}{8}$ 15,875	$\frac{3}{4}$ 19,05	1 25,40	$1\frac{1}{4}$ 31,75	$1\frac{1}{2}$ 38,10	$1\frac{3}{4}$ 44,45	2 50,80	$2\frac{1}{2}$ 63,50	3 76,20
Centre Distance	(mm)	450	600	750	900	1000	1200	1350	1500	1700	1800	2000

Chain length calculation

To find the chain length in pitches (L) for any contemplated centre distance of a two point adjustable drive use the following formula:-

$$\text{Length } (L) = \frac{Z_1 + Z_2}{2} + \frac{2C}{P} + \frac{\left(\frac{Z_2 - Z_1}{2\pi}\right)^2 \times P}{C}$$

The calculated number of pitches should be rounded up to a whole number of even pitches. Odd numbers of pitches should be avoided because this would involve the use of a cranked link which is not recommended.

If a jockey, or tensioner sprocket is used for adjustment purposes, two pitches should be added to the chain length (L).

C is the contemplated centre distance in mm and should generally be between 30–50 pitches. e.g. for a $1\frac{1}{2}$ P chain $C = 1.5 \times 25.4 \times 40 = 1524$ mm.

Selection of wheel materials

Choice of material and heat treatment will depend upon shape, diameter and mass of the wheel.

Pinion/wheel	*Steady*	*Medium impulsive*	*Highly impulsive*
Up to 29T	EN8 or EN9	EN8 or 9 hardened and tempered or case-hardened mild steel	EN8 or 9 hardened and tempered or case-hardened mild steel
30T and over	C.I.	Mild steel meehanite	Hardened and tempered steel or case-hardened mild steel or flame-hardened teeth

Through hardened EN9 should be oil quenched and tempered to give a hardness of 400 VPN nominal.

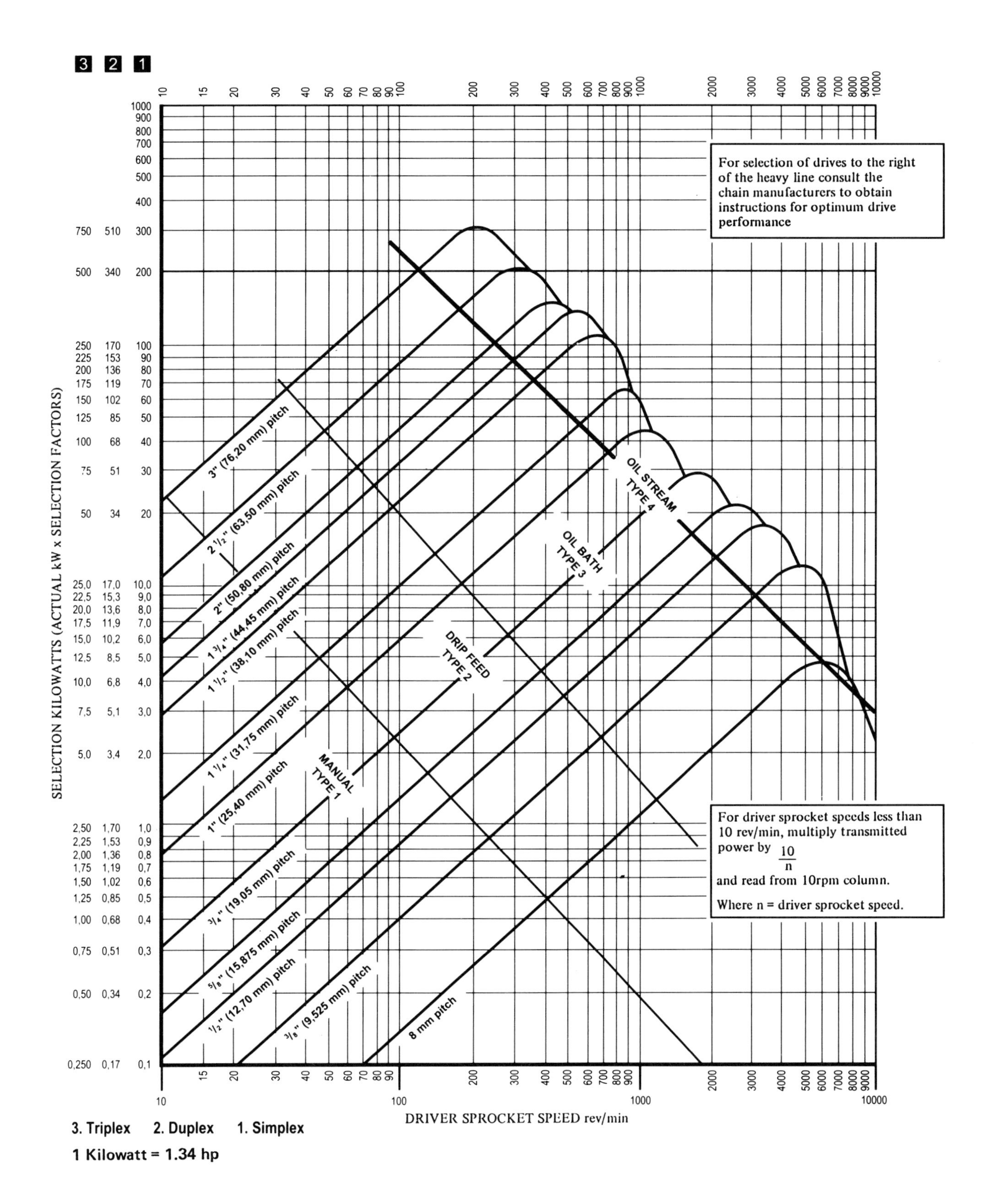

Figure 2.1 BS/DIN Chain drives – Ratings chart using 19T driver sprockets (ISO 606B)

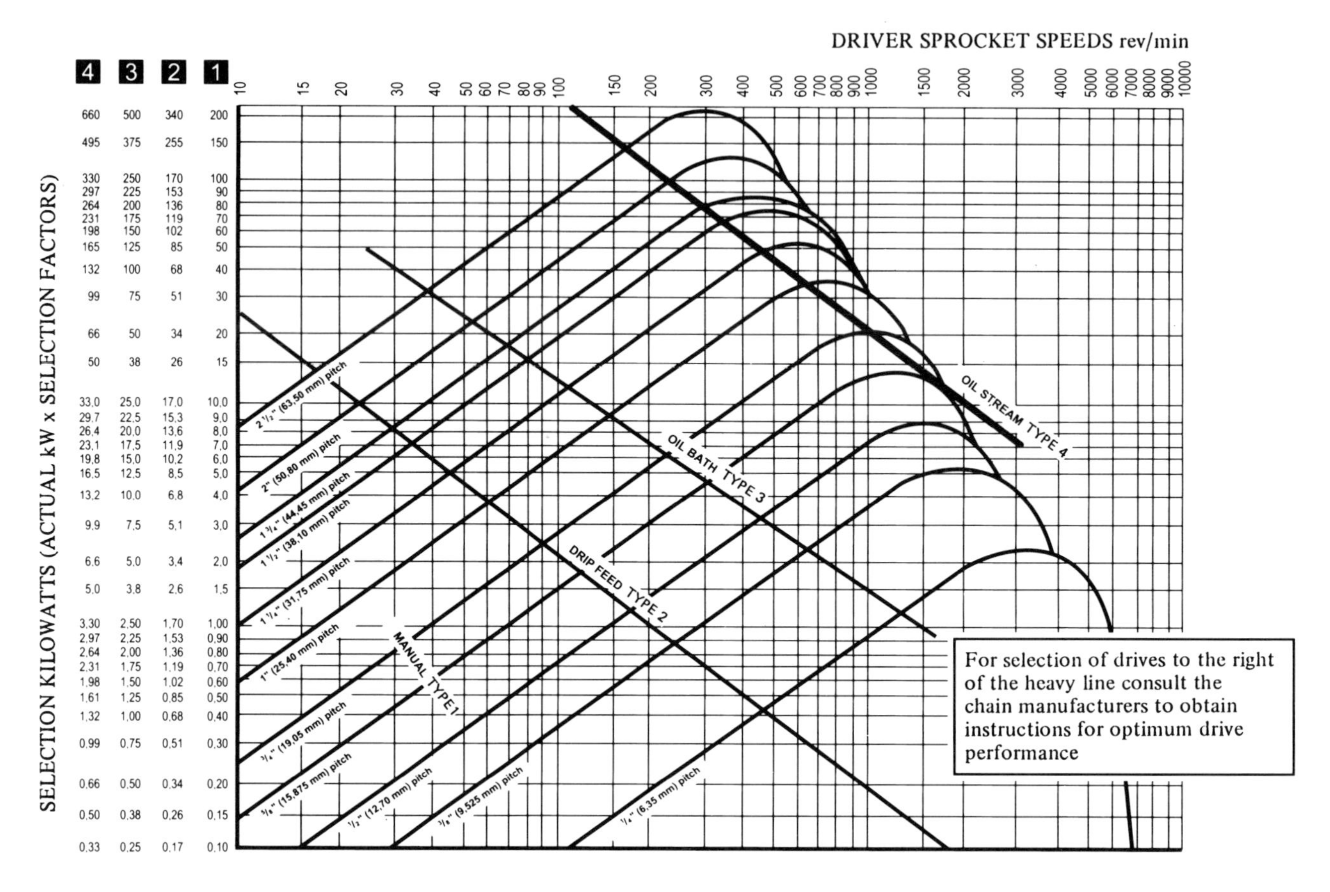

Figure 2.2 ANSI Chain drives – Ratings chart using 19T driver sprockets (ISO 606A)

LUBRICATION

Chain drives should be protected against dirt and moisture and be lubricated with good quality non-detergent petroleum based oil. A periodic change of oil is desirable. Heavy oils and greases are generally too stiff to enter the chain working surfaces and should not be used.

Care must be taken to ensure that the lubricant reaches the bearing area of the chain. This can be done by directing the oil into the clearances between the inner and outer link plates, preferably at the point where the chain enters the wheel on the bottom strand.

The table below indicates the correct lubricant viscosity for various ambient temperatures.

Recommended Lubricants.

Ambient temperature		*Oil viscosity rating*	
°C	*°F (approx)*	*SAE*	*BS 4231*
−5 to +5	20 to 40	20	46 to 68
5 to 40	40 to 100	30	100
40 to 50	100 to 120	40	150 to 220
50 to 60	120 to 140	50	320

For the majority of applications in the above temperature range a multigrade SAE 20/50 oil would be suitable.

Use of grease

As mentioned above, the use of grease is not recommended. However, if grease lubrication is essential the following points should be noted:

(a) Limit chain speed to 4 m/s.

(b) Applying normal greases to the outside of a chain only seals the bearing surfaces and will not work into them. This causes premature failure. Grease has to be heated until fluid and chain are immersed and allowed to soak until all air bubbles cease to rise. If this system is used the chains need regular cleaning and regreasing at intervals depending on power/speed.

Abnormal ambient temperatures

For elevated temperatures up to 250°C dry lubricants such as colloidal graphite or MoS_2 in white spirit or polyalkaline glycol carriers are most suitable.

Conversely, at low temperatures between −5 to −40, special low temperature initial greases and subsequent oil lubricants are necessary. Lubricant suppliers will give recommendations.

LUBRICATION METHODS

There are four basic methods for lubricating chain drives. The recommended methods shown in the ratings charts are determined by chain speed and power transmitted. The use of better methods is acceptable and may be beneficial.

Manual operation

Type 1
Oil is applied periodically with a brush or oil can, preferably once every 8 hours of operation. Volume and frequency should be sufficient to just keep the chain wet with oil and allow penetration of clean lubricant into the chain joints. Applying lubricant by aerosol can be satisfactory under some conditions, but it is important that the aerosol lubricant is of an approved type for the application. An ideal lubricant 'winds in' to the pin/bush/roller clearances, resisting both the tendency to drip or drain when the chain is stationary, and centrifugal 'flinging' when the chain is moving.

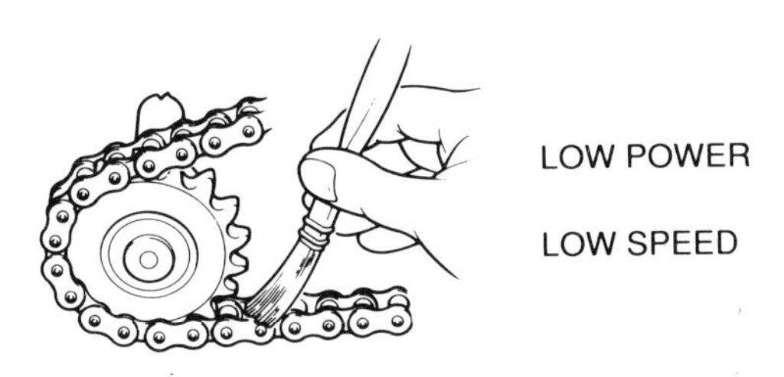

Drip lubrication

Type 2
Oil drips are directed between the link plate edges from a drip lubricator. Volume and frequency should be sufficient to allow penetration of lubricant into the chain joints.

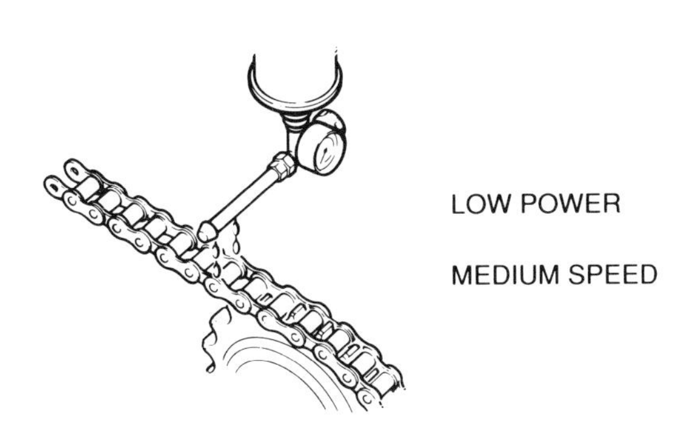

Bath or disc lubrication

Type 3
With oil bath lubrication the lower strand of chain runs through a sump of oil in the drive housing. The oil level should cover the chain at its lowest point whilst operating. With slinger disc lubrication an oil bath is used but the chain operates above the oil level. A disc picks up oil from the sump and deposits it on the chain by means of deflection plates. When such discs are employed they should be designed to have peripheral speeds between the minimum and maximum limits of 180 to 2440 m/min.

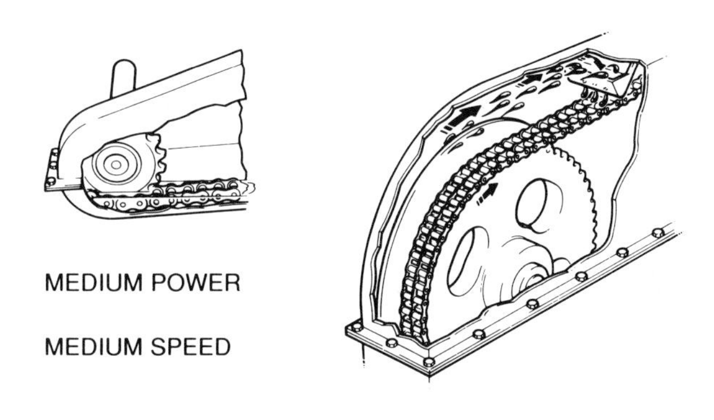

Oil stream lubrication

Type 4
A continuous supply of oil from a circulating pump or central lubricating system is directed onto the chain. It is important to ensure that the spray pipe holes, from which the oil emerges, are in line with the chain plate edges. The spray pipe should be positioned so that the oil is delivered onto the chain just before it engages with the driver wheel. This ensures that the lubricant is centrifuged through the chain and assists in cushioning roller impact on the sprocket teeth. When a chain is properly lubricated a wedge of clean lubricant is formed in the chain joints and metal contact is minimised. Oil stream lubrication also provides effective cooling and impact damping at high speeds. It is, therefore, important that the method of lubrication specified in the ratings chart is closely followed.

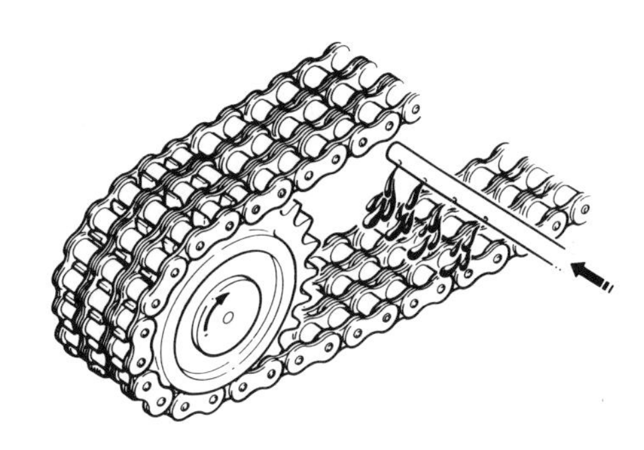

INSTALLATION AND MAINTENANCE

Wheel alignment

Make sure that the shafts are properly supported in bearings. Shaft, bearings and foundations should be suitable to maintain the initial static alignment.

Sprockets should be arranged close to the bearings. Accurate alignment of shafts and sprocket tooth faces provides uniform distribution of the load across the entire chain width and contributes substantially to maximum drive life.

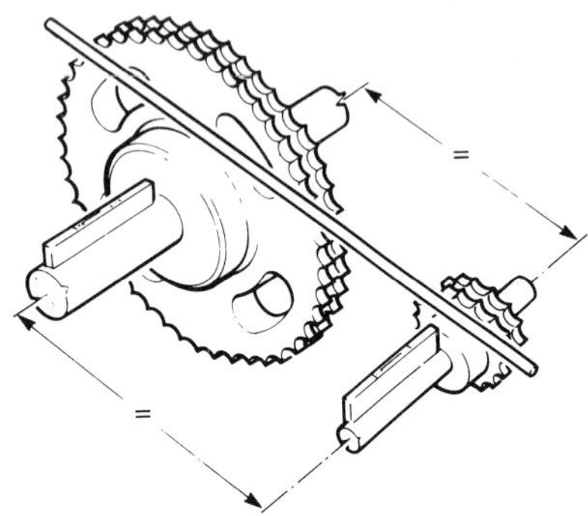

To measure wheel wear

Examination of the tooth flanks will give an indication of the amount of wear which has occurred. Under normal circumstances this will be evident as a polished worn strip about the pitch circle diameter of the sprocket tooth. If the depth of this wear has reached an amount equal to 10% of the 'Y'-dimension (see diagram) then steps should be taken to replace the sprocket. Running new chain on sprockets having this amount of tooth wear will cause rapid chain wear. It should be noted that in normal operating conditions with correct lubrication, the amount of wear at 'X' will not occur until several chains have been used.

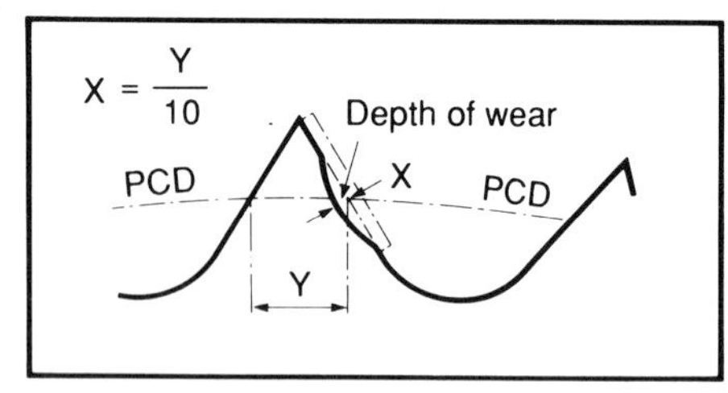

Chain adjustment

The chain should be adjusted regularly so that, with one strand tight, the slack strand can be moved a distance 'A' at mid point (see diagram below). To cater for any eccentricities of mounting, the adjustment of the chain should be tried through a complete revolution of the large sprocket.

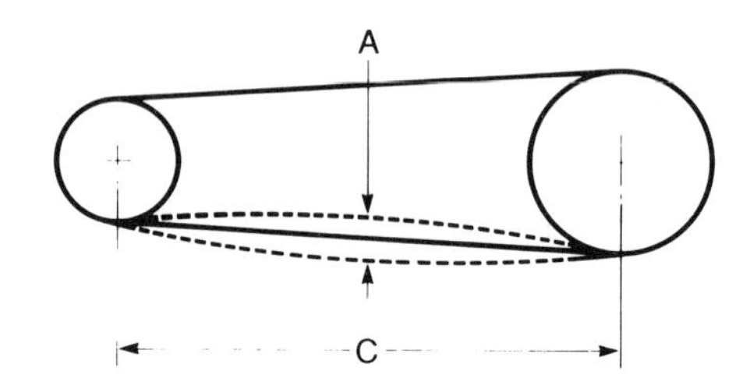

$A = \frac{C}{K}$ Where K = 25 for smooth drives
= 50 for impulsive drives

To measure chain wear

Measure length M in millimetres (see diagram). The percentage extension can then be calculated using the following formula:

$$\text{Percentage extension} = \frac{M - (X \times P)}{X \times P} \times 100$$

where X = number of pitches measured.
P = Pitch in mm

As a general rule, the useful life of the chain is terminated, and the chain replaced, when the percentage extension reaches 2% (1% in the case of extended pitch chains). For drives with no provision for adjustment, the rejection limit is lower, dependent upon speed and layout. A useful figure is between 0.7% and 1% extension.

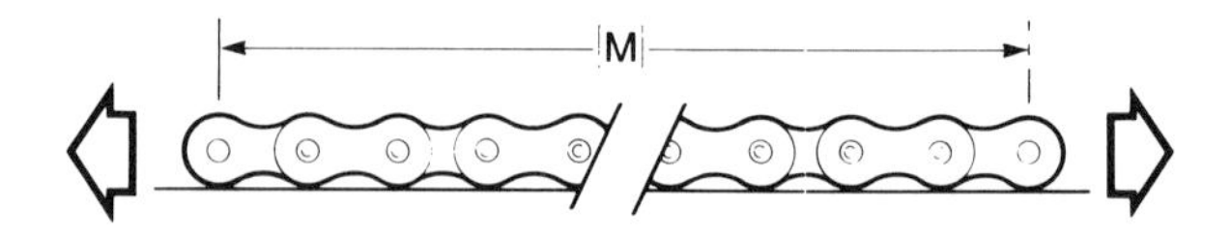

Health and safety

CAUTION
The following precautions must be taken before disconnecting and removing a chain from a drive prior to replacement, repair or length alteration:

1. Always isolate the power source from the drive or equipment.
2. Always wear safety glasses.
3. Always wear appropriate protective clothing, e.g. hats, gloves and safety shoes etc. as warranted by circumstances.
4. Always ensure tools are in good working condition and use in the proper manner.
5. Always ensure that directions for the correct use of any tools are followed.
6. Always loosen tensioning devices.
7. Always support the chain to avoid sudden unexpected movement of chain or components.
8. Never attempt to disconnect or re-connect a chain unless the chain construction is fully understood.
9. Never re-use individual components.
10. Never re-use a damaged chain or chain part.
11. On light duty drives where a spring clip joint (No. 26) is used always ensure that the clip is fitted correctly in relation to direction of travel, with open end trailing.

GEAR TYPES

External spur gears

Cylindrical gears with straight teeth cut parallel to the axes, tooth load produces no axial thrust. Give excellent results at moderate peripheral speeds, tendency to be noisy at high speeds. Shafts rotate in opposite directions.

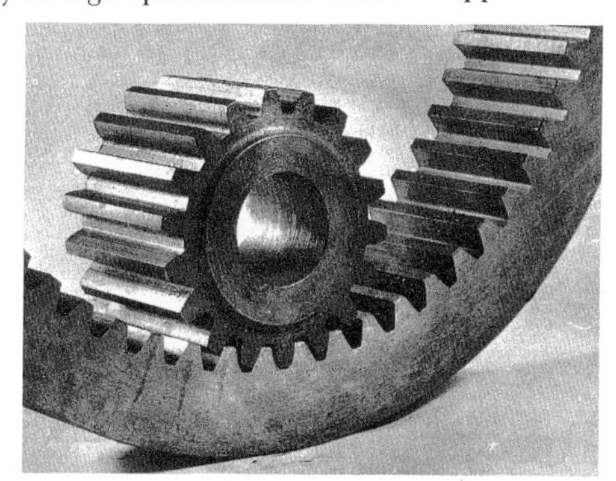

Internal spur gears

Provide compact drive for transmitting motion between parallel shafts rotating in same direction.

Helical gears

Serve same purpose as external spur gears in providing drive between two parallel shafts rotating in opposite directions. Superior in load carrying capacity and quietness in operation. Tooth load produces axial thrust.

Straight bevel gears

Used to connect two shafts on intersecting axes, shaft angle equals angle between the two axes containing the meshing gear teeth. Gear teeth are radial towards apex, end thrust is developed under tooth load tending to separate the gears.

Spiral bevel gears

Used to connect two shafts on intersecting axes same as straight bevels. Have curved oblique teeth contacting each other gradually and smoothly from one end of the tooth to the other. Meshes similar to straight bevel but are smoother and quieter in action. Have better load carrying capacity. Hand of spiral left-hand teeth incline away from axis in anti-clockwise direction looking on small end of pinion or face of gear, right hand teeth incline away from axis in clockwise direction. The hand of spiral of the pinion is always opposite to that of the gear and the hand of spiral of the pinion is used to identify the gear pair. The spiral angle does not affect the smoothness and quietness of operation or the efficiency but does affect the direction of the thrust loads created, a left hand spiral pinion driving clockwise when viewed from large end of pinion creates an axial thrust that tends to move the pinion out of mesh.

Zerol bevel gears

Zerol bevel gears have curved teeth lying in the same general direction as straight bevel gears but should be considered as spiral bevel gears with zero spiral angle.

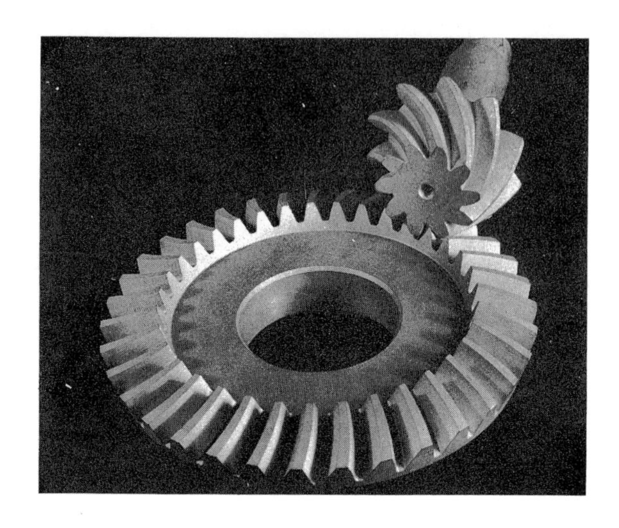

Hypoid bevel gears

Hypoid gears are a cross between spiral bevel gears and worm gears, the axes of a pair of hypoid bevel gears are non-intersecting, the distance between the 'axes' being called the offset. The offset allows higher ratios of reduction than practicable with bevel gears. Hypoid gears have curved oblique teeth on which contact begins gradually and continues smoothly from one end of the tooth to the other.

Worm gears

Worm gears are used to transmit motion between shafts at right angles that do not lie in a common plane. They are also used occasionally to connect shafts at other angles. Worm gears have line tooth contact and are used for power transmission, but the higher the ratio the lower the efficiency.

APPLICATION OF GEARS

Table 3.1 Scope and torque capacity of gears

Relation between shaft axes	*Gear ratio (RG)*	*Max. tooth speed V* (M/Sec)	*Type of tooth*	*Max. wheel torque* (Nm)
Parallel	up to 10 to 1	5	Helical or straight	900×10^4
		25	Helical	220×10^4
			Profile ground straight	22×10^4
		205	Helical	56×10^4
Intersecting	up to 7 to 1	2.5	Spiral bevel or straight bevel	9×10^4
		60	Spiral bevel	4.5×10^4
Non-intersecting at 90°	up to 10 to 1	60	Hypoid bevel	6×10^4
Non-intersecting crossed at 90°	up to 50 to 1	50	Worm and wormwheel	28×10^4
			Crossed helical	17×10^4
Non-intersecting crossed at 80° to 100° but not 90°	up to 50 to 1	50	Worm and wormwheel	11×10^4
			Crossed helical	17×10^4

Note: The above figures are for general guidance only. Any case that approaches or exceeds the quoted limits needs special consideration of details of available gear-cutting equipment.

CHOICE OF MATERIALS

Table 3.2 Allowable stresses on materials for spur, helical, straight bevel, spiral bevel and hypoid bevel gears

Material	S_{CO} N/mm^2	S_{BOS} *(skin)* N/mm^2	S_{BOC} *(core)* N/mm^2	*BHN* *(core)*	*ULT* *(core)* N/mm^2	*YIELD* *(core)* N/mm^2	*VHN* *(skin)* *(min)*	*ULT* *(skin)* N/mm^2	*BS* *spec.*	*Metallurgical condition and treatment*
PB1	5	50	50	70	185	–	–	–	1400PB2	Sand cast
PB2	6	60	60	85	230	–	–	–	1400PB2	Chill cast centrifugal cast
PB3	7	70	70	90	260	–	–	–	1400PB2	
M1	6	76	76	140	310	–	–	–	309W24/8	Malleable iron
C1	7	40	40	165–190	185	–	–	–	1452	Ord. grade
C2	9	52	52	210–220	245	–	–	–	1452	Med. grade
C3	10	70	70	220–230	340	–	–	–	1452	High grade
C5	10	110	110	145–170	540	280	–	–	592	Steel as cast
									BS 970	
1	10.5	145	145	150–180	540–618	280	–	–	080A35	Normalised
2	12.5	160	160	180–230	618–740	430	–	–	150M28	Q
3	16.5	172.5	172.5	200–250	695–850	355	–	–	–	Normalised
4	18.5	186	186	220–270	740–895	585	–	–	708M40	S
5	18.5	186	186	220–270	740–895	585	–	–	817M40	S
6	21.5	215	215	250–300	850–1000	680	–	–	830M31	T
7	26.0	235	235	290–340	970–1130	770	–	–	830M31	V
8	45.0	110	145	150–180	540–618	–	450	1390	080A35	(FH)†
9	45.0	165	145	150–180	540–618	–	450	1390	080A35	(CH)†
10	45.0	130	172	200–250	695–850	–	450	1390	708M40	R(FH)†
11	45.0	195	172	200–250	695–850	–	450	1390	708M40	R(CH)†
12	45.0	145	186	220–270	740–895	–	450	1390	817M40	S(FH)†
13	45.0	207	186	220–270	740–895	–	450	1390	817M40	S(CH)†
14	45.0	240	215	250–300	850–1000	–	450	1390	826M31	T(CH)†
15	45.0	172	227	270–320	925–1005	–	450	1390	826M40	U(FH)†
16	45.0	255	227	270–320	925–1005	–	450	1390	826M40	U(CH)†
17	55.0	130	172	200–250	695–850	–	710	1850	722M24	R. Nitrided
18	62.0	145	186	223–270	740–895	–	710	1850	722M24	S. Nitrided
19	69.0	207	276	375–444	1205–1390	–	710	1850	897M39	Nitrided
20	69.0	255*	172	210–240	695–772	–	710	1850	665M17	
21	76.0	282*	214	250–300	740–1005	–	710	1850	655M13	Case-hardened
22	83.0	345*	262	350–410	1160–1312	–	710	1850	659M15	

Where:

S_{CO} = allowable contact stress
S_{BOS} = allowable skin bending stress
S_{BOC} = allowable core bending stress
CI: Cast iron
CS: Cast steel
BHN: Brinell hardness number
VHN: Vickers hardness number
MI: Malleable iron
PB: Phosphor bronze

*Multiply by 1.8 for very smooth fillets not ground after hardening.

(FH)†: Hardening by flame or induction over the whole working surfaces of the tooth flanks but excludes the fillets – applies to modules larger than 3.5.

(CH)† Hardening by flame or induction over the whole tooth flanks, fillets and connecting root surfaces – applies to modules between 5 and 28.
Spin hardening – applies to modules between 3.5 and 2.0.

Notes:

1. Materials 8 to 22, the basic allowable bending stress (S_{BO}), used in estimating load capacity of gears depends on the ratio of the depth of the hard skin at the root fillets to the normal pitch (circular pitch) of the teeth.

$$S_{BO} = S_{BOS} \text{ or } \frac{S_{BOC}}{[1\text{–}7.5 \text{ (depth of skin)/normal pitch}]}$$

whichever is the less.

2. Materials 8 to 22, values of S_{CO} are reliable only for skin thicker than:

$0.003\, d \times D\ (d + D)$

3. Materials 1 to 8, the value of S_{BO} is approx:

$$S_{BO} = 600 \times \text{Ult Tensile} - \frac{\text{Ult. Tensile}^3}{60}$$

4. Gear cutting becomes difficult if BHN exceeds 270.

Table 3.3 Allowable stresses for various materials used for crossed helicals and wormwheels

S_{CO2} N/mm^2	S_{BO2} N/mm^2	*Wheel material*		*BHN*	*Ultimate tensile strength N/mm^2*
10.5	50	Phosphor	Sand cast	70	185–216
12.5	60	Bronze	Chill cast	82	230–260
15.2	70	BS 1400	Centrifugally cast	90	260–293
		P.B.2			
7	41.4	Cast iron	Ordinary grade	150	185–216
7	51.7	BS 1452	Medium grade	165	245–262
7	70.0		High grade	180	340–370

Note: The pinion or worm in a pair of worm gears should be of steel, materials 3 to 7 or 20 to 22, Table 3.2, and always harder than the material used for the wheel.

Non-metallic materials for gears

To help in securing quiet running of spur, helical and straight and spiral bevel gears fabric-reinforced resin materials can be used. The basic allowable stresses for these materials are approximately S_{CO} = 10.5 N/mm^2 and S_{BO} = 31.0 N/mm^2, but confirmation should always be obtained from the material supplier.

Other plastic materials are also available and information on their allowable stresses should be obtained from the material supplier.

Material combinations

1. With spur, helical, straight and spiral bevel gears, material combinations of cast iron – phosphor bronze, malleable iron – phosphor bronze, cast iron – malleable iron or cast iron – cast iron are permissible.
2. The material for the pinion should preferably be harder than the wheel material.

Where other materials are used:

(a) Where cast steel and materials 1 to 7, Table 3.2 are used, it is desirable that the ultimate tensile strength for the wheel should lie between the ultimate tensile strength and the yield stress of the pinion.

(b) Materials 8 to 22, Table 3.2, may be used in any combination.

(c) Gears made from materials 8 to 22, Table 3.2, to mate with gears made from any material outside this group, must have very smooth finish on teeth.

GEAR PERFORMANCE

A number of methods of estimating the expected performance of gears have been published as Standards. These use a large number of factors to allow for operational and geometric effects, and for new designs leave a lot to the designers' judgement, for the matching of the design to suit a particular application. They are, however, more readily applicable to the development of existing designs.

Early methods

Lewis Formula – Dates back to 1890s and is used to calculate the shear strength of the gear tooth and relate it to the yield strength of the material.

Buckingham Stress Formula – Dates back to mid 1920s and compares the dynamic load with the beam strength of the gear tooth, and a limit load for wear.

British Standard 436 Part 3 1986

Provides methods for determining the actual and permissible contact stresses and bending stresses in a pair of involute spur or helical gears.

Factors covered in this standard include:

Tangential Force	The nominal force for contact and bending stress.
Zone factor	Accounts for the influence of tooth flank curvature at the pitch point on Hertzian stress.
Contact ratio factor:	Accounts for the load sharing influence of the transverse contact ratio and the overlap ratio on the specific loading.
Elasticity factor:	Takes into account the influence of the modulus of elasticity of the material and Poisson's ratio on the Hertzian stress.
Basic endurance limit:	The basic endurance limit for contact takes into account the surface hardness.
Material quality:	This covers the quality of the material used.
Lubricant influence, roughness and speed factor:	The lubricant viscosity, surface roughness and pitch line speed affect the lubricant film thickness which affects the Hertzian stresses.
Work hardening factor:	Accounts for the increase of surface durability due to meshing.
Size factor:	Covers the possible influences of size on the material quality and its response to manufacturing processes.
Life factor:	Accounts for the increase in permissible stress when the number of stress cycles is less than the endurance life.
Application factor:	This allows for load fluctuations from the mean load or loads in the load histogram caused by sources external to the gearing.
Dynamic factor:	Allows for load fluctuations arising from contact conditions at the gear mesh.
Load distribution:	Accounts for the increase in local load due to mal-distribution of load across the face of the gear caused by deflections, alignment tolerances and helix modifications.
Minimum demanded and actual safety factor:	The minimum demanded safety factor is agreed between the supplier and the purchaser. The actual safety factor is calculated.
Geometry factors:	Allow for the influence of the tooth form, the effect of the fillet and the helix angle on the nominal bending stress for application of load at the highest point of single pair tooth contact.
Sensitivity factor:	Allows for the sensitivity of the gear material to the presence of notches, ie: the root fillet.
Surface condition factor:	Accounts for the reduction of endurance limit due to flaws in the material and the surface roughness of the tooth root fillets.

International Standards Organisation I.S.O. 60 'Gears'

Similar in many ways to BS 436 Part 3 1986 but far more comprehensive in its approach. For the average gear design a very complex method of arriving at a conclusion similar to the less complex British Standard. Factors covered in this standard include:

Tangential load:	The nominal load on the gear set.
Application factor:	Accounts for dynamic overloads from sources external to the gearing.
Dynamic factor:	Allows for internally generated dynamic loads, due to vibrations of pinion and wheel against each other.
Load distribution:	Accounts for the effects of non-uniform distribution of load across the face width. Depends on mesh alignment error of the loaded gear pair and the mesh stiffness.
Transverse load distribution factor:	Takes into account the effect of the load distribution on gear-tooth contact stresses, scoring load and tooth root strength.
Gear tooth stiffness constants:	Defined as the load which is necessary to deform one or several meshing gear teeth having 1 mm face width by an amount of 1 μm.
Allowable contact stress:	Permissible Hertzian pressure on gear tooth face.
Minimum demanded and calculated safety factors:	Minimum demanded safety factor agreed between supplier and customer, calculated safety factor is the actual safety factor of the gear pair.
Zone factor:	Accounts for the influence on the Hertzian pressure of the tooth flank curvature at the pitch point.
Elasticity factor:	Accounts for the influence of the material properties, i.e.: modulus of elasticity and Poisson's ratio.
Contact ratio factor:	Accounts for the influence of the transverse contact ratio and the overlap ratio on the specific surface load of gears.
Helix angle factor:	Allows for the influence of the helix angle on surface durability.
Endurance limit:	Is the limit of repeated Hertzian stress that can be permanently endured by a given material.
Life factor:	Takes account of a higher permissible Hertzian stress if only limited durability endurance is demanded.
Lubrication film factor:	The film of lubricant between the tooth flanks influences surface load capacity. Factors include oil viscosity, pitch line velocity and roughness of tooth flanks.
Work hardening factor:	Accounts for the increase in surface durability due to meshing a steel wheel with a hardened pinion with smooth tooth surfaces.
Coefficient of friction:	The mean value of the local coefficient of friction is dependent on several properties of the oil, surface roughness, the 'lay' of surface irregularities, material properties of tooth flanks, tangential velocities, force and size.
Bulk temperature:	Surface temperature.
Thermal flash factor:	Dependent on moduli of elasticity and thermal contact coefficients of pinion and wheel materials, and the geometry of the line of action.
Welding factor:	For different tooth materials and heat treatments.
Geometrical factor:	Defined as a function of the gear ratio and a dimensionless parameter on the line of action.
Integral temperature criterion:	The integral temperature of the gears depends on the oil viscosity and the performance of the gear materials relative to scuffing and scoring.

The figures produced from this standard are very similar to those produced by British Standard 436 Part 3 1986.

British Standard 545, 1982 (Bevel gears)

Specifies tooth form, modules, accuracy requirements, methods of determining load capacity and material requirements for machine-cut bevel gears, connecting intersecting shafts which are perpendicular to each other and having teeth with a normal pressure angle of 20° at the pitch cone, whose lengthwise form may have straight or curved surfaces.
The load capacity of the gears is limited by consideration of both wear and strength, factors taken into account are:

Wear and strength factors: Include the speed, surface stress, zone, pitch, spiral angle overlap ratio and bending stress factors.

Limiting working temperature: The temperature of the oil bath under the specified loading conditions.

Basic stress factors: Given for the various recommended materials.

British Standard 721 Part 2 1983 (Worm gears)

Specifies the requirements for worm gearing based on axial modules. Four classes of gear are specified, which are related to function and accuracy. The standard applies to worm gearing comprising cylindrical involute helicoid worms and wormwheels conjugate thereto. It does not apply to pairs of cylindrical gears connecting non-parallel axes known as crossed helical gears.
The load capacity of the gears is limited by both wear and strength of the wormwheel, factors taken into account include:

Expected life: The strength is calculated to an expected total running life of 26 000 hours. Allows for both steady and variable loads at different running speeds.

Momentary overload capacity: Momentary overload is considered as one whose duration is too short to be defined with certainty but does not exceed 15 seconds.

Efficiency and lubrication: The efficiency, excluding bearing and oil-churning losses for both worm and wormwheel driving.

Basic stress factors: Given for the various recommended materials.

American Gear Manufacturers Association Standards

The American Gear Manufacturers Association Standards are probably the most comprehensive coverage for gear design and are compiled by a committee and technical members representing companies throughout America, both north and south, Australia, Belgium, Finland, France, Great Britain, India, Italy, Japan, Mexico, Sweden, Switzerland and West Germany and are being constantly up-dated.
The standards cover gear design, materials, quality and tolerances, measuring methods and practices, and backlash recommendations.
Gear performance is covered by a series of different standards as follows:

AGMA 170–01 Design guide for vehicle spur and helical gears.

AGMA 210–02 Surface durability (pitting) of spur gear teeth.

AGMA 211–02 Surface durability (pitting) of helical and herringbone gear teeth.

AGMA 212–02 Surface durability (pitting) formulas for straight bevel and zerol bevel gear teeth.

AGMA 215–01 Information sheet for surface durability (pitting) of spur, helical, herringbone and bevel gear teeth.

AGMA 216–01 Surface durability (pitting) formulas for spiral bevel gear teeth.

AGMA 217–01 Information sheet – gear scoring design guide for aerospace spur and helical power gears.

AGMA 220–02 Rating the strength of spur gear teeth.

AGMA 221–02 Rating the strength of helical and herringbone gear teeth.

American Gear Manufacturers Association Standards (Contd.)

AGMA 222–02	Rating the strength of straight bevel and zerol bevel gear teeth.
AGMA 223–01	Rating the strength of spiral bevel gear teeth.
AGMA 225–01	Information sheet for strength of spur, helical, herringbone and bevel gear teeth.
AGMA 226–01	Information sheet – geometry factors for determining the strength of spur, helical, herringbone and bevel gear teeth.
AGMA 2001–B.88	Fundamental rating factors and calculation methods for involute spur and helical gear teeth.
AGMA 2005–B.88	Design manual for bevel gears.

Factors taken into account in these standards include:

Unit load:	Calculated from tangential load, size of gear teeth and face width of gear.
Bending stress factor:	The relation of calculated bending stress to allowable bending stress.
Geometry factor:	The geometry factor evaluates the radii of curvature of the contacting tooth profiles based on the tooth geometry.
Transmitted tangential load:	Represents the tooth load due to the driven apparatus.
Dynamic factors:	Account for internally generated tooth loads induced by non-conjugate meshing action of the gear teeth.
Application factor:	Allows for any externally applied loads in excess of the nominal tangential load.
Elastic coefficient:	Accounts for both the modulus of elasticity of the gears and Poisson's ratio.
Surface condition:	Allows for surface finish on the teeth, residual stress and the plasticity effects (work hardening) of the materials.
Size factor:	Reflects non-uniformity of material properties, tooth size, diameter of gears, ratio of tooth size to diameter, face width, area of stress pattern, ratio of case depth to tooth size and hardenability and heat treatment of materials.
Load distribution factors:	Modifies the rating equations to reflect the non-uniform distribution of the load along the lines of contact.
Allowable stress numbers:	Depend upon the material composition, mechanical properties, residual stress, hardness and type of heat treatment.
Hardness ratio factor:	Covers the gear ratio and the hardness of both pinion and gear teeth.
Life factor:	Adjusts the allowable stress numbers for the required number of cycles of operation.
Reliability factor:	Accounts for the effect of the normal statistical distribution of failures found in materials testing.
Temperatures factor:	Takes into account the temperature in which the gears operate.

Other factors are included in the standards depending upon the actual usage of the gears, e.g. motor vehicles, marine diesels, etc.

Comparison of design standards

From the list of factors given it can be seen that all three standards approach the gear performance problem in a similar manner but due to slight variances in methods used to calculate the factors the stress allowable figures will differ.

British Standard 436 Part 3 1986 is a radical up-date of the original BS 436 and in many ways brings it in line with ISO/TC 60 whilst it can be seen from the list of AGMA Standards that these are constantly reviewed to meet the demands of industry.

Flexible couplings are used to connect the shafts of separate machines and take up any small misalignment that there may be between them.

COUPLING TYPES

Gear couplings

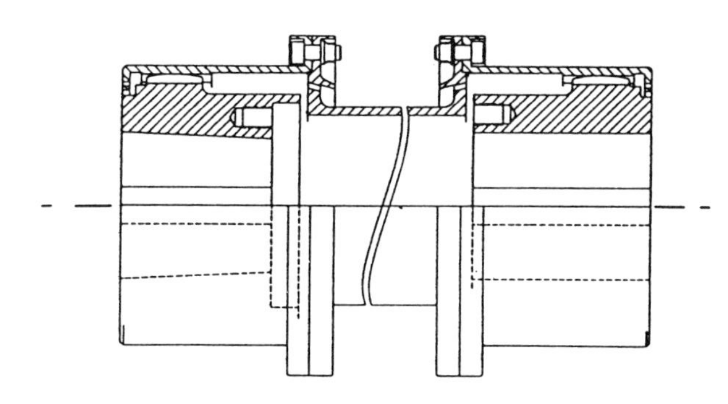

Hubs fitted to the shafts of the coupled machines have gear teeth around their periphery and these mate with internal gear teeth on a sleeve which couples the two hubs together.

The gear meshes transmit the torque between the machines but allow relative movement to accommodate the misalignment.

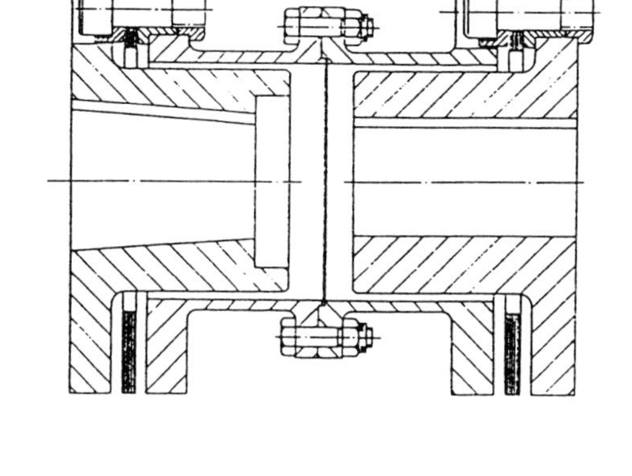

Multiple membrane couplings

The hubs on the shafts of the coupled machines are connected to an intermediate spacer by flexible members. These members are made from stacks of thin laminations so that they are flexible in bending and strong in tension and shear.

Some couplings use the flexible members as tangential links to provide tension connections. Other couplings use radial or disc shaped links in which the torque is transmitted in shear.

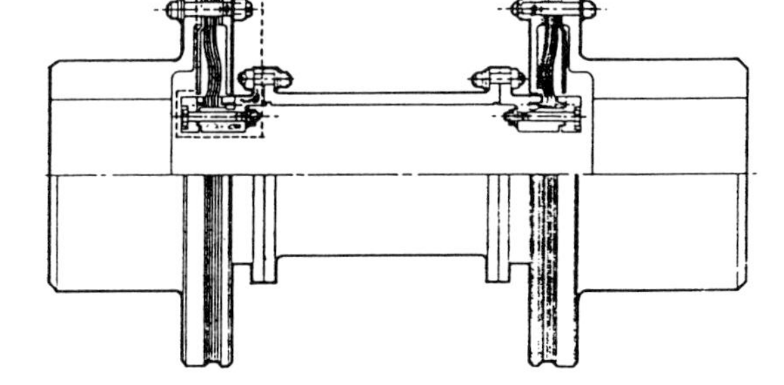

Contoured disc couplings

In this type of coupling the hubs on the machine shafts are coupled to an intermediate shaft by thin discs. These discs have a variable thickness in a radial direction to give a more even stress distribution and are made of high strength material.

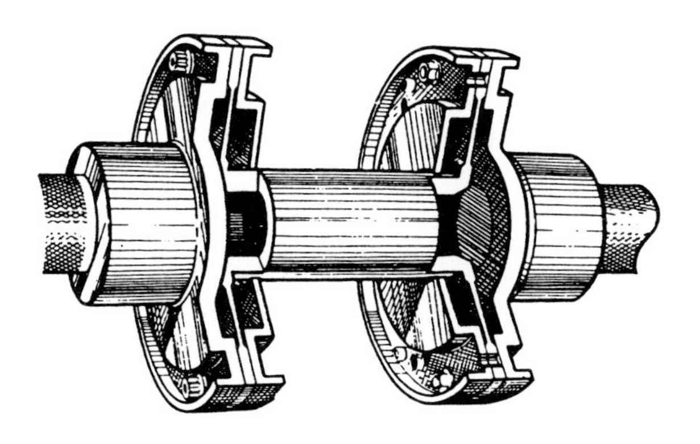

Elastomeric element couplings

There are various designs of coupling which use elastomeric materials to transmit the torque while allowing some flexibility.

The most highly rated couplings use the rubber mainly in compression in the form of rubber blocks located between radial blades. A hub with radial blades on its periphery is fitted to one machine, and a sleeve member with corresponding inwardly extending blades is fitted to the shaft of the other machine.

Convoluted axial spring couplings

The hubs on the coupled machines have a number of contoured blocks around their periphery. A convoluted axial spring is fitted around the hubs and into the slots between the blocks. The driving torque is transmitted by bending and shear in the axial bars of the spring, which can also deflect to take up misalignment.

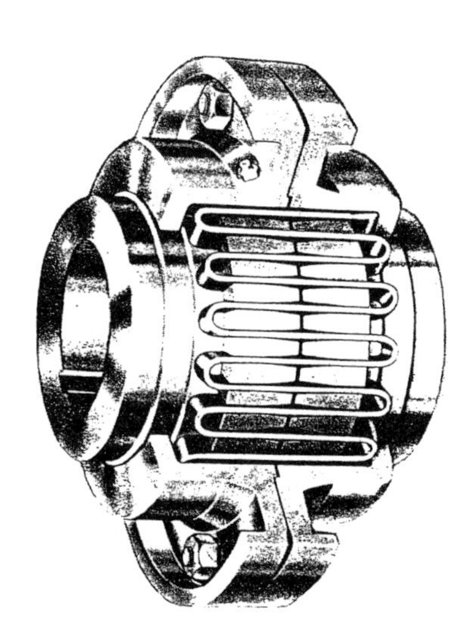

Quill shafts

The machines can also be coupled together by a shaft with a diameter which is just adequate to transmit the maximum torque, and made long enough to give lateral flexibility in order to take up misalignment. Quill shaft couplings do not permit any relative axial movement between the coupled machines.

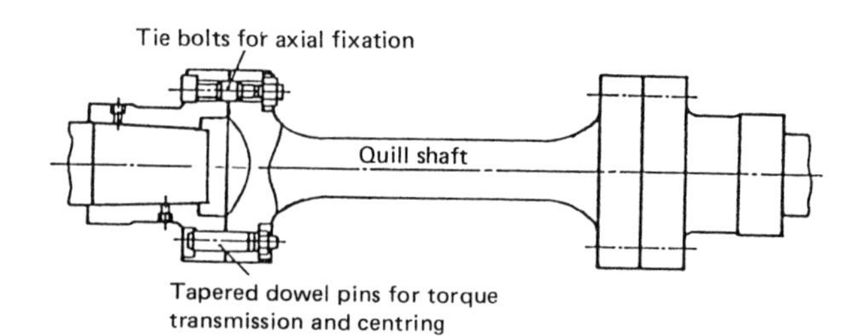

COUPLING PERFORMANCE

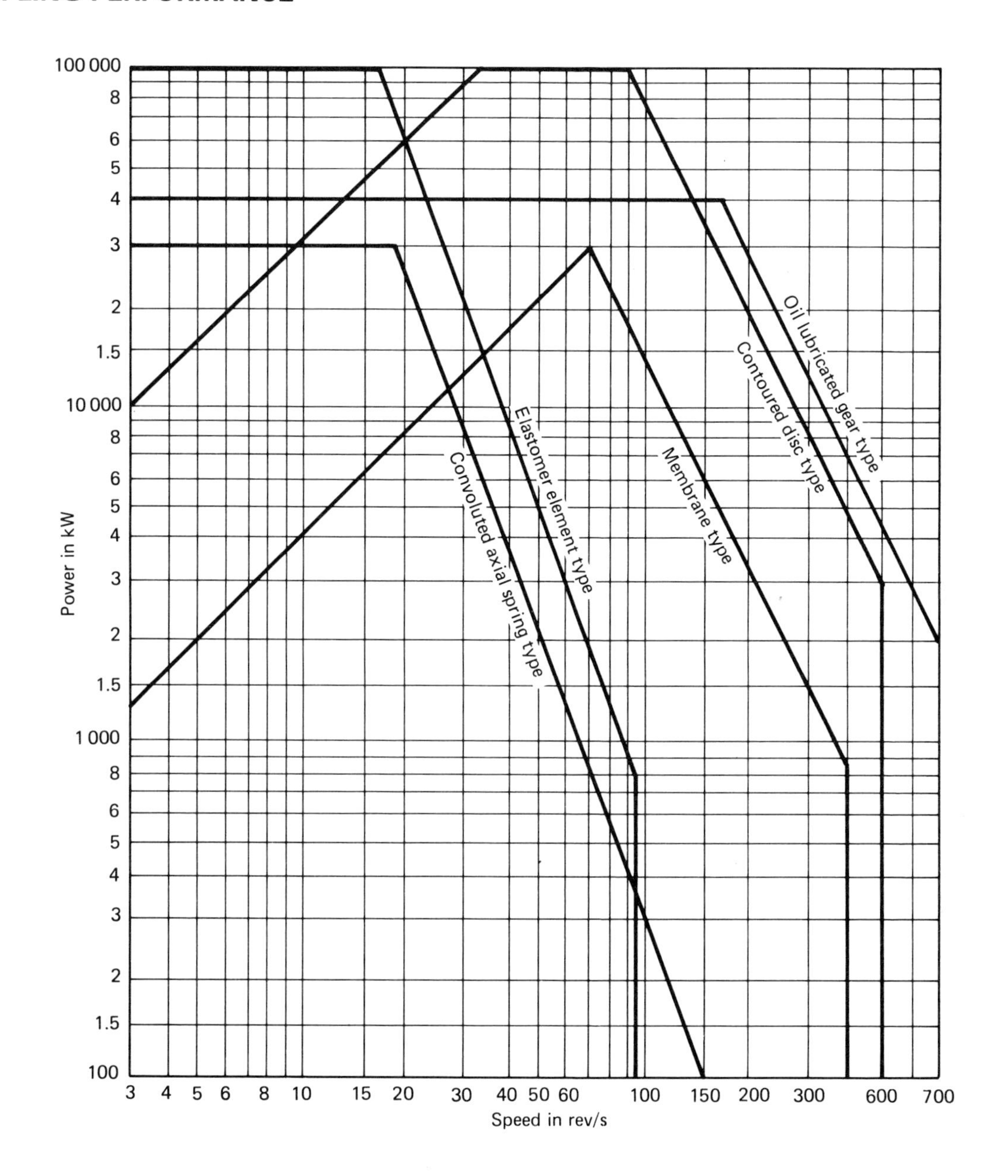

Figure 4.1 The power and speed limits of flexible couplings

In this figure the performance limits at higher speeds are determined largely by centrifugal stresses in the components.

The maximum power limits of gear couplings and elastomeric element couplings arise from contact and compressive stresses.

The lower speed performance limits of disc and membrane couplings arise from the maximum torque that can be transmitted within the stress limits of the material of the flexible members.

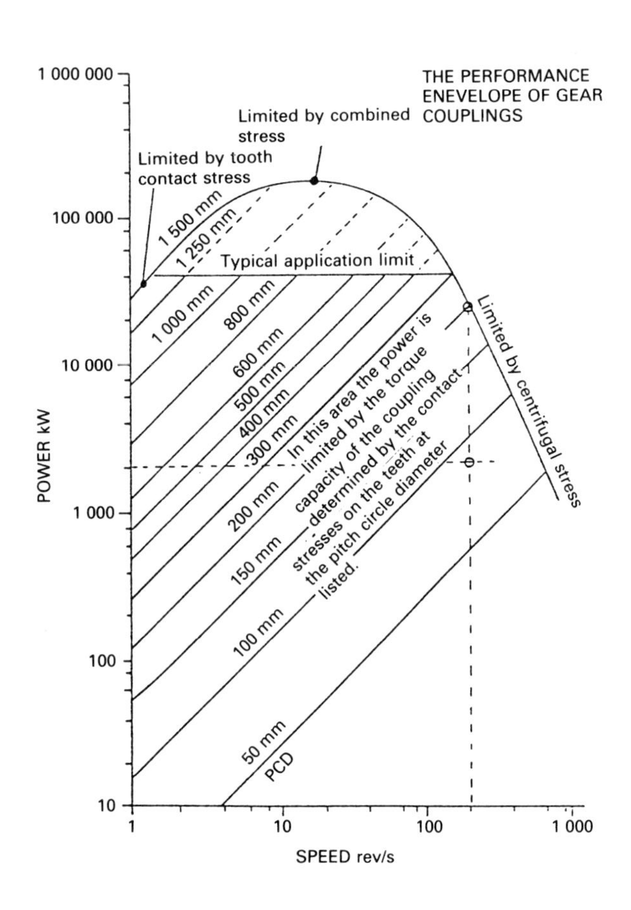

Figure 4.2 The performance envelope for gear couplings

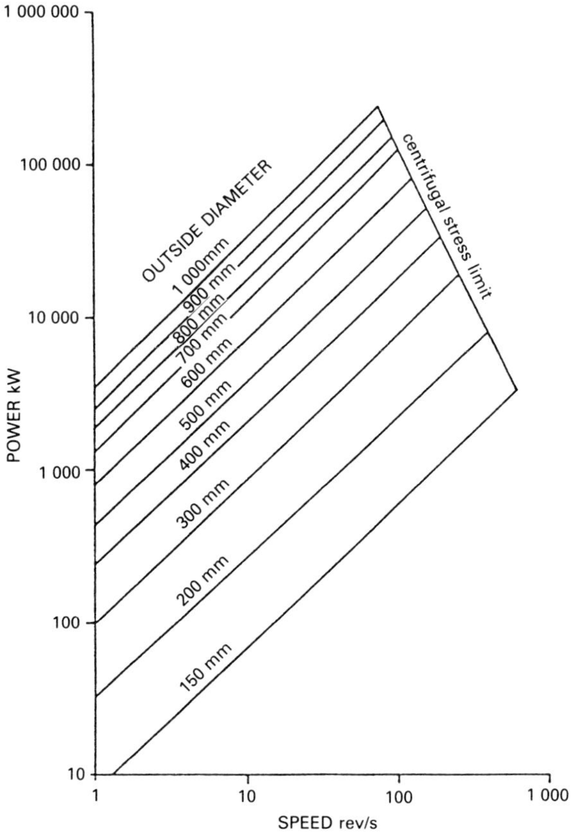

Figure 4.3 Performance envelope for contoured disc couplings

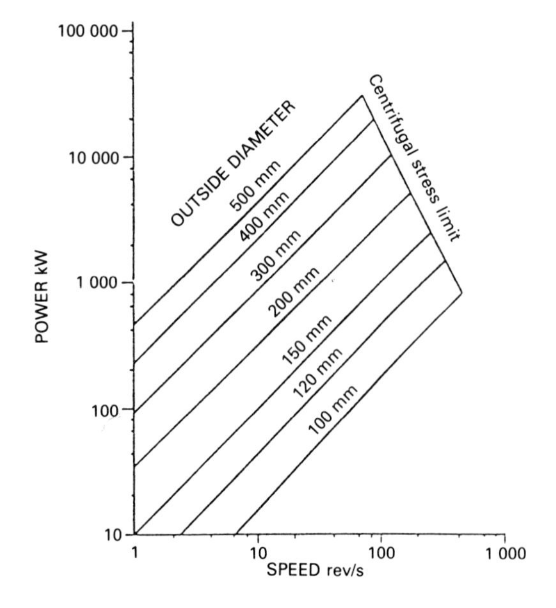

Figure 4.4 Performance envelope for multiple membrane couplings

The parallel inclined lines in these figures correspond to lives of constant torque. In these areas the performance of the coupling is limited by component stresses, which arise directly from the transmitted torque.

Table 4.1 Relative advantages and disadvantages of the various types of coupling

Coupling type	*Advantage*	*Disadvantages*
Gear couplings	Of the couplings that offer a range of flexibilities, they are the most compact and can provide the minimum overhung mass on the machine shaft. They allow the maximum axial movement of any coupling.	Lubrication is essential and while they can be packed with grease for low speed applications a continuous oil feed is required at high speeds. See Figure 4.5 for lubrication limits. They can apply substantial axial and lateral loads to the coupled machines. See Figure 4.7.
Multiple membrane couplings	They require no lubrication or maintenance and once correctly assembled should maintain their balance.	Their relatively high mass can affect the lateral stability of machine rotors. The diaphragm clamping is a critical assembly feature.
Contoured disc couplings	They require no lubrication and have inherently good balance. Their performance is predictable and consistent.	Have a large diameter which can give rise to windage losses and noise.
Elastomeric element couplings	Robust. Can absorb torsional shocks. Can be designed to de-tune torsional resonances in machine systems.	Relatively large diameter which limits their maximum speed capability.
Convoluted axial spring couplings	Robust with some torsional shock absorbing capability. Decoupling is simple, by removing covers and the convoluted spring.	Require grease lubrication which together with balance consistency, limits their maximum speed capability.
Quill shafts	Simplicity. Low mass. Balance retention	No axial movement capability and limited lateral flexibility.

Table 4.2 Approximate maximum misalignments allowable across the various types of coupling

Coupling type	*Lateral misalignment*	*Axial misalignment*
Gear couplings	Typically 0.002 radians per mesh but depends on diameter and rotational speeds, as it is limited by the rubbing speed at the teeth. See Figure 4.6.	Limited only by the axial width of the widest tooth row.
Multiple membrane couplings	Typically up to 0.008 radians/disc but depends on design and operating conditions. High axial displacements reduce the allowable angular misalignment.	Typically up to ±6 mm but depends on design.
Contoured disc couplings	Typically up to 0.010 radians/disc but depends on design and operating conditions.	Typically up to ±6 mm but depends on design.
Electromeric element couplings	Typically 0.008 radians and 1 mm laterally, but depends on design.	Typically up to 4 mm but depends on design.
Convoluted axial spring couplings.	Up to 0.2 mm laterally.	Up to 10 mm approximately.
Quill shafts	Depends on design possibly 0.002 radians along the length.	None unless used in conjunction with a disc coupling at one end.

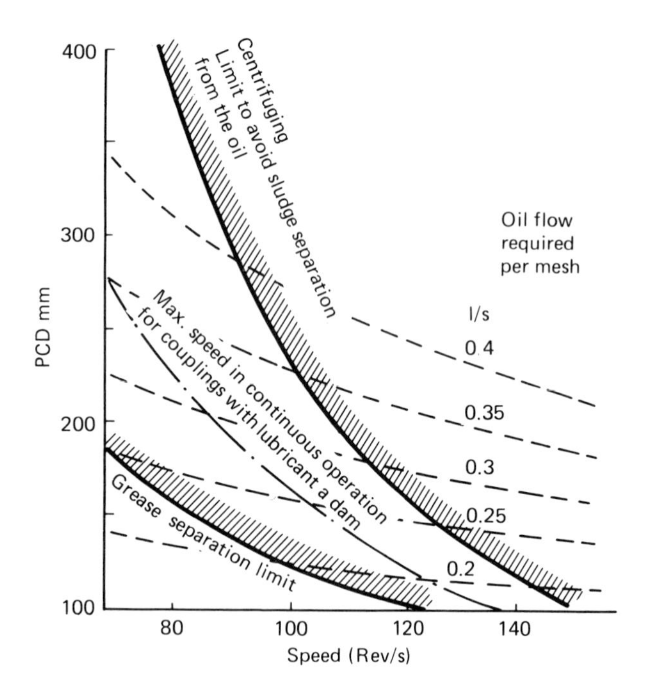

Figure 4.5 The lubrication requirements of gear couplings

Note: Damless coupling designs, with continuous lubricant feed to each tooth must be used for applications above the oil centrifuging limit and it is recommended that they should also be used above the chain dotted line where long periods of unattended reliable operation are required.

Flow rates indicated are for this type of design.

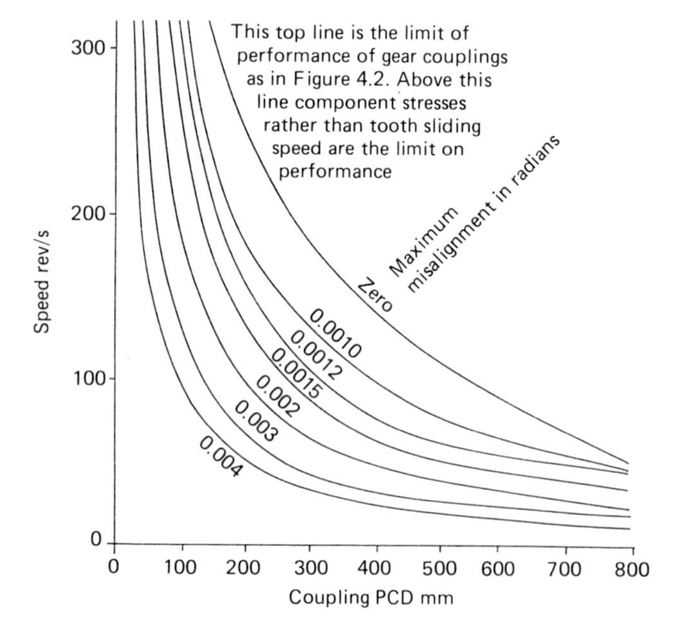

Figure 4.6 The maximum misalignment at a gear coupling mesh to avoid excessive tooth wear

Note

1. 0.001 radians of angular misalignment is equivalent to 0.001 inches per inch or 1 mm per metre misalignment across the mesh.
2. The maximum angular misalignments given are for continuous misaligned operation. If the misalignment is only transient, values up to 1.6 times greater are permissible.
3. The lines are plotted for a constant tooth sliding velocity of 0.12 m/s

COUPLING EFFECTS

Effect on critical speeds

Since couplings are fitted at the end of machine shafts, they constitute an overhung mass. Overhung masses reduce the lateral critical speed of rotors. If a machine is operating near its critical speed the overhung mass of the coupling needs to be considered.

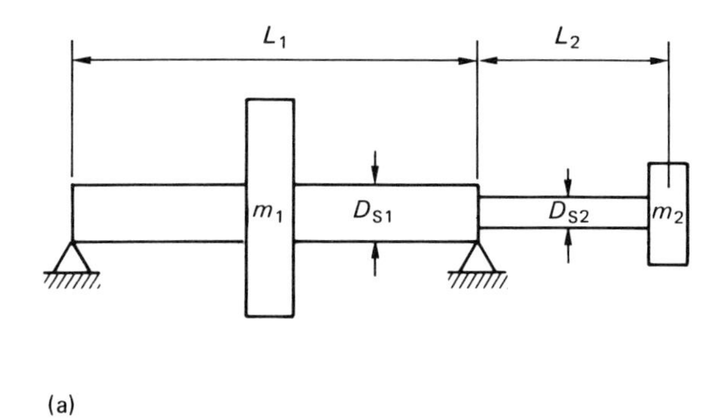

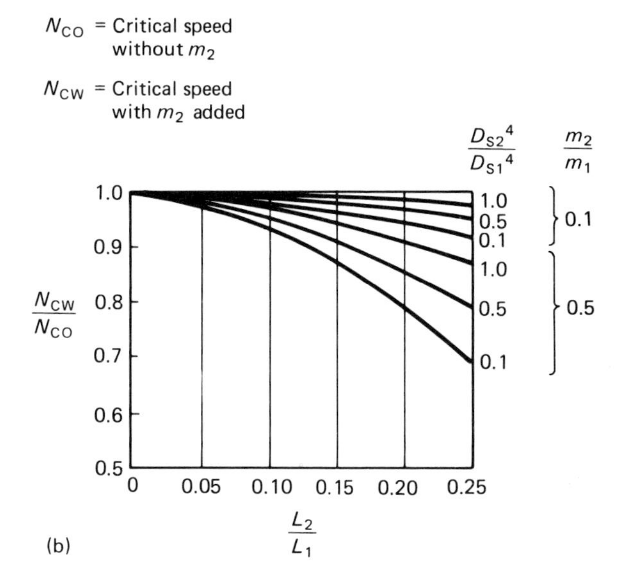

Gear coupling effects

The axial loads that may be applied by gear couplings to the coupled machines can be estimated from the tooth contact forces generated from the torque transmission multiplied by the likely coefficient of friction. This will normally have a value of about 0.15 but if the surface of the teeth becomes damaged could rise to 0.3.

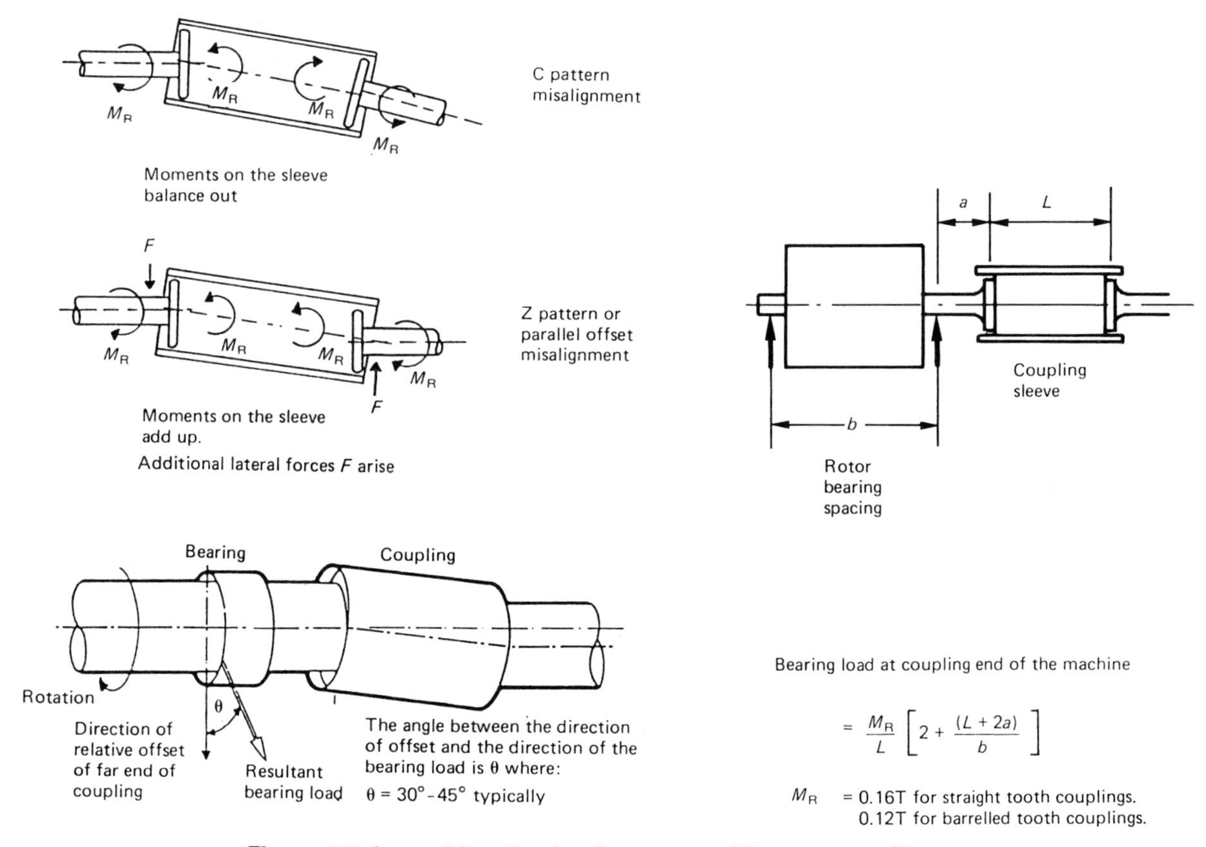

Figure 4.7 Lateral bearing loads generated by gear couplings

DESIGN AND OPERATION

Many machine systems require clutches which can pick up the drive to a machine that is already rotating or alternatively can release the drive when another driver takes over. For high power applications friction clutches tend to be large and generate heat due to slipping before synchronisation. Simple free wheels or sprag clutches are unsuitable for high torques because of their small driving contact areas.

High power overrunning clutches need to provide positive engagement and achieve a large driving surface area by transmitting torque through concentric internal and external teeth, generally gear teeth. When the clutch is disengaged these concentric teeth are separated axially.

A synchronising self-shifting clutch is shown in Figure 5.1. This has a pawl and ratchet mechanism to sense synchronism between the input and output shafts and to align the teeth which are then shifted into mesh by the small torque applied through the pawls to helical splines. Conversely reverse torque on the splines causes the teeth to disengage.

Powers of up to 300 MW have been transmitted by these clutches and the limiting factor in their design is centrifugal stresses in the outer geared rings.

High speeds in both the overrunning and engaged modes can be achieved (up to 15000 rev/min) because the driving teeth of the clutch are separated axially when the clutch is disengaged and because the pawl and ratchet mechanism can be designed for such high relative speeds without wear.

Clutches can be mounted directly between two shafts, in a separate casing or in a gearbox.

Oil lubrication is required for the clutch teeth and pawls to prevent wear and corrosion. Lower power and speed (e.g. 1 MW at 3000 rev/min) clutches are usually self-contained units with an integral oil system. Higher powers and speeds require a force fed oil supply, this is normally arranged to be common to the other machines in the system.

When the clutch is disengaged with the output side at high speed there will be a drag torque on the input shaft due to oil viscosity effects that may tend to keep the input machine rotating continuously at low speed. This can usually be accepted, if not a brake can be fitted to the clutch casing.

When the clutch teeth are engaged there is a strong centring effect between the input and output parts. If the input and output parts are always in good alignment (e.g. in a gearbox) this centring effect is acceptable, a semi-rigid clutch of this type is shown in Figure 5.2.

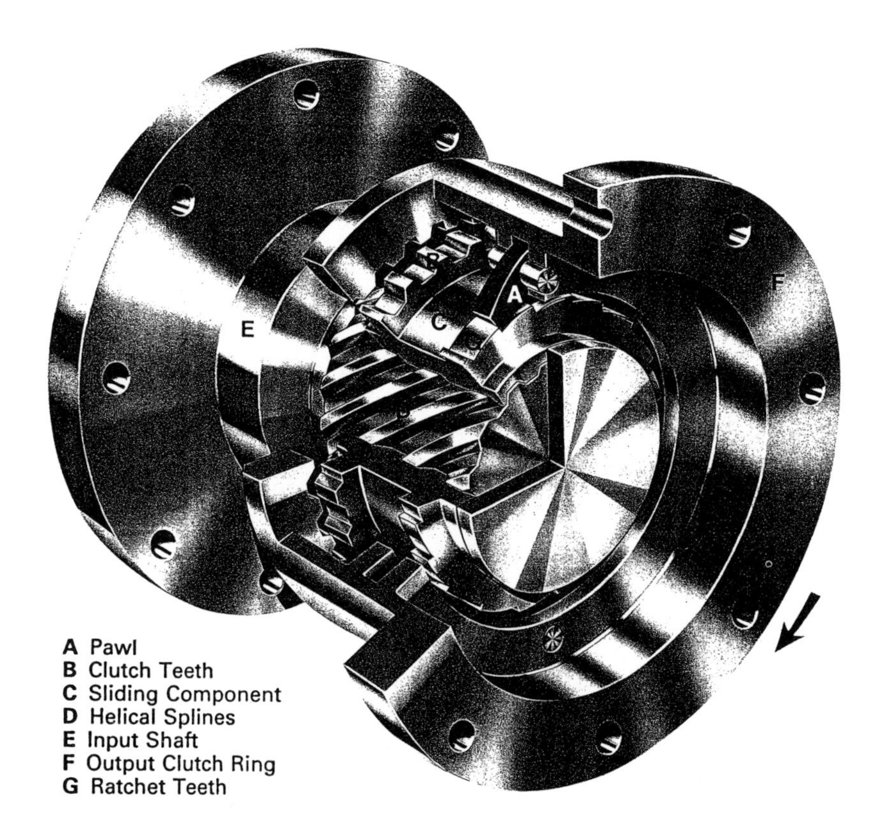

Figure 5.1 A high power clutch with axial engagement activated by helical splines and a ratchet and pawl mechanism

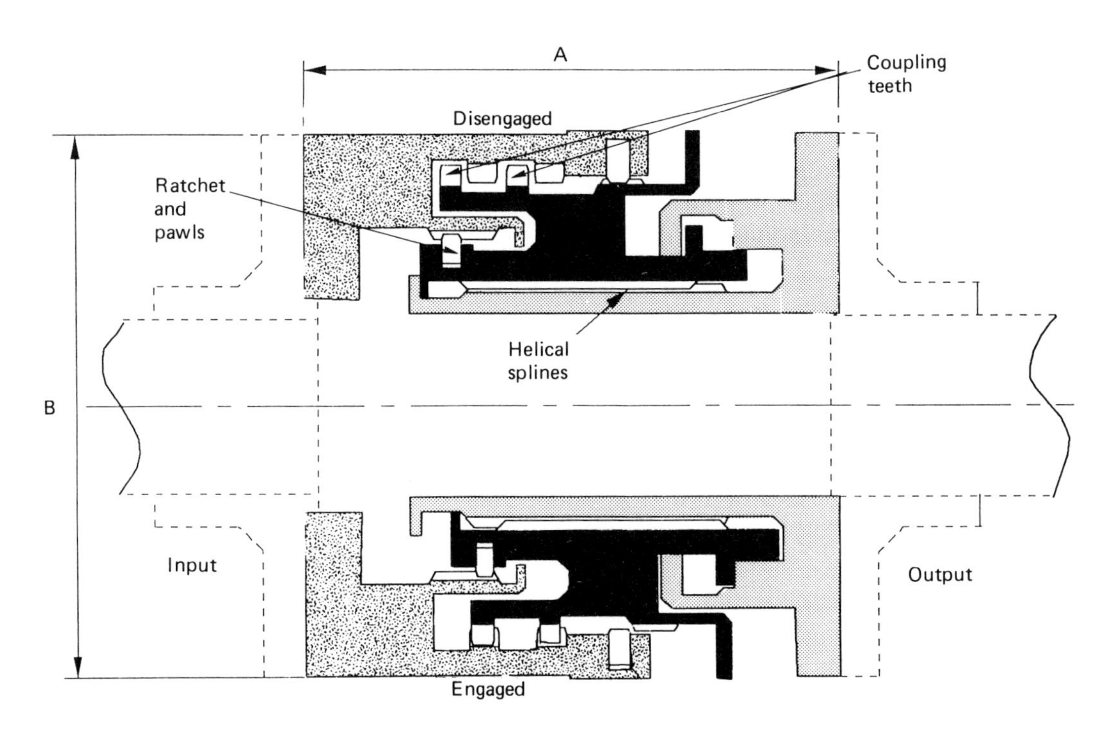

Figure 5.2 A high power semi-rigid clutch in the disengaged and engaged position

Clutches of this type, as shown in Figure 5.2, have typical dimensions and weights as shown in Table 5.1.

Table 5.1 Typical dimensions and weights of synchronising self shifting clutches

Torque capacity	*A*	*B*	*Wt*
40 kNm	360 mm	375 mm	150 kg
60	330	440	210
100	290	510	280
160	370	600	420
380	580	840	1125
1000	720	1000	2750

If there is misalignment between the input and output sides of the clutch (e.g: when a clutch is installed between separate machines such as a fan and a steam turbine) the clutch must be designed to accept this misalignment. Such a clutch acts like a spacer flexible coupling. A spacer clutch of this type is shown in Figure 5.3.

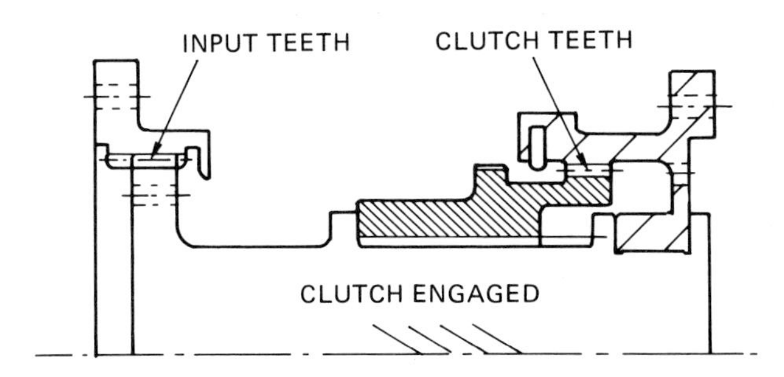

Figure 5.3 A spacer clutch which can also act as a flexible coupling between two shafts

The clutch can also be designed to accept changes in axial length, as may be necessary due to the thermal expansion of a generator system.

In dual drive systems one driver can be stopped and maintenance can be carried out on that machine whilst the driven machine continues to run, driven by the other driver. To ensure the safety of operators the input side of the clutch must be held stationary by a maintenance lock which can be mounted within the clutch casing.

Clutches can be fitted with an internal thrust bearing between the input and output sides so that the clutch is a fixed length and positions the output machine from the input or vice versa. Such a bearing is usually only subjected to high thrust forces when the clutch is engaged and in this condition there is no relative rotation across the bearing so it has a high thrust capacity.

Clutches can be fitted with a lock so that once engaged and locked they do not disengage and hence can transmit reverse torque (e.g. in a marine reversing drive). The lock requires a control system to lock and unlock the clutch and this is normally arranged through a hydraulic servo interlinked with the plant control system. The lock can be preselected before clutch engagement so that the clutch locks immediately it engages.

The clutches can be fitted with a lock-out so that when selected the clutch will not engage (e.g. in a ship where an engine must be tested without driving the propeller). The lock-out again requires a control system.

Clutches can be arranged so that they only engage at high speeds for example in a steam turbine generator where the steam turbine must turn slowly forward for cooling without engaging the clutch and turning the generator.

APPLICATIONS

Table 5.2 Applications and operating conditions of self synchronising clutches

Industrial drives		*Powers*	*Speeds*
Fan drive	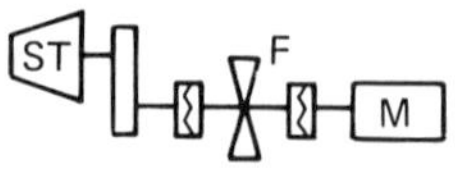 ST F M	500 kW	1500 rev/min
Pump drive	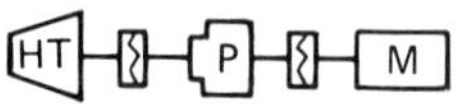 HT P M	1000 kW	3000 rev/min
Compressor drive	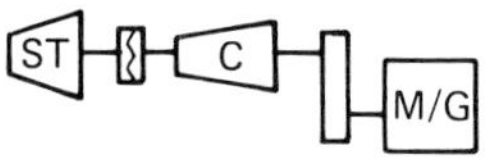 ST C M/G	3000 kW	6000 rev/min
Power generation			
Dual driven generator	 M/G	2 × 50 MW	3000 rev/min
Combined cycle	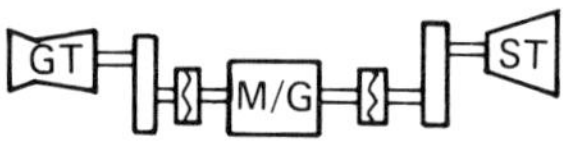 GT M/G ST	100 MW/50 MW	3000 rev/min
Synchronous condensing	 GT M/G	170 MW	3000 rev/min
Auxiliary drives			
Steam turbine barring	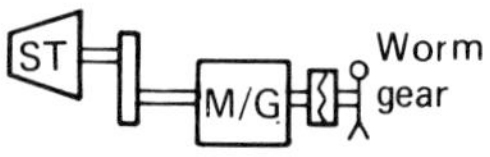 ST M/G Worm gear	20 kW	10 rev/min
Gas turbine starting	 GT M/G Starter	200 kW	5000 rev/min
Marine drives			
Combined diesel or gas turbine propulsion (CODOG)	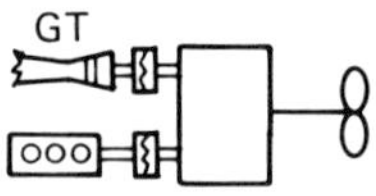 GT	25 MW 5 MW	3600 rev/min 1200 rev/min

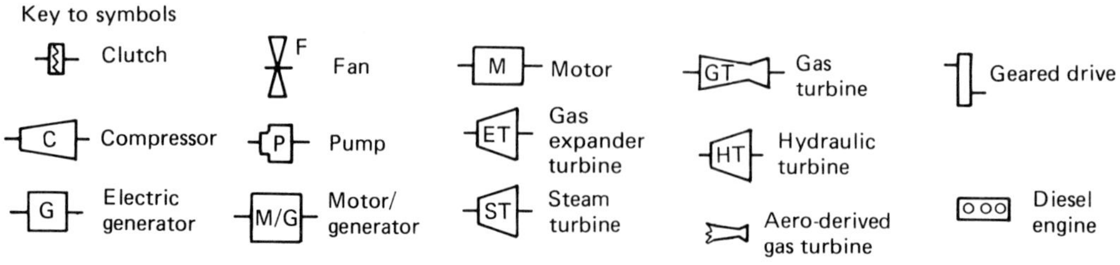

Key to symbols

- Clutch
- F Fan
- M Motor
- GT Gas turbine
- Geared drive
- C Compressor
- P Pump
- ET Gas expander turbine
- HT Hydraulic turbine
- G Electric generator
- M/G Motor/generator
- ST Steam turbine
- Aero-derived gas turbine
- Diesel engine

A freewheel or one-way clutch is a device which transmits torque in one direction, and disengages or freewheels in the other direction. The most familiar example is the bicycle freewheel which enables the cyclist to stop pedalling at will.

Numerous applications occur in small machinery. Two-speed drives can be designed with two motors of different speeds, coupled to the driven machine by one-way clutches. Indexing mechanisms using one-way clutches can convert reciprocating motion into unidirectional rotation. Uphill conveyors can be prevented from rolling back using a one-way clutch as a backstop.

Table 6.1 Types of one-way clutch

Type	*Description*	*Special characteristics*
Ratchet and pawl 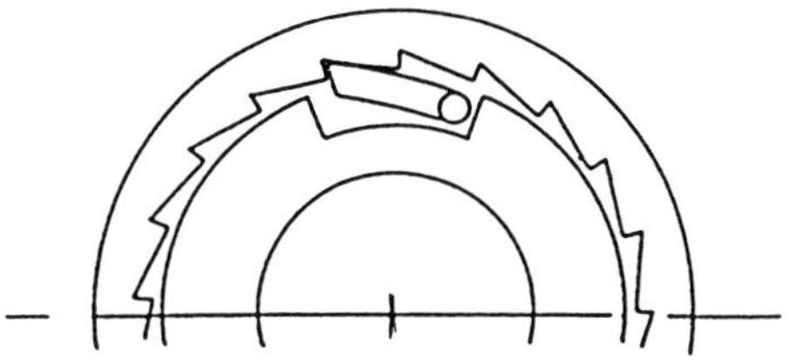	Hard steel spring loaded pawls engage with ratchet teeth. (Springs not shown).	Backlash equal to ratchet tooth spacing. Makes clicking sound and suffers wear during freewheeling. Low precision, low cost. Plastic versions with axial configurations are possible.
Locking roller 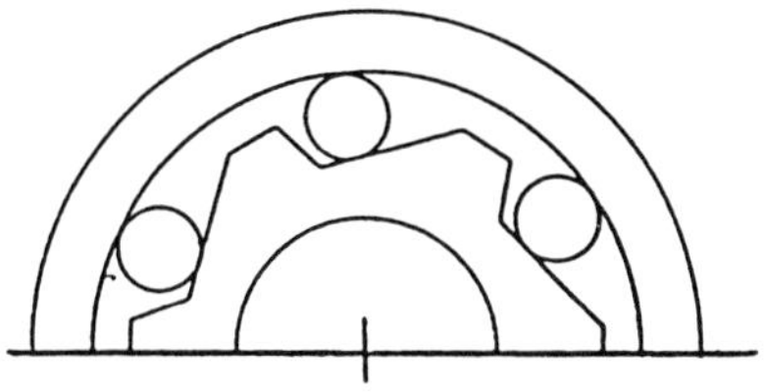	Rollers ride up ramps and drive by wedging in place. Spring bias to engage. (Springs not shown).	Low cost because made by modifying roller bearings.
Locking needle roller 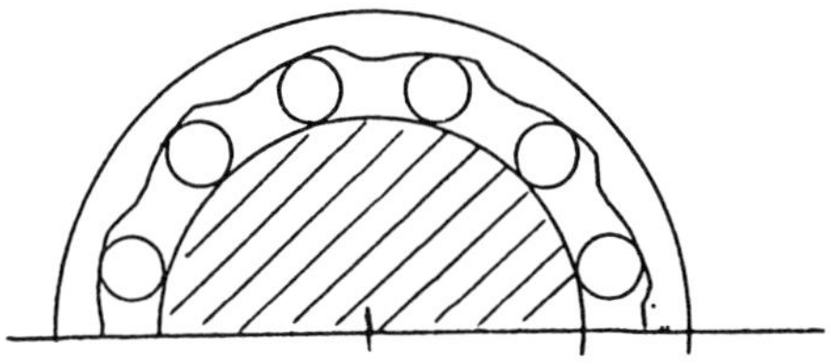	As for locking roller, but ramps are in the outer member. (Springs not shown).	Can operate directly against hardened shaft. Very compact unit, and low cost.
Sprag clutch 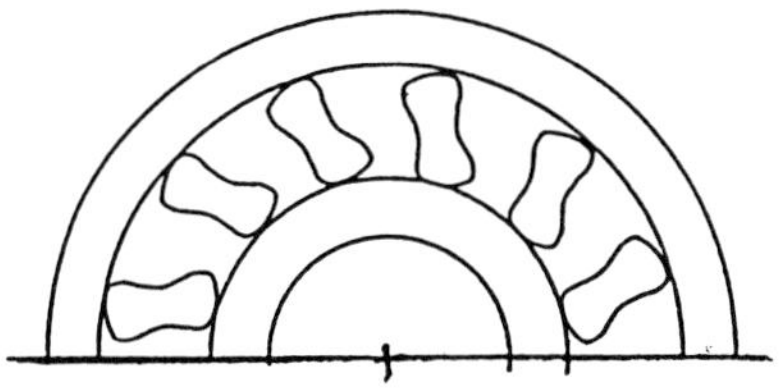	Profiled elements restrained by a cage. Jam in place to drive. Spring bias to engage. (Springs not shown).	Highest torque capability of all types. High cost. Can eliminate wear when freewheeling by means of centrifugal lifting of elements.
Wrap spring 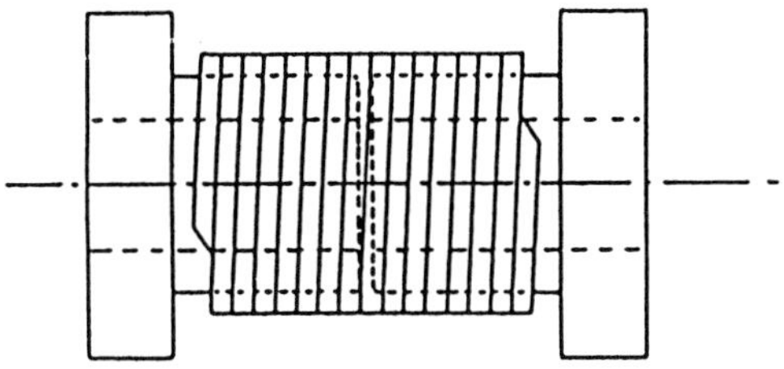	Friction between hubs and coil spring causes spring to tighten onto hubs and drive in one direction. In opposite direction spring unwraps and slips easily.	Can be made very small. Potentially low cost but commercially available units are expensive.

Table 6.2 Characteristics of one-way clutches

Type	*Max torque*	*Max. freewheeling speed*	*Indexing accuracy*	*Cost*
Ratchet and Pawl	10^3 Nm	low	poor	low
Roller clutch	5×10^4 Nm	moderate	good	low
Needle roller clutch	10^2 Nm	high	good	low
Sprag clutch	10^6 Nm	high	good	high
Wrap spring clutch	5×10^2 Nm	low	moderate	low/moderate

The above characteristics are for commercially readily available units. Larger, higher torque versions of all types are possible as special designs. For very high torques and power a better alternative is a self-synchronising clutch, described in section 5.

The graph below shows the approximate relationship between torque capacity and freewheeling speed for the various types.

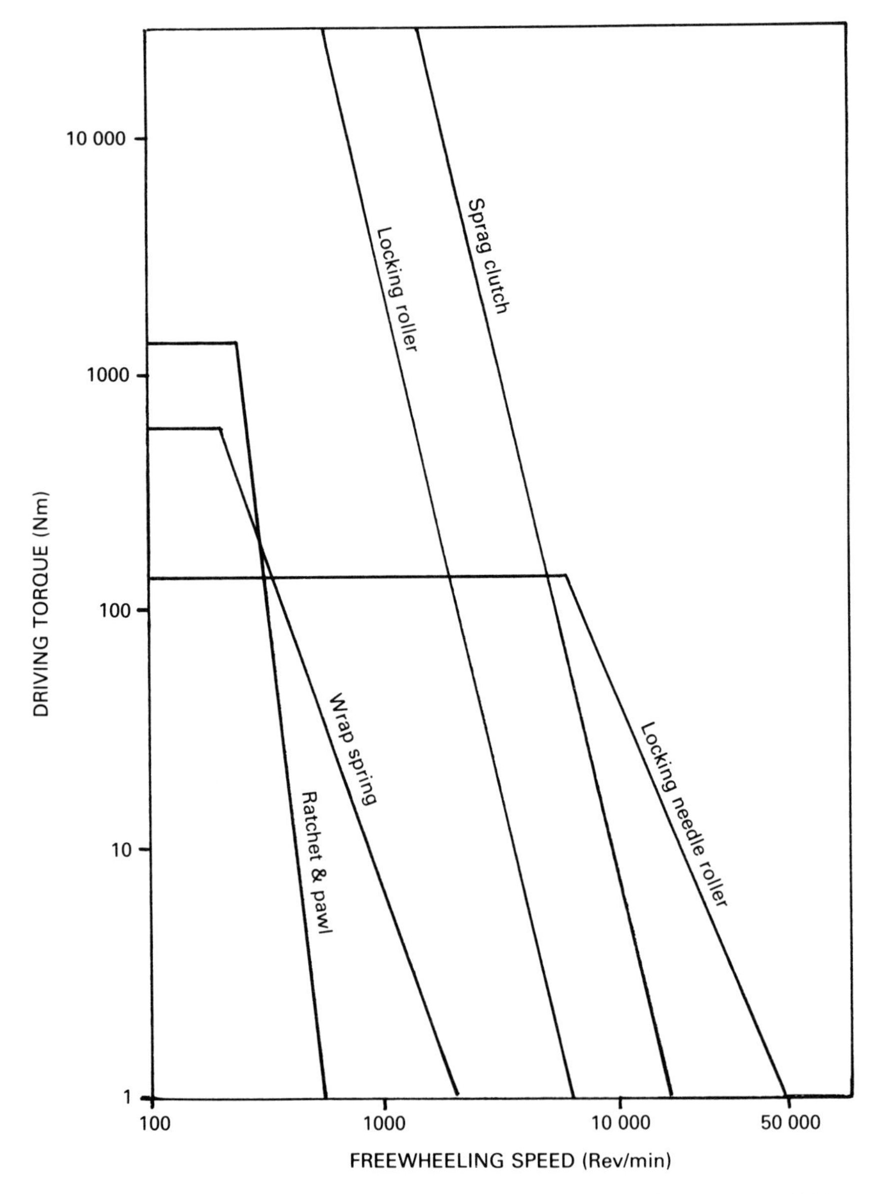

Figure 6.1 Approximate torque and freewheeling speed limitations for commercially available clutches

CLUTCH SELECTION

A clutch is used to transmit motion from a power source to a driven component and bring the two to the same speed. Once full engagement has been made, the clutch must usually be capable of transmitting, without slip, the maximum torque that can be applied to it. The operating characteristics of the various clutch designs, and the requirements of the application, can be used as a guide to the selection of an appropriate type of clutch.

Table 7.1 Types of friction clutch

Type of clutch	*Figure No.*	*Special characteristics*	*Typical applications*
Cone type	7.1, 7.2	Embodies the mechanical advantage of the wedge which reduces the axial force required to transmit a given torque. It also has greater facilities for heat dissipation than a plate clutch of similar size and so may be more heavily rated	In general engineering its use is restricted to more rugged applications such as contractors' plant. Machine-tool applications include feed drives, and bar feed for auto. lathes. (Figure 7.2)
Single-plate (disc)	7.3	Used where the diameter is not restricted. Coil or diaphragm springs usually provide the clamping pressure by forcing the spinner plate against the driving plate. Simple construction, and if of the open type ensures no distortion of the spinner plate by overheating	Wide applications in automobile and other traction drives. Figure 7.5 (a and b) show alternative operating methods
Multi-plate	7.4	Main feature is that the power transmitted can be increased by using more plates, thus allowing a reduction in diameter. The spline friction should be minimised to ensure that clamping losses are small, and that the working rates of all plates are as uniform as possible. If working in oil,* it must be enclosed, whereas a dry plate clutch can often have circulating air to carry away the heat generated	Extensively used in machine tool head-stocks, or in any gearbox drive where space is limited between shaft centres. Figure 7.5(c) shows an operating method
Expanding ring or band	7.6, 7.7	Will transmit high torque at low speed. Centrifugal force augments gripping power, so withdrawal force must be adequate. Both cases show positive engagement	Large excavators. Textile machinery drives. Machine tools where clutch is located in main driving pulley
Centrifugal	7.8, 7.9	Automatic in operation, the torque without spring control increasing as the square of the speed. An electric motor with a low starting torque can commence engagement without shock, the clutch acting as a safety device against stalling and overload. Shoes are often spring-loaded to prevent engagement until 75% of full speed has been reached	Wide applications on all types of electric motor drives, generally reducing motor size and cost. Industrial diesel engine drives
Magnetic	7.10, 7.11, 7.12	Units are compact, and operate by a direct magnet pull through switch engagement. No end thrust is transmitted, and centrifugal force has no effect on drive. Stock clutches are generally wound for 24 V d.c. with powers up to about 37 kW at 100 rev/min. The response curves, (Figure 7.12) show typical torque and voltage time charts	Drives with automatic speed-changing systems requiring remote control. Machine tool gearboxes. N.C. machine tools. Tracer control systems for copying
Hydraulic	7.13	Clutches can be incorporated into automatic machine cycles by remote control of solenoid valves. They give high torque with minimum size, i.e. compared with magnetic clutches identical dimensions result in a torque ratio of 3:1. Maintenance low because of working in oil, and piston stroke compensates for wear	Marine reversing drives. Excavators. Diesel turbine drives. Winch drives. Presses
Pneumatic		Function similar to that of hydraulic clutch, producing axial pressure by cylinder and piston. General design is usually of the multi-disc type	Crank presses. Flying shear. Rolling mill drives. Cranes. Hammers

* Working in oil gives a reduction in friction, but this can be counteracted by higher operating pressures. As long as there is an oil film on the plates, the friction and the engagement torque remain low, but as soon as the film breaks the engagement torque rises rapidly and may lead to rapid acceleration. The friction surface pressure should not exceed 1 MN/m^2 with a sliding speed maximum of 20 m/s for steel on steel. With oil immersed clutches having steel and sintered plates, the relationship between static and dynamic coefficient of friction is more favourable. Friction surface pressure and sliding speed may then be up to 3 MN/m^2 and 30 m/s.

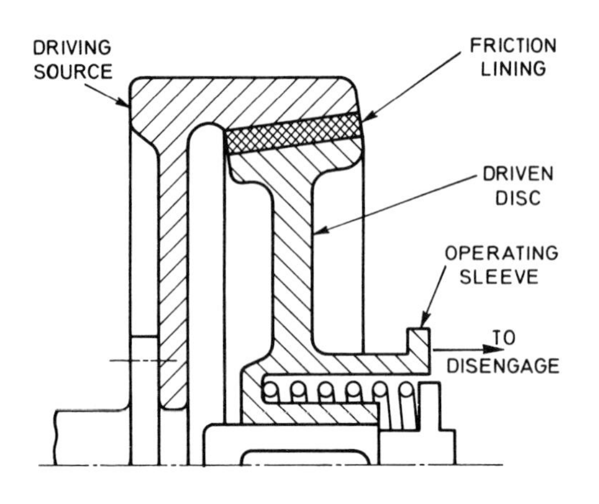

Figure 7.1 Cone-type friction clutch

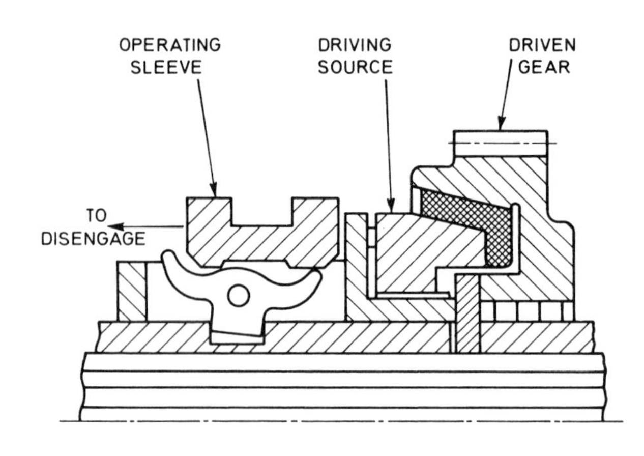

Figure 7.2 Cone clutch used on automatic lathe

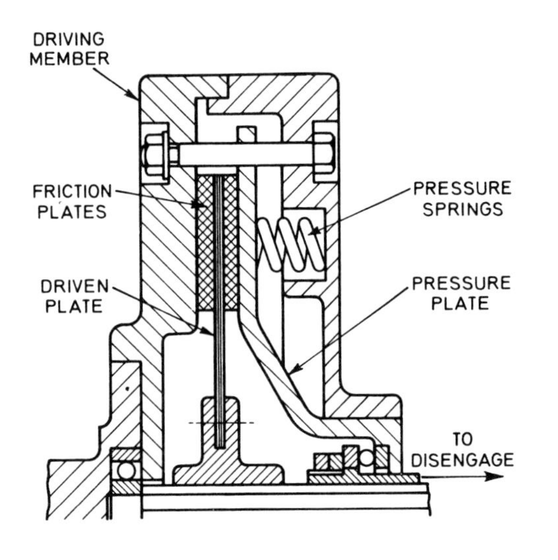

Figure 7.3 Construction of single-plate clutch

PRESSURE TOGGLES

OPERATING SLEEVE

FRICTION PLATES

DRIVEN MEMBER

TO DISENGAGE

DRIVING SLEEVE

Figure 7.4 Design of multi-plate clutch

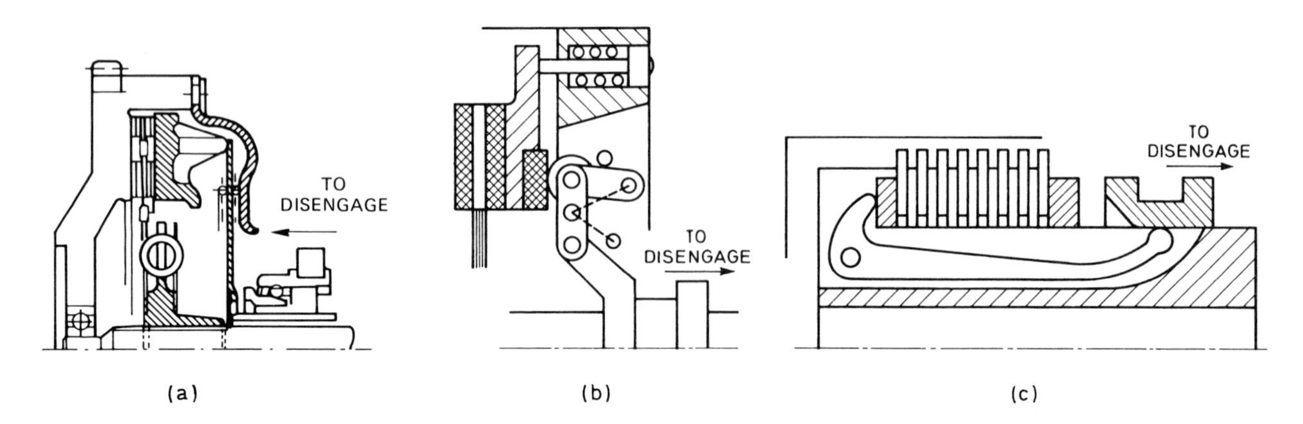

***Figure 7.5 Alternative operating methods**: (a) diaphragm spring clutch; (b) over-centre clutch, suitable for long engagement periods—no external force is required once clutch is engaged; (c) pivoted lever design for engagement of multi-plate clutch*

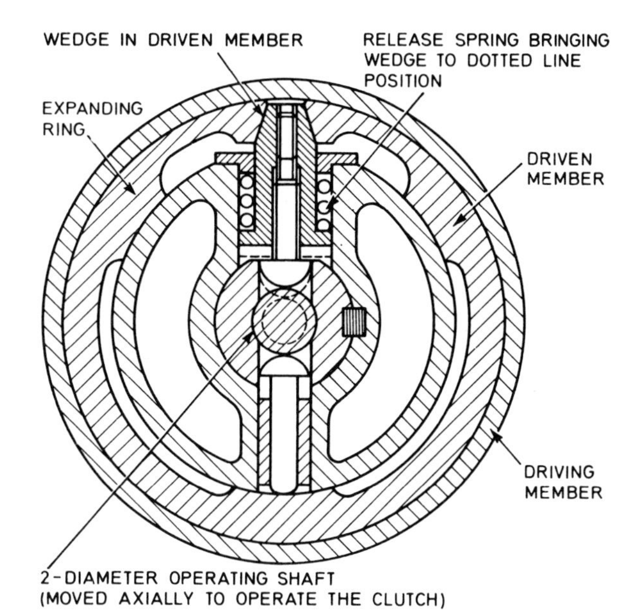

Figure 7.6 Expanding ring type of friction clutch

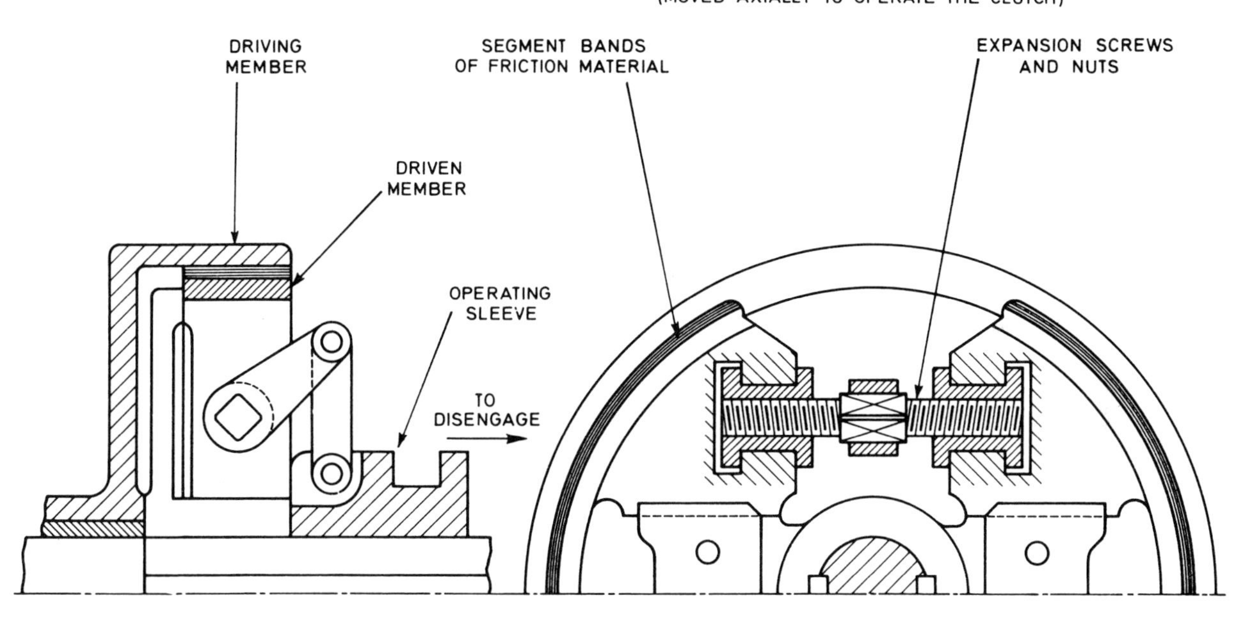

Figure 7.7 Expanding band friction clutch

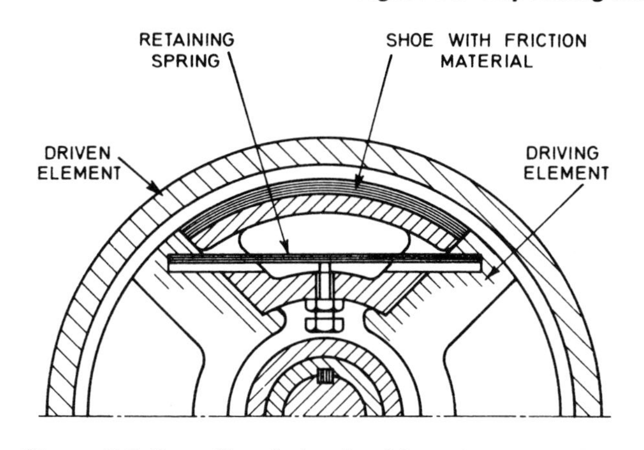

Figure 7.8 Centrifugal clutch with spring control

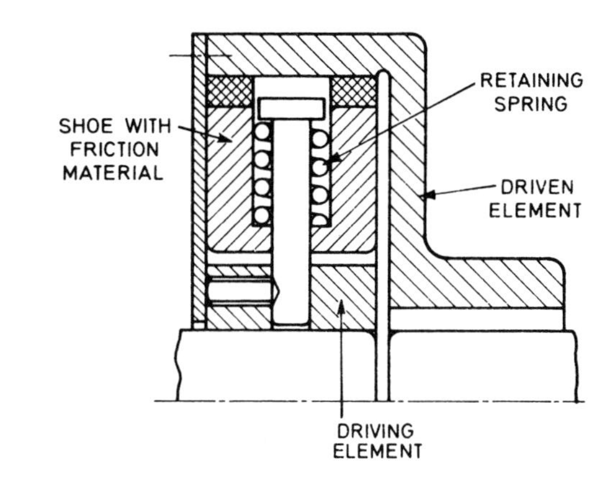

Figure 7.9 Light-duty type of centrifugal clutch

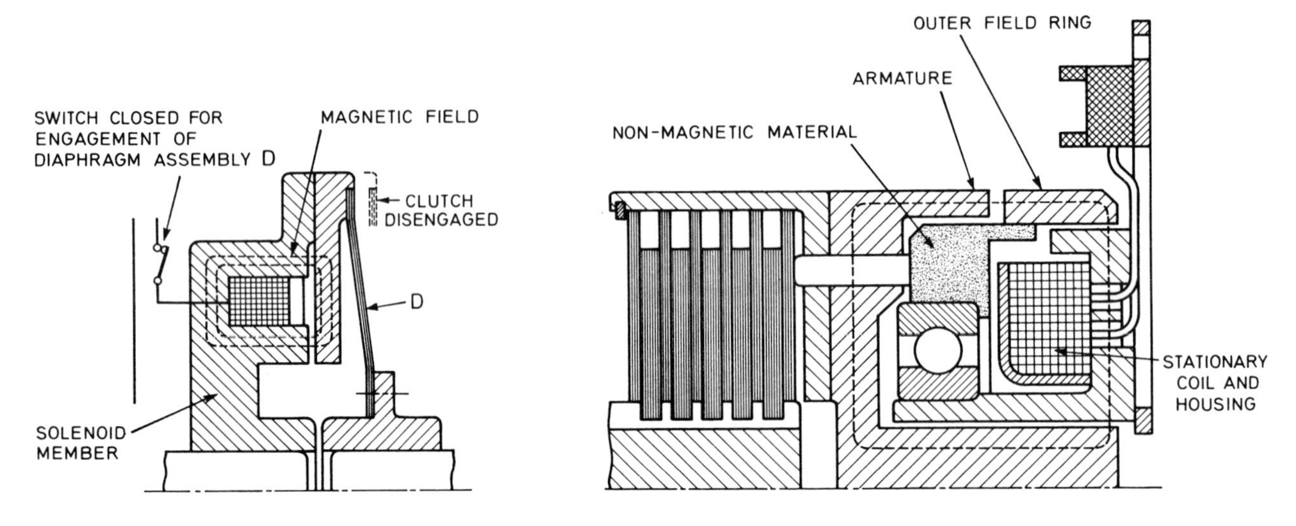

***Figure 7.10** Diaphragm type of magnetic clutch*

***Figure 7.11** Multi-plate type of magnetic clutch*

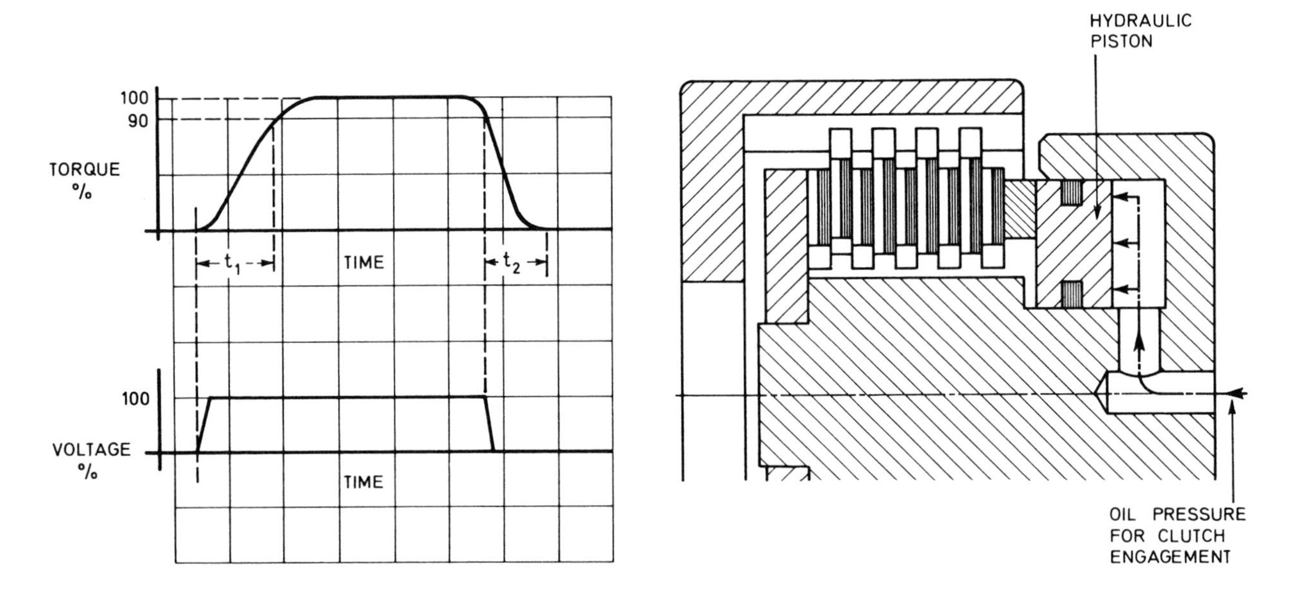

***Figure 7.12** Response curves of magnetic clutch*

***Figure 7.13** Diagram showing operation of hydraulic clutch*

DESIGN OF CLUTCHES

Choice of lining area

The total lining area required, A, may be determined from: $A = H/K_l$, where H is the power to be dissipated, and K_l is the duty rating of the lining material.

The power to be dissipated, H, may be determined from the energy to be imparted to the driven component in the time of engagement or, if the transmitted torque is considered to be constant, from the transmitted torque and the mean relative speed during engagement.

When the total area is known, the detail dimensions of the linings can be readily calculated for plate or drum clutches. For cone clutches, the developed shape of the lining is shown in Figure 7.14 and can be calculated from

$$R = \frac{D}{2 \sin \theta}$$

$$\beta = 360° \sin \theta$$

$$A = 2R \sin \frac{\beta}{2}$$

$$B = R - (R - b) \cos \frac{\beta}{2}$$

The required service life and the wear rate and allowable wear of the lining are also important, particularly in determining the lining thickness.

Typical lining arrangements, contact pressures and suggested duty ratings for lining materials are given in Table 7.2.

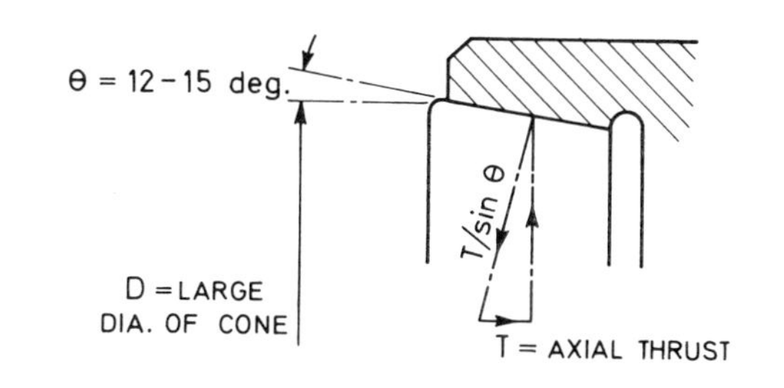

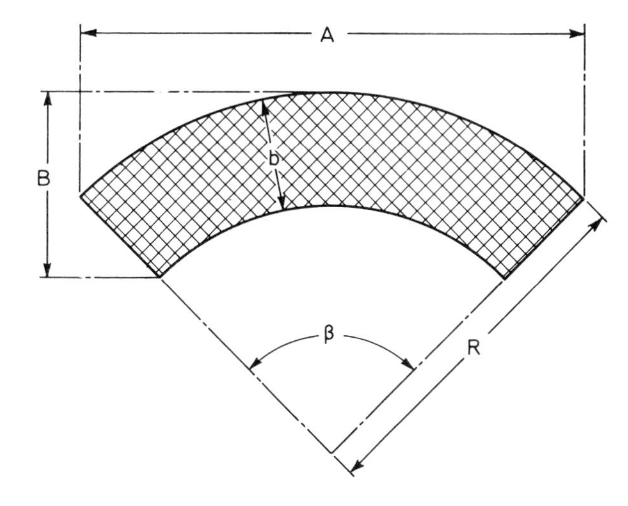

Figure 7.14 Development of liner for cone clutch

Table 7.2 Guide to design and material selection for various types of clutch

Type of friction device	*Self-energisation and sensitivity*	*Angle of lining arc*	*Pressure range* MN/m^2	*Material*	*Severity of the application*	*Suitable duty rating of lining,* K_l W/mm^2	*Coefficient of friction*
Two leading shoe drum clutch	High	100° per shoe	0.14–1.0	Moulded or woven	High	0.24	0.30
					Medium	0.58	0.35
					Low	1.75	0.40
Band clutch operating with drum rotation	Very high	270°	0.07–0.7	Rigid moulded segments, flexible moulded or woven	High	0.24	0.25
					Medium	0.58	0.30
					Low	1.15	0.35
Plate clutch operating under dry conditions	None	360°	0.07–0.4	Rigid woven or moulded	High	0.12	0.30
					Medium	0.24	0.35
					Low	0.58	0.40
Plate clutch operating under oil immersed conditions	None	360°	0.28–2.8	Rigid woven or moulded	High	0.15	0.06
					Medium	0.29	0.08
					Low	0.70	0.10
Cone clutch	None	360°	0.05–0.2	Moulded or woven	High	0.24	0.25
					Medium	0.35	0.30
					Low	0.80	0.35

Lining design for oil-immersed clutches

For working in oil, multi-plate clutches are suitable. Oil acts as a cushion and energy released by heat is carried away by oil. The main disadvantage is a reduction in friction, but this can be counteracted by higher operating pressures. As long as there is an oil film on the plates, the friction characteristic and engagement torque remain low, but as soon as the film breaks the engagement torque rises rapidly and may lead to rapid acceleration. The friction surface pressure should usually not exceed 1 MN/m^2 with a sliding speed maximum of 20 m/s, steel on steel. With oil-immersed clutches having steel and sintered plates the relationship between the static and dynamic coefficient of friction is more favourable. Friction surface pressure and sliding speed may be up to 3 MN/m^2 and 30 m/s.

Facing grooves and anti-distortion slots

The most common facing groove is a single- or multi-lead spiral. This helps to prevent the formation of an oil film, which, if formed, would lower the coefficient of friction. It also provides space for the oil to be dispersed during clutch engagement. Spiral grooves are between 0.6 and 1.5 mm wide and 0.2–1 mm deep, depending on diameter and face thickness. The pitch is between 1.5 and 6.0 mm, depending on the size of the disc.

Figure 7.15 shows an anti-distortion slot in a sintered metal facing disc. Two widths of slots are frequently used, 4 or 5 mm wide. The slots minimise distortion and warping. Thermal expansion of the sintered metal and backing plates from absorption of heat causes expansion and contraction with temperature variations. The slots permit dimensional changes without buckling or dishing the backing plate. External and internal slots are used as in Figure 7.16. They number from 4 to 8 of each.

On moulded and woven fabrics the slots are normally 2.5–3.5 mm wide and 1–1.5 mm deep. Eighteen to twenty-four equi-distant slots or sets of three are usual. Thermal expansion slots are not required for these materials.

Clutch plates (Figure 7.17) are also available, which present a sine wave cross-section along a periphery, i.e. a corrugation, which generates a spring action. During engagement the pressure on the friction surface increases until the sine wave becomes a flat surface. The spring action ensures positive declutching and only line contact results in the disengaged position, thus ensuring minimum torque and heating when idling.

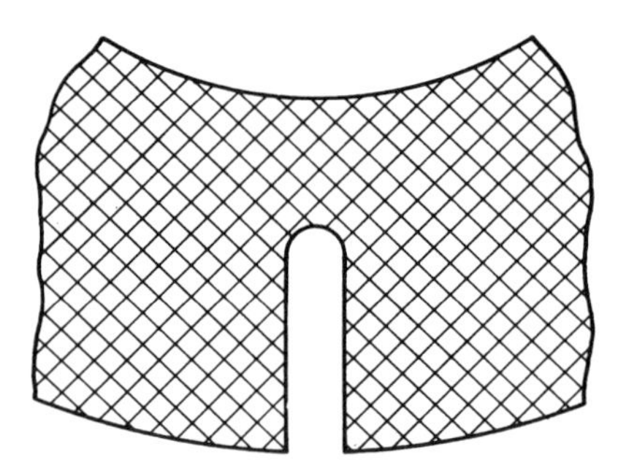

Figure 7.15 Anti-distortion slot in friction plate

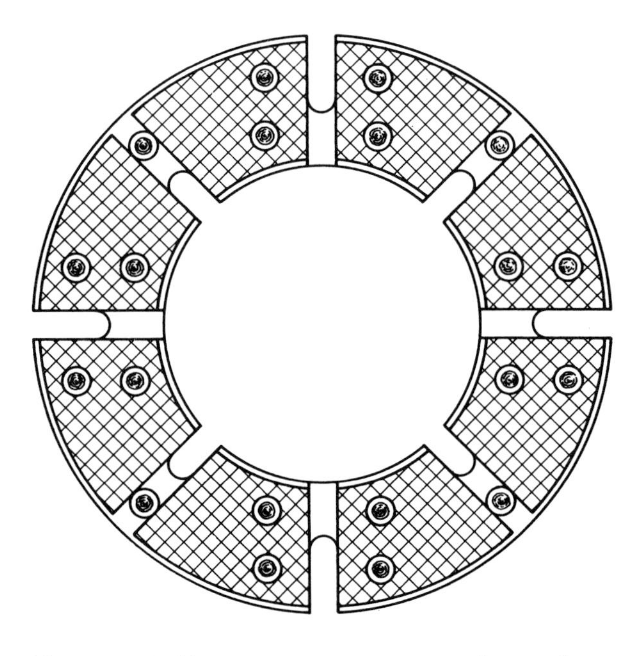

Figure 7.16 Slots to counteract expansion and contraction of plates

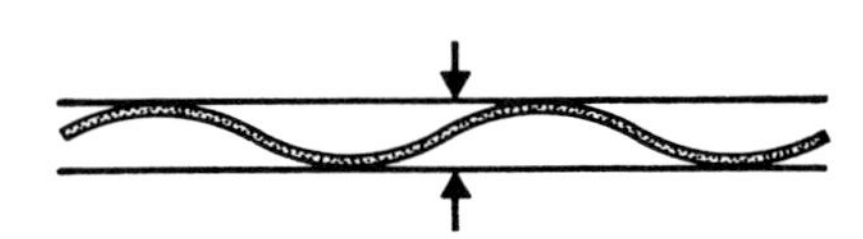

Figure 7.17 Friction plates with sine wave cross-section

FITTING OF FRICTION LININGS

Copper or brass semi-tubular rivets are used for the attachment of the majority of the linings. One manufacturer uses brass containing 70% copper in 150° head semi-tubular rivets, as shown in Figure 7.18. The recommended dimensions and lining area/rivet are as follows:

Lining thickness (mm)	4.8	6.35	9.5	12.7	19.0
Rivet shank dia. (mm)	4.0	4.8	6.35	8.0	9.5
Lining area/rivet (mm^2)	1900	2300	3600	4500	6500

With riveting, some lining area is lost to rivet holes, and up to a third of the thickness is used to accommodate rivet heads, thus reducing wear life. Friction clutch facings, particularly those used on cone and band clutches, can be bonded to the metal carrier using proprietary adhesives and techniques (contact the manufacturers). Bonded facings have the advantage that all the friction material can be worn away.

Some precautions to observe when lining cone clutches are shown in Figure 7.19.

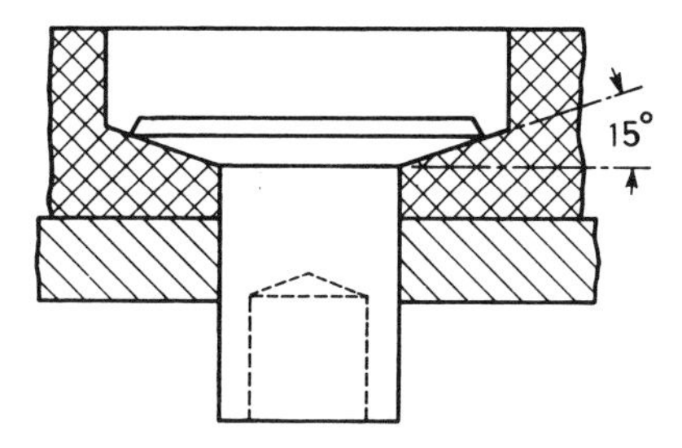

Figure 7.18 Type of rivet and fastening for clutch linings

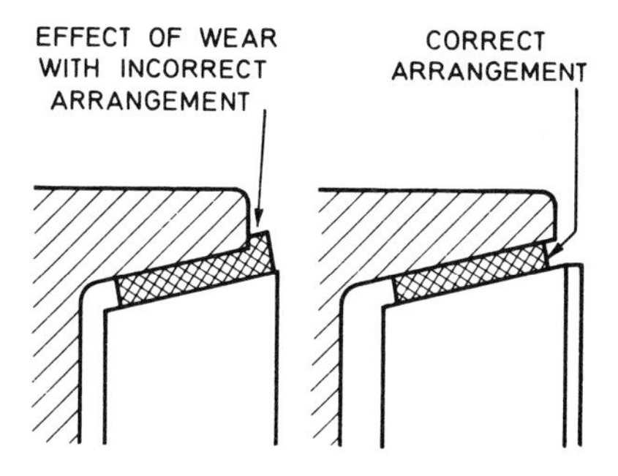

Figure 7.19 Precautions to take when lining cone clutches

LINING MATERIALS

Impregnated woven cotton based linings are used to obtain high friction, but the maximum operating temperature is limited to that at which cotton begins to char (100°C), therefore asbestos and non-asbestos fibres have replaced cotton for applications where greater heat resistance is required. The fibres are woven to produce a fabric which is impregnated with a resin solution and cured. Zinc or copper wire is often introduced to increase thermal conductivity. Asbestos and non-asbestos moulded friction materials consist basically of a cured mix of short asbestos or other fibres and bonding resins and may also contain metal particles.

Asbestos or non-asbestos tape or yarn can be wound into discs and bound together using resin or rubber compounds.

Sintered metals are used for a limited number of friction applications. The metal base is usually bronze, to which is added lead, graphite and iron in powder form. The material is suitable for applications where very high temperatures and pressures are encountered. It is rigid and has a high heat conductivity, but gives low and variable friction.

Information on the various lining materials is given in Tables 7.3, 7.4 and 7.5.

Mating surfaces

The requirements are: (1) requisite strength and low thermal expansion; (2) hardness sufficient to give long wear life and resist abrasion; (3) heat soak capacity sufficient to prevent heat spotting and crazing.

Close-grained pearlitic grey cast iron meets these requirements, a suitable specification being an iron with the following percentage additions: 3.3 carbon, 2.1 silicon, 1.0 manganese, 0.3 chromium, 0.1 sulphur, 0.2 phosphorous, 4.0 molybdenum, 0.5 copper plus nickel. Hardness should ideally be in the range 200–230 BHN.

Table 7.3 Clutch facing materials and their applications

Type	*Uses*
Woven	Industrial band, plate and cone clutches—cranes, lifts, excavators, winches, machine tools and general engineering applications
Asbestos/non asbestos wound tape and yarn	Mainly automotive and light commercial vehicles, agricultural and industrial tractors
Moulded	Automotive, commercial vehicles, agricultural and industrial tractors
Sintered	Tractors, heavy vehicles, road rollers, winches, machine tool applications
Cermet	Heavy earth-moving equipment, crawler tractors, sweepers, trenchers and graders
Oil immersed	
paper	Automotive and agricultural automatic transmissions
woven	Band linings and segments for automatic transmissions
moulded	Industrial transmissions and agricultural equipment
sintered	Power shift transmission, presses, heavy-duty general engineering applications

Facings are available in a wide range of sizes 75–610 mm outside diameter, and the designer should consult with the friction material manufacturer to determine which stock size would best fit his requirements.

Table 7.4 Friction, and allowable operating conditions for various clutch facing materials

Operating dry	μ	*Temperature, °C*		*Working pressure* kN/m^2	*Power rating* W/mm^2
		Maximum	*Continuous*		
LIGHT DUTY					
Woven	0.35–0.40	250	150	175–520	0.3–0.6
Millboard	0.40	250	150	175–700	0.3–0.6
MEDIUM DUTY					
Asbestos/non asbestos wound tape and yarn	0.40	350	200	175–700	0.3–1.2
Moulded	0.35	350	200	175–700	0.6–1.2
HEAVY DUTY					
Sintered*	0.36/0.30‡	500	300	350–2800	1.7
Cermet†	0.40			700–1400	4.0
OIL IMMERSED					
Paper	0.11			700–1750	2.3
Woven	0.08			700–1750	1.8
Moulded	0.04			700–1750	0.6
Moulded (grooved)	0.06				
Sintered	0.10/0.05‡			700–4200	2.3
Sintered (grooved)	0.10/0.06‡			700–4200	2.3

* Facings are sintered on to core plates or backing plates.
† Supplied as buttons in steel cups.
‡ First figure static, second dynamic coefficient.

Table 7.5 Typical physical and mechanical properties of clutch facings

	Resin-based materials	*Sintered metals*
Thermal conductivity	0.80 W/m°C	16 W/m°C
Specific heat	1.25 kJ/kg°C	0.42 kJ/kg°C
Thermal expansion	0.5×10^{-4}/°C	0.13×10^{-4}/°C
Specific gravity	1.6 for woven/wound 2.8 for moulded	6.0
Brinell hardness number	6–15	13
	MN/m^2	MN/m^2
Ultimate tensile strength	20	45
Ultimate shear strength	12	35
Ultimate compressive strength	100	150

If the maximum temperatures in the tables are exceeded the μ may fall badly. Any increase in temperature will increase the wear rate. Figure 7.20 shows how the wear of resin-based materials typically increases with temperature, taking the wear at 100°C as unity.

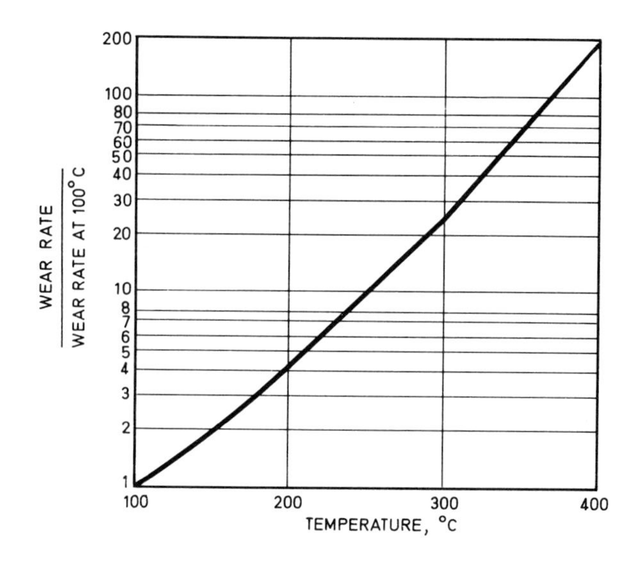

Figure 7.20 The effect of temperature on the wear rate of clutch facing materials

B8 Brakes

A brake has to develop the required torque in a stable and controlled manner, and must not reach temperatures high enough to impair its performance, or damage its components. There are three main types of brake: band brakes, drum brakes and disc brakes.

A brake is characterised by its Brake Factor which is defined as, for band and drum brakes, as the frictional force at the drum radius divided by the actuating force, provided it is the same on both shoes (for drum brakes). For disc brakes the brake factor $= 2\mu$ where μ is the coefficient of friction.

BAND BRAKES

A flexible steel band lined with friction material is tightened against a rotating drum. Because of its self-servo action a band brake can be made very powerful. Positive self servo occurs when the frictional force augments the actuating force so increasing the torque, that is, the brake has a high brake factor. The brake factor increases rapidly with μ and the angle of wrap δ, in the case of simple band brakes, as shown in Table 8.1.

Too much self-servo makes the brake unstable and likely to grab and judder (it is usual to work with $\delta = 270°$ and $\mu = 0.3$–0.4).

The relationship between drum diameter and torque capacity for band brakes of conventional proportions is shown in Figure 8.2.

If the drum rotates in the opposite sense to the actuating force (negative servo) the brake factor is very small, but by suitable design of the actuating mechanism the brake can be made equally effective for both directions of rotation.

As band brakes require small actuating forces they are generally suitable for manual operation, particularly when they are used only intermittently.

In band brakes, the width of the rubbing path is typically one fifth of the diameter.

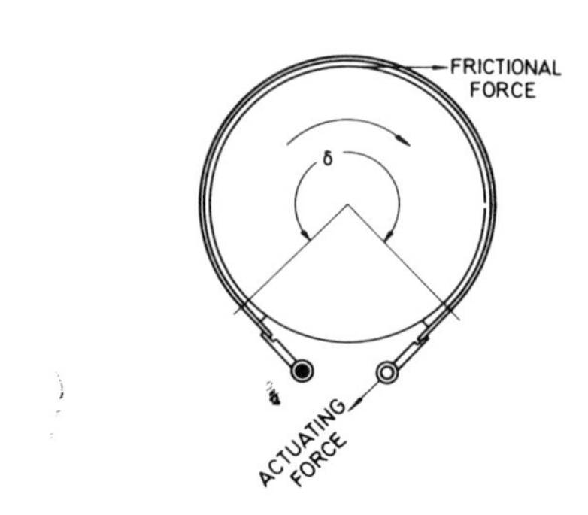

Table 8.1 Brake factor for different δ and μ

δ degrees	μ				
	0.1	0.2	0.3	0.4	0.5
210	0.44	1.08	2.00	3.33	5.25
240	0.52	1.31	2.51	4.34	7.13
270	0.60	1.57	3.11	5.59	9.55
300	0.69	1.85	3.81	7.12	12.72
330	0.78	2.16	4.63	9.00	16.82
360	0.87	2.51	5.59	11.34	22.16

Table 8.2 The various types of band brake

Type	*Brake factor* ($\mu = 0.3$)	*Uses*	*Type*
Simple	3.11 $\delta = 270°$	Winches, hoists, excavators, tractors, etc.	
Reversible	High; depends on lever ratio	As above but where the brake has to be equally effective in either direction	

Table 8.2 (continued)

Type	*Brake factor* (μ = 0.3)	*Uses*	*Type*
Screw-operated (reversible)	1.81 δ = 155° for each lining	Winches, reeling brakes for marine applications	
Differential	Very high; depends on lever ratio	Parking or holding brake only, as the brake is unstable	
Double wound	15.9 δ = 540°	In rugged installations such as oil-well equipment where operation is manual and precise control not required	

RIGID SHOE (DRUM) BRAKES

Shoes are lined with friction material, usually over an arc of 90–110 deg. Shoes may be leading (positive self-servo) or trailing (negative self-servo). By suitable combination of shoes, or by altering the geometry of the brake, the amount of servo required for the actuating force available, and the amount of stability required, can be obtained.

On most drum brakes the actuating force on each shoe is the same and the brake factor B is defined as:

$$B = \frac{\text{Total frictional force at drum radius}}{\text{Actuating force}} = \frac{P + P'}{F}$$

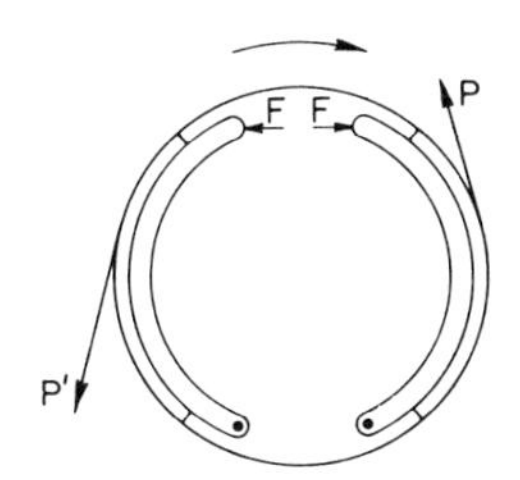

and some values of B are shown plotted against μ for different brakes in Figure 8.1

Knowing the torque, type of brake, its diameter, and the μ of the linings, the actuating force required can be obtained from Figure 8.1 (μ would normally lie between 0.3 and 0.4—a lining with the lower μ would give a more stable brake and would last longer, but a larger actuating force would be required).

Alternatively, knowing the torque required, the type of brake, and the μ of the linings, the actuating force for a given diameter brake, or the diameter of the brake for a given actuating force, can be determined from Figure 8.1.

The rubbing path area must be adequate to keep temperatures within limits so that performance is not impaired and the friction material has a good life. Table 8.5 shows the area required for different types of duty.

Smaller, wider drums of sufficient thermal capacity reduce bending stresses and give more uniform pressure on the linings, and reduce the moment of inertia of the drum. Drums for internal expanding brakes are generally stiffened with external circumferential ribs to avoid 'bell-mouthing' but can 'barrel' if ribs constrain ends but not the centre of the rubbing path. These fins also improve cooling; axial fins may develop high thermal stresses.

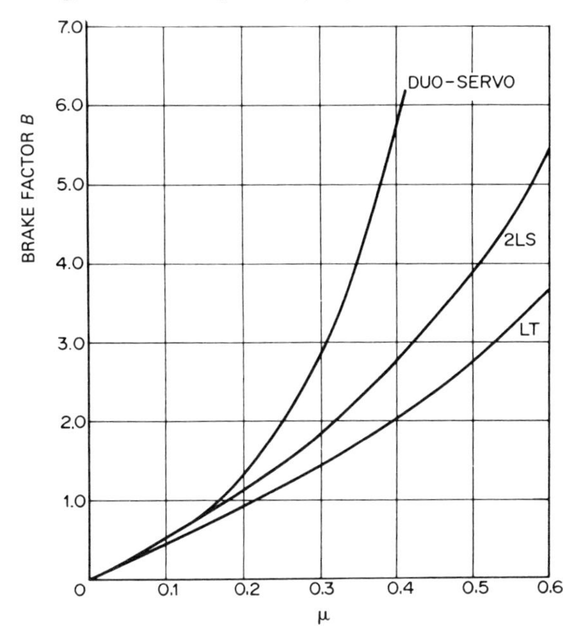

Figure 8.1 Typical variation of Brake Factor B with μ for a LT (leading-trailing shoe), a 2LS (two leading shoe), and a duo-servo drum brake having linings of arc length 100°

Table 8.3 The various types of drum brake

Type	*Brake factor* ($\mu = 0.35$)	*Uses*	*Type*
Post brake (external contracting, leading–trailing shoe)	1.7	Most widely used type of industrial brake. Used on mills, colliery winders, lift equipment, etc.	
Internal expanding, leading—trailing shoe	1.7	Mainly used for automotive purposes	
Internal expanding, two leading shoe	2.3	Mainly used for automotive purposes	
Duo-servo (shoes linked together)	4.0	Mainly used for automotive purposes	

The torque capacity of drum brakes of various sizes in shown in Figure 8.2

DISC BRAKES

A caliper carrying friction pads straddles the rotating disc and the pads are forced against the disc to apply the brake. The brake factor is low and equal to 2μ, but the disc brake is stable and less affected by high temperatures than the band or drum brake. The diameter of the disc and the area of friction material can be obtained from the torque curves of Figure 8.2 or from Table 8.5. The area of the pads of a spot-type disc brake is about one tenth of the area of the rubbing path. The actuating force required to act on each paid is $F = T/\mu D$, where T is the torque, and D the diameter to the midline of the pads.

Table 8.4 The various types of disc brake

Type	*Brake factor* ($\mu = 0.3$)	*Uses*	*Type*
Spot type	0.70	Cars and light commercial vehicles, reel tension brakes for paper mills, cable winding equipment, steel strip mills, forge brakes, marine shaft brakes, wind generators, etc.	
Plate type	0.70 (two rubbing faces)	Machine tools, and other machinery hydro-electric turbines agricultural tractors	

Table 8.5 Required brake areas for various duties

Duty	*Typical application*	*Area/Power* mm^2/w	
		Band and drum brakes	*Disc brake (Spot type)*
Infrequent (time to cool to ambient temperature between applications)	Emergency brakes; safety brakes	0.5	0.17
Intermittent	General-duty applications, winding engines, cranes, lifts	1.7	0.43
Heavy	Excavators, presses, drop stamps, haulage gear	3.4–4.2	0.86

Torque capacity

The torque capacity of band, drum and disc brakes of conventional proportions for medium-duty applications is shown in Figure 8.2

For light-duty work multiply figures on the torque axis by 3 for band and drum brakes, by 2.5 for disc brakes, for brakes of a given diameter.

For heavy-duty work divide figures on the torque axis by 2 for band, drum and disc brakes.

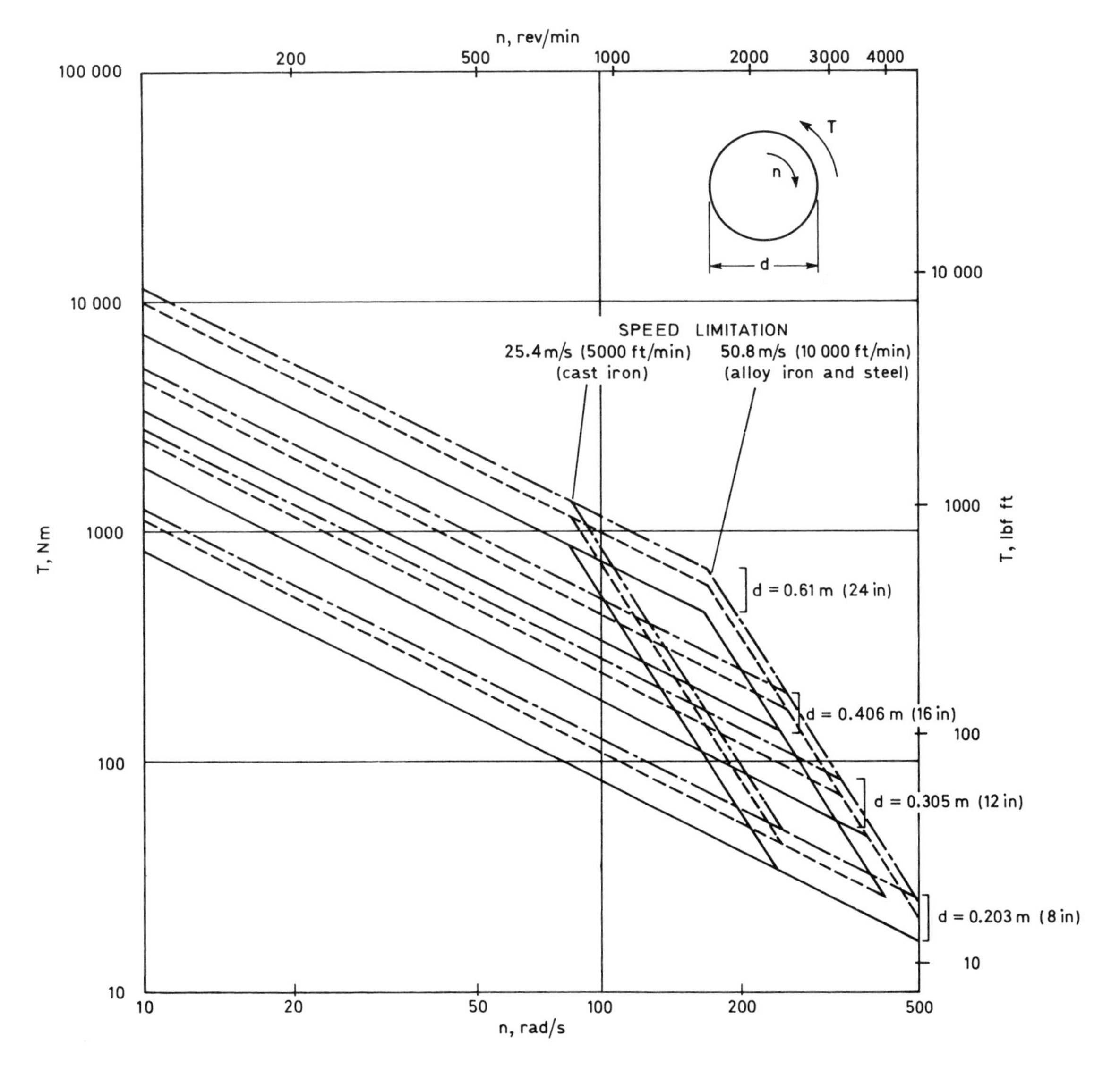

Figure 8.2 The torque capacity of band, drum and disc brakes of conventional proportions for medium-duty applications

Band brakes—The width of lining is $\frac{1}{5}$ the diameter (Rating: 1.7 mm^2/W, 2 in^2/hp) ———

Drum brakes—The width of lining is $\frac{1}{5}$ the diameter (Rating: 1.7 mm^2/W, 2 in^2/hp) — — —

Disc brakes—Length of pad $\frac{1}{10}$ mean diameter, width $\frac{3}{4}$ length (Rating: 6.8 mm^2/W, 8 in^2/hp) — — —

BRAKE SELECTION

Table 8.6 Brake selection for special performance requirements

Type of brake	*Maximum operating temperatures*	*Brake factor*	*Stability*	*Standard parts available*	*Setting up*	*Type*
Positive servo band brakes	Low	Very high	Very low	Some	Straightforward but watch excessive expansion in narrow tracks	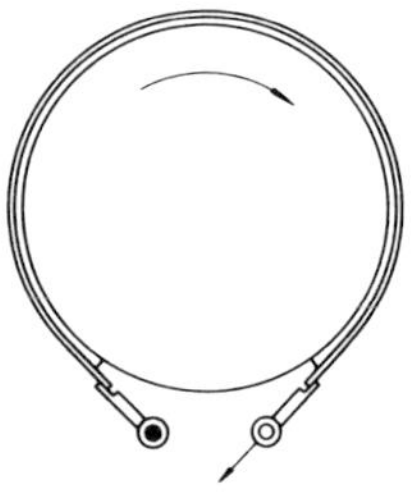
Drum brakes with two leading shoes	Low for external shoes, higher for internal shoes	High	Low	Some	Watch drum ovality and run-out and drum stiffness	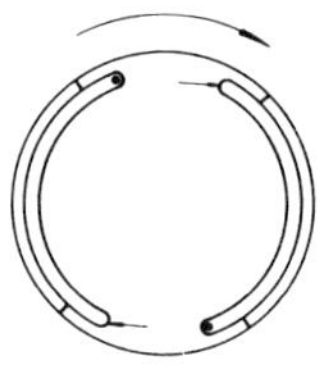
Drum brakes with one leading and one trailing shoe	Low for external shoes, higher for internal shoes	Moderate	Moderate	Some	Watch drum ovality and run-out and drum stiffness	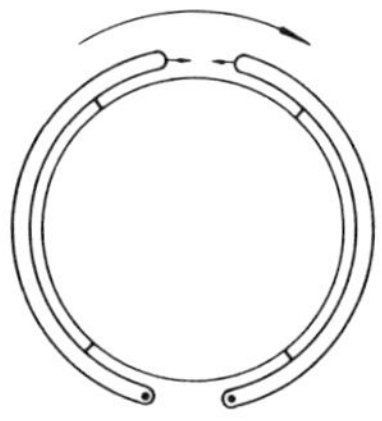
Duo-servo brakes	Low	Very high	Low	Some	Watch drum ovality and run-out	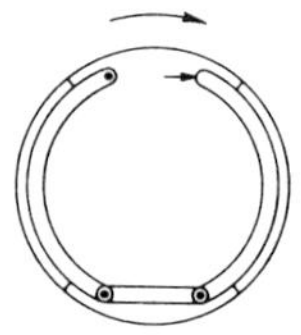
Disc brakes	High	Low	High	Some	Watch disc run-out thick/thin variation round disc, disc stiffness, disc coning.	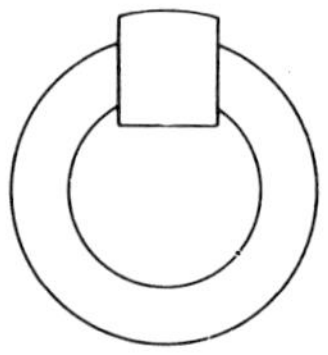

Table 8.7 Brake selection for special environmental conditions

Type of brake	*High temperature*	*Low temperature*	*Wet and humid*	*Dirt and dust*	*External vibration*	*Type*
Band	Good up to temperature limit of friction material	Good but avoid ice formation	Unstable but still effective	Good	Good	
Drum	Good up to temperature limit of friction material	Good but avoid ice formation	Unstable if humid. If very wet complete loss of performance on internal brakes but external brakes are more effective	Very good if sealed	Good	
Disc	Good up to temperature limit of friction material	Good but avoid ice formation	Good	Poor, should be shielded	Good	
General comments	Watch effect of thermal expansion on clearances		Watch corrosion and 'stiction' after prolonged parking in high humidity conditions			

Table 8.8 Methods of actuation

	Advantages	*Disadvantages*	*Points to watch*	*Uses*
Mechanical	Robust. Simple. Manual operation gives good control	Large leverage needed	Frictional losses at pins and pivots	Band brakes Drum brakes Disc brakes
Pneumatic	Large forces available	Compressed air supply needed. Brake chambers may be bulky. Slow response	Length of stroke (particularly if diaphragm type)	Band brakes Drum brakes Disc brakes
Hydraulic	Compact. Large forces available. Quick response and good control	Special fluid needed. Temperatures must not be high enough to vaporise fluid	Seals	Band brakes Drum brakes Disc brakes (spot type)
Electrical	Suitable for automatic control, quick response	On/off operation	Air gap	Band brakes Drum brakes Disc brakes (spot and plate types)

Note: Many brakes fail-safe; powerful springs apply the brake which is held off by one of the above means. Reduction in, for example, hydraulic pressure applies the brake.

FRICTION MATERIALS

A very wide range of friction materials is available, and in many cases materials have been developed for specific applications. The friction material manufacturer should therefore be consulted at an early stage in the design of the brake and should also be consulted concerning stock sizes – standard sizes are much cheaper than non-standard and are likely to be immediately available. The non-asbestos lining materials are normally made in flexible rolls in standard lengths, (e.g. 4 m) and widths (330 mm) and various thicknesses. Linings of the required sizes are slit from the standard sheets. These linings are bonded to the shoes and, by increasing the temperature and time of bonding, the linings can be made more rigid, and able to withstand higher and higher duties.

Industrial disc pads are generally based on automobile and CV pad types. They may be classified as organic-non-asbestos, low steel, or semi-metallic pads. They are based on thermosetting polymers reinforced by inorganic (e.g. glass) or organic fibres, 10–15%, or 50% by weight of steel fibre; and suitable fillers are added to give the pads the required tribological properties. The organic non-asbestos pads are suitable for lighter duties, and the greater the amount of steel fibre the higher the temperature the pads can withstand. The non-metallics tend to give less squeal and groan and cold judder, and less lining and rotor wear at low temperatures; steel fibres give a more stable μ and better high temperature lining life, but they can cause corrosion problems and they allow more heat to pass into the brake assembly instead of into the disc.

Data for typical materials are shown in Tables 8.9, 8.10 and 8.11. These figures are meant as a guide only; materials vary from manufacturer to manufacturer, and any one manufacturer may make up a number of different materials of the one type which may vary somewhat in properties.

Table 8.9 Material types and applications

Type	*Manufacture*	*Typical dimensions*	*Uses*
LININGS*			
Woven cotton	Closely woven belt of fabric is impregnated with resins which are then polymerised	As rolls; thickness 3.2–25.4 mm width up to 304.8 mm and lengths up to 15.2 m	Industrial drum brakes, minewinding equipment, cranes, lifts
Woven asbestos	Open woven belt of fabric is impregnated with resins which are then polymerised. May contain wire to scour the surface	As radiused linings thickness 3.2–12.7 mm width up to 203 mm, minimum radius 76 mm, maximum arc 160°	Industrial band and drum brakes, cranes, lifts, excavators, winches, concrete mixers. Mine equipment
Non-asbestos flexible semi-flexible rigid	Steel, glass or inorganic fibre and friction modifiers mixed with thermosetting polymer and mixture heated under pressure	Linings: thicknesses up to 35 mm Maximum radius about 15–30 times thickness depending upon flexibility	Industrial drum brakes Heavy-duty drum brakes— excavators, tractors, presses
PADS			
Resin-based	Similar to linings but choice of resin not as restricted as flexibility not required	In pads up to 25.4 mm in thickness or on backplate to fit proprietary calipers	Heavy-duty brakes and clutches, press brakes, earth-moving equipment
Sintered metal	Iron and/or copper powders mixed with friction modifiers and the whole sintered		Heavy-duty brakes and clutches, press brakes, earth-moving equipment
Cermets	Similar to sintered metal pad, but large proportion of ceramic material present	Supplied in buttons, cups	As above

* Many lining materials supplied as large pads can be bolted, or riveted, using brass rivets, to the band or shoe; the pads can be moved along the band or shoe as wear occurs and so maximum life obtained from the friction material despite uneven wear along its length. Alternatively, and particularly with weaker materials, the friction material can be bonded to the metal carrier using proprietary adhesives and techniques (contact the manufacturer). On safety-critical applications the friction material should be attached by both bonding and riveting.

Table 8.10 Performance and allowable operating conditions for various materials

Materials	μ	Temperatures, °C		Working pressures	Maximum pressure
		Maximum	Maximum operating	kN/m^2	MN/m^2
LININGS					
Woven cotton	0.50	150	100	70–700	1.5
Woven asbestos	0.40	250	125	70–700	2.1
Non-asbestos					
light duty (flexible)	0.38	350	175	70–700	2.1
medium duty (semi-flexible)	0.35	400	200	70–700	2.8
heavy duty (rigid)	0.35	500	225	70–700	3.8
PADS					
Resin based	0.32	650	300	350–1750	5.5
Sintered	0.30	650	300	350–3500	5.5
Cermet	0.32	800	400	350–1050	6.9

Table 8.11 Typical mechanical properties

Materials	Ultimate strength			Rivet-holding capacity	Specific Gravity
	Tensile	Shear	Compressive		
	MN/m^2	MN/m^2	MN/m^2	MN/m^2	
LINING					
Woven cotton	20.7	12.4	96.5	69	1.0
Woven asbestos	24.1	13.8	103.4	83	1.5–2.0
Non-asbestos					
light duty	8.2	8.2	41.3	103	1.7
medium duty	10.3	8.2	96.5	152	1.7
heavy duty	13.8	9.0	103.4	172	2.0
PAD					
Resin-based	—	9.0	103.4	—	2.0
Sintered metals	48.2	68.9	151.6	—	6.0

Mating surfaces

Woven cotton or asbestos linings, and those with steel and inorganic or organic fibre reinforcement, should run against fine-grained pearlitic cast-iron or alloy cast-iron of Brinell Hardness 180–240 or steel cold-rolled or forged with a Brinell Hardness greater than 200. The surface should be fine-turned or ground to a finish of at least 2.5 μm CLA. Cast steel and non-ferrous materials are not recommended. Some friction materials are very sensitive to trace amounts of titanium (and some other elements) in the cast iron rotor, and these trace elements can considerably reduce μ, though they also tend to increase the life of the friction material.

Sintered metals should run against fine-grained pearlitic cast iron or alloy irons, Brinell Hardness 180–250. High carbon steel such as EN6 for moderately loaded, and EN42 for heavy-duty thin counterplates in multidisc clutches. Minimum Brinell Hardness 200 for heavy duty. The surface finish should be 0.9–1.5 μm CLA.

Cermets should run against similar cast irons with Brinell Hardness greater than 200. High carbon steels with a hardness between 200 and 300 are acceptable. The surface finish should again be 0.9–1.5 μm CLA.

Screws are used as linear actuators or jacks and can generate substantial axial forces. They can operate with an external drive to either the screw or the nut, and the driving system often incorporates a worm gear in order to obtain a high reduction ratio.

TYPES OF SCREW

Plain screws

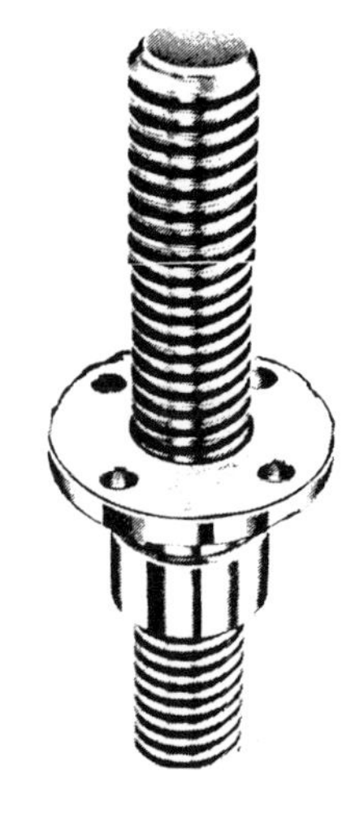

In these screws the load is transmitted by direct rubbing contact between the screw and the nut.

These are the simplest and inherently the most robust. The thread section may be of a square profile or more commonly is of the acme type with a trapezoidal cross section.

Their operating friction is relatively high but on larger diameter screws can be reduced to very low levels by incorporating hydrostatic pads into the operating surfaces of the nut. This is usually only justified economically in special screws such as the roll adjustment screws on large rolling mills.

Ball screws

In these screws the load is transmitted by close packed balls, rolling between the grooves of the screw and the nut.

These provide the lowest friction and are used particularly for positioning screws in automatically controlled machines. The nuts need to incorporate a system for re-circulating the balls. The load capacity is less than in other types of screw and is limited by the contact stresses between the balls and the screw.

Planetary roller screws

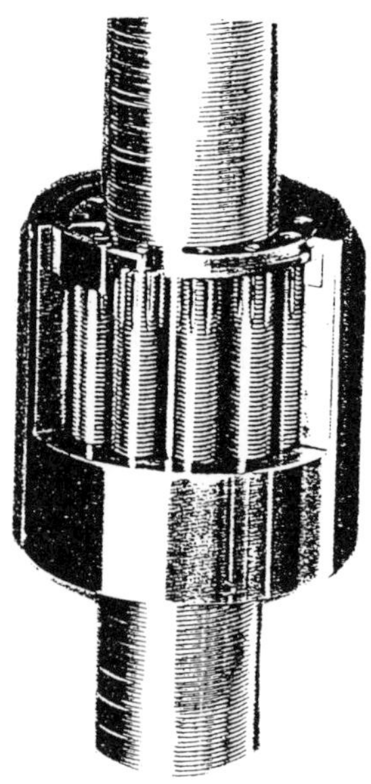

In these screws a number of rollers are positioned between the screw and the nut and rotate between them, around the screw, with a planetary motion.

Those with the highest load capacity have helical threads on the rollers and nut, matching the pitch of the screw. The whole space between the screw and the nut can be packed with rollers but these need to have synchronised rotation by a gear drive to ensure that they retain their axial position.

Alternative types are available in which the rollers and nut have simple parallel ribs matching the pitch of the screw. The screw however has to be multistart because the number of rollers that can be fitted equals the number of starts on the thread. Also the nut cannot be used as the driver if synchronised external movement is required, because of the possibility of slip between the rollers and the nut. In these cases the screw or the planetary roller carrier has to be driven, but not the nut.

PERFORMANCE

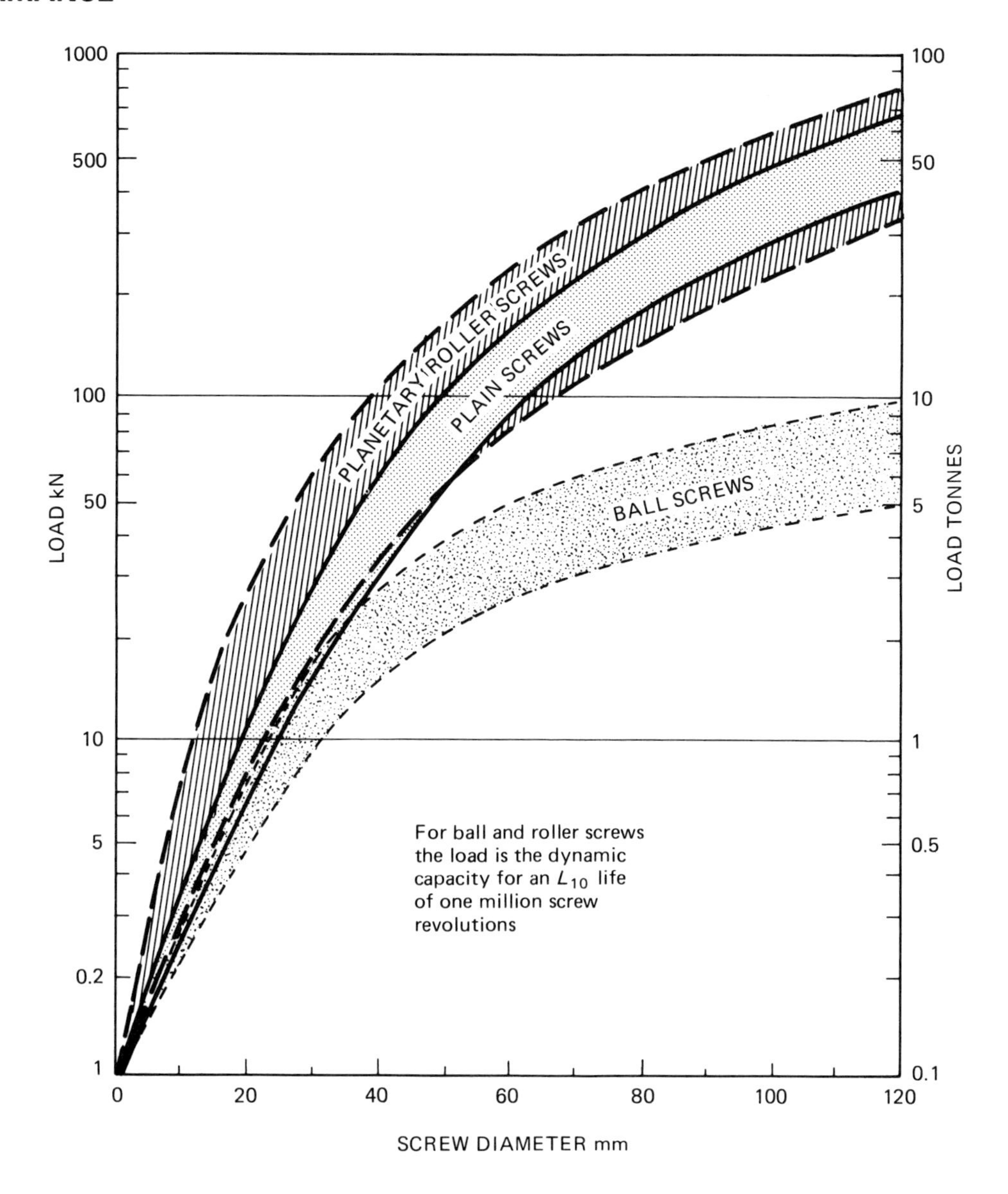

Figure 9.1 The axial load capacity of various types of screw

Mechanical efficiencies of screws

Plain screws	30%–50% approximately, with larger diameter screws tending to have the lower values.
Roller screws	65%–85%
Ball screws	75%–90%

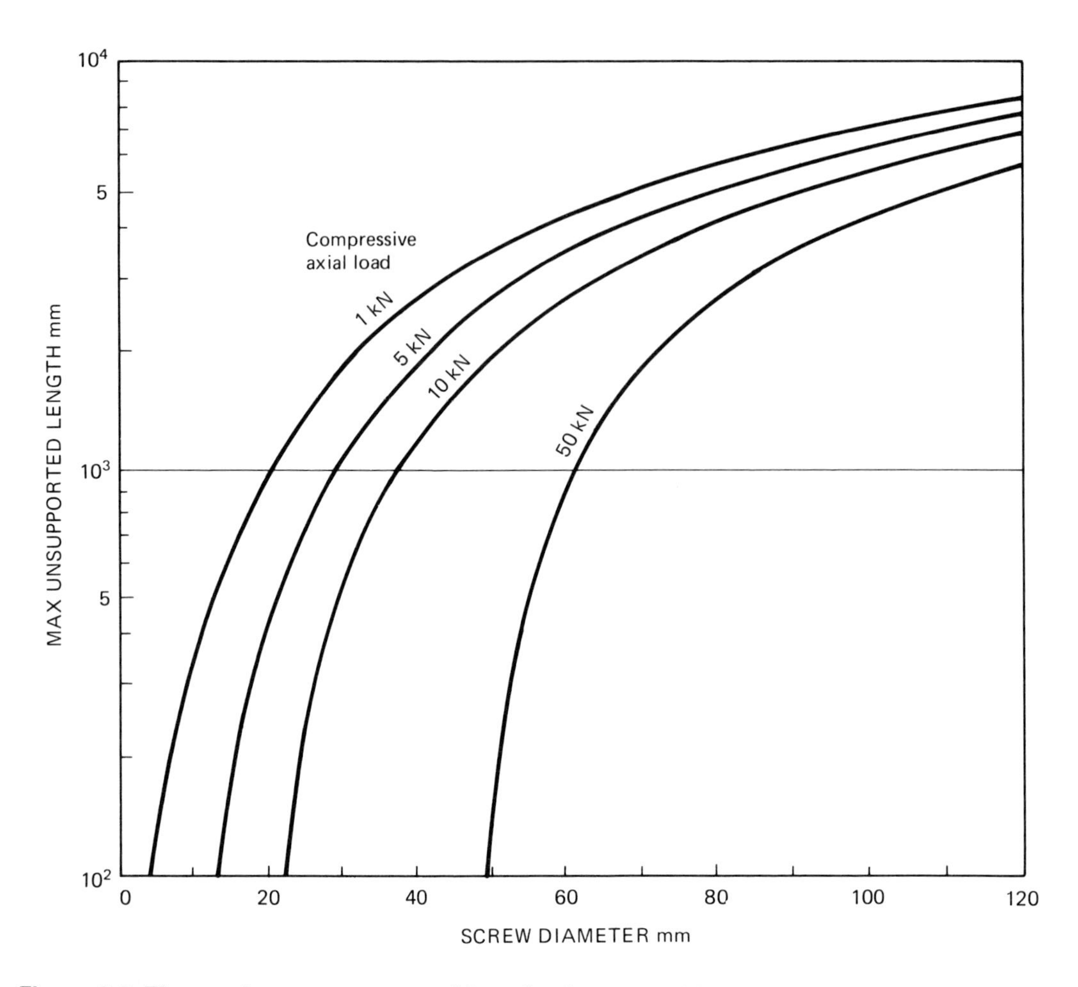

Figure 9.2 The maximum unsupported length of screw, with one end free, to avoid buckling

Installation

The performance of all screws will be reduced when subject to misalignment and sideways loads. Ball screws are particularly sensitive to these effects.

All screws require lubrication, either by regular greasing or by operation within an enclosure with oil or fluid grease.

For precision installations the screws and nuts need protection from external contamination and flexible convoluted gaiters are commonly used for their protection.

COMMON MODES OF FAILURE

Three main forms of cam and tappet failures occur. These are pitting, polish wear and scuffing. Failure may occur on either the cam or tappet, often in differing degree on both.

Pitting

This is the failure of a surface, manifested initially by the breaking-out of small roughly triangular portions of the material surface. This failure is primarily due to high stresses causing fatigue failure to start at a point below the surface where the highest combined stresses occur. After initiation a crack propagates to the surface and it may be that the subsequent failure mechanism is that the crack then becomes filled with lubricant, which helps to lever out a triangular portion of material.

Heavily loaded surfaces will continue to pit with increasing severity with time. Figure 10.1 shows some pitted cam followers.

Polish wear

This is the general attrition of the contacting surfaces. When conditions are right this will be small, but occasionally very rapid wear can occur, particularly with chilled and hardened cast iron flat-faced tappets. Often a casual look will suggest that the surfaces are brightly polished and in good condition but dimensional checks reveal that considerable wear has occurred. Polish wear appears to be an intermediate case between pitting and scuffing assisted by some form of chemical action involving the oil – certainly surfaces which develop a bloom after running do not normally give 'polish wear'.

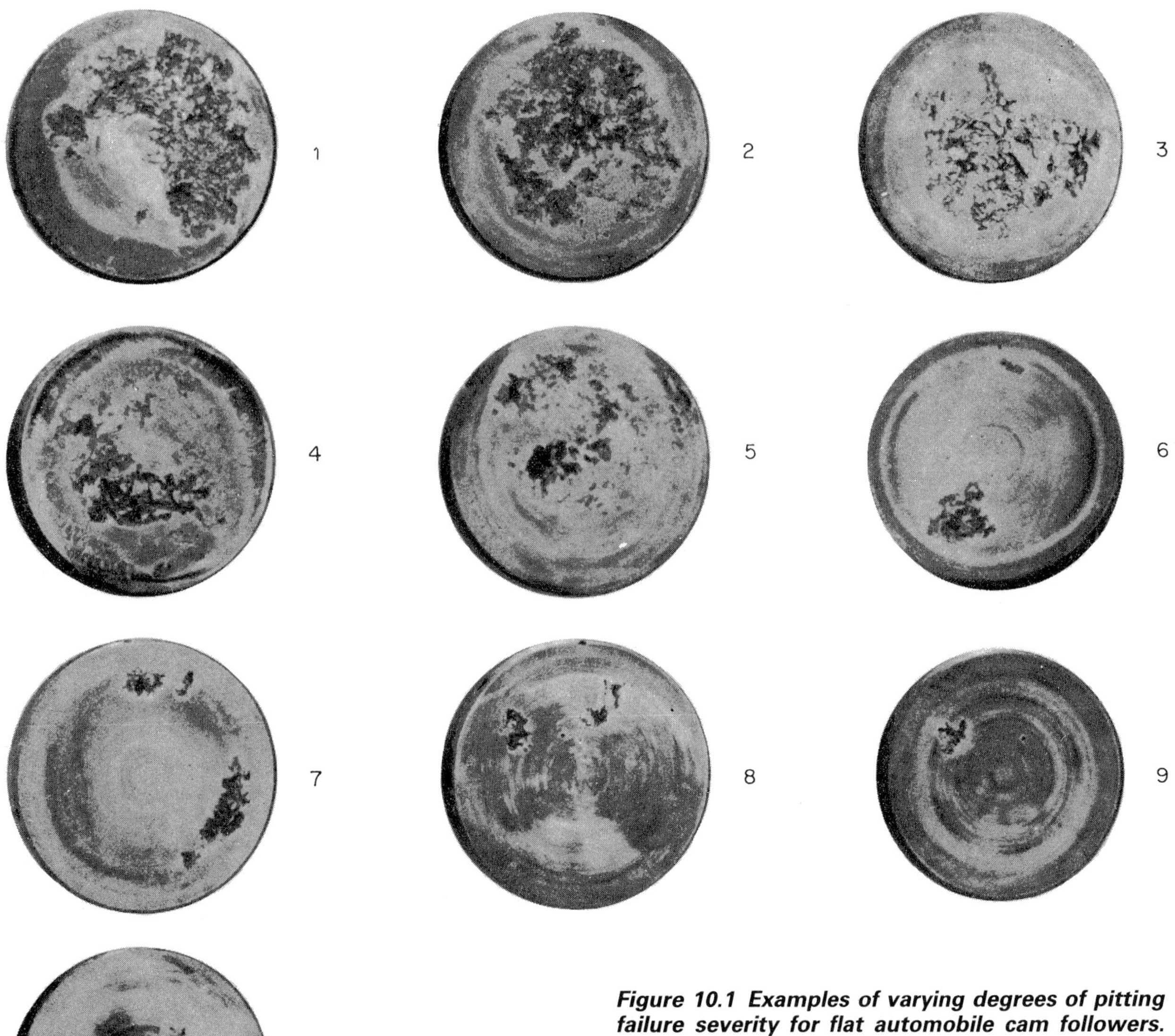

Figure 10.1 Examples of varying degrees of pitting failure severity for flat automobile cam followers. *Numbers indicate awards for ratings for lack of damage during oil standardisation tests (courtesy Orobis Ltd.)*

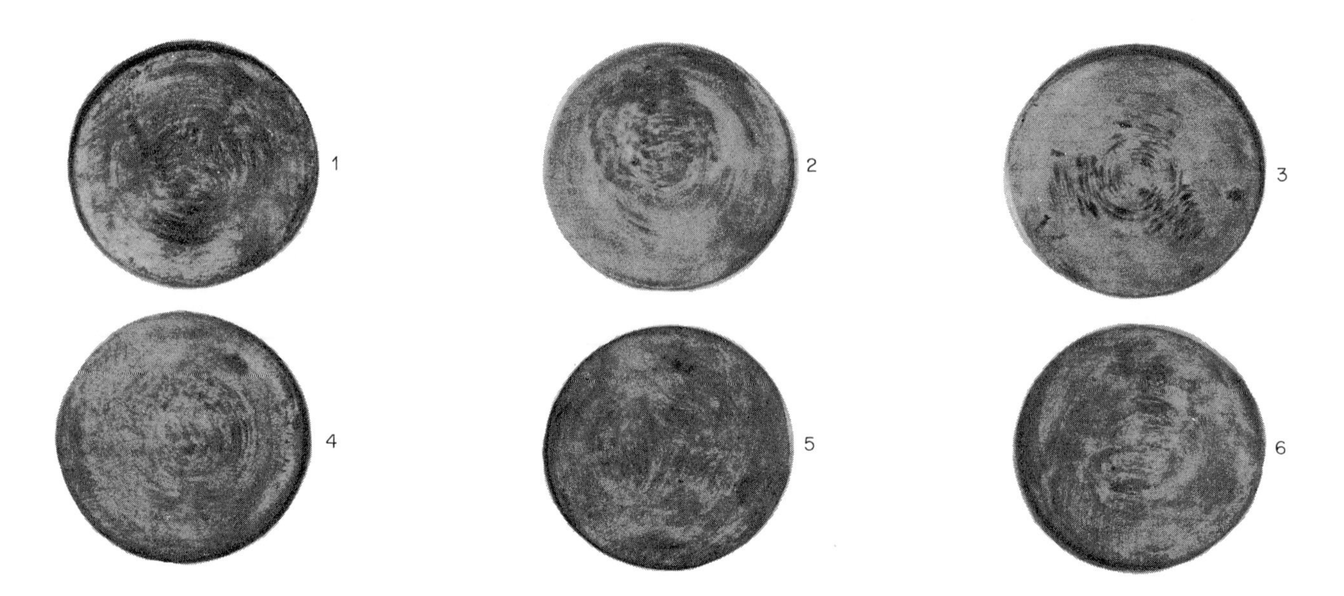

Figure 10.2 Examples of varying severity of cam follower, scuffing for flat automotive tappets *(courtesy:* Orobis Ltd.)

Scuffing

This is the local welding together of two heavily loaded surfaces, particularly when a high degree of relative sliding occurs under poor lubrication conditions, followed by the tearing apart of the welded material. It is particularly likely to start from high spots, due to poor surface finish, during early running of new parts.

CHECKING THE TRIBOLOGICAL DESIGN

It is usual to assess cam/tappet designs on the basis of the maximum contact stress between the contacting cam and tappet, with some consideration of the relative sliding velocity. This requires the determination of the loads acting between the cam and tappet throughout the lift period (at various speeds if the mechanism operates over a speed range), the instantaneous radius of curvature for the cam throughout the lift period, and the cam follower radius of curvature. Figure 10.3 shows the relationship between these various quantities for a typical automotive cam. In addition it is possible to assess the quality of lubrication at the cam/tappet interface by calculating the elastohydrodynamic (EHL) film thickness and relating this to the surface roughness of the components. An approximate method for the calculation is given later in this section.

Calculation of the instantaneous radius of curvature of the cam

Where the cam is made up of geometric arcs and tangents the appropriate values for the radii of curvature can be read from the drawing. Many cams are now generated from lift ordinates computed from a mathematical law incorporating the desired characteristics, so it is necessary to calculate the instantaneous radius of cam curvature around the profile. At any cam angle the instantaneous radius of curvature at that angle is given by the following:

For flat followers (tappets)

$$R_c = R_{base} + y + 3282.8\,y''$$

where R_{base} = base circle radius in mm
y = cam lift at desired angle in mm
y'' = cam acceleration at chosen angle in mm/deg^2
R_c = radius of curvature in mm

For curved followers

$$R_c = \left\{ \frac{[(R_b + R_F + y)^2 + V^2]^{3/2}}{(R_b + R_F + y)^2 + 2V^2 - (R_b + R_F + y)A} \right\} - R_F$$

where R_b = cam base circle radius in mm
R_F = follower radius in mm
y = cam lift at chosen angle in mm
V = follower velocity at chosen angle in mm/rad
= 57.29 × velocity in mm/deg
A = follower acceleration mm/rad^2
= 3282.8 × acceleration (mm/deg^2)

The value for R_c will be positive for a convex cam flank and negative for a concave (i.e. hollow) flank.

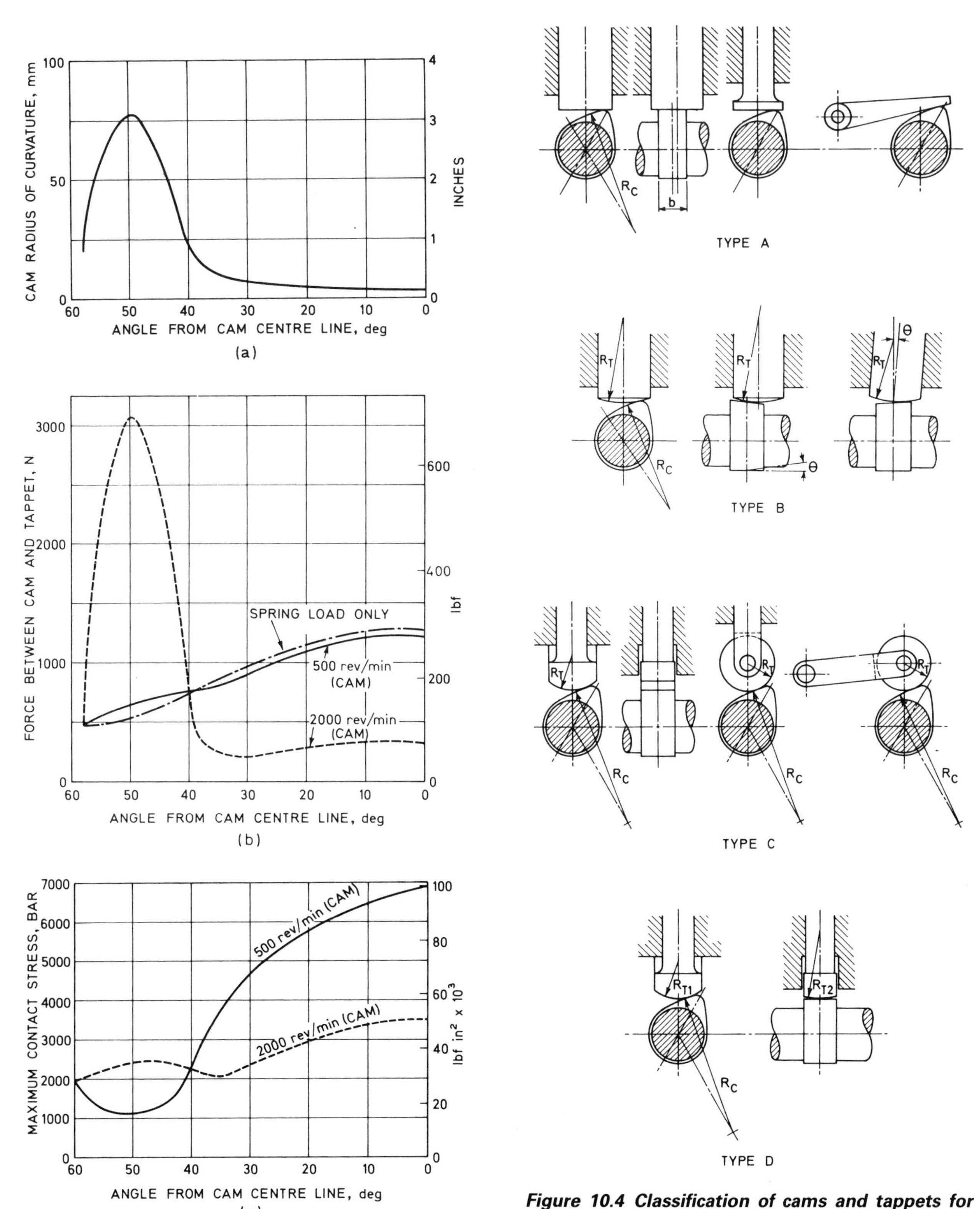

Figure 10.3 Typical variation for an automotive cam of: *(a) instantaneous radius of curvature; (b) cam/tappet force; (c) maximum contact stress*

Figure 10.4 Classification of cams and tappets for determination of contact stresses. *Type A: flat follower faces. Type B: spherical faced tappets. Type C: curved and roller followers with flat transverse faces. Type D: curved tappets with transverse radius of curvature*

Calculation of contact (Hertzian) stress

It is now necessary to calculate the Hertzian stresses between the cam and tappet. Most tappets and cams can be classified into one of the forms shown in Figure 10.4. The appropriate formulae for the Hertzian stress are listed below.

The following symbols and units are used:

W = load between cam and tappet (N)

b = width of cam (mm)

R_c = cam radius of curvature at point under consideration (mm)

R_T = tappet radius curvature (mm)

R_{T1} = tappet radius of curvature in plane of cam (mm)

R_{T2} = tappet radius of curvature at right angles to plane of cam (mm)

f_{max} = peak Hertzian stress at point under consideration (N/mm^2)

Type A: Flat tappet face on cam

$$f_{max} = K \left[\frac{W}{R_c \, . \, b} \right]^{1/2}$$

Material combination	K
Steel on steel	188
Steel on cast iron	168
Cast iron on cast iron	153

The centre line of the tappet is often displaced slightly axially from the centre line of the cam to promote rotation of the tappet about its axis. This improves scuffing resistance but is considered by some to slightly reduce pitting resistance.

Type B: Spherical faced tappet

Since the theoretical line contact of Type A tappets on the cam is often not achieved, due to dimensional inaccuracies including asymmetric deflection of the cam on its shaft, edge loading occurs. To avoid this a large spherical radius is often used for the tappet face. Automotive engines use a spherical radius of between 760 to 2540 mm (30 to 100 in). To promote tappet rotation the tappet centre line is displaced slightly from the axial centre line of the cam and the cam face tapered (10–14 min of arc with 760 mm tappet radius and 4–7 min for 1500–2540mm tappet radius). Alternatively the longitudinal tappet axis is tilted by a corresponding amount to the camshaft axis. The theoretical point contact extends into an elongated ellipse under load to give a better contact zone than with the nominally flat face.

$$f_{max} = X \, . \, K \left[\frac{2}{R_T} + \frac{1}{R_c} \right]^{2/3} . \, W^{1/3}$$

K is obtained from Figure 10.5 after evaluating

$$\left[1 + \frac{2R_c}{R_T} \right]$$

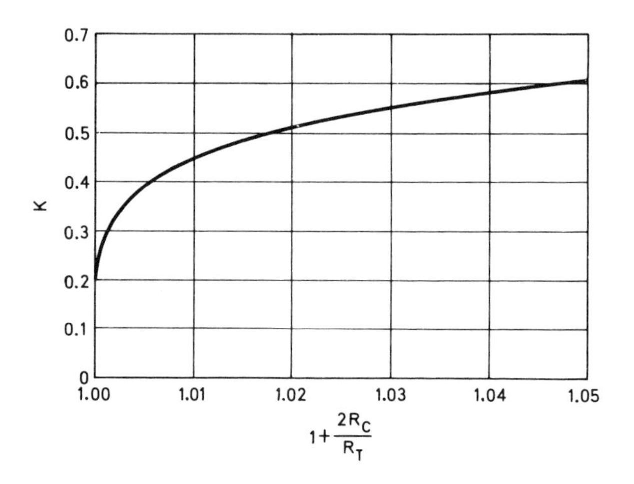

Figure 10.5 Constant for the determination of contact stresses with spherical-ended tappets

Material combination	X
Steel on steel	838
Steel on cast iron	722
Cast iron on cast iron	640

Type C: Curved and roller tappets with flat transverse face

$$f_{max} = K \left[\left(\frac{1}{R_c} + \frac{1}{R_T} \right) \frac{W}{b} \right]^{1/2}$$

Where K is the same as for type A, flat tappet face on cam.

Type D: Curved tappet with large transverse curvature (crowning)

The large transverse radius of curvature has values similar to those used in Type B.

$$f_{max} = X \,.\, K\left[\frac{1}{R_{T1}} + \frac{1}{R_{T2}} + \frac{1}{R_c}\right]^{2/3} . \; W^{1/3}$$

X values for material combinations as for Type B.
K is obtained from Figure 10.5 after evaluating

$$\left[\frac{\dfrac{1}{R_{T1}} + \dfrac{1}{R_{T2}} + \dfrac{1}{R_c}}{\dfrac{1}{R_{T1}} - \dfrac{1}{R_{T2}} + \dfrac{1}{R_c}}\right]$$

and using scale labelled

$$\left(1 + \frac{2R_c}{R_T}\right).$$

Allowable design values for contact stress

Safe values for contact stress (Hertzian stress) are dependent on a number of factors such as the combination of materials in use; heat treatment and surface treatment; quality of lubrication.

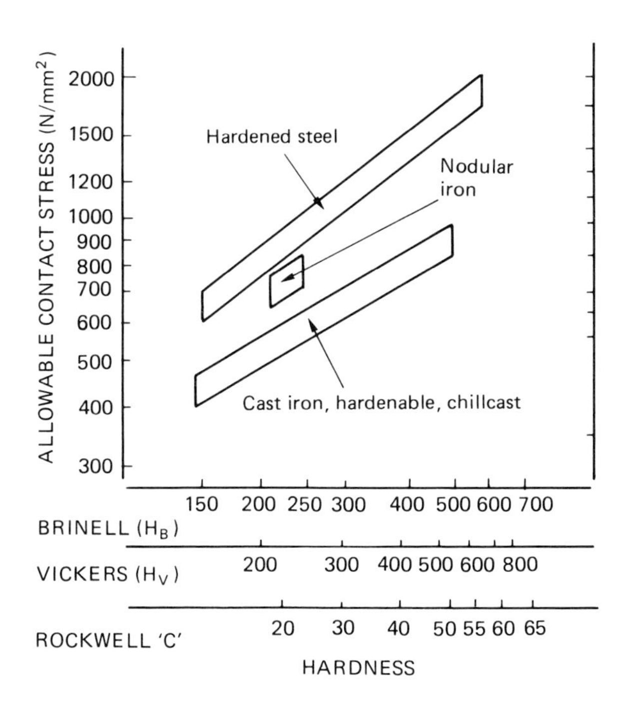

Figure 10.6 Typical allowable contact stresses under good lubrication conditions

Figure 10.6 gives allowable contact stress for iron and steel components of various hardnesses. These values can only be applied if lubrication conditions are good, and this needs to be checked using the assessment method below.

ASSESSMENT OF LUBRICATION QUALITY

Calculation of film thickness

The lubrication mechanism in non-conformal contacts such as in ball bearings, gears, and cams and followers, is Elastohydrodynamic lubrication or EHL. This mechanism can generate oil films of thicknesses up to the order of 1 μm. There is a long formula for accurately calculating the film thickness, but a simple formula is given below which gives sufficient accuracy for assessing the lubrication quality of cams and followers. This formula applies only to iron or steel components with mineral oil lubrication.

$$h \;= 5 \times 10^{-6} \times (\eta \, u \, R_r)^{0.5}$$

where:

h = EHL film thickness (mm)
η = lubricant viscosity at working temperature (Poise)
u = entrainment velocity (mm/s).
– for evaluation of u, see below
R_r = relative radius of curvature (mm)
– for flat tappets $R_r = R_c$
– for curved tappets
$R_r = (1/R_c + 1/R_T)^{-1}$
– for spherical or barrelled roller tappets assume $R_T = R_{T_1}$

Evaluation of entrainment velocity u

The entrainment velocity u can vary enormously through the cam cycle, reversing in sign, and in some cases remaining close to zero for part of the cycle. This last condition leads to very thin or zero thickness films.

For roller followers, u can be taken as being approximately the surface speed of the cam. Calculation of the EHL film is only required at the cam nose and on the base circle. Roller followers usually have good lubrication conditions at the expense of high contact stress for a given size.

For plain tappets the entrainment velocity, u at any instant is the mean of the velocity of the cam surface relative to the contact point and the velocity of the follower surface relative to the contact point.

On the base circle therefore, where the contact point is stationary, u is half the cam surface speed.

At all other parts of the cycle, the contact point is moving. The entrainment velocity u can be calculated from the following equation.

$$u = \omega\left[\frac{R_b}{2} + \frac{y}{2} - R_c\right]$$

ω = cam speed in rad/s
R_b = base circle radius (mm)
y = cam lift (mm)
R_c = cam radius at point of contact (mm)

This applies for flat tappets, and for curved tappets with a radius much larger than the cam radius it can be used as a reasonable approximation.

Ideally the values for u and R_r should be calculated for all points on the cycle, but as a minimum they should be calculated for the base circle and the maximum lift position.

For cams with curved sliding contact followers the equation for u is very complex. However, to check the value of u at the maximum lift position only, the following approximate formula can be used.

$$u = \omega\left[\frac{3280y'' R_F}{(R_F + R_b + y)} + \frac{(R_b + y)}{2}\right]$$

(curved follower, max lift position only)

where

y = max cam lift (mm)
y'' = max cam acceleration at nose (mm/deg^2) which is a negative value
R_F = follower radius (mm)
R_b = base circle radius (mm).

Evaluation of mode of lubrication

Once a value for the film thickness has been calculated, the mode of lubrication can be determined by comparing it with the effective surface roughness of the components. The effective surface roughness is generally taken as the combined surface roughness R_{qt}, defined as

$$R_{qt} = (R_{q1}{}^2 + R_{q2}{}^2)^{0.5}$$

R_{q1} and R_{q2} are the RMS roughnesses of the cam and tappet respectively, typically 1.3 times the R_a (or CLA) roughness values.

If the EHL film thickness h is greater than R_{qt} then lubrication will be satisfactory.

If the EHL film thickness is less than about 0.5 R_{qt} then there will be some solid contact and boundary lubrication conditions apply. Under these circumstances, surface treatments and surface coatings to promote good running-in will be desirable, and anti-wear additives in the oil may be necessary.

Alternatively, it may be appropriate to improve surface finishes, or to change the design to an improved profile giving better EHL films, or to use roller followers which are inherently easier to lubricate.

SURFACE FINISH

Extremely good surface finishes are desirable for successful operation, as the EHL lubrication film is usually very thin.

Typical achievable values are 0.4 μm R_a for the cam, and 0.15 μm R_a for the tappets.

SURFACE TREATMENTS

Some surface treatment and heat treatment processes which can be used with cams and tappets are given below:

Phosphating	Running-in aid. Retains lubricant.
'Tufftride'	Running-in aid. Scuff resistant
'Noscuff'	Greater depth than Tufftride. Less hard.
'Sulf. B.T.'	Low distortion. Anti-scuffing
Flame hardening, Induction hardening	Can give distortion
Laser hardening	Low distortion. 0.25 to 1 mm case depth
Carburising	0.5 mm case depth typical
Nitriding	Depth 0.3 mm. Hardening and scuff resistance.
Plasma Nitriding	As nitriding, but low distortion
Sulfinuz	Good scuff resistance
Boriding	Good wear resistance

OIL AND ADDITIVES

The oil type is frequently constrained by requirements of other parts of the machine. However, for best lubrication of the cam and tappet (i.e.: thickest EHL film), the viscosity of the lubricant at the working temperature should be as high as possible. Often the best way of achieving this is to provide good cooling at the cams, by means of a copious supply of oil.

Trends in vehicle engine design such as overhead camshafts, and higher underbonnet temperatures, have led to high camshaft temperatures and low lubricant viscosities. Some cam wear problems may be partially attributed to this.

Oil additives, principally ZDDP (zinc-dialkyldithiophosphate) and similar, are used in vehicle engine oils, and are beneficial to cam and tappet wear. There is evidence that these additives can promote pitting at high temperatures, due to their chemical effects. Additives should therefore be used with care, and are certainly not an appropriate alternative to good design.

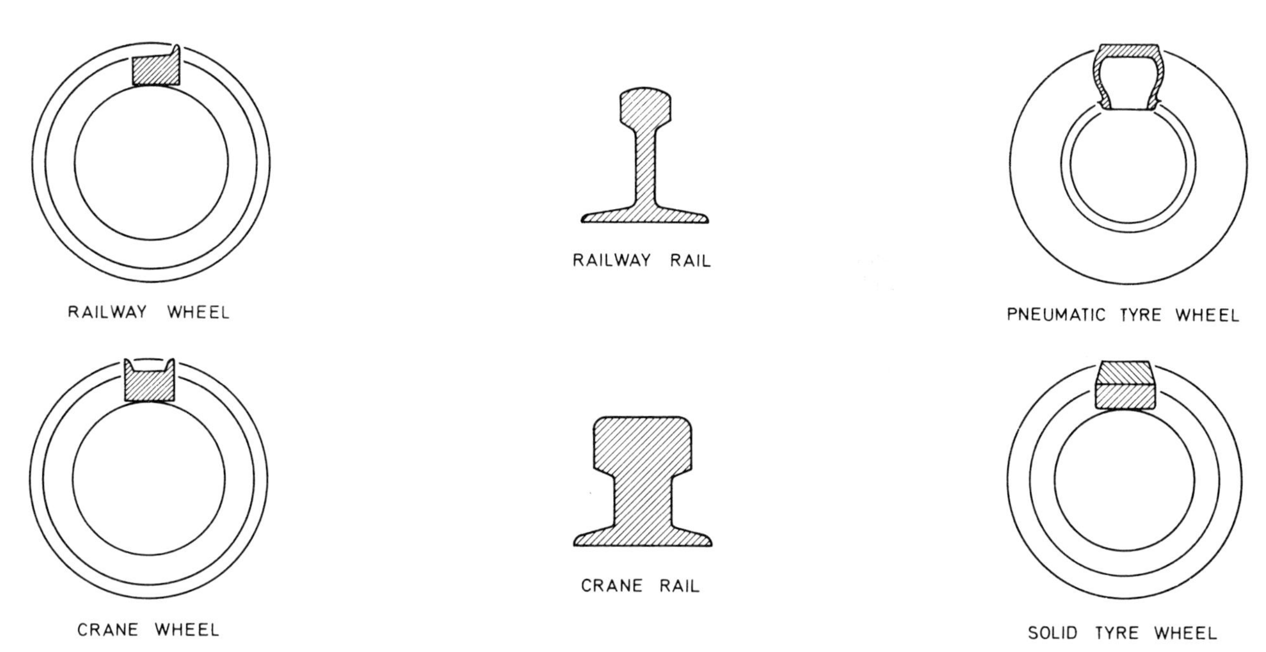

Figure 11.1 Cross sections of typical wheels and rails

Table 11.1 The effect of various factors on wear

	Load (W) lbf *or* N	*Speed*	*Elevated temp.*	*Water*	*Wheel dia.* in *or* mm	*Material composition*	*Mating surface*	*Other*
Steel rails (vertical wear)	See Figure 11.2. Wear α W^{β} where $\beta = \frac{1}{2}$ to 1	Little overall effect except on side or flange wear. Increased wear at high speeds counteracted by increased wear due to more acceleration, gradient climbing, and braking at low speeds	Increased corrosive wear at temperatures above 200°C, depending on the atmosphere	Increased wear due to corrosion. Hence the slope of the lines in Figure 11.2	Rail wear reduced by increasing wheel size	Increased hardness and corrosion resistance reduce wear substantially. Steel should have a fine pearlitic microstructure	Wear is increased by a rough or distorted mating surface, or by sanding the track	Atmospheric pollution increases wear substantially
Steel tyres	Similar to rail wear. See Figure 11.2		As for rails	Little effect	Wear α (diameter)$^{\beta}$ where $\beta = 2$ to 3	Increased hardness and toughness reduce wear	Corrugated rail surfaces cause wear and noise. Curves in the track increase wear	
Pneumatic rubber tyres	See Figure 11.5	See Figure 11.5. Wear due mainly to the effect of braking and cornering. One right-angled turn equivalent to 3 stops	Maximum temperature for continuous use 120°C	Wear reduced to 20–50% of dry wear	Wear α (diameter)$^{\beta}$ where $\beta \not< 2$	Radial ply construction reduces wear by about 40%. The wear resistance of different rubber varies widely	Harshness more important than roughness in road surfaces, by a factor of about 3. See Figure 11.5	For effect of inflation pressure, see Figure 11.5
Solid tyres	Proportional to W^{γ} where $\gamma > 1$	Use limited to low speeds due to poor heat dissipation	Important limiting factor: dependent on tyre material	Reduced wear	Wear decreases with increasing size	Materials should be chosen to suit the conditions	Rough, sharp surfaces cause rapid wear	

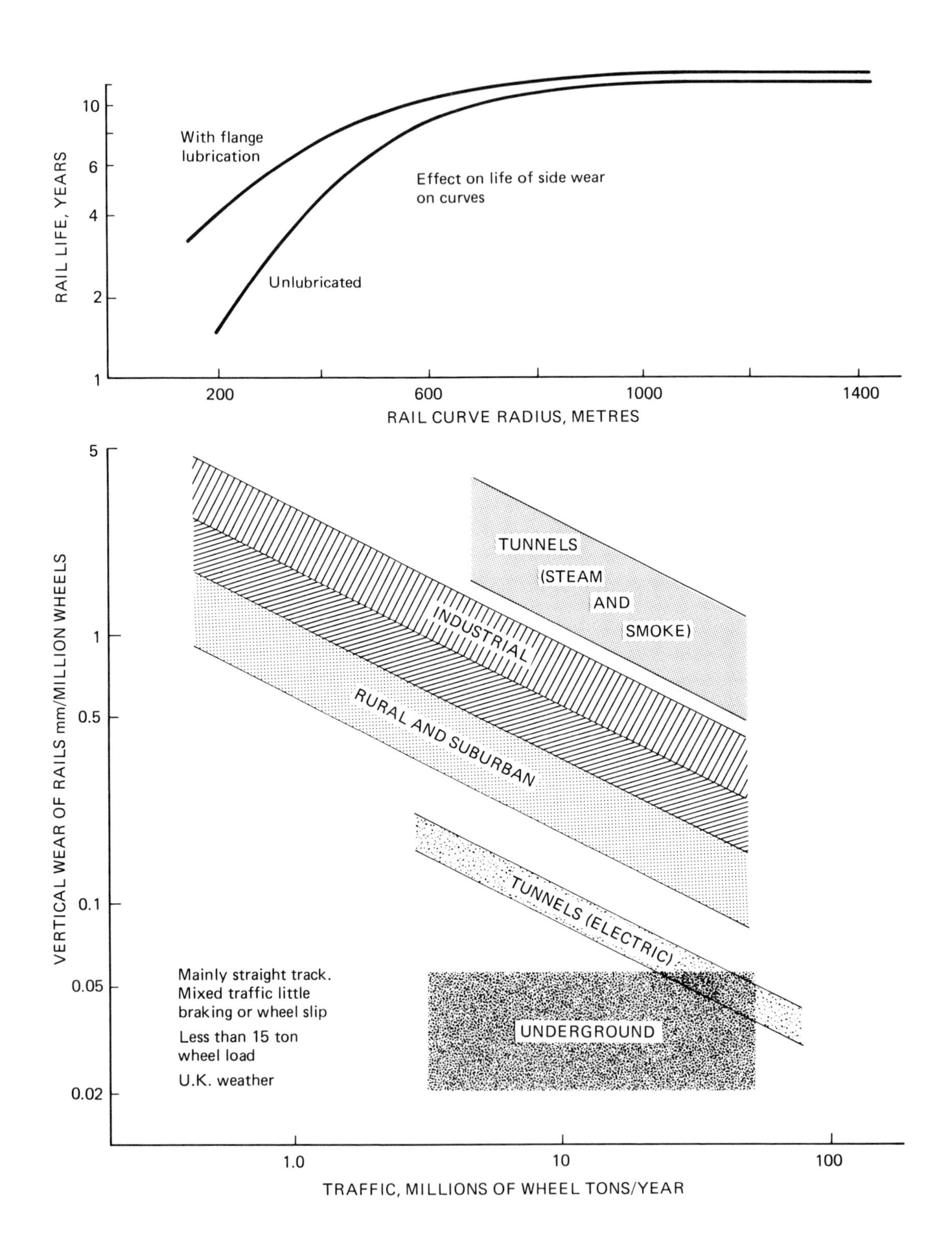

Figure 11.2 The effect of traffic density on rail wear is plotted to show the influence of the environment and type of traffic. *Level of atmospheric pollution is an important controlling factor. Rail side wear is one to three times greater than vertical wear on straight track, and increases as the radius of the curve diminishes below 1000 metres*

Table 11.2 The effect of various factors on load capacity

	Speed (V)	*Elevated temp.*	*Wheel dia.*	*Material composition*	*Inflation pressure*	*Wear*	*Other factors*
Steel tyres on steel rails	Small effect only	Not important except in the extreme. High temperatures generated during braking can cause thermal fatigue of tyre treads	See Figure 11.4	Hardened steels or increased 'carbon equivalent' content increase load capacity	Not applicable	Wear will reduce the section modulus and hence the load capacity of the rail	
Pneumatic rubber tyres	The effect of speed on load capacity can be reduced by increasing inflation pressure, see Figure 11.5	Load capacity must be limited to prevent excessive temperature rise. Maximum temperature 120°C. Increased inflation pressure reduces running temperature	See Figure 11.1. Outside diameter and tread width can vary widely for a particular rim size	Load capacity increases with tyre ply rating. Materials with low internal friction are more suitable for consistent high-speed use	Inflation pressure is a vital factor affecting load capacity, see Figure 11.1 and other columns	Within the legal limits wear has no effect provided the tyre has been correctly used and is undamaged	

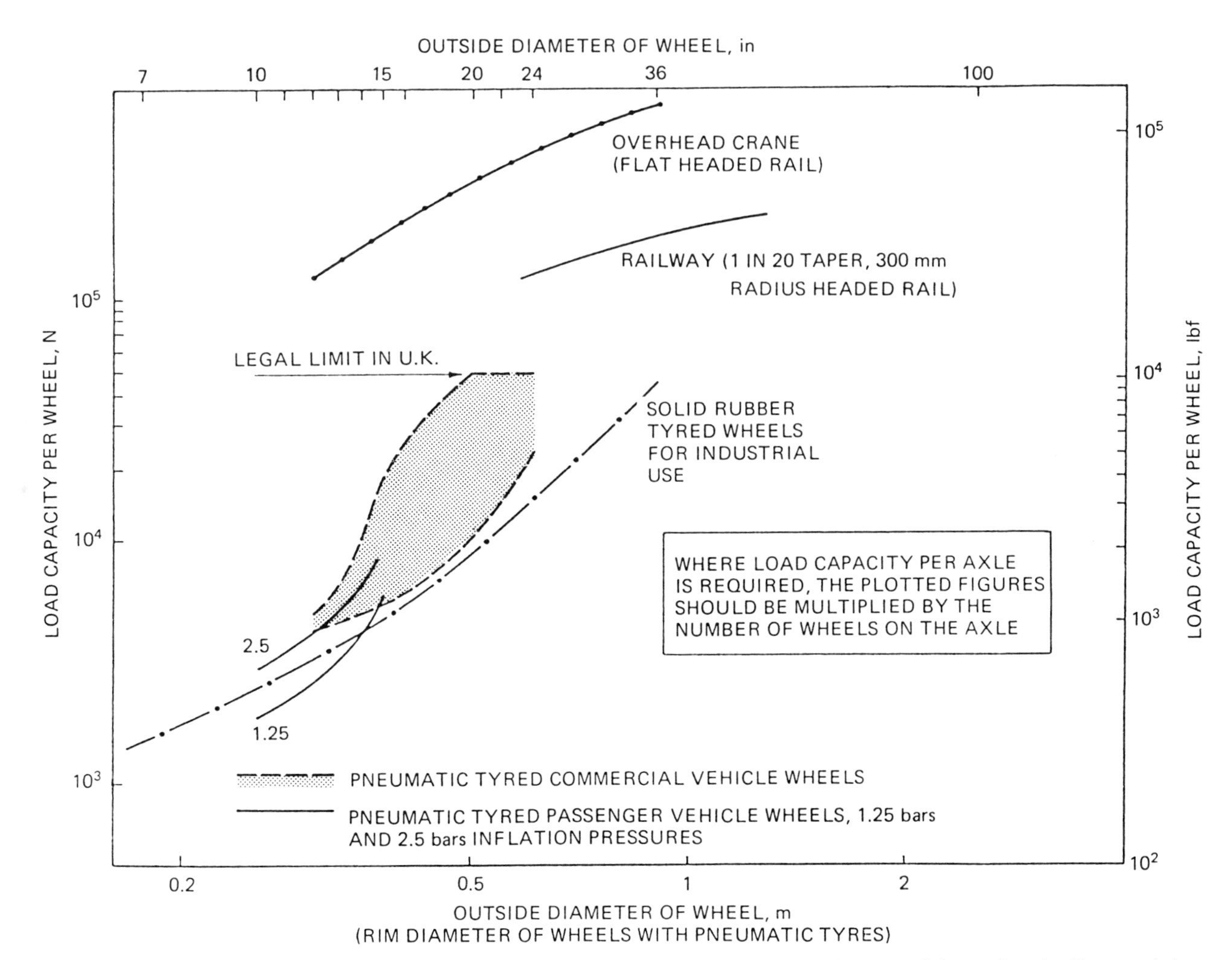

Figure 11.3 Load capacity is plotted against wheel size for typical UK operating conditions. *Steel rail material to ISO 5003 (0.45% C 1% Mn). Steel tyre material to BS 970 Pt 2 1988. Pneumatic tyres are of cross ply construction to ISO 4251 and 4223. Solid rubber tyres are to American Tyre and Rim Association formula taking width and thickness as $\frac{1}{4}$ and $\frac{1}{16}$ outside diameter respectively*

Table 11.2 *(continued)*

	Speed (V)	*Elevated temp.*	*Wheel dia.*	*Material composition*	*Inflation pressure*	*Wear*	*Other factors*
Solid tyres	Load capacity roughly inversely proportional to speed with rubber tyres	Important limiting factor. Maximum dependent on tyre material, e.g. 120°C for rubber tyres	Load capacity proportional to (diameter)$^{\beta}$ where $\beta = 0.5$ to 1.25	Load capacity is dependent on material, physical and mechanical properties, such as yield strength and Young's modulus	Not applicable	Tyre damage, i.e. bond failure, cracking, cutting and tearing limit load capacity	Load capacity reduced if wheel is driven

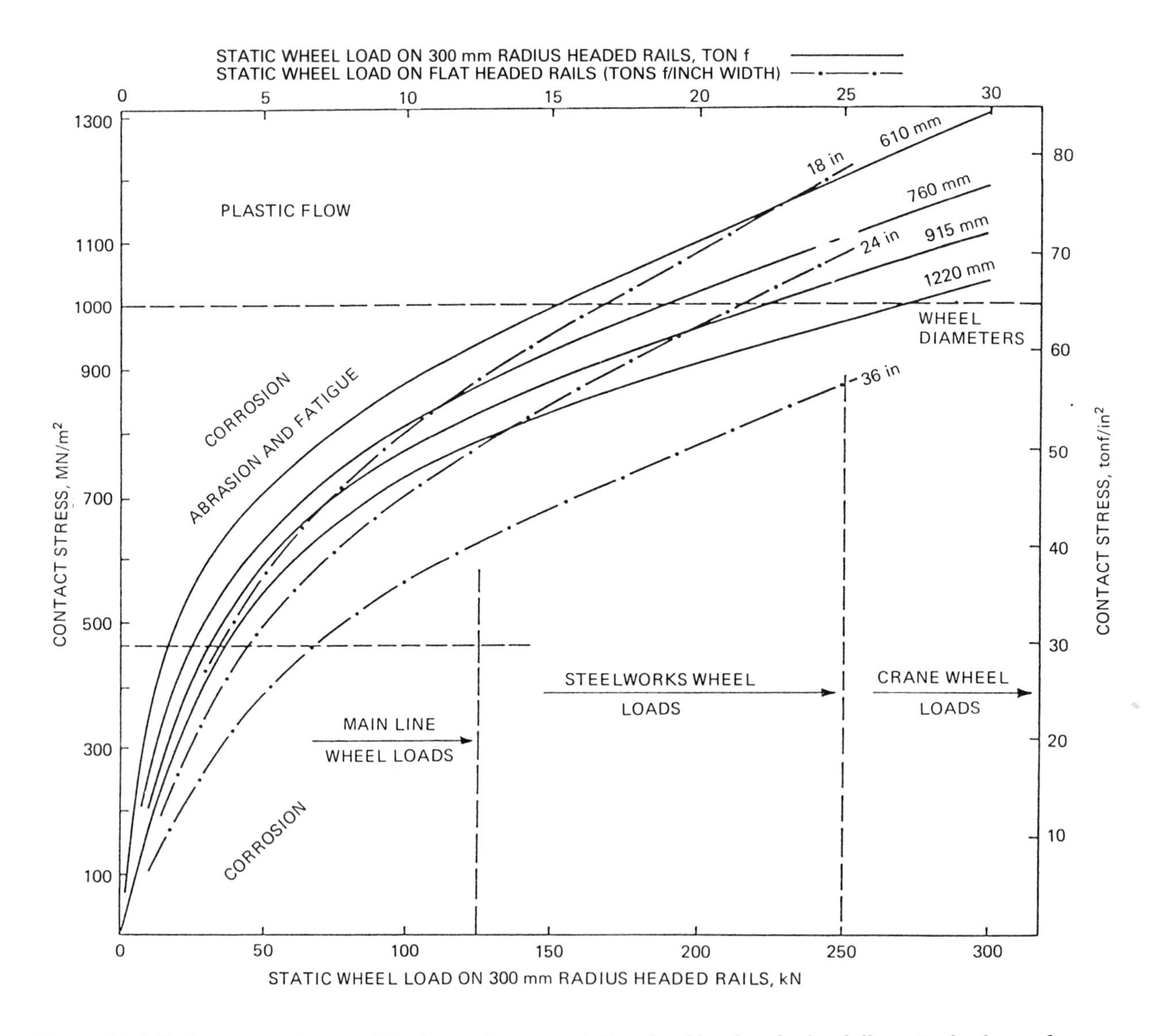

Figure 11.4 Rail contact stress and its dependence on static wheel load and wheel diameter is shown for flat-headed crane wheels and a typical main-line rail with 300 mm head radius. *The predominant wear mechanisms over the ranges of stress are shown. Wear of main-line rails typically takes the form of corrosion followed by abrasion. Fatigue cracking and plastic deformation become important where load and traffic density are high.*

Table 11.3 The effect of various factors on adhesion or traction (T), skidding (S) and rolling resistance (RR)

	Load (W)	Speed	Elevated Temp.	Water	Wheel dia. (D)	Wheel width (ω)	Material	Mating surfaces	Others
Steel tyres	$RR \alpha W^{0.9}$ $T \alpha W$ as a first approximation	Adhesion decreases slightly with speed, see Figure 11.6. Increased resistance with speed results mainly from the effect of rail joints and suspension characteristics	Little effect, except where very high local temperatures are applied to burn off adhesion-limiting surface contaminants on wheel and rail	Light rain has a marked deleterious effect on adhesion. Continuous heavy rain can improve adhesion by cleaning the rail surface	$RR \alpha D^{-\gamma}$ where $\gamma = 0.5$ to 1	Little effect	Little effect	Adhesion can be increased by removing contaminants from the surface of the rail or by sanding	Diesel- or electric-driven wheels have greater adhesion than steam because of the smoother torque
Pneumatic rubber tyres	*RR* increases with load unless inflation pressure is also increased. $T \alpha W$ as a first approximation	See Figure 11.6 Adhesion reduced by the effect of sideways forces α (speed)^2 in cornering	Rubber friction coefficient decreases with temperature – about 10% per 15°C change	Large reduction in friction, see Figure 11.6. Some rubbers are less affected than others	Adhesion increases, while rolling resistance decreases with increase in diameter	Rolling resistance decreases slightly with width with modern tyre designs. Traction increases with width	Radial ply construction gives improvements in all three coefficients. Rubber composition is more important than tread pattern	On wet roads harshness is more important than roughness except at high speeds, see Figure 11.6. Road surface is a more important factor than tyre condition	Aquaplaning speed α (inflation pressure)$^{1/2}$. Wear down to 2 mm tread depth has little effect on adhesion: smooth treads give as little as half the adhesion of unworn patterned treads on poor, wet surfaces
Solid tyres	$RR \alpha W^{4/3}$ $T \alpha W$ as a first approximation	Not suitable for moderate high speeds (above 40 km/h) unless specially designed	Dependent on properties of tyre material	Affected in a similar manner to pneumatic tyres, but more so	$RR \alpha D^{-0.6}$	$RR \alpha \omega^{-0.3}$	$RR \alpha E^{1/2}$ (*E* is Young's modulus)	Solid tyres only suitable for smooth surfaces	

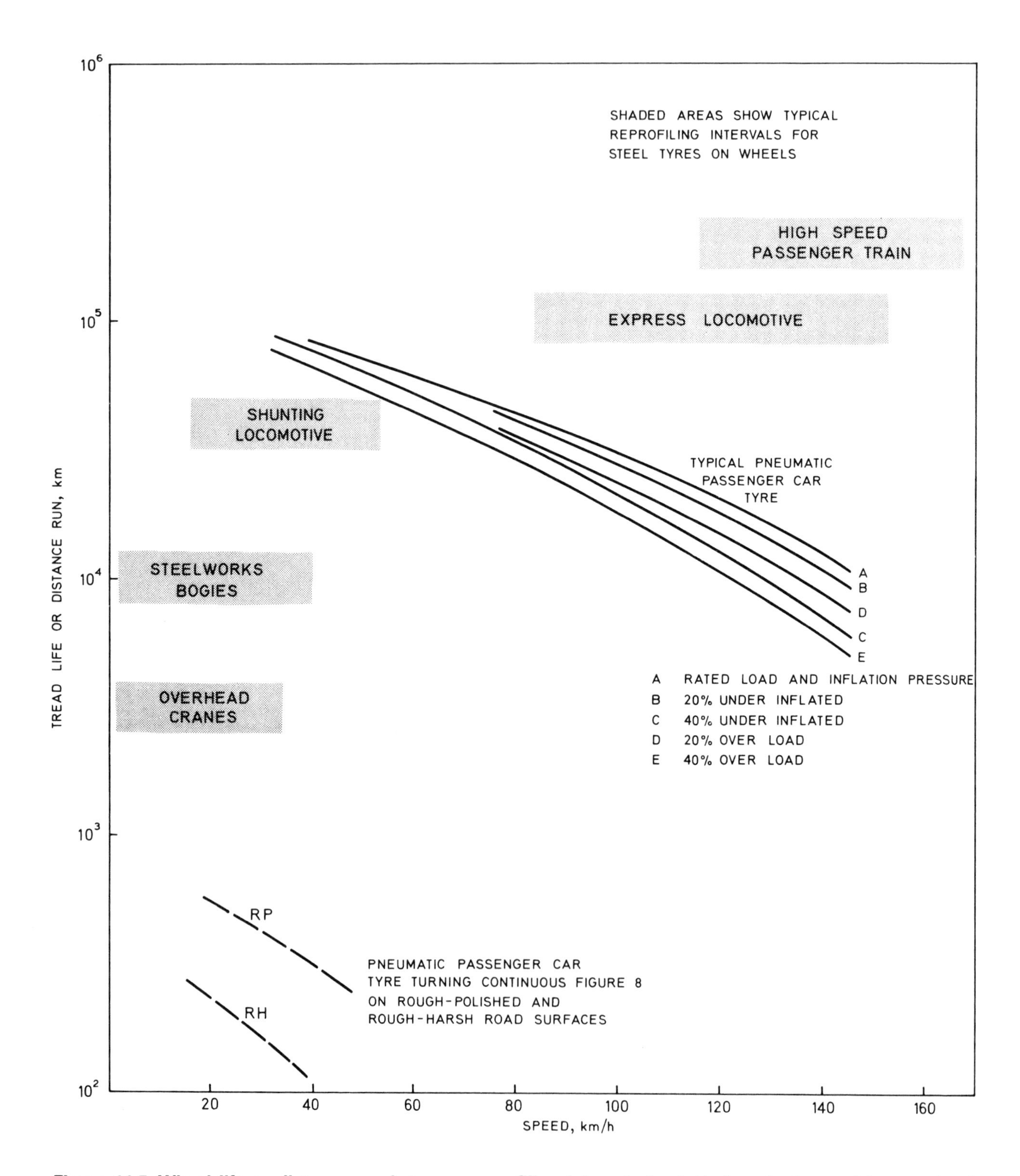

Figure 11.5 Wheel life or distance run between reprofiling intervals is plotted against speed for overhead cranes, different types of railway rolling stock, and passenger-car pneumatic tyres

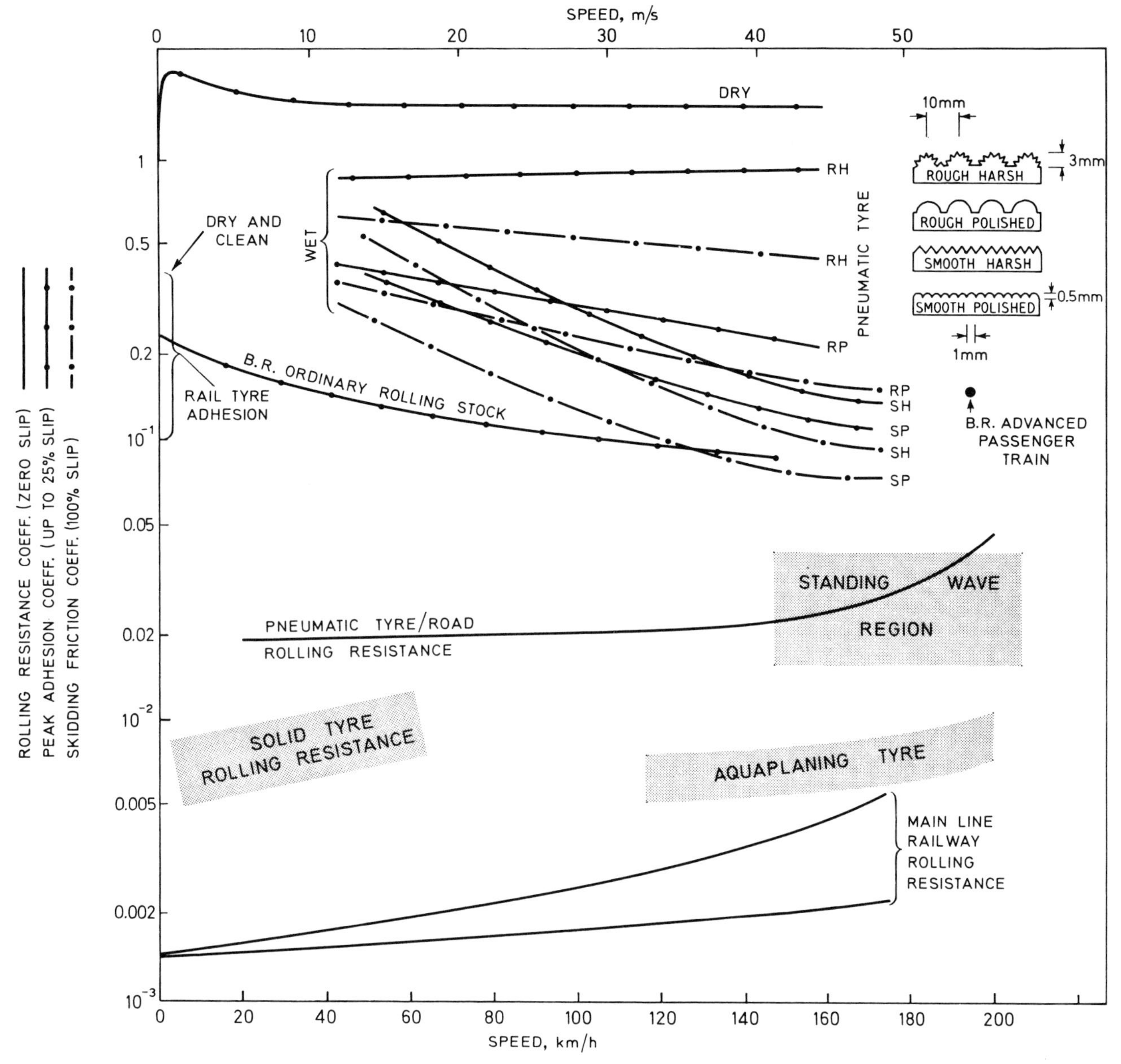

Figure 11.6 Rolling resistance, peak adhesion or traction, and skidding friction coefficients are plotted against speed for passenger-car tyres on wet and dry road surfaces of various typical characteristics, for solid tyres and for main-line railway rolling stock

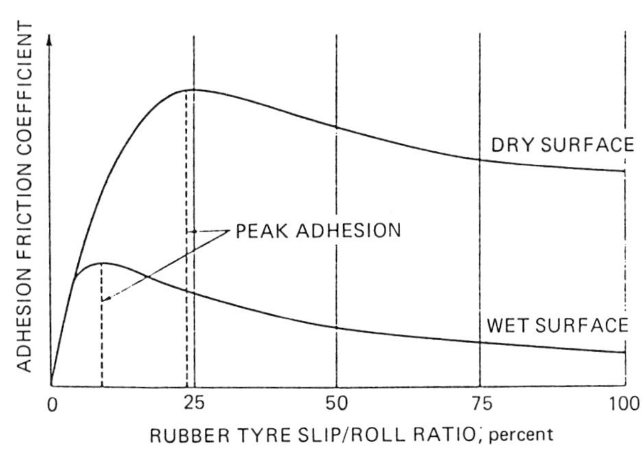

Figure 11.7 Typical adhesion versus tyre slip/roll ratio for treaded rubber tyres on wet and dry surfaces. 0 per cent signifies pure rolling, 100 per cent pure sliding

Capstans and drums are employed for rope drives. The former are generally friction drives whilst the latter are usually direct drives with the rope attached to the drum. The roles, however, may be reversed. Friction drives control the rope motion by developing traction between the driving sheave and the rope and might be preferred to direct drives for reasons of economy (smaller drive sheave), safety (slippage possible) or necessity (e.g. endless haulage).

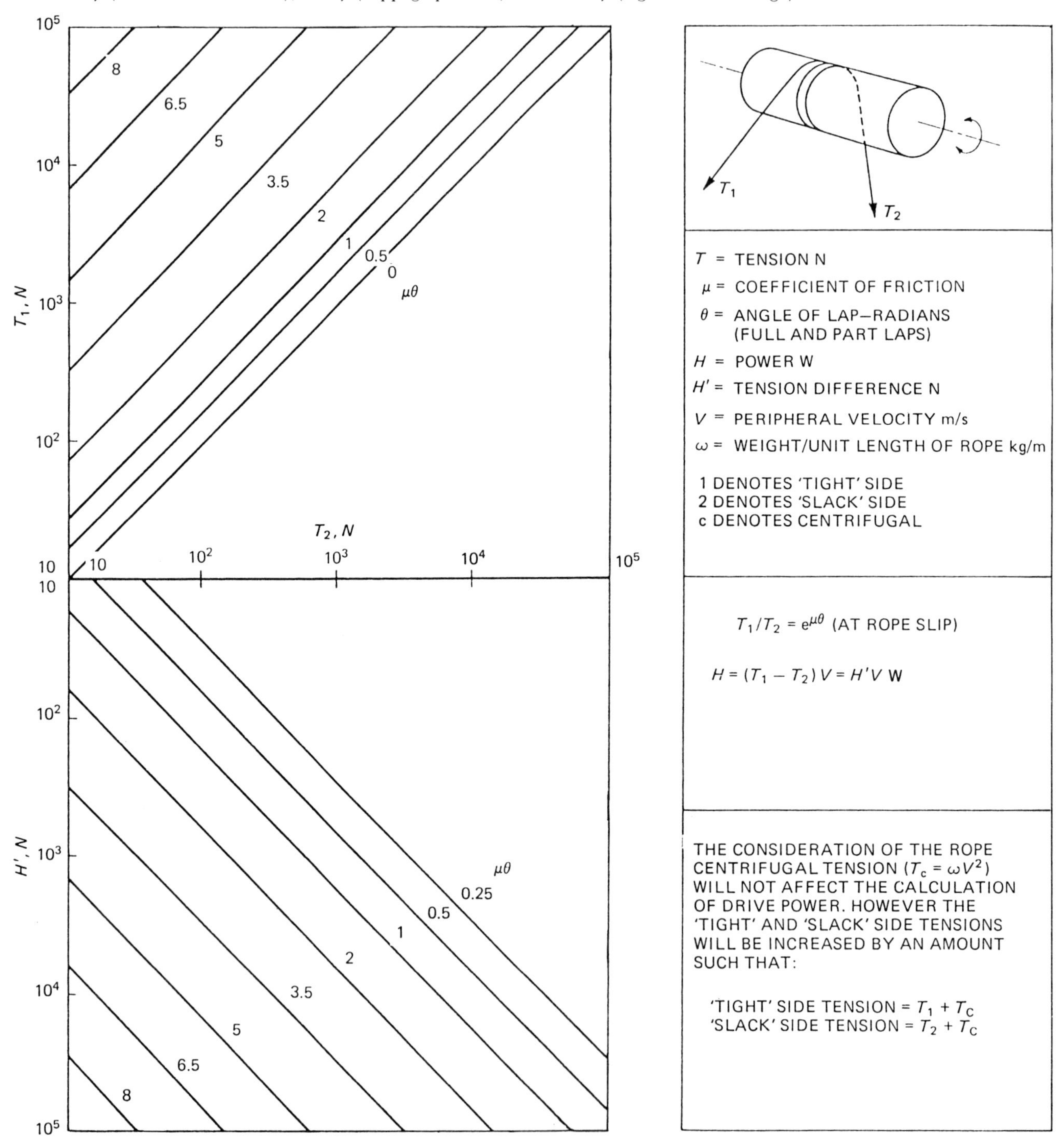

FRICTION DRIVES

The figure shows the determination of the approximate performance of friction drives at the rope slip condition. In contrast to the belt and pulley situation the required 'tight' or 'slack' side tension is usually known. Capstans are widely used with vegetable, animal or man-made fibre ropes, but more rigorous conditions demand wire rope.

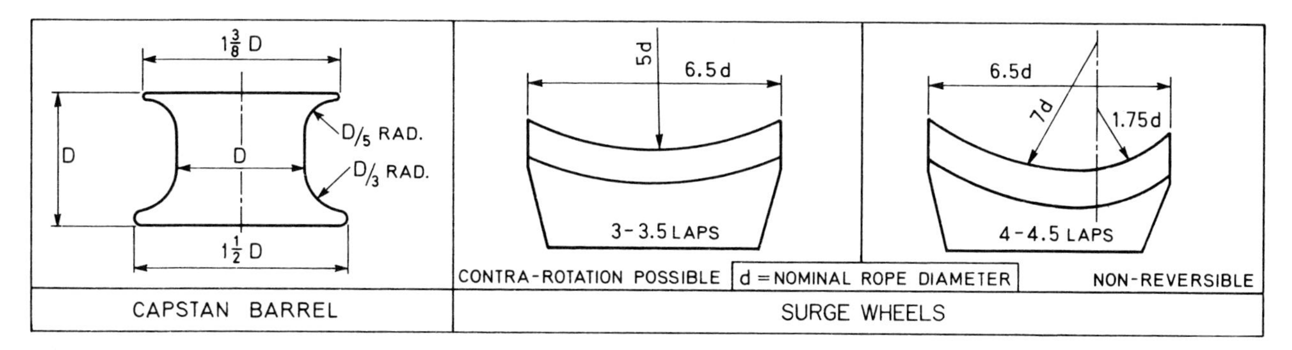

The figure shows the typical profiles of a capstan barrel and surge wheels. Grooving is not appropriate. Capstans are often employed when relatively low rope tensions are involved and hence must have a large flare, which whilst ensuring free movement does not allow disengagement of the rope. Surge wheels are used on endless haulage systems with wire ropes. The large rope tensions involved mean that only a moderate flare is necessary. The laps slip or surge sideways across the surface as the rope moves on and off the wheel, hence the term surge wheel. This movement necessitates differing wheel shapes depending upon the rotational requirements.

	PLAIN	U	V	UNDERCUT U
GROOVE PROFILE			α (TYPICALLY 40°)	α
APPARENT COEFFICIENT OF FRICTION μ'	μ	$4\mu/\pi$	$\mu/\sin\frac{\alpha}{2}$	$\frac{4\mu(1-\sin\frac{\alpha}{2})}{\pi-\alpha-\sin\alpha}$

The figure shows grooving in friction drives to increase tractive effort. The apparent coefficient of friction, μ', replaces the normal coefficient in calculations.

Rope material	*Drive sheave or sheave liner material*	*Friction coefficient at slip (dry conditions)*
Wire	Iron or steel	0.12
	Wood	0.24
	Rubber or leather	0.50
Nylon	Aluminium	0.28
	Iron or steel	0.25
Woven cotton	Iron or steel	0.22
Leather	Iron or steel	0.50

For wet conditions reduce the friction coefficient by 25%, for greasy conditions 50%

The table gives an approximate guide to the value of friction coefficient at slip for various rope and driver material combinations. A factor of safety reducing the values shown and appropriate to the application is usually incorporated.

DIRECT DRIVES

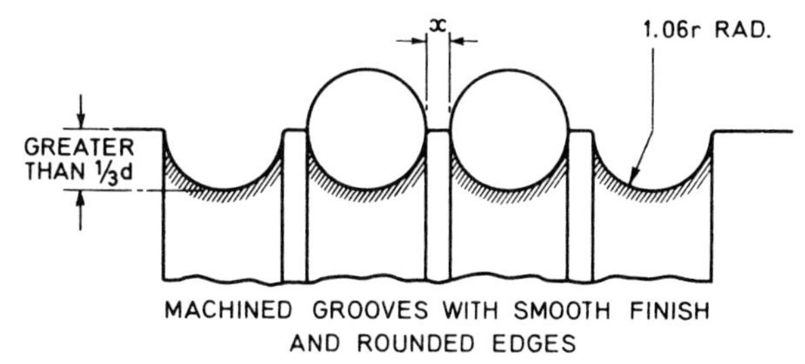

MACHINED GROOVES WITH SMOOTH FINISH AND ROUNDED EDGES

ROPE NOMINAL DIAMETER d	CLEARANCE BETWEEN TURNS x
UP TO 13 mm	1.6 mm
UP TO 28 mm	2.4 mm
UP TO 38 mm	3.2 mm

d = 2r = NOMINAL ROPE DIAMETER
x = CLEARANCE BETWEEN TURNS

The figure shows typical grooving of a wire rope drum drive with the rope attached to the drum. Performance is unaffected by frictional considerations. Pinching of the rope is avoided where grooves are employed for guidance purposes. Drum grooves are normally of cast iron, carbon steel or alloy steel and reduce wear of both drum and rope.

SELECTION OF LOAD CARRYING WIRE ROPES

Table 13.1 Ropes for industrial applications

Application	Conveyors, small hoists, small stays, trawl warps*	Excavators (drag), lifts,* sidelines (dredging), mast stays, pendants, crane ropes (small)	Lifts,* drilling lines, scrapers, trip ropes, skip hoists, pile driving	Crane hoists (multi-fall), grabs, slings	Tower cranes (hoist), mobile cranes (small), drop ball cranes, boatfalls	Deck cranes (single fall), mobile cranes (large)
Construction	6 × 7(6/1) *IWRC*	6 × 19(9/9/1) *IWRC*	6 × 19 (12/6 + 6F/1) *IWRC*	6 × 36 (14/7 and 7/7/1) *IWRC*	17 × 7(6/1) *FC*	12 × 6 over 3 × 24
MBL kN	$d^2 \times 0.633$	$d^2 \times 0.632$	$d^2 \times 0.643$	$d^2 \times 0.630$	$d^2 \times 0.562$	$d^2 \times 0.530$
Wt kg/100 m	$d^2 \times 0.382$	$d^2 \times 0.398$	$d^2 \times 0.408$	$d^2 \times 0.406$	$d^2 \times 0.372$	$d^2 \times 0.362$
E N/mm^2	68 600	64 700	64 700	64 700	53 900	53 900

d = nominal diameter of rope
FC = Fibre Core
E = Modulus of Elasticity of the rope (differs from elastic modulus of wires in rope due to helical formation of latter)
MBL = Minimum Breaking Load
$IWRC$ = Independent Wire Rope Core

* These applications have Fibre Main Cores

Notes: Minimum Breaking Loads based on a tensile strength of 1.8 kN/mm^2

With Fibre Cores: Reduce *MBL* by 8% approximately.
Reduce weight by 9% approximately.

Construction and fatigue performance

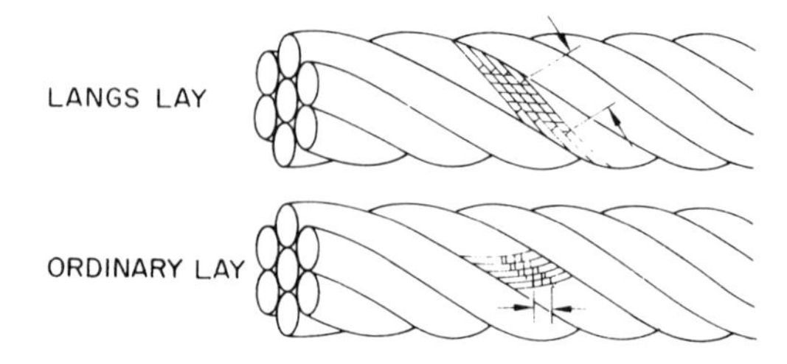

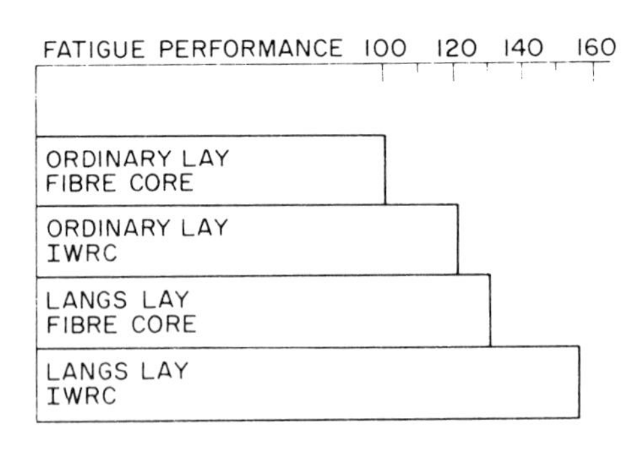

Figure 13.1 Comparison of ropes in fatigue based on a factor of safety of six

Wires in a Langs Lay rope are laid in the same direction as the strands, thus rope in this construction can permit a greater reduction in volume of steel from the rope surface for the same loss in breaking load than is possible with an ordinary lay rope.

Langs Lay ropes have greater resistance to abrasion, and are also superior to ordinary lay ropes in fatigue resistance. However, they develop high torque, and cannot be used when this factor is of importance.

Table 13.2 Ropes for mining applications

Application	Haulage (general), conveyors, winding (small sizes), cableway haulage	Winding	Winding—drum hoists, friction hoists	Shaft guides	Balance for drum and friction hoists	Aerial carrying ropes, cableways
Construction	6 × 8(7/1)Δ flattened strand 160 Grade	6 × 25 12/12/1 Br 9/3 flattened Strand 180 Grade	Locked coil—winding	Locked coil—guide	'Superflex' 20 × 6/17 × 6/ 13 × 6/6 × 19 110 Grade	Locked coil—aerial
MBL kN	$d^2 \times 0.565$	$d^2 \times 0.613$	$d^2 \times 0.851$	$d^2 \times 0.500$	$d^2 \times 0.368$	$d^2 \times 0.809$
Wt kg/100 m	$d^2 \times 0.407$	$d^2 \times 0.413$	$d^2 \times 0.563$	$d^2 \times 0.550$	$d^2 \times 0.385$	$d^2 \times 0.568$
E N/mm^2	61 800	58 800	98 100	117 700	53 900	117 700

Loading and performances

Static load $(T_S) = W_C + W_L + W_R$

Static factor of safety $= MBL: T_S$ (i.e. $FOS = 6.5:1$)

for cranes etc. W_R is not considered

Dynamic load $(T_D) = T_S \cdot a$

$$\text{Bending load } (T_B) = \frac{E\Delta A}{D}$$

Total load $(T_{max}) = T_S + T_D + T_B$

$$\text{Percentage stress} = \frac{T_{max} \times 100}{MBL}$$

(See Figure 13.2 for stress reversals based on percentage stress.)

$$\text{Elastic stretch} = \frac{T_S \cdot L}{EA}$$

W_C = Weight of conveyance and attachments
W_L = Payload
W_R = Weight of suspended rope
a = Acceleration
g = Gravitational constant
E = Elastic Modulus (based on rope area)
Δ = Diameter of rope at centre of outer wires
A = Area of rope
D = Diameter of drum or sheave
L = Rope length

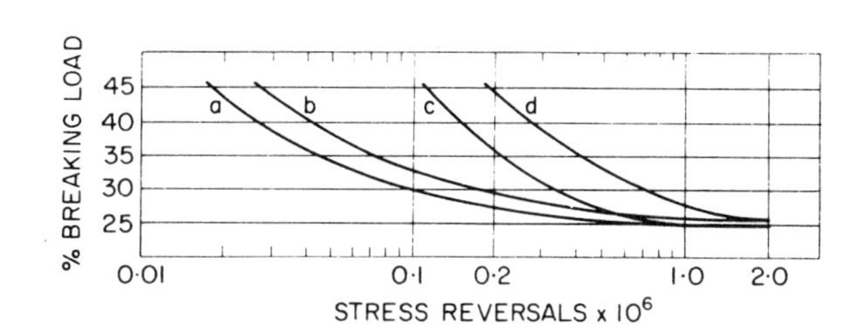

Figure 13.2 Fatigue tests on wire ropes *(a) flattened strand, winding (b) locked coil, winding (c) 6 × 25 RS IWRC ordinary lay (d) 6 × 36 IWRC ordinary lay*

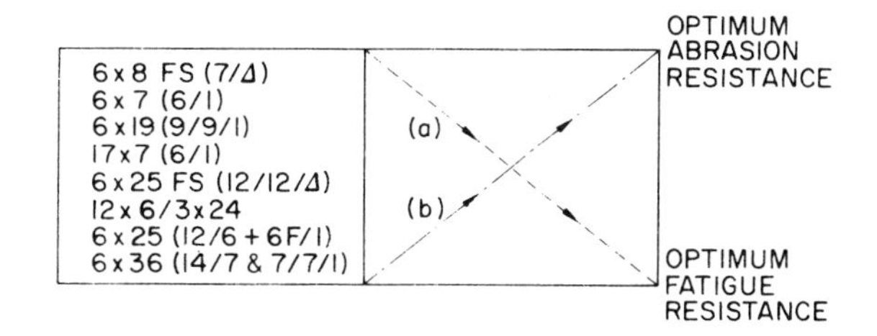

Figure 13.3 Comparison of construction for *(a) resistance to bending fatigue (b) resistance to abrasion and crushing*

SELECTION OF CONTROL CABLES AND WIRES

The form of cables and wires which can be used for control purposes can be defined as follows:

Wire—a single 'solid' circular section i.e. piano or music wire.

Single strand— an assembly of wires of appropriate shape and dimensions, spun helically in one or more layers around a core.

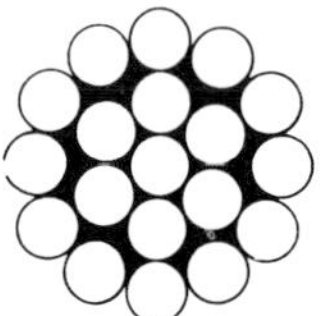

19 WIRES

Multi strand—an assembly of strands spun helically in one or more layers around a core.

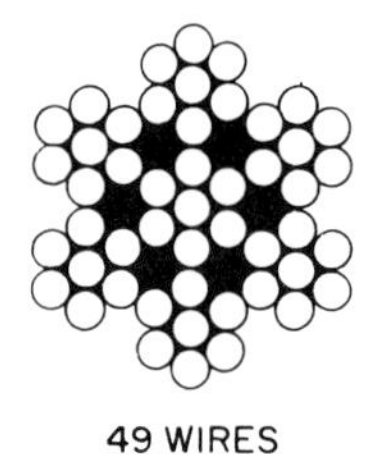

49 WIRES

Wire can be used in push and pull conditions inside suitable conduits (outer casings). Single strand cable should be used inside conduits for the transmission of tension loads. Rope constructions are used around pulleys. Single strand cable reinforced with a helically close wound steel tape around the periphery of the strand provides sufficient resistance to buckling for use within suitable conduits to transmit both push and pull loads.

Load capacity

Minimum breaking loads of cable taking into account the helical effects of their construction are approximately 85% of the calculated aggregate breaking load. Considerations of safety factors to cater for possible intermittent increased load conditions modify considerably the load capacity of cables from their minimum breaking load values to practical load values for good design. The table gives a recommended typical selection of single strands in relation to their use within conduit.

Table 14.1 Performance of stranded cables

Strand construction	*Overall diameter* mm	*Minimum breaking load* N	*Conduit type*	*Overall diameter conduit* mm	*Conduit minimum bend radius* mm	*Recommended maximum input load* N
19 WIRES	1.53	2220	Single wire section	4.45	54	220
	2.00	3340	Twin wire section	8.10	127	510
	2.80	8000	Twin wire section	9.80	150	1780
	3.50	12 900	Twin wire section	11.40	170	3380
36 WIRES	5.50	28 900	Twin wire section	16.40	250	8400

Notes: Overall diameter of strand excludes thermoplastic covering. Conduits incorporate appropriate thermoplastic liners. Minimum bend radii for conduits may well vary from the above values when the conduit construction comprises long lay wires.

Pulley size and cable fatigue

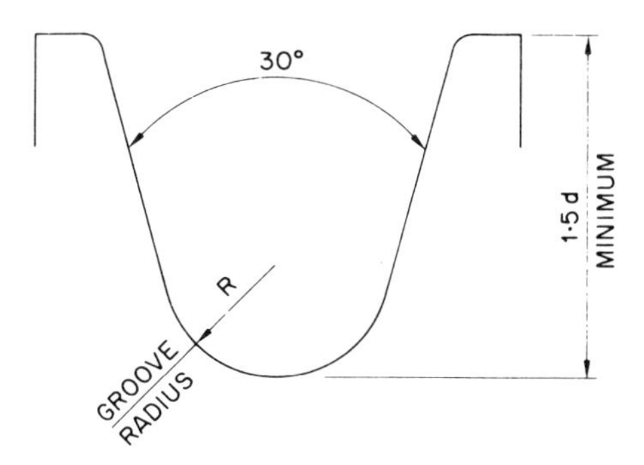

Figure 14.1 Pulley groove form for control purposes. *Groove radius R = 0.53d where d = cable diameter inclusive of any thermoplastic covering*

Although Bowden cables were conceived to eliminate the use of ropes and pulleys, control configurations using steel wire ropes and pulleys are in use.

For very small rope below 2 mm diameter, pulley diameter (over groove) should not be less than thirty times rope diameter. For small rope used as controls (2 to 6 mm diameter), pulley diameter over groove should not be less than thirty-five times actual rope diameter (exclude the effect of any plastic covering). Nylon or acetal moulded pulleys give excellent life, resist corrosion and reduce friction and wear between rope and pulley. Prediction of fatigue life of ropes around pulleys is difficult and involves many factors. Practical tests have shown that provided the minimum recommended pulley diameters are used and that the tension load in the rope does not exceed 10% of the minimum breaking load of the rope, long life can be expected. Angles of wrap between 10° and 30° should be avoided; reverse bends should be avoided wherever possible.

The following table shows the effect on fatigue life with varying

$$\frac{\text{pulley dia. over groove } (D)}{\text{rope dia. } (d)}$$

on a single pulley. 7 × 7 rope 2.60 mm (d), nylon covered to 3.60 mm. Pulley—nylon; axle mild steel, grease lubricated. 90° angle of wrap. Tension load in rope 625 N. Minimum breaking load of rope 6250 N.

Table 14.2 Effect of diameter ratio on fatigue life

D/d	*Cycles to failure*
20	56 000
25	170 000
30	330 000
37.5	after 650 000 no failure

Friction in Bowden cables

The friction of wires and cables inside conduits (Bowden cables) depends on the materials and construction of the cable and conduit and on the bend radii obtained when the system is installed. Control runs employing flexible conduit should take a natural path between the conduit grounding points.

The materials used may be:

Metal cable in a metal conduit with grease lubrication.
Metal cable in a plastic lined conduit.
Plastic coated metal cable in a conduit lined with a dissimilar plastic.

The last combination gives low friction and high efficiency compared with the first, as shown in the figure below.

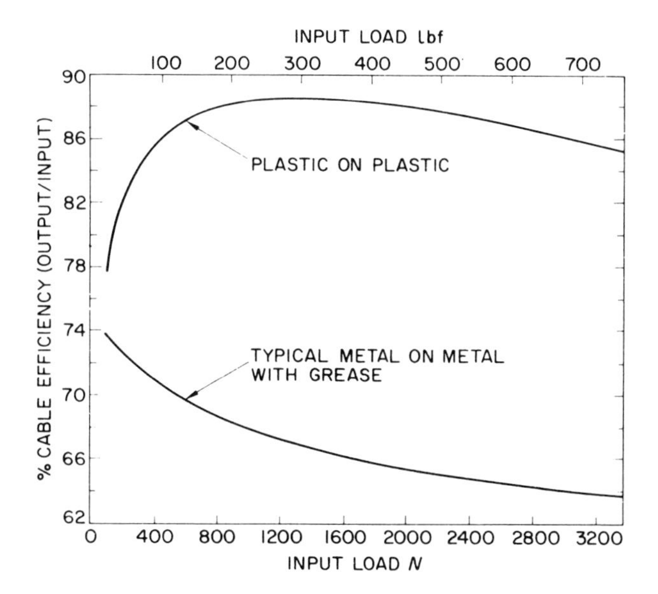

Figure 14.2 Variation of cable efficiency with input load. *Configuration: 180° 'U' bend 210 mm radius and conduit length 800 mm. For plastic on plastic cables with 90° bend add 2% to the above values of efficiency. With 270° bends deduct 5% from the above values of efficiency*

GENERAL CHARACTERISTICS

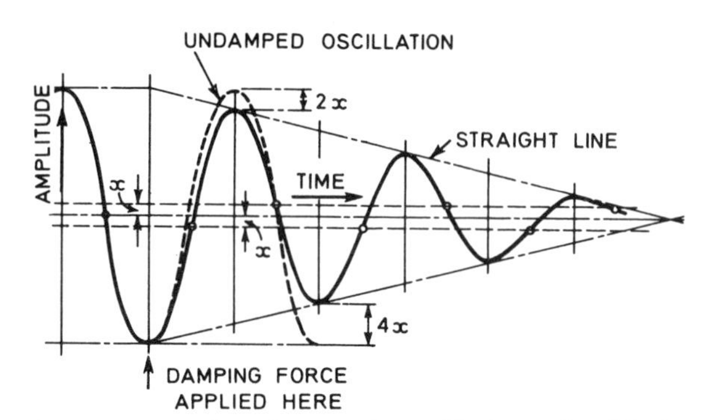

Figure 15.1 The decay of oscillations by friction damping

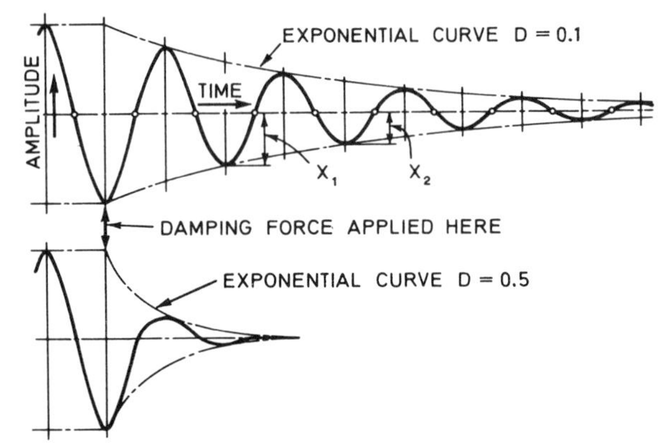

Figure 15.2 The decay of oscillations by viscous hydraulic damping

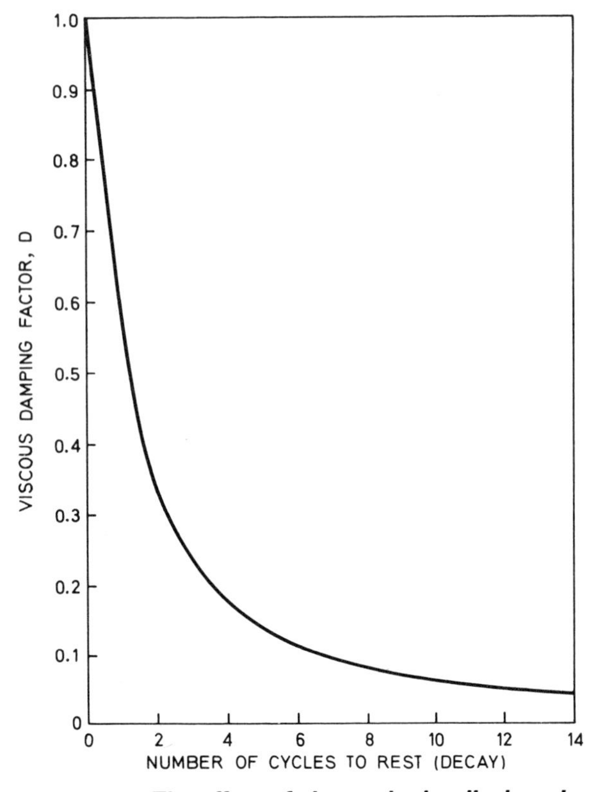

Figure 15.3 The effect of viscous hydraulic damping factor on the number of cycles required to come to rest after a single impact

Damping devices are used to provide forces to resist relative motion and oscillation. The two main types are friction dampers, using friction between solid components and hydraulic dampers using mainly viscous effects.

If an oscillation is damped with a friction damper, the oscillation decay will follow a straight line (Figure 15.1). The amplitude of each successive oscillation will be reduced by $4x$

where $x = F/k$

and F = damping force (newtons)

k = system spring rate (newtons/metre)

With a hydraulic damper, the decay of the oscillation follows an exponential curve (Figure 15.2).

The diagram of the damping action of a hydraulic damper shows the effect of different damping factors. These are related to the number of cycles that it takes for the system to come to rest after an impact displacement. This relationship can be used to make an experimental measurement of the damping factor of an existing system.

The relationship between damping force and velocity varies with the type of damper and can conveniently be described by the formula

$$F = cV^n$$

where c = a constant which depends on factors such as the size of the damper

V = the displacement velocity

n = a constant which depends on the working principle of the damper

Type of damper	*Value of n*	*Force characteristic*
Friction damper	0	F is constant for all values of V
Hydraulic damper with constant area flow passages	2	F is proportional to V^2
Hydraulic damper with valves to control the flow	1	F is proportional to V

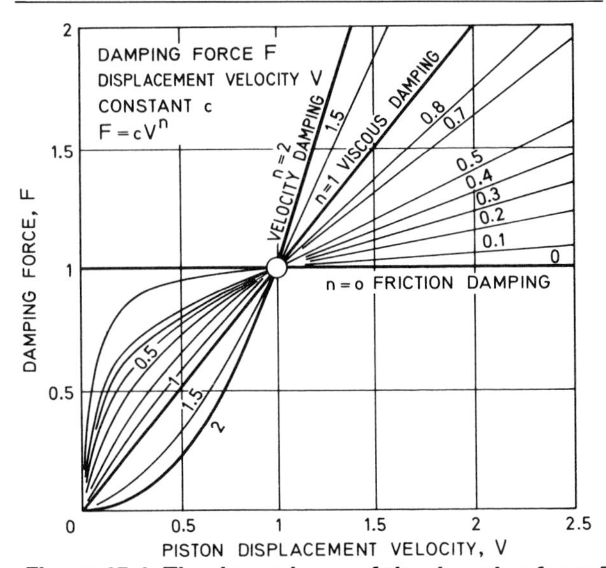

Figure 15.4 The dependence of the damping force F on the piston velocity V at various values of the exponent n

PERFORMANCE OF HYDRAULIC DAMPERS FITTED WITH VALVES

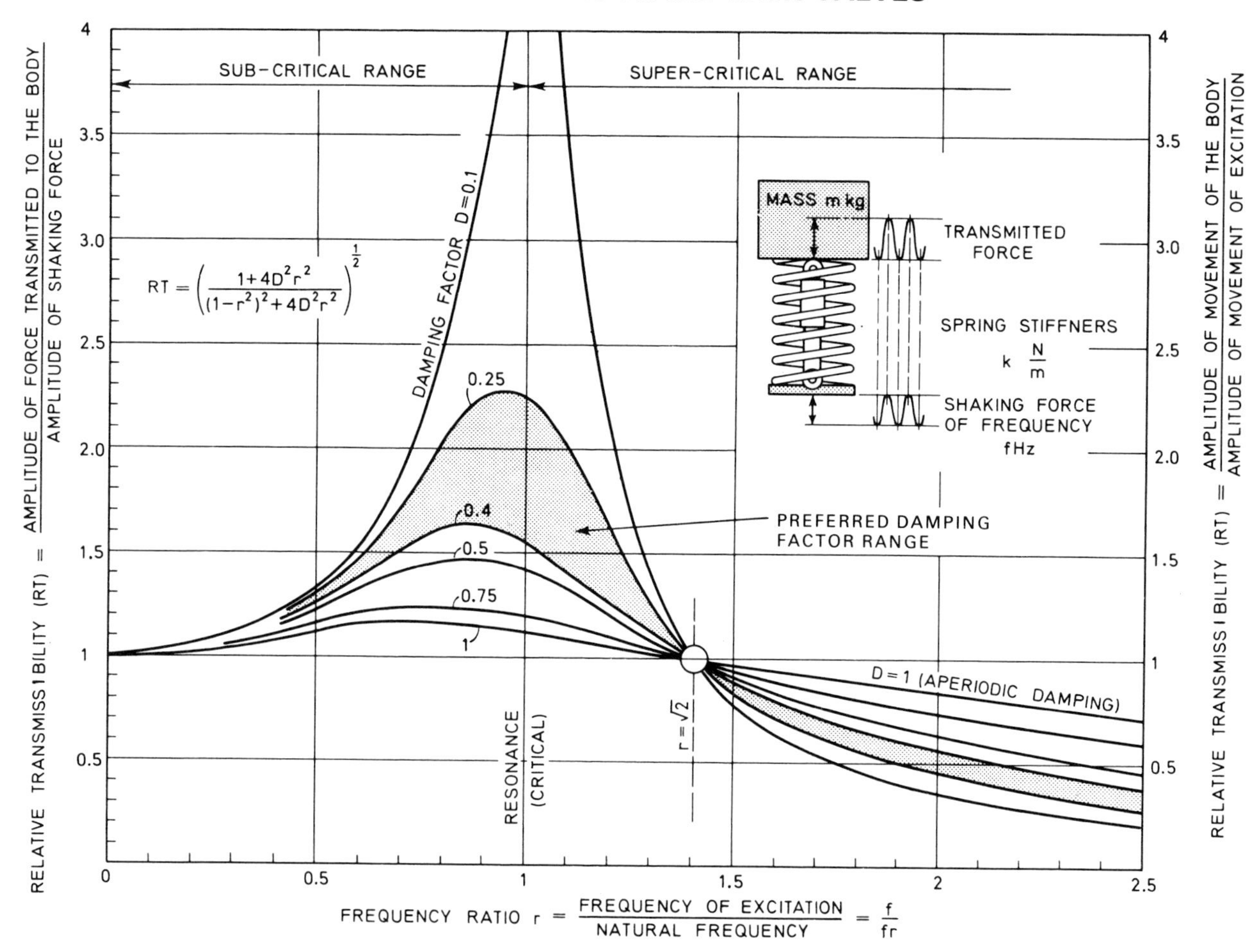

Figure 15.5 Curves of relative transmissibility for a number of representative viscous damping factors

If two components are connected by a spring system, such as a wheel and the body of a vehicle, the forces and oscillating movement transmitted by one to the other, will depend on the frequency of the oscillations being transmitted in relation to the natural frequency of the system, and the damping factor of any damping device which is incorporated.

In a viscous hydraulic damper, in which F is proportional to V, the damping factor D is given by

$$D = \frac{C_1}{4\pi\, m f_r}$$

where C^1 = damping force per unit velocity N/ms^{-1}
m = mass of the critical component kg
f_r = natural (resonant) frequency of the system

$$\frac{1}{2\pi}\left(\frac{k}{m}\right)^{1/2} \text{ Hz}$$

This damping factor is not affected by velocity for a damper of this kind, and consequently the relative transmissibility of forces and movements will depend mainly on the frequency ratio r, of the excitation frequency to the resonant frequency.

For a system of this kind with resonance at $r = 1$, damping is desirable over the range $r = 0$ to $r = 1.4$. If r is normally greater than 1.4 low D values become more desirable. For conditions extending over a wide range of r, values of $D = 0.25$ to 0.4 are found to give the best results.

Viscous hydraulic dampers of good design are available with adjustable valves to give controlled relationships between damping force and damper velocity.

POINTS TO NOTE IN DAMPER SELECTION AND DESIGN

1. Friction dampers can potentially absorb the most energy, but are unsuitable for systems subject to steady oscillations at or near resonance, unless the friction dampers are large enough to provide a damping force of at least four times any excitation force.
2. To give satisfactory performance with acceptable wear rates, the contact pressures in friction pad dampers should not exceed 250 kN/m^2 (34 p.s.i.).
3. With hydraulic dampers, damping factors in the range 0.25 to 0.4 are generally suitable, but at frequency ratios above $r = 2.5$ the transmitted forces are higher than with friction dampers.
4. With viscous hydraulic dampers giving straight line force/velocity characteristics it is important to specify the velocity at which the force should be levelled to a constant value by a cut-off valve. A valve of adequate capacity is needed to prevent instantaneous velocity peaks from bursting the housing or causing damage to the working valves or the end mountings. The maximum allowable pressure at cut-off is usually of the order of 3.5 MN/m^2 (500 p.s.i.).
5. If damping devices are likely to be operated near the resonant frequency of the system, the rigidity of the rubber pads or mounting bushes need to be checked for adequate stiffness.

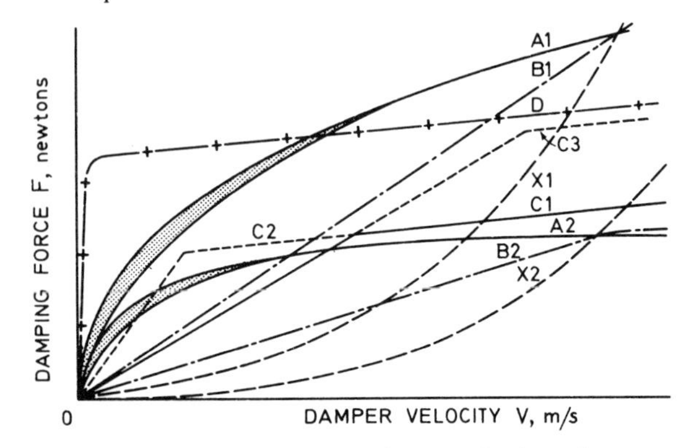

Figure 15.6 The range of force-velocity characteristics attainable with a damper of the type illustrated. *These characteristics are obtained as follows:*

A1 *with one adjustable valve and a small diameter passage*
A2 *as above, but a larger diameter passage*
B1 *as A1 but a smaller orifice in the adjustable valve*
B2 *as A2 but a larger diameter orifice in the adjustable valve*
C1 *with two adjustable valves, the first valve controlling up to the cut-off point along lines of B1 and B2, and the characteristic above cut-off being determined by the second adjustable valve. The bore in the seat of the second adjustable valve is larger than that of the first*
C2 *as above, but with a different bore and orifice in the first valve*
C3 *as C1 but with the second valve set more closely*
D *with two or more adjustable valves, all set to identical values*

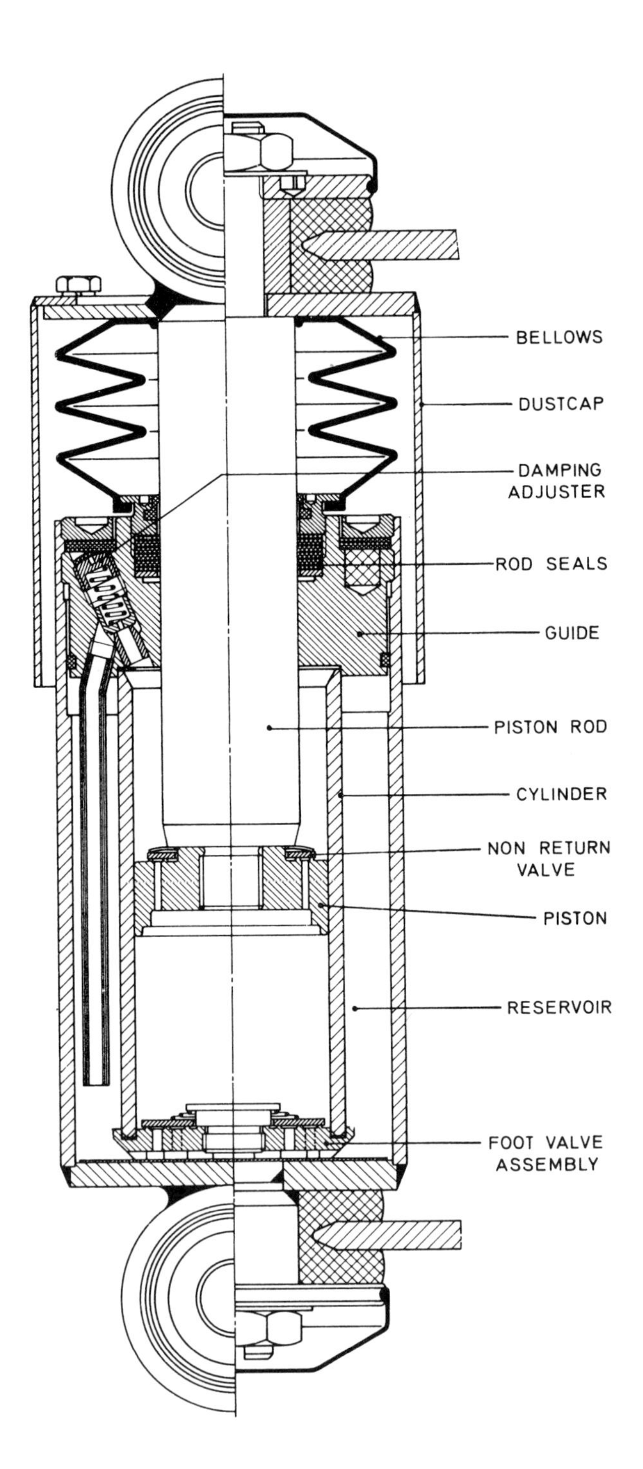

Figure 15.7 A viscous hydraulic damper. *The adjustable valve provides a simple method of changing the slope of the force-velocity characteristic. The damper is intended to deal with vertical oscillation and shows alternative types of end fixing*

HYDRAULIC PISTONS

Type A—Semi-static operation (pressure unidirectional)

Piston may be made from aluminium or cast iron. Operation may be by a push rod or a piston rod bolted through the piston. Sealing is usually by a cup washer fitted against a flat end face.

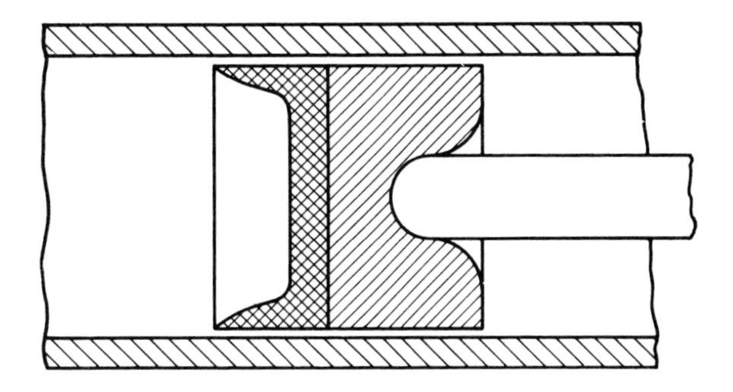

Type B—Double-acting hydraulic actuators

Pistons of aluminium, cast iron or steel, may be fitted with lip seals ('U' packings) for operating pressures up to 21 MN/m^2 (3000 p.s.i.), or PTFE or metallic rings.

Hydraulic piston with 'U' packings

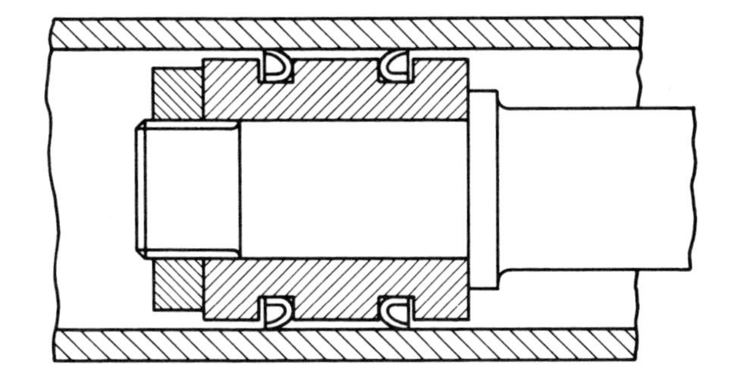

Hydraulic piston with PTFE sealing rings and bearer bands

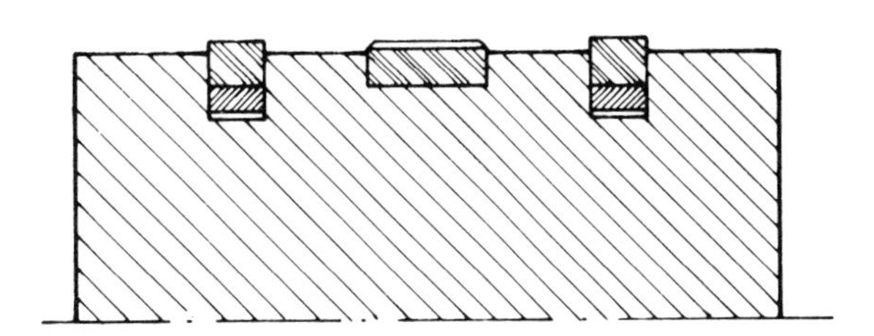

Scarf-step sealing ring arrangement, hydraulic or pneumatic: (a) unidirectional; (b) double-acting

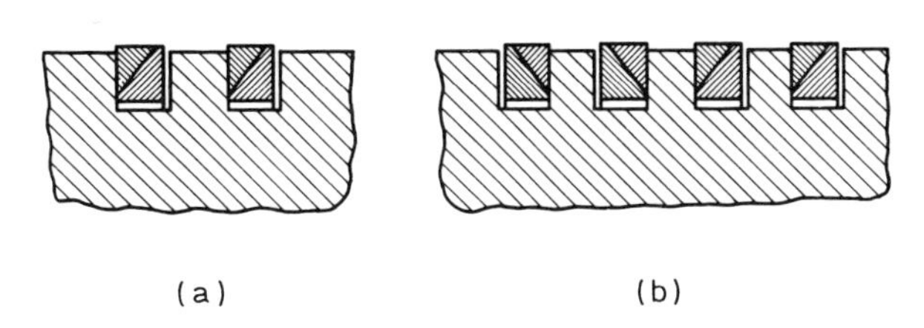

Figure 16.1 Various types of hydraulic pistons

CRANK OPERATED HYDRAULIC PUMP PISTONS

Piston may be made from aluminium or cast iron, with sealing usually by metallic piston rings. The minimum strength of the piston parts should be as for IC engines.

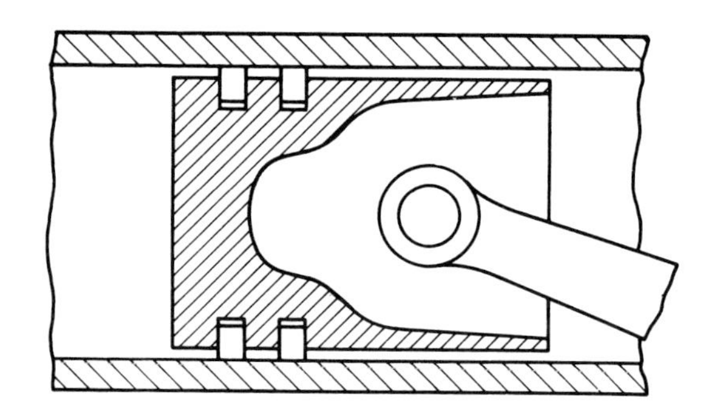

Some simple robust low-speed designs use a plain, solid cylindrical piston operating through a gland bush.

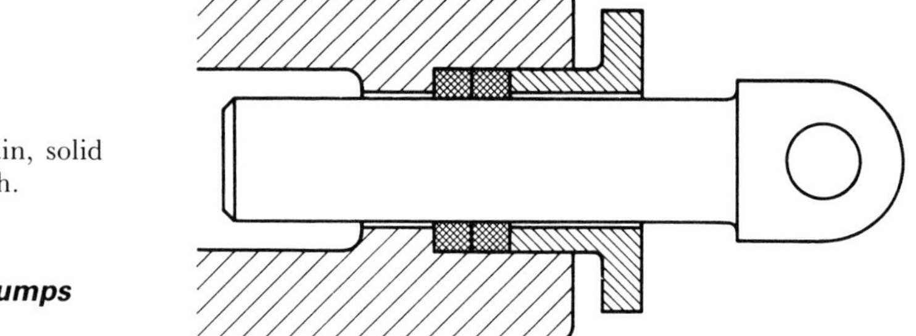

Figure 16.2 Types of pistons for hydraulic pumps

AIR AND GAS COMPRESSOR PISTONS

Piston materials may be aluminium or cast iron. Some common arrangements are shown below. Minimum strength of piston parts to be as for IC engines.

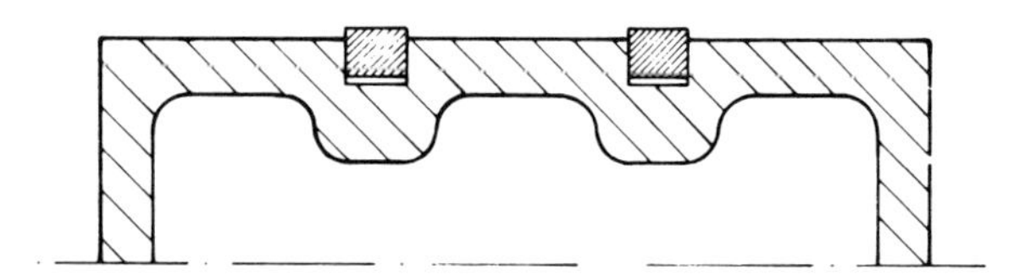

Double-acting compressor piston (lubricated) with plain piston rings

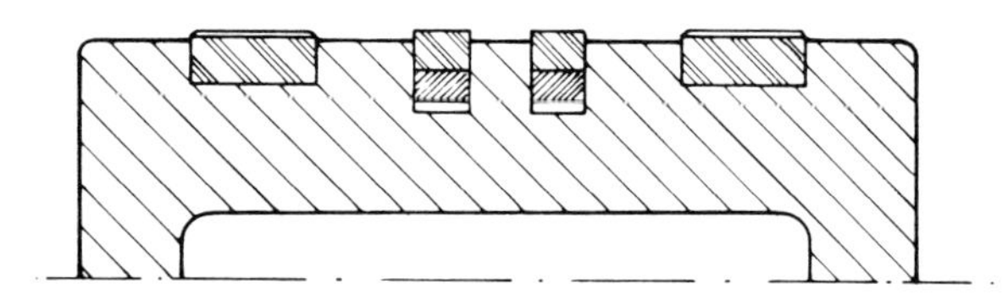

Double-acting compressor piston (non-lubricated) with PTFE sealing rings and bearer bands

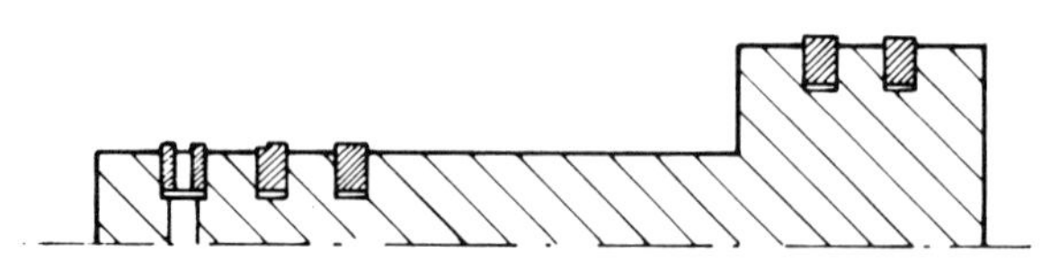

Two-stage compressor piston (lubricated)

Two-stage compressor piston (non-lubricated) with PTFE sealing rings and bearer bands

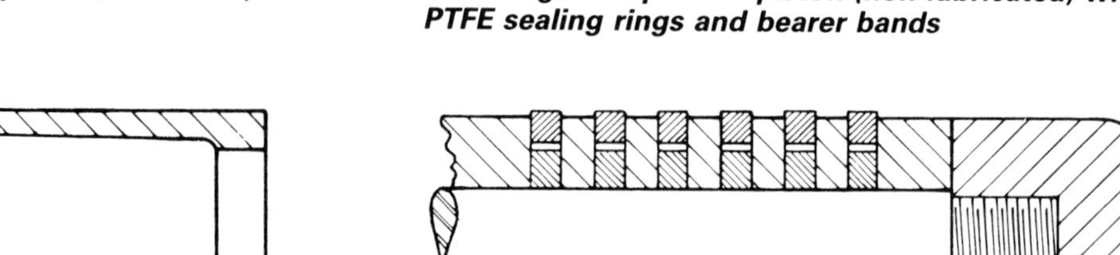

Simple single-stage piston (lubricated)

Built-up piston for high-pressure compressor

Figure 16.3 Pistons for air and gas compressors

GASOLINE ENGINE PISTONS

The key features of a piston for a gasoline engine are shown in Figure 16.4.

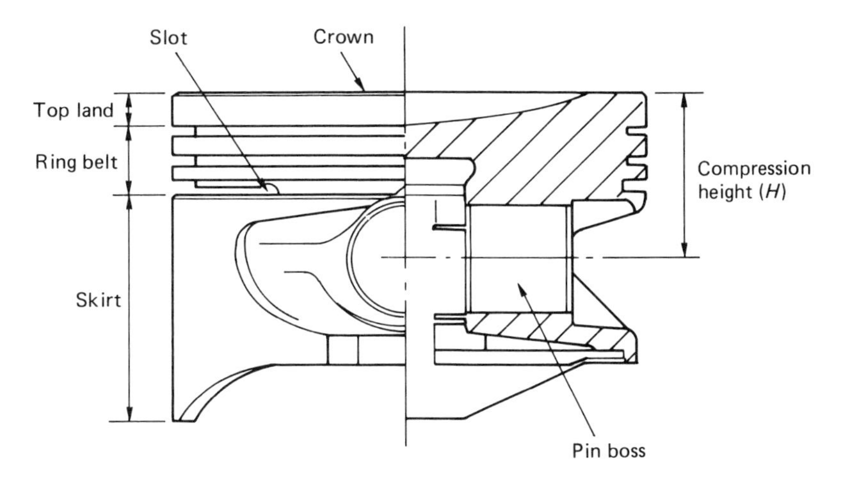

Figure 16.4 Features of a gasoline engine piston

The skirt guides the piston and must be machined slightly out-of-round to compensate for the thermal expansion of the piston and to facilitate hydrodynamic lubrication between the piston and the cylinder bore.

In gasoline pistons in engines of less than 65 bhp/litre slots may be used around part of the circumference of the piston to act as a thermal break between the skirt and the crown, and to introduce more flexibility into the skirt. This allows the running clearance of the skirt to be minimised so as to improve piston stability.

The compression height of the piston (H) is governed by the top land height, the ring belt height and the piston pin diameter. The compression height of gasoline pistons is kept to a minimum in order to reduce reciprocating mass and give improved powertrain refinement.

Piston weight may be characterised by the apparent density (K), defined as follows:-

$$K = \frac{M}{D^3} \text{gcm}^{-3}$$

where M = mass in grams of bare piston (no rings or pin)

D = bore diameter in centimetres.

The relationship between apparent density and compression height for lightweight designs of gasoline pistons is shown in Figure 16.5.

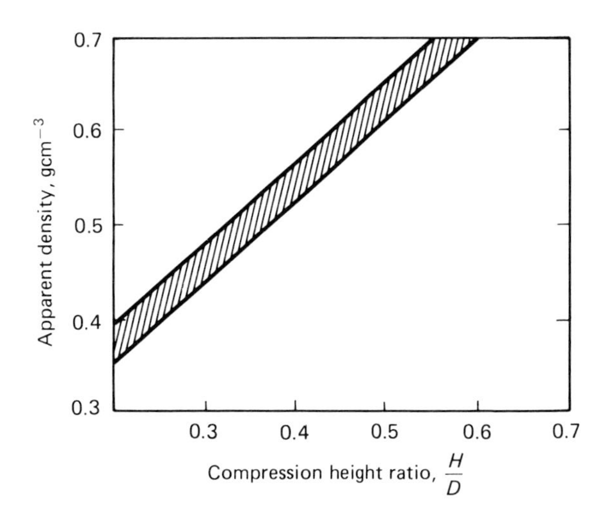

Figure 16.5 The variation of apparent piston density with compression height for a lightweight gasoline piston.

DIESEL ENGINE PISTONS

The higher loads experienced by diesel pistons means that additional features are commonly necessary. These are shown below:-

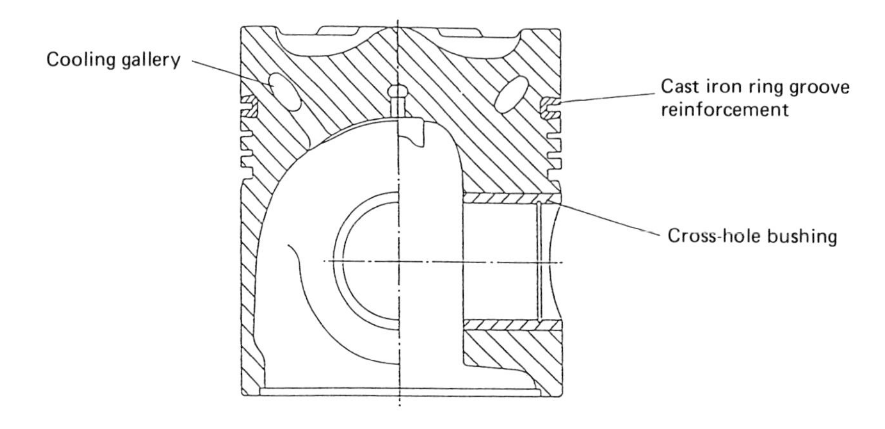

Figure 16.6 A typical highly rated diesel engine piston

The reinforcement of the top ring groove is usually by means of an austenitic cast iron insert, integrally joined to the parent material during the piston casting process.

At engine ratings above 3.5 MW/m^2 of piston area internal oil cooling galleries are often adopted. The gallery is formed by a soluble salt core which is removed after the casting has been produced.

Two-piece pistons are usually specified for cylinder sizes in excess of 300 mm, particularly for engines burning heavy fuel. They are also required at engine ratings above 5.0 MW/m^2 of piston area. The most prevalent combinations are a steel crown with a forged aluminium body and a steel crown with a nodular cast iron body.

Typical bolted crown designs are shown in Figure 16.7.

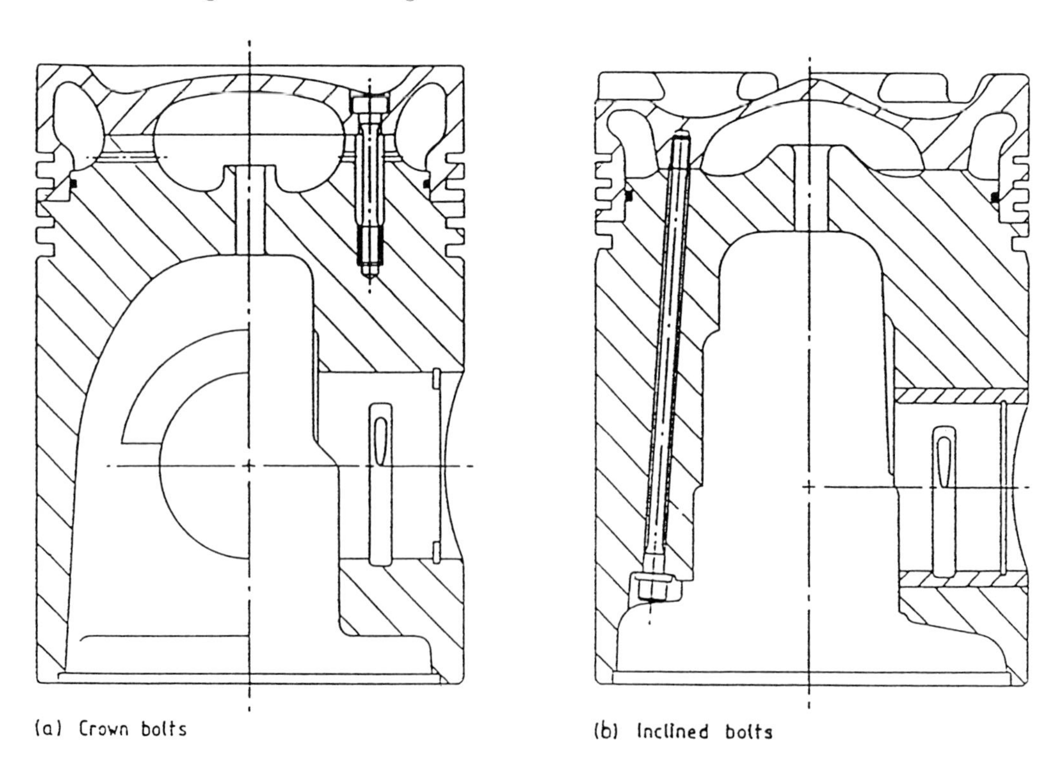

Figure 16.7 Piston designs with bolted crowns

I.C. ENGINE PISTON DESIGN

Piston materials

The % composition of typical eutectic piston alloys is shown below:-

	Gasoline pistons %	*Diesel pistons* %
Silicon	10.0–13.0	11.0–12.5
Copper	0.7–1.5	0.7–1.5
Magnesium	0.8–1.5	0.7–1.3
Iron	1.0 max	0.5 max
Manganese	0.5 max	0.25 max
Zinc	0.5 max	0.1 max
Aluminium	remainder	remainder

Ring arrangements

Typical ring arrangements for the four main types of IC engines are shown in Figure 16.8.

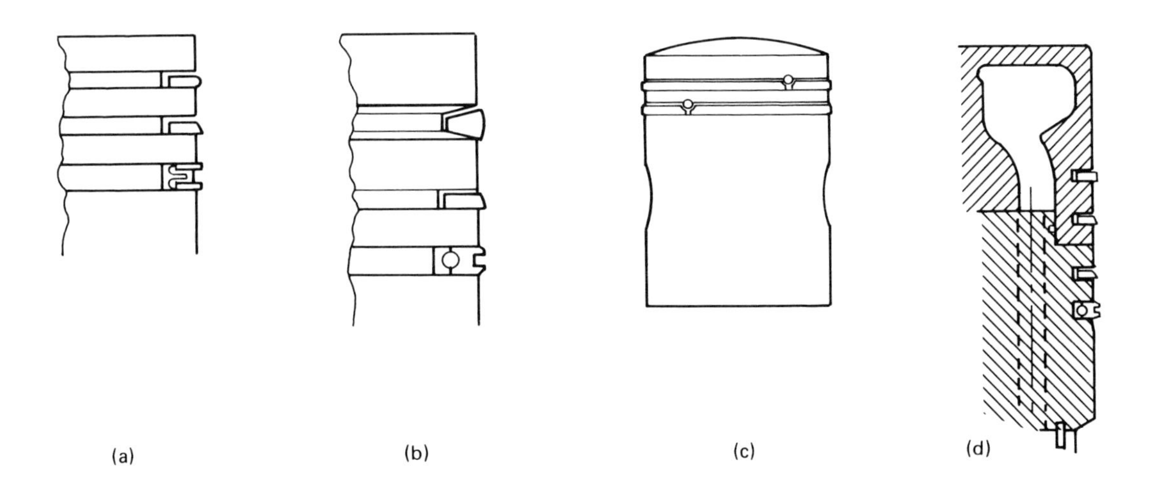

(a) Typical four-stroke petrol engine ring layout:
(b) Typical four-stroke diesel engine ring layout:
(c) Two-stroke petrol engine piston showing ring layout:
(d) Four-stroke medium speed diesel piston showing ring layout

Figure 16.8 Typical ring arrangements for the four main types of IC engines

Piston dimensions

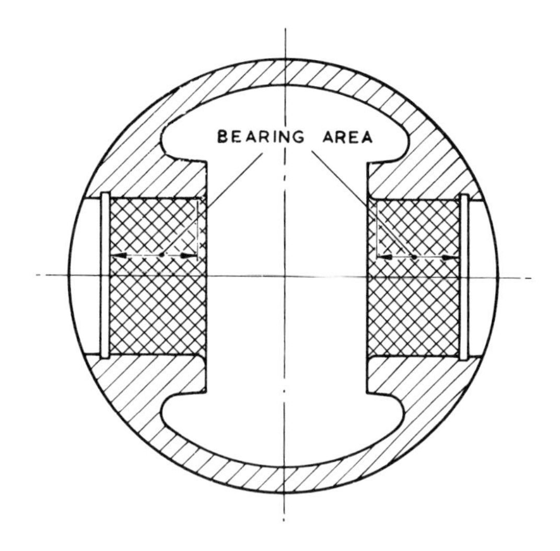

Gudgeon pin bearing area

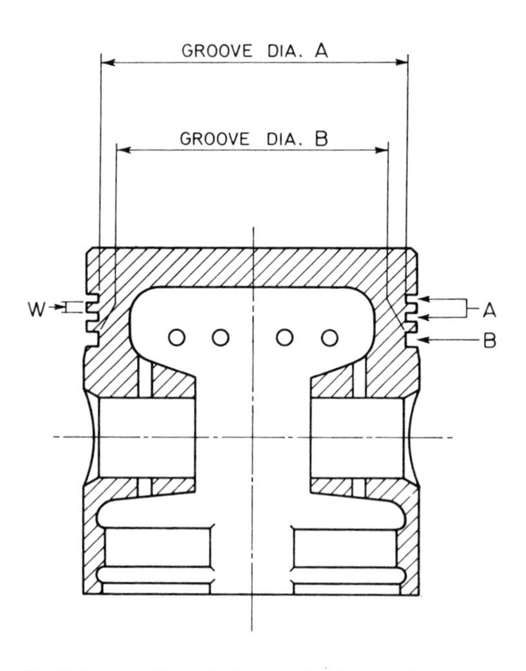

r = MAXIMUM GROOVE ROOT FILLET RADIUS
D = NOMINAL CYLINDER BORE DIA.
t = MAXIMUM RING RADIAL DEPTH

DIA. A = $D-(2t + 0.006\,D + 0.2 + 2r)$ mm
DIA. B = $D - 2\,(t + r)$ mm

Gudgeon pin diameter to be decided so that piston boss bearing area will support total pressure loading on piston, with bearing pressure not exceeding 69 MN/m^2 (10 000 lbf/in^2) in aluminium. Gudgeon pin thickness is determined from Figure 16.10 and fatigue strength from Figure 16.11 to ensure that suitable steel is used (DIN 17210 [1984] - 17 Cr 3 or 16 Mn Cr 5 where higher fatigue strength required).

Minimum ring side clearance

	DIESEL	PETROL
Top Ring	0.064 mm	0.04 mm
2nd Ring	0.050 mm	0.03 mm
Oil Ring	0.038 mm	0.038 mm

Minimum intermediate land thickness: calculated from ring groove depth and operating pressure – see Figure 16.9

NB: For turbo-charged designs increase calculated width by 29%

Minimum piston/bore clearances: internal combustion engines per each mm of cylinder bore diameter.

Piston dimensions

	Solid skirt mm	*Top land* mm
Lo-Ex silicon aluminium alloy (LM13 type)	0.00025	0.006

Surface finishes on wearing surfaces	μm Ra
Ring groove side faces	0.6 max.
Skirt bearing surface (aluminium-diamond turn)	3.2 – 4.8
Gudgeon pin o.d. 80 mm	0.1 max.
Gudgeon pin o.d. 80 – 140 mm	0.2 max.
Gudgeon pin bore in piston	0.3 max.

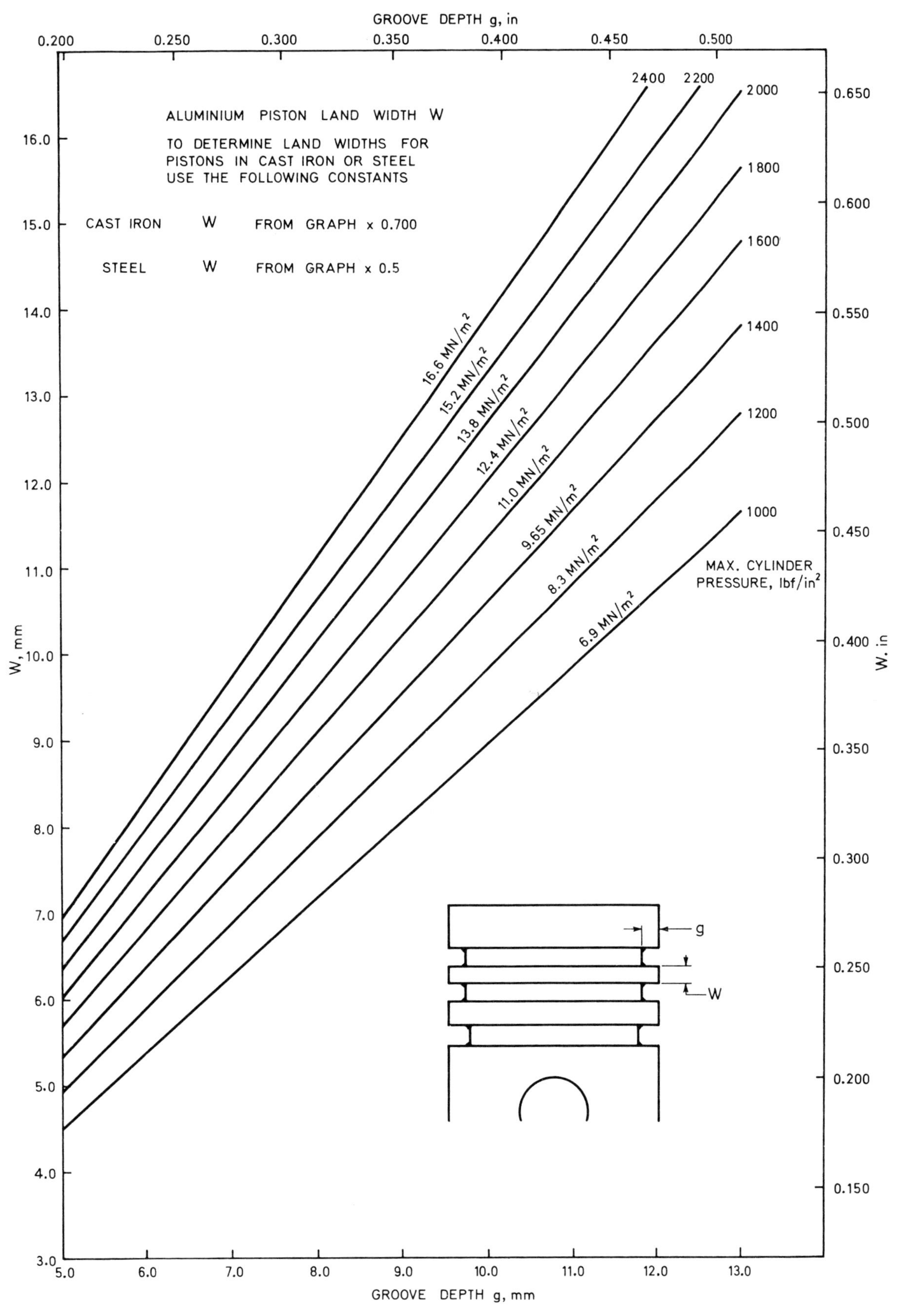

Figure 16.9 The land width required for various groove depths and maximum cylinder pressures

Gudgeon pin dimensions

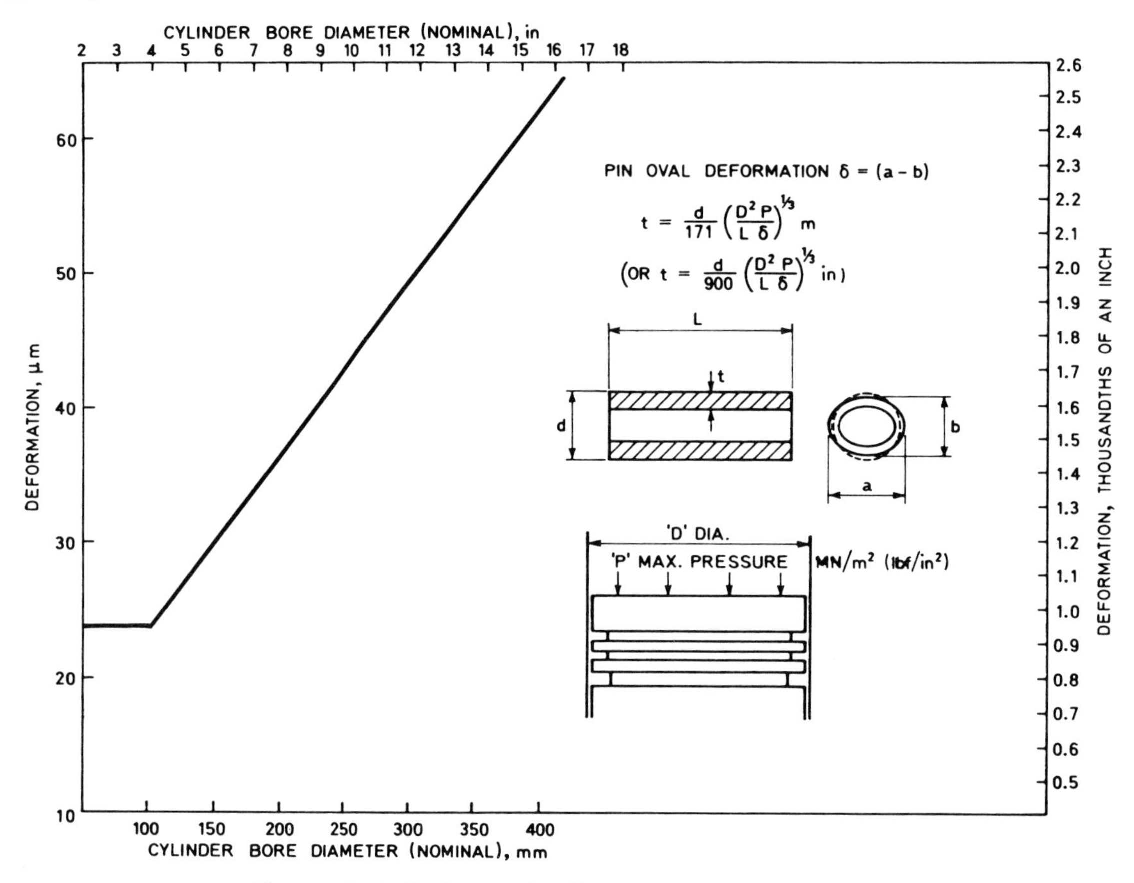

Figure 16.10 Gudgeon pin allowable oval deformation

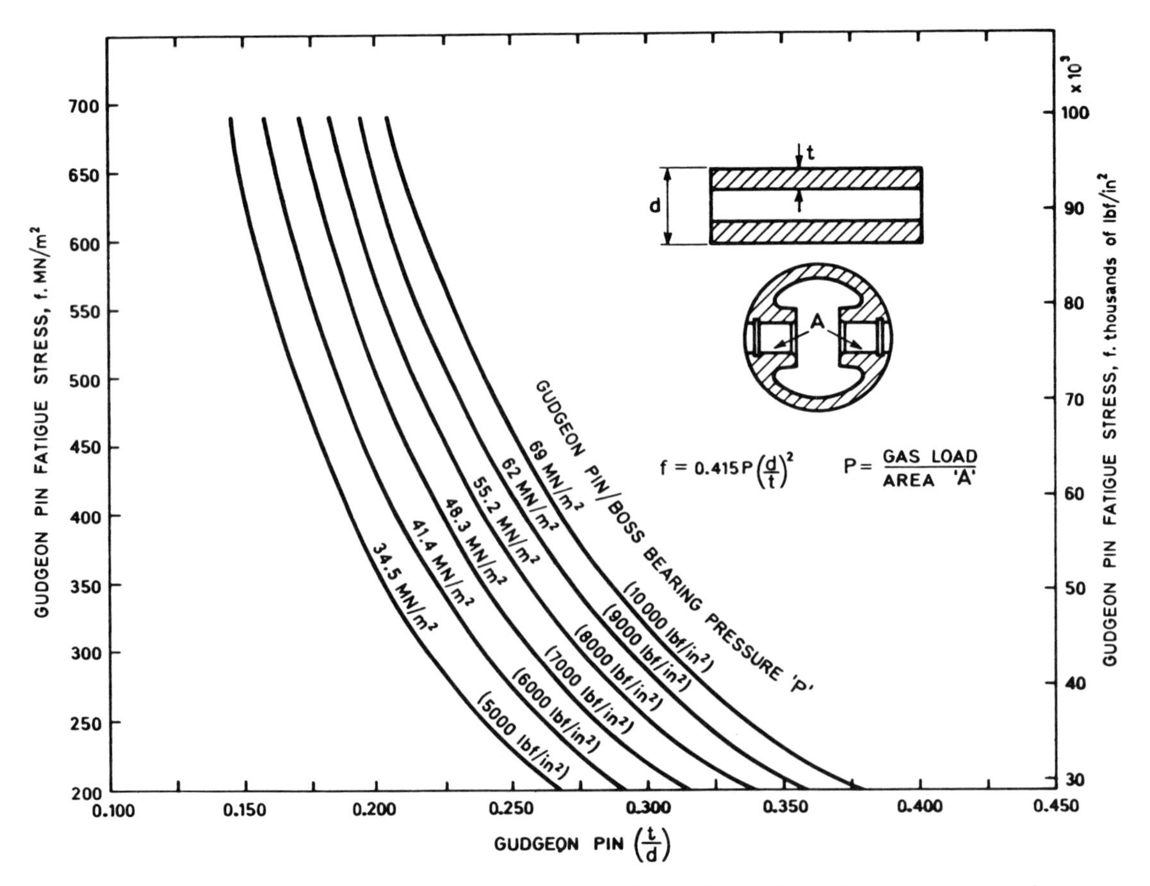

Figure 16.11 Fatigue stress in gudgeon pins for various pin and piston geometries

METALLIC LUBRICATED RINGS

Selection of type and materials

Table 17.1 Types of piston rings and guidance on selection*

Ring type		Description
COMPRESSION RINGS		
STRAIGHT-FACED RECTANGULAR	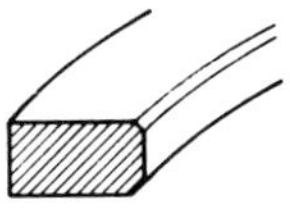 1	The most simple ring can be chromium-plated on peripheral face to give longer life
BARREL-FACED CHROMIUM-PLATED	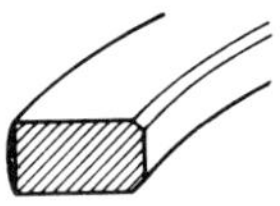 2	Widely used as top compression ring in diesel and petrol engines. Gives quick bed-in, good scuff resistance and long life. Has neutral oil control characteristic
RECTANGULAR INLAID	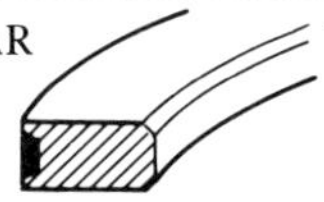 3a	Various low-wear-rate scuff-resistant materials, such as electroplated chrome, sprayed chrome, molybdenum, etc. are set into ring periphery. Outer lands give edge protection to deposited material
SEMI-INLAID	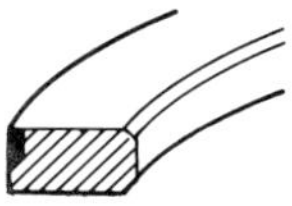 3b	
KEYSTONE	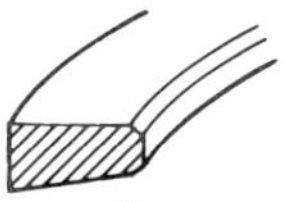 4a	A common top ring in diesel engines prevents sticking due to carbon formation fitted in groove with similar taper
HALF KEYSTONE	 4b	
TAPER FACED	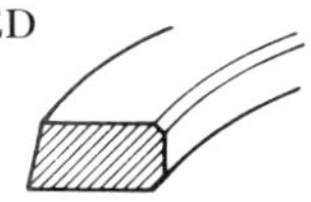 5	Normally between $\frac{1}{2}^\circ$ to $1\frac{1}{2}^\circ$, gives quick bed-in and combines gas sealing and oil control features. A witness land at periphery is often added
DYKES PRESSURE BACKED	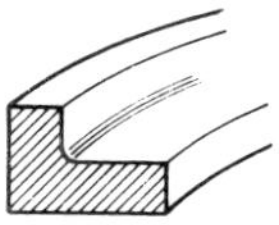 6	Mainly used in high-speed racing applications to prevent blow-by due to 'flutter' under high inertia loading. Fitted in a groove of similar shape
INTERNALLY STEPPED POSITIVE TWIST TYPE	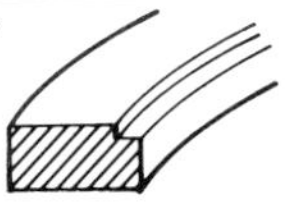 7a	Step or bevel relief on inner edge causes ring to dish when fitted, giving bottom edge contact and good oil control
NEGATIVE TWIST TYPE	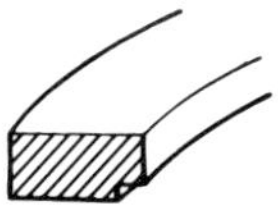 7b	
GROOVED INLAID	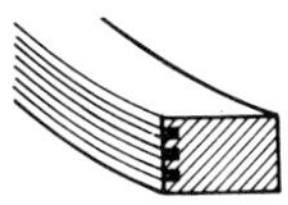 8	Materials such as chrome, bronze and ferrox are inlaid in multi-groove configurations, providing good scuff resistance
DUAL PURPOSE		
EXTERNALLY-STEPPED COMPRESSION AND SCRAPER	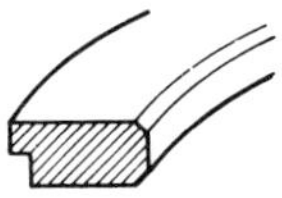 9	Combines gas sealing and oil control functions, the step giving the ring a torsional twist when fitted
OIL CONTROL RINGS		
NAPIER	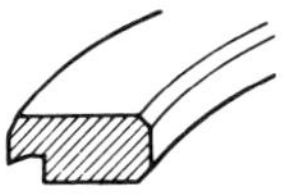 10	Variation of externally stepped ring; hooked relief gives sharp scraping edge with good oil control
COIL SPRING LOADED SLOTTED OIL CONTROL	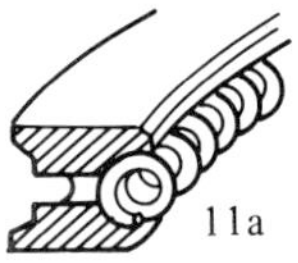 11a	Very popular ring particularly in high-speed diesels. Main wall pressure is derived from butting helical coil expander. Chromium-plated for wall pressures above 700 kN/m^2
BEVELLED EDGE TYPE	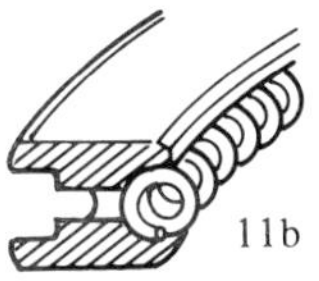 11b	
SLOTTED OIL CONTROL	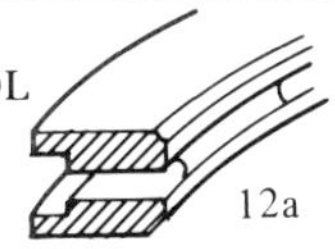 12a	A common form of bulk oil scraper with two scraping lands separated by drainage slots or holes
BEVELLED EDGE TYPE	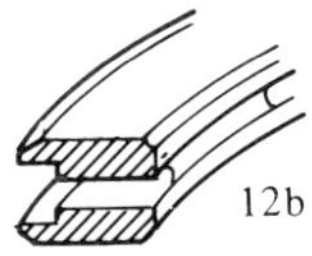 12b	
STEEL RAIL MULTIPIECE (*a*)	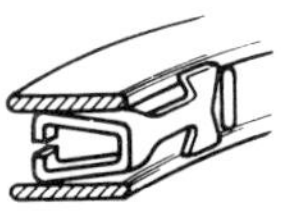	(*a*) Combined spacer expander (*b*) Separate spacer expander These rings allow very high scraping pressures to be applied with good conformability, and have chromium-plated rails
(*b*)	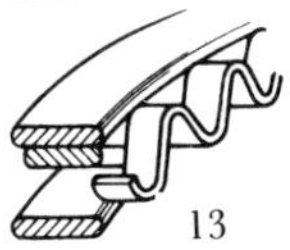 13	

* Basic ring types only shown above. Several features often combined in one ring design, e.g. 4, 7 and 2, giving internally stepped, barrel-faced, chromium-plated, keystone ring.

Table 17.2 Properties of typical ring materials

	Modulus of elasticity E_n GN/m²	*Tensile strength* MN/m²	*Hardness BHN*	*Fatigue rating*	*Wear rating*	*Scuff compatibility rating*
Grey irons	83–124	230–310	210/310	Fair/good	Good	Very good/excellent. Good on chrome
Carbidic malleable irons	140–160	400–580	250/320	Good/very good	Excellent	Good. Very good on chrome
Malleable/nodular irons	155–165	540–820	200/440	Excellent	Poor. Usually chromium-plated	Poor. Usually chromium-plated
Sintered irons	120	250–390	130/150	Good	Good	Very good

Table 17.3 Ring coatings

Coating	*Ring wear*	*Scuff/compatibility*	*Bore wear*	*Comment*
Chromium—electroplated	Excellent	Very good	Very good	Most widely used coating
Chrome sprayed	Very good	Very good	Fair	Wide variation in bore wear performance
Molybdenum sprayed	Fair	Excellent	Fair	Suffers from temperature/time break-up, especially above 250°C
Tungsten-carbide sprayed	Excellent	Good	Good	—
Iron oxide (ferrox)	Fair	Very good	Good	—
Phosphates (Parko-lubrising)	Fair	Running-in scuff very good	—	Mainly used for running-in
Copper plating	Poor	Running-in scuff excellent	Very good	Mainly used for running-in. Can also be applied to chromium-plate

Ring design and performance

Ring design uses well established elastic bending theory and is based on a careful compromise between opening stress when fitting the ring on to the piston, and working stress when the ring is in the cylinder.

The uniform elastic wall pressure P is given by:

$$P = \frac{E_n L}{7.07D(D/t - 1)^3}$$

where E_n = nominal modulus of elasticity
L = free gap
D = external ring diameter
t = radial thickness.

Depending on application, the ring free shape is sometimes modified to give high or low pressure at the horns. This is termed positive or negative ovality and is measured by means of a flexible band placed round the ring and closed to the bore diameter (Figure 17.1). Typically, negative ovality is used in two-stroke applications where ports have to be crossed and highly rated top ring applications where compensation is required to offset thermal distortion effects. Positive ovality limits the onset of flutter in high-speed applications.

To obtain a uniform pressure distribution, the ring free shape is given by:

Radial ordinate

$$R_c = R + U + \delta u$$

where

$$U = \frac{FR^4}{E_n I}(1 - \cos\alpha + \tfrac{1}{2}\alpha\sin\alpha)$$

$$\delta u = \frac{R}{2}\left(\frac{FR^3}{E_n I}\right)^2(\alpha - \tfrac{1}{2}\alpha\cos\alpha - \tfrac{1}{2}\sin\alpha)(3\sin\alpha + \alpha\cos\alpha)$$

F = mean wall pressure × ring axial width
R = radius at neutral axis, when in the cylinder
I = moment of inertia

(see Figure 17.2)

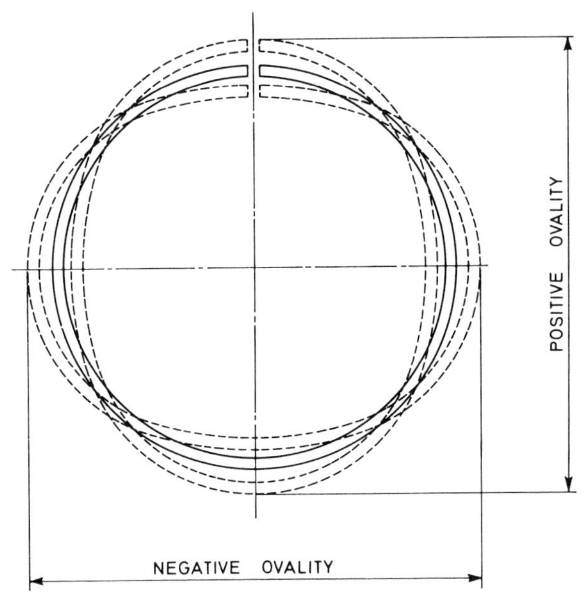

Figure 17.1 Definition of positive and negative ovality

Figure 17.2 Relation between free and closed ring shape for constant pressure distribution

$$\frac{\text{fitting stress}}{E_n} = \frac{4(8t - L + 0.004D)}{3\pi t\left(\frac{D}{t} - 1\right)^2}$$

$$\frac{\text{working stress}}{E_n} = \frac{4(L - 0.004D)}{3\pi t\left(\frac{D}{t} - 1\right)^2}$$

$$\frac{\text{fitting + working stress}}{E_n} = \frac{32}{3\pi\left(\frac{D}{t} - 1\right)^2}$$

which is independent of free gap L

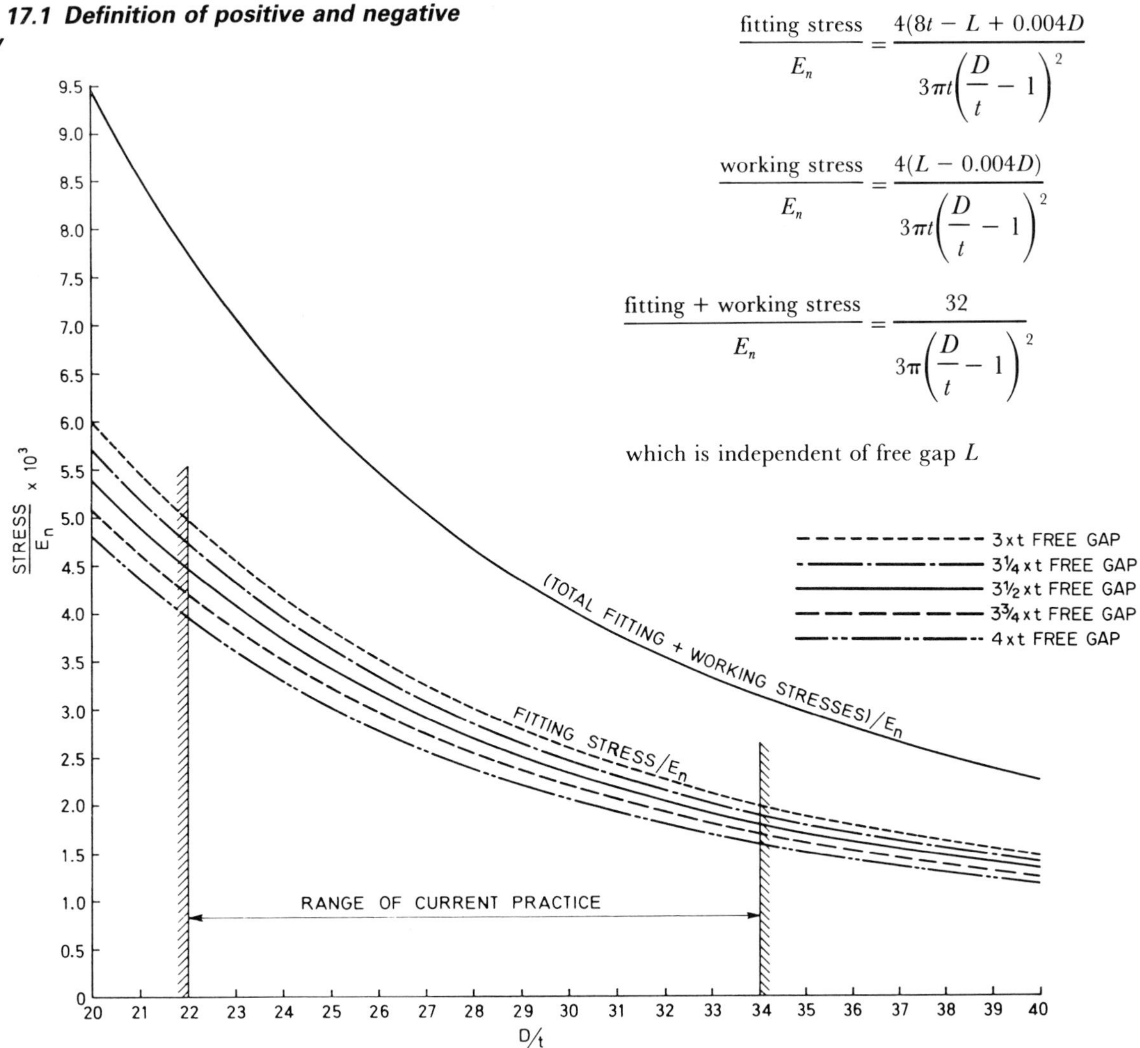

Figure 17.3 The effect of the ring radial thickness on the fitting and working stresses

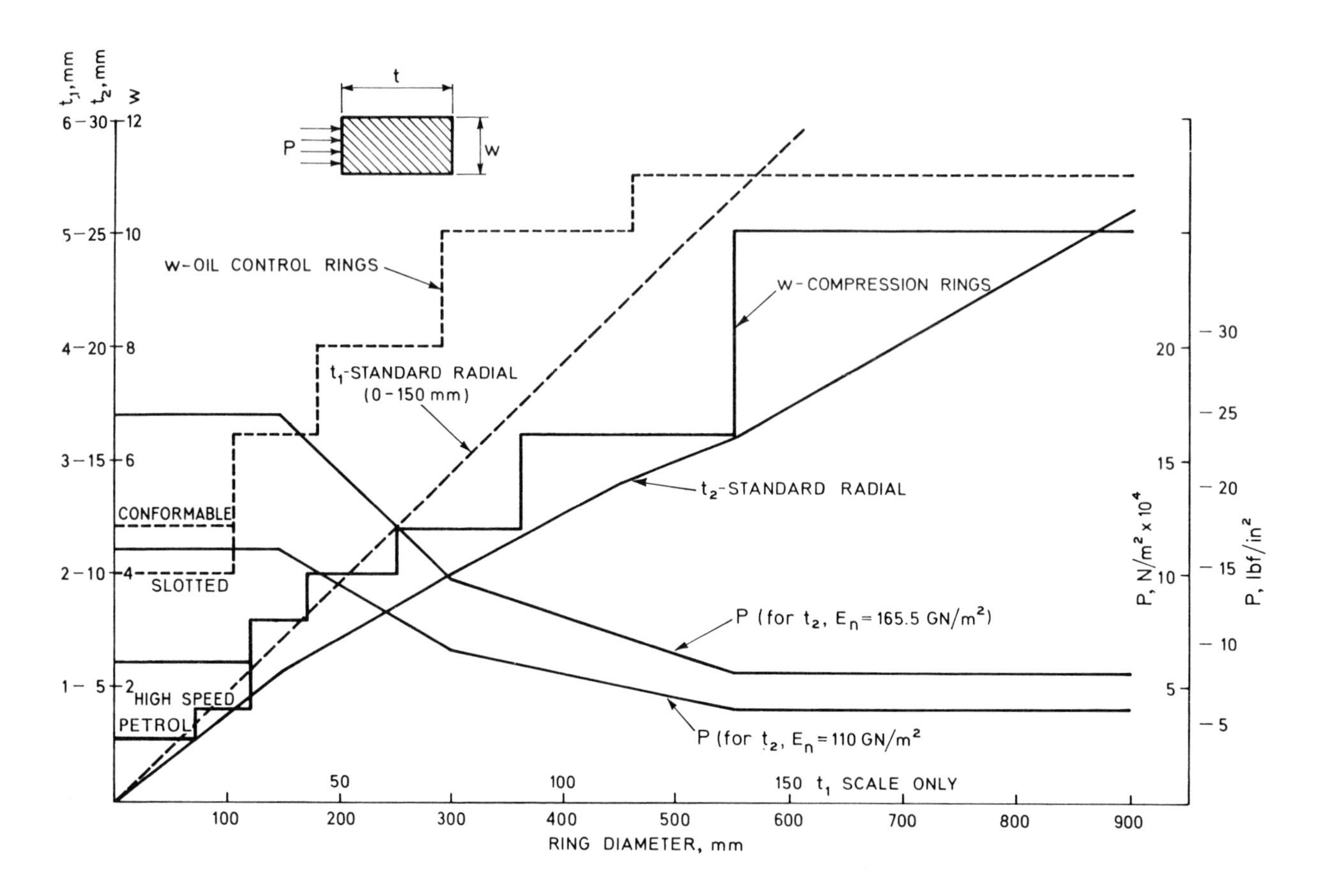

Figure 17.4 Typical ring dimensions

Table 17.4 Minimum side clearance

Application	*Bore size* (mm)	*Minimum groove side clearance* (mm) *(on dimension W – Figure 17.4)*	
		Top compression ring	2nd and 3rd compression rings Oil control rings
Petrol	50–100	0.035	0.035
	100–150	0.06	0.06
Diesel	75–175	0.06	0.04
	175–250	0.08	0.06
	250–400	0.10	0.08
	400–600	0.15	0.13
	Over 600	0.15	0.13
Compressors	50–150	0.025	0.025
	150–300	0.04	0.04
	300–450	0.05	0.05

Ring pack arrangements

Narrower rings are commonly used in gasoline engines. A typical gasoline ring pack is shown in Figure 17.5.

The top ring may be manufactured from steel, and may be surface hardened by a nitriding process, or may be chrome plated or molybdenum-inlaid.

In gasoline engines of less than 2.0 litres, the oil-control ring may be of a multi-piece design.

Typical steel specifications are as follows:-

Nitriding steel – DIN X65 CrMo14

Hardened tempered steel – SAE 9254

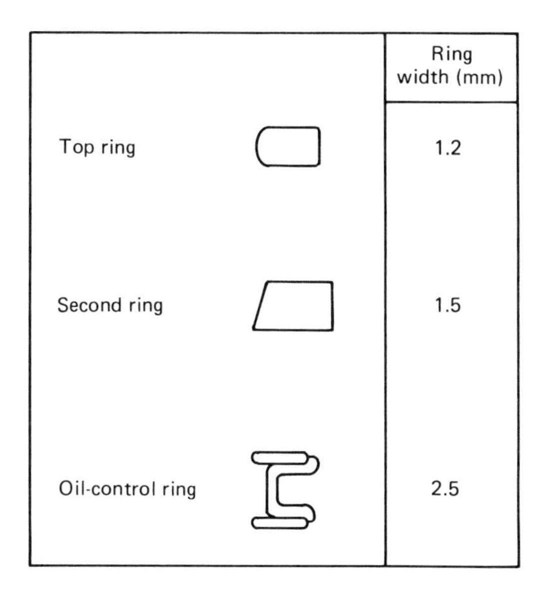

Figure 17.5 A typical gasoline engine ring pack

In both high and medium speed four-stroke diesel engines, three rings are now common, and typical ring packs are as shown.

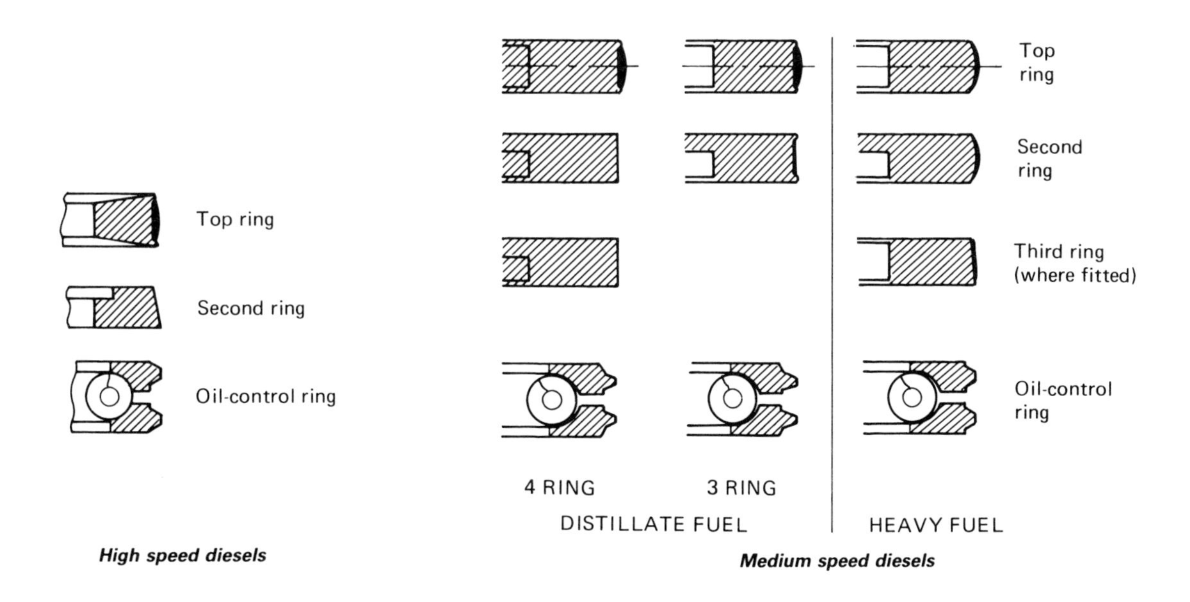

Figure 17.6 Typical ring pack arrangements on high speed and medium speed diesel engines

NON METALLIC PISTON RINGS

Metallic piston rings require lubrication for satisfactory operation. There are, however, many applications where lubricants would be considered a contaminant or even a fire hazard, e.g. in food-processing equipment. For these applications, piston rings can be made from self-lubricating materials. These materials can also be used in lubricated applications where there is a risk of lubricant breakdown.

Ring materials

Table 17.5 Typical properties of ring materials

Material	*Tensile strength* MN/m^2	*Specific gravity*	*Typical coefficients of expansion* × 10^{-6}/°C
Carbon-filled PTFE	10	2.05	55
Glass-filled PTFE	17	2.26	80
Graphite/MoS_2 filled PTFE	20	2.20	115
Bronze-filled PTFE	13	3.90	118
Resin-bonded PTFE	29	1.75	30
Resin-bonded fabric	110	1.36	22.5/87.5*
Carbon	43	1.8	43
Resin-bonded carbon	20	1.9	20

* Material is anisotropic, thus the lower expansion is parallel to, and the alternative figure is normal to, the plane of pressing.

Table 17.6 Suggested operating conditions for various materials

Material		*Terminal pressure bars*	*Maximum speed* m/s	*Maximum temp.* °C	*Average coefficient of friction (dry)*	*Minimum humidity* p.p.m.	*Minimal lubrication*
Matrix	*Filler*						
PTFE	Carbon	200	6.0	250	0.1/0.15	3	Very good
	Glass	200	6.0	200	0.1/0.15	3	Very good
	Graphite/MoS_2	200	6.0	200	0.12/0.18	3	Very good
	Bronze	100	4.0	200	0.15/0.2	40	Good
Resin-bonded PTFE		200	4.0	200	0.15	10	Very good
Resin-bonded fabric		100	3.0	150	0.15/0.2	40	Poor
Carbon		60	3.8	350	0.2/0.25	40	Poor
Resin-bonded carbon		100	4.5	180	0.2	40	Poor

Ring design

Table 17.7 Preferred number of rings

Differential pressure bars	0–9	10–14	15–24	25–29	30–49	50–99	100–200
Minimum number of rings	2	3	4	5	6	7	8

Table 17.8 Suggested sizes of rings

Units	*Cylinder diameter (b)*	Piston ring		*Groove side clearance (C)*
		Axial width	*Radial thickness*	
Millimetres	25–75	3–5	3–6	$C = 0.025 \times$ axial width
	76–150	5–10	5–10	
	151–230	6–12	6–12	(N.B.—Groove tolerance H7)
	230–400	10–19	10–19	
	400–800	12–25	12–25	
Inches	1–3	0.125–0.187	0.125–0.218	
	3–6	0.187–0.375	0.187–0.437	
	6–9	0.250–0.500	0.250–0.500	
	9–16	0.375–0.750	0.375–0.750	
	16–30	0.500–1.000	0.500–1.000	

Table 17.9 Types of joints

Butt	*Scarf*	*Lap or step*
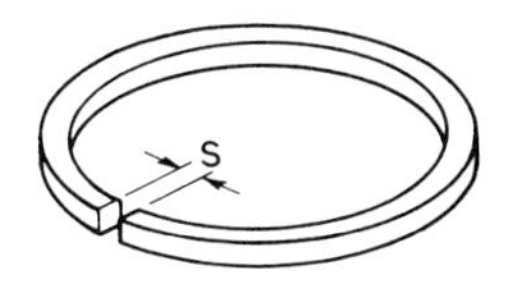	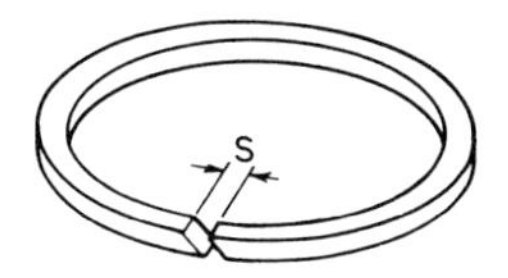	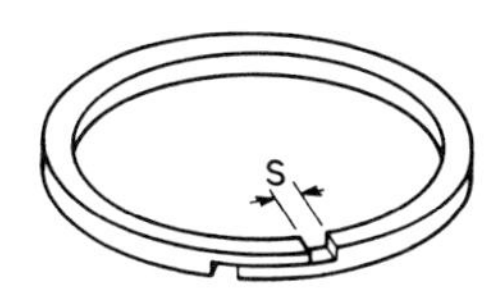
Suitable for all pressures	Suitable for all pressures	Not recommended where pressure differential exceeds 10 atmospheres
Circumferential clearance (S)	$S = \pi \times D \times \alpha_p \times T$ where D = cylinder diameter, α_p = Coeff. of expansion of piston ring material, T = Operating temperature	

Cylinder materials and finishes

Table 17.10 Typical cylinder materials

Material	*Type*	*Remarks*
Cast iron	ISO RI85 220 grade	Suitable only for continuous operation
Ni-Resist	ISO 2892 AUS101. ASTM A436/1	Preferred to cast iron—less danger of corrosion
Stainless steel	ISO 683/1 316S16. AISI 316	Used for machines where long shutdowns occur
Meehanite cast iron	CR or CRS	Similar to Ni-Resist

A suitable surface finish for these cylinder liners is 0.4 to 0.6 μm R_a or 2.4 to 3.6 μm R_{max}.

MATERIALS AND DESIGN

Table 18.1 Choice of materials for cylinders and cylinder liners

Application	*Type of construction*	*Comment*	*Material*		*Comment*	*Surface finish and treatment*
			Block	*Liner*		
I.C. engines	Monobloc	Most petrol engines. Some oil engines ('siamesed' cylinders used for cheaper and low specific power engines or where space is at a premium)	Grey C.I. (low phosphorus)	—	Simplest and cheapest method of building mass-production engines	Most applications use an untreated cross-hatched honed finish—See Note 3
			Aluminium alloy (high silicon) Aluminium with nickel plate containing silicon carbide particles	—	Gives maximum reduction in weight but poses special problems in material compatibility with mating component, i.e. piston and rings	For greater scuff resistance of C.I. a phosphate treatment can be used. For greater wear resistance bores may be hardened on surface, through- or zone-hardened, or hard chromium-plated on cast iron or steel liner. Surface porosity (by reverse plating) is necessary with chromium plating to give scuff resistance. Porous coatings aid oil retention reducing scuffing and wear. Silicon carbide impregnation can be used to combat bore polishing.
	Dry liner	Oil engines and petrol engines	Grey C.I. (low phosphorus)	Grey C.I. (low-to-medium phosphorus)	Liner normally pressed-in but may be slip fit. Much improved wear. Can pose cooling problems. Used for engine reconditioning in monobloc system	
			Aluminium alloy	Grey C.I. (low-to-medium phosphorus)	Liner normally cast-in or pressed-in	Aluminium alloys (high silicon) require special surface finish to allow free silicon to stand out from the matrix. Nickel plate with silicon carbide particles is the most common solution for aluminium bores. Some cheaper aluminium alloys may be used for 'throw-away' engines. Piston skirts may be electroplated with iron or chromium.
	Wet liner	High-performance petrol and most oil engines	Grey C.I.	Grey C.I. (low-to-medium phosphorus) Grey C.I. with silicon carbide impregnation Austenitic cast iron	Wet liner requirement for long life, good cooling and ease of maintenance	
			Aluminium alloy	Grey C.I. austenitic C.I. Aluminium alloy (high silicon) Aluminium with nickel plate containing silicon carbide particles	High-performance petrol engines to reduce weight	Costs rise significantly from the basic monobloc cast-iron cylinder block. Care must be taken to ensure the minimum specification consistent with technical requirements
Compressors	Monobloc	Small size and low pressure (up to 3–4 in dia. and 100 p.s.i.)	Grey C.I. (low phosphorus)	—	As in i.c. engines	As in i.c. engines except where plastics piston rings are used in which case honed 'mirror-finish' is desirable
	Wet liner	Heavy duty long life high reliability units of all sizes and operating pressures	As in i.c. engines		As in i.c. engines	
Hydraulic actuators and fluid piston pumps	To suit design requirements	—	Grey cast iron Bronze Aluminium alloy Steel		Material depends on environmental requirements of pressure, duty, reliability and fluid in use	Fine turned or honed to mirror finish Hardened steel bores usually ground or lapped

Table 18.2 Cylinder/cylinder liner tolerances

	Ovality mm	*Concentricity* mm
Monoblocs	0.025 FIM max	
Press fit dry type cylinder liners	0.150 FIM max	0.150 FIM max
Slip fit dry type cylinder liners*	0.050 FIM max	0.100 FIM max
Wet type cylinder liners*	0.025 FIM max	0.100 FIM max

* It is also vital that the flange be parallel and square to the major axis of the liner within 0.050 mm.

Table 18.3 Interference fits

Diameter mm	*1 Cast iron liners in cast iron blocks* *2 Aluminium liners or austenitic iron liners in aluminium blocks* mm	*Grey cast iron in aluminium blocks* mm
Up to 40	0.025 min	0.075 min
over 40 to 50	0.040 min	0.100 min
over 50 to 75	0.050 min	0.115 min
over 75 to 100	0.075 min	0.125 min
over 100 to 155	0.100 min	0.140 min

Notes:

1. Choice of construction and material is dependent on market being catered for: i.e. cost, power output or delivery requirement, life requirement, size and intended application.
2. Choice of material is also dependent on material used for pistons and rings and on any surface coatings given to these. Also, but to a lesser extent, on the surface treatment.
3. Honing specifications generally satisfactory; lies in the range 20 to 40 micro-inches, with a horizontal included angle of cross-hatch of 30/60° and a 60% plateau area. Surfaces must be free from folds, tears, burrs and burnished areas (see illustration). Suitable surface conditions can be most easily accomplished with silicon carbide hones. Diamond hones can be used but are best confined to roughing-cuts. Finishing can then be performed with silicon carbide honing stones or for more critical applications with silicon carbide particles in a soft matrix such as cork. Control of production tools and machines is vital for satisfactory performance in series production.
4. Sealing of wet liners is of great importance—see BS 4518 for Sealing Rings. Proprietary sealant/adhesive materials are available for assisting in sealing and in fixing liners.

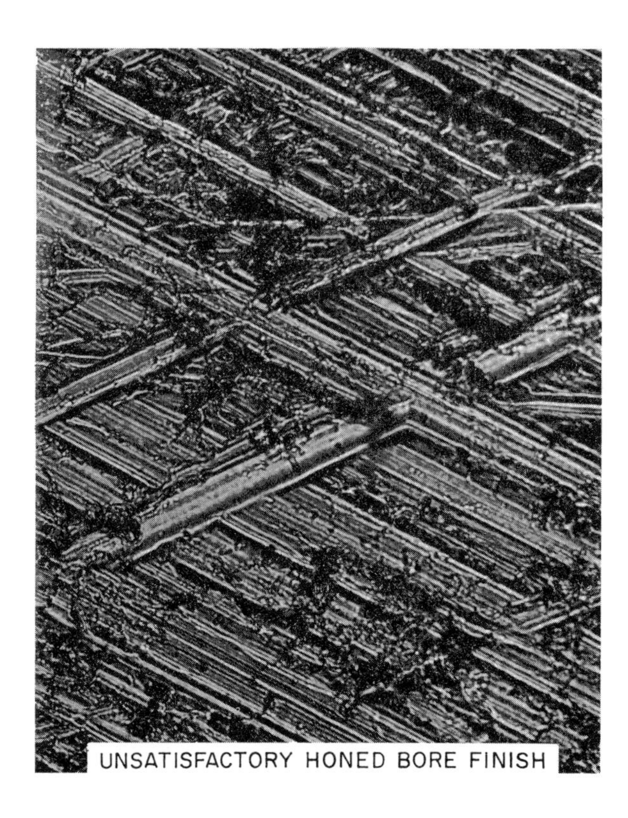
UNSATISFACTORY HONED BORE FINISH

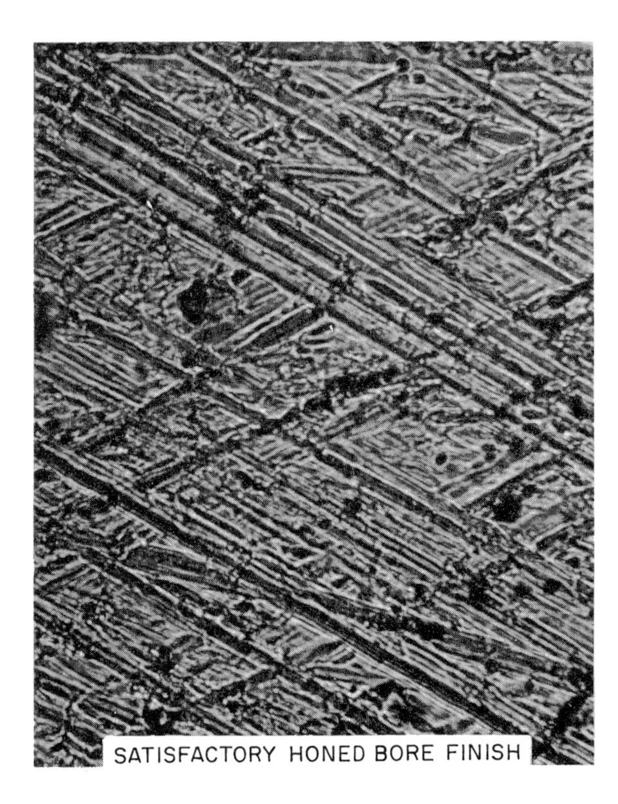
SATISFACTORY HONED BORE FINISH

Table 18.4 Materials, compositions and properties

Type	*Typical composition (%) (Remainder is F_e)*								*Typical properties*		
	C	*Si*	*S*	*P*	*Mn*	*Ni*	*Cr*	*Others*	*Coeff. of thermal expansion 20°-200°C*	*U.T.S*	*BHN*
Sand-cast blocks and barrels	3.3	2.1	0.1	0.15	0.6	0.3	0.2		11.0×10^{-6}/°C.0 ×10	220MN/m^2	200
Sand-cast liners	3.3	1.8	0.1	0.25	0.8	—	0.4		11.0×10^{-6}/°C	230MN/m^2	200
Centrifugally-cast grey iron liners	3.4	2.3	0.06	0.5	0.8	—	0.4		11.0×10^{-6}/°C	260MN/m^2	250
Centrifugally-cast alloy iron liners	3.3	2.2	0.06	0.2	0.8	—	0.4	Ni and Cu 0.5–1.5 Mo/Va 0.4	10.5×10^{-6}/°C	320MN/m^2	280
Austenitic iron liners	2.9	2.0	0.06	0.3	0.8	14.0	2.0	Cu 7.0	19.3×10^{-6}/°C	190MN/m^2	180

Table 18.5 Microstructures required

Type	*Microstructure*
Sand-cast blocks and barrels	Flake graphite, pearlitic matrix, no free carbides, phosphide eutectic network increases with phosphorus content, minimum of free ferrite desirable to minimise possibility of scuffing but less important with increasing phosphide
Sand-cast liners	As for sand cast but with finer graphite tending towards rosette or undercooled. Matrix martensitic/bainitic if linear hardened and tempered
Centrifugally-cast grey iron liners	Compact graphite or quasi-flakes, pearlitic matrix, islands of wear-resistant alloy carbides distributed throughout (approx. 5% by volume) matrix. Phosphide exists as ternary eutectic with carbides. Minimum of free ferrite ideal, but not important in presence of carbides
Centrifugally-cast alloy iron liners	Uniformly distributed fine flake graphite along with some undercooled graphite (ASTM Types A along with some D and E, Sizes 5–8. Fine-grained cored austenite matrix. Complex carbides and ternary phosphide eutectic are present in controlled amounts in a broken, non-continuous network
Austenitic iron liners	

BASIC SEAL TYPES AND THEIR CHARACTERISTICS

Dynamic seal

Sealing takes place between surfaces in sliding contact or narrowly separated.

Static seal

Sealing takes place between surfaces which do not move relative to each other.

Pseudo-static seal

Limited relative motion is possible at the sealing surfaces, or the seal itself allows limited motion; e.g. swivel couplings for pipes, flexible diaphragms.

Exclusion seal

A device to restrict access of dirt, etc., to a system, often used in conjunction with a dynamic seal.

Table 19.1 Characteristics of dynamic seals

	Contact seals		*Clearance seals*
Sealing interface	Surfaces loaded together: *(i)* Hydrodynamic operation (normal loads, speeds and viscosities) LOAD ABOUT 1 μm FLUID FILM (a)	*(ii)* Boundary lubrication (high loads, low speeds, low viscosities) LOAD MOLECULAR FILM (b)	Predetermined separation ABOUT 25 μm PRESET GAP (c)
Leakage	*(i)* Low to very low or virtually zero	*(ii)* As *(i)*	High, except for viscoseal and centrifugal seal at design optimum
Friction	Moderate	High	Low
Life	Moderate to good	Short	Indefinite
Reliability	Moderate to good	Poor	Good

Table 19.2 Types of dynamic and static seals

Dynamic seals				
Contact seals		*Clearance seals*		*Static seals*
Rotary	*Reciprocatory oscillatory*	*Rotary*	*Reciprocatory*	
Lip seal (Figure 19.1) Mechanical seal (Figure 19.2) Packed gland (Figure 19.3) 'O' ring* Felt ring	'U' ring, etc. (Figure 19.4) Chevron (Figure 19.5) 'O' ring (Figure 19.6) Lobed 'O' ring (Figure 19.7) Coaxial PTFE seal (Figure 19.8) Packed gland (Figure 19.3) Piston ring Bellows (Figure 19.9) Diaphragm	Labyrinth† (Figure 19.10a) Viscoseal (Figure 19.10b) Fixed bushing (Figure 19.10d) Floating bushing (Figure 19.10e) Centrifugal seal (Figure 19.10c) Polymeric bushing (Figure 19.10f)	Labyrinth† (Figure 19.10a) Fixed bushing (Figure 19.10d) Floating bushing (Figure 19.10e)	Bonded fibre sheet Spiral wound gasket Elastomeric gasket Plastic gasket Sealant, setting Sealant, non-setting 'O' ring Inflatable gasket Pipe coupling Bellows

* Only for very slow speeds.
† Usually for steam or gas.

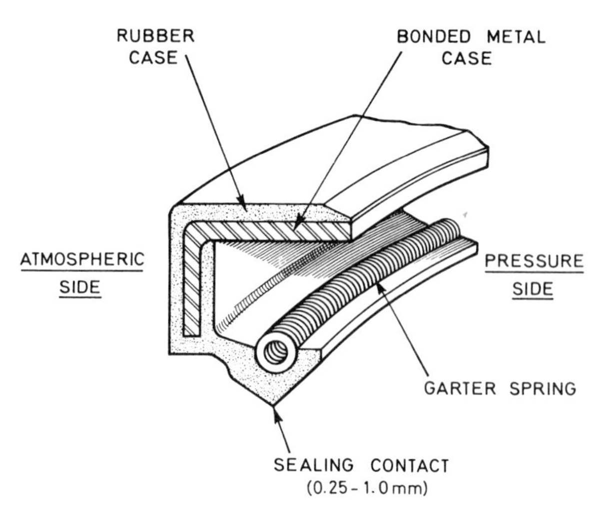

Figure 19.1 Rotary lip seal

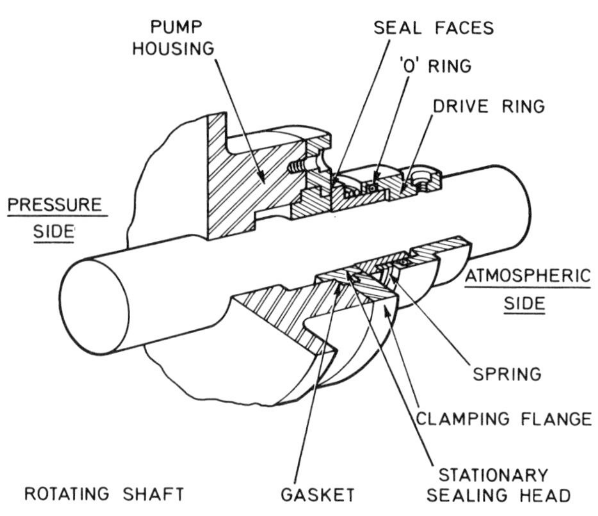

Figure 19.2 Mechanical seal

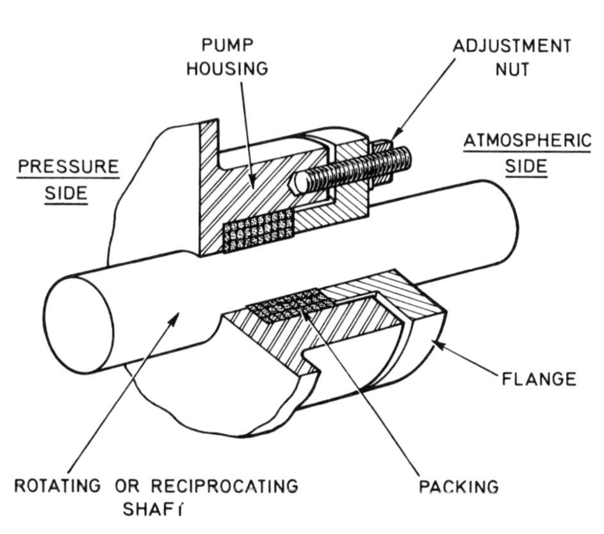

Figure 19.3 Packed gland

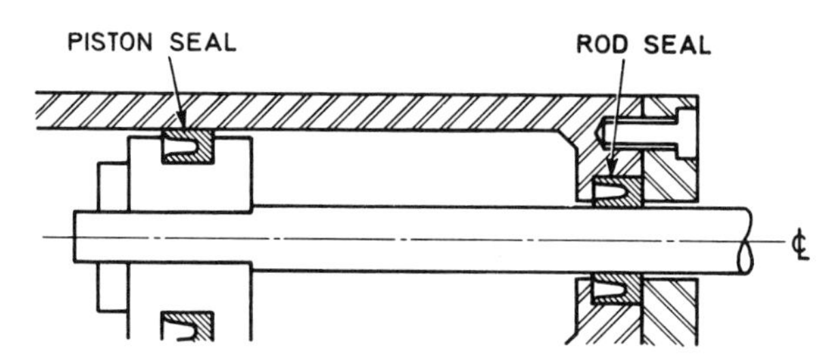

Figure 19.4 Square-backed 'U' seals as piston and rod seals in a hydraulic cylinder

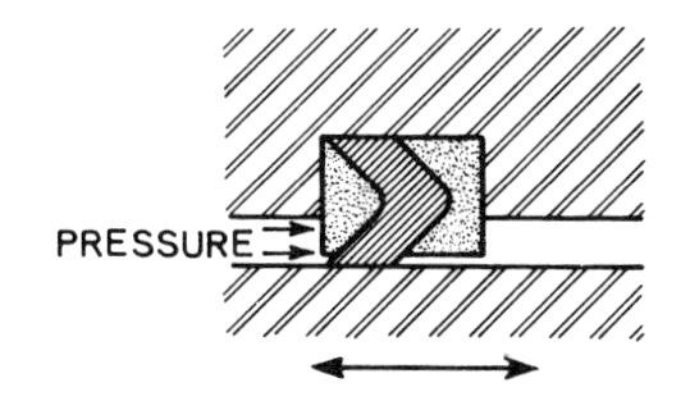

Figure 19.5 Chevron seal with shaped support rings

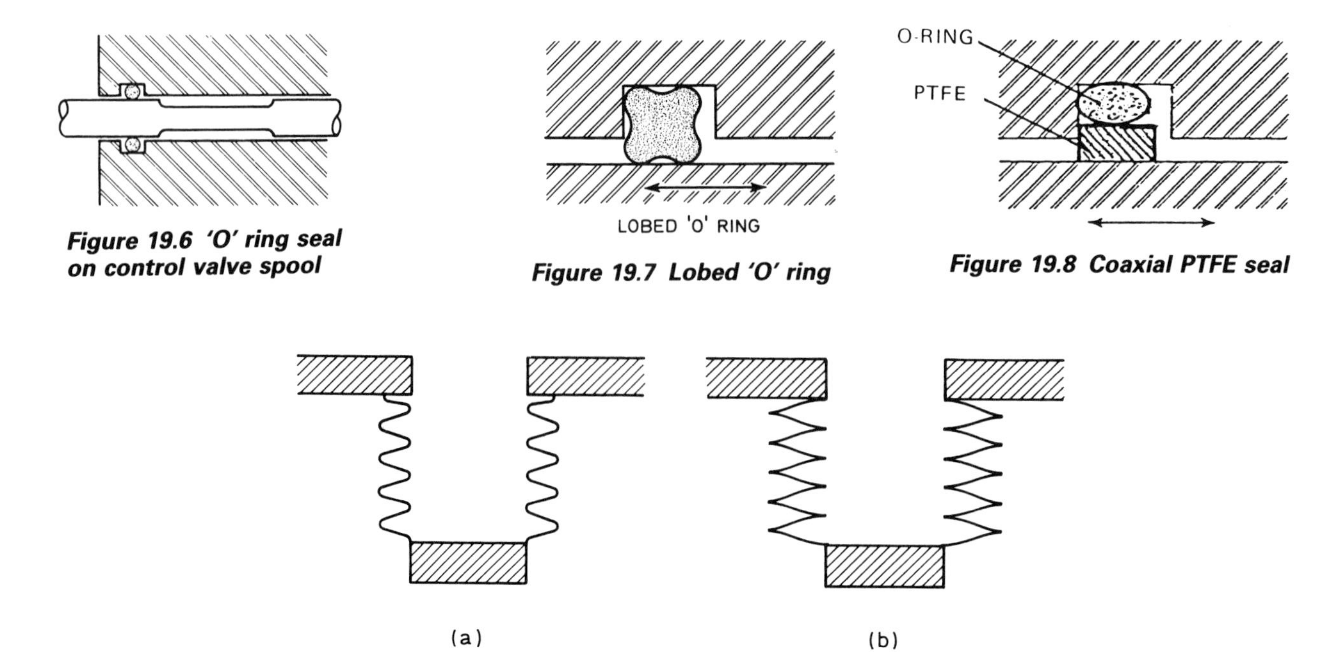

Figure 19.6 'O' ring seal on control valve spool

Figure 19.7 Lobed 'O' ring

Figure 19.8 Coaxial PTFE seal

Figure 19.9 Metal bellows: *(a) formed; (b) welded*

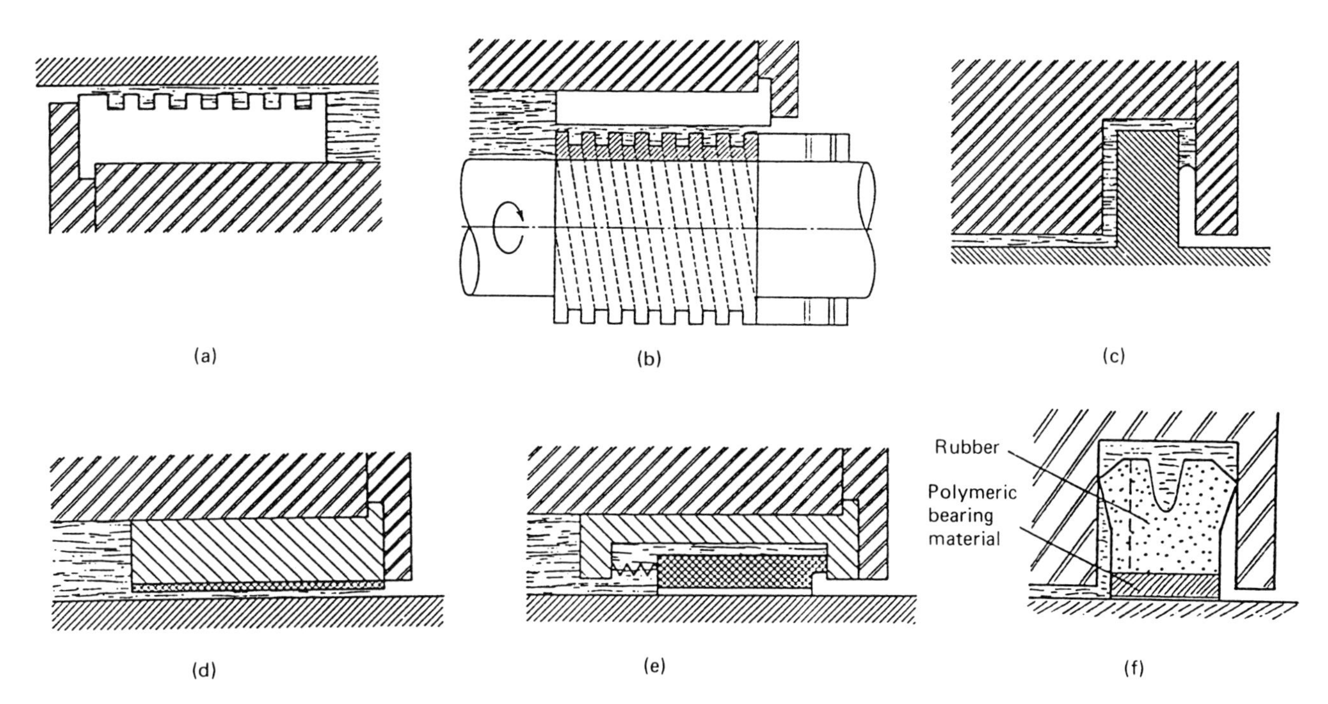

***Figure 19.10 Examples of clearance seals:** (a) labyrinth; (b) viscoseal; (c) centrifugal seal; (d) fixed bushing; (e) floating bushing; (f) polymeric-bushing seal*

Multiple seals

One seal or several in series may be used, depending on the severity of the application. Table 19.3 shows six basic dynamic sealing problems where two fluids have to be separated. Since contact seals rely on the sealed fluids for lubrication of the sliding parts it is essential that the seal(s) chosen should be exposed to a suitable lubricating liquid. Where this is not already so, a second seal enclosing a suitable 'buffer' liquid must be used. Multiple seals are also used where the pressure is so large that it must be broken down in stages to comply with the pressure limits of the individual seals, or where severe limitations on contamination exist. Table 19.3 lists the procedures for dealing with these various situations. Where a buffer fluid is used, care should be taken to ensure proper pressure control, especially when exposed to temperature variation. The pressure drop across successive seals will not be identical unless positive control is provided.

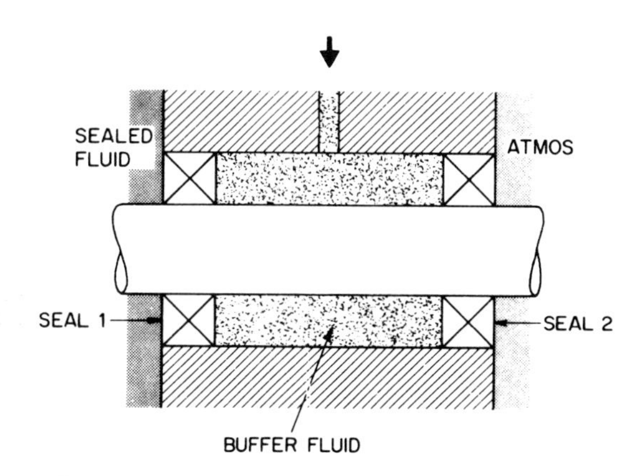

Figure 19.11 Multiple seals, with buffer fluid

Terminology:

'Tandem seals' multiple seals facing same direction, used to stage the pressure drop of the system. Inter-stage pressures progressively lower than sealed pressure.

'Double seals' pair of seals facing opposite directions, used to control escape of hazardous or toxic sealed fluid to environment, or to permit liquid lubrication of the inner seal. The buffer pressure is normally higher than the sealed pressure.

SEAL SELECTION

Table 19.3 The use of dynamic contact seals in the six dynamic sealing situations

Configuration (see diagram)	*Single seal*	*Multiple seal*
(a)	Satisfactory unless: *(i)* no contamination permissible *(ii)* $\lvert p_1 - p_2 \rvert$ large *(iii)* liquids both poor lubricants *(iv)* abrasive present	Buffer fluid = gas or vacuum: $p_B > p_1, p_2$ or $p_B \ll p_1, p_2$ Buffer fluid = liquid 1 or 2: $p_B \simeq (p_1 + p_2)/2$ Buffer fluid = good lubricant: $p_B > p_1, p_2$ Buffer fluid = clean liquid: $p_B > p_1$ or p_2, subject to abrasive location
(b)	Satisfactory unless: *(i)* no contamination permissible *(ii)* $\lvert p_1 - p_2 \rvert$ large *(iii)* the liquid is a poor lubricant *(iv)* abrasive present	Buffer fluid = gas or vacuum: $p_B > p_1$ or $p_B > p_2$ or $p_B > p_1, p_2$ or $p_B \ll p_1, p_2$ Buffer fluid = liquid: $p_B \simeq (p_1 + p_2)/2$ Buffer fluid = good lubricant: $p_B > p_1, p_2$ Buffer fluid = clean liquid: $p_B > p_1$ or p_2, subject to abrasive location
(c)	Satisfactory unless: *(i)* no contamination of vacuum permissible *(ii)* $p_2 \gg p_2$ *(iii)* the liquid is a poor lubricant *(iv)* abrasive present in liquid	Buffer fluid = compatible liquid or gas; alternatively use clearance seal(s) and evacuate buffer zone $p_B \geqslant p_1$ Buffer fluid = clean liquid: $p_B > p_1$
(d)	Unsatisfactory	Buffer fluid = compatible liquid lubricant: $p_B > p_1, p_2$
(e)	Unsatisfactory	Buffer fluid = compatible liquid lubricant: $p_B > p_1$
(f)	Unsatisfactory	Buffer liquid = compatible liquid lubricant: $p_B > p_1, p_2$

LIQUID 1 P_1 LIQUID 2 P_2 (a)

LIQUID P_1 GAS P_2 (b)

LIQUID P_1 VACUUM P_2 (c)

GAS 1 P_1 GAS 2 P_2 (d)

GAS P_1 VACUUM P_2 (e)

VACUUM 1 P_1 VACUUM 2 P_2 (f)

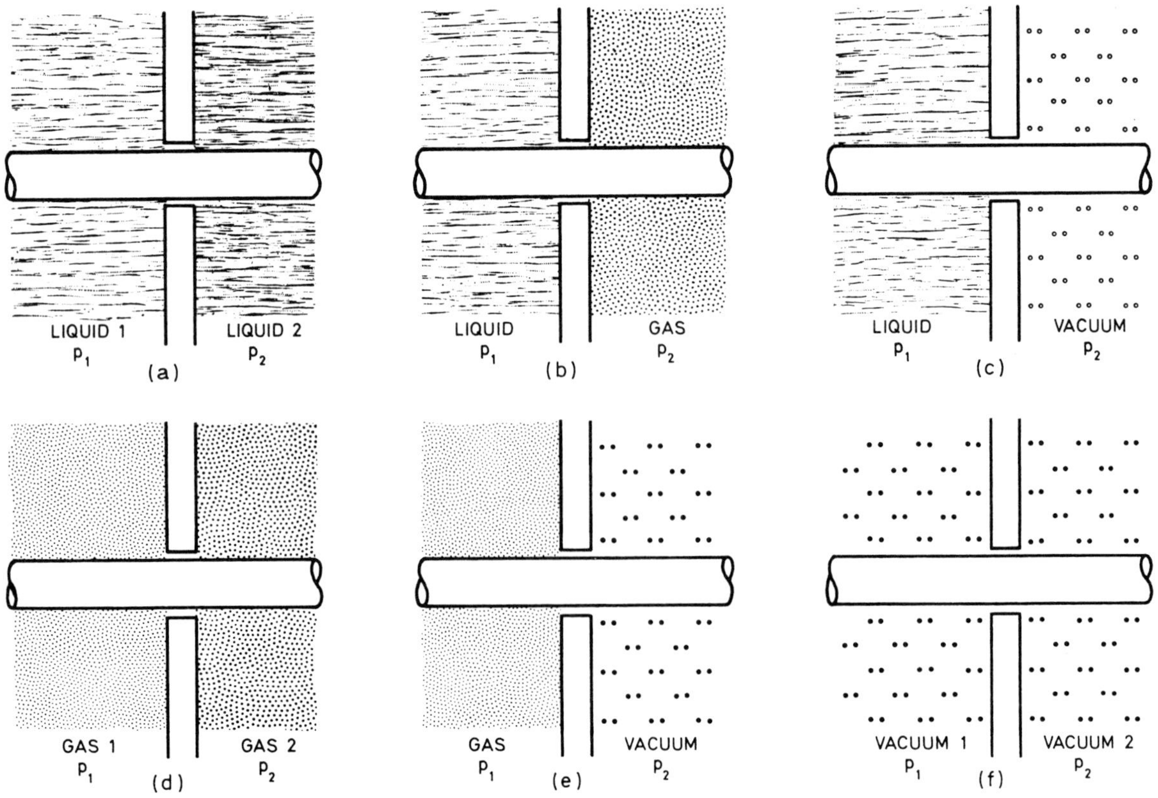

Check-list for seal selection

Temperature (see Figure 19.12): seals containing rubber, natural fibres or plastic (which includes many face seals) may have severe temperature limitations, depending on the material, for example:

Natural rubber	−50 to +80°C
Nitrile rubber	−40 to +130
Fluorocarbon rubber	−40 to +200
Perfluorocarbon rubber	−10 to +300
PTFE, plastic	−100 to +280

At low temperatures, certain of the fluoroelastomers may become less 'rubbery' and may seal less well at high pressure.

Speed (see Figure 19.13)
Pressure (see Figure 19.13)
Size (see Figure 19.14)
Leakage (see Figure 19.15)

After making an initial choice of a suitable type of seal, the section of this handbook which relates to that type of seal should be studied. Discussion with seal manufacturers is also recommended.

Fluid compatibility: check all materials which may be exposed to the fluid, especially rubbers.

Abrasion resistance: harder sliding contact materials are usually better but it is preferable to keep abrasives away from the seal if at all possible, for example by flushing with a clean fluid.

Polyurethane and natural rubber are particularly abrasion resistant polymers. Where low friction is also necessary filled PTFE may be considered.

Vibration: should be minimised, but rubber seals are likely to function better than hard seals.

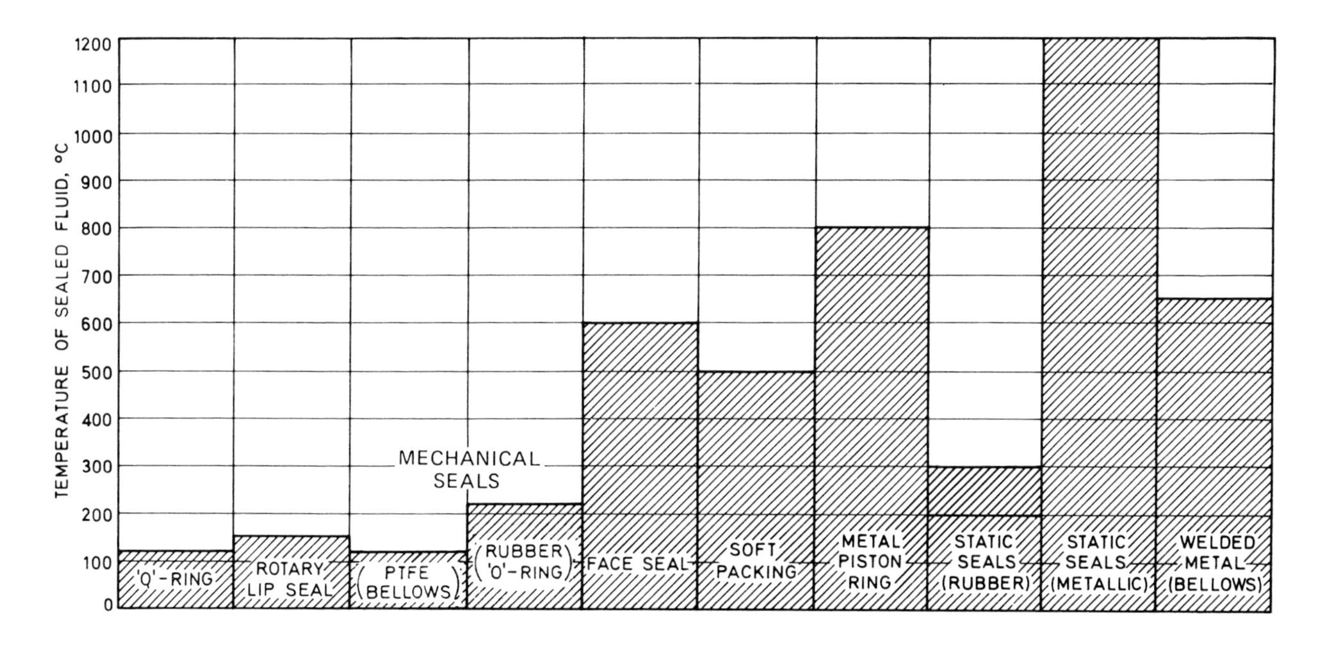

Figure 19.12 Approximate upper temperature limits for seals

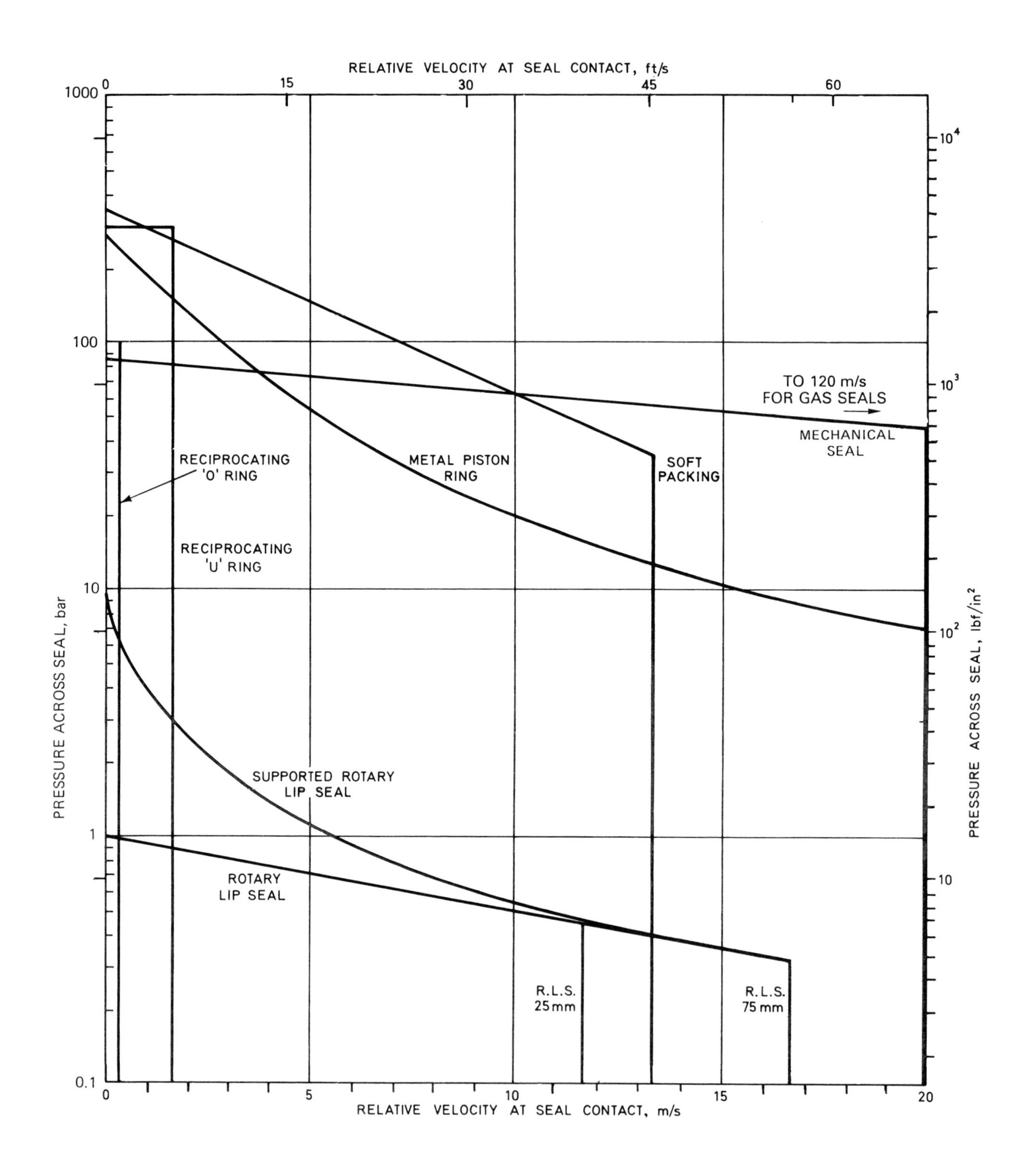

Figure 19.13 Limits of pressure and rubbing speed for various types of seal

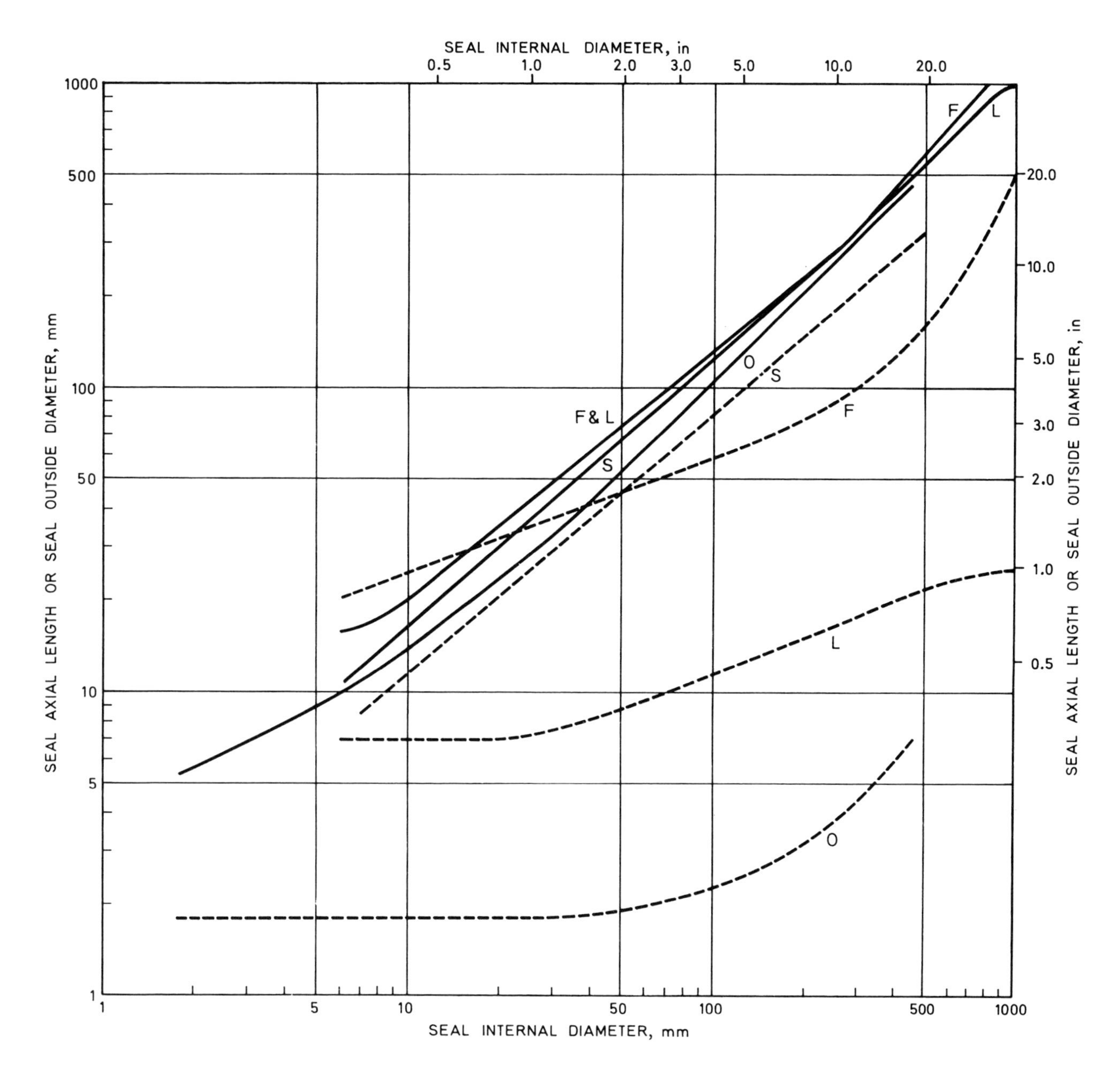

***Figure 19.14** Normal minimum seal sizes (———, outside dia.; – – – – –, length; F, mechanical seal; L, lip seal; O, 'O' ring; S, soft packing)*

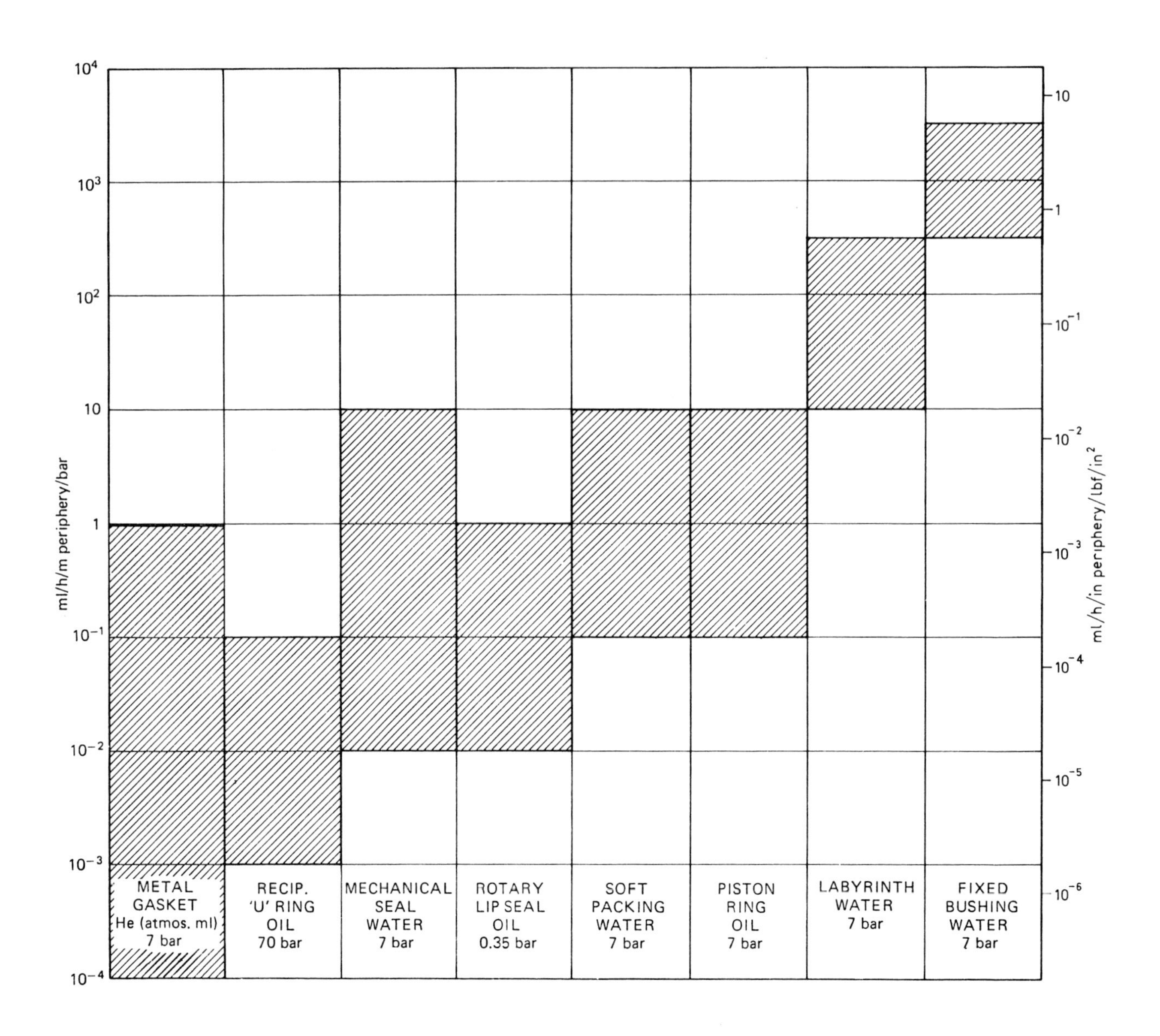

Figure 19.8 Approximate leakage rates for various types of seal

When operating in dirty and dusty conditions, the reliability of equipment depends almost entirely on the amount of abrasive material present. Natural soils contain abrasive materials in amounts varying from 98% down to 20% by weight.

Table 20.1 The source, nature and effect of contaminants

Source	*Nature of contaminant*	*Operating conditions, effect on reliability and basic requirements*		
		Wet (more than 15% by weight of water)	*Dry*	*Wet and dry*
Contact with soils (high silicon)	Sharp faceted grains predominantly silica (SiO_2) 98% by weight occurs frequently	Severe, highly abrasive. Considerable loss of reliability unless extensive sealing provided	Some loss of reliability. Machinery ingesting air requires efficient air cleaning	Severest, maximum abrasive effect. Very good air cleaning and sealing required
(low silicon)	Predominantly grains of Calcium (CaO). Silica less than 25% by weight.	Clogging rather than abrasive. Some loss of reliability. Sealing required	Little loss of reliability. Good air cleaners and sealing required	Poor conditions. Good air cleaning and sealing required
Airborne dust	Any finely divided material in the dry state picked up in air currents	—	Reduction in reliability dependent on dust concentration. Very large air cleaners required for highest concentrations	—

DESIGN OF SEALING SYSTEMS

Design to reduce the effects of dirt and dust

1. Keep to a minimum the number of rotary or sliding parts exposed to bad conditions.
2. Provide local clean environments for bearings and reciprocating hydraulic mechanisms by means of separate housings or sealing arrangements.
3. Provide adequate space in the sealing arrangement for oil lubrication.
4. Do not use grease lubrication for bearings, unless design for oil becomes uneconomic.
5. Provide adequate means for replenishment of lubricant; easily accessible.
6. Protect lubrication nipples locally to avoid erosion or fracture from stones and soil.
7. Provide positive means for checking amount of lubricant in housing.
8. Never use a common hydraulic fluid system for such mechanisms as reciprocating hydraulic motors and exposed hydraulic rams for earth moving equipment. Abrasive material is bound to enter the ram system which will be highly destructive to precision mechanisms. Provide independent fluid systems.
9. For mechanisms relatively crude in function where lubricant retention of any sort is either too costly or impracticable, load carrying bearings and reciprocating parts may be made in material with very hard or work hardening contact surfaces. Austenitic manganese steels have work hardening properties, but are not readily machinable. The shape of parts must be arranged to be used as cast or with ground surfaces.
10. Arrange the position of air cleaner intakes to avoid locally induced dust clouds from the motion of the mechanism.

Table 20.2 Sealing of rotary parts

Method	*Description*	*Comments*
Type A 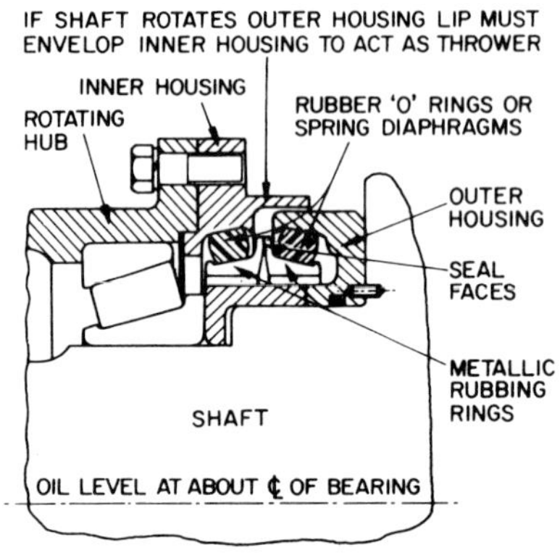 IF SHAFT ROTATES OUTER HOUSING LIP MUST ENVELOP INNER HOUSING TO ACT AS THROWER INNER HOUSING ROTATING HUB RUBBER 'O' RINGS OR SPRING DIAPHRAGMS OUTER HOUSING SEAL FACES METALLIC RUBBING RINGS SHAFT OIL LEVEL AT ABOUT ℄ OF BEARING	Metallic rubbing rings mounted between rubber 'O' rings, spring diaphragms or rubber housings. Contact faces 2–3 mm radial width. Surface finish not greater than 3 μm R_a Axial pressure between contact faces 140–210 kN/m^2 (20–30 lbf/in^2). Hardness of contact faces not less than 800 VPN	Very high level of protection and durability, wet or dry. Satisfactory when submerged in sea water to at least 3 m. Rubbing speeds of up to at least 3 m/s, but essential to use oil lubrication. If rubbing speed is restricted to not more than 0.1 m/s grease may be used. Standard parts available up to 250 mm dia. Rings of special size readily obtainable as precision castings which require only the contact faces to be ground and finished
	Material of rubbing rings: 1st choice Stellite or similar	Highly abrasion and corrosion resistant. Use for worst conditions of operation
	2nd choice (a) Highly alloyed cast irons (proprietary mixes) (b) Hard facings applied to rings of cheaper steels	Much less corrosion resistant
Type B 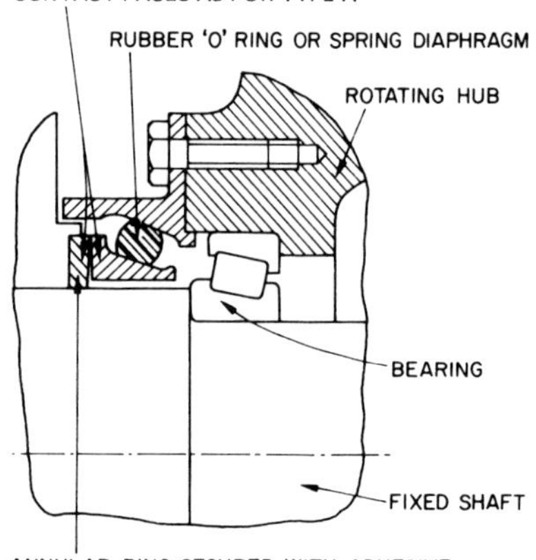 CONTACT FACES AS FOR TYPE A RUBBER 'O' RING OR SPRING DIAPHRAGM ROTATING HUB BEARING FIXED SHAFT ANNULAR RING SECURED WITH ADHESIVE OR BY SOME MECHANICAL DEVICE	Similar to Type A, only one rubbing ring flexibly mounted	Occupies less volume but level of protection as for Type A. Requires more careful mounting and fitting of fixed annular ring
Type C 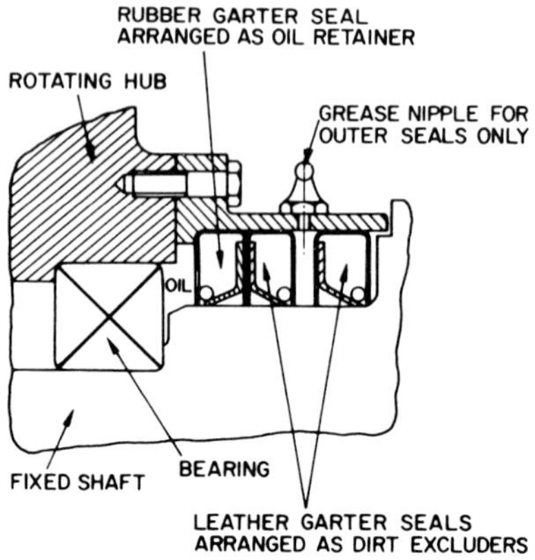 RUBBER GARTER SEAL ARRANGED AS OIL RETAINER ROTATING HUB GREASE NIPPLE FOR OUTER SEALS ONLY OIL FIXED SHAFT BEARING LEATHER GARTER SEALS ARRANGED AS DIRT EXCLUDERS	Three garter seals arranged as oil retainer and dust excluders, with either rubber or leather sealing elements. The rubber lipped seal is for oil retention only, the adjacent leather lipped seal prevents contaminated grease from entering the bearing cavity and the outer leather seal allows fresh grease to escape carrying contaminated material with it	Level of protection much lower than either Type A or B, but less costly. Standard seals more easily obtainable. Directions in which lips of seals are mounted are critical. Leather sealing elements are not abraded away so fast as rubber by dirt and mud, and must be used for dirt excluders. Not suitable for total immersion in any depth longer than a few minutes unless oil is replaced and fresh grease is applied immediately after coming out of water. Limiting speeds are as for general practice when using seals of this type. In worst environment grease replenishment required daily. Pump in until grease is seen to exude from outer housing

Table 20.3 Sealing of reciprocating parts

Method	*Description*	*Comments*
A relay system as in Figure 20.1	The hydraulic device is built into a housing and a relay system converts the reciprocating into rotary motion. The rotating parts are sealed as shown in Table 20.2	High level of protection. The primary hydraulic seal functions in clean conditions
A flexible cover system as in Figure 20.2	A flexible cover is mounted over the main hydraulic seals and means provided for breathing clean air through piping from the inside of the cover to a clean zone or through an air cleaner. The cover material is highly oil resistant and preferably reinforced with fabric	At maximum speed of operation provide piping to limit pressure difference to 35 kN/m^2 (5 lbf/in^2) between inside and outside of cover. The hydraulic system is dependent on the cover and although small pin-holes will not seriously reduce protection in dry conditions they will cause early failure if present when operating under water or in wet soil conditions
Standard chevron seals, 'O' rings etc.	Non metallic reciprocating seals used singly or in groups. All sliding parts through and adjacent to the seals highly corrosion resistant	Suitable where some loss of hydraulic fluid is not critical. An adequate reserve of hydraulic fluid must be provided to keep system full
Flexible metallic or non-metallic diaphragm	Hydraulic system sealed off completely	Very high level of reliability but restricted to small usable movements depending on diameter of diaphragm

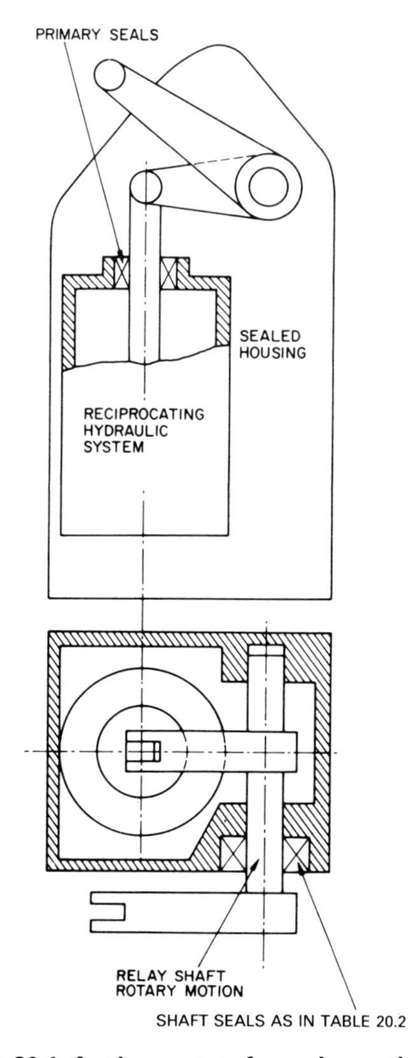

Figure 20.1 A relay system for reciprocating motion

FLEXIBLE COVER
OIL RESISTANT
REINFORCED WITH FABRIC
RECIPROCATING ROD
HYDRAULIC DEVICE
PRIMARY HYDRAULIC SEALS
BREATHER PIPE TO CLEAN ZONE OR AIR CLEANER

Figure 20.2 A vented flexible cover system

Table 20.4 Sealing with limited rotary or axial movement

Method	*Description*	*Comments*
Elastomeric deflection	Annular elastomer bushes either single or multi-layered, bonded or fastened to the adjacent parts.	Very high level of reliability in all environments at low cost. Elastomer must be matched to local contaminants. All motion either torsional or axial must occur in the elastomer. Usually, bushes made specially to suit load requirements

RECIPROCATING ENGINE BREATHING

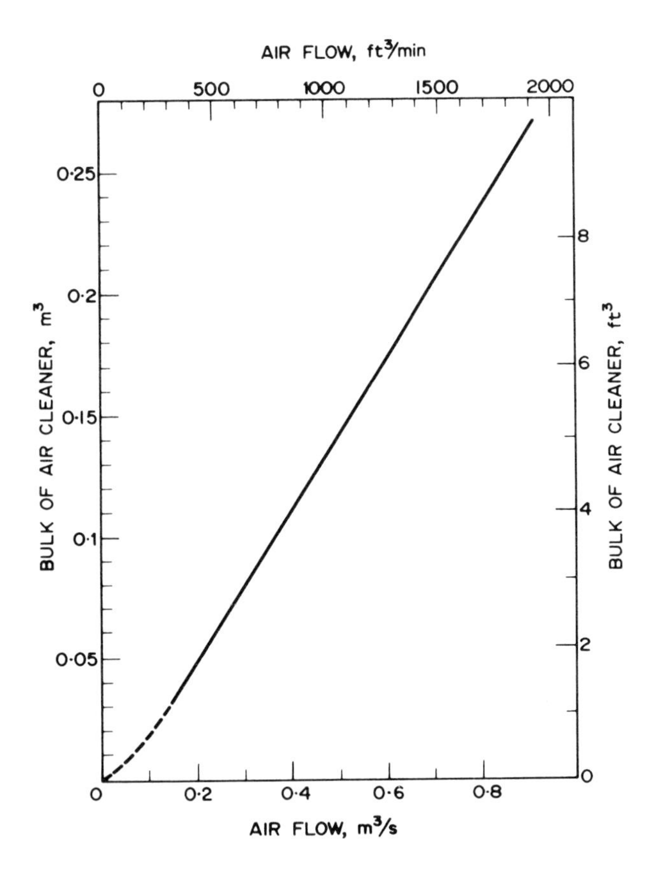

Figure 20.3 Air cleaner requirements for reciprocating engine breathing; 20–100 h maintenance periods for max. dust concentration of 0.0015 kg/m^3, restriction 15–25 in w.g.

Dust concentration up to 0.0015 kg/m^3.

Type of cleaner: 2-stage, primary centrifuge with fabric secondary stage.

Fabric required for 2nd stage:
0.1 m^2 × 150 mm thick/37 kW.

Approx. relationship between air flow and bulk volume of complete cleaner shown in Figure 20.3.

Oil issuing from a bearing as end leakage will travel along a shaft for a finite distance before centrifugal dispersal of the film takes place. Many clearance seals will permit oil leakage from the bearing housing if they are situated within the shaft oil-film regime. Flinger rings and drain grooves can prevent the oil reaching the seal.

GENERAL PROPORTIONS

The natural dispersal length of the oil film along the shaft varies with the diameter and the speed as shown in Figure 21.1.

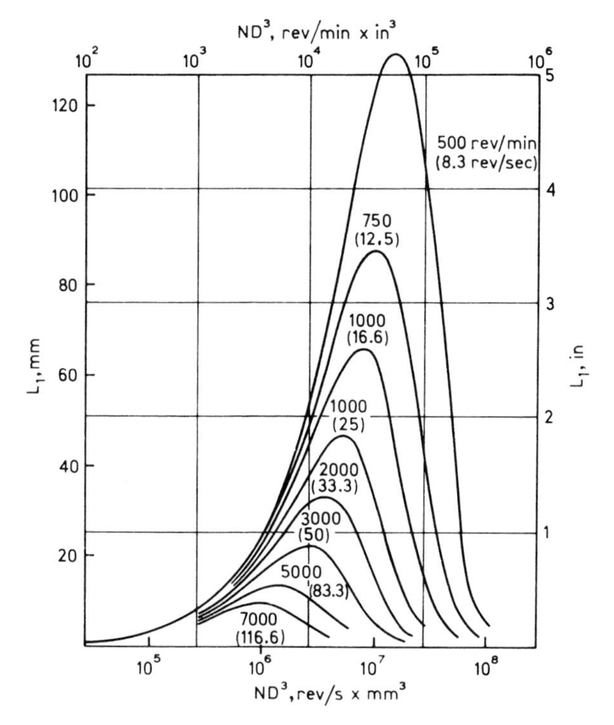

Figure 21.1

Notation:

L_1 = natural dispersal length of oil film—in (mm)
L_2 = distance of oil thrower from end of bearing—in (mm)
D = shaft diameter—in (mm)
D_0 = outside diameter of oil thrower—in (mm)
N = shaft speed—rev/min (rev/s)

Using the value of L_1 corresponding to the design value of ND^3 in Figure 21.1, the oil thrower diameter should be derived from:

$$D_0 = \sqrt{\frac{C}{N^2 D} \log_e \frac{L_1}{L_2} + D^2} \qquad (1)$$

where C has the value

30×10^6 for inch rev/min units.
and 136×10^6 for millimetre, rev/s units.

In general, high-speed shafts require small throwers and low-speed shafts require large ones, particularly if the thrower is close to the bearing.

Where shafts must operate at any speed within a speed range, flingers should be designed by the foregoing methods using the minimum range speed.

Where shafts are further wetted by oil splash and where oil can drain down the inside walls of the bearing housing on to the thrower itself, larger thrower diameters than given by equation (1) are frequently employed. Figure 21.2 gives a guide to 'safe' thrower proportions to meet this condition.

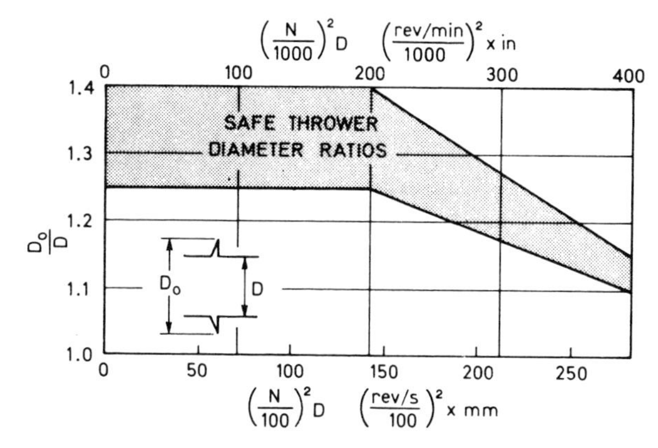

Figure 21.2

TYPICAL THROWERS

Scale details of some well-proven throwers are given in Figures 21.3, 21.4 and 21.5. Relevant values of D_0/D for the originals are given in each case. The application of each type may be assessed from Figure 21.2.

Type 1

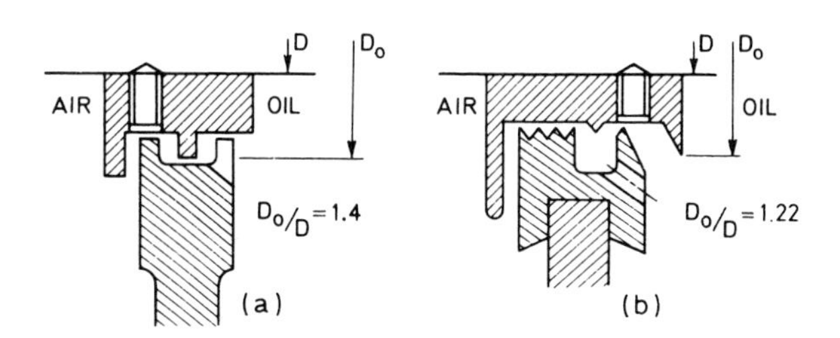

Figure 21.3 Throwers for slow/medium speeds

These are simple throwers of the slip-on type. Mild steel is the usual thrower material while a self-lubricating material such as leaded bronze is preferred for the split housing.

Note:

1. The drain groove from the annulus in (a) and the drain hole in (b).
2. The chamfer at the outer periphery of the (b) split housing to drain away oil washing down the walls of the bearing housing.
3. The chamfer at the back of the main thrower of (b) and the mating chamfer on the housing.

The above features are also common to the other types shown in Figures 21.4 and 21.5.

Type 2

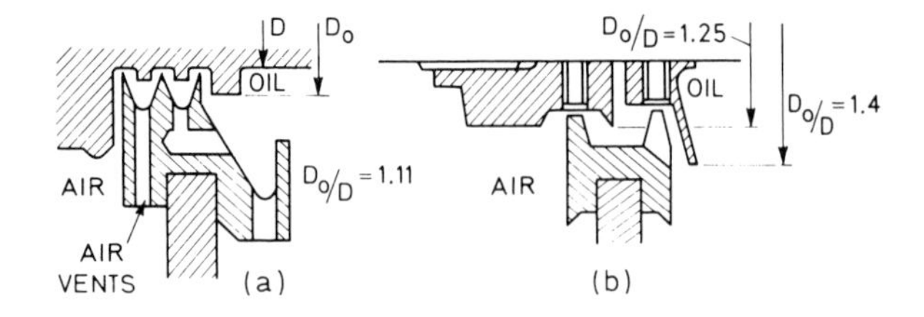

Figure 21.4 Throwers for medium/high speeds

Note how the shaft enlargement on (a) has necessitated the introduction of a second annular space, vented to the atmosphere. Such enlargements, coupling hubs, etc. can create pressure depressions which can pull oil mist through the seal. Note the two-piece construction of (b) which gives a good sized secondary thrower. The shaped primary thrower is perhaps overlarge for a high-speed machine, but this is a good fault!

Type 3

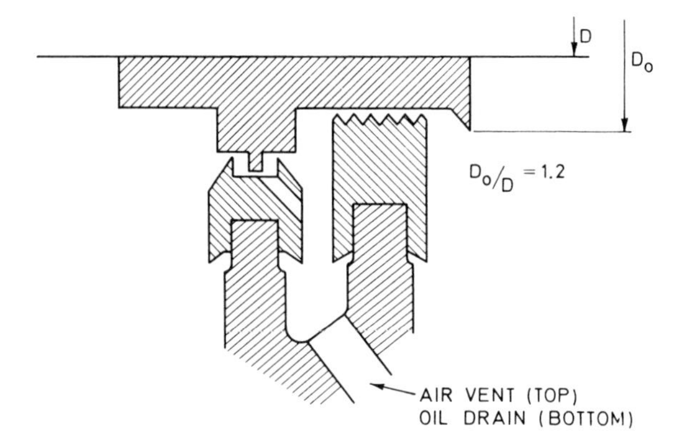

Figure 21.5 A medium/high speed two piece thrower

As an alternative to Type 2, a two-piece arrangement can be used if space permits. The primary seal can be of the visco seal or windback type. The secondary seal can be of the simple Type 1 variety. A substantial air vent is provided between the seals to combat partial vacuum on the air side.

DETAIL DIMENSIONS

Drain hole/oil groove sizing

Hole/groove area $\geqslant k \times$ thrower annular clearance area, corresponding to maximum design tolerances.

Suggested variation of k with shaft speed is given in Figure 21.6.

The individual diameter of the several drain holes making up the above area should not be less than 5 mm ($\frac{3}{16}$) in say.

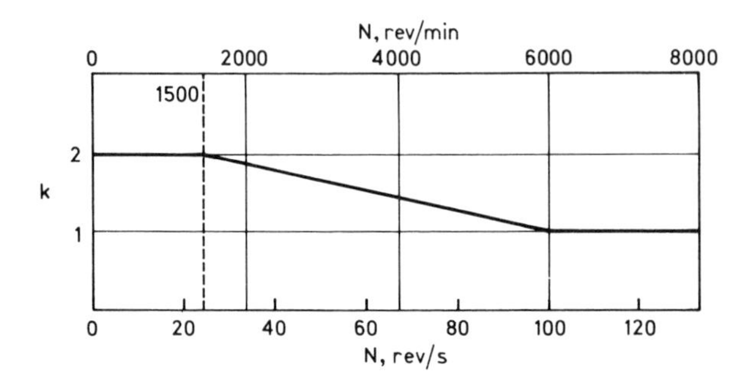

Figure 21.6

Internal clearances

These are a matter of judgement. Suggested values for diametral clearance are:

Low-speed shafts: 1.25 × max. design bearing diametral clearance.

High-speed shafts: D/250 or 2 × max. design bearing diametral clearance, whichever is greater.

PLAIN BUSH SEALS

Fixed bush seal

Leakage is limited by throttling the flow with a close-fitting bush—Figure 22.1.

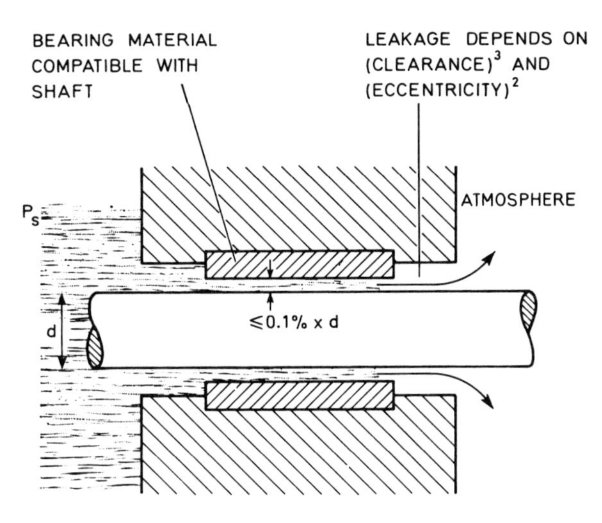

Figure 22.1 Typical fixed bush

Rigid floating bush seal

Alignment of a fixed bush can be difficult but by allowing some radial float this problem can be avoided (Figure 22.2)

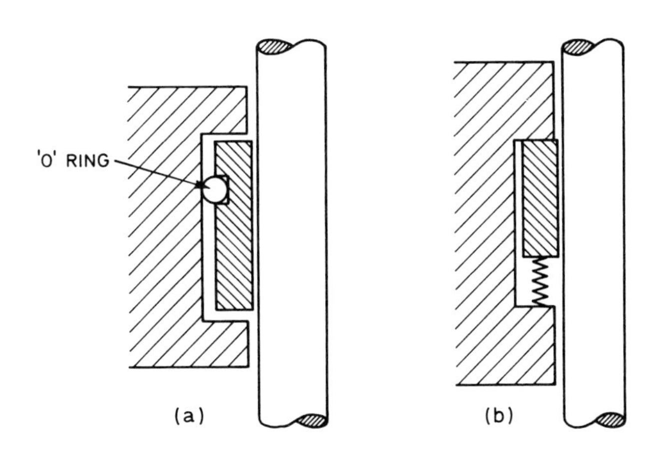

Figure 22.2 Bush seals with radial float

Leakage calculation

The appropriate formula is indicated in Table 22.1 for laminar flow conditions. For an axial bush with an incompressible fluid, Figure 22.3 can be used in both laminar and turbulent regions.

Table 22.1 Bush seal volumetric leakage with laminar flow

$q \equiv$ volumetric flow rate/unit pressure gradient/unit periphery $\eta \equiv$ absolute viscosity	*Fluid incompressible*	*Fluid compressible**
$\left(\varepsilon = \frac{\delta}{c}\right)$ AXIAL BUSH (l, Q, P_s, V, δ, a, c, P_a)	$Q = \dfrac{2\pi a(P_s - P_a)}{l} \cdot q$ m³/s or in³/s	
	† $q = \dfrac{c^3}{12\eta} \cdot (1 + 1.5\epsilon^2)$	$q = \dfrac{c^3}{24\eta} \cdot \dfrac{(P_s + P_a)}{P_a}$
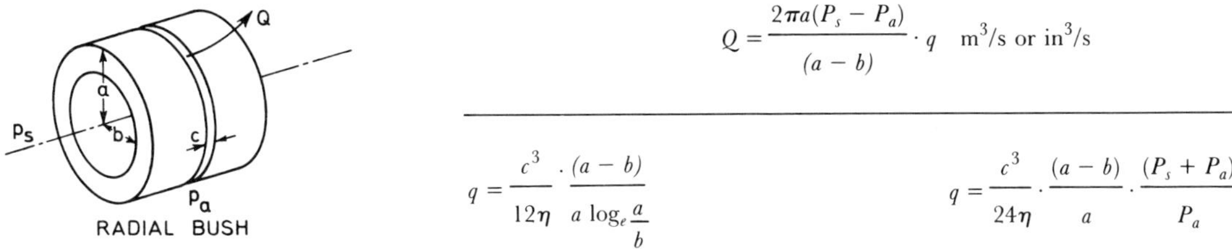 RADIAL BUSH (Q, a, b, c, P_s, P_a)	$Q = \dfrac{2\pi a(P_s - P_a)}{(a - b)} \cdot q$ m³/s or in³/s	
	$q = \dfrac{c^3}{12\eta} \cdot \dfrac{(a - b)}{a \log_e \frac{a}{b}}$	$q = \dfrac{c^3}{24\eta} \cdot \dfrac{(a - b)}{a} \cdot \dfrac{(P_s + P_a)}{P_a}$

* For Mach number < 1.0, i.e. fluid velocity < local velocity of sound.

† If shaft rotates, onset of Taylor vortices limits validity to $\dfrac{V_c}{\nu}\sqrt{\dfrac{c}{a}} < 41.3$ (where ν = kinematic viscosity).

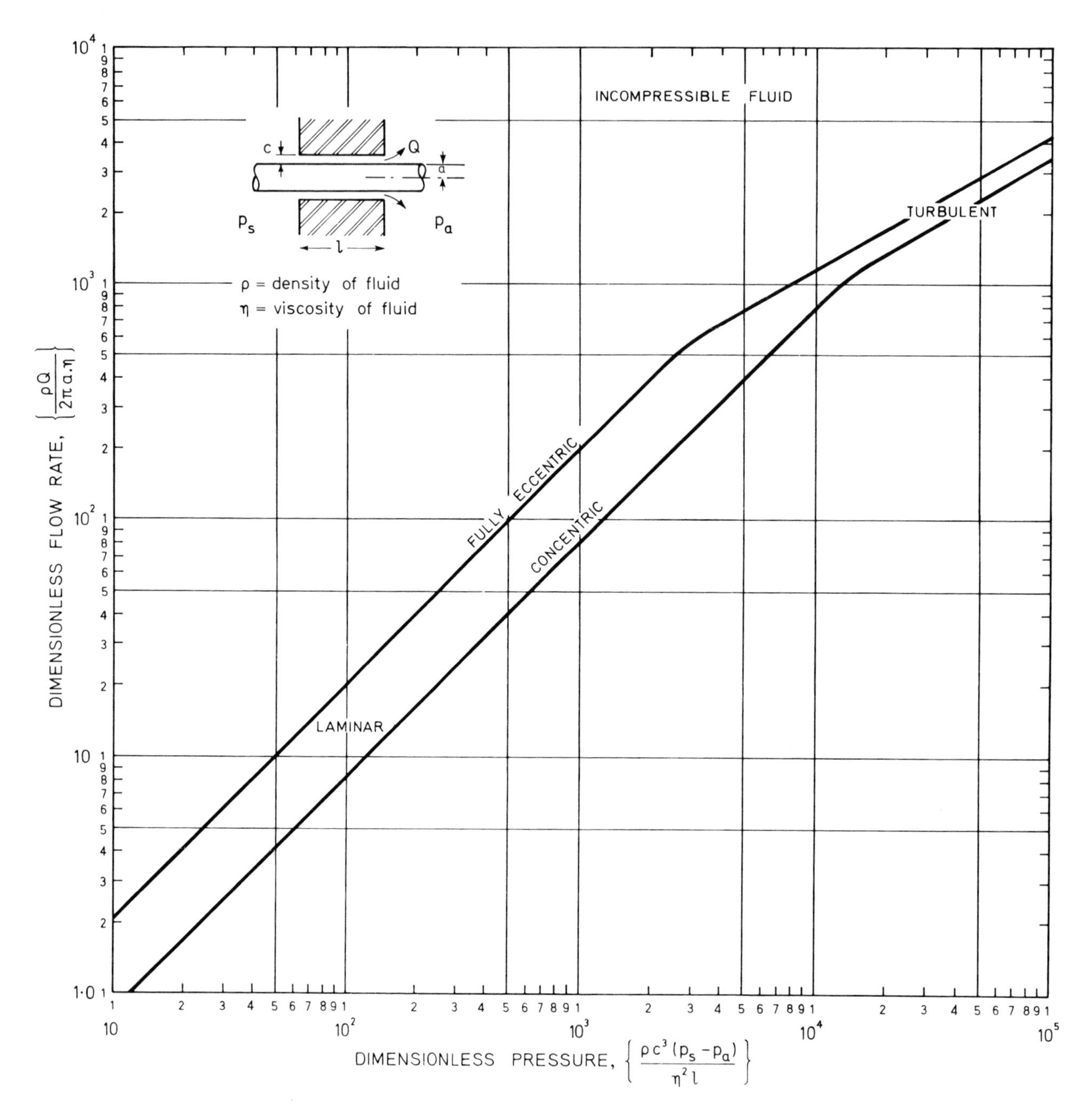

Figure 22.3 Leakage flow of an incompressible fluid through an axial bush at various shaft eccentricities

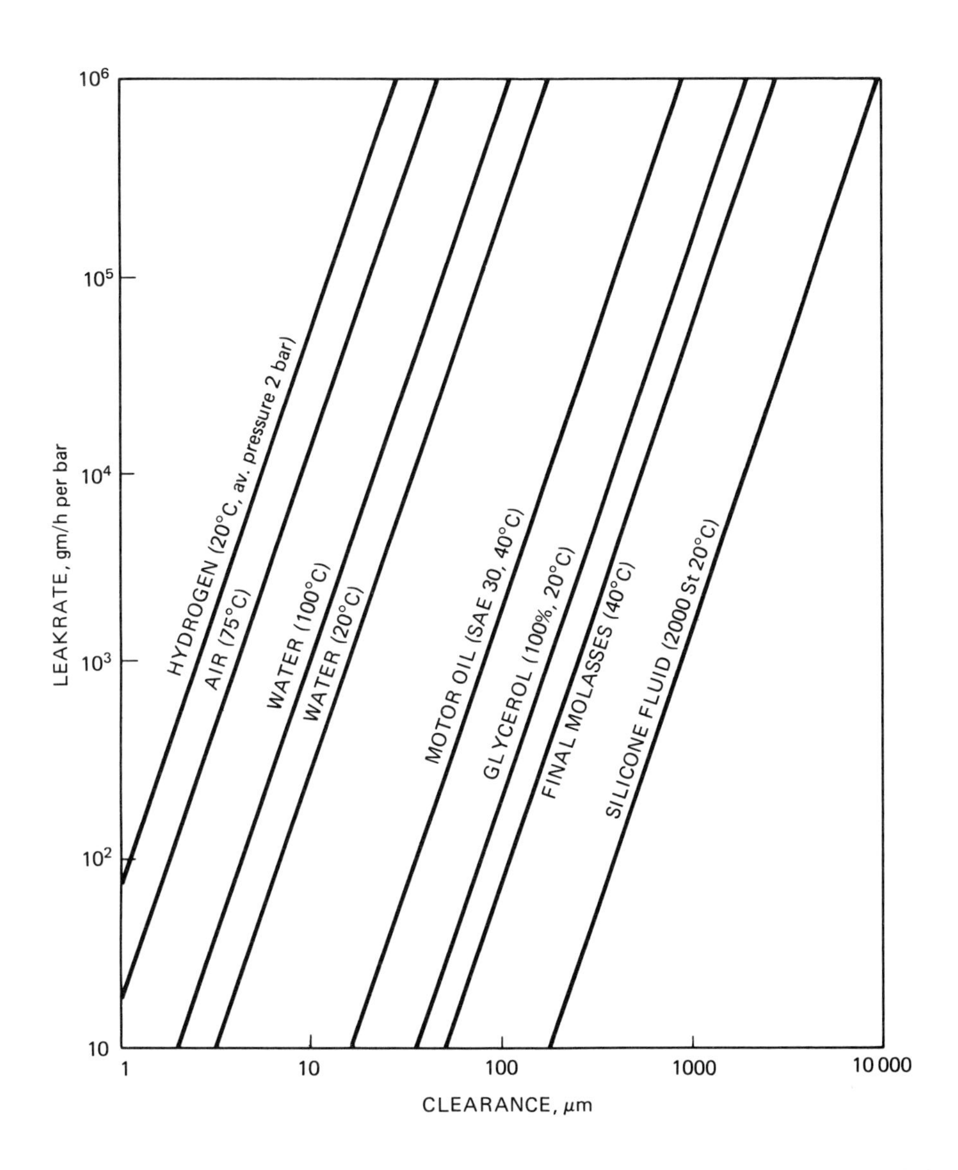

Figure 22.4 Effect of clearance and viscosity on mass flow leakage from a bush seal with length equal to perimeter (if perimeter = k × length then multiply leakrate by k)

Mass flow rate (M) through the clearance of a bush seal:

$$M = 9.4 \times 10^{-8} \frac{\rho h^3 \Delta_p D}{\eta L} \text{gm/h}$$

where ρ = density, kg/m^3
h = clearance, μm
Δ_p = pressure differential, bar
D = mean diameter, mm
η = dynamic viscosity, $\frac{\text{Ns}}{\text{m}^2}$
L = leakage path length, mm

Materials

Bush and shaft may sometimes rub; therefore compatible bearing materials should be selected. Segmented bushes are commonly made of carbon–graphite and may run on a shaft sleeve of centrifugally cast bronze (90% Cu, 10% Sn) when sealing water. The life of such a combination could be 5–10 years if abrasives are not present. Other suitable materials for the sleeve or shaft include carbon steels, for more severe conditions nitriding or flame hardening is recommended, and stainless steel. In the latter case a compatible babbit may be used to line the bush.

BRUSH SEALS

Brush seals are an alternative for labyrinths in gas turbine engine applications, reducing leakage by a factor up to five or tenfold, although relatively expensive. The brush seal comprises a bundle of metal filaments welded at the base. The filaments are angled circumferentially at about 45 degrees, filament length is chosen to give an interference of 0.1–0.2 mm with the sealing counterface. Filaments are typically about 0.7 mm diameter and manufactured from such materials as high temperature alloys of nickel or cobalt. Suitable counterface materials include hardfacings of chromium carbide, tungsten carbide or alumina

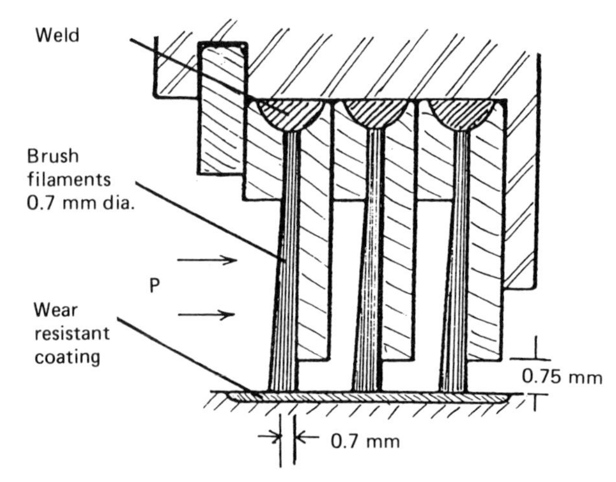

Figure 22.5 Brush seal with typical dimensions

LABYRINTHS

A labyrinth [Figure 22.6] can reduce leakage below that of a bush seal by about half. This is because eddies formed in the grooves between the vanes increase the flow resistance. Labyrinths are commonly used for gas and steam turbines.

Typically, the vane axial spacing might be twenty times the vane lip clearance. If the spacing is too small eddy formation is inefficient, if too large then the seal becomes unduly long.

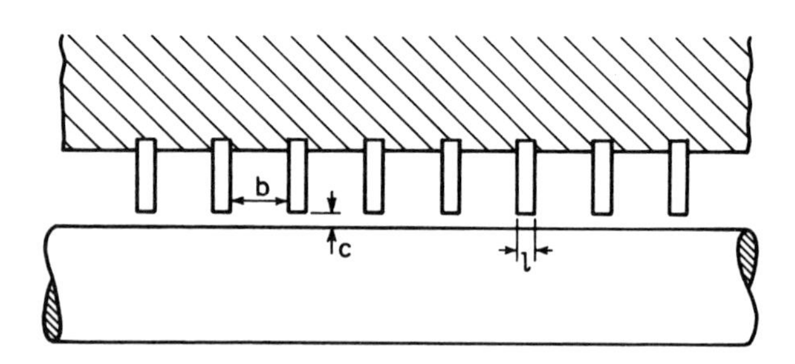

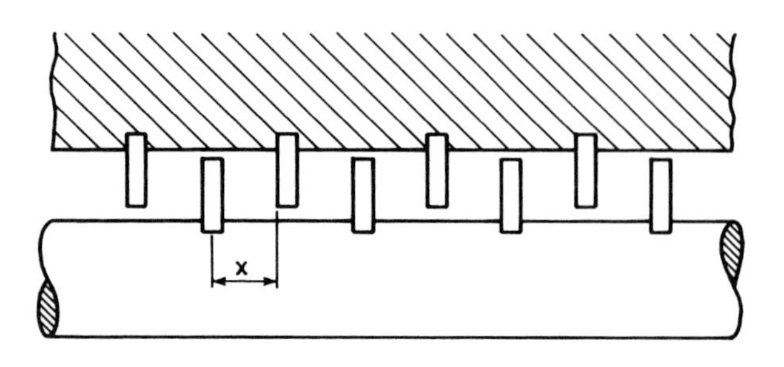

Figure 22.6 Typical labyrinth arrangements

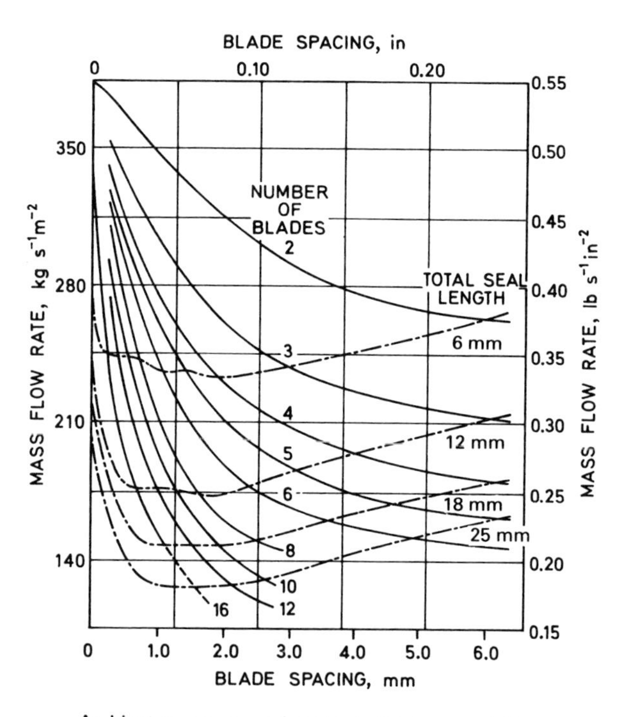

Ambient pressure = 1 bar; blade thickness = 0.14 mm radial clearance = 0.127 mm pressure ratio = 0.551; temperature 310 K

Figure 22.7 Performance of a typical labyrinth seal

Materials

Vane and rotor must be compatible bearing materials, in case of rubbing contact, but in turbines stresses are high and creep is a critical factor. Metal foil honeycomb is a convenient form of material for labyrinths since its integrity is retained if rubbing occurs, and yet it deforms readily. Suitable honeycomb is produced in 18/8 stainless steel and Nimonic 75, 80a and 90. Another convenient combination comprises metal fins and a carbon bush. The latter can be segmented if necessary for large diameters.

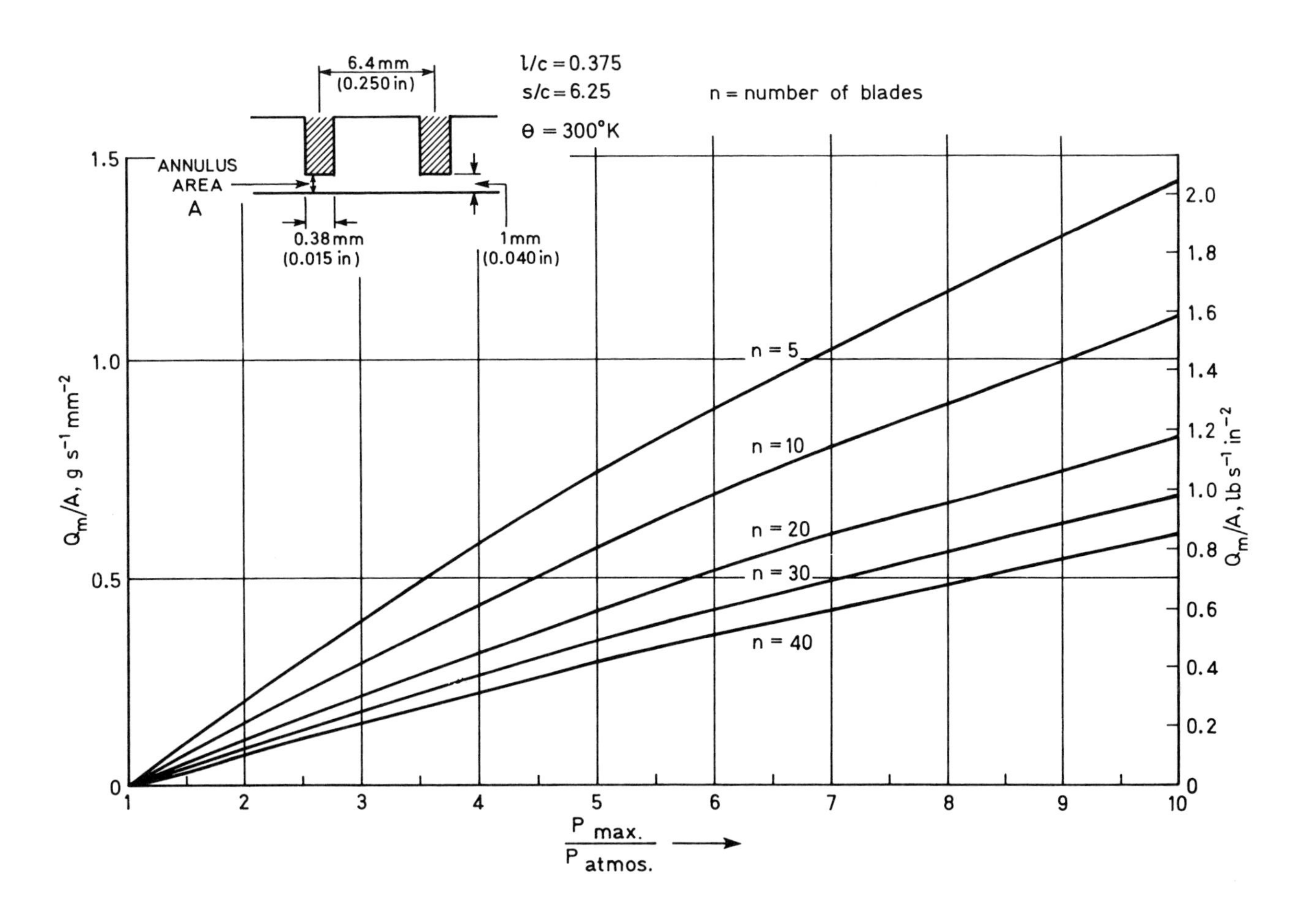

TEMPERATURE

For other temperatures multiply Q_m/A by $\frac{17.3}{\sqrt{\theta_k}}$

GEOMETRY

For other values of s/c or l/c multiply Q_m/A by g:
(see below left)

STEPPED LABYRINTH

For a stepped labyrinth use an 'effective' n value, kn:
(see below right)

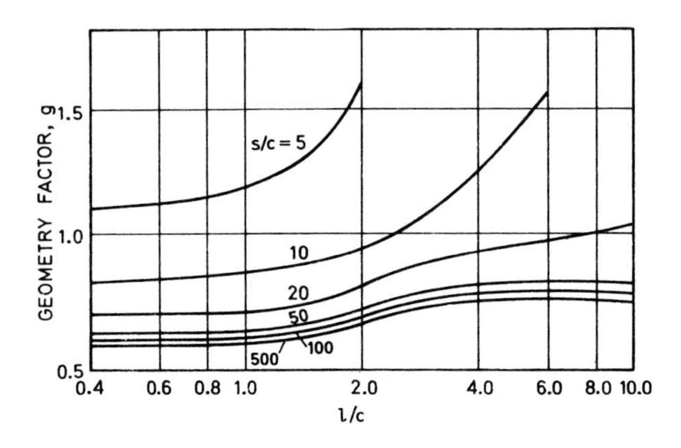

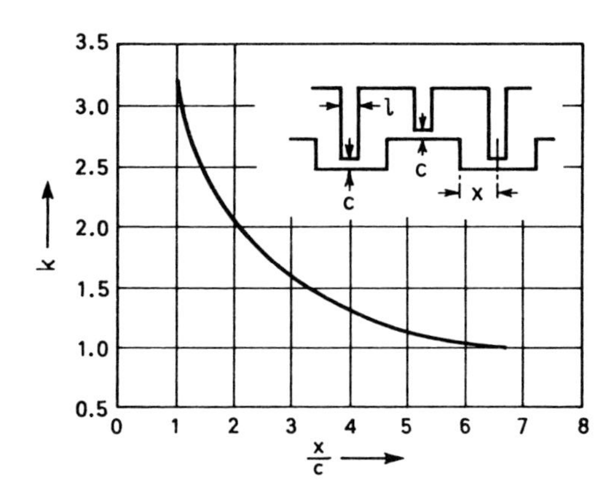

Figure 22.8 Calculation of mass flow rate of labyrinth

VISCOSEAL
(Also called a screw seal or wind back seal)

Resembles a bush seal in which a helical groove has been cut in the bore of the bush or on the shaft (Figure 22.9). As the shaft rotates the helix pumps any leaking fluid back into the sealed system. There is no *sealing action* if the shaft is not rotating: an auxiliary seal can be fitted to prevent static leakage and may be arranged to lift off automatically when the shaft rotates, thereby reducing wear.

Viscoseals are used with *viscous fluids*, or at *high rotational speeds*, to seal low or moderate presures. When the pumping action just balances the leakage flow with the helix full of liquid there is no net leakage—this is the *sealing pressure*. If the system pressure exceeds the sealing pressure the seal leaks. At lower pressures the helix runs partially dry.

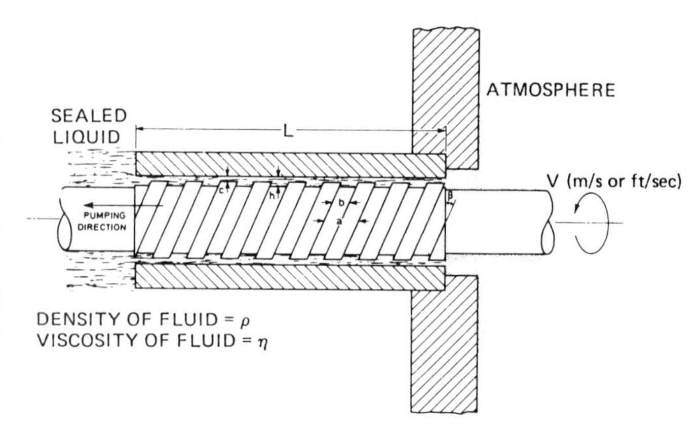

Figure 22.9 Viscoseal or wind-back seal

Optimum design and performance prediction

Helix angle, β = 15.7°
Clearance ratio, h/c = 3.7
Land ratio b/a = 0.5
Groove profile rectangular

The 'sealing pressure' for a seal of these dimensions, is given by:

$$p_s = \frac{6\eta VL}{c^2} \cdot k \quad (\eta = \text{absolute viscosity})$$

(i) $k = 0.1$ if $(Vc\rho/\eta) < 500$ (i.e. laminar flow)
or (ii) k is given by Figure 22.10 if $(Vc\rho/\eta) > 500$ (i.e. turbulent flow).

The optimum design varies with Reynolds number $[(Re) = Vc\rho/\eta]$ for turbulent flow; in particular a smaller helix angle than the optimum laminar value is better.

Eccentricity will lower the sealing pressure with laminar flow but may raise it for turbulent flow.

A problem may arise due to 'gas ingestion' when running in the sealing condition. This is due to instability at the liquid/air interface and is not easily prevented. The effect is to cause air to be pumped into the sealed system.

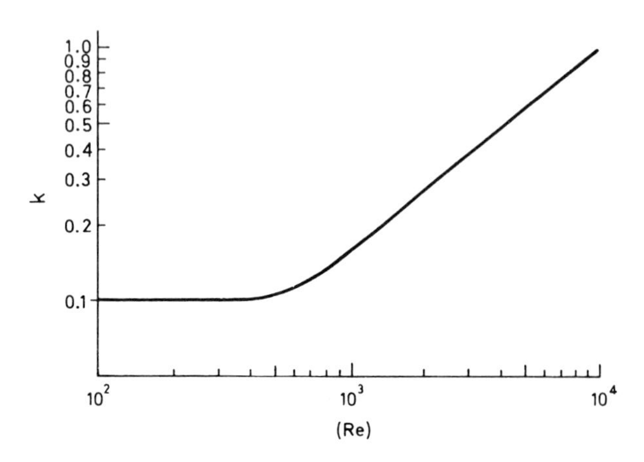

Figure 22.10 Variation of k with Reynolds Number

Typical performance

A 25 mm dia. seal with a clearance of 0.025 mm, length 25 mm, sealing water with a shaft speed of 2.5 m/s would be expected to seal at pressures up to 0.67 atm.

BARRIER VISCOSEAL

By installing a pair of conventional viscoseals back to back, so that pressure is built up between them, a pressure barrier forms to prevent the sealed fluid escaping (Figure 22.11). The sealed fluid must not be miscible with the barrier fluid. Typically the former is a gas and the latter a liquid or grease.

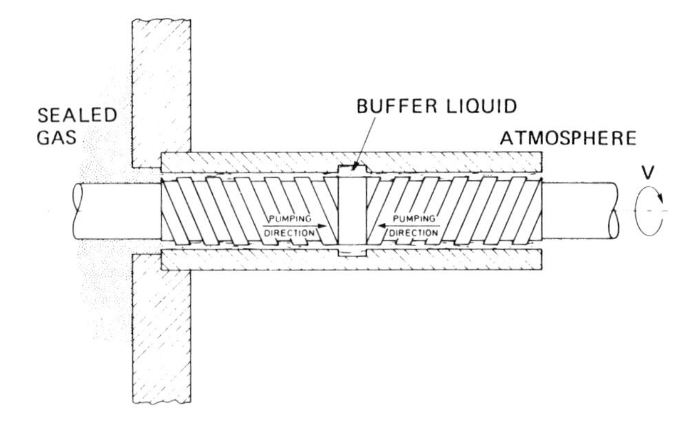

Figure 22.11 Barrier viscoseal

Typical performance

In tests using grease as the barrier fluid, gas pressures up to 10 atm have been sealed by a 13 mm dia. seal at 1000 rev/min.

SEALS FOR ROTATING SHAFTS

Design variations

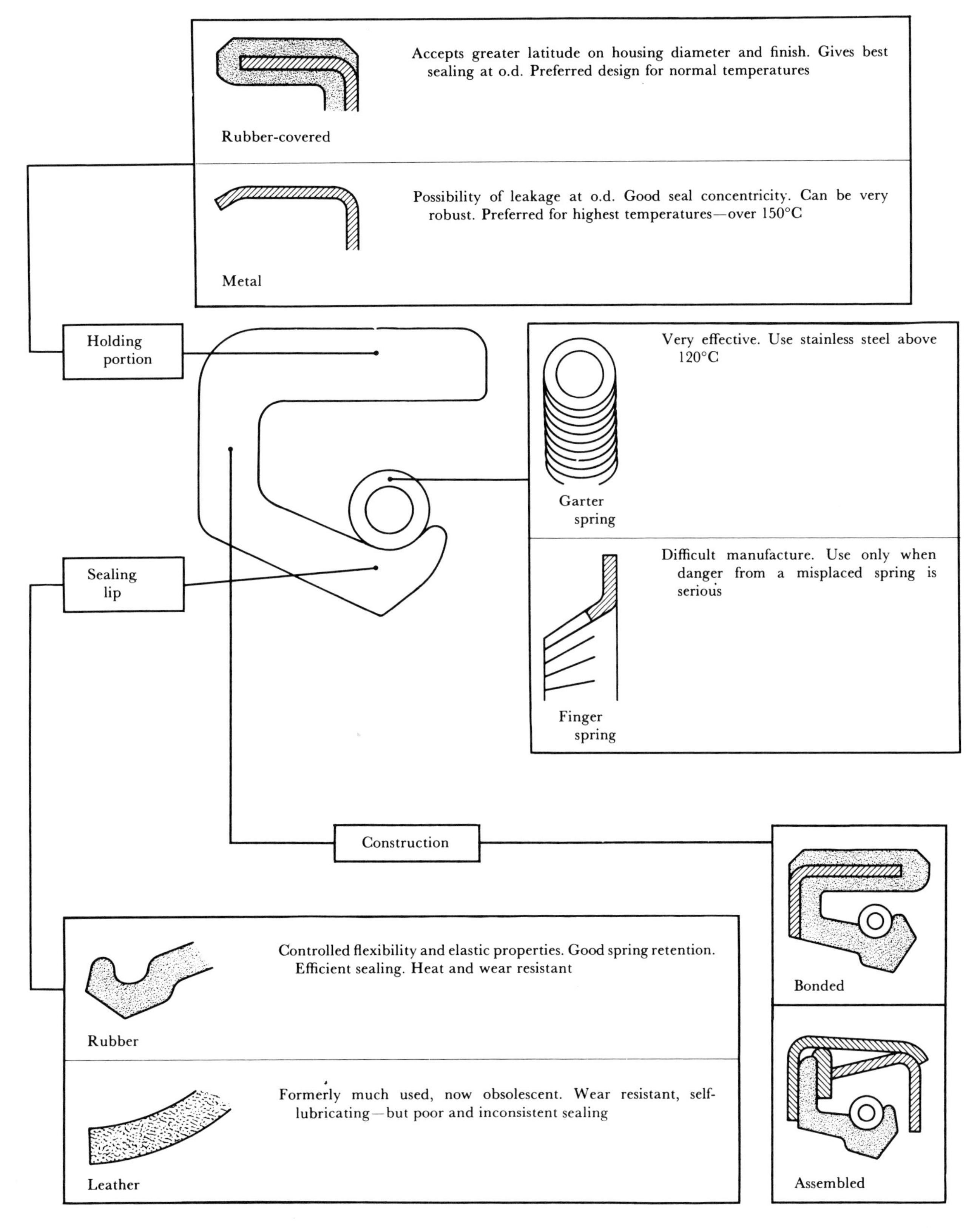

The above variants are the important ones to the seal user or machine designer, apart from special seals for dirty conditions. There are multitudinous detail modifications introduced by seal manufacturers, but for normal applications these can be ignored.

Operating conditions

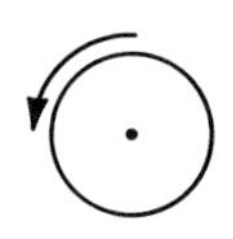 Max. speed	Up to 35 mm ($1\frac{1}{4}$ in) dia. 75 mm (3 in) dia. Over 75 mm (3 in) dia.	Approx. 8000 rev/min Approx. 4000 rev/min Approx. 15 m/s (50 ft/s) peripheral speed
 Max. fluid pressure	Up to 75 mm (3 in) dia. Over 75 mm (3 in) dia. 	Approx. 0.6 bar (10 p.s.i.) Approx. 0.3 bar (5 p.s.i.) By using a profiled backing washer to support the lip, pressures up to 6 bar (100 p.s.i.) can be accommodated
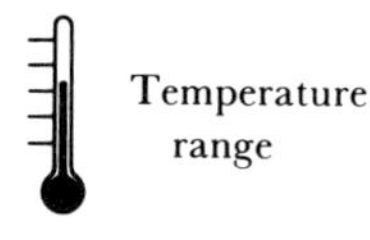 Temperature range	See table of rubber materials	Permissible oil temperatures are set by the sealing lip material. Do not ignore low-temperature conditions
 Eccentricity	Housing	Better than 0.25 mm (0.010 in) total indicator reading when clocked from shaft
	Shaft	Depends on speed. Aim for better than 0.025 mm (0.001 in) total indicator reading when rotated in its own bearings
Surface finish	Housing	Fine turned. Provide lead-in chamber
	Shaft	Grind and polish to better than 0.5 μm R_a. Surface must be free from all defects greater than 0.0025 mm (0.0001 in) deep. Use cardboard protection sleeve during manufacture
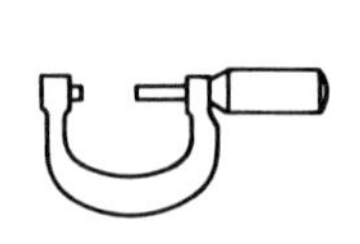 Machining tolerances	Housing	Up to 100 mm (4 in) ±0.025 mm (0.001 in) 100–175 mm (4–7 in) ±0.037 mm (0.0015 in) Over 175 mm (7 in) ±0.05 mm (0.002 in)
	Shaft	Up to 50 mm (2 in) ±0.025 mm (0.001 in) 50–100 mm (2–4 in) ±0.037 mm (0.0015 in) 100–200 mm (4–8 in) ±0.05 mm (0.002 in) Over 200 mm (8 in) ±0.125 mm (0.005 in)

Sealing dirt and grit

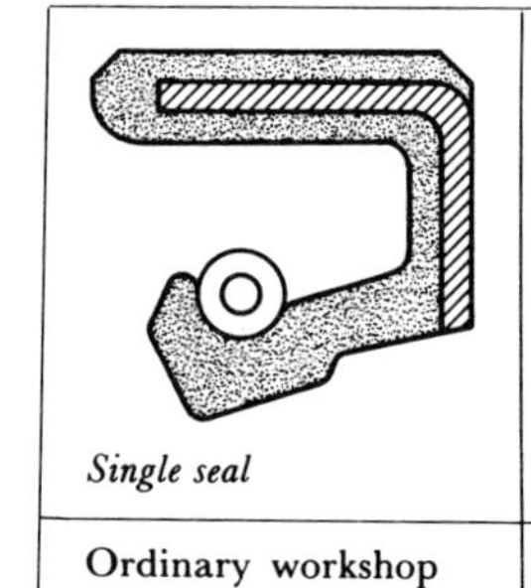 *Single seal*	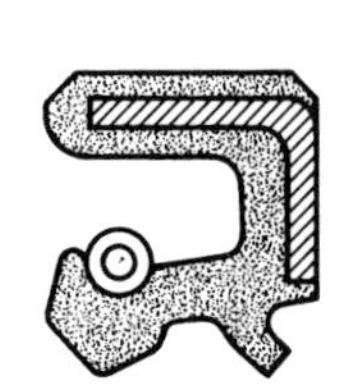  *Auxiliary sealing lip*	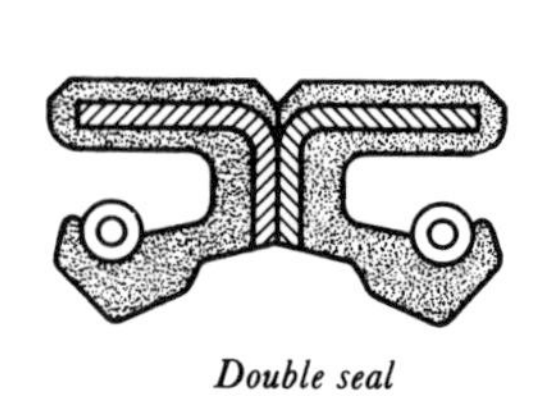 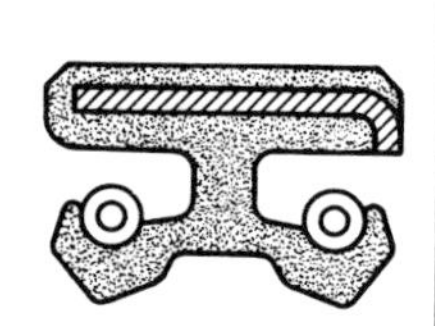 *Double seal*
Ordinary workshop or road conditions	Only slightly better than a single seal. Not worth the extra cost	For arduous conditions—short of being submerged in wet mud

Sealing lip material (rubber)

Type of rubber	*Typical trade names*	*Working temperature range, °C*	*Resistance to:*		*Relative cost of seal*
			Mineral oil	*Chemical fluids*	
Nitrile	Hycar Polysar	−40 to +100	Excellent	Fair	1
Acrylate	Krynac Cyanacryl	−20 to +130	Excellent	Fair	2
Fluoropolymer	Viton	−30 to +200	Excellent	Excellent	10
Polysiloxane	Silastomer	−70 to +200	Fair	Poor	4

Nitrile synthetic rubber is the universal choice for sealing oil or grease at temperatures below 100°C. For more extreme temperature conditions the choice is normally one of the other materials shown in the table—with some penalty in other directions. When in doubt consult the seal manufacturer.

Positive-action seals

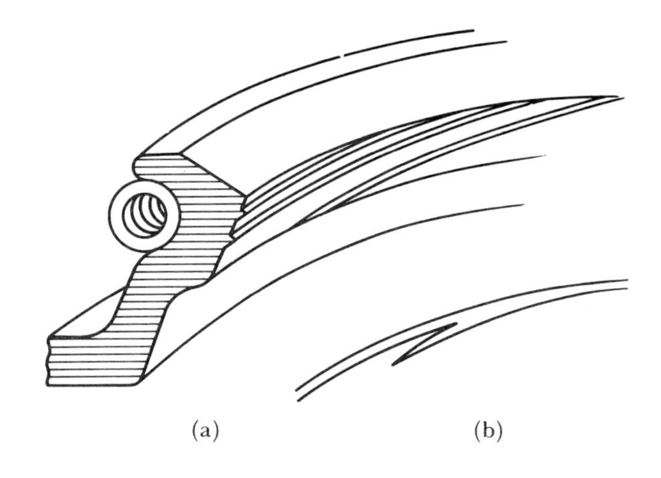

Figure 23.1
(a) Section of a positive-action seal, showing the helical ridges on the air side.

(b) View of contact band through a glass shaft, showing one of the thread run-outs.

The 'positive-action' feature improves sealing performance and reliability. It is essential in conditions where eccentricity or vibration is beyond the limit for a normal seal.

Storage and fitting

1. Store in a cool place in manufacturer's package.
2. Lubricate before installing. Handle carefully.
3. Use a sleeve on the shaft to protect seal from damage by sharp edges, keyways, etc.
4. Press home squarely into housing, using a proper tool.

SEALS FOR RECIPROCATING SHAFTS

Packing types

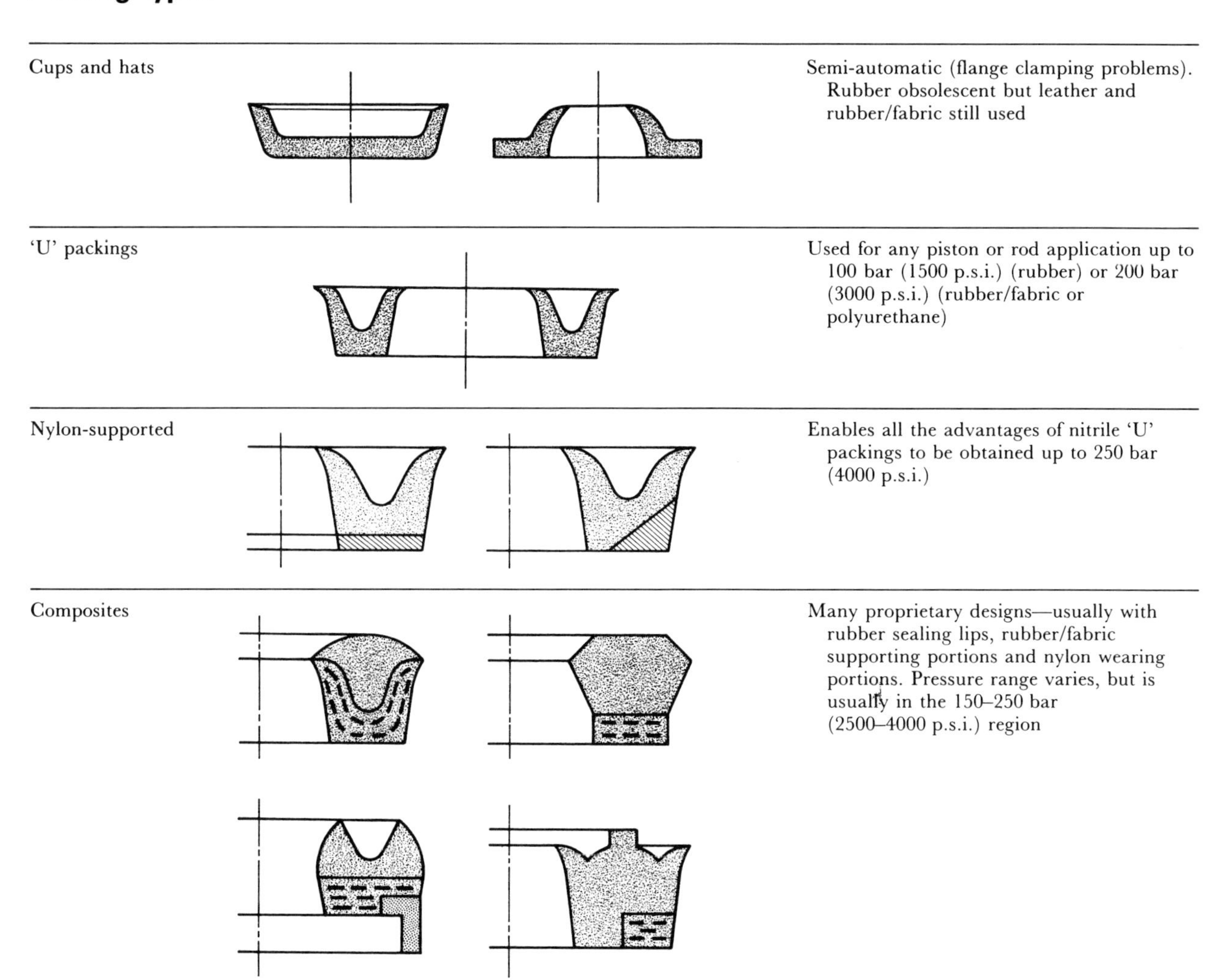

Cups and hats	Semi-automatic (flange clamping problems). Rubber obsolescent but leather and rubber/fabric still used
'U' packings	Used for any piston or rod application up to 100 bar (1500 p.s.i.) (rubber) or 200 bar (3000 p.s.i.) (rubber/fabric or polyurethane)
Nylon-supported	Enables all the advantages of nitrile 'U' packings to be obtained up to 250 bar (4000 p.s.i.)
Composites	Many proprietary designs—usually with rubber sealing lips, rubber/fabric supporting portions and nylon wearing portions. Pressure range varies, but is usually in the 150–250 bar (2500–4000 p.s.i.) region

Materials

Material	Properties
Rubber (nitrile)	Highest sealing efficiency. Easily formed to shape. Low cost. Limited pressure capability, 100 bar (1500 p.s.i.). Excellent wear resistance. Poor extrusion resistance
Rubber-impregnated fabric	Great toughness, resistance to cutting and extrusion. Wear resistance inferior to plain rubber
Polyurethane rubber	Best toughness and wear resistance. Used for flexible packings at the highest pressures
Leather	Good wear and extrusion resistance. Limited shaping capability. Poor resistance to permanent set
Nylon	Combines well with rubber to resist extrusion and to provide good bearing surfaces

Extrusion clearance—mm(in)

	Up to 100 bar (1500 p.s.i.)		100–200 bar (1500–3000 p.s.i.)		Over 200 bar (3000 p.s.i.)	
	Normal	*Short life*	*Normal*	*Short life*	*Normal*	*Short life*
Rubber	0.25 (0.010)	0.5 (0.020)	—	—	—	—
Rubber/fabric leather	0.4 (0.015)	0.6 (0.025)	0.25 (0.010)	0.5 (0.020)	0.1 (0.005)	0.25 (0.010)
Polyurethane	0.4 (0.015)	0.6 (0.025)	0.25 (0.010)	0.5 (0.020)	0.1 (0.005)	0.25 (0.010)
Nylon support	—	—	0.25 (0.010)	1.0 (0.040)	0.1 (0.005)	0.5 (0.020)

Design of metal parts

	Cylinders	*Piston rods*
Preferred materials	Steel Cast iron	Steel
Heat treatment	Not required	Harden if possible
Plating	Not required	Chrome plate if possible
Surface finish	Grind or hone 0.5 μm R_a max.	Grind and polish 0.5 μm R_a max.
Machining tolerances	Fixed by extrusion clearance—see previous table	

Friction

Friction varies considerably with working conditions but for preliminary design purposes it can be assumed that it will be between 0.5% and 3% of the load which would be produced on a piston of the same diameter by the fluid pressure involved. For more accurate values the seal manufacturer must be consulted.

Points the designer should watch

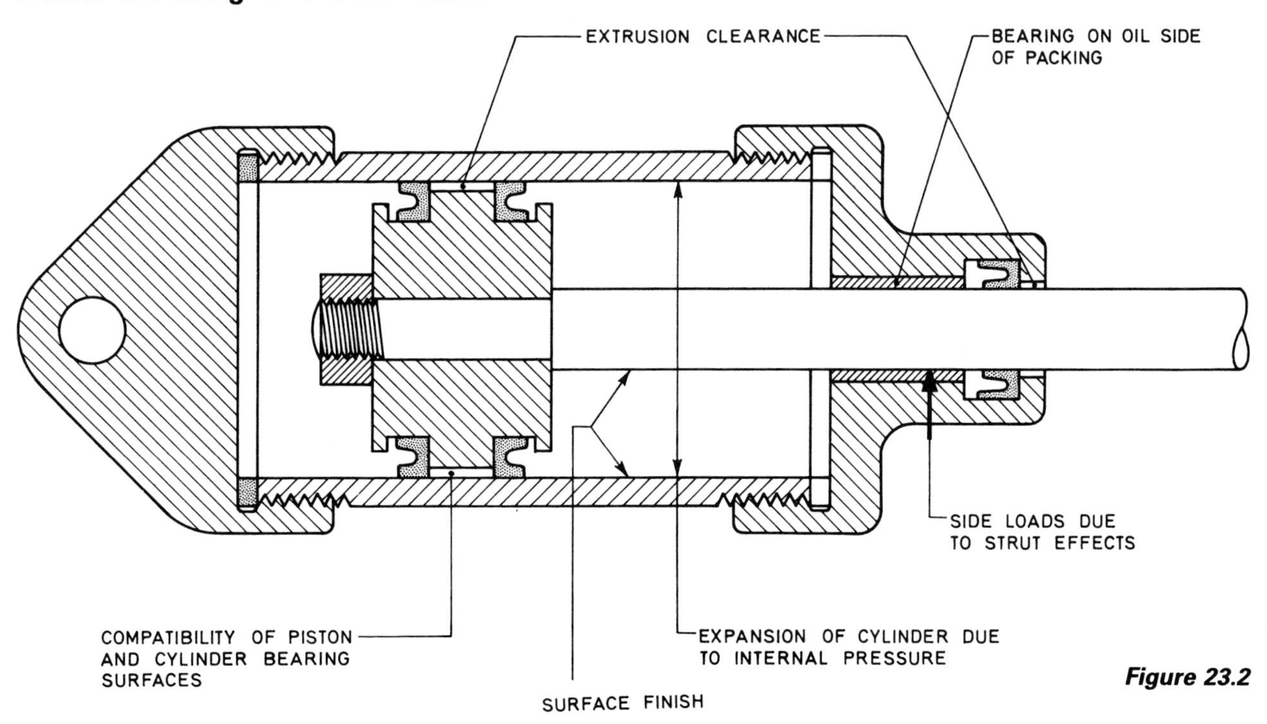

Figure 23.2

TYPES OF MECHANICAL SEAL

Table 24.1 Standard types of mechanical seal

Type of seal	*Maximum allowable sealed pressure*	*Remarks*
UNBALANCED SEALS 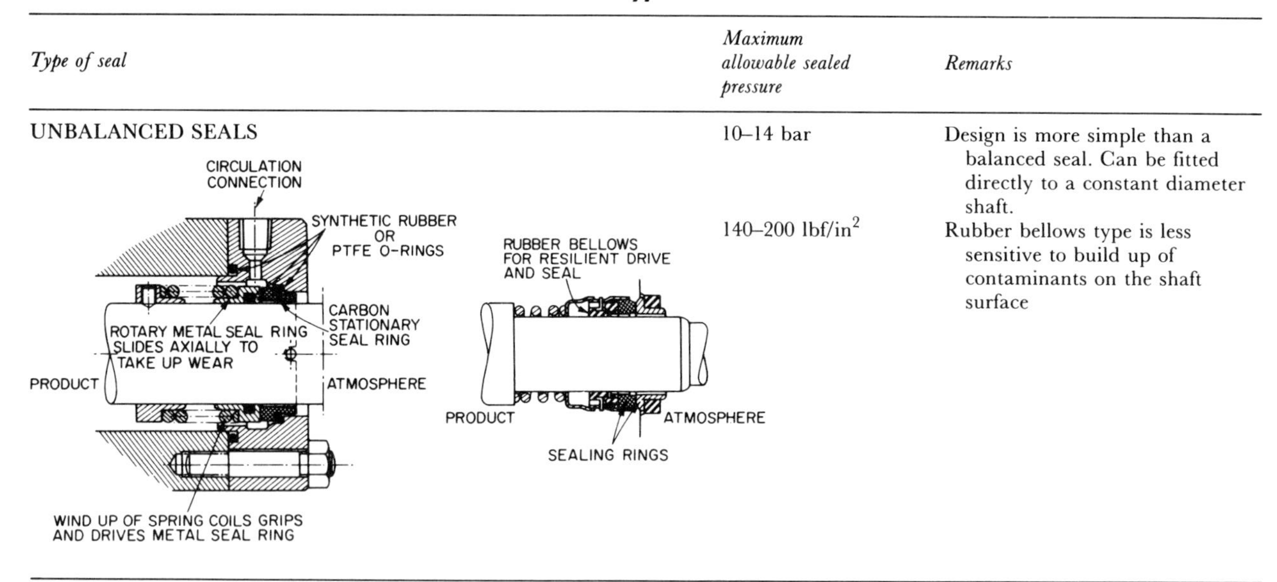	10–14 bar 140–200 lbf/in^2	Design is more simple than a balanced seal. Can be fitted directly to a constant diameter shaft. Rubber bellows type is less sensitive to build up of contaminants on the shaft surface
BALANCED SEALS 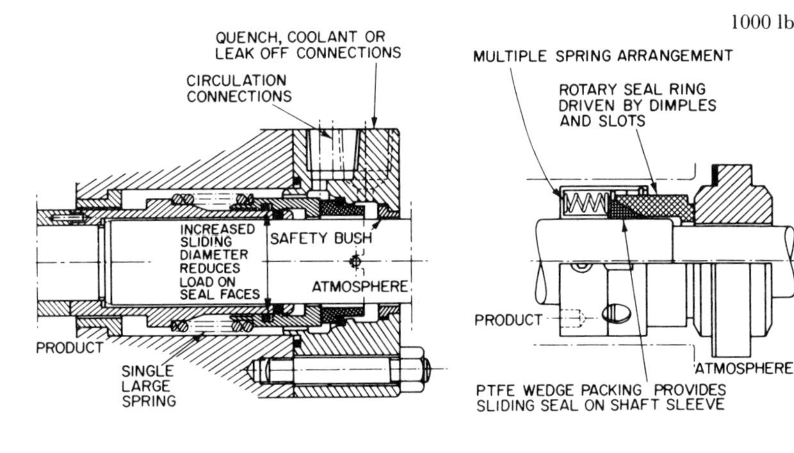	70 bar 1000 lbf/in^2	Balanced design reduces load on seal faces and enables higher pressures to be sealed. Design is more complex and shaft sleeve is essential
BACK TO BACK SEAL 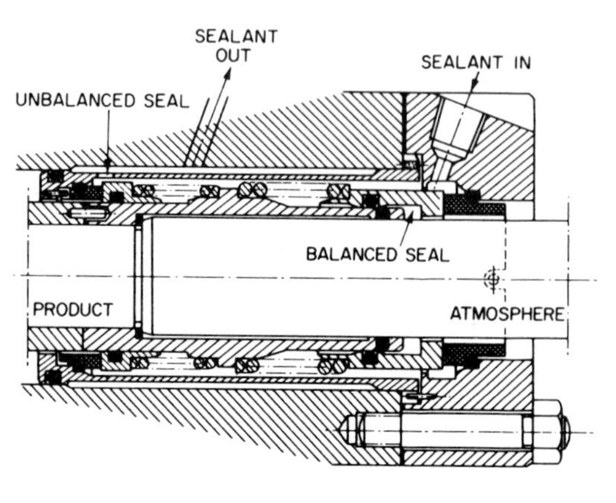	As above	If fluid to be sealed is: a gas, a liquid near its boiling point, abrasive, a poor lubricant, this arrangement enables a separate liquid sealant (such as oil) to be used. It is, however, rather complex. Environmental awareness and safety requirements make the use of this seal more necessary.

Table 24.2 Special seals and additional design features

Feature	*Reason for inclusion*	*Remarks*
CERAMIC/PTFE SEAL 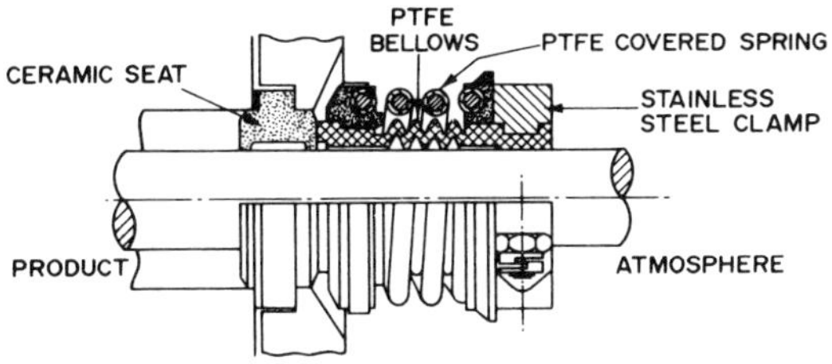	Excellent corrosion resistance of components in contact with the fluid to be sealed	Particularly suitable for chemical applications
SPLIT SEAL 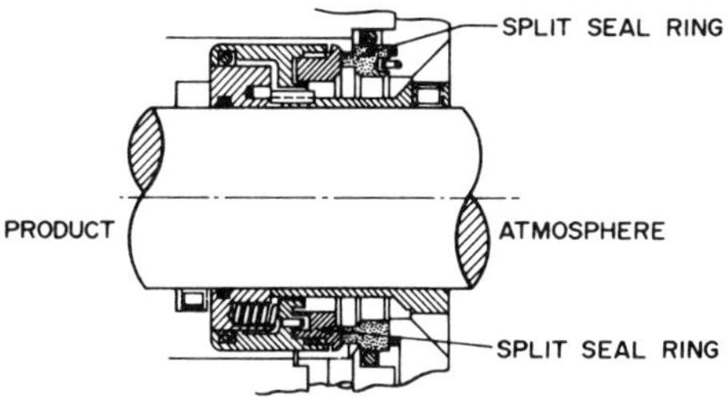	Ease of replacement of wearing components	The splitting of both sealing components reduces allowable performance and gives some leakage
FLUSH 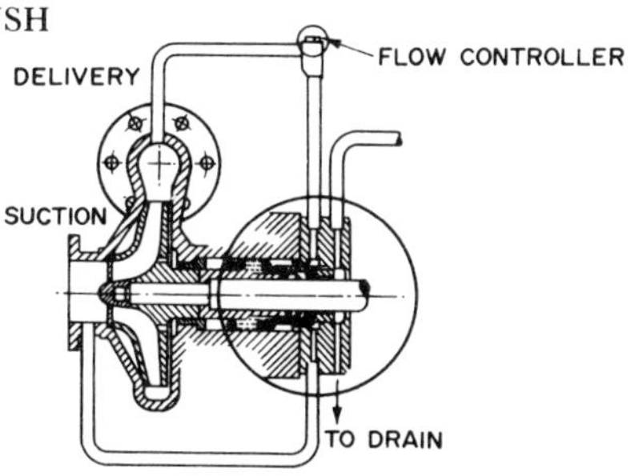 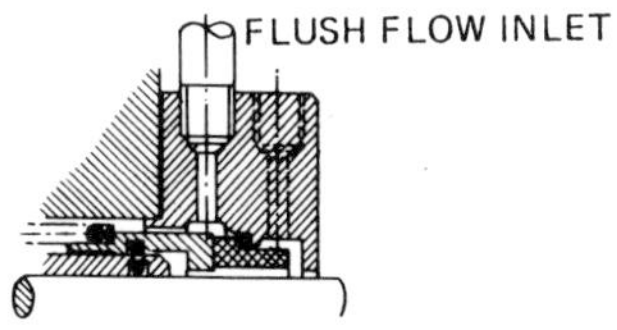	To flush sediments and other debris from the sealing zone and to remove frictional heat. To keep seals handling hot products cool. To keep seals handling cold products warm	On pumps with the seal on suction pressure, a small flow of product from pump discharge is piped to the seal, and returns to pump suction through the balance holes in the impeller. If the differential head exceeds 15 m a flow controller is needed to limit the flow rate. This is only suitable if the product is free of particles above 50 μm. With dirty products a cyclone separator should be used instead, to provide clean product to the seal and return dirty product to the suction
QUENCH 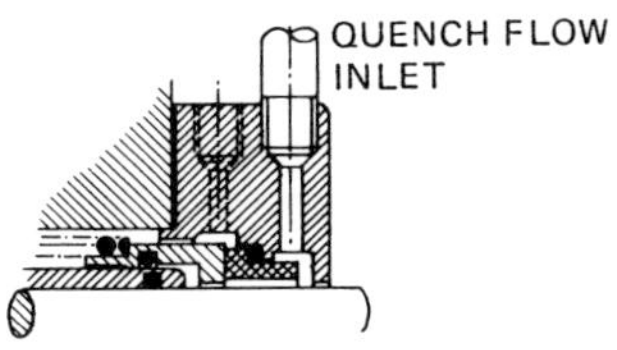	To remove or prevent the build up of crystals or decomposition products on the atmospheric side of the seal	Water flushing will generally remove crystal deposits and low pressure steam will remove 'coke' from hydrocarbons
SAFETY BUSH (illustrated on diagram of balanced seal)	To give extra protection in the event of seal failure by reducing the pressure of the escaping product	The bush is usually made of non-ferrous material to reduce risk of sparking or seizure
AUXILIARY PACKING/back up seal	Used instead of a safety bush when conditions are so hazardous that a second or emergency safeguard must be provided	Usually consists of two turns of soft packing in an auxiliary stuffing box, or a lip seal of PTFE based material

SEAL PERFORMANCE

The main factors which determine the operating limits of mechanical seals are:-

1. The stability (boiling etc.) of the fluid film between the seal faces.
2. The wear life of the seal face materials.
3. The compatibility of the materials, from which the seal is made with the sealing environment.
4. Temperature of operation.

When operating correctly the sealing faces are separated by a very thin fluid film of 0.25–8 μm thick.

Rubbing contact occurs during starting and stopping, and occasionally at normal speeds. The material combinations used, therefore, need to be selected for adequate friction and wear performance, and Table 24.3 gives some general guidance.

PV values for mechanical seals are calculated in a different way from the more usual load per unit projected area times rubbing velocity.

In a mechanical seal P is taken as the pressure drop across the seal in bar and V is the mean sliding velocity at the interface in metres per second.

With unbalanced seals 100 to 140% of the sealed pressure acts on the seal faces. Balanced seals have part of the sealed pressure hydraulically relieved to reduce the force applied to the seal faces.

Table 24.3 Typical PV limits for seal face materials corresponding to a seal life of 8000 h

Face material combination	*Product*	*PV bar x m/s*	
		Unbalanced	*Balanced*
Stainless steel carbon*	Water	9	Never
	Oil	30	
Lead bronze carbon*	Water	20	Never
	Oil	35	
Stellite carbon*	Water	35	100 non-
	Oil	100	700 preferred
Tungsten carbide carbon**	Water	100	250
	Oil	150	1000
Silicon carbide carbon*	Water	150	500
	Oil	200	1500
Alumina/carbon**	Water	100	
Tungsten carbide[×] tungsten carbide	Water	60	
	Oil	100	
Silicon carbide[×] silicon carbide	Water	100	
	Oil	150	
Tungsten carbide[×] silicon carbide	Water	100	
	Oil	300	

* metal impregnated carbon.
** resin impregnated carbon (which gives improved corrosion resistance).
[×] for fluids containing abrasive solids.

The materials chosen must also have good corrosion resistance, particularly since one seal face is usually carbon-graphite which generates a high electrolytic potential with most metals, when in an electrolyte.

For the components other than the seal faces, such as seal chambers, springs and shaft sleeves, the basic material is austenitic stainless steel (18/8/3) progressing as increasing chemical resistance is required to Hastelloy B (Ni 61/Co 2/Mo 28/Fe 5) and Hastelloy C (Ni 5/Co 2.5/Cr 15.5/Mo 16/W 4/Fe5). It is an advantage to coat the shaft sleeve under the sliding packing with a hard facing to reduce abrasion and corrosion.

Temperature has a major effect on the choice of packing/secondary sealing materials, as shown in Table 24.4.

Table 24.4a The effect of temperature on secondary sealing materials and design

Below −75°C	Vertical shafts with double seal arrangement (Seals warm gas or vapour instead of cold liquid on horizontal shaft layout)	Use synthetic rubber or PTFE packings
−75°C 30°C	Vertical or horizontal shafts with double seal arrangement	Use synthetic rubber or PTFE packings
−30°C to Ambient	Vertical or horizontal shafts with balanced seal	Use synthetic rubber or PTFE packings
Ambient to 100°C	Balanced or unbalanced seals	Use synthetic rubber or PTFE packings
100 to 250°C	Balanced or unbalanced fluoroelostomer or perfluoroelastomer rings	Use PTFE packings or glass
250 to 250°C	Balanced seal,	Use filled PTFE wedge or packings
	or double seal with cooling by intermediate sealant perfluoroelastomer or graphite foil rings	Use PTFE packings

Table 24.4b Allowable temperature ranges for various secondary sealing materials

Fluorosilicone	−60 to 100°C
Ethylene/Propylene	−50 to 140°C
Butyl	−50 to 100°C
Neoprene	−45 to 110°C
Nitrile	−30 to 120°C
Fluoroelastomer	−25 to 180°C
Perfluoroelastomer	−10 to 250°C
PTFE (spring loaded 'O' rings)	−100 to 250°C
PTFE (glass filled wedge)	−100 to 250°C
Graphite foil	0 to 480°C

Table 24.5 Thermal properties of seal face materials

Material	*Expansion coefficient* $\times 10^6$/k	*Conductivity* W/mK	*Diffusivity* $\times 10^6 m^2/s$	*Specific heat* J/kg K	*Maximum temperature* °C continuous
Stainless steel (18/8)	16	16	4	510	600
Ni-resist	17	40	12	460	400
Alumina (99%)	8	25	7	1000	1500
Tungsten carbide (6% cobalt)	5	90	31	195	600
Silicon carbide (reaction bonded)	4	150	44	1100	1350
Silicon carbide (converted)	4	46	29	840	450
Carbon graphite					
(resin impregnated)	5	18	13	750	250
(antimony impregnated)	5	22	13	750	400
PTFE + glass	100	0.3	0.2	900	200
Stellite (cobalt based)	12	25	7	430	800
Bronze (leaded)	17	150	48	340	300

Thermal diffusivity is a measure of the rate of heat diffusion through a material (for copper the value is 112).

It equals $\dfrac{\text{thermal conductivity}}{\text{specific heat} \times \text{density}}$

High diffusivity increases the performance of a mechanical seal.
Low diffusivity can lead to damaging thermal cycling behaviour.

Table 24.6 Chemical pH tolerance of materials

Material	pH	Material	pH
Tungsten carbide (cobalt binder)	7–13	Silicon carbide (sintered alpha)	1–14
Tungsten carbide (nickel binder)	6–13	Filled PTFE	0–14
Silicon carbide (reaction bonded)	5–12	High alumina ceramic	0–14
Tungsten carbide (nickel chrome moly bonded)	2–13		

Spring arrangements

The spring arrangements which can be used depend on the size, operating speed and product sealed: guidance is given in Table 24.7.

Rubber or metal bellows can be used in place of springs.

Table 24.7 Spring arrangements suitable for various sizes and speeds

Speed rev/min	*Shaft dia mm*	*Spring arrangement*			
		Rotary		*Stationary*	
		Single	*Multi*	*Single*	*Multi*
Up to 3000	Up to 100	Yes	Yes	Yes	Yes
Up to 3000	Over 100	No	Yes	Yes	Yes
Up to 4500	Up to 75	Yes	Yes	Yes	Yes
Over 4500	Up to 100	No	No	Yes	Yes
Over 4500	Over 100	No	No	No	Yes

SELECTION PROCEDURE

In order to make the correct seal selection it is necessary to know:

The product to be sealed
Pressure to be sealed
Shaft speed
Shaft or sleeve dia.
Temperature of product in the seal area

It is assumed that by knowing the product, its boiling point at the pressure to be sealed is also known.
In the procedure described here, products to be sealed are divided into aqueous products and hydrocarbons.

1. Consult Table 24.8 for Aqueous Products or Table 24.9 Hydrocarbons and note:

   The vapour pressure curve number, and possible seal face materials, seal packing and other seal component materials, for compatibility with the product, together with any special remarks for satisfactory operation.

2. Refer to the central diagram in the left-hand column of the General Seal Selection Graphs to determine for the particular shaft speed and shaft diameter an appropriate seal configuration with regard to the spring layout. Note the parameter, peripheral speed.
3. Depending whether the seal pressure is above or below 1 MN/m^2 (140 p.s.i.); refer to the Face Material Selection curves—balanced at the top of the page unbalanced at the bottom. For the particular seal pressure and peripheral speed, note the face material choices available.
4. Move to the right to the temperature and pressure curves. From the centre figure on the right, transpose from the temperature scale for the particular product corresponding to the vapour pressure curve no., to the top for a balanced seal and to the bottom for an unbalanced seal, choosing a pressure line relating to the diffusivity of the face materials (see Table 24.5). The curves then represent a stability line in terms of sealing pressure and operating temperature for particular speeds (m/sec). The area to the left of the curve is one of stable operation: the area between the curve and the saturation curve is one of instability. If the operating point falls within this latter region, either some cooling is required or the seal pressure must be changed, to bring the operating point to the left of the stability curve.
5. Compare the face material selected from corrosion and compatibility considerations from Tables 24.6, 24.8 and 24.9 with those from stability considerations as in step 4 above and decide on the appropriate combination.
6. Decide what cooling (or seal pressure change) is required; decide on the circulation system and what additional design features are required by way of quench, safety bush etc.

Selection is now complete.

Example

Water at 10 bar 3000 rev/min, 45 mm shaft dia., 175°C temperature at seal.

1. From the Aqueous Products list:
   Vapour pressure curve no. is 21
   Face material choice is stainless steel 18/8
   lead bronze, Stellite, tungsten carbide, alumina, or silicon carbide
   Seal packings are high nitrile rubber up to 100°C or ethylene propylene up to 150°C (note temperature at seal is 175°C)
   Other seal components, 17/2 stainless steel.
   No special remarks are listed.
2. From the Seal Type Selection curve either a balanced or unbalanced seal may be used (at a peripheral speed of 7 m/sec).
3. For Face Material Selection consider:
   (a) a balanced seal when tungsten carbide/carbon and silicon carbide/carbon are possibilities.
   (b) an unbalanced seal when silicon carbide/carbon is the only possible combination.
4. From the temperature and pressure curves consider
   (a) balanced seal and transposing from curve no. 21, note that at 10 bar and 7 m/sec the maximum operating temperature range is 160°C to 130°C, depending on seal face
   (b) unbalanced seal diffusivity maximum operating temperature is 70°C
5. Comparison of possible selections

Balanced seal	Cooling required from 175°C to 160°C or 130°C depending on the diffusivity of the chosen face seal.
Unbalanced seal	Cooling required from 175°C to 70°C and high diffusivity face seal.

From these possible selections, the preferred one is the balanced seal using silicon carbide running against a carbon stationary seal ring. The minimum of cooling is required and silicon carbide is chemically resistant.

Note that nitrile rubber packings are unsuitable. PTFE or graphite are suitable however. Other metal seal components are 17/2 or 18/8 stainless steel.

Table 24.8 Selection of materials for aqueous products

Product	vapour pressure curve no	Suitable seal ring materials for corrosion resistance*							seal packings −30 to 100°C	Other seal component materials	Remarks
		stainless steel 18/8	silicon carbide	stainless 17/2	lead bronze	Stellite	tungsten carbide	alumina			
Acetic acid	27	*	*			*			P	18/8 stainless steel	Up to 100°C only
Ammonia liquid	9	*	*			*	*	*	E	17/2 stainless steel	Use double seal indoors
Ammonium carbonate	21	*	*			*	*	*	N	17/2 stainless steel	Avoid crystallising conditions at seal
Ammonium chloride	21	*	*			*	*	*	N	18/8 stainless steel	Avoid crystallising conditions at seal. Use Monel spring
Ammonium hydroxide	21	*	*			*	*	*	N	18/8 stainless steel	
Beer	21	*	*	*	*	*	*	*	N	17/2 stainless steel	Use V packings above 100°C
Brine (calcium chloride)	21	*	*		*	*	*	*	N	18/8 stainless steel	Use Monel spring and V packings above 100°C
Brine (sodium chloride)	21	*	*		*	*	*	*	N	18/8 stainless steel	Use V packings above 100°C
Calcium carbonate	21	*	*		*	*	*	*	N	17/2 stainless steel	Avoid crystallising conditions at seal
Calcium chloride up to 100°C	21	*	*			*	*	*	N	18/8 stainless steel	Avoid crystallising conditions at seal. Use Monel spring
Carbon disulphide	20	*	*			*	*	*	V	18/8 stainless steel	Use V packings above 100°C
Citric acid up to 50% conc.	21	*	*			*			N	18/8 stainless steel	
Citric acid above 50% conc.	21								N	17/2 stainless steel	Use Hastelloy seal ring
Copper sulphate	21	*	*			*	*	*	N	18/8 stainless steel	Avoid crystallising conditions at seal Use V packings above 100°C
Dye liquors	21	*	*			*	*	*	V	18/8 stainless steel	May require P packings on occasion
Hydrogen peroxide	29		*						E	18/8 stainless steel	Use Hastelloy seal ring
Lime slurries	21	*	*			*	*	*	N	18/8 stainless steel	Use clean injection
Lye (caustic)	21	*	*			*	*	*	E	18/8 stainless steel	
Milk	21	*	*	*	*	*	*	*	N	17/2 stainless steel	
Paper stock	21	*	*		*	*	*	*	N	18/8 stainless steel	Use clean water injection with neck bush restriction
Phosphoric acid 0–20% conc.	21	*	*			*			N	18/8 stainless steel	
Phosphoric acid 20–45% conc.	21	*	*			*			E	18/8 stainless steel	
Phosphoric acid 45–100% conc.	21	*				*			P	18/8 stainless steel	
Potassium carbonate	21	*	*	*	*	*	*	*	N	17/2 stainless steel	Avoid crystallising conditions at seal
Potassium dichromate	21	*	*		*	*	*	*	E	18/8 stainless steel	Avoid crystallising conditions at seal
Potassium hydroxide up to 30% conc.	21	*	*			*	*	*	N	18/8 stainless steel	Avoid crystallising conditions at seal
Potassium hydroxide 30% to 100% conc.	21	*				*	*	*	E	18/8 stainless steel	Avoid crystallising conditions at seal
Sewage	21	*	*			*	*	*	N	18/8 stainless steel	Avoid solidification at seal. Use clean injection
Sodium bicarbonate	21	*	*	*	*	*	*	*	N	17/2 stainless steel	Avoid crystallising conditions at seal
Sodium carbonate	21	*	*			*	*	*	N	18/8 stainless steel	Avoid crystallising conditions at seal
Sodium chloride	21	*	*			*	*	*	N	18/8 stainless steel	Avoid crystallising conditions at seal. Use Monel spring
Sodium hydroxide up to 10% conc.	21	*	*			*	*	*	E	18/8 stainless steel	Avoid crystallising conditions at seal
Sodium hydroxide 10% conc. and above	21								P	Hastelloy	Use Hastelloy seal ring avoid crystallising conditions at seal
Sodium sulphate	21	*	*			*	*	*	N	18/8 stainless steel	Avoid crystallising conditions at seal
Sulphur dioxide liquid	13	*	*			*	*	*	E	18/8 stainless steel	
Water	21	*	*	*	*	*	*	*	N	17/2 stainless steel	E packings up to 150°C
Water (demineralised)	21	*	*		*	*	*	*	N	18/8 stainless steel	
Water (boiler feed)	21	*	*		*	*	*	*	N	18/8 stainless steel	
Water (sea)	21	*	*		*	*	*	*	N	18/8 stainless steel	

Seal Packing code
E = Ethylene propylene synthetic rubber
N = High Nitrile synthetic rubber
V = Viton A fluorocarbon elastomer
P = PTFE or perfluoro-elastomer or graphite

Unless otherwise stated in remarks column PTFE can be used as the seal packing between −100°C and 250°C.
To avoid crystallising conditions at seal use the quench feature or clean solvent injection into the seal chamber through the circulation connections.

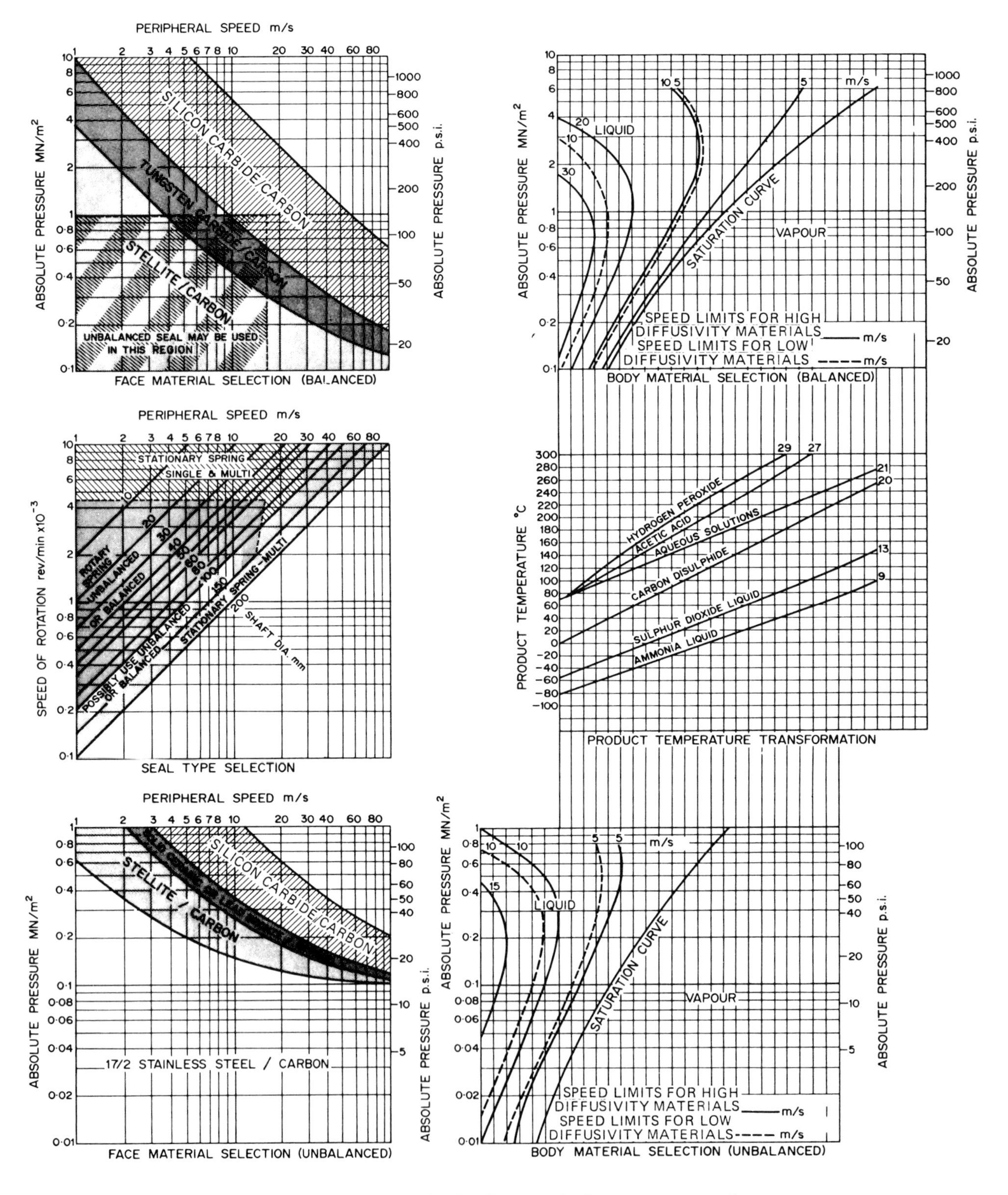

Figure 24.1 General seal selection graphs for aqueous products

Table 24.9 Selection of materials for hydrocarbons

Product	vapour pressure curve no	*Suitable seal ring materials for corrosion resistance** stainless steel 18/8	silicon carbide	stainless steel 17/2	lead bronze	Stellite	tungsten carbide	alumina	seal packings −30 to 100°C	*Other seal component materials*	*Remarks*
Acetone	22	*	*	*	*	*	*	*	E	17/2 stainless steel	
Arcton 9	18	*	*	*	*	*	*	*	N	17/2 stainless steel	
Benzene	24	*	*	*	*	*	*	*	P	17/2 stainless steel	
Butadiene	15	*	*	*	*	*	*	*	N	17/2 stainless steel	
Butane	16	*	*	*	*	*	*	*	N	17/2 stainless steel	
Butyl alcohol	26	*	*	*	*	*	*	*	N	17/2 stainless steel	
Butylene	15	*	*		*	*	*	*	V	18/8 stainless steel	
Cumene	30	*	*			*	*	*	V	18/8 stainless steel	
Cyclohexane	24	*	*	*	*	*	*	*	N	17/2 stainless steel	
Dimethyl ketone	22	*	*	*	*	*	*	*	E	17/2 stainless steel	
Ethane	3	*	*	*	*	*	*	*	N	17/2 stainless steel	
Ethylene	1	*	*	*	*	*	*	*	N	17/2 stainless steel	
Ethylene glycol	32	*	*	*	*	*	*	*	N	17/2 stainless steel	
Formaldehyde	13	*	*	*	*	*	*	*	N	17/2 stainless steel	
Freon 11	18	*	*	*	*	*	*	*	N	17/2 stainless steel	
Freon 12	12	*	*	*	*	*	*	*	N	17/2 stainless steel	
Freon 22	10	*	*	*	*	*	*	*	E	17/2 stainless steel	
Furfural	30	*	*	*	*	*	*	*	E	17/2 stainless steel	Use double seals above 121°C, use permanent quench below 121°C
Hexane	22	*	*	*	*	*	*	*	N	17/2 stainless steel	
Iso butane	14	*	*	*	*	*	*	*	N	17/2 stainless steel	
Iso pentane	18	*	*		*	*	*	*	N	17/2 stainless steel	
Iso propyl alcohol	24	*	*	*	*	*	*	*	E	17/2 stainless steel	
Methyl alcohol	22	*	*	*	*	*	*	*	N	17/2 stainless steel	
Methyl chloride	12	*	*	*	*	*	*	*	P	17/2 stainless steel	Use Monel spring
Methyl ethyl ketone	24	*	*	*	*	*	*	*	E	17/2 stainless steel	
Pentane	19	*	*	*	*	*	*	*	N	17/2 stainless steel	
Phenol	32	*	*		*	*	*	*	E	18/8 stainless steel	Avoid solidification at seal
Propane	10	*	*	*	*	*	*	*	N	17/2 stainless steel	
Propyl alcohol	25	*	*	*	*	*	*	*	N	17/2 stainless steel	
Propylene	9	*	*	*	*	*	*	*	N	17/2 stainless steel	
Soap solutions	21	*	*		*	*	*		N	18/8 stainless steel	Avoid solidification at seal
Styrene	29	*	*		*	*	*	*	P	18/8 stainless steel	
Toluene	26	*	*	*	*	*	*	*	P	17/2 stainless steel	
Urea (ammonia contaminated)	21	*	*			*	*		P	18/8 stainless steel	Avoid crystallising conditions at seal
Urea (ammonia free)	21	*	*	*	*	*	*	*	P	17/2 stainless steel	Avoid crystallising conditions at seal
Vinyl chloride	14	*	*			*	*	*	P	18/8 stainless steel	

Seal Packing code
E = Ethylene propylene synthetic rubber
N = High Nitrile synthetic rubber
V = Viton A fluorocarbon elastomer
P = PTFE or perfluoro-elastomer or graphite

Unless otherwise stated in remarks column PTFE can be used as the seal packing between −100°C and 250°C.
To avoid solidification at the seal on high viscosity products, heat the seal area and circulation lines 30 minutes before start-up.
To avoid crystallising conditions at seal use the quench feature or clean solvent injection into the seal chamber through the circulation connections.

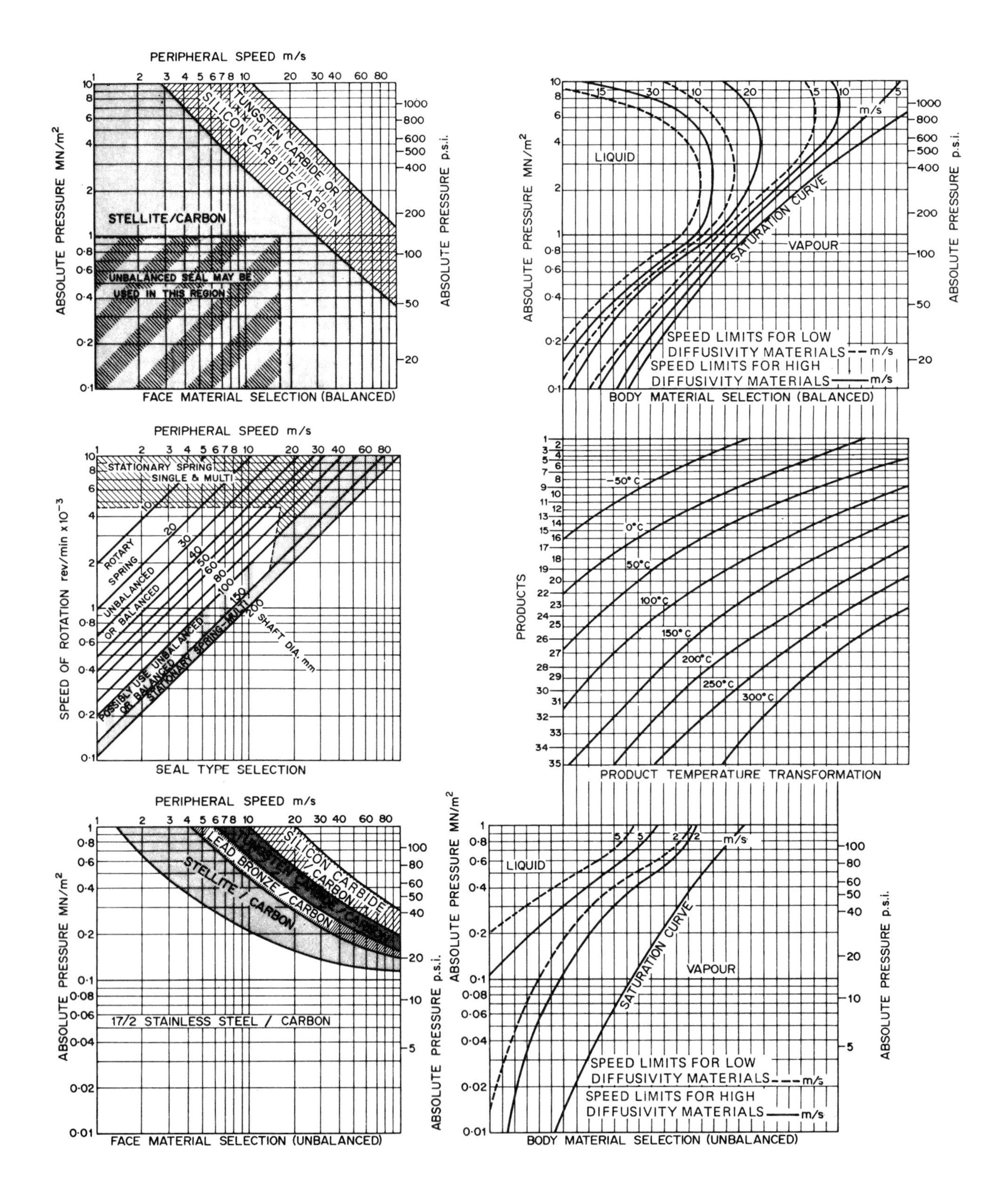

Figure 24.2 General seal selection graphs for hydrocarbons

Power absorption and starting torque

The coefficient of friction between the seal faces varies from a maximum dry value at zero product pressure to a minimum value in the region 0.7–1 MN/m^2 (100–140 p.s.i.). It increases imperceptibly at higher product pressures. Over the normal range of pump speeds, power absorption is proportional to speed.

The starting torque is normally about 5 times the running torque, but when a pump has been standing and the liquid film has been squeezed from between the seal faces, starting torque may be doubled. If when the pump has been standing for a long time and corrosion (chemical or electrolytic) has formed a chemical bond at the periphery of the seal rings, which has to be broken before rotation can begin, the break-out torque may increase to many times (over 5 times) the normal starting torque. Chemically inert seal faces—ceramic, glass-filled PTFE etc.—should then be used.

Power absorption and starting torque for aqueous solutions, light oils and medium hydrocarbons, use values in Figure 24.3. For light hydrocarbons use $\frac{2}{3}$ of these values. For heavy hydrocarbons use $1\frac{1}{3}$ of these values. Allow ±25% on all values.

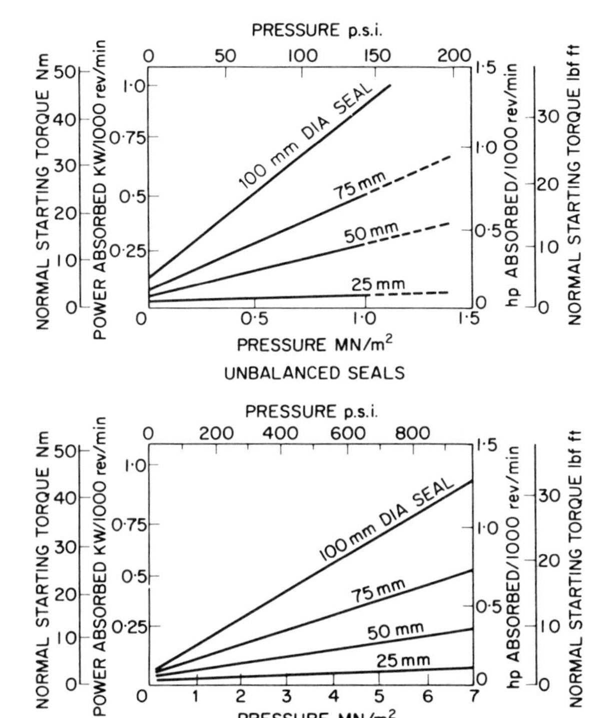

Figure 24.3 The starting torque for various seals

Installation and allowable malalignments

Shaft tolerances should be within ±0.05 mm (±0.002 in) and where PTFE wedge packings are used must be round within 0.01 mm (0.0005 in). Shaft surface finish should be better than 0.8 μm R_a (32 μin cla) except where PTFE wedge packings are used which require better than 0.4 μm R_a (16 μin cla) and preferably a corrosion resistant surface (Stellite) on which to slide. Where the sliding packing passes over the shaft a 10° chamfer and lead-in, free from burrs should be provided.

Seal face flatness is usually better than two or three helium light bands (0.6 or 0.9 μm) per 25 mm of working face diameter.

Seal face surface finish depends very much on the material – about 0.25 μm R_a (10 μ in cla) for carbon/graphite to about 0.05 μm R_a (2 μin cla) for hard faces.

Permissible eccentricity for O-ring and wedge fitted seals varies from 0.10 mm (0.004 in) TIR at 1000 rev/min to 0.03 mm (0.001 in) TIR at 3000 rev/min: rubber bellows fitted seals accommodate 2 to 3 times these values. Axial run out for the non-sliding seal ring is about the same order as the above but for speeds higher than 3000 rev/min a useful guide is 0.0002 mm/mm of dia (0.0002 in/inch of dia).

Axial setting is not very critical for the single spring design of seal ±2.5 mm (±0.100 in): for the multi spring design however, it is limited to about ±0.5 mm (±0.020 in).

The main applications of packed glands are for sealing the stems of valves, the shafts of rotary pumps and the plungers of reciprocating pumps. With a correct choice of gland design and packing material they can operate for extended periods with the minimum need for adjustment.

VALVE STEMS

Valve stem packings use up to 5 rings of packing material as in Figure 25.1. For high temperature/high pressure steam, moulded rings of expanded graphite foil material are commonly used. This gives low valve stem friction.

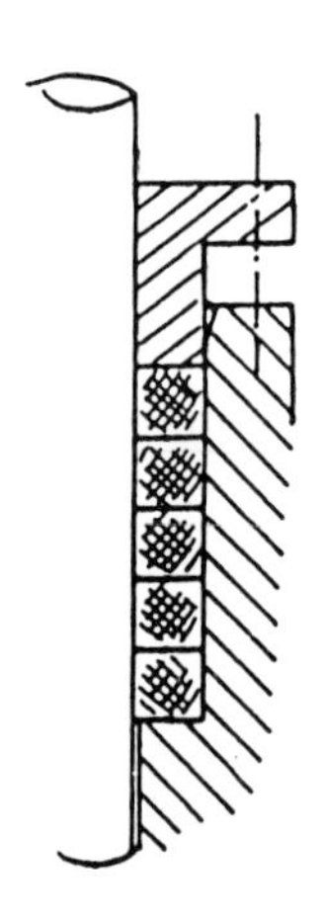

Figure 25.1 A typical valve stem packing

To reduce the risk of extrusion of the lamellar graphite during frequent valve operation, the end rings of the packing can be made from graphite/yarn filament.

Materials of this type only compress in service by a small amount and can provide a virtually maintenance free valve packing if used with live loading as shown in Figure 25.2.

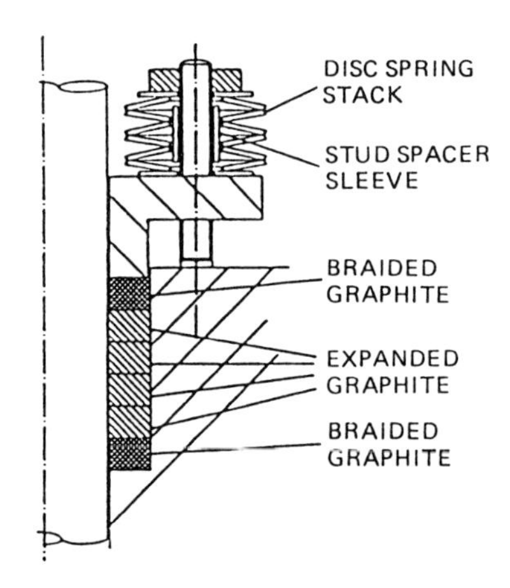

Figure 25.2 A valve stem packing using spring loading to maintain compression of the valve packing and avoid leakage

ROTARY PUMPS

Rotary pump glands commonly use up to 5 rings of packing material. For most applications up to a PV of 150 bar m/sec (sealed pressure × shaft surface speed) a simple design as in Figure 25.3 is adequate. In most pumps the pressure at the gland will be 5 bar or less and those with pressures over 10 bar will be exceptional.

At PV values over 150 bar m/sec direct water cooling or jacket cooling are usually necessary and typical arrangements are shown in Figure 25.4 and 25.5.

When pumping abrasive or toxic fluids there may be a need to provide a flushing fluid entry at the fluid end of the glands, as in Figure 25.6, or a high pressure barrier fluid which is usually injected near the centre of the gland as in Figure 25.7.

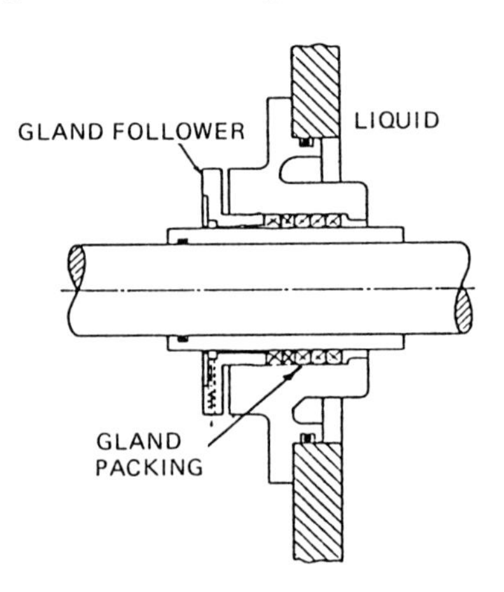

Figure 25.3 A general duty rotary pump gland

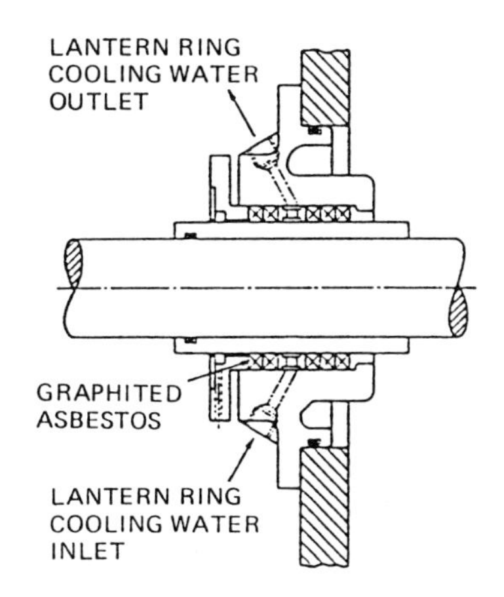

Figure 25.4 A gland packing with direct cooling via a lantern ring

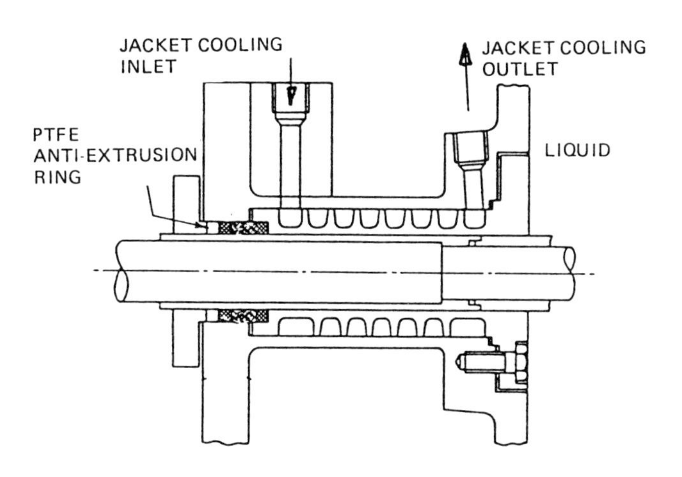

Figure 25.5 A gland packing with a cooling jacket for high temperature applications

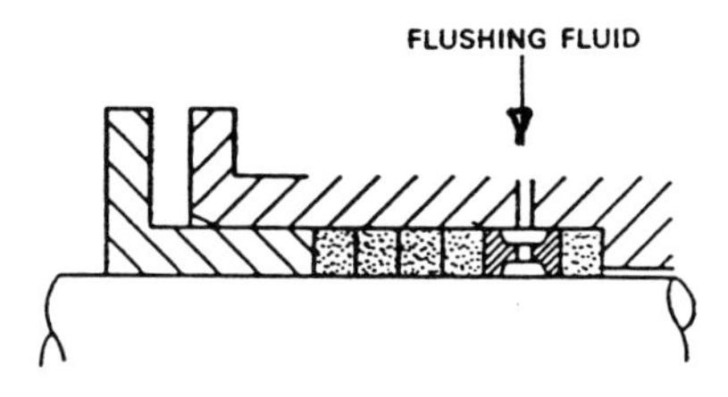

Figure 25.6 A packing gland with a flushing fluid system

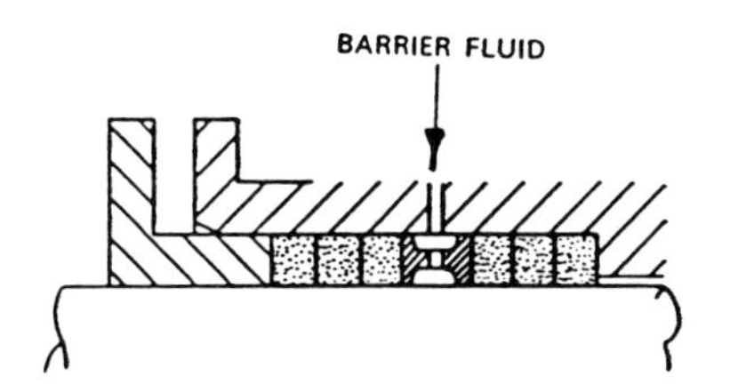

Figure 25.7 A packed gland with a barrier fluid system

RECIPROCATING PUMPS

Reciprocating pumps also use typically 5 packing rings. However due to the increased risk of extrusion of the packing due to the combination of high pressure and reciprocating movement, anti extrusion elements are usually incorporated in the gland.

Self adjusting glands can be used on reciprocating pumps but the spring loading for compression take up must act in the same direction as the fluid pressure loading, as shown in Figure 25.10.

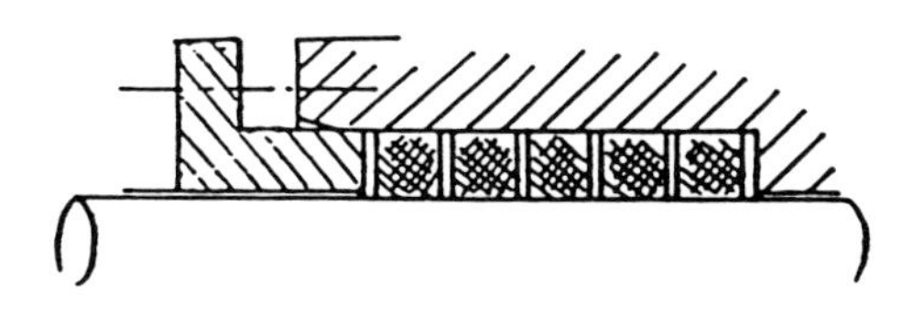

Figure 25.8 A reciprocating pump gland with PTFE anti-extrusion washers between the packing rings

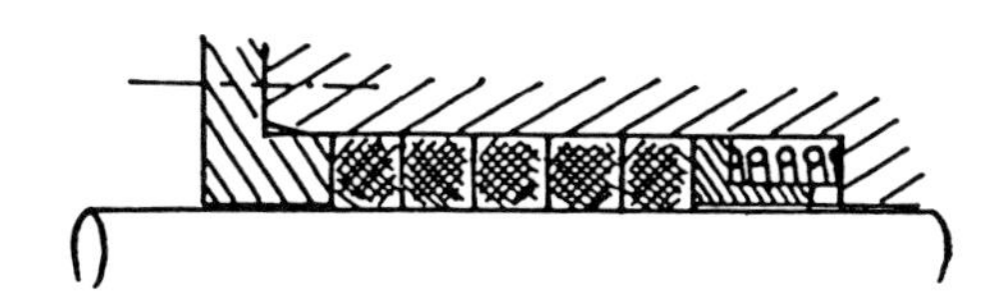

Figure 25.10 A reciprocating pump gland with internal spring loading to maintain compression of the packing

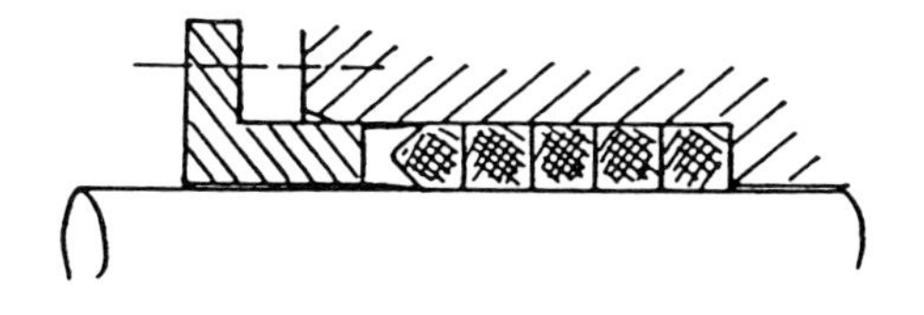

Figure 25.9 A reciprocating pump gland with an anti-extrusion moulded hard fabric lip seal

PACKING MATERIALS

Table 25.1 Materials for use in packed glands

Material	*Maximum operating temperature* °C	*Special properties*	*Typical applications*
Expanded graphite foil	550°C 2500°C in non-oxidising environments	Low friction, self lubrication, low compression set and contains no volatile constituents. Available as rings	Valve stems
Graphite/yarn filament	550°C	Available as cross plaited square section lengths. Resistant to extrusion	Valve stems
Aramid (Kevlar) fibre	250°C	Tough and abrasion resistant	Valve stems and pumps
PTFE filament	250°C	Low friction and good chemical resistance	Valve stems Pumps at surface speeds below 10 m/s
Hybrid graphite/PTFE yarn	250°C	Particularly suitable for high speed rotary shafts. Close bush clearances needed to reduce risk of extrusion. Good resistance to abrasives.	Pump shafts for speeds of the order of 25 m/s
Ramie	120°C	Good water resistance	Rotary and reciprocating water pumps

Note: Many of these packings can be provided with a central rubber core which can increase their elasticity and thus assist in maintaining the gland compression. Their application depends on the temperature and chemical resistance of the type of rubber used.

PACKING DIMENSIONS AND FITTING

Typical gland dimensions are shown in Figure 25.11 and packing sizes in Table 25.2.

Table 25.2 Typical radial housing widths in relation to shaft diameters. All dimensions in mm

All packings except expanded graphite		*Expanded graphite*	
Shaft diameter	*Housing radial width*	*Shaft diameter*	*Housing radial width*
Up to 12	3	up to 18	3
above 12 to 18	5	above 18 to 75	5
18 to 25	6.5	75 to 150	7.5
25 to 50	8	150 and above	10
50 to 90	10		
90 to 150	12.5		
150	15		

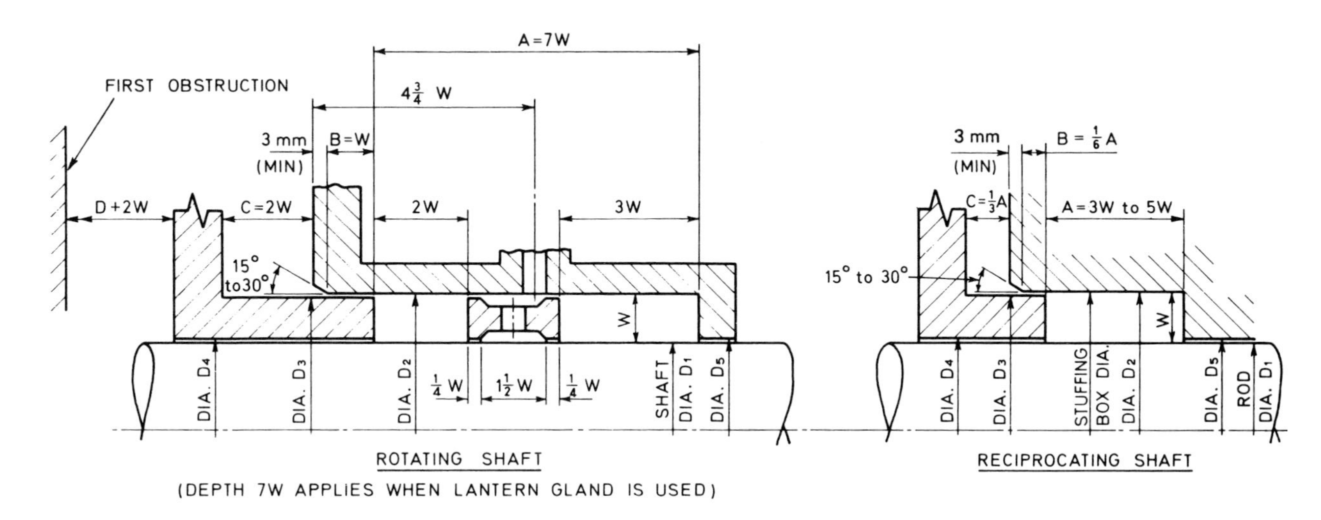

Figure 25.11 Typical gland dimensions for rotating and reciprocating shafts

Pump shafts, valve stems and reciprocating rams should have a surface finish of better than 0.4 μm Ra. Their hardness should not be less than 250 Brinell.

Rings should be cut with square butt joins and each fitted individually with joins staggered at a minimum of 90°. After applying a small degree of compression to the complete set, gland nuts must be slackened off to finger tight prior to start up. Once running, any excessive leakage can then be gradually reduced by repeated small degrees of adjustment. The major cause of packing failure is excessive compression, particularly at the initial fitting stage.

Further advice may be obtained from packing manufacturers.

The figure shows a typical general arrangement of a mechanical rod packing assembly. The packing (sealing) rings are free to move radially in the cups and are given an axial clearance appropriate to the materials used (see Table 26.2). The back clearance is in the range of 1 to 5 mm. ($\frac{1}{32}$ to $\frac{3}{16}$ in). The diametral clearance of the cups is chosen to prevent contact with the rod; it lies typically in the range 1 to 5 mm ($\frac{1}{32}$ to $\frac{3}{16}$ in). The sealing faces on the rings and cups are accurately ground or lapped.

The case material can be cast iron, carbon steel, stainless steel or bronze to suit the chemical conditions. It may be drilled to provide lubricant feed to the packing, to vent leakage gas or to provide water cooling.

The rings are held in contact with the rod by spring pressure; sealing action however, depends on gas forces which hold the rings radially in contact with the rod and axially against the next cup.

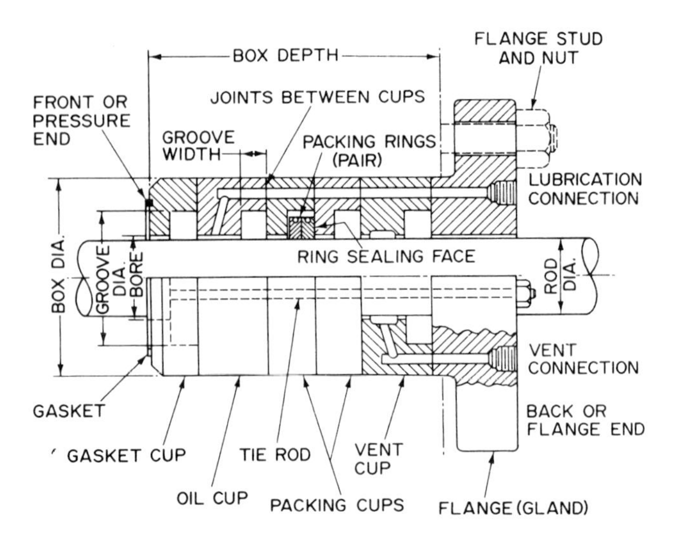

Figure 26.1 General arrangement of a typical mechanical piston rod packing assembly

SELECTION OF TYPE OF PACKING

1 Pressure breaker

Description

Three-piece ring with bore matching rod. Total circumferential clearance 0.25 mm. Garter spring to ensure contact with rod.

Applications

Used in first one or two compartments next to high pressure, when sealing pressure above 35 bar (500 p.s.i.) to reduce pressure and pressure fluctuations on sealing rings.

2 Radial cut/Tangential cut pair

Description

The radial cut ring is mounted on the high pressure side. (Two tangential cut rings can be used when there is a reversing pressure drop.) The rings are pegged to prevent the radial slots from lining up. Garter springs are fitted to ensure rod contact. Ring bores match the rod.

Applications

The standard design of segmental packing. Used for both metallic and filled PTFE packings.

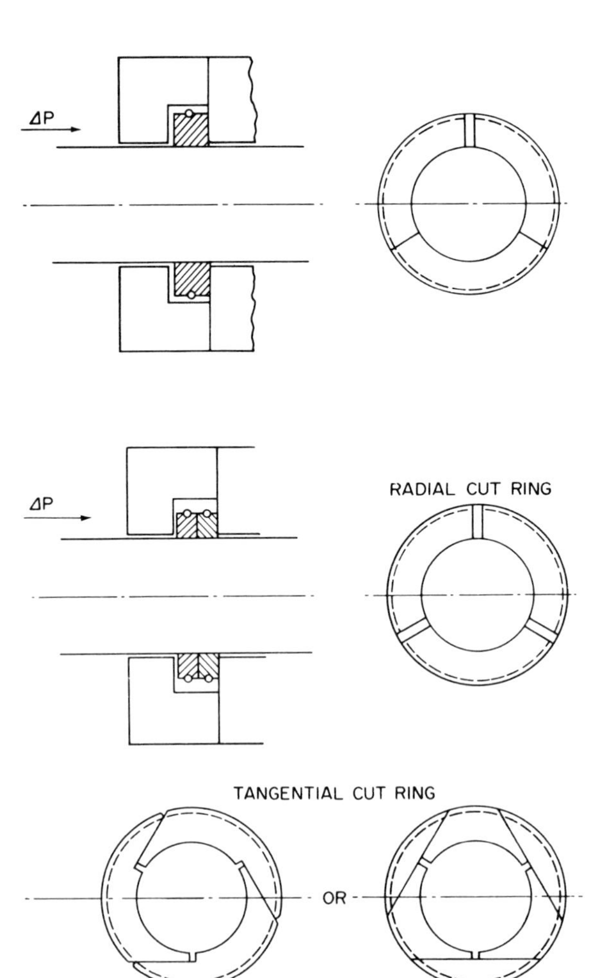

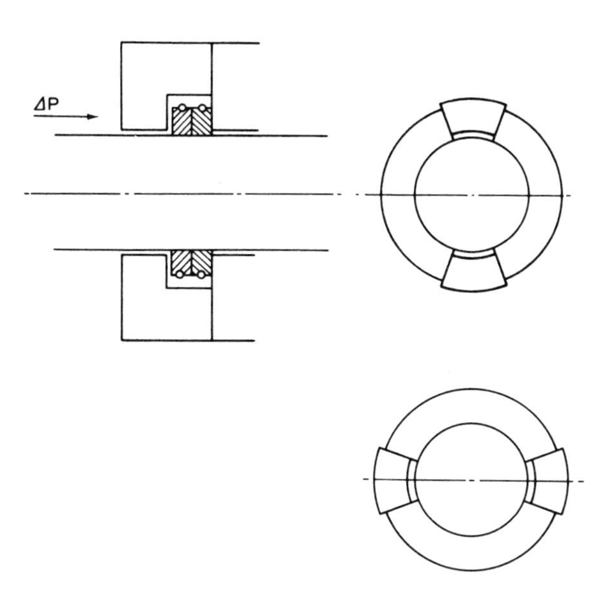

3 Unequal segment ring

Description

The rings are pegged to prevent the gaps lining up. Garter springs are fitted to ensure rod contact. The bore of the larger segment matches the rod.

Applications

Rather more robust than tangentially cut rings (2) and hence more suitable for carbon-graphite packings.

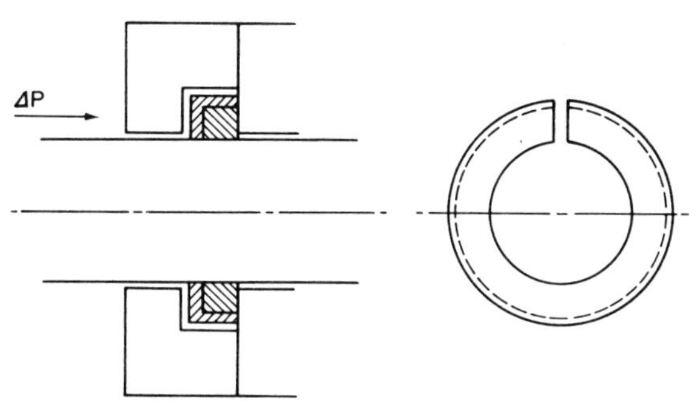

4 Contracting rod packing

Description

Cast iron L-ring with bronze or white metal inner ring or three piece packing with filled PTFE and metallic back-up ring. Contact with rod maintained by ring tension. Rings pegged to prevent the gaps from lining up. *Note*: this style of packing has to be assembled over the end of the rod.

Applications

Used for both metallic and filled PTFE packings.

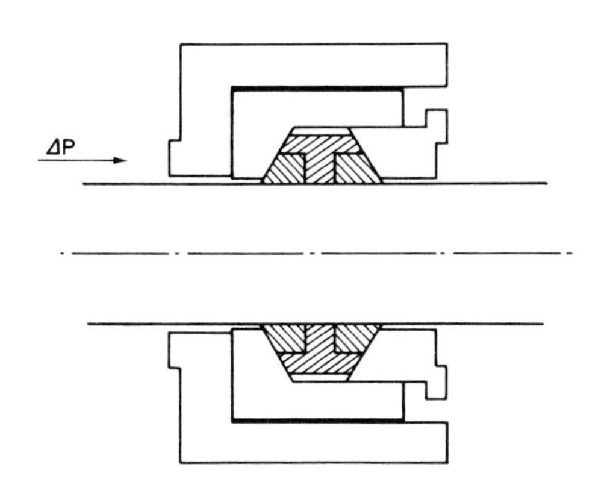

5 Cone ring

Description

Three ring seal—each ring in three segments with bore, matching rod. Cone angle ranging from 75° at pressure end to 45° at atmosphere end.

Applications

Used for both metallic and filled PTFE packings.

DESIGN OF PACKING ARRANGEMENT

Number of sealing rings

There is no theoretical basis for determining the number of sealing rings. Table 26.1 gives values that are typical of good practice:

Table 26.1 The number of sealing rings for various pressures

Pressure	*No. of sets of sealing rings*
up to 10 bar (150 p.s.i.)	3
10–20 bar (150–250 p.s.i.)	4
20–35 bar (250–500 p.s.i.)	5
35–70 bar (500–1000 p.s.i.)	6
70–150 bar (1000–2000 p.s.i.)	8
above 150 bar (2000 p.s.i.)	9–12

Notes: 1 With Type 4 packings increase number of sealing rings by 50–100%.
2 With Type 5 packings four sets of sealing rings should be adequate.
3 Above 35 bar (500 p.s.i.) use one or two pressure breakers (Type 1) in addition, on the pressure side.

Piston rods

Rod material is chosen for strength or chemical resistance. Carbon, low alloy and high chromium steels are suitable. For the harder packings (lead bronze and cast iron) hardened rods should be used; treatment can be flame or induction hardening, or nitriding. Chrome plating or high chromium steel is used for chemical resistance.

Surface finish

Metal and filled PTFE packing 0.2–0.4 μm R_a (8–16 μin cla).
Carbon/graphite and metal/graphite sinters 0.1–0.2 μm R_a (4–8 μin cla)

Dimensional tolerances

Diameter	+0.05 mm (+0.002 in) −0.05 mm (−0.002 in)
Taper over stroked length	0.01 mm (0.0005 in)
Out-of-roundness	0.025 mm (0.001 in)

Packing materials

Table 26.2 The types of packing material and their applications

Material	*Rod hardness*	*Axial clearance*	*Applications*
(1) Lead-bronze	250 BHN min	0.08–0.12 mm (0.003–0.005 in)	Optimum material with high thermal conductivity and good lubricated bearing properties. Used where chemical conditions allow. Suitable for pressures up to 3000 bar (50 000 p.s.i.)
(2) Flake graphite grey cast iron	400 BHN min	0.08–0.12 mm (0.003–0.005 in)	Cheaper alternative to (1); bore may be tin coated to assist running in. Suitable up to 70 bar (1000 p.s.i.) for lubricated operation
(3) White metal (Babbitt)	not critical	0.08–0.12 mm (0.003–0.005 in)	Used where (1) and (2) not suitable because of chemical conditions. Preferred material for high chromium steel and chrome-plated rods. Max. pressure 350 bar (5000 p.s.i.). Max. temperature 120°C
(4) Filled PTFE	400 BHN min	0.4–0.5 mm (0.015–0.020 in)	Suitable for unlubricated and marginally lubricated operation as well as fully lubricated. Very good chemical resistance. Above 25 bar (400 p.s.i.) a lead bronze backing ring (0.1/0.2 mm) clear of rod should be used to give support and improved heat removal
(5) Reinforced p.f. resin	not critical	0.25–0.4 mm (0.010–0.015 in)	Used with sour hydrocarbon gases and where lubricant may be thinned by solvents in gas stream
(6) Carbon-graphite	400 BHN min	0.03–0.06 mm (0.001–0.002 in)	Used with carbon-graphite piston rings. Must be kept oil free. Suitable up to 350°C
(7) Graphite/metal sinter	250 BHN min	0.08–0.12 mm (0.003–0.005 in)	Alternative to (4) and (6)

FITTING AND RUNNING IN

1. Cleanliness is essential so that cups bear squarely together and to prevent scuffing or damage at start up.
2. Handle segments carefully to avoid damage during assembly.
3. Check packings float freely in cups.
4. With lubricated packings, check that plenty of oil is present before starting to run-in. Oil line must have a check valve between the lubricator and the packing. Manually fill the oil lines before starting. Use maximum lubrication feed rate during run-in.
5. If the temperature of the rod rises excessively (say above 100°C) during run-in, stop and allow to cool and then re-start run-in.
6. Run in with short no-load period.

SELECTION AND DESIGN

Table 27.1 Guidance on the selection of basic types

Type name	*Distributor*	*'U'*	*Cup*	*'O' ring*
External—fitted to piston, sealing in bore			CONTROLLED COMPRESSION	
Internal—fitted in housing, sealing on piston or rod				
Simple housing design	Good	Good	Poor	Very good
Low wear rate	Very good	Good	Good	Poor
High stability (resistance to roll)	Good	Fair	Very good	Poor
Low friction	Fair	Fair	Fair	Good
Resistance to extrusion	Good	Good	Good	Fair
Availability in small sizes	Fair	Good	Poor	Very good
Availability in large sizes	Good	Fair	Good	Good
Bidirectional sealing	Single-acting only. Use in pairs back-to-back for double-acting. 'Non-return' valve action can be useful			Effective but usually used in pairs
Remarks	Do not allow heel to touch mating surface except under high pressure. Use correct fits and guided piston, etc. If seal too soft for pressure, lip may curl away from surface			Avoid parting line flash on the sealing surfaces. Unsuitable for rotational movement

Application notes

Long lips take up wear better and improve stability but increase friction. Use plastic back-up rings to reduce extrusion at high pressures. The use of a thin oil will reduce wear but may increase friction. For pneumatic assemblies use light grease which may contain colloidal graphite or MoS_2. Choose light hydraulic oil for mist lubrication.

Avoid metal-to-metal contact due to side loading or piston weight. If seals will not maintain concentricity use acetal resin, nylon, PTFE, glass fibre/PTFE or metal bearings.

To prevent mixing of unlike fluids, e.g. aeration of oil, use two seals and vent the space in between to atmosphere.

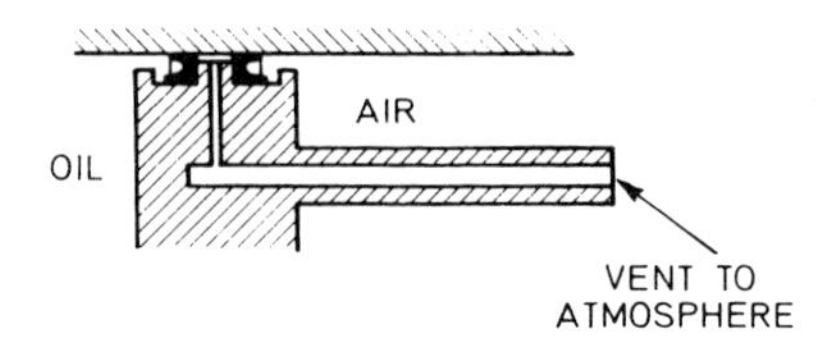

Table 27.2 Seals derived from basic types

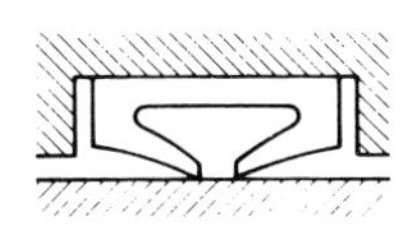	Double-acting, one-piece, narrow width, but pressure can be trapped between lips and seal may jam. Needs composite piston
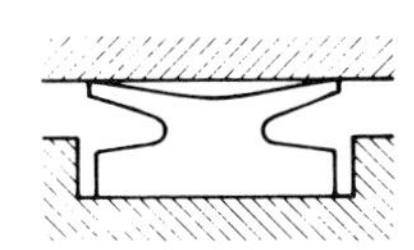	Similar, but no pressure trap and can be fitted to one-piece piston
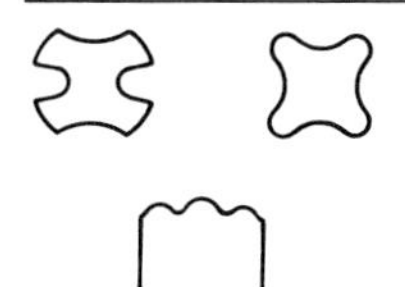	Derived from 'O' ring. Less tendency to roll. Improved and multiple sealing surfaces. Sealing forces reduced and parting line flash removed from working surface
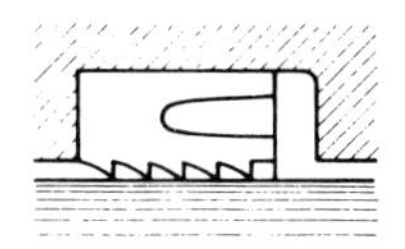	Multiple sealing lips to obviate leakage due to curl
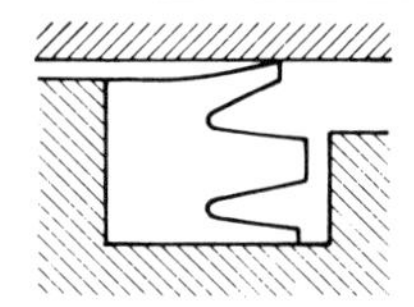	'W' section. Good for hydraulic applications and high pressures. Can be used internally or externally

Table 27.3 Special seals

		Material
	Static seal in body sections Dynamic seal on piston	Rubber
	Register between body sections Static seal in body sections Dynamic seal on piston	Polyethylene
	Piston head seal	Polyethylene
SPRING	Fits 'O' ring groove. Use internally or externally. Suitable for rotational movement.	Usually PTFE

Table 27.4 Mating surface materials

Materials	*Type*	*Finish*		*Remarks*
		0.6 μm max. 0.2 to 0.4 μm (8 to 16 μin) preferred	0.2 μm max. 0.05 to 0.1 μm (2 to 4 μin) preferred	
Brass	As drawn	✓		Best untreated materials. Improve with use. High cost.
Copper	As drawn	✓		
Aluminium alloy	As drawn	✓		Liable to scuff and corrode. Low cost
	Polished		✓	
	Anodised	✓		Short life
	Hard anodised		✓	Abrasive, therefore polish
Mild steel	As drawn	✓		Corrodes. Low cost.
	Honed	✓		Corrodes readily
	Hard chrome plated		✓	Very abrasive, polish before and after plating. High cost.
	Ground		✓	Used mostly for piston rods
Stainless steel	Ground		✓	

Notes: Anodising and plating can be porous to air causing apparent seal leakage. The finish on the seal housing can be 0.8 μm. Use rust prevention treatment for mild steel in storage.

INSTALLATION

Table 27.5 Assembly hazards

Problem	*Suggested solution*
Multiple seal grooves 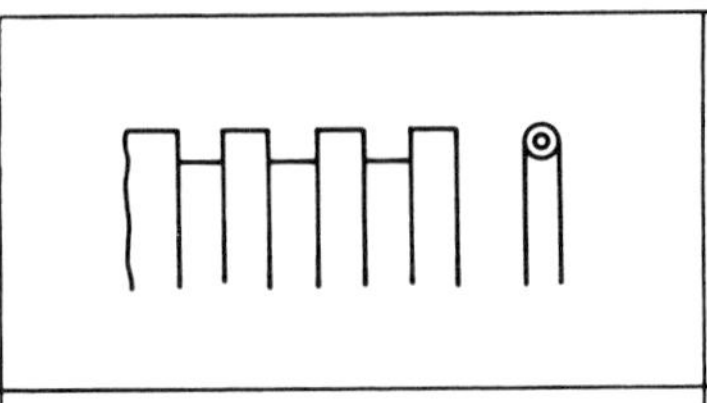 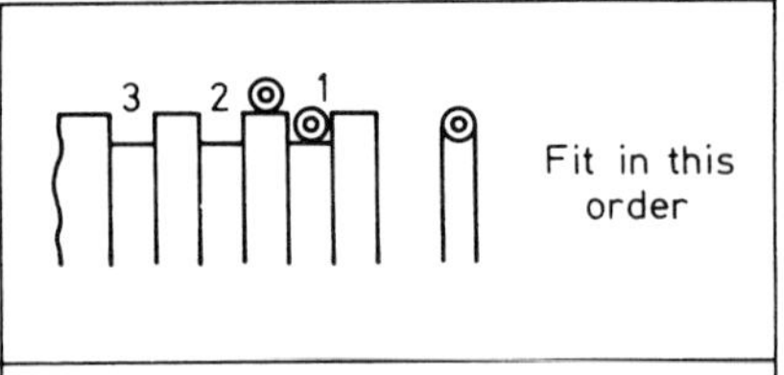	3 2 1 Fit in this order
External grooves	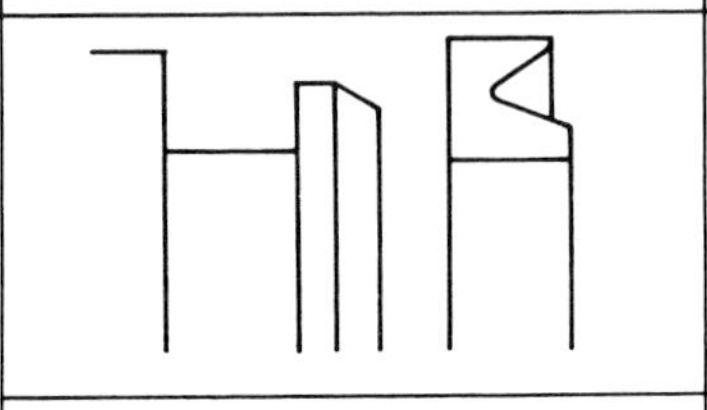 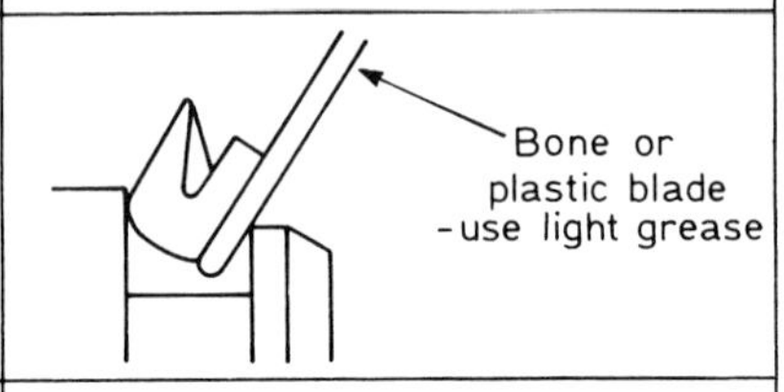 Bone or plastic blade -use light grease
Internal grooves	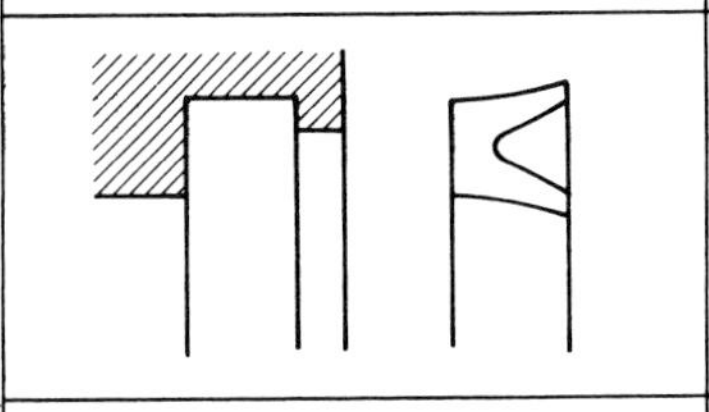 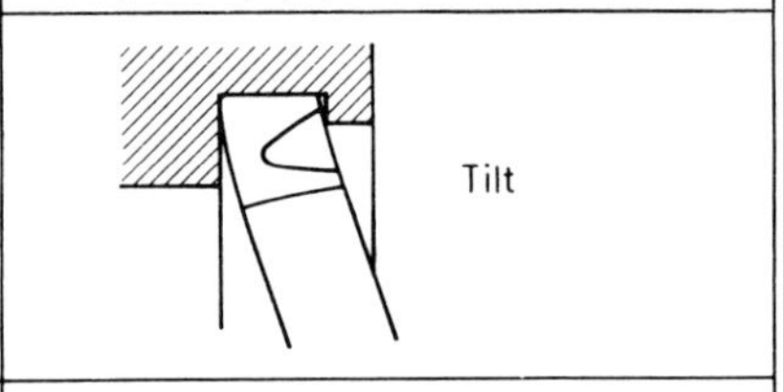 Tilt
Crossing ports	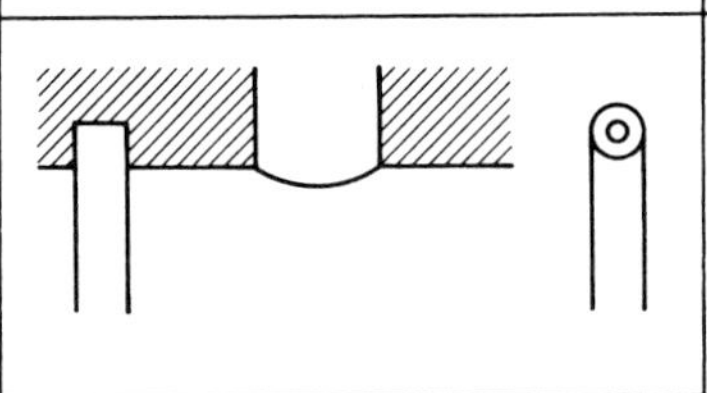 Deburr, chamfer, use assembly sleeve or temporary plug in port
Crossing threads 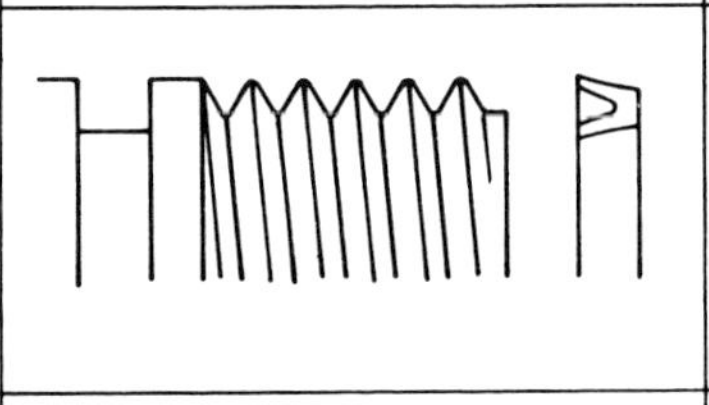 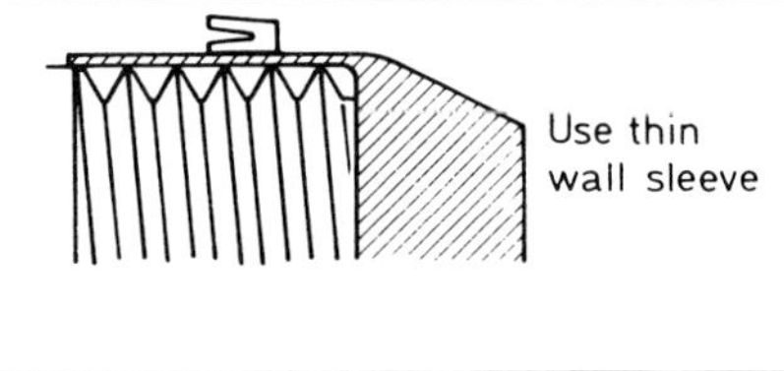	Use thin wall sleeve
Crossing edges and circlip grooves 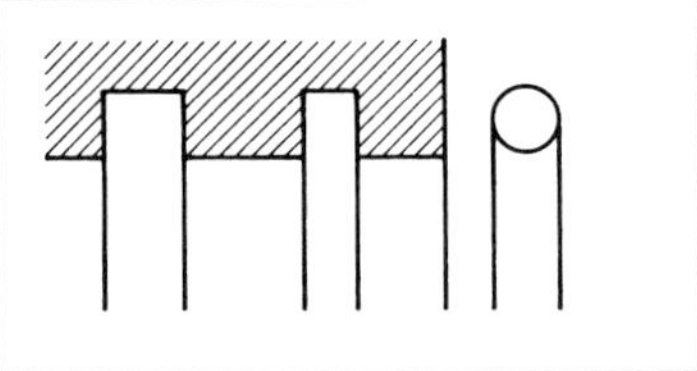 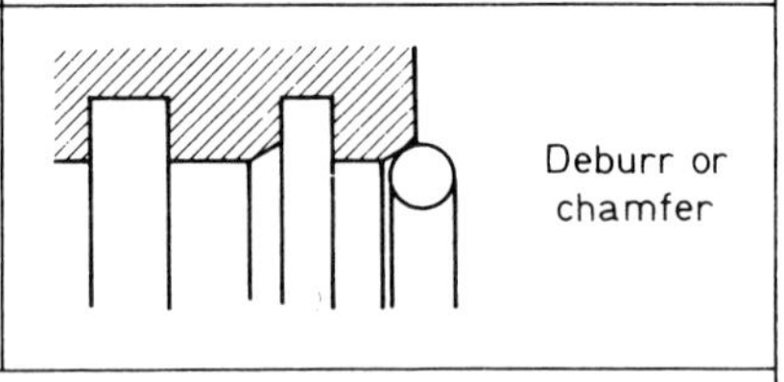	Deburr or chamfer
Fitting piston assemblies to bores	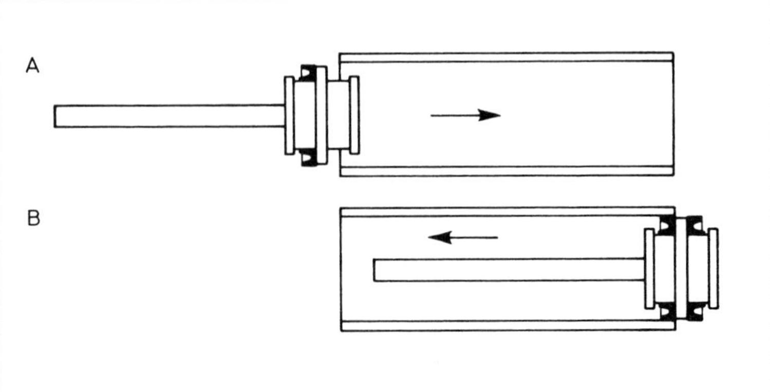 A B

Table 27.6 'O' ring fits

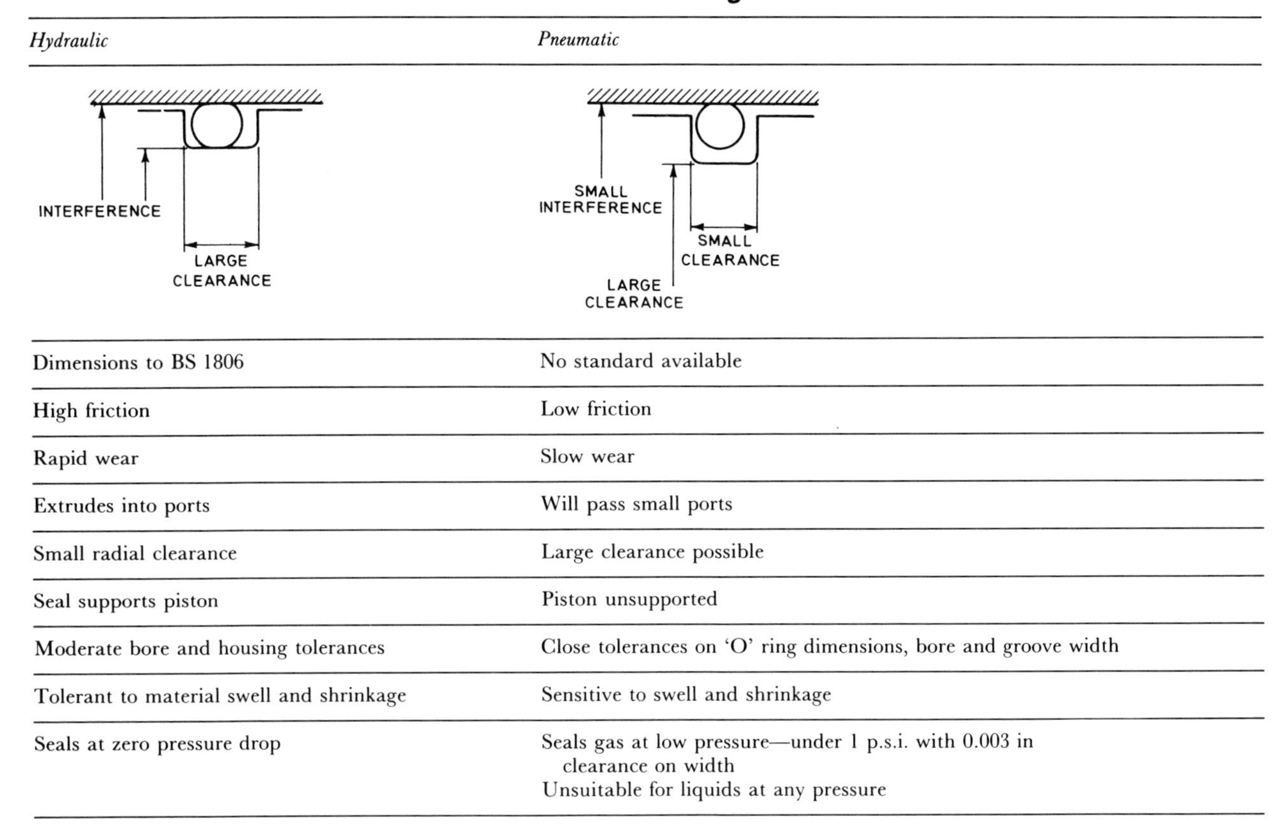

Hydraulic	*Pneumatic*
Dimensions to BS 1806	No standard available
High friction	Low friction
Rapid wear	Slow wear
Extrudes into ports	Will pass small ports
Small radial clearance	Large clearance possible
Seal supports piston	Piston unsupported
Moderate bore and housing tolerances	Close tolerances on 'O' ring dimensions, bore and groove width
Tolerant to material swell and shrinkage	Sensitive to swell and shrinkage
Seals at zero pressure drop	Seals gas at low pressure—under 1 p.s.i. with 0.003 in clearance on width Unsuitable for liquids at any pressure

FAILURE

Table 27.7 Types and causes of failure

Type	*Usual symptom*	*Cause*
Channelling (fluid cutting)	Small, straight grooves across the sealing surface	Fluid leaking across seal at high velocity
Abrasive wear	Flat on 'O' ring Circumferential groove on lip seal Sharp sealing edge on lip seal	Pressure too high or abrasive mating surface
Extrusion	Surface broken Slivers of rubber	Pressure too high or too much clearance
Chemical attack	Softening or hardening—may break up	Incompatible fluid
Temperature effects	Hardening and breaking up Breaking up	Too hot and/or excess friction Too cold

Notes: Symptoms of contamination by solid particles are similar to channelling but the grooves are less regular. Uneven distribution of wear suggests eccentricity or side loading. 'O' ring rolling produces variation in shape and size of section.

Table 1.1 Importance of lubricant properties in relation to bearing type

Lubricant property \ Type of component	Plain journal bearing	Rolling bearing	Closed gears	Open gears, ropes, chains, etc.	Clock and instrument pivots	Hinges, slides, latches, etc.
1. Boundary lubricating properties	+	++	+++	++	++	+
2. Cooling	++	++	+++	—	—	—
3. Friction or torque	+	++	++	—	++	+
4. Ability to remain in bearing	+	++	—	+	+++	+
5. Ability to seal out contaminants	—	++	—	+	—	+
6. Temperature range	+	++	++	+	—	+
7. Protection against corrosion	+	++	—	++	—	+
8. Volatility	+	+	—	++	++	+

Note: The relative importance of each lubricant property in a particular class of component is indicated on a scale from + + + = highly important to − = quite unimportant.

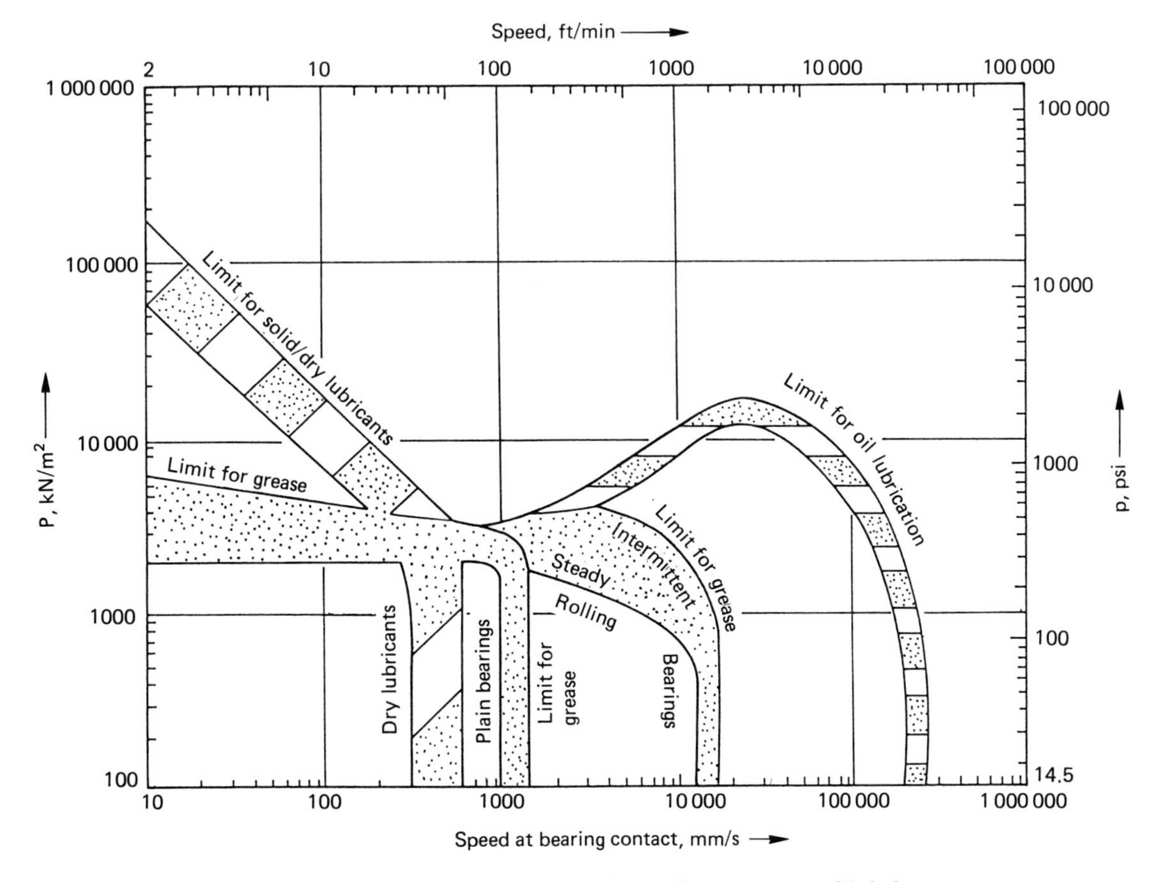

Figure 1.1 Speed/load limitations for different types of lubricant

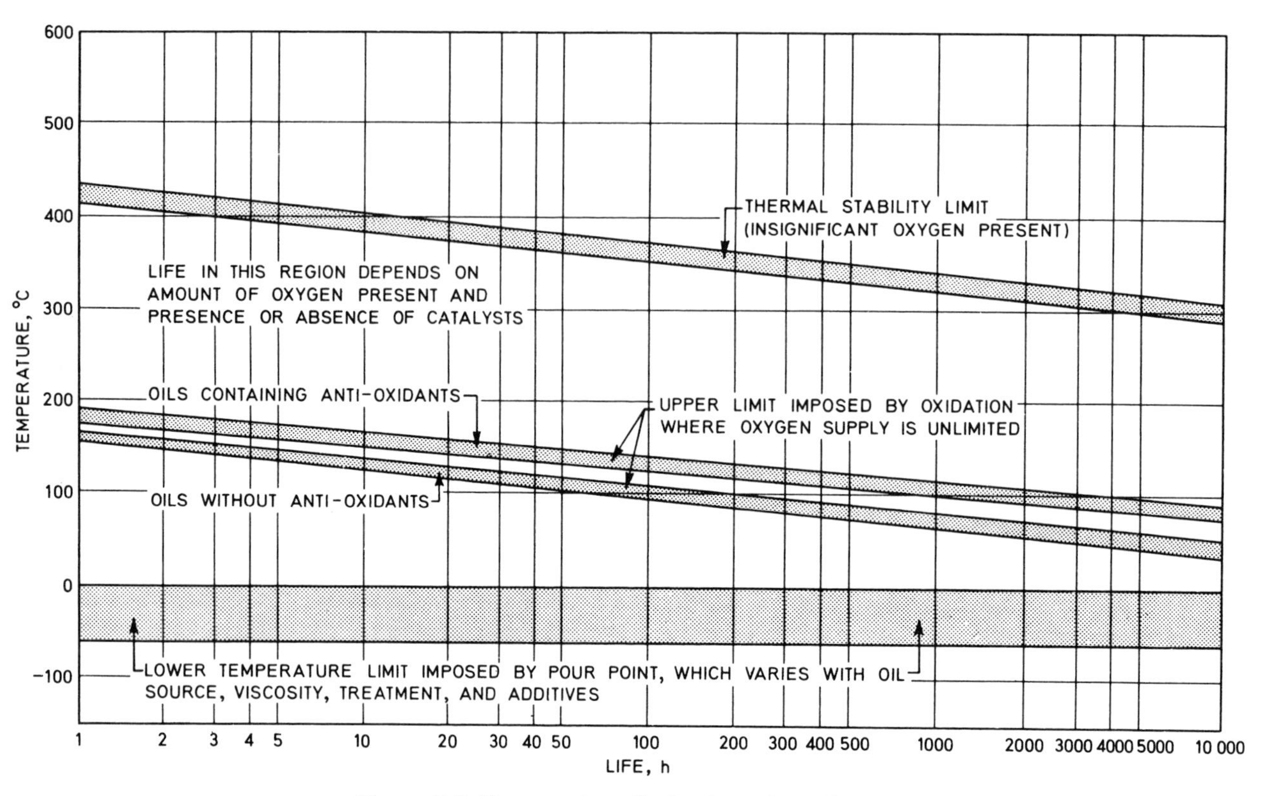

Figure 1.2 Temperature limits for mineral oils

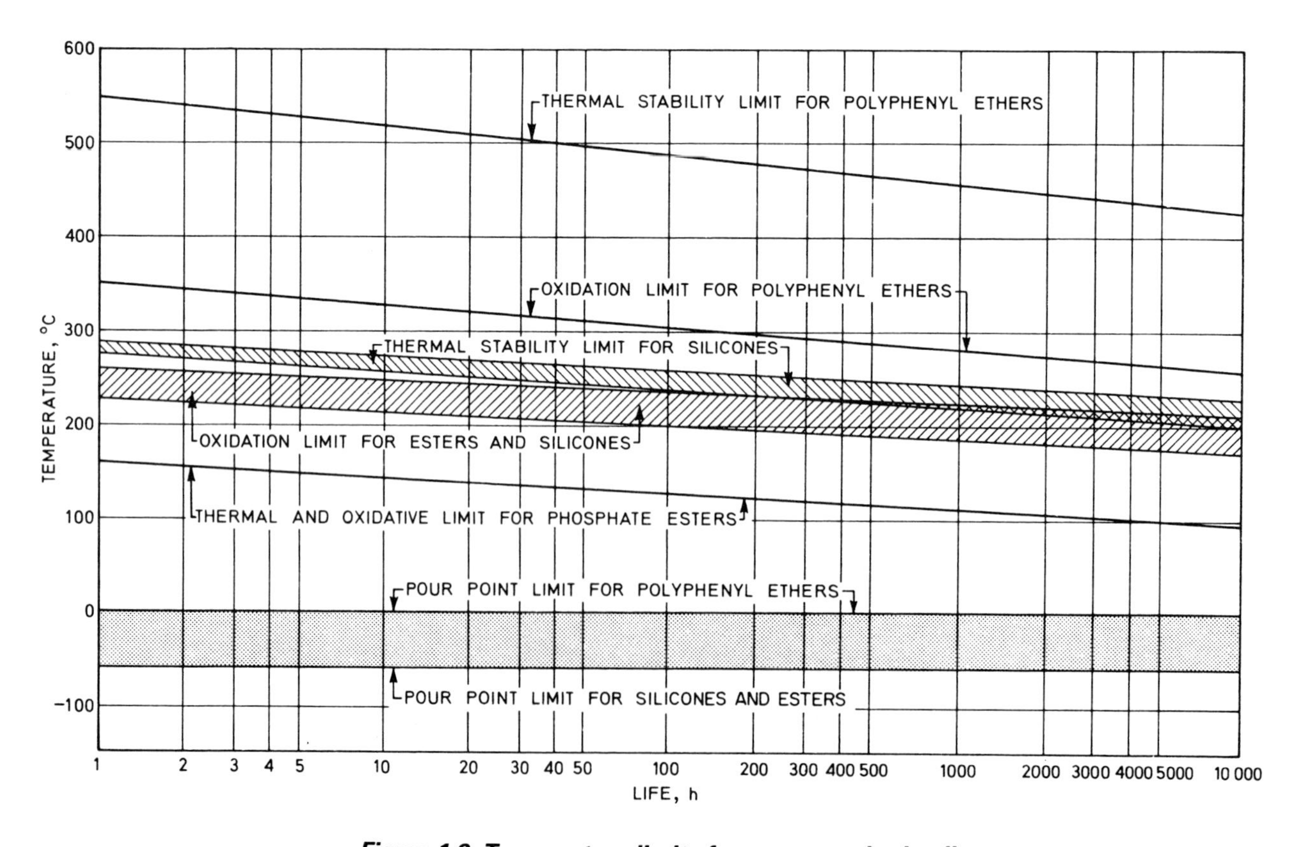

Figure 1.3 Temperature limits for some synthetic oils

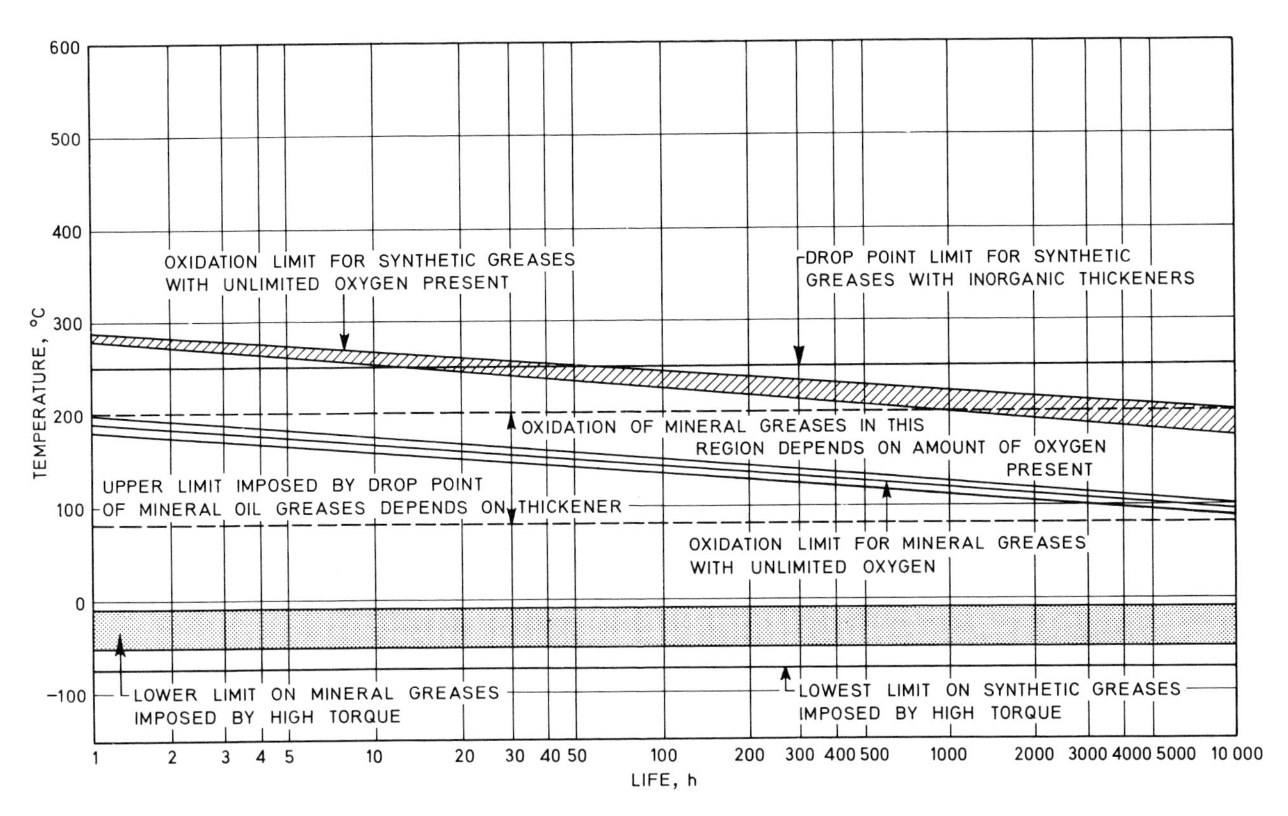

Figure 1.4 Temperature limits for greases. *In many cases the grease life will be controlled by volatility or migration. This cannot be depicted simply, as it varies with pressure and the degree of ventilation, but in general the limits may be slightly below the oxidation limits*

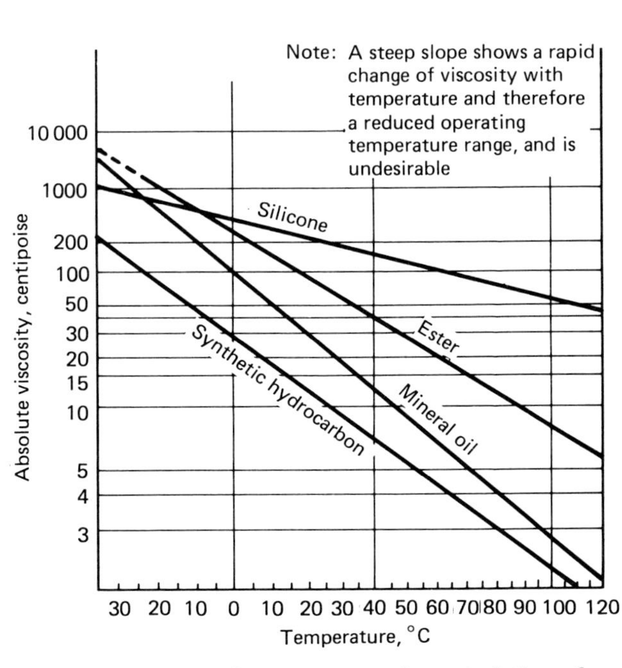

Figure 1.5 Viscosity/temperature characteristics of various oils

The effective viscosity of a lubricant in a bearing may be different from the quoted viscosity measured by a standard test method, and the difference depends on the shear rate in the bearing.

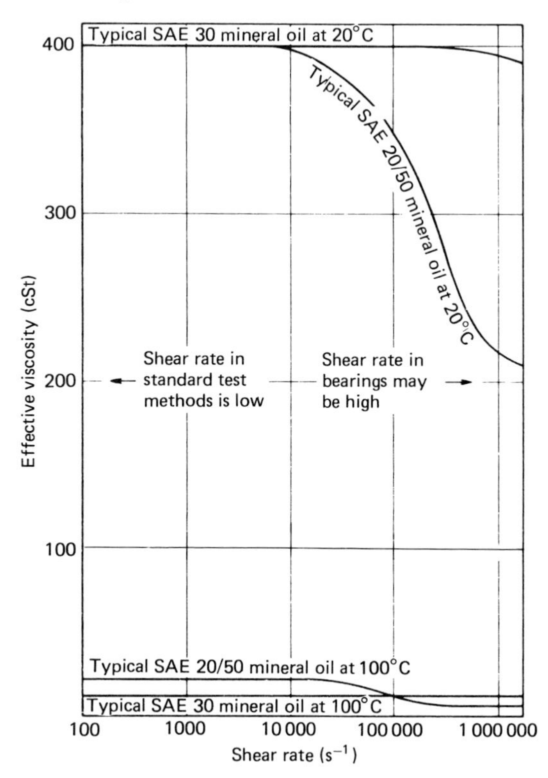

Figure 1.6 Variation of viscosity with shear rate

Table 1.2 Examples of specific mechanisms and possible lubricants and systems

	Lubricant	*Lubricating system*	*Maintenance cost*	*Investment cost*	*Rate of heat removal by lubricant*	*Remarks*
Journal bearings	Oil	By hand	High	Low	Small	Only for light duty
		Circulating system	Low	High	High	Necessary oil flow must be ensured
		Ring lubrication	Low	Low	Moderate	Only for moderate circumferential speeds
		Porous bearings	Low	Low	Small	Only for moderate circumferential speeds and low specific pressures
	Grease	By hand	High	Low	Nil	Only for light duty
		Centralised system	Low	High	Nil	Good pumpability of grease required if long lines to bearings
Rolling bearings	Oil	Oil mist	Low	High	Small	If compressed air in necessary quantity and cleanliness available, investment costs are moderate
		Circulating system	Low	High	High	Oil feed jets must be properly designed and positioned to ensure optimum lubrication and heat removal
		Bath	Low	Low	Small	Careful design and filling required to avoid excessive churning
		Splash	Low	Low	Moderate	Careful design of housing and other components (e.g. gears) necessary to ensure adequate oil supply
	Greas	Packed	Low	Low	Nil	Overfilling must be avoided. Maintenance costs are only low if re-lubrication period not too short
		Centralised system	Low	High	Nil	Possibility for used grease to escape must be provided. Shield delivery lines from heat
Gears	Oil	Bath	Low	Low	Moderate	Careful design of housing required to ensure adequate oil supply to all gears and to avoid excessive churning
		Circulating system	Low	High	High	Jets have to be properly designed and placed to ensure even oil distribution and heat removal
	Grease	By hand	High	Low	Nil	Only for light duty
		Housing filled	Low	Low	Nil	Principally for small, low-speed gears, otherwise, use stiffer greases to avoid slumping and overheating

CLASSIFICATION

Mineral oils are basically hydrocarbons, but all contain thousands of different types of varying structure, molecular weight and volatility, as well as minor but important amounts of hydrocarbon derivatives containing one or more of the elements nitrogen, oxygen and sulphur. They are classified in various ways as follows.

Types of crude petroleum

Paraffinic Contains significant amounts of waxy hydrocarbons and has 'wax' pour point (see below) but little or no asphaltic matter. Their naphthenes have long side-chains.

Naphthenic Contains asphaltic matter in least volatile fractions, but little or no wax. Their naphthenes have short side-chains. Has 'viscosity' pour point.

Mixed base Contains both waxy and asphaltic materials. Their naphthenes have moderate to long side-chains. Has 'wax' pour point.

Viscosity index

Lubricating oils are also commonly classified by their change in kinematic viscosity with temperature, i.e. by their kinematic viscosity index or KVI. Formerly, KVIs ranged between 0 and 100 only, the higher figures representing lower degrees of viscosity change with temperature, but nowadays oils may be obtained with KVIs outside these limits. They are generally grouped into high, medium and low, as in Table 2.1.

Table 2.1 Classification by viscosity index

Group	*Kinematic viscosity index*
Low viscosity index (LVI)	Below 35
Medium viscosity index (MVI)	35–80
High viscosity index (HVI)	80–110
Very high viscosity index (VHVI)	Over 110

It should be noted, however, that in Table 2.5 viscosity index has been determined from dynamic viscosities by the method of Roelands, Blok and Vlugter,[1] since this is a more fundamental system and allows truer comparison between mineral oils. Except for low viscosity oils, when DVIs are higher than KVIs, there is little difference between KVI and DVI for mineral oils.

Traditional use

Dating from before viscosity could be measured accurately, mineral oils were roughly classified into viscosity grades by their typical uses as follows:

Spindle oils Low viscosity oils (e.g. below about 0.01 Ns/m^2 at 60°C,) suitable for the lubrication of high-speed bearings such as textile spindles.

Light machine oils Medium viscosity oils (e.g. 0.01–0.02 Ns/m^2) at 60°C, suitable for machinery running at moderate speeds.

Heavy machine oils Higher viscosity oils (e.g. 0.02–0.10 Ns/m^2) at 60°C, suitable for slow-moving machinery.

Cylinder oils Suitable for the lubrication of steam engine cylinder; viscosities from 0.12 to 0.3 Ns/m^2 at 60°C

Hydrocarbon types

The various hydrocarbon types are classified as follows:

(*a*) Chemically saturated (i.e. no double valence bonds) straight and branched chain. (Paraffins or alkanes.)

(*b*) Saturated 5- and 6-membered rings with attached side-chains of various lengths up to 20 carbon atoms long. (Naphthenes.)

(*c*) As (*b*) but also containing 1, 2 or more 6-membered unsaturated ring groups, i.e. containing double valence bonds, e.g. mono-aromatics, di-aromatics, polynuclear aromatics, respectively.

A typical paraffinic lubricating oil may have these hydrocarbon types in the proportions given in Table 2.2.

Table 2.1 Hydrocarbon types in Venezuelan 95 VI solvent extracted and dewaxed distillate

Hydrocarbon types		*% Volume*
Saturates (KVI = 105)	Paraffins	15
	Naphthenes	60
Aromatics	Mono-aromatics	18
	Di-aromatics	6
	Poly-aromatics	1

The VI of the saturates has a predominant influence on the VI of the oil. In paraffinic oils the VI of the saturates may be 105–120 and 60–80 in naphthenic oils.

Structural group analyses

This is a useful way of accurately characterising mineral oils and of obtaining a general picture of their structure which is particularly relevant to physical properties, e.g. increase of viscosity with pressure. From certain other physical properties the statistical distribution of carbon atoms in aromatic groups ($\%C_A$), in naphthenic groups ($\% C_N$), in paraffinic groups ($\% C_P$), and the total number (R_T) of naphthenic and aromatic rings (R_N and R_A) joined together. Table 2.3 presents examples on a number of typical oils.

Table 2.3 Typical structural group analyses (*courtesy:* Institution of Mechanical Engineers)

Oil type	*Specific gravity at 15.6°C*	*Viscosity Ns/m² at 100°C*	*Mean molecular weight*	$\% C_A$	$\% C_N$	$\% C_P$	R_A	R_N	R_T
LVI spindle oil	0.926	0.0027	280	22	32	46	0.8	1.4	2.2
LVI heavy machine oil	0.943	0.0074	370	23	26	51	1.1	1.6	2.7
MVI light machine oil	0.882	0.0039	385	4	37	59	0.2	2.1	2.3
MVI heavy machine oil	0.910	0.0075	440	8	37	54	0.4	2.7	3.1
HVI light machine oil	0.871	0.0043	405	6	26	68	0.3	1.4	1.7
HVI heavy machine oil	0.883	0.0091	520	7	23	70	0.4	1.8	2.2
HVI cylinder oil	0.899	0.0268	685	8	22	70	0.7	2.3	3.0
Medicinal white oil	0.890	0.0065	445	0	42	58	0	2.8	2.8

REFINING

Distillation

Lubricants are produced from crude petroleum by distillation according to the outline scheme given in Figure 2.1.

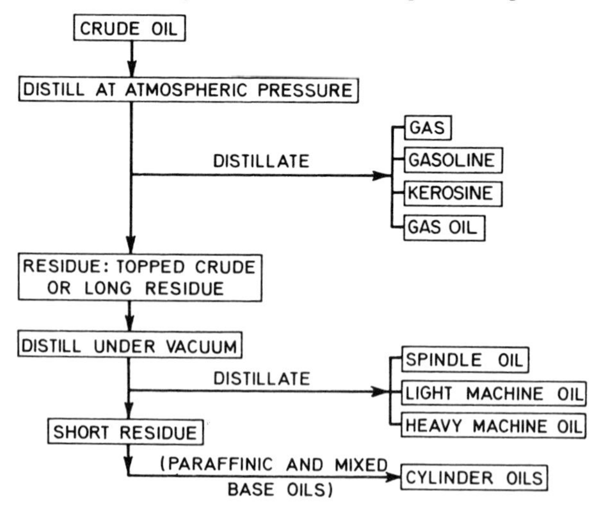

Figure 2.1 (*courtesy:* Institution of Mechanical Engineers)

The second distillation is carried out under vacuum to avoid subjecting the oil to temperatures over about 370°C, which would rapidly crack the oil.

The vacuum residues of naphthenic crudes are bitumens. These are not usually classified as lubricants but are used as such on some plain bearings subject to high temperatures and as blending components in oils and greases to form very viscous lubricants for open gears, etc.

Refining processes

Table 2.4 Refining processes (*courtesy:* Institution of Mechanical Engineers)

Process	*Purpose*
De-waxing	Removes waxy materials from paraffinic and mixed-base oils to prevent early solidification when the oil is cooled to low temperatures, i.e. to reduce pour point
De-asphalting	Removes asphaltic matter, particularly from mixed-base short residues, which would separate out at high and low temperatures and block oil-ways
Solvent extraction	Removes more highly aromatic materials, chiefly the polyaromatics, in order to improve oxidation stability
Hydrotreating	Reduces sulphur content, and according to severity, reduces aromatic content by conversion to naphthenes
Acid treatment	Now mainly used as additional to other treatments to produce special qualities such as transformer oils, white oils and medicinal oils
Earth treatment	Mainly to obtain rapid separation of oil from water, i.e. good demulsibility

The distillates and residues are used to a minor extent as such, but generally they are treated or refined both before and after vacuum distillation to fit them for the more stringent requirements. The principal processes listed in Table 2.4 are selected to suit the type of crude oil and the properties required.

Elimination of aromatics increases the VI of an oil. A lightly refined naphthenic oil may be LVI but MVI if highly refined. Similarly a lightly refined mixed-base oil may be MVI but HVI if highly refined. Elimination of aromatics also reduces nitrogen, oxygen and sulphur contents.

The distillates and residues may be used alone or blended together. Additionally, minor amounts of fatty oils or of special oil-soluble chemicals (additives) are blended in to form additive engine oils, cutting oils, gear oils, hydraulic oils, turbine oils, and so on, with superior properties to plain oils, as discussed below. The tolerance in blend viscosity for commercial branded oils is typically $\pm 4\%$ but official standards usually have wider limits, e.g. $\pm 10\%$ for ISO 3448.

PHYSICAL PROPERTIES

Viscosity-Temperature

Figure 2.4 illustrates the variation of viscosity with temperature for a series of oils with kinematic viscosity index of 95 (dynamic viscosity index 93). Figure 2.2 shows the difference between 150 Grade ISO 3448 oils with KVIs of 0 and 95.

Viscosity–Pressure

The viscosity of oils increases significantly under pressure. Naphthenic oils are more affected than paraffinic but, very roughly, both double their viscosity for every 35 MN/m^2 increase of pressure. Figure 2.3 gives an impression of the variation in viscosity of an SAE 20W ISO 3448 or medium machine oil, HVI type, with both temperature and pressure.

In elastohydrodynamic (ehl) formulae it is usually assumed that the viscosity increases exponentially with pressure. Though in fact considerable deviations from an exponential increase may occur at high pressures, the assumption is valid up to pressures which control ehl behaviour, i.e. about 35 MN/m^2. Typical pressure viscosity coefficients are given in Table 2.5, together with other physical properties.

Pour point

De-waxed paraffinic oils still contain 1% or so of waxy hydrocarbons, whereas naphthenic oils only have traces of them. At about 0°C, according to the degree of de-waxing, the waxes in paraffinic oils crystallise out of solution and at about −10°C the crystals grow to the extent that the remaining oil can no longer flow. This temperature, or close to it, when determined under specified conditions is known as the pour point. Naphthenic oils, in contrast, simply become so viscous with decreasing temperature that they fail to flow, although no wax crystal structure develops. Paraffinic oils are therefore said to have 'wax' pour points while naphthenic oils are said to have 'viscosity' pour points.

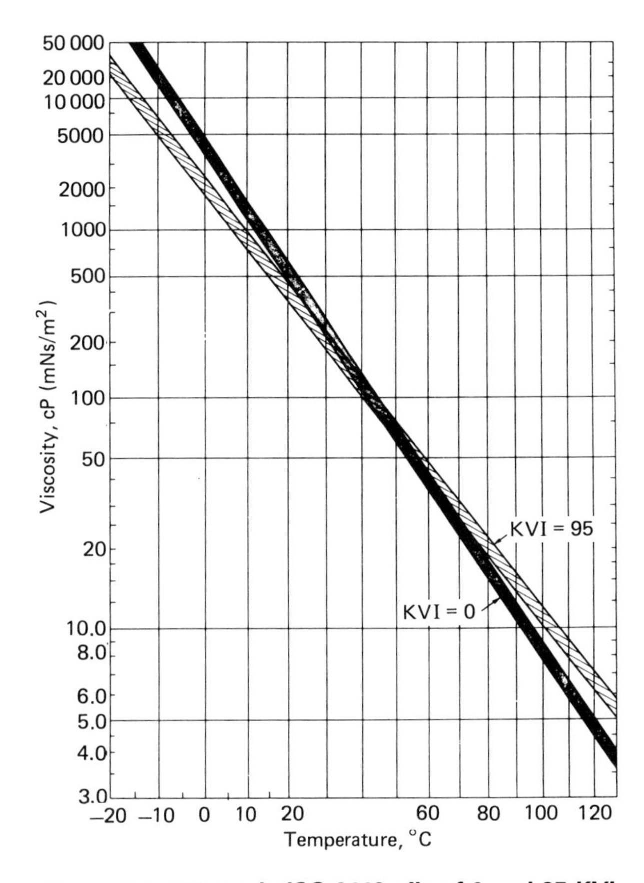

Figure 2.2 150 grade ISO 3448 oils of 0 and 95 KVI

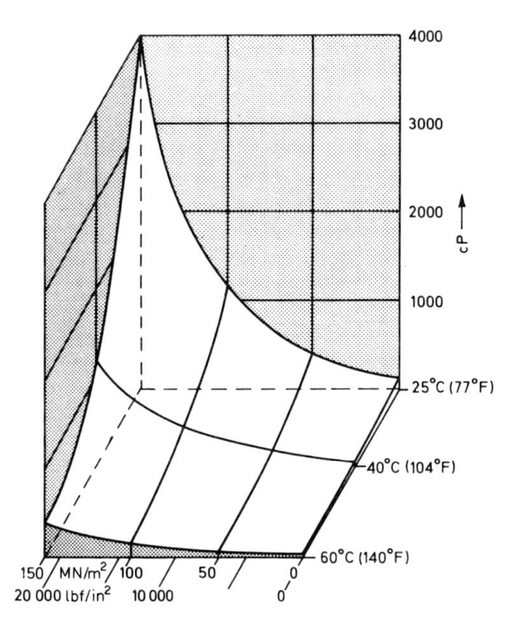

Figure 2.3 Variation of viscosity with temperature and pressure of an SAE 20W (HVI) oil *(courtesy:* Institution of Mechanical Engineers)

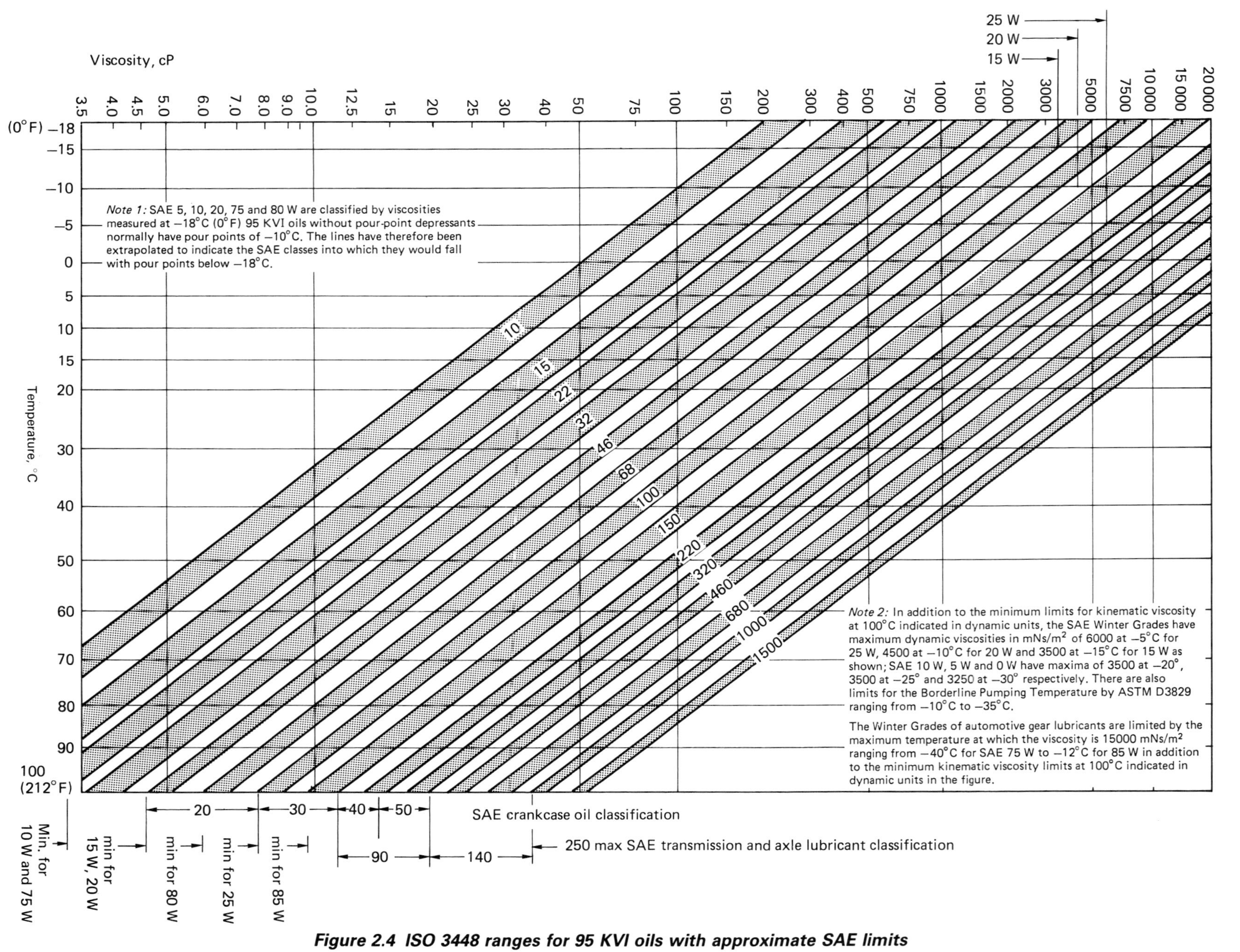

Figure 2.4 ISO 3448 ranges for 95 KVI oils with approximate SAE limits

Thermal properties

Table 2.5 Typical physical properties of highly refined mineral oils (*courtesy:* Institution of Mechanical Engineers)

		Naphthenic oils			*Paraffinic oils*		
		Spindle	*Light machine*	*Heavy machine*	*Light machine*	*Heavy machine*	*Cylinder*
Density (kg/m^3) at	25°C	862	880	897	862	875	891
Viscosity (mNs/m^2) at	30°C	18.6	45.0	171	42.0	153	810
	60°C	6.3	12.0	31	13.5	34	135
	100°C	2.4	3.9	7.5	4.3	9.1	27
Dynamic viscosity index		92	68	38	109	96	96
Kinematic viscosity index		45	45	43	98	95	95
Pour point, °C		−43	−40	−29	−9	−9	−9
Pressure—viscosity coefficient ($m^2/N \times 10^8$) at	30°C	2.1	2.6	2.8	2.2	2.4	3.4
	60°C	1.6	2.0	2.3	1.9	2.1	2.8
	100°C	1.3	1.6	1.8	1.4	1.6	2.2
Isentropic secant bulk modulus at 35 MN/m^2 and	30°C	—	—	—	198	206	—
	60°C	—	—	—	172	177	—
	100°C	—	—	—	141	149	—
Thermal capacity (J/kg °C) at	30°C	1880	1860	1850	1960	1910	1880
	60°C	1990	1960	1910	2020	2010	1990
	100°C	2120	2100	2080	2170	2150	2120
Thermal conductivity (Wm/m^2 °C) at	30°C	0.132	0.130	0.128	0.133	0.131	0.128
	60°C	0.131	0.128	0.126	0.131	0.129	0.126
	100°C	0.127	0.125	0.123	0.127	0.126	0.123
Temperature (°C) for vapour pressure of 0.001 mmHg		35	60	95	95	110	125
Flash point, open, °C		163	175	210	227	257	300

DETERIORATION

Lubricating oils can become unfit for further service by: oxidation, thermal decomposition, and contamination.

Oxidation

Mineral oils are very stable relative to fatty oils and pure hydrocarbons. This stability is ascribed to the combination of saturated and unsaturated hydrocarbons and to certain of the hydrocarbon derivatives, i.e. compounds containing oxygen, nitrogen and sulphur atoms—the so-called 'natural inhibitors'.

Factors influencing oxidation

Temperature	Rate doubles for every 8–10°C temperature rise.
Oxygen access	Degree of agitation of the oil with air.
Catalysis	Particularly iron and copper in finely divided or soluble form.
Top-up rate	Replenishment of inhibition (natural or added).
Oil type	Proportions and type of aromatics and especially on the compounds containing nitrogen, oxygen, sulphur.

Table 2.6 Effects of oxidation and methods of test

Types of product produced by oxidation	*Factors involved*	*Methods of test*
Organic acids which are liable to corrode cadmium, lead and zinc and thereby to promote the formation of emulsions Lightly polymerised materials which increase the viscosity of the oil Moderately polymerised materials which become insoluble in the oil, especially when cold. When dispersed these also promote emulsification and increase of viscosity. When settled out they clog filter screens and block oil-ways Highly polymerised coke-like materials formed locally on very hot surfaces where they may remain	The relative proportions of the various types of products depend on the conditions of oxidation and the type of oil The degree of oxidation which can be tolerated depends on the lubrication system: more can be tolerated in simple easily cleaned bath systems without sensitive metals, less in complex circulation systems	Total acid number or neutralisation value, which assesses the concentration of organic acids, and is therefore an indication of the concentration of the usually more deleterious polymerised materials, is the most convenient and precise test to carry out. Limits vary between 0.2 mg KOH/g and 4.0 or more With many additive oils proof of the continued effective presence of the necessary additives, e.g. anti-oxidant, is more important

Thermal decomposition

Mineral oils are also relatively stable to thermal decomposition in the absence of oxygen, but at temperatures over about 330°C, dependent on time, mineral oils will decompose into fragments, some of which polymerise to form hard insoluble products.

Table 2.7 Thermal decomposition products

Product	*Effect*
Light hydrocarbons	Flash point is reduced; viscosity is reduced
Carbonaceous residues	Hard deposits on heater surfaces reduce flow rates and accentuate overheating

Some additives are more liable to thermal decomposition than the base oils, e.g. extreme pressure additives; and surface temperature may have to be limited to temperatures as low as 130°C.

Contamination

Contamination is probably the most common reason for changing an oil. Contaminants may be classified as shown in Table 2.8.

Table 2.8 Contaminants

Type	*Example*
Gaseous	Air, ammonia
Liquid	Water, oil of another type or viscosity grade or both
Solid	Fuel soot, road dust, fly ash, wear products

Where appropriate, oils are formulated to cope with likely contaminants, for example turbine oils are designed to separate water and air rapidly, diesel engine oils are designed to suspend fuel soot in harmless finely divided form and to neutralise acids formed from combustion of the fuel.

Solid contaminants may be controlled by appropriate filtering or centrifuging or both. Limits depend on the abrasiveness of the contaminant and the sensitivity of the system.

Oil life

Summarising the data given under the headings Oxidation and Thermal decomposition, above, Figure 2.5 gives an indication of the time/temperature limits imposed by thermal and oxidation stability on the life of a well-refined HVI paraffinic oil.

ADDITIVE OILS

Plain mineral oils are used in many units and systems for the lubrication of bearings, gears and other mechanisms where their oxidation stability, operating temperature range, ability to prevent wear, etc., are adequate. Nowadays, however, the requirements are often greater than plain oils are able to provide, and special chemicals or additives are 'added' to many oils to improve their properties. The functions required of these 'additives' gives them their common names listed in Table 2.9.

Table 2.9 Types of additives

Main type	*Function and sub-types*
Acid neutralisers	Neutralise contaminating strong acids formed, for example, by combustion of high sulphur fuels or, less often, by decomposition of active EP additives
Anti-foam	Reduces surface foam
Anti-oxidants	Reduce oxidation. Various types are: oxidation inhibitors, retarders; anti-catalyst metal deactivators, metal passivators
Anti-rust	Reduces rusting of ferrous surfaces swept by oil
Anti-wear agents	Reduce wear and prevent scuffing of rubbing surfaces under steady load operating conditions; the nature of the film is uncertain
Corrosion inhibitors	Type (*a*) reduces corrosion of lead; type (*b*) reduces corrosion of cuprous metals
Detergents	Reduce or prevent deposits formed at high temperatures, e.g. in ic engines
Dispersants	Prevent deposition of sludge by dispersing a finely divided suspension of the insoluble material formed at low temperature
Emulsifier	Forms emulsions; either water-in-oil or oil-in-water according to type
Extreme pressure	Prevents scuffing of rubbing surfaces under severe operating conditions, e.g. heavy shock load, by formation of a mainly inorganic surface film
Oiliness	Reduces friction under boundary lubrication conditions; increases load-carrying capacity where limited by temperature rise by formation of mainly organic surface films
Pour point depressant	Reduces pour point of paraffinic oils
Tackiness	Reduces loss of oil by gravity, e.g. from vertical sliding surfaces, or by centrifugal force
Viscosity index improvers	Reduce the decrease in viscosity due to increase of temperature

Table 2.10 Types of additive oil required for various types of machinery

Type of machinery	*Usual base oil type*	*Usual additives*	*Special requirements*
Food processing	Medicinal white oil	None	Safety in case of ingestion
Oil hydraulic	Paraffinic down to about −20°C, naphthenic below	Anti-oxidant Anti-rust Anti-wear Pour point depressant VI improver Anti-foam	Minimum viscosity change with temperature; minimum wear of steel/steel
Steam and gas turbines	Paraffinic or naphthenic distillates	Anti-oxidant Anti-rust	Ready separation from water, good oxidation stability
Steam engine cylinders	Unrefined or refined residual or high-viscosity distillates	None or fatty oil	Maintenance of oil film on hot surfaces; resistance to washing away by wet steam
Air compressor cylinders	Paraffinic or naphthenic distillates	Anti-oxidant Anti-rust	Low deposit formation tendency
Gears (steel/steel)	Paraffinic or naphthenic	Anti-wear, EP Anti-oxidant Anti-foam Pour point depressant	Protections against abrasion and scuffing
Gears (steel/bronze)	Paraffinic	Oiliness Anti-oxidant	Reduce friction, temperature rise, wear and oxidation
Machine tool slideways	Paraffinic or naphthenic	Oiliness; tackiness	Maintains smooth sliding at very low speeds. Keeps film on vertical surfaces
Hermetically sealed refrigerators	Naphthenic	None	Good thermal stability, miscibility with refrigerant, low flow point
Diesel engines	Paraffinic or naphthenic	Detergent Dispersant Anti-oxidant Acid neutraliser Anti-foam Anti-wear Corrosion inhibitor	Vary with type of engine thus affecting additive combination

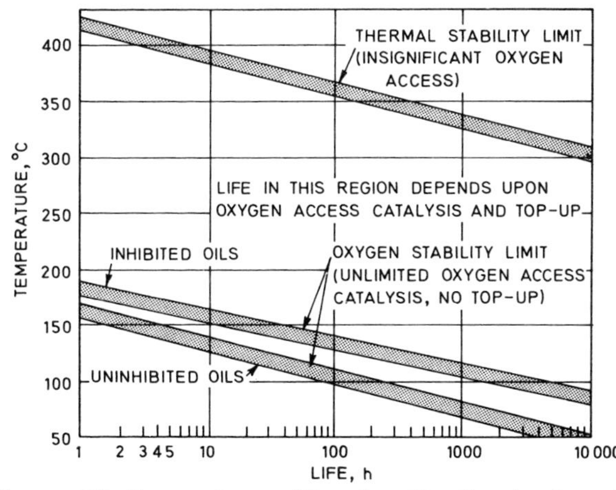

Figure 2.5 Approximate life of well-refined mineral oils (*courtesy:* Institution of Mechanical Engineers)

Selection of additive combinations

Additives and oils are combined in various ways to provide the performance required. It must be emphasised, however, that indiscriminate mixing can produce undesired interactions, e.g. neutralisation of the effect of other additives, corrosivity and the formation of insoluble materials.

Indeed, some additives may be included in a blend simply to overcome problems caused by other additives. The more properties that are required of a lubricant, and the more additives that have to be used to achieve the result, the greater the amount of testing that has to be carried out to ensure satisfactory performance.

Synthetic oils

Application data for a variety of synthetic oils are given in the table below. The list is not complete, but most readily available synthetic oils are included.

Table 3.1

Property \ Fluid	*Di-ester*	*Inhibited Esters*	*Typical Phosphate Ester*	*Typical Methyl Silicone*	*Typical Phenyl Methyl Silicone*	*Chlorinated Phenyl Methyl Silicone*	*Polyglycol (inhibited)*	*Perfluorinated Polyether*
Maximum temperature in absence of oxygen (°C)	250	300	110	220	320	305	260	370
Maximum temperature in presence of oxygen (°C)	210	240	110	180	250	230	200	310
Maximum temperature due to decrease in viscosity (°C)	150	180	100	200	250	280	200	300
Minimum temperature due to increase in viscosity (°C)	−35	−65	−55	−50	−30	−65	−20	−60
Density (g/ml)	0.91	1.01	1.12	0.97	1.06	1.04	1.02	1.88
Viscosity index	145	140	0	200	175	195	160	100–300
Flash point (°C)	230	255	200	310	290	270	180	
Spontaneous ignition temperature	Low	Medium	Very high	High	High	Very high	Medium	Very high
Thermal conductivity (W/M °C)	0.15	0.14	0.13	0.16	0.15	0.15	0.15	
Thermal capacity (J/kg°C)	2000	1700	1600	1550	1550	1550	2000	
Bulk modulus	Medium	Medium	Medium	Very low	Low	Low	Medium	Low
Boundary lubrication	Good	Good	Very good	Fair but poor for steel on steel	Fair but poor for steel on steel	Good	Very good	Poor
Toxicity	Slight	Slight	Some toxicity	Non-toxic	Non-toxic	Non-toxic	Believed to be low	Low
Suitable rubbers	Nitrile, silicone	Silicone	Butyl, EPR	Neoprene, viton	Neoprene, viton	Viton, fluoro-silicone	Nitrile	Many
Effect on plastics	May act as plasticisers		Powerful solvent	Slight, but may leach out plasticisers	Slight, but may leach out plasticisers	Slight, but may leach out plasticisers	Generally mild	Mild
Resistance to attack by water	Good	Good	Fair	Very good	Very good	Good	Good	Very good
Resistance to chemicals	Attacked by alkali	Attacked by alkali	Attacked by many chemicals	Attacked by strong alkali	Attacked by strong alkali	Attacked by alkali	Attacked by oxidants	Very good
Effect on metals	Slightly corrosive to non-ferrous metals	Corrosive to some non-ferrous metals when hot	Enhance corrosion in presence of water	Non-corrosive	Non-corrosive	Corrosive in presence of water to ferrous metals	Non-corrosive	Removes oxide films at elevated temperature
Cost (relative to mineral oil)	4	6	6	15	25	40	4	500

The data are generalisations, and no account has been taken of the availability and property variations of different viscosity grades in each chemical type.

Table 3.1 *continued*

Property \ Fluid	*Chlorinated Diphenyl*	*Silicate Ester or Disiloxane*	*Polyphenyl Ether*	*Fluorocarbon*	*Mineral Oil (for comparison)*	*Remarks*
Maximum temperature in absence of oxygen (°C)	315	300	450	300	200	For esters this temperature will be higher in the absence of metals
Maximum temperature in presence of oxygen (°C)	145	200	320	300	150	This limit is arbitrary. It will be higher if oxygen concentration is low and life is short
Maximum temperature due to decrease in viscosity (°C)	100	240	150	140	200	With external pressurisation or low loads this limit will be higher
Minimum temperature due to increase in viscosity (°C)	−10	−60	0	−50	0 to −50	This limit depends on the power available to overcome the effect of increased viscosity
Density (g/ml)	1.42	1.02	1.19	1.95	0.88	
Viscosity index	−200 to +25	150	−60	−25	0 to 140	A high viscosity index is desirable
Flash point (°C)	180	170	275	None	150 to 200	Above this temperature the vapour of the fluid may be ignited by an open flame
Spontaneous ignition temperature	Very high	Medium	High	Very high	Low	Above this temperature the fluid may ignite without any flame being present
Thermal conductivity (W/m °C)	0.12	0.15	0.14	0.13	0.13	A high thermal conductivity and high thermal capacity are desirable for effective cooling
Thermal capacity (J/kg °C)	1200	1700	1750	1350	2000	
Bulk modulus	Medium	Low	Medium	Low	Fairly high	There are four different values of bulk modulus for each fluid but the relative qualities are consistent
Boundary lubrication	Very good	Fair	Fair	Very good	Good	This refers primarily to anti-wear properties when some metal contact is occurring
Toxicity	Irritant vapour when hot	Slight	Believed to be low	Non-toxic unless overheated	Slight	Specialist advice should always be taken on toxic hazards
Suitable rubbers	Viton	Viton nitrile, fluorosilicone	(None for very high temperatures)	Silicone	Nitrile	
Effect on plastics	Powerful solvent	Generally mild	Polyimides satisfactory	Some softening when hot	Generally slight	
Resistance to attack by water	Excellent	Poor	Very good	Excellent	Excellent	This refers to breakdown of the fluid itself and not the effect of water on the system
Resistance to chemicals	Very resistant	Generally poor	Resistant	Resistant but attacked by alkali and amines	Very resistant	
Effect on metals	Some corrosion of copper alloys	Non-corrosive	Non-corrosive	Non-corrosive, but unsafe with aluminium and magnesium	Non-corrosive when pure	
Cost (relative to mineral oil)	10	8	100	300	1	These are rough approximations, and vary with quality and supply position

A grease may be defined as solid to semi-fluid lubricant consisting of a dispersion of a thickening agent in a lubricating fluid. The thickening agent may consist of e.g. a soap, a clay or a dyestuff. The lubricating fluid is usually a mineral oil, a diester or a silicone.

Tables 4.1, 4.2 and 4.3 illustrate some of the properties of greases containing these three types of fluid. All values and remarks are for greases typical of their class, some proprietary grades may give better or worse performance in some or even all respects.

TYPES OF GREASE

Table 4.1 Grease containing mineral oils

Soap base or thickener	*Min. drop pt. °C (°F) ASTM D 556 IP 132*	*Min. usable temp. °C (°F)*	*Max. usable temp. °C (°F)*	*Rust protection*	*Available with extreme pressure additive*	*Use*	*Cost*	*Official specification*
Lime (calcium)	90 (190)	−20 (0)	60 (140)	—	Yes	General purpose	Low	BS 3223
Lime (calcium) heat stable	99.5 (210) or 140 (280)	−20 (0) −55 (−65) *	80 (175) *	(Sometimes)	Yes	General purpose and rolling bearings	Low	BS 3223 DEF STAN 91–17 (LG 280) DEF STAN 91–27 (XG 279) MIL-G-10924B
Sodium (conventional)	205 (400)	0 (32)	150–175 (300–350)	Yes	No	Glands, seals low–medium speed rolling bearings	Low	—
Sodium and calcium (mixed)	150 (300)	−40 (−40)	120–150 (250–300)	Yes	No	High-speed rolling bearings	Medium	DEF 2261A (XG 271) MIL-L-7711A
Lithium	175 (350)	−40 (−40)	150 (300)	Yes	Yes	All rolling bearings	Medium	DEF STAN 91–12 (XG 271) DEF STAN 91–28 (XG 274) MIL-L-7711A
Aluminium Complex	200 (390)	−40 (−40)	160 (320)	Yes	Mild EP	Rolling bearings	Medium/ high	—
Lithium Complex	235 (450)	−40 (−40)	175 (350)	Yes	Mild EP	Rolling bearings	Medium	—
Clay	None	−30 (−20)	205 (400)*	Poor	Yes	Sliding friction	Medium	—

* Depending on conditions of service.

Although mineral oil viscosity and other characteristics of the fluid have been omitted from this table, these play a very large and often complicated part in grease performance. Certain bearing manufacturers demand certain viscosities and other characteristics of the mineral oil, which should be observed. Apart from these requirements, the finished characteristics of the grease, as a whole, should be regarded as the most important factor.

Table 4.2 Grease containing esters

Soap base or thickener	*Min. drop pt. °C (°F) ASTM D 566 IP 132*	*Min. usable temp. °C (°F)*	*Max. usable temp. °C (°F)*	*Rust protection*	*Available with extreme pressure additive*	*Use*	*Cost*	*Official specification*
Lithium	175 (350)	−75 (−100)	120 (250)	Yes	Yes	Rolling bearings	High	DTD 5598 (XG 287) MIL G-23827A
Clay	None	−55 (−65)	Not applicable†	Yes	Yes	Rolling bearings	High	DTD 5598 (XG 287) MIL-G-23827A
Dye	260 (500)	−40 (−40)	175 (350)†	Yes	No	Rolling bearings	Very high	DTD 5579 (XG 292) MIL-G-25760A

† Upper limit will depend on type of ester used and conditions of service.

Table 4.3 Grease containing silicones

Soap base or thickener	*Min. drop pt. °C (°F)*	*Min. usable temp. °C (°F)*	*Max. usable temp. °C (°F)*	*Rust protection*	*Available with extreme pressure additive*	*Use*	*Cost*	*Official specification*
Lithium	175 (350)	−55 (−65)	205 (400)	Yes	No	Rolling bearings	High	—
Dye	260 (500)	−75 (−100)	260 (500)	Yes	Yes	Rolling bearings/ miniature	Very high	DTD 5585 (XG 300) MIL-G-25013D
Silicone soap	205 (400)	−55 (−65)	260 (500)	Yes	No	Miniature bearings	Extremely high	—

CONSISTENCY

The consistency of grease depends on, amongst other things, the percentage of soap, or thickener in the grease. It is obtained by measuring in tenths of a millimetre, the depth to which a standard cone sinks into the grease in five seconds at a temperature of 25°C (77°F) (ASTM D 217–IP 50). These are called 'units', a *non* dimensional value which *strictly should not be regarded as* tenths of a millimetre. It is called Penetration.

Penetration has been classified by the National Lubricating Grease Institute (NLGI) into a series of single numbers which cover a very wide range of consistencies. This classification does not take into account the nature of the grease, nor does it give any indication of its quality or use.

Table 4.4 NLGI consistency range for greases

Description	*NLGI no.*	*Penetration range (ASTM D 217–IP 50)*	*Types generally available*	*Some common uses*
Semi-fluid	000	445–475	Not dye	Centralised systems Total loss systems
Semi-fluid	00	400–430	Not dye	
Very soft	0	355–385	Not dye	
Soft	1	310–340	All	Rolling bearings General purpose
Medium soft	2	265–295	All	
Medium	3	220–250	All	
Stiff	4	175–205	Na or Ca only	Plain bearings so called 'Block' or 'Brick' greases
Very stiff	5	130–160	Na or Ca only	
Very stiff	6	85–115	Na or Ca only	

The commonest consistencies used in rolling bearings are in the NLGI 2 or 3 ranges but, since modern grease manufacturing technology has greatly improved stability of rolling bearing greases, the tendency is to use softer greases. In centralised lubrication systems, it is unusual to use a grease stiffer than NLGI 2 and often a grease as soft as an NLGI 0 may be found best. The extremes (000, 00, 0 and 4, 5, 6) are rarely, if ever, used in normal rolling bearings (other than 0 in centralised systems), but these softer greases are often used for gear lubrication applications.

GREASE SELECTION

When choosing a grease consideration must be given to circumstances and nature of use. The first decision is always the consistency range. This is a function of the method of application (e.g. centralised, single shot, etc.). This will in general dictate within one or two NLGI ranges, the grade required. Normally, however, an NLGI 2 will be found to be most universally acceptable and suitable for all but a few applications.

The question of operating temperature range comes next. Care should be taken that the operating range is known with a reasonable degree of accuracy. It is quite common to overestimate the upper limit: for example, if a piece of equipment is near or alongside an oven, it will not necessarily be at that oven temperature—it may be higher due to actual temperature-rise of bearing itself, or lower due to cooling effects by convection, radiation, etc.

Likewise, in very low-temperature conditions, the ambient temperature often has little effect after start-up due to internal heat generation of the bearing. It is always advisable, if possible, to measure the temperature by a thermocouple or similar device. A measured temperature, even if it is not the true bearing temperature, will be a much better guide than a guess. By using Tables 4.1, 4.2 and 4.3 above, the soap and fluid can be readily decided.

Normally, more than one type of grease will be found suitable. Unless it is for use in a rolling bearing or a heavily-loaded plain bearing the choice will then depend more or less on price, but logistically it may be advisable to use a more expensive grease if this is already in use for a different purpose. For a rolling bearing application, speed and size are the main considerations; the following Table 4.5 is intended as a guide only for normal ambient temperature.

Table 4.5 Selection of greases for rolling bearings

	SPEED				
	Very slow (under 500 r.p.m.)	*Slow (500 r.p.m.)*	*Average (1000 r.p.m.)*	*Fast (2000 r.p.m.)*	*Very fast (over 2000 r.p.m.)*
BEARING SIZE					
Micro (1–5 mm)	A specially selected ultra-clean grease required				
Miniature (under 10 mm)	Normally aviation greases are used, e.g. XG 287 type				
Small (20–40 mm) Medium (65 mm)	Calcium (LG 280 type)	Calcium or lithium (XG 271 type)	Lithium (XG 271 type)	(Lithium (XG 271 type)) (Lithium (XG 274 type))	Soda–calcium oil type
Large (100 mm)	Lithium/oil (XG 274 or XG 271 type)	Lithium (XG 274 type)	Lithium (XG 274 type)	Lithium (XG 274 type)	—
Very large (200 mm or more)	Calcium or lithium	Lithium (XG 274 type)	Lithium (XG 274 type)	Lithium or soda–calcium (XG 274 type)	—

Note: for definition of these 'types' refer to Table 4.1

If the bearing is heavily loaded for its size, i.e. approaching the maker's recommended maximum, or is subject to shock loading, it is important to use a good extreme-pressure grease. Likewise a heavily-loaded plain bearing will demand a good EP grease.

In general it is advisable always to have good anti-rust properties in the grease, but since most commercial greases available incorporate either additives for the purpose or are in themselves good rust inhibitors, this is not usually a major problem.

Table 4.6 Uses of greases containing fillers

Filler	*Graphite*	*Molybdenum disulphide*	*Metallic powder**
GREASE			
Calcium	Plain bearings Sliding friction and/or localised high temperatures	Not commonly used see lithium	High temperature thread lubricant Electrically conducting†
Lithium	As for calcium but not commonly used	Sliding friction Anti fretting Knuckle or ball joint Reciprocal motion Rolling bearings Care must be taken with localised high temperatures	Not commonly used
Clay	As for calcium	As for calcium	As for calcium

* Metal powder is normally zinc dust, copper flake or finely-divided lead.
† Useful where a current must be passed through rolling bearing from axle to outer race.

Note: filled greases should not normally be used in rolling bearings. Finally, it should be remembered that all of the above remarks refer to typical greases of their type. It is quite possible by special manufacturing techniques, use of specialised additives, etc., to reverse many of the properties either by accident or design. The grease supplier will always advise on any of these abnormal properties exhibited by his grease.

C5 Solid lubricants and coatings

SOLID LUBRICANTS REQUIRED WHEN FLUIDS ARE:			
Undesirable	Contaminate product	—	Food machinery, electrical contacts
	Maintenance difficult	—	Inaccessibility, storage problems
Ineffective	Hostile environments	—	Corrosive gases, dirt and dust
	High temperatures	—	Metalworking, missiles
	Cryogenic temperatures	—	Missiles, refrigeration plant
	Radiation	—	Reactors, space
	Space/vacuum	—	Satellites, X-ray equipment
	Fretting conditions	—	General, often used with oils
	Extreme pressures	—	General, often used with oils

A TYPES OF SOLID LUBRICANT

Materials are required which form a coherent film of low shear strength between two sliding surfaces.

	T_{max} (°C)	*Special features*
	Lamellar solids	
MoS_2	350	Stable to > 1150 °C in vacuum
WS_2	400	Oxidative stability > MoS_2
Graphite	500	Ineffective in vacuum/dry gases
TaS_2	550	Low electrical resistivity
CaF_2/BaF_2	1000	Ineffective below 300 °C
	Polymers	
UHMWPE	100	Exceptionally low wear
FEP	210	Chemically inert; useful at cryogenic temperatures
PTFE	275	Chemically inert; useful at cryogenic temperatures
Polyimides	300	Friction > PTFE
Polyurethanes	100	Useful for abrasion resistance but friction relatively high.
Nylon 11	150	Useful for abrasion resistance but friction relatively high.
	Soft metals	
Pb, Au, Ag, Sn, In		Useful in vacuum.
	Oxides	
MoO_3, PbO/SiO_2, B_2O_3/PbS		Effective only at high temperatures
	Miscellaneous	
$AsSbS_4$, $Sb(SbS_4)$, $Ce_2(MoS_4)$		Oil & grease additives.
$Zn_2P_2O_7/Ca(OH)_2$		White lubricant additive for grease.
	Plasma-sprayed coatings	
$Ag/Ni\text{-}Cr/CaF_2/Glass$		Wide temperature range lubricant formulations, 20–1000 °C.
$Ag/Cr_2C_3/Ni\text{-}Al/BaF_2\text{-}CaF_2$		Wide temperature range lubricant formulations, 20–1000 °C.

B METHODS OF USE

General

Powder — Rubbed on to surfaces to form a 'burnished film', 0.1–10 μm thick. See subsection C.

Dispersion with resin in volatile fluids — Sprayed on to surfaces and cured to form a 'bonded coating', 5–25 μm thick. See subsection D.

Dispersion in non-volatile fluids — Directly as a lubricating medium, or as an additive to oils and greases. See subsection E.

Specialised

As lubricating additives to metal, carbon and polymer bearing materials.

As proprietary coatings produced by vacuum deposition, plasma spraying, particle impingement, or electrophoresis.

C BURNISHED FILMS

Effects of operational variables

Results obtained from laboratory tests with a ball sliding on a film-covered disc. Applicable to MoS_2, WS_2 and related materials, but not to PTFE and graphite.

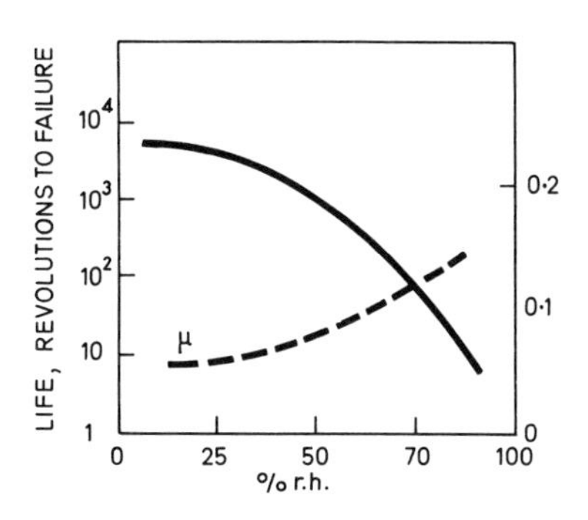

RELATIVE HUMIDITY

Acidity may develop at high r.h. Possibility of corrosion or loss of film adhesion.

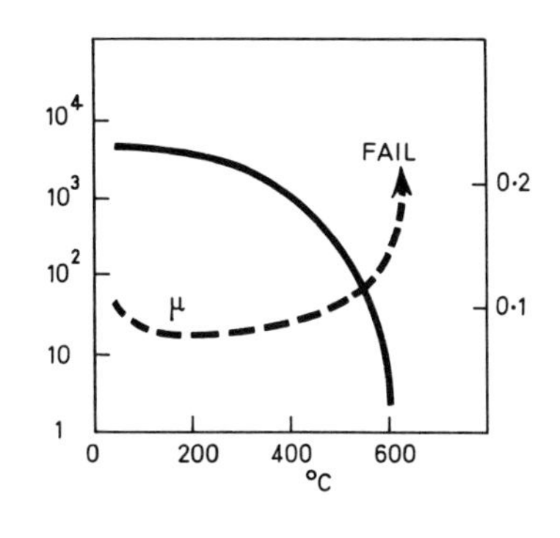

TEMPERATURE

Life limited by oxidation, but MoO_3 is not abrasive. Temperature lowers humidity.

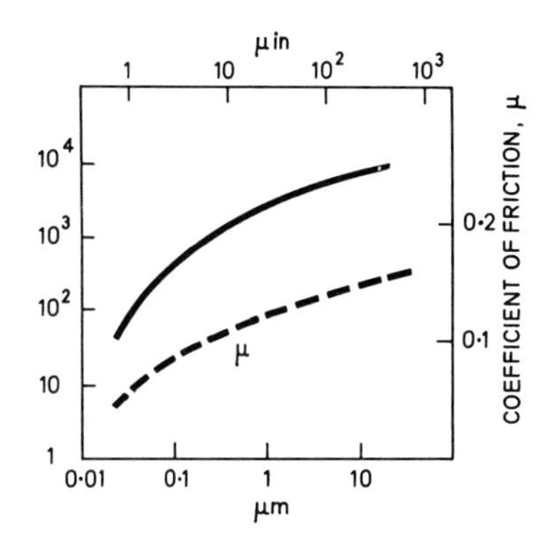

FILM THICKNESS

Difficult to produce films >10 μm thick except at high r.h.

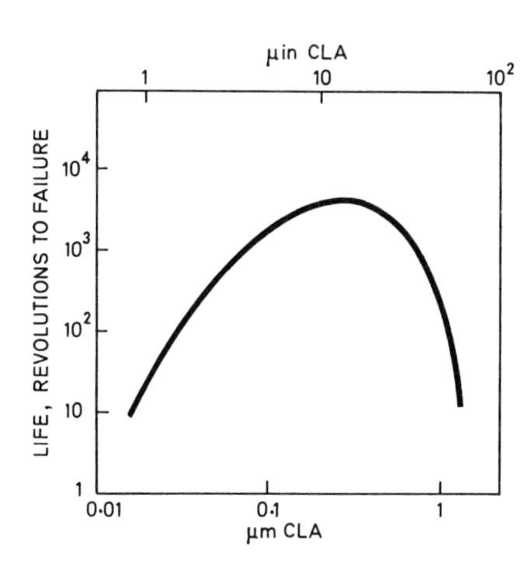

SURFACE ROUGHNESS

Optimum depends on method of surface preparation. Abrasion, 0.5 μm; grit-blast, 0.75 μm; grinding, 1.0 μm; turning, 1.25 μm.

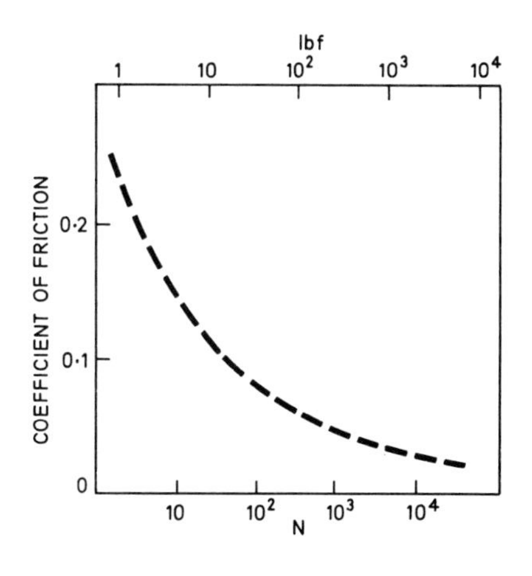

LOAD

As load increases, a greater proportion is supported by the substrate. A similar trend occurs as the substrate hardness increases.

No well-defined trend exists between film life and substrate hardness. Molybdenum is usually an excellent substrate for MoS_2 films. Generally similar trends with film thickness and load also apply to soft metal films.

D BONDED COATINGS

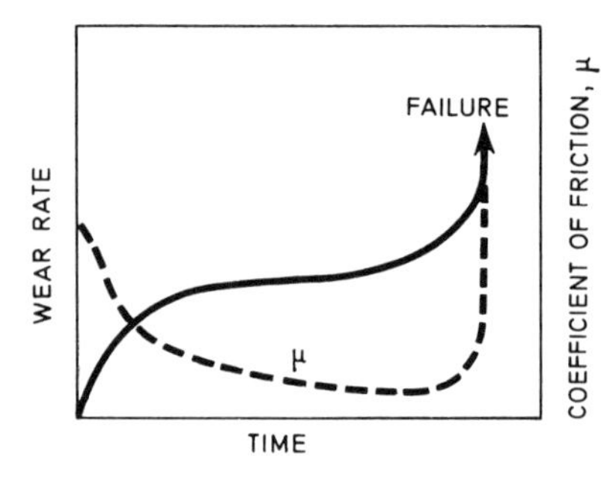

MoS_2 resin coatings show performance trends broadly similar to those for burnished films but there is less dependence of wear life upon relative humidity.

Both the coefficient of friction and the wear rate of the coating vary with time.

Laboratory testing is frequently used to rate different coatings for particular applications. The most common tests are:

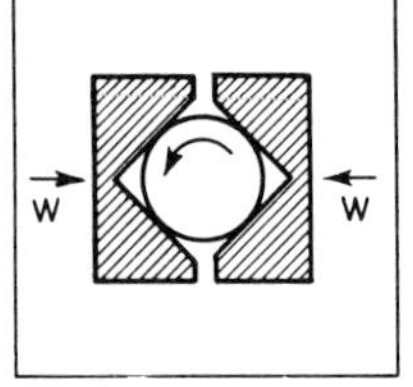

Falex (adopted for specification tests)

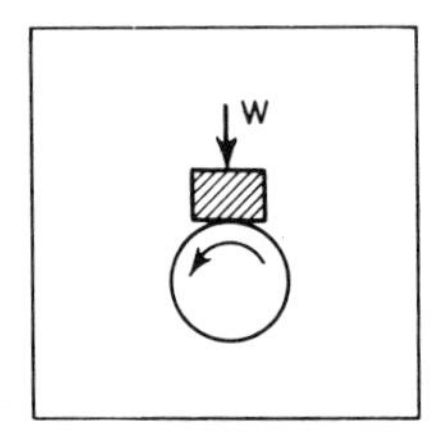

Pad-on-ring (Timken and LFW-1

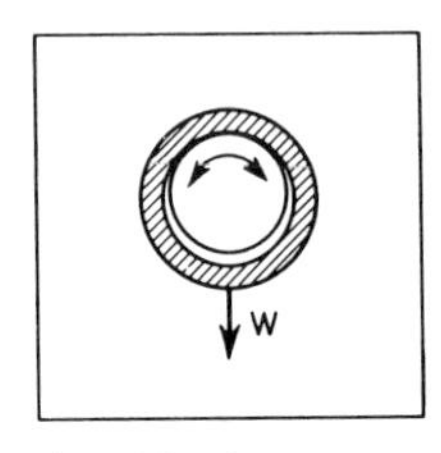

Oscillating plain bearing

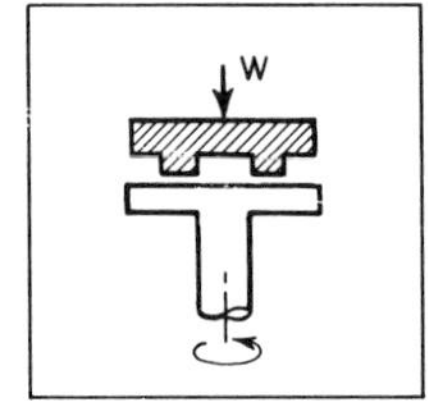

Thrust washer

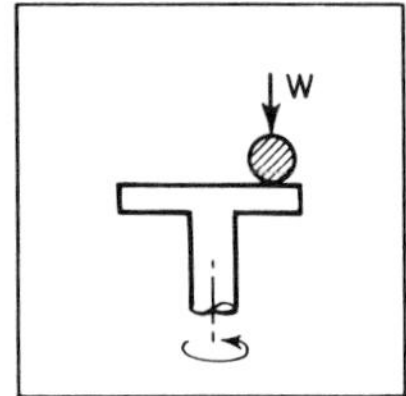

Ball on flat (may be 3 balls)

It is essential to coat the moving surface. Coating both surfaces usually increases the wear life, but by much less than 100% (≏ 30% for plain bearings, ≏ 1% for Falex tests). Considerable variations in wear life are often found in replicate tests (and service conditions).

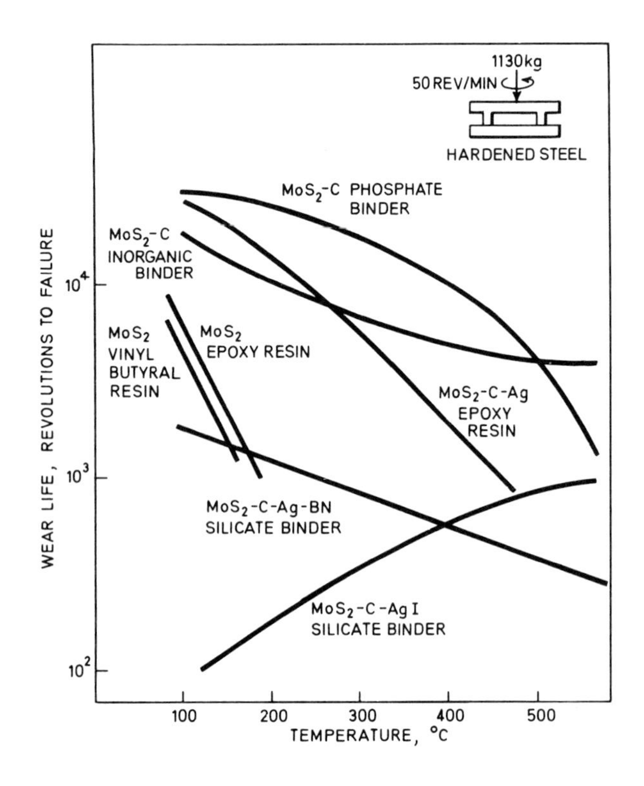

Performance of MoS_2 bonded coatings at elevated temperatures is greatly dependent on the type of resin binder and on the presence of additives in the formulation. Typical additives include graphite, soft metals (Au, Pb, Ag), lead phosphite, antimony trioxide, and sulphides of other metals.

General characteristics of MoS_2 films with different binders

Type	*Curing temp. for 1 hr (°C)*	*Max. temp. (°C)*	*Adhesion to substrate*	*Relative wear life**
Acrylic	20	65	Fair	7
Cellulose	20	65	Fair	25
Alkyd	120	95	Good	1
Phenolic	150	150	Good	75
Epoxy	200	200	Excellent	18
Silicone	250	300	Fair	7
Silicates	20	450	Fair	50
Vitreous	300–600	550	Good	50

* Based on simplified laboratory tests.

Points to note in design

1. Wide variety of types available; supplier's advice should always be sought.
2. Watch effect of cure temperature on substrate.
3. Use acrylic binders on rubbers, cellulose on wood and plastics.
4. Substrate pretreatment essential.
5. Fluids usually deleterious to life.

Preparation of coatings

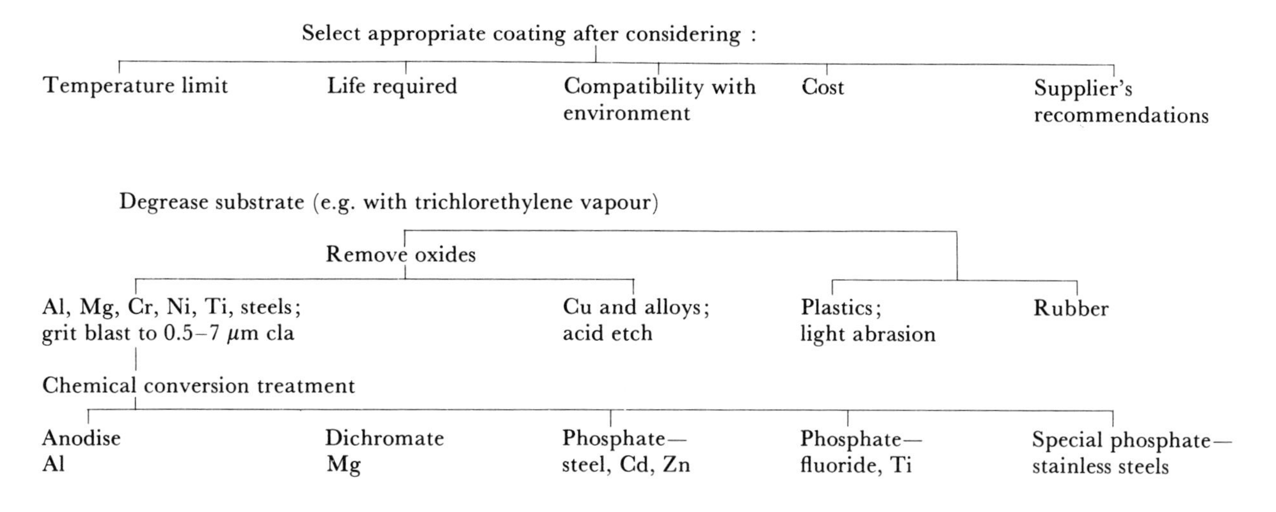

Spray, or dip, to give 2–10 μm film after curing (trials usually needed)

Cure coating; as detailed by supplier

Specifications for solid film bonded coatings

US-MIL-L23398	Lubricant, solid film, air-drying
UK-DEF-STAN 91-19/1 US-MIL-L-8937	Lubricant, solid film, heat-curing
US-MIL-L-46010	Lubricant, solid film, heat cured, corrosion inhibited
US–MIL–L-81329	Lubricant, solid film, extreme environment

Other requirements

Satisfactory appearance

Limits on { Curing time/temperature; Film thickness }

Adhesion – tape test

Thermal stability – resistance to flaking/cracking at temperature extremes

Fluid compatibility – no softening/peeling after immersion

Performance { Wear life; Load carrying capacity }

Storage stability of dispersion

Corrosion – anodised aluminium or phosphated steel

E DISPERSIONS

Graphite, MoS_2 and PTFE dispersions are available in a wide variety of fluids: water, alcohol, toluene, white spirit, mineral oils, etc.

In addition to uses for bonded coatings, other applications include:

Additives to oils and greases	Improving lubricant performance Running-in Assembly problems Wire-rope and chain lubrication Grease thickeners	Compatibility with other additives may pose problems
Parting and anti-stick agents	Moulds for plastics Die-casting moulds Saucepans, etc. Cutting-tool treatments Impregnation of grinding wheels	Use PTFE if product discoloration is a problem
Anti-seize compounds	High-temperature thread lubricants Metalworking generally	Formulations generally proprietary and complex

Specifications for solid lubricant dispersions in oils and greases

Paste

UK-DTD-392B US-MIL-T-5544	Anti-seize compound, high temperatures (50% graphite in petrolatum)
UK-DTD-5617	Anti-seize compound, MoS_2 (50% MoS_2 in mineral oil)
US-MIL-A-13881	Anti-seize compound, mica base (40% mica in mineral oil)
US-MIL-L-25681C	Lubricant, MoS_2, silicone (50% MoS_2 – anti-seize compound)

Grease

US-MIL-G-23549A	Grease, general purpose (5% MoS_2, mineral oil base)
UK-DTD-5527A US-MIL-G-21164C	Grease, MoS_2, low and high temperature (5% MoS_2, synthetic oil base)
US-MIL-G-81827	Grease, MoS_2, high load, wide temperature range (5% MoS_2)
UK-DEF-STAN 91-18/1	Grease, graphite, medium (5% in mineral oil base)
UK-DEF-STAN 91-8/1	Grease, graphite (40% in mineral oil base)

Oil

UK-DEF-STAN 91-30/1 US-MIL-L-3572	Lubricating oil, colloidal graphite (10% in mineral oil)

There is a wide variety of liquids with many different uses and which may interact with tribological components. In these cases, the most important property of the liquid is usually its viscosity. Viscosity values are therefore presented for some common liquids and for some of the more important process fluids.

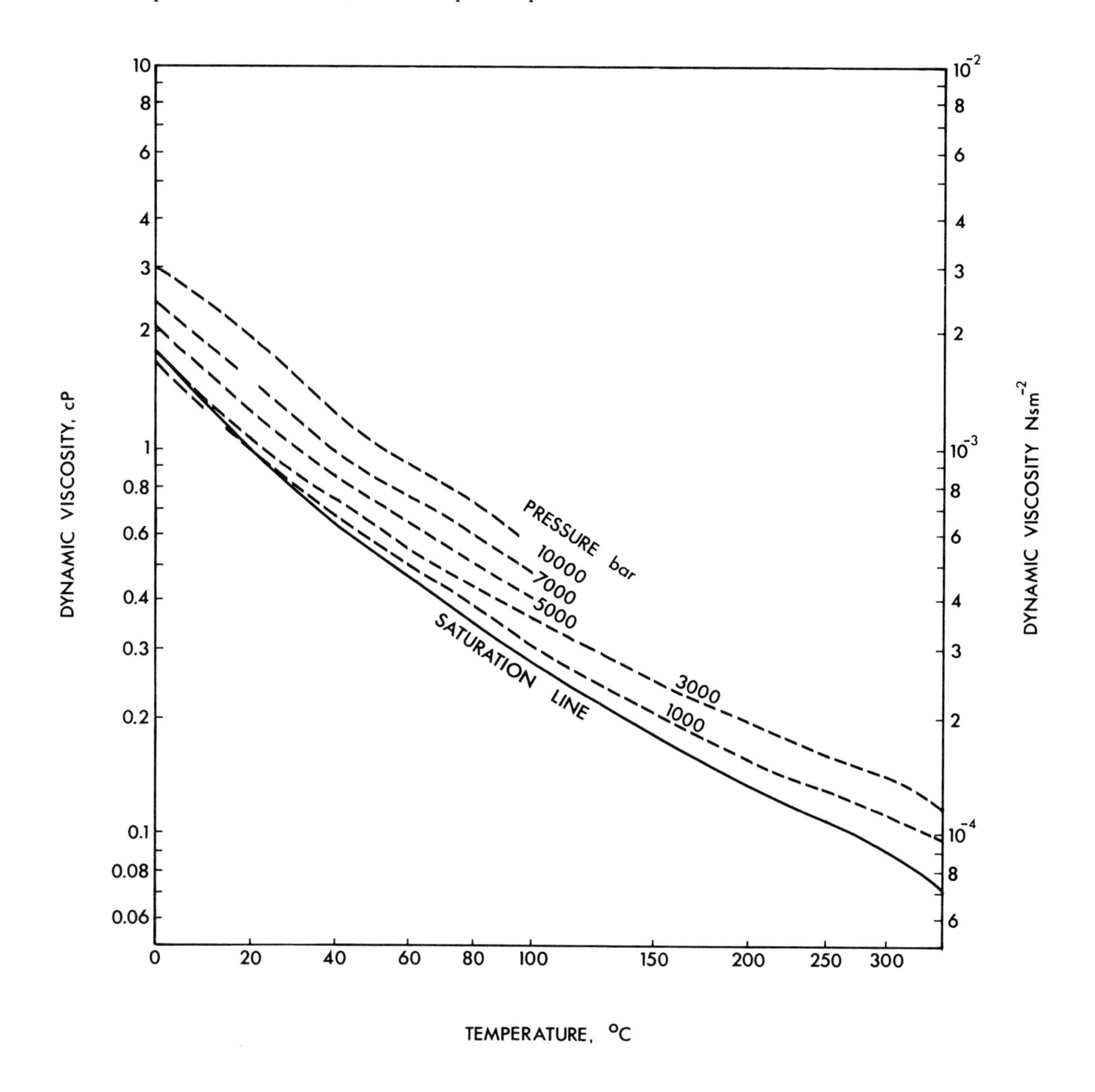

Figure 6.1 The viscosity of water at various temperatures and pressures

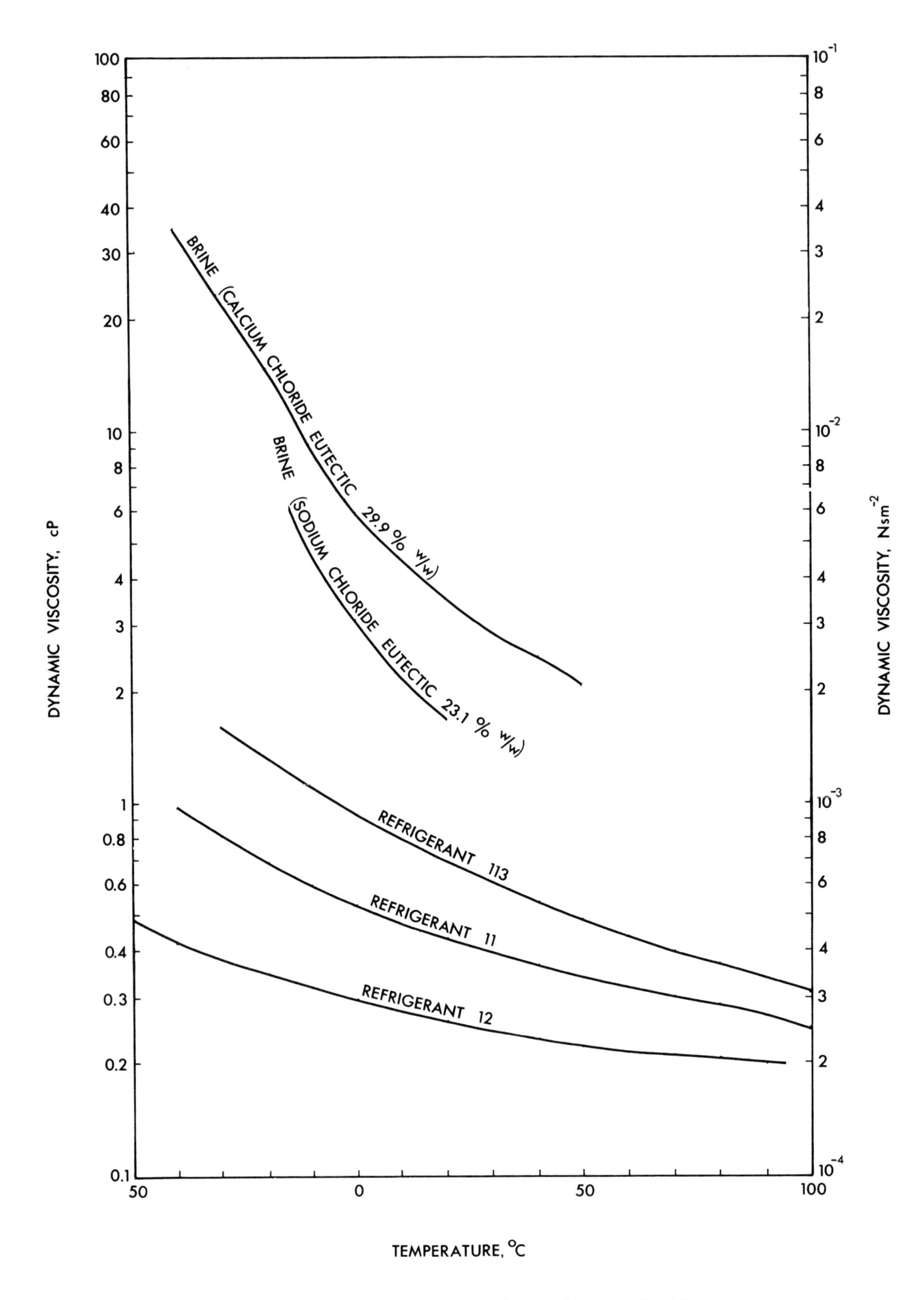

Figure 6.2 The viscosity of various refrigerant liquids

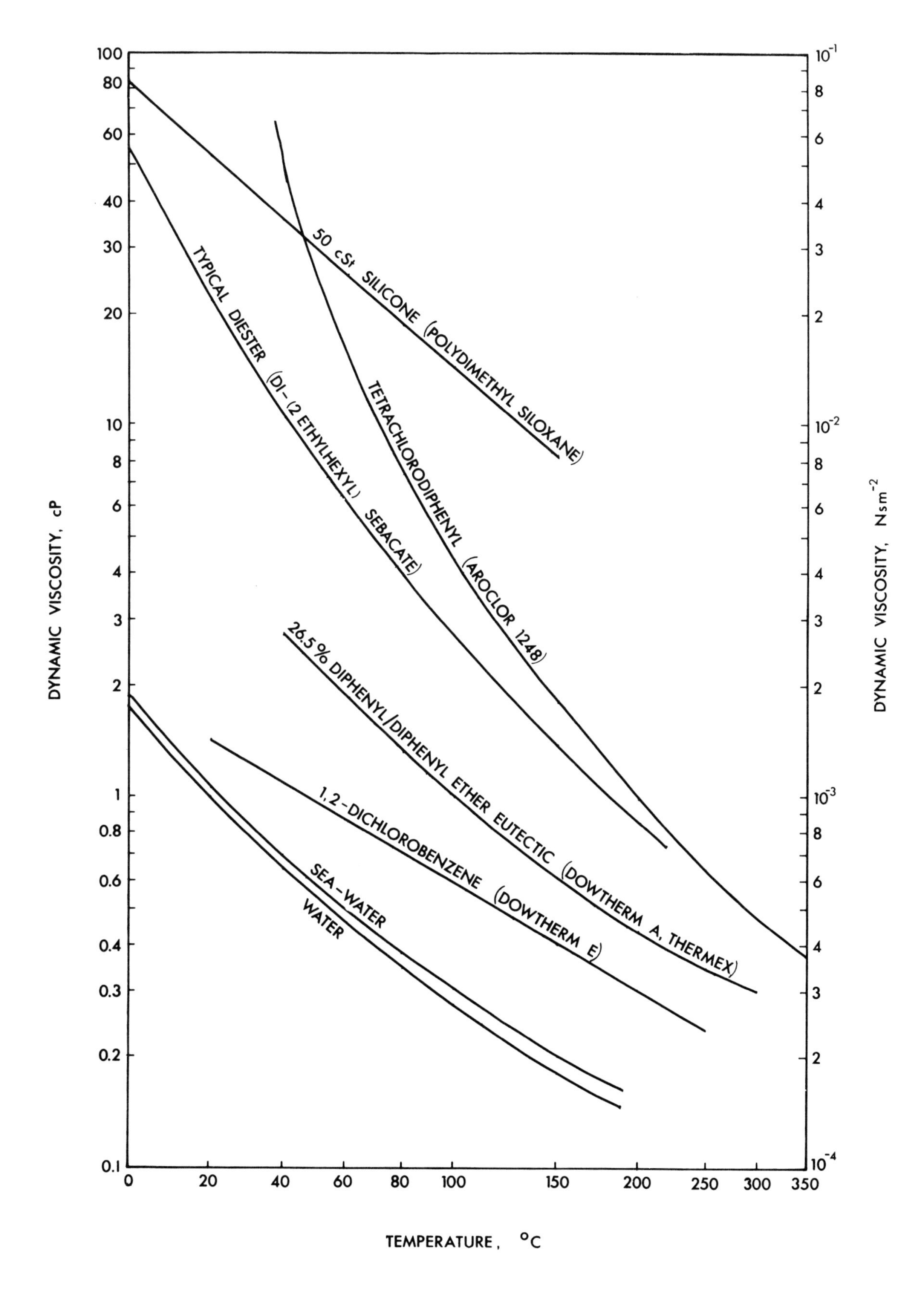

Figure 6.3 The viscosity of various heat transfer fluids

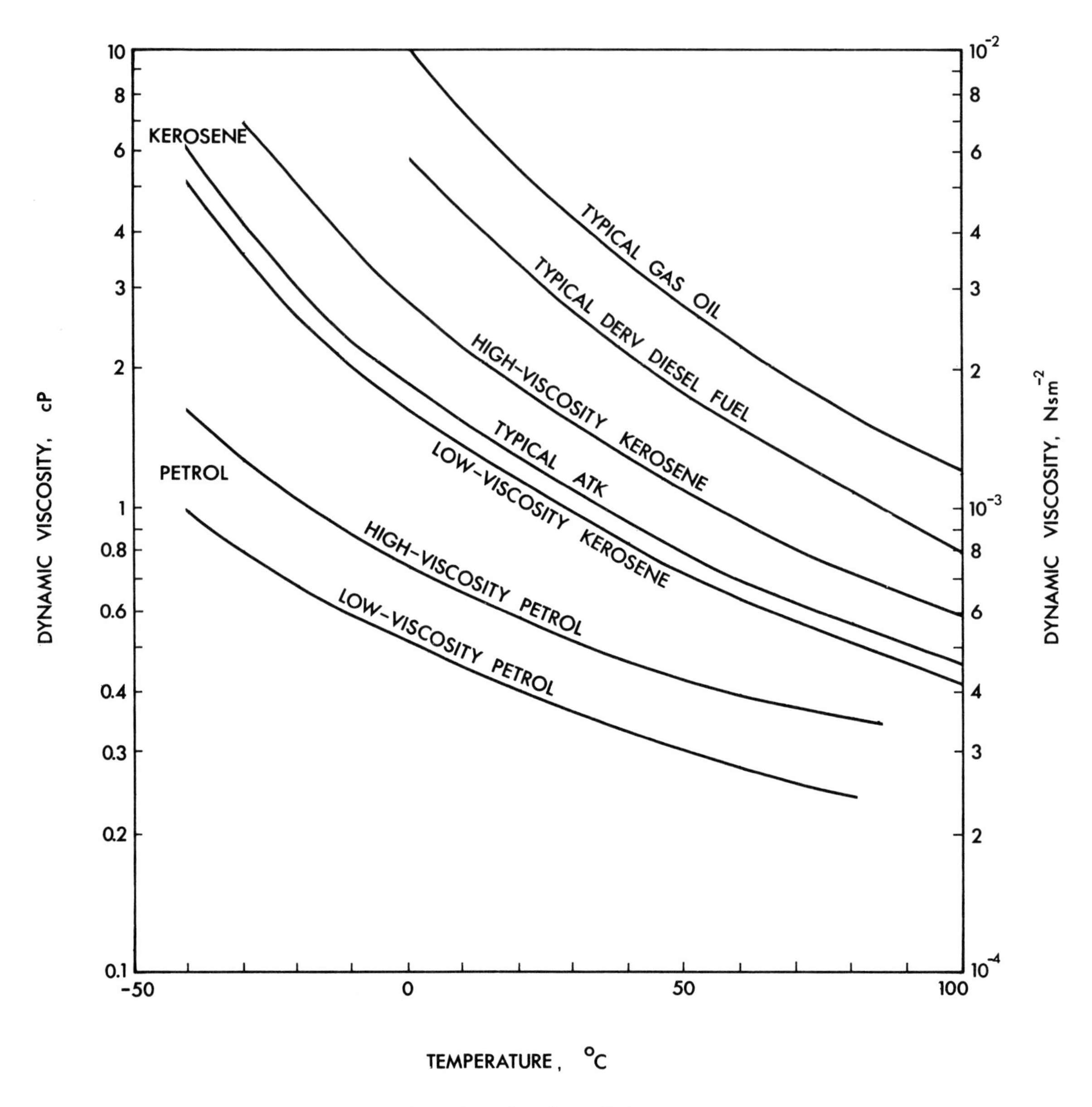

Figure 6.4 The viscosity of various light petroleum products

Petroleum products are variable in composition and so only typical values or ranges of values are given.

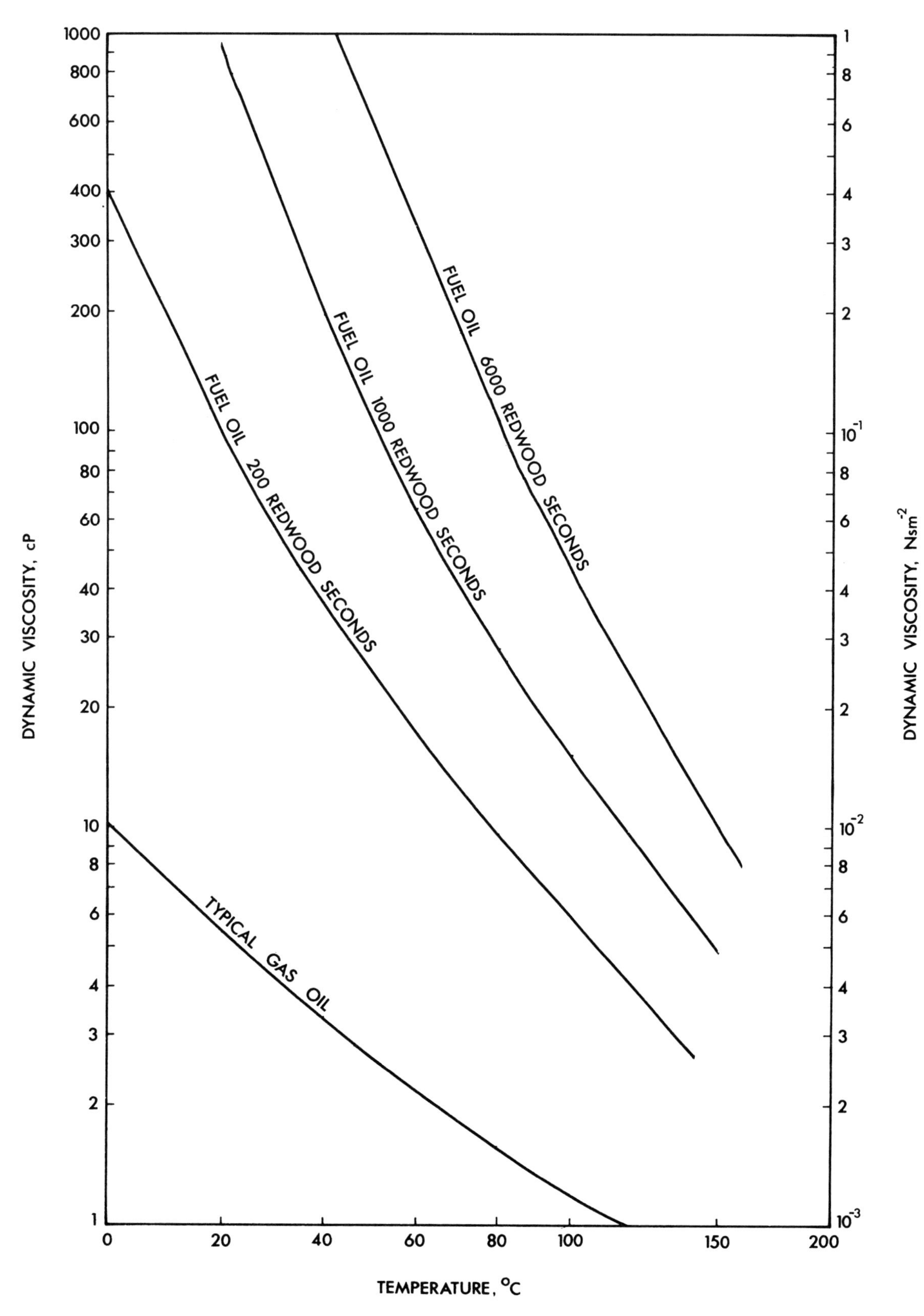

Figure 6.5 The viscosity of various heavy petroleum products

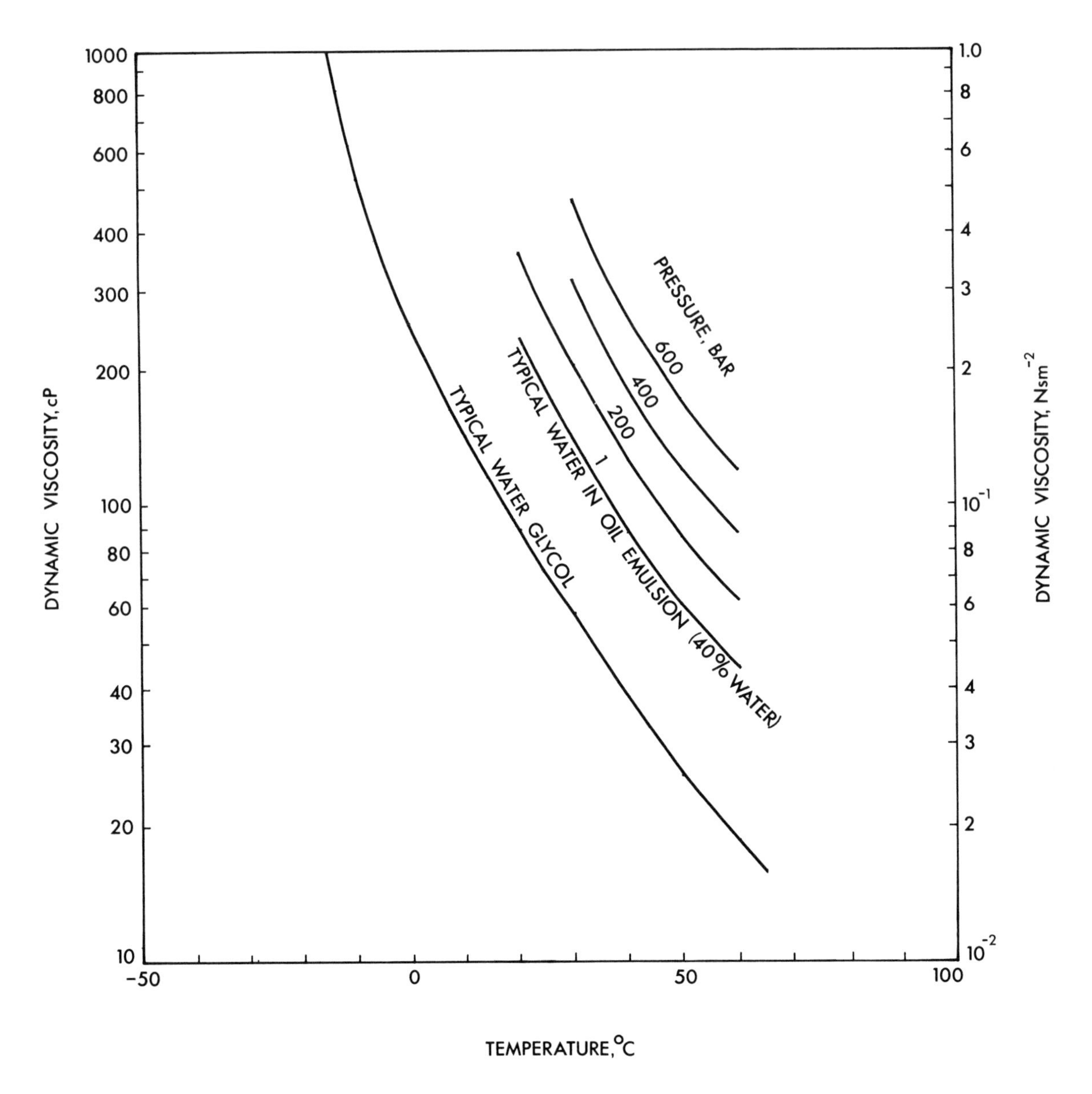

Figure 6.6 The viscosity of various water-based mixtures

For all practical purposes the above fluids may be classed as Newtonian but other fluids, such as water-in-oil emulsions, are non-Newtonian. The viscosity values given for the typical 40% water-in-oil emulsion are for very low shear rates. For this emulsion the viscosity will decrease by 10% at shear rates of about $3000\,s^{-1}$ and by 20% at shear rates of about $10\,000\,s^{-1}$.

Mineral oils and greases are the most suitable lubricants for plain bearings in most applications. Synthetic oils may be required if system temperatures are very high. Water and process fluids can also be used as lubricants in certain applications. The general characteristics of these main classes of lubricants are summarised in Table 7.1.

Table 7.1 Choice of lubricant

Lubricant	*Operating range*	*Remarks*
Mineral oils	All conditions of load and speed	Wide range of viscosities available. Potential corrosion problems with certain additive oils (e.g. extreme pressure) (see Table 7.9)
Synthetic oils	All conditions if suitable viscosity available	Good high and low temperature properties. Costly
Greases	Use restricted to operating speeds below 1 to 2 m/s	Good where sealing against dirt and moisture necessary and where motion is intermittent
Process fluids	Depends on properties of fluid	May be necessary to avoid contamination of food products, chemicals, etc. Special attention to design and selection of bearing materials

The most important property of a lubricant for plain bearings is its viscosity. If the viscosity is too low the bearing will have inadequate load-carrying capacity, whilst if the viscosity is too high the power loss and the operating temperature will be unnecessarily high. Figure 7.1 gives a guide to the value of the minimum allowable viscosity for a range of speeds and loads. It should be noted that these values apply for a fluid at the mean bearing temperature. The viscosity of mineral oils falls with increasing temperature. The viscosity/temperature characteristics of typical mineral oils are shown in Figure 7.2. The most widely used methods of supplying lubricating oils to plain bearings are listed in Table 7.2

The lubricating properties of greases are determined to a large extent by the viscosity of the base oil and the type of thickener used in their manufacture. The section of this handbook on greases summarises the properties of the various types.

Additive oils are not required for plain bearing lubrication but other requirements of the system may demand their use. Additives and certain contaminants may create potential corrosion problems. Tables 7.3 and 7.4 give a guide to additive and bearing material requirements, with examples of situations in which problems can arise.

Table 7.2 Methods of liquid lubricant supply

Method of Supply	*Main characteristics*	*Examples*
Hand oiling	Non automatic, irregular. Low initial cost. High maintenance cost	Low-speed, cheap journal bearings
Drip and wick feed	Non automatic, adjustable. Moderately efficient. Cheap	Journals in some machine tools, axles
Ring and collar feed	Automatic, reliable. Efficient, fairly cheap. Mainly horizontal bearings	Journals in pumps, blowers, large electric motors
Bath and splash lubrication	Automatic, reliable, efficient. Oil-tight housing required High initial cost	Thrust bearings, bath only. Engines, process machinery, general
Pressure feed	Automatic. Positive and adjustable. Reliable and efficient. High initial cost	High-speed and heavily loaded journal and thrust bearings in machine tools, engines and compressors

Notes

Pressure oil feed: This is usually necessary when the heat dissipation of the bearing housing and its surroundings is not sufficient to restrict its temperature rise to 20°C or less

Journal bearings: Oil must be introduced by means of oil grooves in the bearing housing. Some common arrangements are shown in Figure 7.3

Thrust bearings: These must be lubricated by oil bath or by pressure feed from the centre of the bearing

Cleanliness: Cleanliness of the oil supply is essential for satisfactory performance and long life

Table 7.3 Principal additives and contaminants

Problem	*Occurs in*	*Requirements*
Oxidation of lubricant	IC engines Steam turbines Compressors High-speed gearboxes	Antioxidant additives
Scuffiing	Gearboxes Cam mechanisms	Extreme-pressure additive
Deposit formation	IC engines Compressors	Dispersant additives
Excessive wear of lubricated surfaces	General	Antiwear additives
Water contamination	IC engines Steam turbines Compressors	Good demulsification properties. Turbine-quality oils may be required
Dirt particle contamination	IC engines Industrial plant	Dispersant additives
Weak organic acid contamination	IC engines	Acid neutraliser
Strong mineral acid contamination	Diesel engines Process fluids	Acid neutraliser
Rusting	IC engines Turbines Industrial plant General	Rust inhibitor

Plain journal bearings

Surface speed, $u = \pi dn$, ms^{-1}

Mean pressure, $\bar{p} = \frac{W}{ld}$, kNm^{-2}

where
n = shaft speed, s^{-1}
l = bearing width, m
d = shaft diameter, m
W = load, kN

Minimum allowable viscosity $\eta_{min.}$, cP, may be read directly

Plain thrust bearings

Surface speed, $u = \pi Dn$, ms^{-1}

Mean pressure $\bar{p} = \frac{0.4W}{lD}$, kNm^{-2}

where
n = shaft speed, s^{-1}
l = width of bearing ring, m
D = mean pad diameter, m
W = thrust load, kN

Minimum allowable viscosity $\eta_{thrust} = \eta_{min.}\left(\frac{D}{l}\right)$

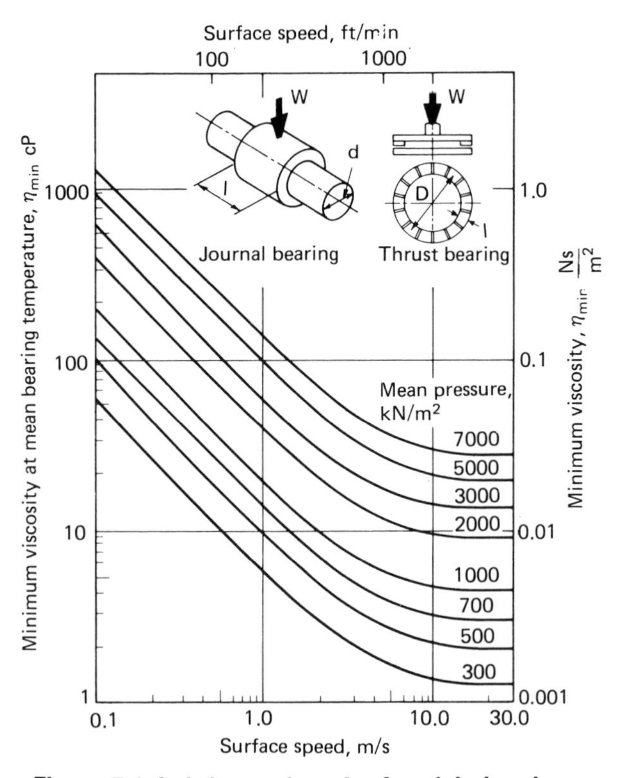

Figure 7.1 Lubricant viscosity for plain bearings

Table 7.4 Resistance to corrosion of bearing metals

	Maximum operating temperature, °C	*Additive or contaminant*				
		Extreme-pressure additive	*Antioxidant*	*Weak organic acids*	*Strong mineral acids*	*Synthetic oil*
Lead-base white metal	130	Good	Good	Moderate/poor	Fair	Good
Tin-base white metal	130	Good	Good	Excellent	Very good	Good
Copper–lead (without overlay)	170	Good	Good	Poor	Fair	Good
Lead–bronze (without overlay)	180	Good with good quality bronze	Good	Poor	Moderate	Good
Aluminium–tin alloy	170	Good	Good	Good	Fair	Good
Silver	180	Sulphur-containing additives must not be used	Good	Good—except for sulphur	Moderate	Good
Phosphor–bronze	220	Depends on quality of bronze. Sulphurised additives can intensify corrosion	Good	Fair	Fair	Good
Copper–lead or lead–bronze with suitable overlay	170	Good	Good	Good	Moderate	Good

Note: corrosion of bearing metals is a complex subject. The above offers a general guide. Special care is required with extreme-pressure lubricants; if in doubt refer to bearing or lubricants supplier.

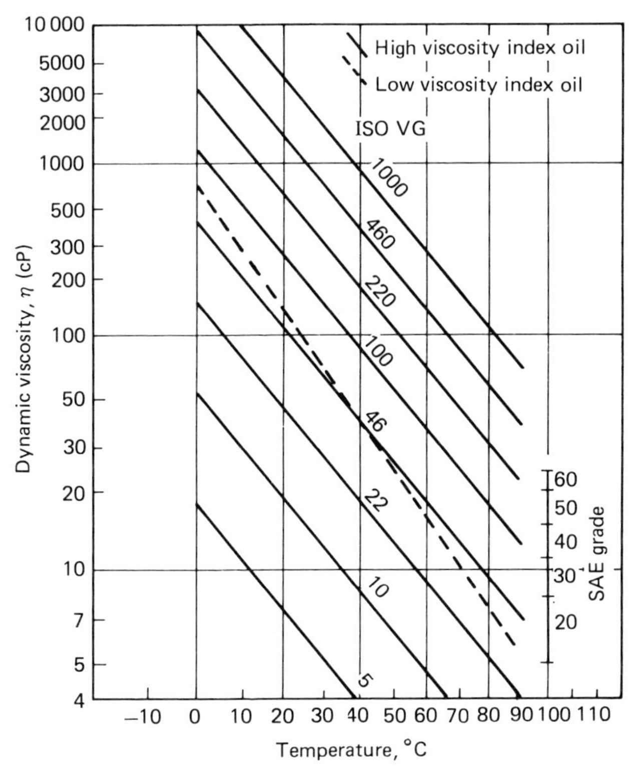

Figure 7.2 Typical viscosity/temperature characteristics of mineral oils

Bearing temperature

Lubricant supply rate should be sufficient to restrict the temperature rise through the bearing to less than 20°C. A working estimate of the mean bearing temperature, $\theta_{bearing}$, is given by

$$\theta_{bearing} = \theta_{supply} + 20, \text{ °C}$$

Dynamic and Kinematic Viscosity

Dynamic Viscosity, η (cP)

= Density × Kinematic Viscosity (cSt)

Viscosity classification grades are usually expressed in terms of Kinematic Viscosities.

Rotation	*Housing*	*Direction of load*	
		Fixed	*Variable*
Unidirectional	Unit	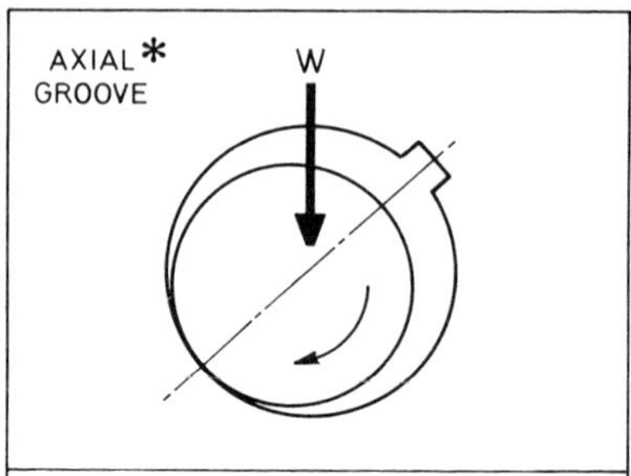 AXIAL* GROOVE W	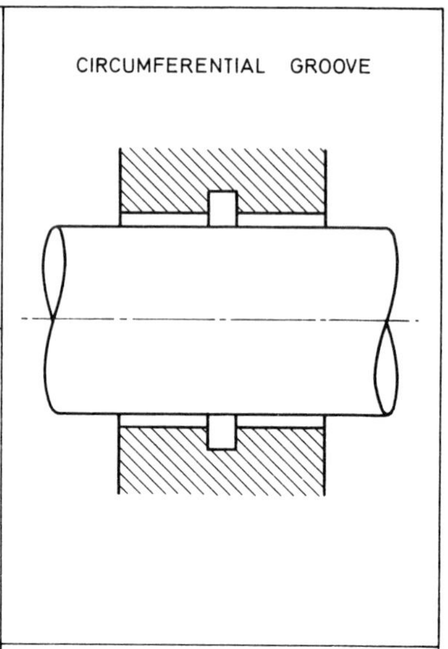 CIRCUMFERENTIAL GROOVE
	Split	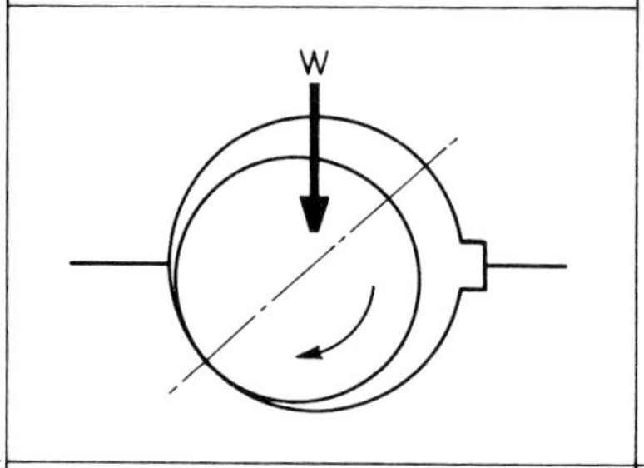 W	
Reversible	Unit/split	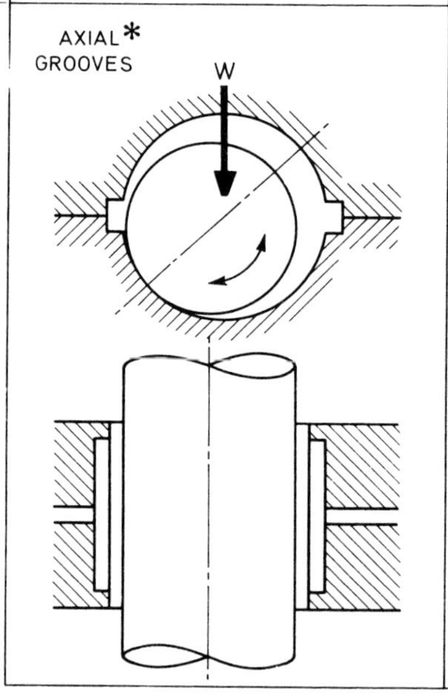 AXIAL* GROOVES W	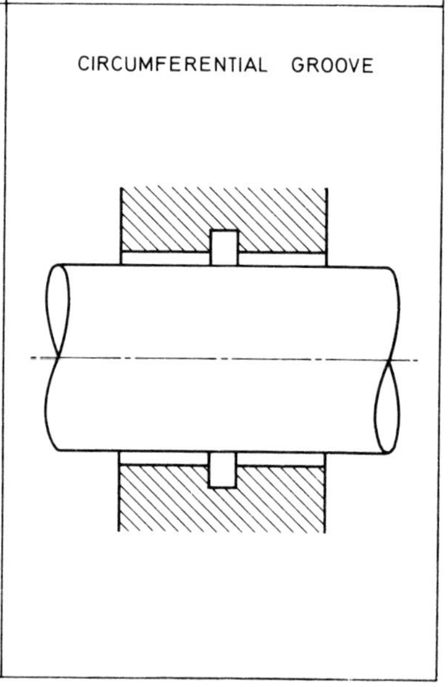 CIRCUMFERENTIAL GROOVE

* At moderate speeds oil holes may be substituted if l/d does not exceed 1.

Note: the load-carrying capacity of bearings with circumferential grooves is somewhat lower than with axial grooves owing to the effect of the groove on pressure generation.

Figure 7.3 Oil grooves in journal bearings

SELECTION OF THE LUBRICANT

Table 8.1 General guide for choosing between grease and oil lubrication

Factor affecting the choice	*Use grease*	*Use oil*
Temperature	Up to 120°C—with special greases or short relubrication intervals up to 200/220°C	Up to bulk oil temperature of 90°C or bearing temperature of 200°C—these temperatures may be exceeded with special oils
Speed factor*	Up to *dn* factors of 300 000/350 000 (depending on design)	Up to *dn* factors of 450 000/500 000 (depending on type of bearing)
Load	Low to moderate	All loads up to maximum
Bearing design	Not for asymmetrical spherical roller thrust bearings	All types
Housing design	Relatively simple	More complex seals and feeding devices necessary
Long periods without attention	Yes, depends on operating conditions, especially temperature	No
Central oil supply for other machine elements	No—cannot transfer heat efficiently or operate hydraulic systems	Yes
Lowest torque	When properly packed can be lower than oil on which the grease is based	For lowest torques use a circulating system with scavenge pumps or oil mist
Dirty conditions	Yes—proper design prevents entry of contaminants	Yes, if circulating system with filtration

* *dn* factor (bearing bore (mm) × speed (rev/min)).

Note: for large bearings (> 65 mm bore) use nd_m (d_m is the arithmetic mean of outer diameter and bore (mm)).

GREASE LUBRICATION

Grease selection

The principle factors governing the selection of greases for rolling bearings are speed, temperature, load, environment and method of application. Guides to the selection of a suitable grease taking account of the above factors are given in Tables 8.2 and 8.3.

The appropriate maximum speeds for grease lubrication of a given bearing type are given in Figure 8.1. The life required from the grease is also obviously important and Figure 8.2 gives a guide to the variation of grease operating life with percentage speed rating and temperature for a high-quality lithium hydroxystearate grease as derived from Figure 8.1. (These greases give the highest speed ratings.)

When shock loading and/or high operating temperatures tend to shake the grease out of the covers into the bearing, a grease of a harder consistency should be chosen, e.g. a no. 3 grease instead of a no. 2 grease.

Note: it should be recognised that the curves in Figures 8.1 and 8.2 can only be a guide. Considerable variations in life are possible depending on precise details of the application, e.g. vibration, air flow across the bearing, clearances, etc.

Table 8.2 The effect of the method of application on the choice of a suitable grade of grease

System	*NLGI grade no.*
Air pressure	0 to 2 depending on type
Pressure-guns or mechanical lubricators	Up to 3
Compression cups	Up to 5
Centralised lubrication	2 or below
(*a*) Systems with separate metering valves	Normally 1 or 2
(*b*) Spring return systems	1
(*c*) Systems with multi-delivery pumps	3

Table 8.3 The effect of environmental conditions on the choice of a suitable type of grease

Type of grease	*NLGI grade no.*	*Speed maximum (percentage recommended maximum for grease)*	*Environment*	*Typical service temperature*				*Base oil viscosity (approximate values)*	*Comments*
				Maximum		*Minimum*			
				°C	°F	°C	°F		
Lithium	2	100 75	Wet or dry	100 135	210 275	−25	−13	Up to 140 cSt at 100°F	Multi-purpose, not advised at max. speeds or max. temperatures for bearings above 65 mm bore or on vertical shafts
Lithium	3	100 75	Wet or dry	100 135	210 275	−25	−13		For max. speeds recommended where vibration loads occur at high speeds
Lithium EP	1	75	Wet or dry	90	195	−15	5	14.5 cSt at 210°F	Recommended for roll-neck bearings and heavily-loaded taper-roller bearings
Lithium EP	2	100 75	Wet or dry	70 90	160 195	−15	5		
Calcium (conventional)	1, 2 and 3	50	Wet or dry	60	140	−10	14	140 cSt at 100°F	
Calcium EP	1 and 2	50	Wet or dry	60	140	−5	25	14.5 cSt at 210°F	
Sodium (conventional)	3	75/100	Dry	80	175	−30	−22	30 cSt at 100°F	Sometimes contains 20% calcium
Clay		50	Wet or dry	200	390	10	50	550 cSt at 100°F	
Clay		100	Wet or dry	135	275	−30	−22	Up to 140 cSt at 100°F	
Clay		100	Wet or dry	120	248	−55	−67	12 cSt at 100°F	Based on synthetic esters
Silicone/lithium		75	Wet or dry	200	390	−40	−40	150 cSt at 25°C	Not advised for conditions where sliding occurs at high speed and load

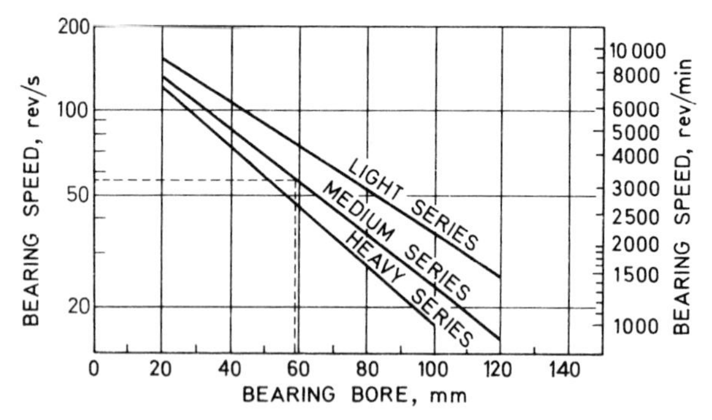

Figure 8.1 Approximate maximum speeds for grease lubrication. (Basic diagram for calculating bearing speed ratings)

	Bearing type	*Multiply bearing speed from Figure 8.1 by this factor to get the maximum speed for each type of bearing*
Ball bearings and cylindrical roller bearings	Cage centred on inner race	As Figure 8.1
	Pressed cages centred on rolling elements	1.5–1.75
	Machined cages centred on rolling elements	1.75–2.0
	Machined cages centred on outer race	1.25–2.0
	Taper- and spherical- roller bearings	0.5
	Bearings mounted in adjacent pairs	0.75
	Bearings on vertical shafts	0.75
	Bearings with rotating outer races and fixed inner races	0.5

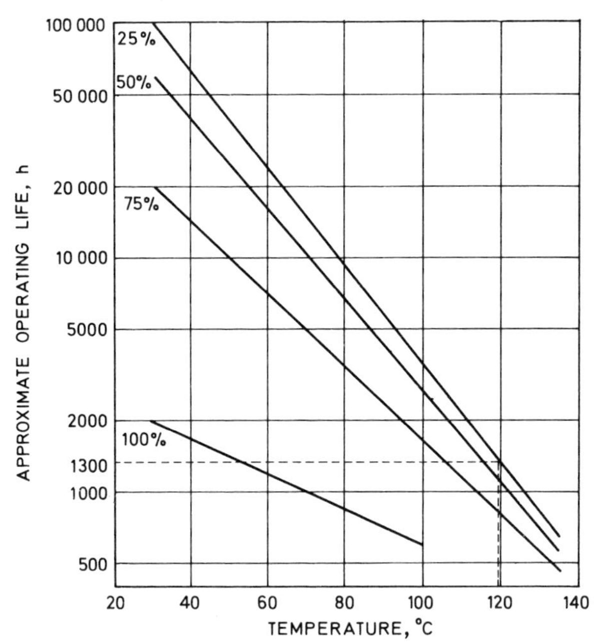

Figure 8.2 Variation of operating life of a high-quality grade 3 lithium hydroxystearate grease with speed and temperature

Calculation of relubrication interval

The relubrication period for ball and roller bearings may be estimated using Figures 8.1 and 8.2. The following is an example in terms of a typical application:

Required to know	: Approximate relubrication period for the following:
Bearing type	: Medium series bearing 60 mm bore.
Cage	: Pressed cage centred on balls.
Speed	: 950 rev/min.
Temperature	: 120°C [The bearing temperature (not merely the local ambient temperature) i.e. either measured or estimated as closely as possible.]
Position	: Vertical shaft.
Grease	: Lithium grade 3.
Duty	: Continuous.
From Figure 8.1	: 60 mm bore position on the lower edge of the graph intersects the medium series curve at approximately 3100 rev/min.

Factor for pressed cages on balls is about 1.5; thus $3100 \times 1.5 = 4650$ rev/min.

Factor for vertical mounting is 0.75. Thus $4650 \times 0.75 = 3488$ rev/min.

This is the maximum speed rating (100%).

Now actual speed = 950 rev/min; therefore percentage of maximum $= \frac{950}{3488} \times 100 = 27\%$ (say 25% approximately).

In Figure 8.2 the 120°C vertical line intersects the 25% speed rating curve for the grade 3 lithium grease at approximately 1300 hours, which is the required answer.

Method of lubrication

Rolling bearings may be lubricated with grease by a lubrication system as described in other sections of the handbook or may be packed with grease on assembly.

Packing ball and roller bearings with grease

(*a*) The grease should not occupy more than one-half to three-quarters of the total available free space in the covers with the bearing packed full.

(*b*) One or more bearings mounted horizontally—completely fill bearings and space between, if more than one, but fill only two-thirds to three-quarters of space in covers.

(*c*) Vertically-mounted bearings—completely fill bearing but fill only half of top cover and three-quarters of bottom cover.

(*d*) Low/medium speed bearings in dirty environments—completely fill bearing and covers.

Relubrication of ball and roller bearings

Relubrication may be carried out in two ways, depending on the circumstances:

(*a*) Replenishment, by which is meant the addition of fresh grease to the original charge.

(*b*) Repacking, which normally signifies that the bearing is dismounted and all grease removed and discarded, the bearing then being cleaned and refilled with fresh grease. An alternative, if design permits, is to flush the bearing with fresh grease *in situ*. (Grease relief valves have been developed for this purpose.)

The quantity required per shot is an arbitrary amount. Requirement is only that sufficient grease is injected to disturb the charge in the bearing and to displace same through the seals, or grease relief valves.

A guide can be obtained from

$$W = \frac{D \times w}{200}$$

where W is quantity (g)
D is outside diameter (mm)
and w is width (mm)

If grease relief valves are not fitted, the replenishment charge should not exceed 5% of the original charge. After grease has been added to a bearing, the housing vent plug (if fitted) should be left out for a few minutes after start-up in order to allow excess grease to escape. A better method, if conditions allow, is to push some of the static grease in the cover back into the bearing to redistribute the grease throughout the assembly. This method is likely to be unsatisfactory when operating temperatures exceed about 100°C.

OIL LUBRICATION

Oil viscosity selection

Generally, when speeds are moderate, the following minimum viscosities at the operating temperatures are recommended:

	cSt
Ball and cylindrical-roller bearings	12
Spherical-roller bearings	20
Spherical-roller thrust bearings	32

The oils will generally be HVI or MVI types containing rust and oxidation inhibitors. Oils containing extreme pressure (EP) additives are normally only necessary for bearings where there is appreciable sliding, e.g. taper-roller or spherical-roller bearings, operating under heavy or shock loads, or if required for associated components, e.g. gears. The nomogram, Figure 8.3, shows how to select more precisely the viscosity needed for known bore and speed when the operating temperatures can be estimated. If the operating temperature is not known or cannot be estimated then the manufacturer's advice should be sought.

To use Figure 8.3, starting with the right-hand portion of the graph for the appropriate bearing bore and speed, determine the viscosity required for the oil at the working temperature. The point of intersection of the horizontal line, which represents this oil viscosity, and the vertical line from the working temperature shows the grade of oil to be selected. If the point of intersection lies between two oils, the thicker oil should be chosen.

Examples: Bearing bore $d = 60$ mm, speed $n = 5000$ rev/min (viscosity at working temperature = 6.8 cSt), with working temperature = 65°C. *Select oil S 14 (14 cSt at 50°C approx.)*

Bearing bore $d = 340$ mm, speed $n = 500$ rev/min (viscosity at working temperature = 13.2 cSt), with working temperature = 80°C. *Select oil S 38 (38 cSt at 50°C approx.)*

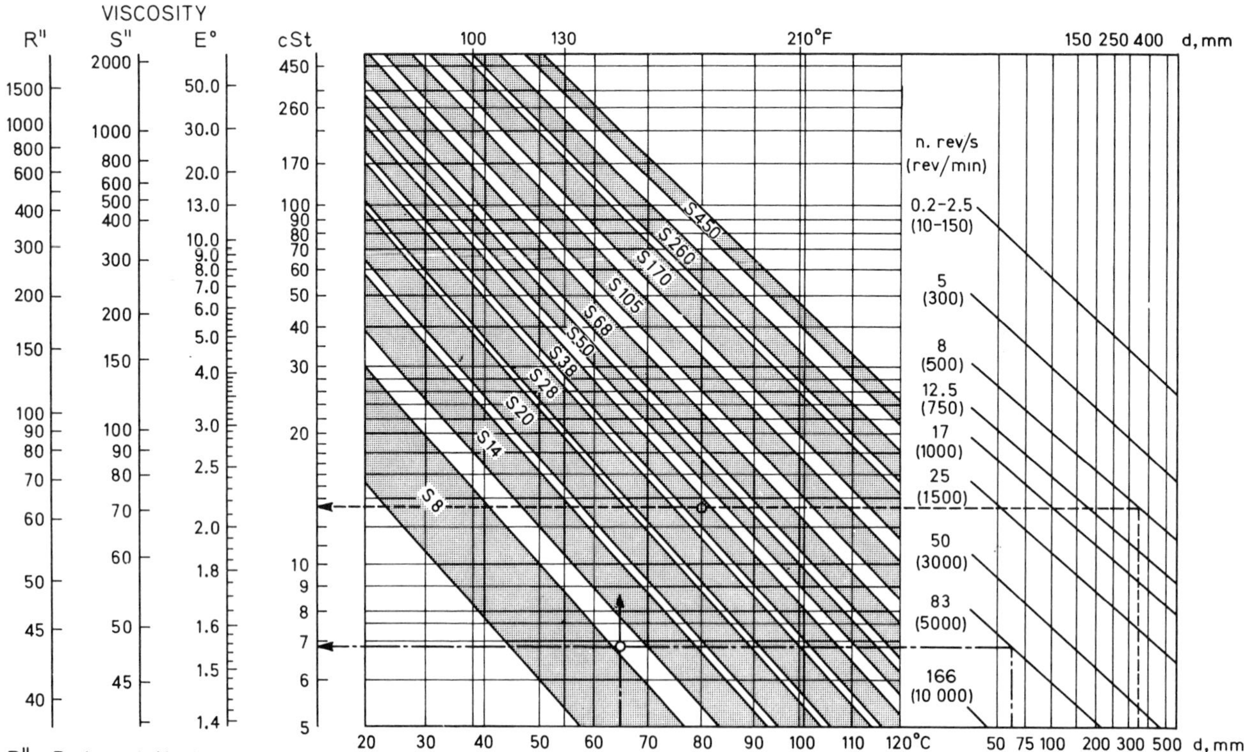

R" = Redwood No. 1 seconds;
S" = Saybolt Universal seconds, SSU
E° = degrees Engler
CSt = centistokes

Figure 8.3 Graph for the selection of oil for roller bearings *(permission of the* Skefko Ball Bearing Co. Ltd). *The graph has been compiled for a viscosity index of 85, which represents a mean value of the variation of the viscosity of the lubricating oil with temperature. Differences for 95 VI oils are negligible*

Application of oil to rolling bearings

System	*Conditions*	*Oil levels/oil flow rates*	*Comments*
Bath/splash	Generally used where speeds are low A limit in *dn* value of 100 000 is sometimes quoted, but higher values can be accommodated if churning is not a problem	Bearings on horizontal and vertical shafts, immerse half lowest rolling element Multi-row bearings on vertical shafts, fully immerse bottom row of elements	
Oil flingers, drip feed lubricators, etc.	Normally as for bath/splash	Flow rate dictated by particular application; ensure flow is sufficient to allow operation of bearing below desired or recommended maximum temperature—generally between 70°C and 90°C	Allows use of lower oil level if temperature-rise is too high with bath/splash
Pressure circulating	No real limit to *dn* value Use oil mist where speeds are very high	As a guide, use:* 0.6 $cm^3/min\ cm^2$ of projected area of bearing (o.d. × width)	The oil flow rate has generally to be decided by consideration of the operating temperature
Oil mist	No real limit to *dn* value Almost invariably used for small bore bearings above 50 000 rev/min, but also used at lower speeds	As a guide, use:* 0.1 to 0.3 × bearing bore (cm/2.54) × no. of rows—cm^3/hour Larger amounts are required for preloaded units, up to 0.6 × bearing bore (cm/2.54) × no. of rows—cm^3/hour	In some cases oil-mist lubrication may be combined with an oil bath, the latter acting as a reserve supply which is particularly valuable when high-speed bearings start to run

* It must be emphasised that values obtained will be very approximate and that the manufacturer's advice should be sought on the performance of equipment of a particular type.

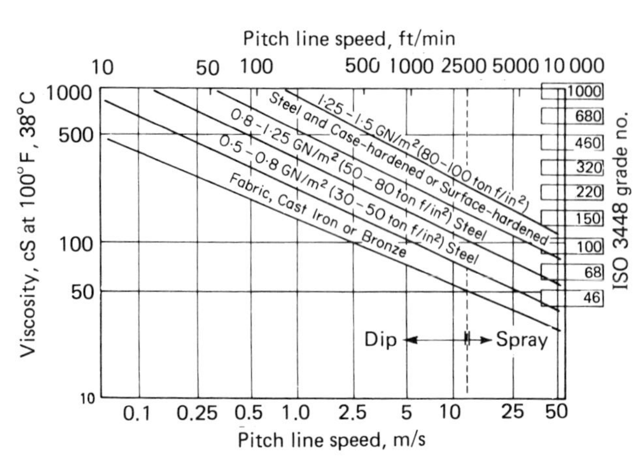

Figure 9.1 Selection of oil for industrial enclosed gear units

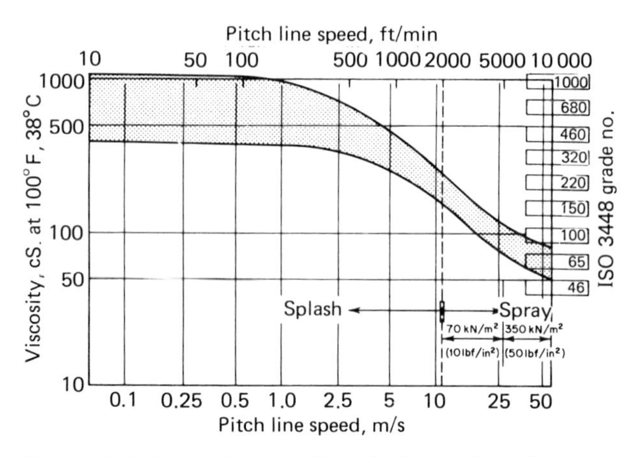

Figure 9.2 Selection of oil for industrial enclosed worm gears

Figure 9.1 is a general guide only. It is based on the criterion: $Sc\ HV/(Vp + 100)$

where Sc = Surface stress factor

$$= \frac{\text{Load/inch line of contact}}{\text{Relative radius of curvature}}$$

and HV = Vickers hardness for the softer member of the gear pair

Vp = Pitch line velocity, ft/min

The chart applies to gears operating in an ambient temperature between 10°C and 25°C. Below 10°C use one grade lower. Above 25°C use one grade higher. Special oils are required for very low and very high temperatures and the manufacturer should be consulted.

With shock loads, or highly-loaded low-speed gears, or gears with a variable speed/load duty cycle, EP oils may be used. Mild EPs such as lead naphthanate should not be used above 80°C(170°F) running temperature. Full hypoid EP oils may attack non ferrous metals. Best EP for normal industrial purposes is low percentage of good quality sulphur/phosphorus or other carefully inhibited additive.

Spray lubrication

Quantity	*Speed* m/sec	10	25	50	100	150
$85 \times 10^{-6} \times$ kW m³/sec	*Pressure* kN/m²	10	100	140	180	210

Quantity	*Speed* ft/min	2500	5000	10 000	20 000	30 000
$0.0085 \times$ hp gpm	*Pressure* lbf/in²	10	15	20	25	30

Suitable lubricants for worm gears are plain mineral oils of a viscosity indicated in Figure 9.2. It is also common practice, but usually unnecessary, to use fatty additive or leaded oils. Such oils may be useful for heavily-loaded, slow-running gears but must not be used above 80°C(170°F) running temperature as rapid oxidation may occur, resulting in acidic products which will attack the bronze wheel and copper or brass bearing-cages.

Worm gears do not usually exceed a pitch line velocity of 2000 ft/min, but if they do, spray lubrication is essential. The sprayed oil must span the face width of the worm.

Quantity	*Speed* m/s	10	15	20	25
$75 \times 10^{-6} \times C^*$ m³/sec	*Pressure* kN/m²	100	170	270	340

C* = centre distance in metres.

Quantity	*Speed* ft/min	2000	3000	4000	5000
$Q = C/4$ gpm	*Pressure* lbf/in²	15	25	40	50

Where C = centre distance, inches.

Recent developments in heavily loaded worm gear lubrication include synthetic fluids which:

(*a*) have a wider operating temperature range
(*b*) reduce tooth friction losses
(*c*) have a higher viscosity index and thus maintain an oil film at higher temperatures than mineral oils
(*d*) have a greatly enhanced thermal and oxidation stability, hence the life is longer

Even more recent developments include the formulation of certain soft synthetic greases which are used in 'lubricated-for-life' worm units. Synthetic lubricants must not be mixed with other lubricants.

AUTOMOTIVE LUBRICANTS

SAE classification of transmission and axle lubricants

SAE visco-sity No.	Centistokes				Redwood seconds			
	0°F – 18°C		210°F 99°C		0°F – 18°C		210°F 99°C	
	Min.	Max.	Min.	Max.	Min.	Max.	Min	Max.
75	—	3250	—	—	—	13 100	—	—
80	3250*	21 700	—	—	13 100	87 600	—	—
90	—	—	14	25	—	—	66	107
140	—	—	25	43	—	—	107	179
250	—	—	41	—	—	—	179	—

Note: * The min. viscosity at 0°F may be waived if the viscosity is not less than 7 cSt at 210°F.

These values are approximate and are given for information only.

Selection of lubricants for transmissions and axles

Almost invariably dip-splash.

The modern tendency is towards universal multipurpose oil.

	Manual gear boxes	Automatic gear boxes	Rear axles (hypoids)	Rear axles (spiral bevel and worms)
Cars	SAE 80 (EP) or multi-purpose	Automatic transmission fluid (ATF)	Highly active SAE 90 EP or multi-purpose	—
Heavy vehicles	SAE 80 EP or SAE 90 (EP) or multi-purpose	SAE 80 EP for semi-automatics ATF fluid for autos	—	SAE 140 or multi-purpose

Notes: (1) Above only to be used where supplier's recommendations are not available.
(2) Above are suitable for normal conditions. In cold conditions (< 0°C) use one SAE grade less. In hot conditions (> 40°C) use one SAE grade higher.
(3) In most cases (except hypoids), straight oils are acceptable. The above EPs are given for safety if supplier's recommendations are not known.
(4) Some synthetic (polyglycol) oils are very successful with worm gears. They must not be mixed with any other oils. ATF fluids must not be mixed with others.
(5) Change periods: (only if manufacturer's recommendations not known). Rear axles—do not change. Top up as required. All manual and automatic gearboxes—change after 20 000 miles. Before that top up as required.

ROLLER CHAINS

Type of lubricant: Viscosity grade no. 150 (ISO 3448). For slow-moving chains on heavy equipment, bituminous viscous lubricant or grease can be used. Conditions of operation determine method of application and topping-up or change periods. Refer to manufacturer for guidance under unusual conditions.

Method of application	Speed limitation m/s(ft/min)	Quantity	Comments
Dip	< 10 (<2000)	—	1.5 x ROLLER PITCH; OIL LEVEL
Slow drip	0–3 (0–600)	5–10 drops/min.	APPLIED TO IN-GOING SIDE OF CHAIN
Fast drip	3–7.5 (600–1500)	> 20	
Spray	> 7·5 (>1500)	Depends upon speed and other conditions	SUMP; PUMP DRIVEN SEPARATELY OR FROM CHAIN SPROCKET

OPEN GEARS

Applies to large, slow-running gears without oil-tight housings.

Requirements of lubricant	Types of lubricant	Methods of application
Must form protective film	Generally bituminous and tacky.	Hand, brush, paddle
Must not be squeezed out	Sometimes cut back by volatile diluent	Dip-shallow pool
Must not be thrown off	Can use grease or heavy EP oils	Drip-automatic
Must be suitable for prevailing ambient conditions		Spray-continuous or intermittent

Temperature °C	Viscosity of open gear lubricant cS at 38°C(100°F)		
	Spray		Drip
	Mild EP oil	Residual compound	Mild EP oil
−10 to 15	—	200–650	—
5 to 35	100–120	650–2000	100–120
25 to 50	180–200	650–2000	180–200

Slide lubrication

Slides are used where a linear motion is required between two components. An inherent feature of this linear motion is that parts of the working surfaces must be exposed during operation. The selection of methods of slide lubrication must therefore consider not only the supply and retention of lubricant, but also the protection of the working surfaces from dirt contamination.

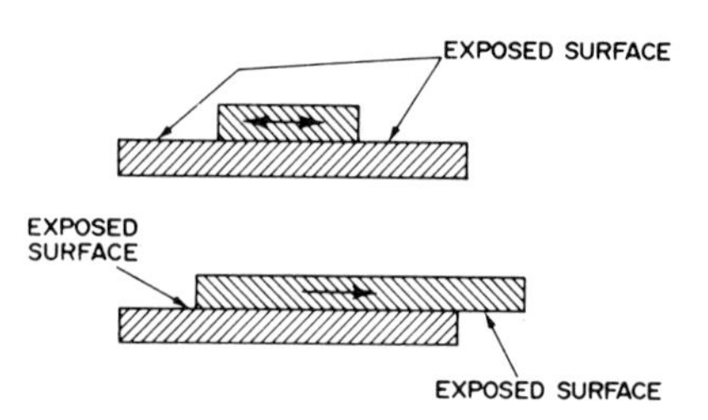

Figure 10.1 Slide movements exposes the working surfaces to contamination

Table 10.1 The lubrication of slides in various applications

Application	*Function of lubricant*	*Lubricant commonly used*	*Remarks*
Slides and linear bearings on machine tools	To minimise the wear of precision surfaces. To avoid any tendency to stick slip motion	Depends on the type of slide or linear bearing. See next Table	Swarf must be prevented from getting between the sliding surfaces
Slides and linear bearings on packaging machines, textile machines, mechanical handling devices	To reduce friction and wear at the moving surfaces without contaminating the material being handled	Greases and solid lubricants are commonly used but air lubrication may also be possible	The sliding contact area should be protected from dirt by fitting scraper seals at each end if possible
Crossheads on reciprocating engines and compressors	To give low friction and wear by maintaining an adequate film thickness to carry the impact loads	The same oil as that used for the bearings	Oil grooves on the stationary surface are desirable to help to provide a full oil film to carry the peak load

Figure 10.2 Typical wick lubricator arrangement on a machine tool

Figure 10.3 Typical roller lubricator arrangement on a machine tool

Table 10.2 The lubrication of various types of linear bearings on machine tools

Type of linear bearing	*Suitable lubricant*	*Method of applying the lubricant*	*Remarks*
Plain slide ways	Cutting oil	By splash from cutting area	Only suitable in machine tool applications using oil as a cutting fluid
	Mineral oil containing polar additives to reduce boundary friction	By wick feed to grooves in the shorter component	Requires scraper seals at the ends of the moving component to exclude swarf
		By rollers in contact with the bottom face of the upper slide member, and contained in oil filled pockets in the lower member	Only suitable for horizontal slides Requires scraper seals at the ends of the moving component to exclude swarf
		By oil mist	Air exhaust keeps the working surfaces clear of swarf
	Grease	By grease gun or cup, to grooves in the surface of the shorter component	Particularly suitable for vertical slides, with occasional slow movement
Hydro-static plain slideways	Air or any other conveniently available fluid	Under high pressure via control valves to shallow pockets in the surface of the shorter member	Gives very low friction and no stick slip combined with high location stiffness. Keeps working surfaces clear of swarf
Linear roller bearings	Oil	Lower race surface should be just covered in an oil bath	Not possible in all configurations. Must be protected from contamination
	Grease	Packed on assembly but with grease nipples for replenishment	Must be protected from contamination

FILLED COUPLINGS (GEAR, SPRING-TYPE, CHAIN)

Table 11.1 Recommendations for the lubrication of filled couplings

Lubricant type	*Limiting criteria* — *Centrifugal Effects* — *Pitch-line acceleration* $d\omega^2/2$ *(m/sec²)*	*Range in practical units* Dn^2 *(ft/sec²)*	*Heat Dissipation* Pn	*Lubricant change period*	*Remarks*
No 1 Grease (mineral oil base)	0.15×10^3 0.5×10^3	25 max 25–80	– –	2 years 12 months	Soft grease preferred to ensure penetration of lubricant to gear teeth
No 3 Grease (mineral oil base)	1.5×10^3 5.0×10^3 12.5×10^3	80–250 250–850 850–2000	– – –	9 months 6 months 3 months	Limitation is loss of oil causing hardening of grease; No 3 grease is more mechanically stable than No 1.
Semi-fluid polyglycol grease or mineral oil	45.0×10^3	3000–5000	230×10^3 max	2 years	Sealing of lubricant in coupling is main problem

d = pcd, m; D = pcd, ft; ω = rads/sec; n = rev/sec; P = hp transmitted

Limits

Grease lubrication, set by soap separation under centrifuging action. Semi-fluid grease lubrication, set by heat dissipation.

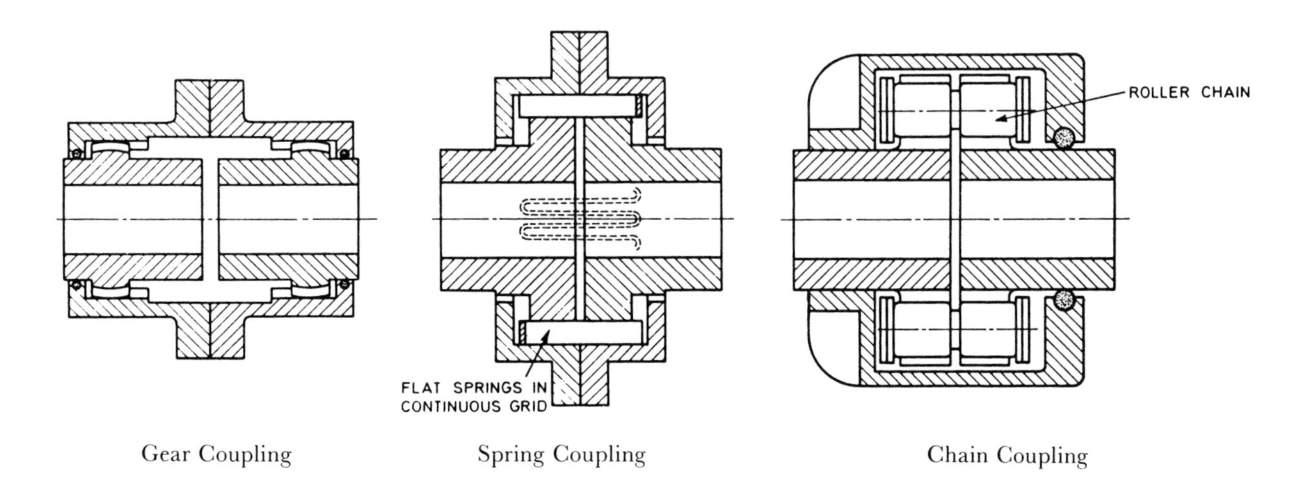

Figure 11.1 Types of filled couplings

CONTINUOUSLY-LUBRICATED GEAR COUPLINGS

Lubrication depends on coupling type

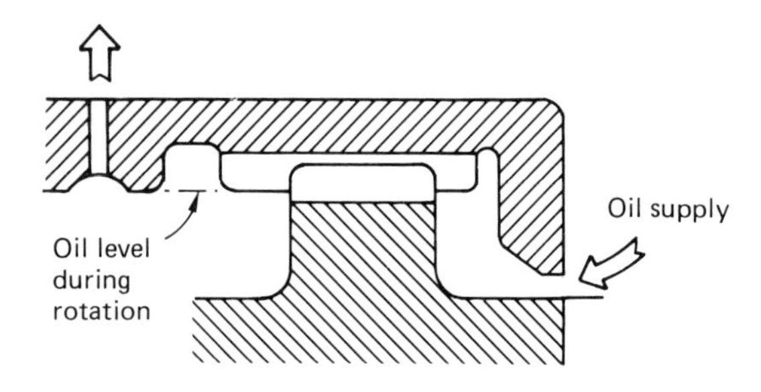

Figure 11.2 Dam-type coupling

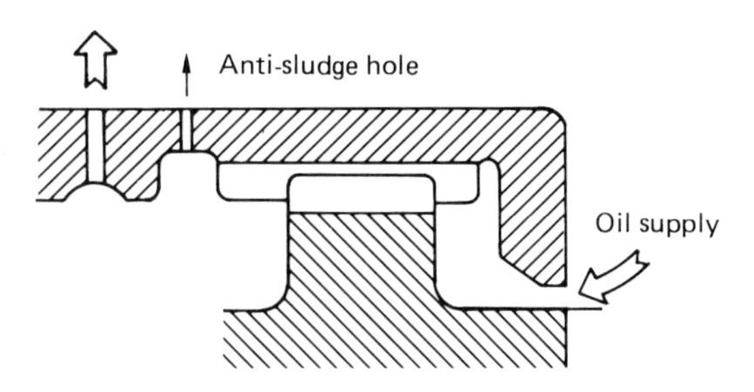

Figure 11.3 Dam-type coupling with anti-sludge holes

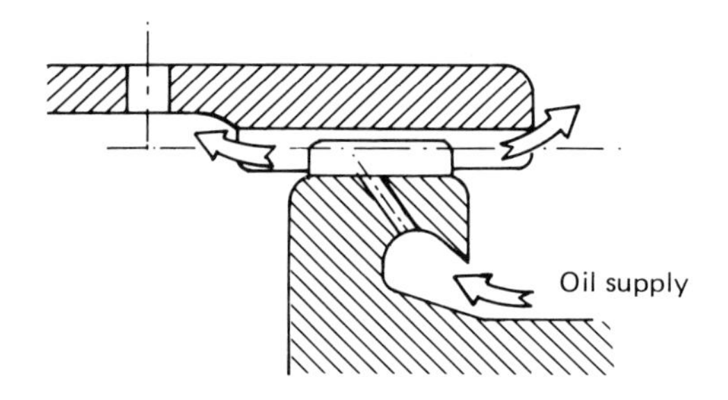

Figure 11.4 Damless-type coupling

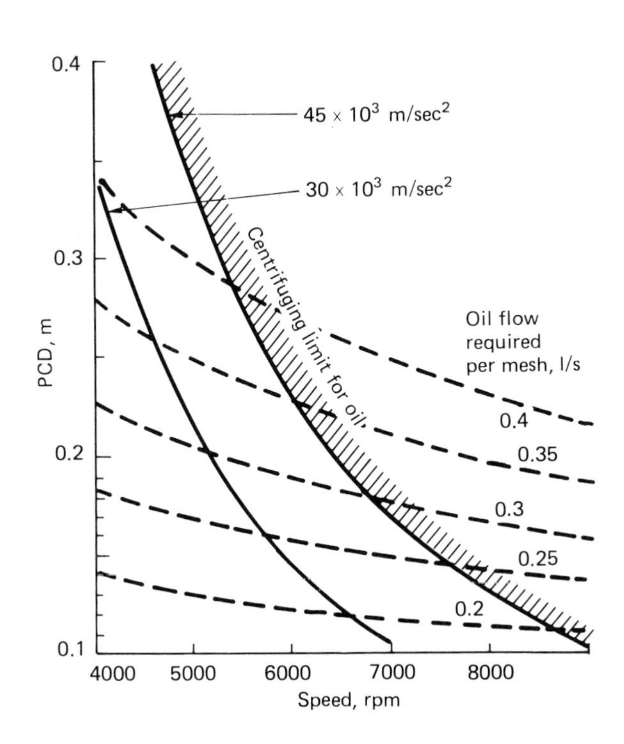

Figure 11.5 Lubrication requirements of gear couplings

Limits:

set by centrifuging of solids or sludge in oil causing coupling lock:

damless-type couplings	45×10^3 m/sec^2
dam-type couplings	30×10^3 m/sec^2

Lubricant feed rate:

damless-type couplings	Rate given on Figure 11.5
dam-type coupling with sludge holes	50% of rate on Figures 11.5
dam-type coupling without sludge holes	25% of rate on Figures 11.5

Lubricant:

Use oil from machine lubrication system (VG32, VG46 or VG68)

THE ADVANTAGES OF LUBRICATION

Increased fatigue life

Correct lubricants will facilitate individual wire adjustment to equalise stress distribution under bending conditions. An improvement of up to 300% can be expected from a correctly lubricated rope compared with a similar unlubricated rope.

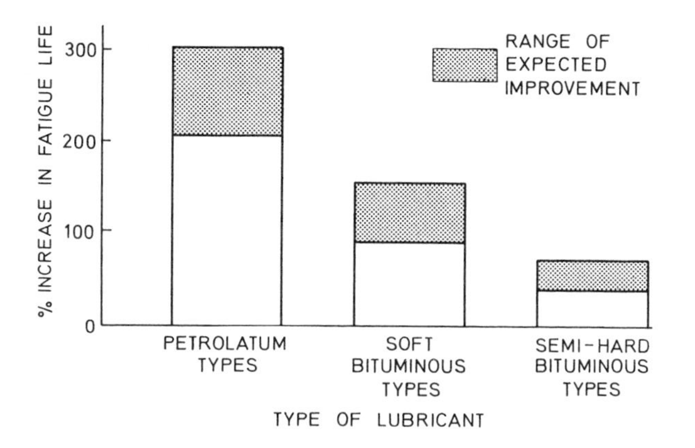

Figure 12.1 Percentage increases in fatigue life of lubricated rope over unlubricated rope

Increased corrosion resistance

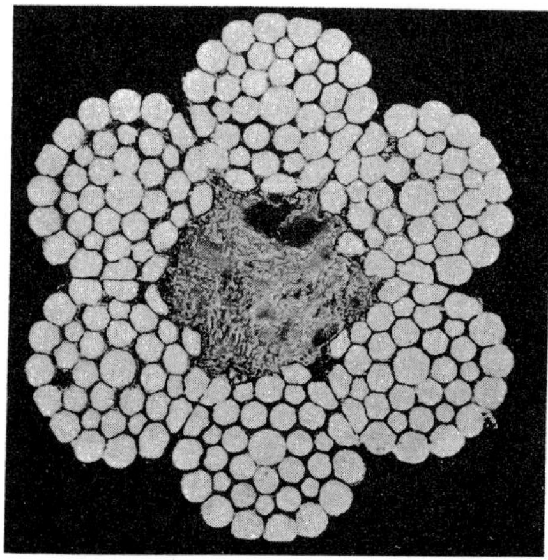

***Figure 12.2 Typical effect of severe internal corrosion**. Moisture has caused the breakdown of the fibre core and then attacked the wires at the strand/core interface*

Figure 12.3 Typical severe corrosion pitting associated with 'wash off' of lubricant by mine water

Increased abrasion resistance

Figure 12.4 Typical abrasion condition which can be limited by the correct service dressing

LUBRICATION DURING MANUFACTURE

The Main Core Fibre cores should be given a suitable dressing during their manufacture. This is more effective than subsequent immersion of the completed core in heated grease.

Independent wire rope cores are lubricated in a similar way to the strands.

The Strands The helical form taken by the individual wires results in a series of spiral tubes in the finished strand. These tubes must be filled with lubricant if the product is to resist corrosive attack. The lubricant is always applied at the spinning point during the stranding operation.

The Rope A number of strands, from three to fifty, will form the final rope construction, again resulting in voids which must be filled with lubricant. The lubricant may be applied during manufacture at the point where the strands are closed to form the rope, or subsequently by immersion through a bath if a heavy surface thickness is required.

Dependent on the application the rope will perform, the lubricant chosen for the stranding and closing process will be either a petrolatum or bituminous based compound. For certain applications the manufacturer may use special techniques for applying the lubricant.

Irrespective of the lubrication carried out during rope manufacture, increased rope performance is closely associated with adequate and correct lubrication of the rope in service.

LUBRICATION OF WIRE ROPES IN SERVICE

	(1)	(2)	(3)	(4)	(5)
Operating conditions	Ropes working in industrial or marine environments	Ropes subject to heavy wear	Ropes working over sheaves where (1) and (2) are not critical	As (3) but for friction drive applications	Standing ropes not subject to bending
Predominant Cause of rope deterioration	Corrosion	Abrasion	Fatigue	Fatigue—corrosion	Corrosion
Typical applications	Cranes and derricks working on ships, on docksides, or in polluted atmospheres	Mine haulage, excavator draglines, scrapers and slushers	Cranes and grabs, jib suspension ropes, piling, percussion and drilling	Lift suspension, compensating and governor ropes, mine hoist ropes on friction winders	Pendant ropes for cranes and excavators. Guys for masts and chimneys
Dressing requirements	Good penetration to rope interior. Ability to displace moisture. Internal and external corrosion protection. Resistance to 'wash off'. Resistance to emulsification	Good antiwear properties. Good adhesion to rope. Resistance to removal by mechanical forces	Good penetration to rope interior. Good lubrication properties. Resistance to 'fling off'	Non slip property. Good penetration to rope interior. Ability to displace moisture. Internal and external corrosion protection	Good corrosion protection. Resistance to 'wash off'. Resistance to surface cracking
Type of lubricant	Usually a formulation containing solvent leaving a thick (0.1 mm) soft grease film	Usually a very viscous oil or soft grease containing MoS_2 or graphite. Tackiness additives can be of advantage	Usually a good general purpose lubricating oil of about SAE 30 viscosity	Usually a solvent-dispersed temporary corrosion preventative leaving a thin, semi-hard film	Usually a relatively thick, bituminous compound with solvent added to assist application
Application technique	Manual or mechanical	Manual or mechanical	Mechanical	Normally by hand	Normally by hand
*Frequency of applications**	Monthly	Weekly	10/20 cycles per day	Monthly	Six monthly/ 2 years

* The periods indicated are for the general case. The frequency of operation, the environmental conditions and the economics of service dressing will more correctly dictate the period required.

APPLICATION TECHNIQUES

Ideally the lubricant should be applied close to the point where the strands of the rope tend to open when passing over a sheave or drum.

The lubricant may be applied manually or mechanically.

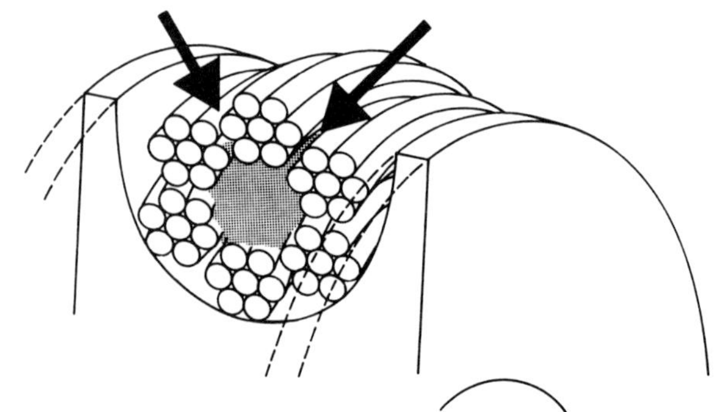

Figure 12.5 Opening of rope section during passage over sheave or drum. *Arrows indicate the access points for lubricant*

Manual – By can or by aerosol

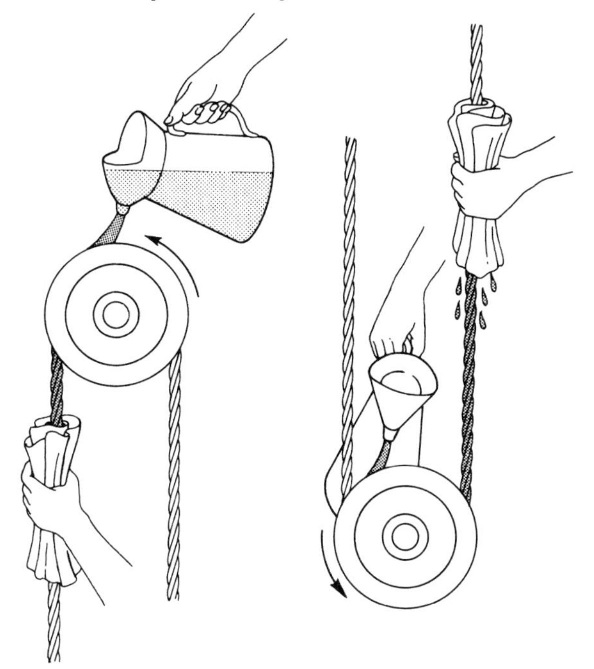

Figure 12,6 Manual application by can

Mechanical—By bath or trough.
By drip feed. By mechanical spray

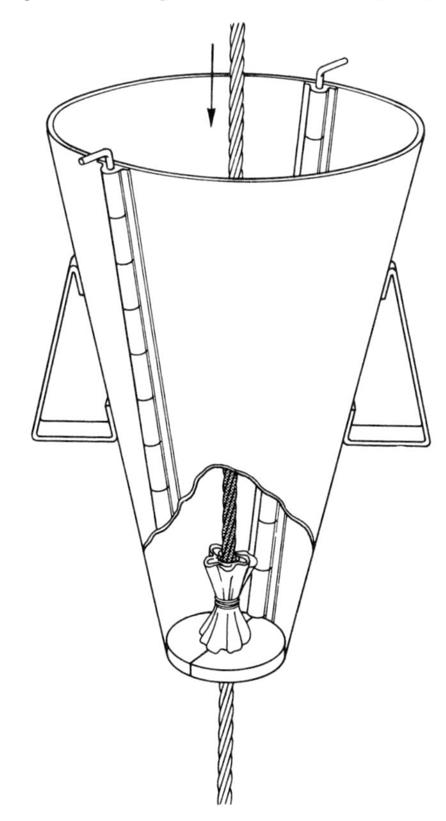

Figure 12.7 Mechanical application by trough

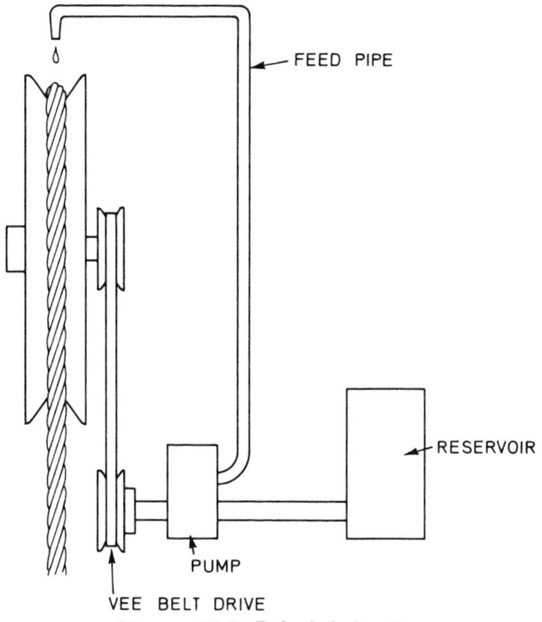

Figure 12.8 Drip lubrication

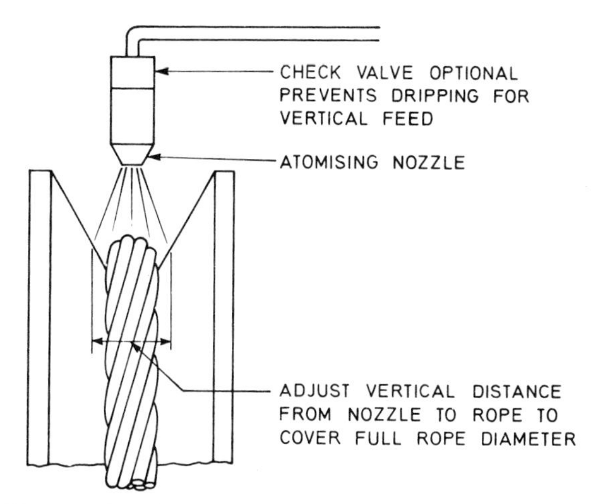

Figure 12.9 Sheave application by spray using fixed nozzle

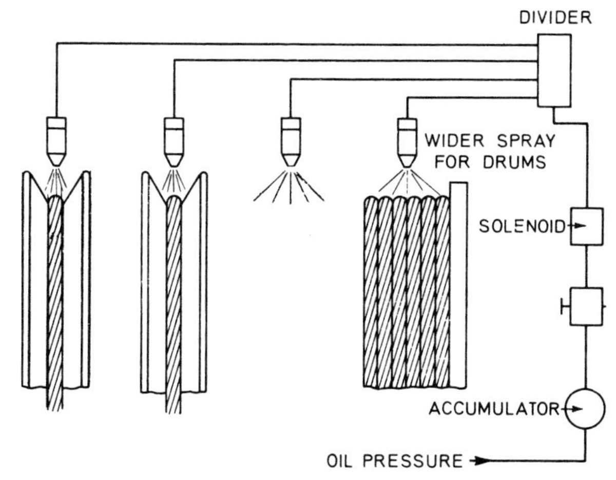

Figure 12.10 Multisheave or drum application by spray

C13 Selection of lubrication systems

For brevity and convenience the vast array of lubrication systems have been grouped under nine headings. These are each more fully discussed in other Sections of the Handbook.

Type	*Title*
A	Total loss grease
B	Hand greasing
C	Centralised greasing
D	Total loss oil
E	Wick/pad oil
F	Ring and disc oil
G	General mist and splash
H	Pressure mist
J	Circulating oil

TYPES OF LUBRICATION SYSTEM

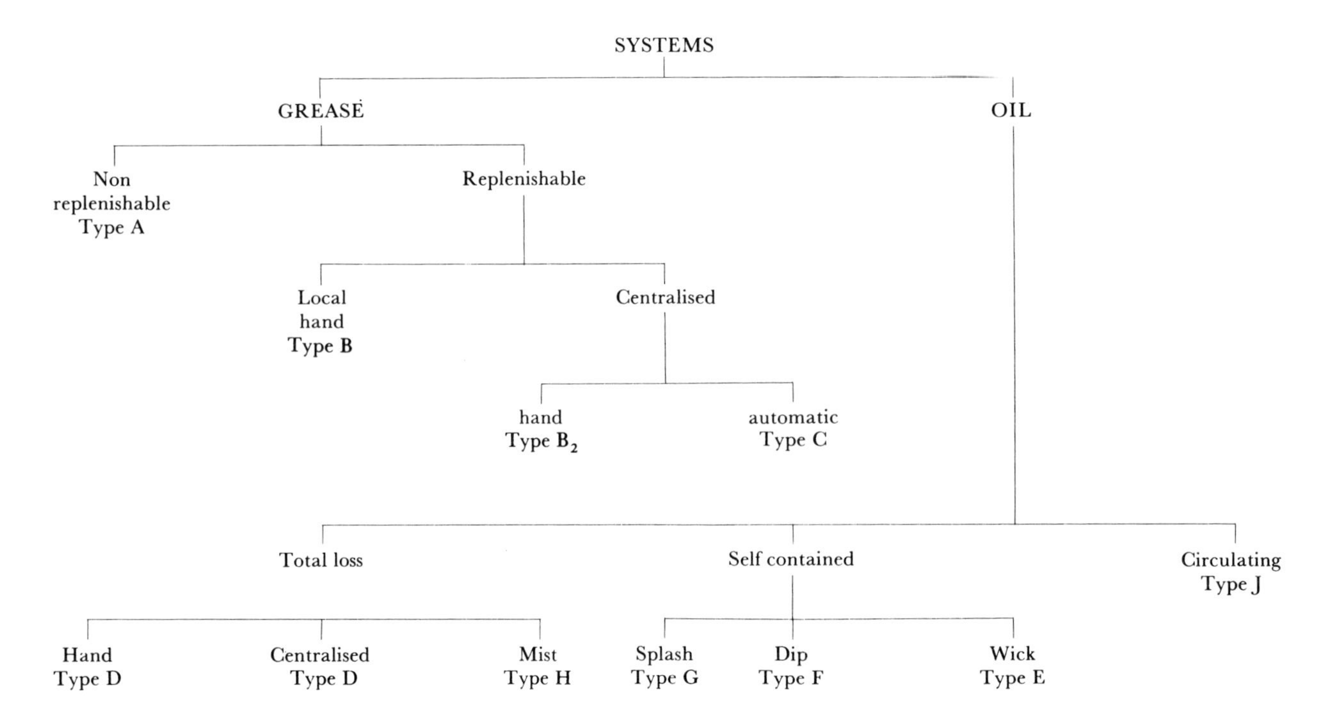

Type letters refer to subsequent tables

METHODS OF SELECTION

Table 13.1 Oil systems

	Type	Characteristics	Application
TOTAL LOSS	Hand Type D	Oil can	Simple bearings, Small numbers, low duty, easy access
TOTAL LOSS	Centralised Type D	Intermittent feed	Small heat removal, difficult access, large numbers
TOTAL LOSS	Mist Type H	Aerosol spray.	Rolling bearings, mechanisms
SELF CONTAINED	Splash Type G	Lubrication by mist	Enclosed mechanisms, gearboxes
SELF CONTAINED	Dip Type F	Ring or disc systems. Oil circulates	Plain bearings—slow or moderate duty
SELF CONTAINED	Wick Type E	Pad or wick feed from reservoir—total loss or limited circulation	Plain bearings, low duty. Light loaded mechanisms
	Circulating Type J	Oil from tank or sump fed by pressure pump to bearings or sprays	Almost all applications where cost is justified

Table 13.2 Grease systems

Type	Characteristics	Applications
Non replenishable Type A	Bearings packed on assembly. Refilling impossible without stripping	Rolling bearings, some plastic bushings
Local—hand Type B_1	Grease nipple to each bearing	Small numbers, easy, access, cheap
Centralised—hand Type B_2	Feed pipes brought to manifold or pump	Reasonable numbers. Inaccessible bearings
Centralised—automatic Type C	Grease pump feed to bearing and sets of bearings from automatic pump	Large numbers, important bearings, great distances. Where frequent relubrication is required

Table 13.3 Relative merits of grease and oil systems

Component or performance requirement	Grease	Oil
Removing heat	No	Yes. Amount depends on system
Keep out dirt	Yes. Forms own seal	No. But total loss systems can flush
Operate in any attitude	Yes. Check design of grease valves	No. Unless specially designed
Plain bearings	Limited (slow speed or light load)	Extensive
Rolling bearings	Extensive	Extensive
Sealed rolling bearings	Almost universal	No
Very low temperatures	No. Except for some special greases	Yes (check pour point of oil)
Very high temperatures	No. Except for some special greases	Yes. With correct system
Very high speeds	No. Except small rolling bearings	Yes
Very low or intermittent speeds	Yes	Hydrodynamic bearings—limited. Hydrostatic bearings—yes
Rubbing plain bearings	Yes	Yes. With limited feed

Table 13.4 Selection by heat removal

System	Effect
Total loss grease or oil	Will not remove heat
Dip or splash	Will remove heat from hot spots. Will remove some heat from the body of the components
Mist	Will remove considerable heat
Recirculating. Oil	May be designed to remove almost any amount of heat

Table 13.5 Selection by type of component to be lubricated

System		*Rolling bearings*	*Fluid film plain bearings*	*Rubbing plain bearings Gears*
Non-replenishable grease Type A	For general use	Very light duty	General	Light duty or slow speeds
Hand grease feed Type B	Heavier loads. Higher temperatures	Slow speed	General for high loads and temperatures	Rarely
Centralised grease feed Type C	Heaviest loads. Higher temperatures. Large sizes	Slow speed. Higher temperatures	General for high loads and temperatures	Slow, heavy duty
Total loss oil Type D	Most applications	All where small heat removal	Yes	Open gears
Wick/Pad, oil Type E	Limited (light duty)	Slow speed	Some	Small gears
Ring or disc, oil Type F	NA	Medium duty	NA	NA
General mist and splash Type G	Most	Light duty	Light duty	Wide
Pressure mist Tvpe H	Almost all. Excellent	Rare	NA	Rare
Circulating oil Type J	All	All	NA	All

Table 13.6 Selection by economic considerations

Lubrication system	*Initial purchase*	*Maintenance of system*	*First fill*	*Subsequent lubricant costs*	*Subsequent labour costs*	*Notes*
Non-replenishable grease Type A	Very cheap	Nil	Cheap	Nil	Nil	Life of bearing is life of lubricant. Expensive if relubricated
Grease feed hand Type B	Cheap	Cheap	Cheap moderate if long liner	Moderate but can be expensive	Expensive	Regular attendance is vital. Neglect can be very expensive
Grease feed automatic Type C	Moderate to expensive	Moderate	Moderate to expensive	Moderate	Moderate	Needs comparatively skilled labour. Costs increase with complications
Total loss oil Type D	Cheap	Cheap	Cheap	Moderate but can be expensive	Moderate	Periodic refilling required. Neglect can be very expensive
Wick or pad Type E	Cheap	Cheap	Cheap	Cheap	Cheap	Also need topping-up but not so often and wick gives some insurance
Ring or disc Type F	Cheap	Nil to cheap	Cheap	Cheap	Cheap	Needs very little attention
General splash Type G	Cheap	Nil	Moderate	Cheap to moderate	Moderate	Oil level must be watched
Pressure mist Type H	Moderate to expensive	Expensive	Cheap	Cheap	Moderate to expensive	Needs comparatively skilled labour. Requires compressed air supply
Circulating oil Type J	Expensive	Expensive	Moderate to expensive	Cheap	Moderate	Simple system requires little attention. Costs increase with complications

Table 13.7 General selection by component. Operating conditions and environment

Type of component	*High temp. (over 150°C)*	*Normal temp. (−10 to 120°C)*	*Low temp. (below −20°C)*	*High speed*	*Medium speed*	*Low speed*	*Dust and dirt*	*Wet and humid*	*Vacuum (special lubricants)*
Rolling bearings	A (special grease), B, C, H*, J*	A, B, C, D, G, H*, J*	A*, D, G, H*, J	A (small), D, H*, J	A, B, C, D, E, H*, J*	A, B, C*, D, E, G, H, J	A, B, C*, D, J	A, B, C*, J	A*, E, Also dry
Fluid film, plain bearings	D, H, J*	A (slow), B, C, D, E, F, G, J	A (slow), D, E, F, G, J	D, G (small sizes), H (small), J	D, E, F, G, H, J*	A (light), B, C, D, E, F, G, J	A, B, C*, D, J	A, B, C, J*	A*, E, J (possible)
Rubbing plain bearings	B, C*, J*	A, B, C*, D, E, F	A, D, G	—	—	A, B, C*, D, E, G, Also dry	A, B, C*, D, Also dry*	A, B, C*, Also dry	A*, Also dry
Slides	C*, G, J*	A, B*, C, D, E, J	A, D, E, G, J	A (light), B, C, D, G, J	A, B, C, D, E (small), F, G, J*	A, B, C*, D, E, J	A, B, C*, D	A, B, C*, D, J	A*, E
Screws	B, G, J*, C	A, B*, C, D, E, G, J	A, D, G, J	A (light), D, G, H, J	A (light), B, C, D, G, H, J	A, B, C*, D, G, J	A, B, C*	A, B, C*, D, E, J	A*, E
Gears	B, C, G, J*	A, B, C, E, H, J*	A, D, G, H, J	D, G, H, J*	A—V (small), C, D, G, H, J	A, B, C, D, E, G, J	A (special greases), J	A, B, C, D, J	A*

* Preferred systems.

Selection of gear lubrication systems

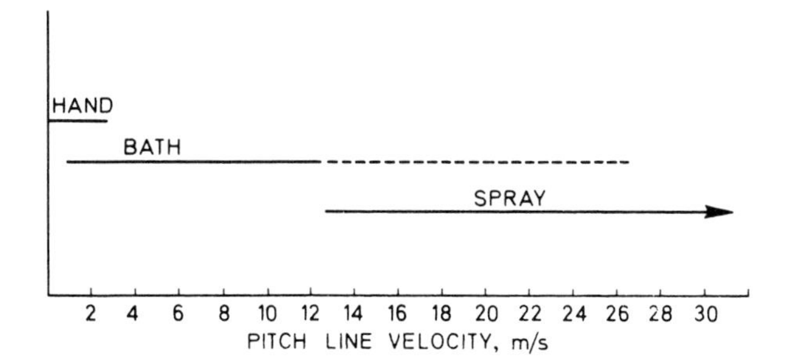

C14 Total loss grease systems

TYPES OF TOTAL-LOSS GREASE SYSTEMS AVAILABLE

Description	*Diagram*	*Operation*	*Drive*	*Suitable grease NLGI No.*	*Typical line pressures*	*Adjustability and typical limiting pipe lengths**
DIRECT FEED Individual piston pumps	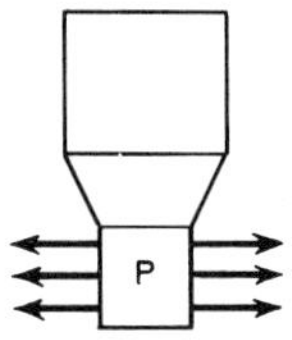	Rotating cam or swash plate operates each piston pump in turn	Motor Machine Manual	0–2	700–2000 kN/m^2 (100–300 lbf/in^2)	By adjustment of stroke at each outlet 9–15 m (30–50 ft)
Distributing piston valve system	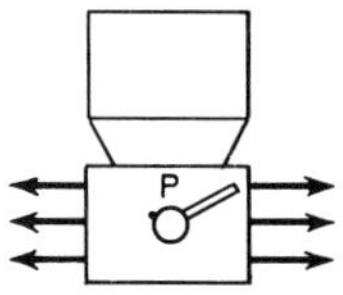	Valve feeds output of a single piston pump to each line in turn	Motor Machine Manual	0–3 0–2	700–2000 kN/m^2 (100–300 lbf/in^2)	None. Output governed by speed of pump 25–60 m (80–200 ft)
Branched system	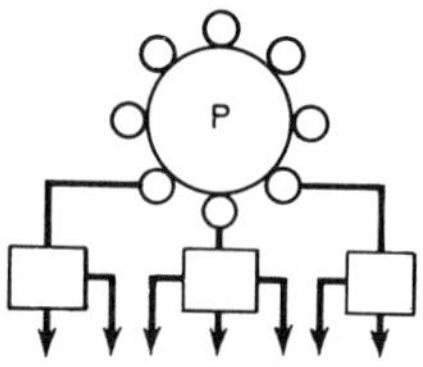	Outputs of individual pumps split by distributing valves	Motor Machine	0–3	700–2800 kN/m^2 (100–400 lbf/in^2)	Adjustment at each outlet/meter block 18–54 m (60–180 ft) } pump to divider 6–9 m (20–30 ft) } divider to bearing
INDIRECT FEED PROGRESSIVE Single line reversing	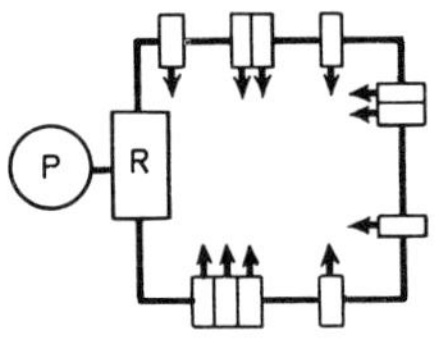	Valves work in turn after each operation of reversing valve *R*				
Single line	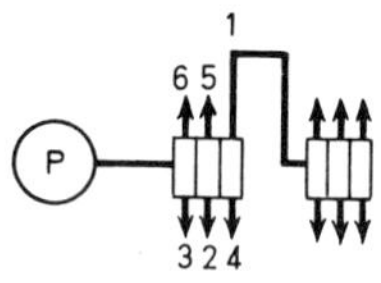	First valve block discharges in the order 1, 2, 3, . . ., etc. One unit of this block is used as a master to set the second block in operation. Second and subsequent blocks operate sequentially	Motor Machine Manual	0–2	14 000–20 000 kN/m^2 (2000–3000 lbf/in^2)	Normally none. Different capacity meter valves available—otherwise adjustment by time cycle Main lines up to 150 m (500 ft) depending on grease and pipe size. Feeder lines to bearings 6–9 m (20–30 ft)
Double line	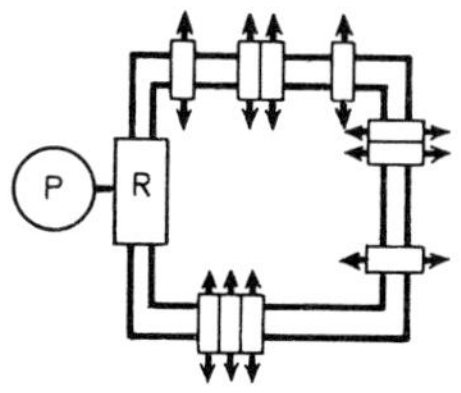	Grease passes through one line and operates half the total number of outlets in sequence. Valve *R* then operates, relieves line pressure and directs grease to the other line, operating remaining outlets				

Description	Diagram	Operation	Drive	Suitable grease NLGI No.	Typical line pressures	Adjustability and typical limiting pipe lengths*
INDIRECT FEED PARALLEL Single line	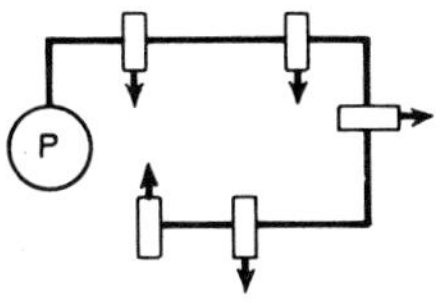	The line is alternately pressurised and relieved by a device on the pump. Two variations of this system exist in which 1 injection is made by the line pressure acting on a piston within the valves 2 injection is made by spring pressure acting on a piston within the valves	Motor Manual	0–1	Up to 17 000 kN/m² (2500 lbf/in²) Up to 8000 kN/m² (1200 lbf/in²)	Operation frequency adjustable (some makes). Output depends on nature of grease 120 m (400 ft)
Single line oil or air actuated	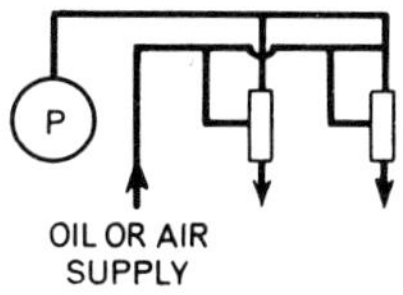 OIL OR AIR SUPPLY	Pump charges line and valves. Oil or air pressure operates valves	Motor with cycle timer	0–3	Up to 40 000 kN/m² (6000 lbf/in²)	Full adjustment at meter valves and by time cycle 600 m (2000 ft)
Double line	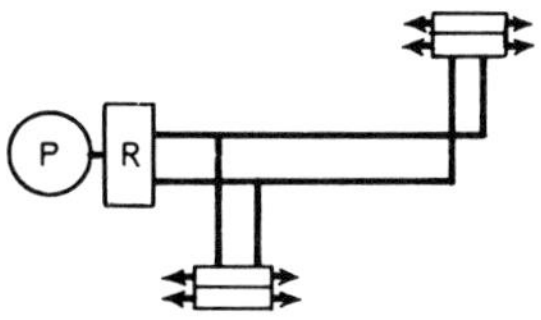	Grease pressure in one line operates half the total number of outlets simultaneously. Valve *R* then relieves line pressure and directs grease to the other line, operating the remaining outlets	Motor Manual	0–2	Up to 40 000 kN/m² (6000 lbf/in²)	As above. 60 m (200 ft) } Manual 120 m (400 ft) } Automatic

* Lengths shown may be exceeded in certain specific circumstances.

Considerations in selecting type of system

	Required performance of system	
Type of bearing	*Grease supply requirements*	*Likely system requirements*
Plain bearings, high loads	Near-continuous supply	Direct feed system
Rolling bearings	Intermittent supply	Indirect feed

PIPE-FLOW CALCULATIONS

To attempt these it is necessary that the user should know:

(*a*) The relationship between the apparent viscosity (or shear stress) and the rate of shear, at the working temperature;

(*b*) The density of the grease at the working temperature.

This information can usually be obtained, for potentially suitable greases, from the lubricant supplier in graphical form as below (logarithmic scales are generally used).

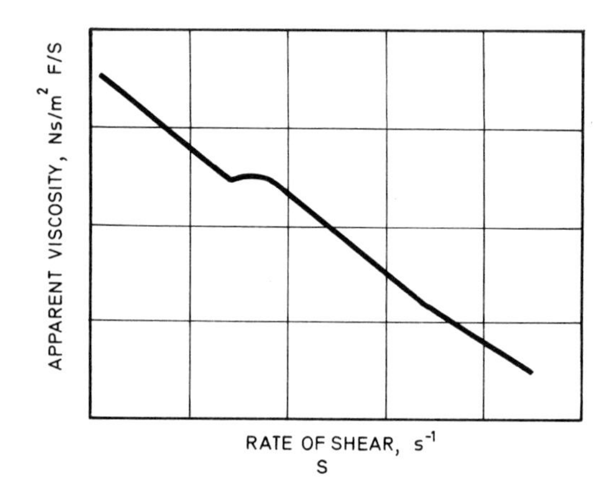

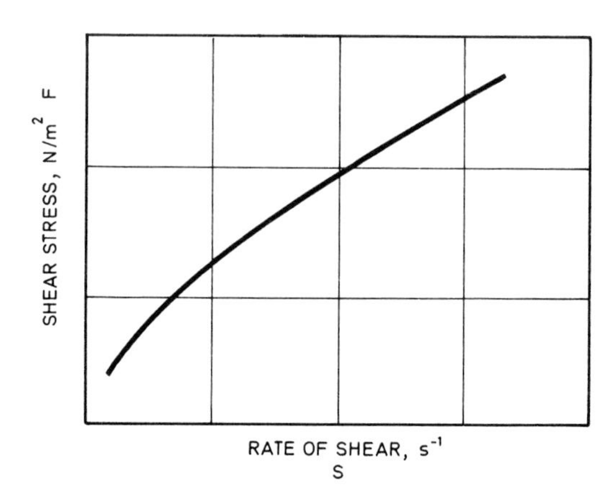

Given	*To estimate*	*Procedure*
Permissible pressure loss P/L, over a known length of pipe	Pipe inside diameter D, to give a certain rate of flow Q	(*a*) Calculate the value of $2.167\sqrt[3]{Q}\,P/L$ (*b*) Plot graph of $(\sqrt[3]{S})\,F$ against F from supplier's data (if convenient on log–log paper) (*c*) Find point on $(\sqrt[3]{S})\,F$ scale equivalent to the value found in (*a*) above (*d*) Note value of F (*e*) Pipe diameter $D = 4LF/P$
Available pumping pressure P, length L and inside diameter D of pipe	Rate of grease flow Q	(*f*) Plot curve F against S from manufacturers' data (*g*) Calculate $F = PD/4L$ from the given P, D and L conditions (*h*) Find corresponding value of S from curve (*i*) Then $Q = \pi D^3 S/32$
Rate of flow Q and inside diameter D of pipe to be used	Pressure gradient P/L to sustain flow rate Q	(*j*) Calculate $S = \dfrac{32Q}{\pi D^3}$ for given conditions (*k*) Find corresponding value of S from the graph of F against S (*l*) Then $P/L = 4F/D$

Notes:

1 These formulae are correct for SI units, P in N/m², L in m, Q in m³/s. F in N/m², S in S^{-1}.

2 D is inside diameter.

Typical pipe sizes used in grease systems

Material	Standard No.	Bore mm	Bore in	Wall thickness mm	Wall thickness in	Outside diameter mm	Outside diameter in
STEEL		**15***	$\frac{1}{2}$	3.25	0.128	21.5	0.756
		20	$\frac{3}{4}$	3.25	0.128	26.5	1.006
Heavyweight grade	ISO 65	**25**	**1**	4.05	0.160	33.1	1.320
		40	**1**	4.05	0.160	48.1	1.820
		50	**2**	4.50	0.176	59.0	2.350
Cold drawn, seamless, fully annealed	ISO 3304	3.35	0.132	0.70	0.028	4.80	$\frac{3}{16}$
		4.50	0.178	0.90	0.036	6.40	$\frac{1}{4}$
		6.10	0.240	0.90	0.036	7.90	$\frac{5}{16}$
		7.10	0.279	1.22	0.048	9.50	$\frac{3}{8}$
		10.30	0.404	1.22	0.048	12.70	$\frac{1}{2}$
COPPER	ISO 196	3.35	0.132	0.70	0.028	4.80	$\frac{3}{16}$
		4.50	0.178	0.70	0.028	6.40	$\frac{1}{4}$
As drawn (M) quality		6.10	0.240	0.90	0.036	7.90	$\frac{5}{16}$
Annealed (O) quality		7.10	0.303	0.90	0.036	9.50	$\frac{3}{8}$
	ISO 274	2.80	0.110	0.60	0.024	**4**	0.158
		3.40	0.130	0.80	0.032	**5**	0.197
		4.00	0.160	1.00	0.039	**6**	0.236
		6.00	0.240	1.00	0.039	**8**	0.315
		7.60	0.300	1.20	0.047	**10**	0.394
BRASS	ISO 196	3.35	0.132	0.70	0.028	4.8	$\frac{3}{16}$
		4.50	0.178	0.90	0.036	6.4	$\frac{1}{4}$
Half-hard or normalised		6.10	0.240	0.90	0.036	7.9	$\frac{5}{16}$
		7.10	0.279	1.22	0.048	9.5	$\frac{3}{8}$
NYLON	ISO 7628	3.10	0.122	0.80	0.033	4.8	$\frac{3}{16}$
		4.10	0.165	1.10	0.043	6.4	$\frac{1}{4}$
		5.40	0.213	1.30	0.050	7.9	$\frac{5}{16}$
		6.20	0.245	1.80	0.070	9.5	$\frac{3}{8}$
		2.40	0.090	0.80	0.032	**4**	0.158
		3.30	0.130	0.85	0.034	**5**	0.197
		4.00	0.160	1.00	0.039	**6**	0.236
		5.40	0.210	1.30	0.051	**8**	0.315
		6.50	0.260	1.75	0.069	**10**	0.394

* Figures in bold type indicate nominal bore or outside diameters used in ordering.

Typical data for flexible hoses used in grease systems

Type		Sizes mm	Sizes in		Typical working pressures at 20°C kN/m²	Typical working pressures at 20°C lbf/in²	Sizes mm	Sizes in		Typical working pressures at 20°C kN/m²	Typical working pressures at 20°C lbf/in²
SYNTHETIC RUBBER Reinforced with cotton, terylene or wire braid or spiral wrapped	Single-wire braid	6	$\frac{1}{4}$	Nominal bore	23 000	3300	50	2	Nominal bore	3500	500
	Multi-spiral wrapped				70 000	10 000				27 000	4000
PLASTICS Simple extruded for low pressure to braided high pressure	High pressure single braid double braid				24 500 28 000	3500 4000	12.5	$\frac{1}{2}$	Nominal bore	17 500 35 000	2500 5000
	Unreinforced	4	$\frac{1}{8}$	o.d.	3500	500	28	$1\frac{1}{8}$	o.d.	1750	250
POLYMER AND METAL Inner core of polymeric material. Stainless steel outer braid		3	$\frac{1}{8}$	Nominal bore	7000	1000					

CONSIDERATIONS IN STORING, PUMPING AND TRANSMITTING GREASE AND GENERAL DESIGN OF SYSTEMS

	Considerations
Storing grease	TEMPERATURE at which grease can be pumped CLEANLINESS of grease PACKAGE sizes available. Ease of handling RESERVOIRS: Exclusion of dirt; internal coating; positive pump prime; provision of strainer, follower plate and scraper; pressure relief; level indicator; emptying and cleaning access; connections for remote feed
Pumping grease	ACCESSIBILITY of pump for maintenance CORRECT ALIGNMENT of pump and motor COUPLING between pump and motor to be adequate size for horse power transmitted MOUNTING to be rigid and secure ACTUATING ROD (if any) to be free from deflection NAME PLATE on pump to be clearly visible and kept clean DIRECTION of rotation to be marked
Transmitting grease	PIPE SIZING: Pressure in lines; length of runs; volume of lubricants; number of bends; temperature fluctuation
Work conditions	INGRESS of water, scale, dust POSSIBLE DAMAGE to piping ACCESS to piping in onerous locations
Human factors	AVAILABILITY of labour RELIABILITY of labour Trade union practices
General design points	CORRECT positioning of lubrication points PRODUCTION requirements QUALITY OF PIPING and materials— use only the best AVOID splitting unmetered grease lines ENSURE that grease holes and grooves communicate directly with faces to be lubricated

GENERAL

Most total loss systems available from manufacturers are now designed to deliver lubricants ranging from light oils to fluid greases of NLGI 000 consistency.

Fluid grease contains approximately 95% oil and has the advantage of being retained in the bearing longer than oil, thus reducing the quantity required whilst continuing to operate satisfactorily in most types of system.

The main applications for total loss systems are for chassis bearings on commercial vehicles, machine tools, textile machinery and packaging plant.

Because of the small quantity of lubricant delivered by these systems, they are not suitable for use where cooling in addition to lubrication is required, e.g. large gear drives.

Fluid grease is rapidly growing in popularity except in the machine tool industry where oil is preferred.

All automatic systems are controlled by electronic or electric adjustable timers, with the more sophisticated products having the facility to operate from cumulated impulses from the parent machine.

Individual lubricant supply to each bearing is fixed and adjustment is effected by changing the injector unit. However, overall lubrication from the system is adjusted by varying the interval time between pump cycles.

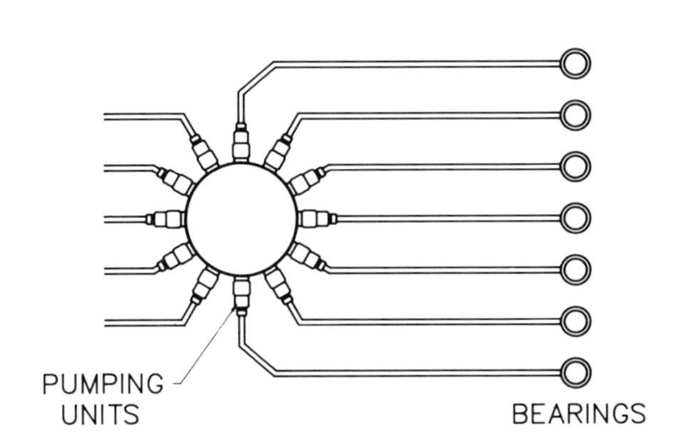

Figure 15.1 Schematic

Multi-outlet – electric or pneumatic

Operation. An electric or pneumatic motor drives cam-operated pumping units positioned radially on the base of the pump. The pump is cycled by an adjustable electronic timer or by electrical impulses from the parent machine, e.g. brake light operations on a commercial vehicle.

Individual 4 mm OD nylon tubes deliver lubricant to each bearing.

Applications: Commercial vehicles, packaging machines and conveyors.

Specification:

Outlets: 1–60 (0.01–1.00 ml)
Pressure: To 10 MN/m^2
Lubricants: 60 cSt oil to NLGI 000 grease (NLGI 2 pneumatic).
Failure warning: Pump operation by light or visual movement.
Cost factor: Low (electric), Medium (pneumatic).

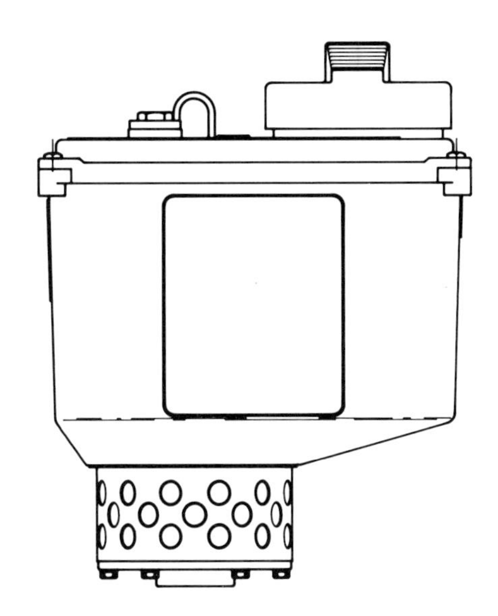

Figure 15.2 Pump

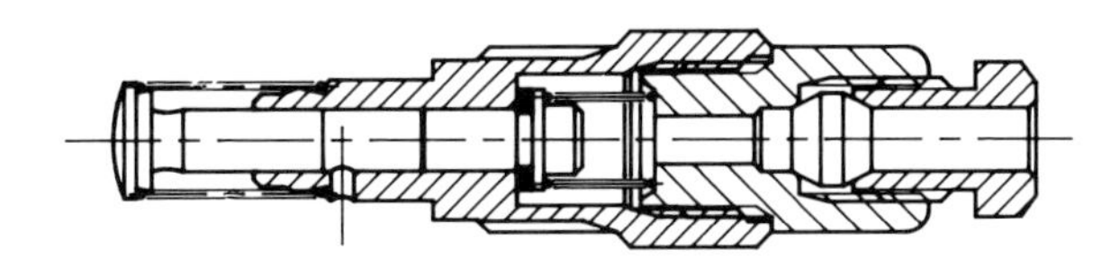

Figure 15.3 Pumping unit

Single line – volumetric injection

Operation. The pump delivers lubricant under pressure to a single line main at timed intervals. When the pressure reaches a predetermined level, each injector or positive displacement unit delivers a fixed volume of lubricant to its bearing through a tailpipe.

When full line pressure has been reached the pump stops and line pressure is reduced to a level at which the injectors recharge with lubricant ready for the next cycle.

Pumps are generally electric gear pumps or pneumatic piston type. All automatic systems are controlled by adjustable electronic timers but hand operated pumps are available.

Main lubricant pipework is normally in 6 or 12 mm sizes and tailpipes in 4 or 6 mm depending the size of system.

Applications: All types of light to medium sized manufacturing plant and commercial vehicles.

Specification:

Outlets: 1–500 (0.005–1.5 ml)
Pressure: 2–5 MN/m^2
Lubricants: 20 cSt oil to NLGI 000 grease.
Failure warning: Main line pressure monitoring.
Cost factor: Medium

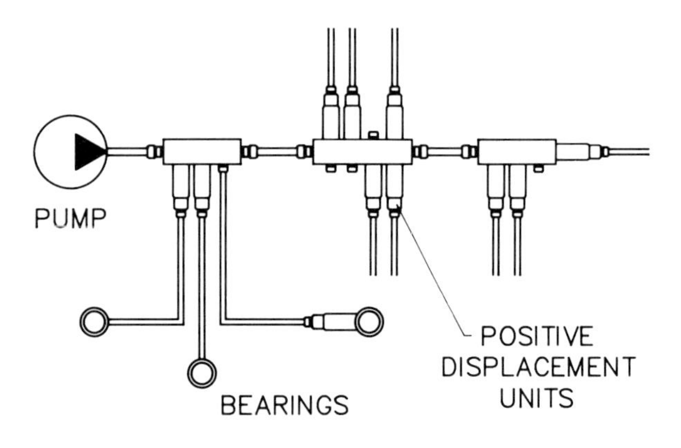

Figure 15.4

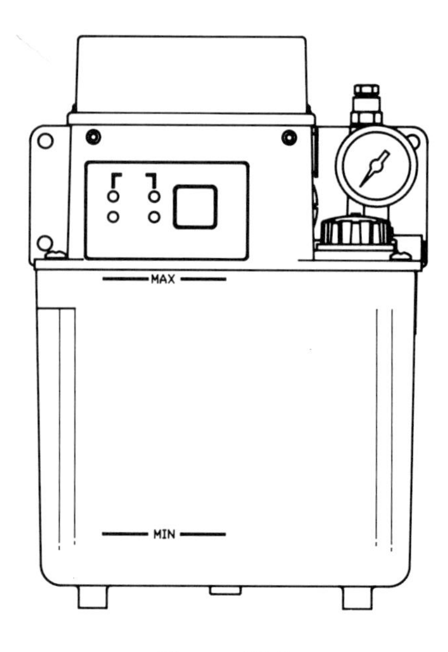

Figure 15.5

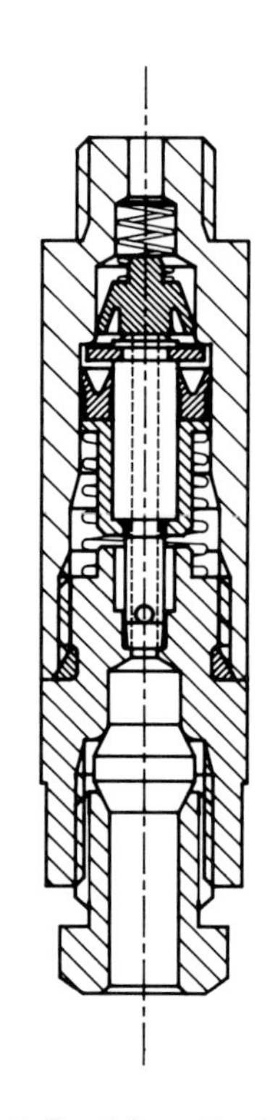

Figure 15.6 Positive displacement unit

Single line-resistance (oil only)

Operation: A motor driven piston pump discharges a predetermined volume of oil at controlled intervals to a single main line. Flow units in the system proportion the total pump discharge according to the relative resistance of the units.

Care needs to be taken in the selection of components to ensure each bearing receives the required volume of lubricant; as a result, this type of system is used predominantly for original equipment application.

Applications: Machine tools and textile machinery.

Specification:

Outlets: 1–100 (0.01–1 ml)
Pressure: 300–500 kN/m^2
Lubricants: 10–1800 cSt (Oil only)
Failure warning: line pressure monitoring.
Cost factor: Low.

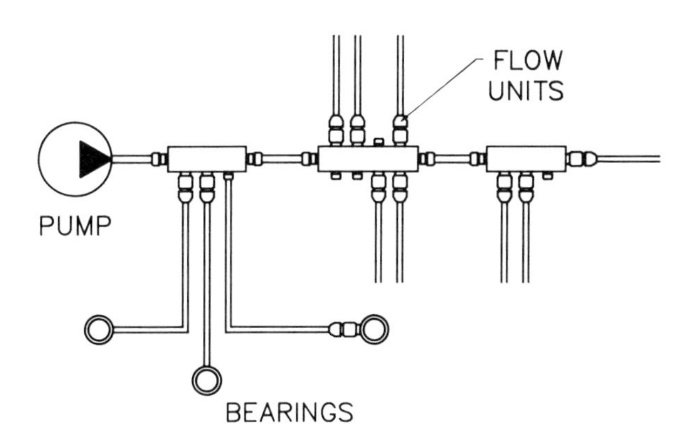

Figure 15.7 Schematic

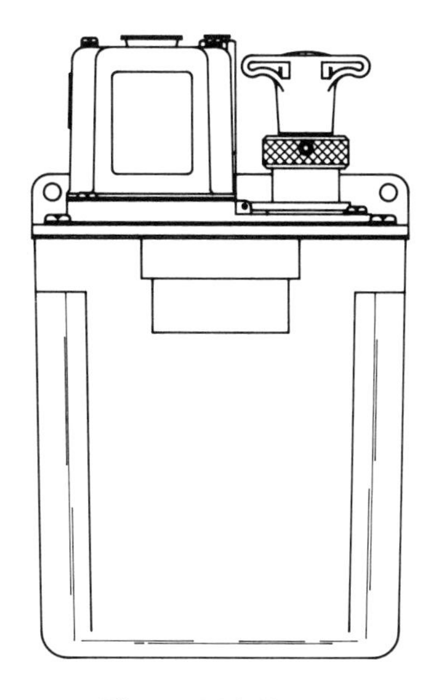

Figure 15.8 Pump

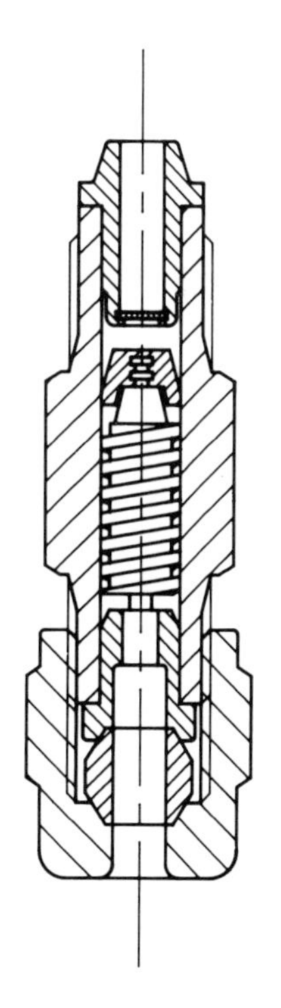

Figure 15.9 Flow unit

Single line progressive

Operation: An electric or pneumatic piston pump delivers lubricant on a timed or continuous basis to a series of divider manifolds. The system is designed in such a way that if a divider outlet fails to operate, the cycle will not be completed and a warning device will be activated. Due to this feature, the system is widely used on large transfer machines in the automobile industry.

Applications. Machine tools and commercial vehicles.

Specification:

Outlets: 200 (0.01–2 ml).
Pressure: 10 MN/m^2
Lubricants: 20 cSt oil to NLGI 2 grease.
Failure warning: Failure of individual injector can activate system alarm.
Cost factor: Medium to high.

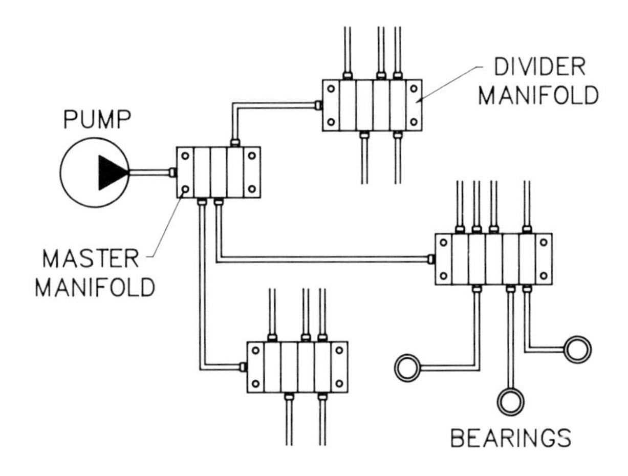

Figure 15.10 Schematic

CHECK LIST FOR SYSTEM SELECTION AND APPLICATION

Most economic method of operation – electric, pneumatic, manual, etc.

Cost installed – Max 2% of parent machine.

Degree of failure warning required – warning devices range from low lubricant level to individual bearing monitoring. Cost of warning devices must be balanced against cost of machine breakdown.

Check pressure drop in main lines of single line systems. Large systems may require larger pipe sizes.

Ensure adequate filtration in pump – Some types of systems are more sensitive to dirt in the lubricant

Is the system protected against the operating environment? e.g.: High/low temperature, humidity, pressure steam cleaning, vibration/physical damage, electrical interference etc.

If using flexible tubing, is it ultra violet stabilised?

If air accidentally enters the system, is it automatically purged without affecting the performance?

Check for overlubrication – particularly in printing and packaging applications.

Is the lubricant suitable for use in the system? (Additives or separation.)

Reservoir capacity adequate? Minimum one month for machinery, three months for commercial vehicles.

Accessibility of indicators, pressure gauges, oil filler caps, etc.

Should system be programmed for a prelube cycle on machine start up?

SPUR, BEVEL AND HELICAL GEARS

All gears, except very slow running ones, require complete enclosure. In general, gears dip into oil for twice tooth depth, to provide sufficient splash for pinions, bearings, etc. and to reduce churning loss to a minimum.

Typical triple reduction helical gear unit

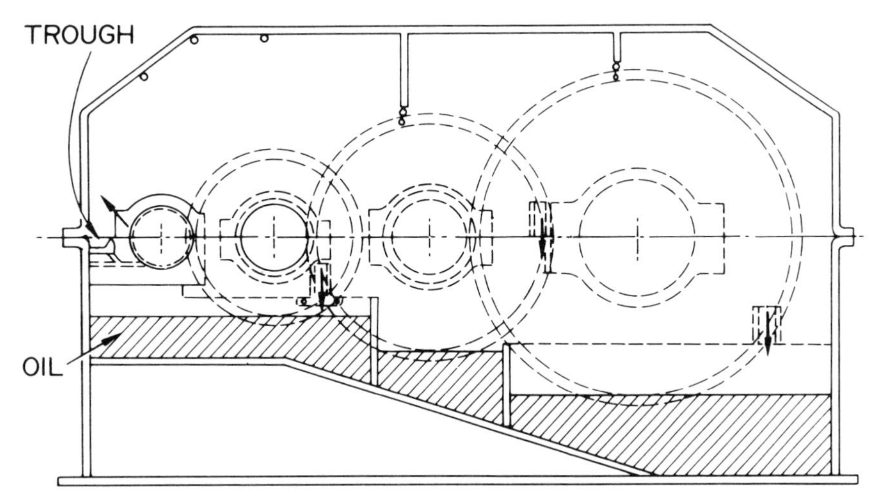

Figure 16.1

This has guards and tanks for individual gears. Bearings are fed by splash and from a trough round the walls of the case. Suitable for up to 12.5 m/s (2500 ft/min) peripheral speed.

Typical single reduction bevel gear unit

Suitable for up to 12.5 m/s (2500 ft/min).

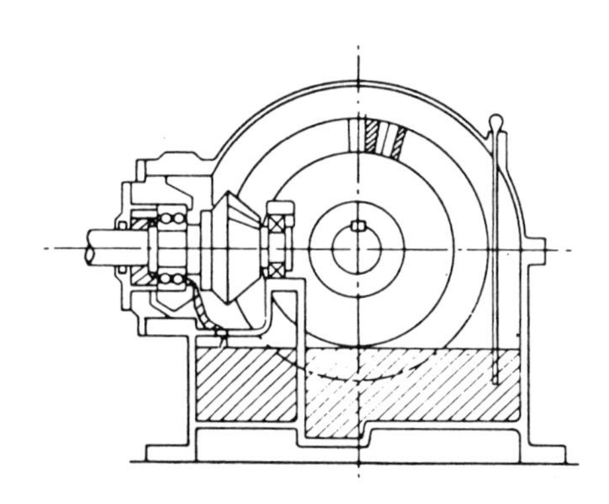

Figure 16.2 Bevel unit with double row bearings on pinion shaft

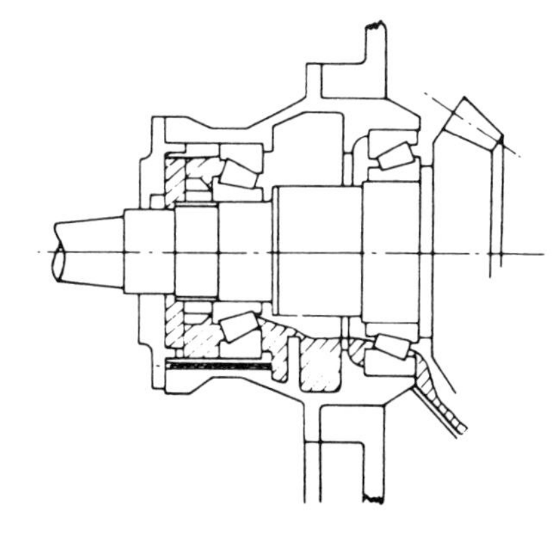

Figure 16.3 Typical bearing lubrication arrangement with taper roller bearings

Typical high-speed gear unit guard

Normally satisfactory up to 25 m/s (5000 ft/min). With special care can be used up to 100 m/s (20 000 ft/min).

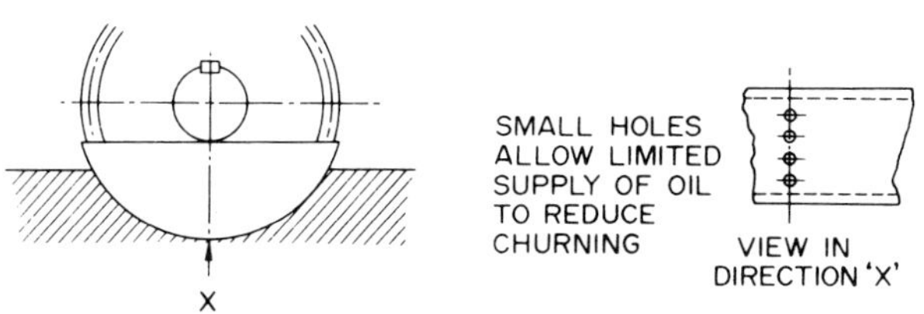

Figure 16.4

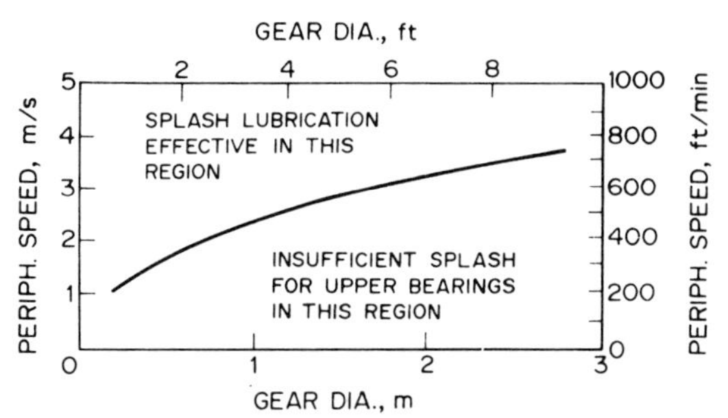

Figure 16.5 Peripheral speed against gear diameter for successful splash lubrication to upper bearings in dip-lubricated gear units

WORM GEARS

Typical under-driven worm gear unit

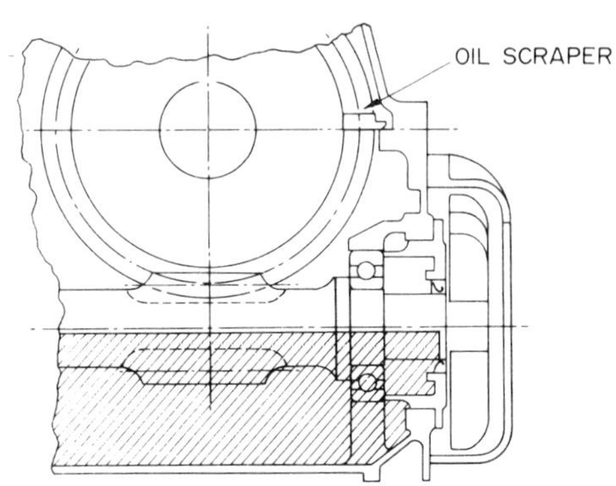

Figure 16.6

Oil is churned by the worm and thrown up to the top and sides of the case. From here it drips down via the wheel bearings to the sump.

A simple lip seal on a hard, ground shaft surface, prevents leakage.

Oil level generally just below worm centre-line.

An oil scraper scrapes oil into a trough to feed the wheel bearings.

Typical over-driven worm gear unit

Similar to under-driven worm gear unit except that the worm is over the wheel at the top of the unit, and the oil level varies in depth from just above wheel tooth depth to almost up to the centre line of the wheel, depending upon speed. The greater the speed, the higher the churning loss, therefore the lower should be the oil level. At low speeds, the churning loss is small and a large depth of oil ensures good heat-transfer characteristics.

GENERAL DESIGN NOTES

Gears

In dip-splash systems, a large oil quantity is beneficial in removing heat from the mesh to the unit walls and thence to the atmosphere.

However, a large quantity may mean special care has to be paid to sealing, and churning losses in gears and bearings may be excessive. It is necessary to achieve a balance between these factors.

Other applications

The cylinders and small-end bearings of reciprocating compressors and automotive internal combustion engines are frequently splash lubricated by oil flung from the rotating components. In these applications the source of the oil is usually the spill from the pressure-fed crankshaft bearings. In some small single-cylinder compressors and four-stroke engines, the cap of the connecting rod may be fitted with a dipper which penetrates up to 10 mm into the oil in the sump and generates splash lubrication as a result. In lightly loaded applications the big-end bearings may also be splash lubricated in this way, and in some cases the dipper may be in the form of a tube which scoops the oil directly into the big-end bearing. In small domestic refrigeration compressors, a similar system may also be used to scoop oil into the end of the crankshaft, in order to lubricate all the crankshaft bearings.

Mist systems, generically known as aerosol systems, employ a generator supplied with filtered compressed air from the normal shop air main, to produce a mist of finely divided oil particles having little tendency to wet a surface. The actual air pressure applied to the inlet of the generator is controlled and adjusted to provide the desired oil output.

The mist must be transmitted at a low velocity below 6 m/s and a low pressure usually between 25 and 50 mbar gauge through steel, copper or plastic tubes. The tubes must be smooth and scrupulously clean internally.

At the lubrication point the mist is throttled to atmospheric pressure through a special nozzle whose orifice size controls the total amount of lubricant applied and raises the mist velocity to a figure in excess of 40 m/s. This causes the lubricant to wet the rubbing surfaces and the air is permitted to escape to atmosphere. Empirical formulae using an arbitrary unit—the 'Lubrication Unit', are used to assess the lubricant requirements of the machine, the total compressed air supply required and the size of tubing needed.

DESIGN

The essential parameters of components are indicated in Table 17.2 and the load factors for bearings are given in Table 17.1.

Table 17.1 Load factors

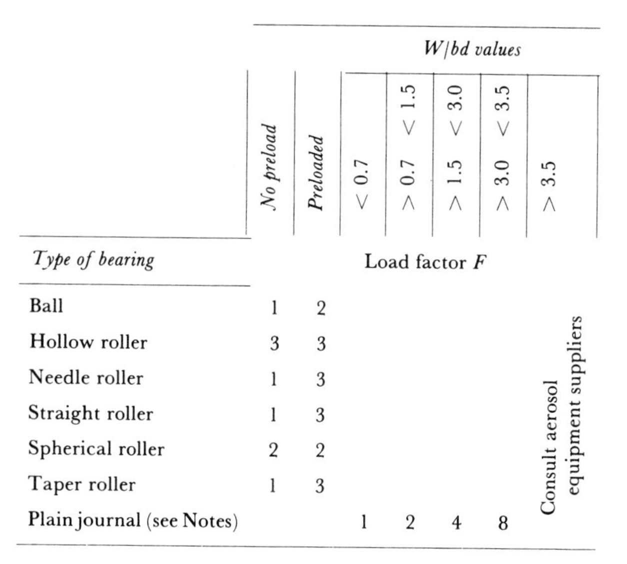

Type of bearing	No preload	Preloaded	W/bd values < 0.7	> 0.7 < 1.5	> 1.5 < 3.0	> 3.0 < 3.5	> 3.5
	Load factor F						
Ball	1	2					Consult aerosol equipment suppliers
Hollow roller	3	3					
Needle roller	1	3					
Straight roller	1	3					
Spherical roller	2	2					
Taper roller	1	3					
Plain journal (see Notes)			1	2	4	8	

Notes: for explanation $\frac{W}{bd}$ see Table 17.2. For plain journal bearings with high end losses or made of white metal, double these load factors.

Table 17.2 Information required for the calculation of lubricant flow rates

Essential parameters	*Units*	*Formulae symbols*	*Ball and roller bearing*	*Ball nuts*	*Plain (journal) bearing*	*Gear pair*	*Worm and gear*	*Gear train*	*Rack and pinion*	*Cam and follower*	*Moving slide*	*Roller chain drive*	*Inverted tooth chain*	*Conveyor chain*
			Type of component											
Contact length of bearing surface	mm	b			X						X			
Shaft diameter	mm	d	X		X									
Load factor (see Table 17.1)		F	X		X									
Load on bearing	N	W			X									
Length of chain	m	l												X
Maximum diameter	mm	D								X				
Number of rows of balls or rollers or chain strands		N	X	X								X		
Pitch circle diameters —Gears—Worm—Sprocket or nut	mm	P		X		X	X	X	X			X	X	X
Chain link pitch	mm	C										X		
Speed of rotation	rev/s	n										X	X	
Contact width of slide or gear or chain width	mm	w				X	X	X	X	X	X		X	X

Note: 'X' denotes information that is required for each type of component

The Lubrication Unit (LU) rating of each component should be calculated from the formulae in Table 17.3, using the values in Tables 17.1 and 17.2.

Table 17.3 Lubrication unit rating

Type of component	*Formulae for lubrication unit rating*
Rolling bearings	$4(d\,.\,F\,.\,N)\,.\,10^{-2}$
Ball nuts	$4P\,[(N-1)+10]\,.\,10^{-3}$
Journal bearings	$2(b\,.\,d\,.\,F)\,.\,10^{-4}$
Gears (see Note)	
Gear pair	$4w\,(P_1+P_2)\,.\,10^{-4}$
Gear train	$4w\,(P_1+P_2+P_3\,\ldots\,+P_n)\,.\,10^{-4}$
Worm and gear	$4\,[(P_M\,.\,w_M)+(P_w\,.\,w_w)]\,.\,10^{-4}$
Rack and pinion	$12\,(P\,.\,w)\,.\,10^{-4}$
Cams	$2(D\,.\,w)\,.\,10^{-4}$
Slides and ways	$8(b\,.\,w)\,.\,10^{-5}$
Chains (see Note)	
Roller chain drive	$(N\,.\,P\,.\,C)\,(n)^{\frac{3}{2}}\,.\,10^{-4}$
Inverted tooth chain	$5(P\,.\,w)\,(n)^{\frac{3}{2}}\,.\,10^{-5}$
Conveyor chains	$5w\,(25l+P)\,.\,10^{-4}$

Notes: if gears are reversing double the calculated LU rating. If the pitch diameter of any gear exceeds twice that of a mating gear, i.e. $P_1 > 2P_2$, consider $P_1 = 2P_2$. For chain drives if $n < 3$ use $n = 3$.

Total the LU ratings of all the components to obtain the total Lubrication Unit Loading (LUL). This is used later for estimating the oil consumption and as a guide for setting the aerosol generators.

Distribution piping

When actual nozzle sizes have been decided, the actual nozzle loadings (measured in Lubrication Units) can be totalled for each section of the pipework, and this determines the size of pipe required for that section. The actual relationship is given in Table 17.4 and Figures 17.2. Where calculated size falls between two standard sizes use the larger size. Machined channels of appropriate cross-sectional area may also be used as distribution manifolds.

Nozzle sizes

Select standard nozzle fitting or suitable drilled orifice size from Figure 17.1 for each component using its calculated LU rating. Where calculated LU rating falls between

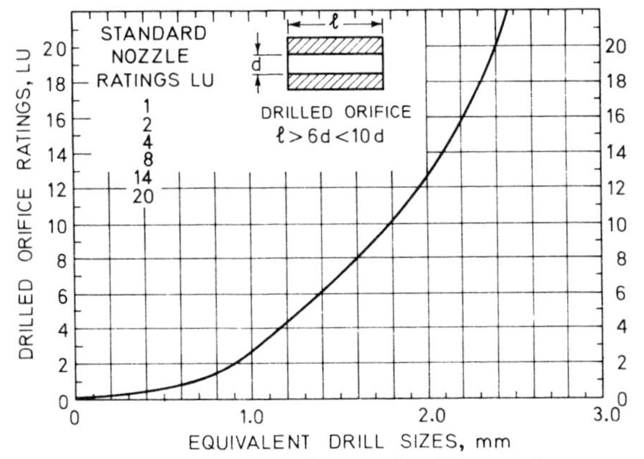

Figure 17.1 Drill size and orifice ratings

two standard fitting or drill sizes, use the larger size fitting or drill. Multiple drillings may be used to produce nozzles with ratings above 20 LU.

Maximum component dimensions (Table 17.1) for a single nozzle.

b = 150 mm
w = 150 mm for slides, 12 mm for chains, 50 mm for other components.

Where these dimensions are exceeded and for gear trains or reversing gears use nozzles of lower LU rating appropriately sized and spaced to provide correct total LU rating for the component.

Table 17.4 Pipe sizes

	Distribution pipe size	
Nozzle loading, LU	*Copper tubing to* ISO 274 *OD* (mm)	*Medium series steel pipe to* ISO 65, *nominal bore* (mm)
10	6	—
15	8	6
30	10	8
50	12	10
75	16	—
100	20	15
200	25	20
300	32	25
500	40	32
650	50	40
1000	63	50

ISO 274 corresponds to BS 2871
ISO 65 corresponds to BS 1387

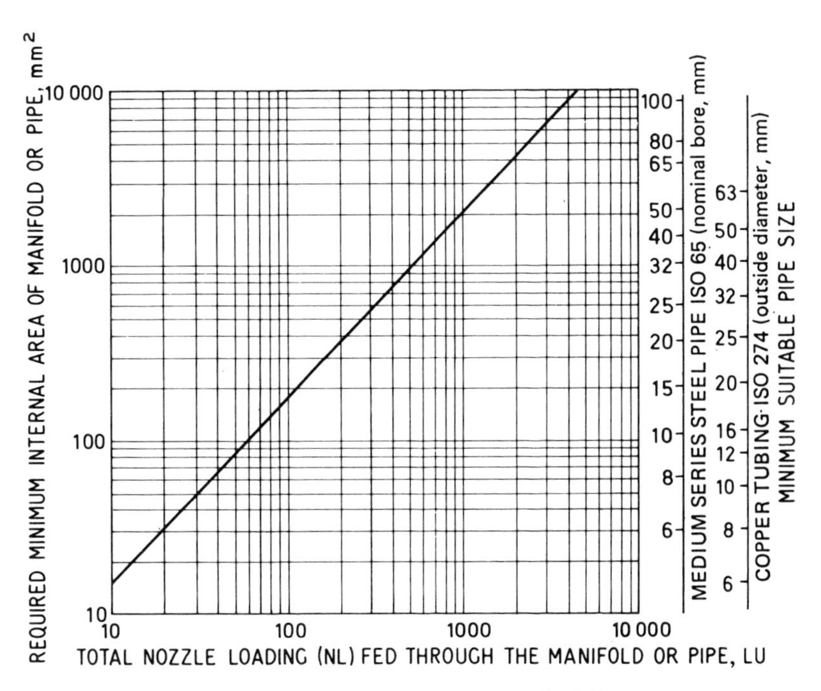

Figure 17.2 Sizing of manifolds and piping

Generator selection

Total the nozzle ratings of all the fittings and orifices to give the total Nozzle Loading (NL) and select generator with appropriate LU rating based on Nozzle Loading. Make certain that the minimum rating of the generator is less than NL.

Air and oil consumption

Air consumption is a function of the total nozzle loading (NL) of the system. Oil consumption depends on the concentration of oil in the air and can be adjusted at the generator to suit the total Lubrication Unit Loading (LUL).

Air consumption

Using the total Nozzle Loading (NL), the approximate air consumption can be calculated in terms of the volume of free air at atmospheric pressure, from:

$$\text{Air consumption} = 0.015\ (\text{NL})\ \text{dm}^3/\text{s}$$

Oil consumption

Using the total Lubrication Unit Loading (LUL), the approximate oil consumption can be calculated from:

$$\text{Oil consumption} = 0.25\ (\text{LUL})\ \text{ml/h}$$

INSTALLATION

Locate nozzle ends between 3 mm and 25 mm from surface being lubricated. Follow normal practice in grooving slides and journal bearings. The positioning of the nozzles in relationship to the surface being lubricated should be similar to that used in circulation systems. See Table 5.

Appropriate vents with hole diameters at least 1.5 times the diameter of the associated supply nozzle must be provided for each lubrication point. If a single vent serves several nozzles, the vent area must be greater than twice the total area of the associated nozzles.

Follow instructions of aerosol generator manufacturer in mounting unit and connecting electrical wiring. Avoid sharp bends and downloops in all pipework. Consult BS 4807: 1991 'Recommendations for Centralised Lubrication as Applied to Plant and Machinery' for general information on installation.

Select appropriate grade of lubricant in consultation with lubricant supplier and generator manufacturer.

Table 17.5 Nozzle positioning

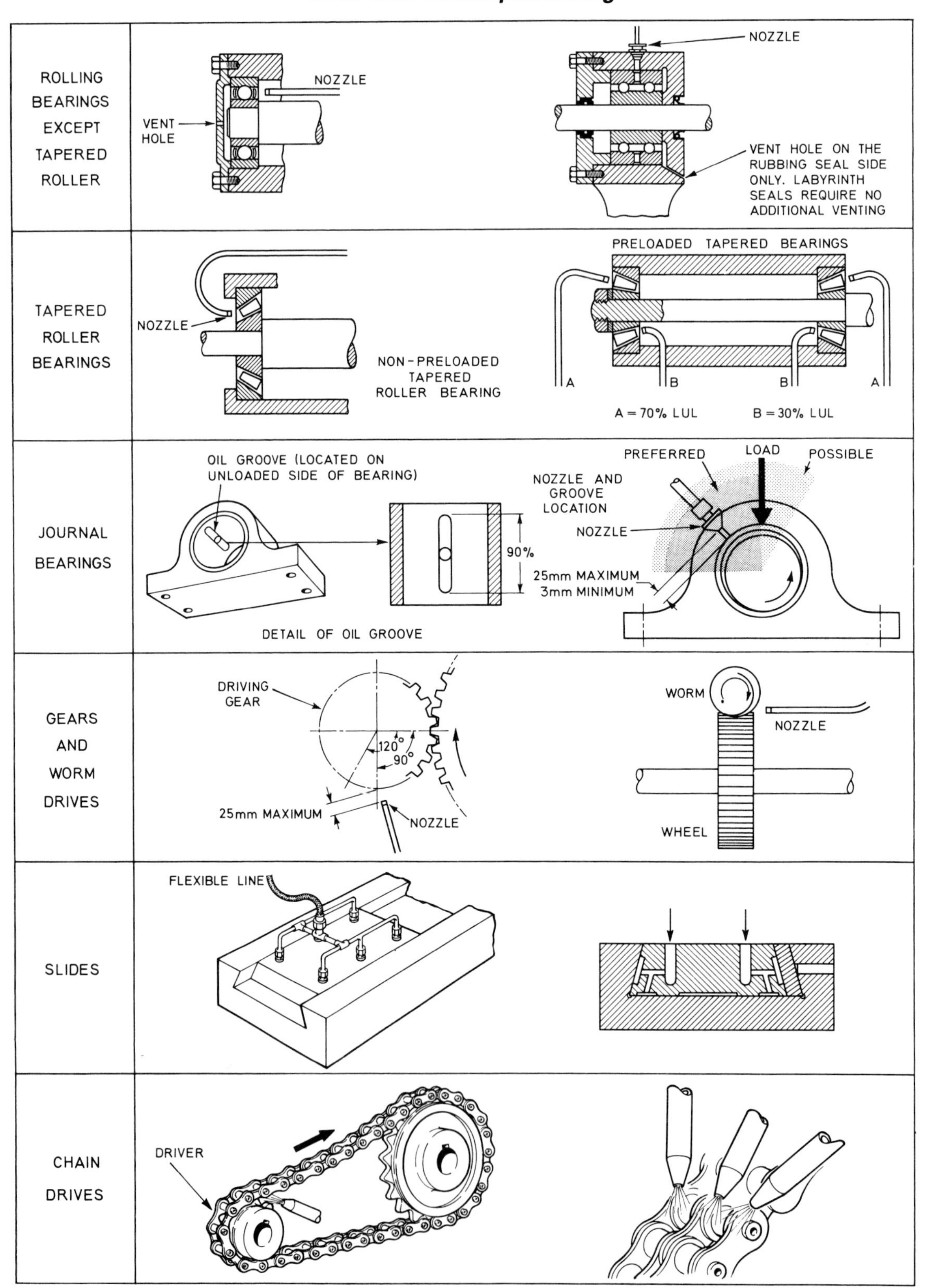

A circulation system is defined as an oil system in which the oil is returned to the reservoir for re-use. There are two groups of systems: *group 1*, lubrication with negligible heat removal; and *group 2*, lubrication and cooling.

GROUP 1 SYSTEMS

Virtually any form of mechanically or electrically driven pump may be used, including piston, plunger, multiplunger, gear, vane, peristaltic, etc. The systems are comparatively simple in design and with low outputs. Various metering devices may be used.

Multiplunger pump systems

These systems utilise the plunger-type oil lubricators of the rising or falling drop type, employing a separate pumping element for each feed, giving individual adjustment and a positive feed to each lubrication point. Generally, the lubricators have up to some 32 outlets with the discharges being adjustable from zero to maximum.

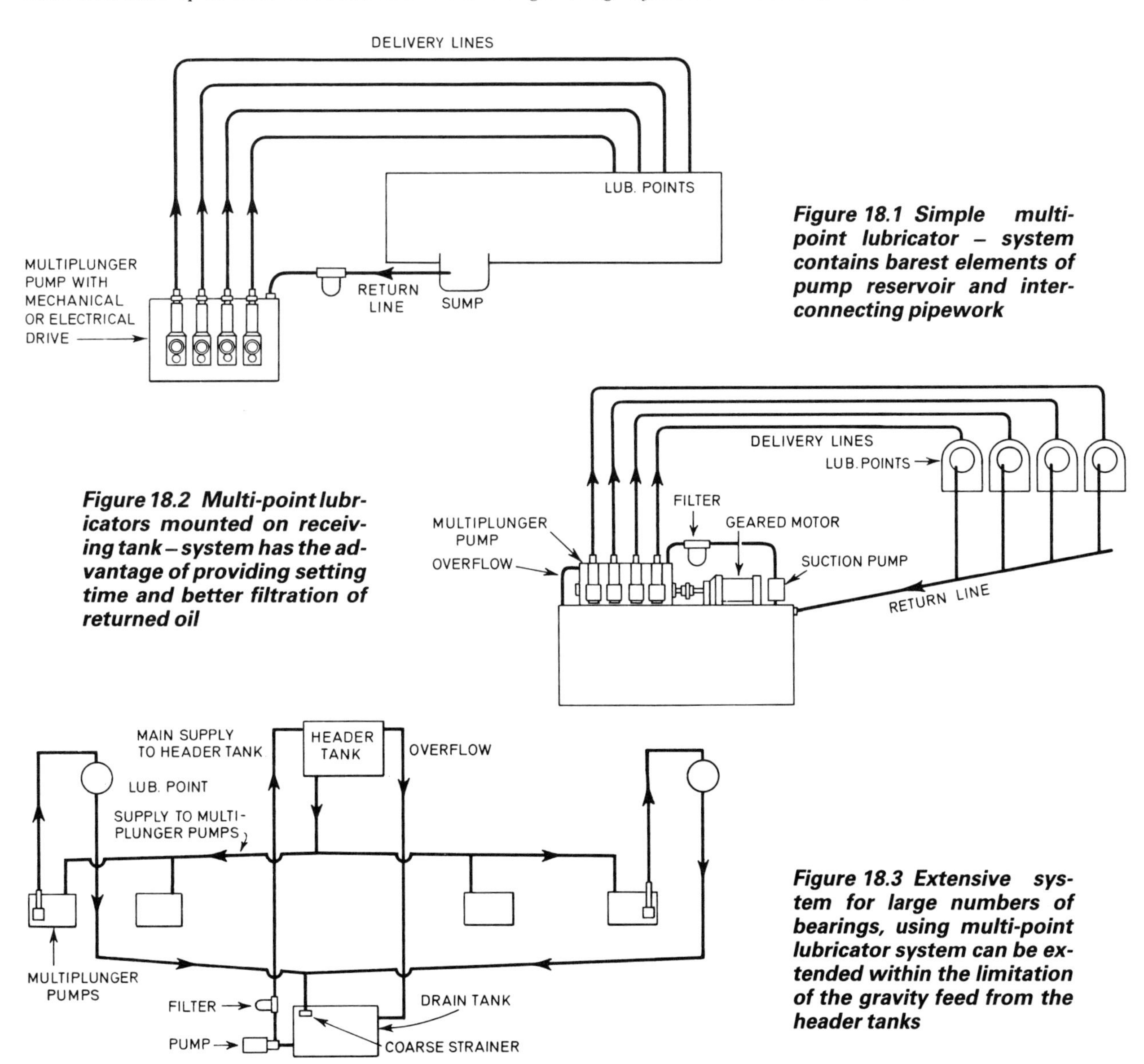

***Figure 18.1** Simple multi-point lubricator – system contains barest elements of pump reservoir and inter-connecting pipework*

***Figure 18.2** Multi-point lubricators mounted on receiving tank – system has the advantage of providing setting time and better filtration of returned oil*

***Figure 18.3** Extensive system for large numbers of bearings, using multi-point lubricator system can be extended within the limitation of the gravity feed from the header tanks*

Typical applications

Paper machines, large kilns, calenders, and general machinery with a large number of bearings requiring a positive feed with feed adjustment.

Positive-split systems

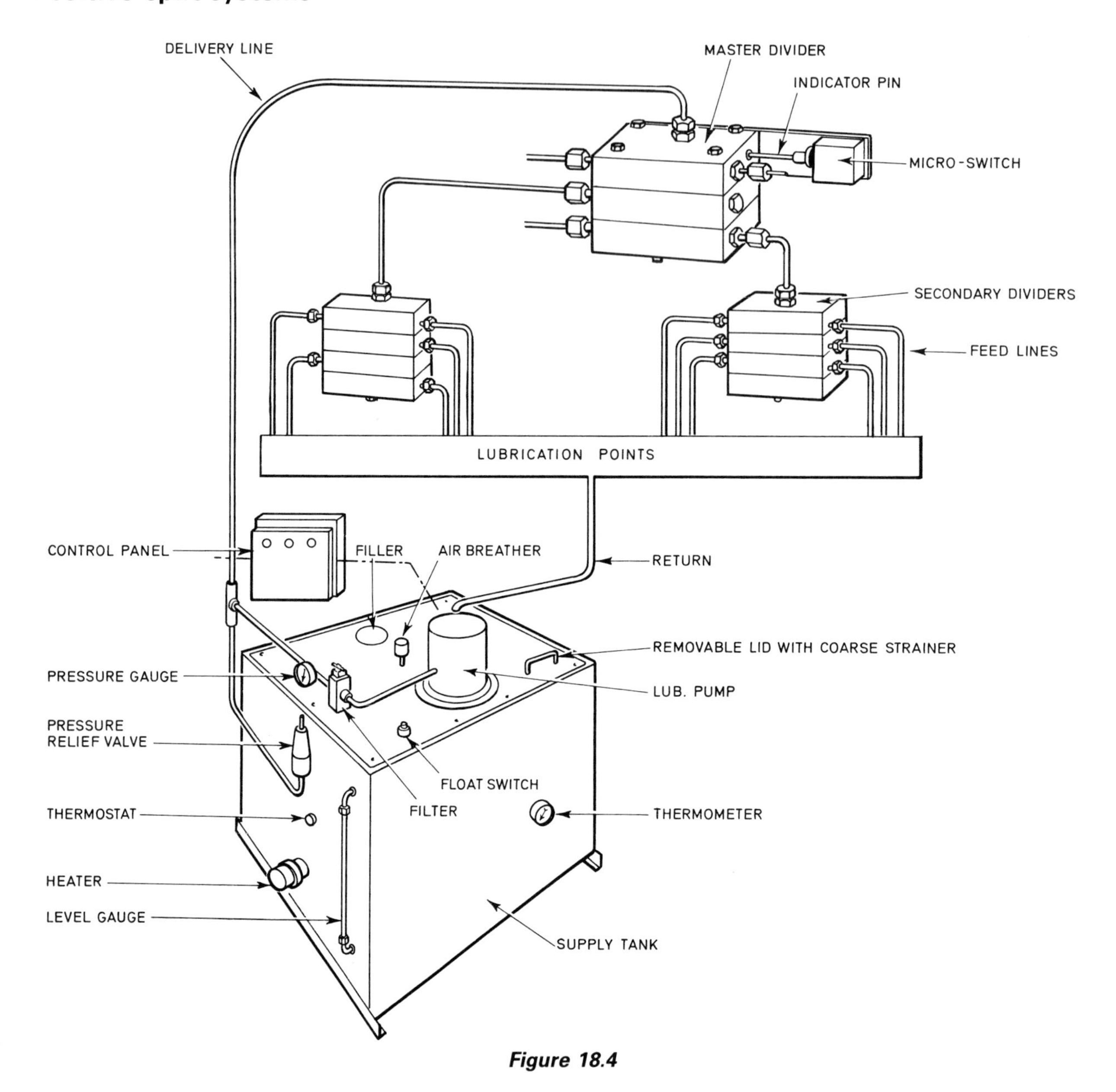

Figure 18.4

In general these systems deliver a larger quantity of lubricant than multiplunger systems. They comprise a small high-pressure pump with or without stand-by, fitted with integral relief valve supplying lubricant to the bearings via positive dividers. These dividers then deliver the oil to the bearings in a predetermined ratio of quantities. By use of a microswitch operated by the indicator pin on the master divider (or single divider if the number of points is small), either timed automatic or continuous operation is available. The microswitch can also be used to give a warning of failure of the system.

Typical applications

Machine tools, sugar industry, gearboxes, printing machines, and special-purpose machinery.

Double-line systems

Double-line elements can be used in conjunction with a reversing valve and piston or gear pump to lubricate larger numbers of points spread over longer distances geographically. These elements and their operation are similar to those previously described under grease systems.

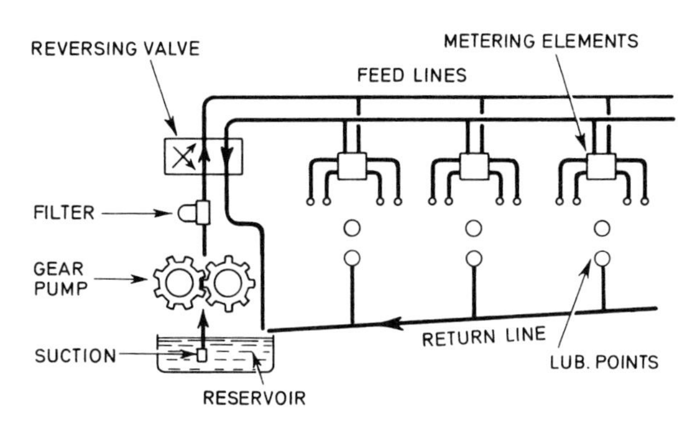

Figure 18.5

Typical applications

Machine tools, textile plant, and special-purpose machinery.

Simple low-pressure systems

The simple form illustrated uses a gear pump feeding the points from connections from a main feed line through needle valves with or without sight glasses.

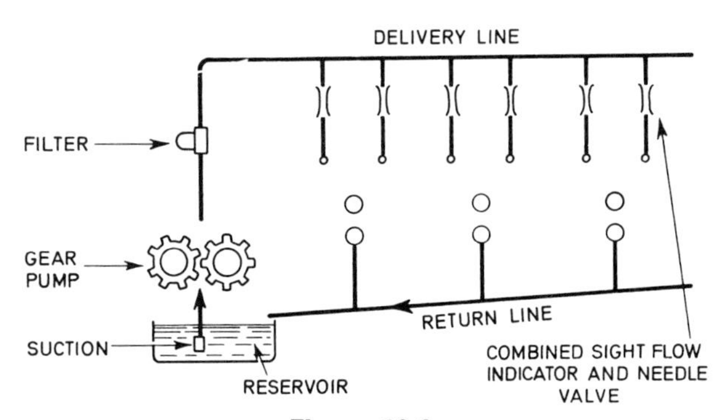

Figure 18.6

Typical applications

Special-purpose machinery and machine tools.

Gravity-feed systems

Gravity-feed systems consist of a header tank, piped through to one or more lubrication points. The level in the header tank is maintained by a gear or other pump with relief valve and filter mounted at the collection tank.

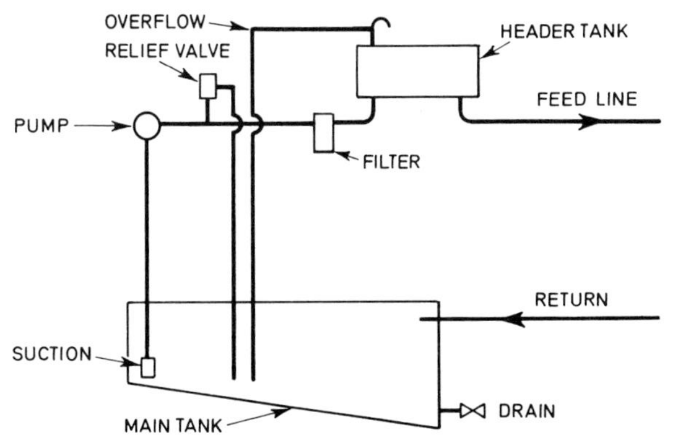

Figure 18.7

This may be used as a back-up for a forced-feed system where important bearings have a long run-down period after removal of the power source, e.g. large air fans.

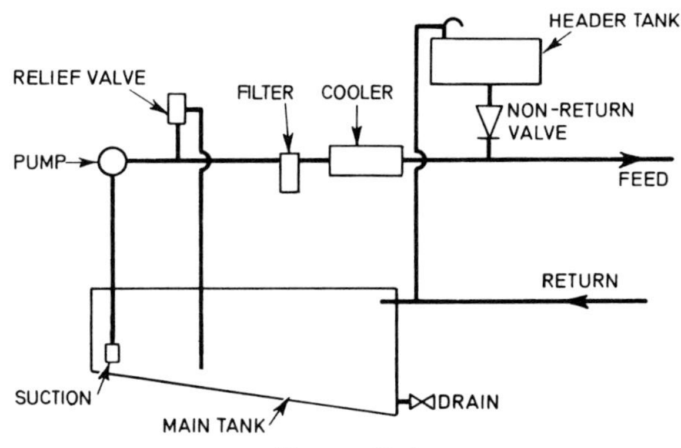

Figure 18.8

GROUP 2 SYSTEMS

The larger and usually more complex type of oil-circulatory system, used for both lubrication and cooling, falls into two distinct classes. The first type, known as the self-contained system, is usually limited in size by the weights of the components. For this reason the storage capacity of this type does not usually exceed 1000 gal. The second type covering the larger systems has the main components laid out at floor level, e.g. in the oil cellar. The detailed design considerations of the main components are discussed elsewhere, but in laying out the system the possible need for the equipment in Table 18.1 should be considered.

Self-contained systems

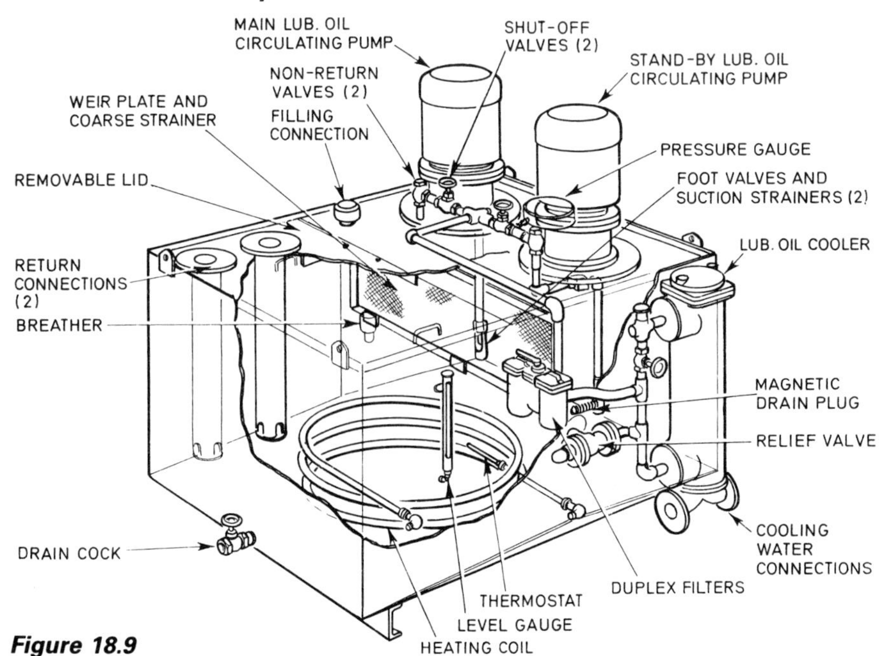

A typical self-contained oil-circulatory system, incorporating a 200 gal tank. These types of system may be used, if required, with a pressure vessel which would be mounted as a separate unit

Figure 18.9

Large oil-circulatory systems

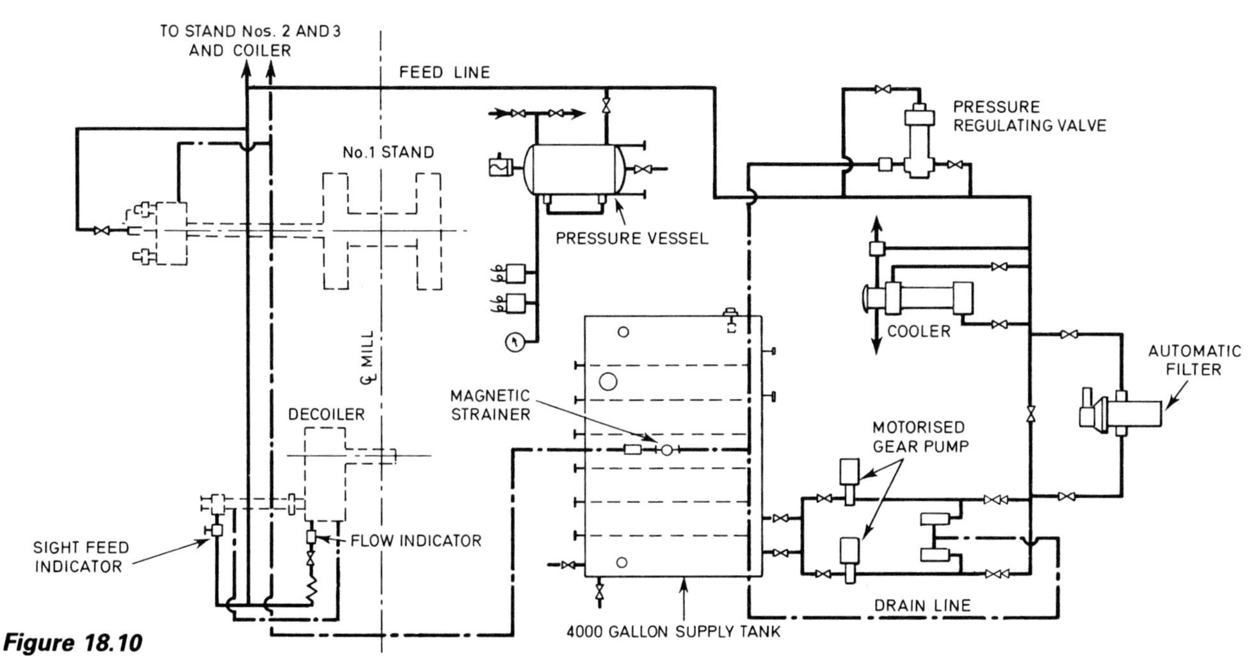

Figure 18.10

The large oil-circulatory systems typical of those in use in steelworks, marine applications and power stations are illustrated diagrammatically above.

Table 18.1 Main components of group 2 systems

Storage tank	One or more storage tanks, dependent on water contamination, will be required with a capacity of between 20 and 40 times the throughput per minute of the system
Tank heating	Electric, steam or hot-water heating are used for raising the temperature of the lubricant in the storage tank
Pumps	One or more main pumps and a stand-by pump are required. The main pumps should have a capacity of at least 25% in excess of basic system requirement
Pressure control valve	System pressure will be maintained by the use of pressure control valves
Non-return valves	A non-return valve is required after each pump unit
Self-cleaning strainer	Either manual or automatic, single or duplex self-cleaning strainers will be used for cleaning the oil
Magnetic strainer	Particularly in the case of gear lubrication, magnetic strainers may be fitted whether in the supply line and/or on the return oil connection to the tank
Pressure vessel	Pressure vessels may be required in order to maintain a flow of oil in the event of a power failure, to allow the run-down of machinery or the completion of a machine operation
Cooler	A cooler may be used to extract the heat taken away from the equipment to the lubricant and maintain the viscosity in the system prior to the supply to the lubrication points
Pressure-reducing stations	On extensive systems the lubrication points can be split into groups for controlling the flow rate by means of a pressure-reducing station, followed by either orifice plates or simple pipe sizing flow control
Water/sludge trap	Where water contamination is likely a water/sludge trap should be fitted in the return line immediately before the tank
Valving	All major equipment should be capable of being shut off by the use of gate valves for maintenance purposes. In the case of filters and coolers a bypass arrangement is necessary
Instrumentation	Normal instrumentation will cater for the specific system requirements, including pressure gauges, thermometers, thermostats, pressure switches, recording devices, etc.

CONTROL OF LUBRICANT QUANTITIES

The quantity fed to the lubrication point can be controlled in a number of ways; typical examples are shown below:

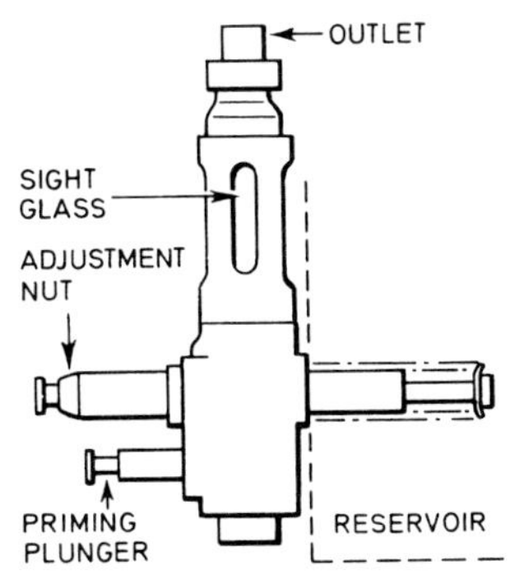

Figure 18.11

The output of a metering pump is itself adjustable by some form of manual adjustment on each pump unit. Sight glasses, of the rising or falling drop type, or of the plug and taper tube type, are normally fitted.

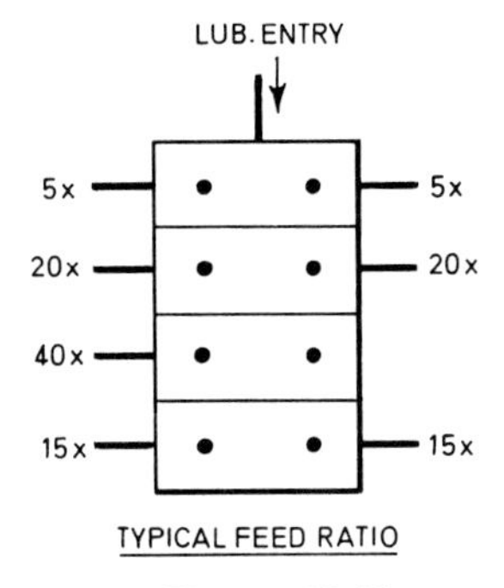

Figure 18.12

The positive dividers may have sections which have different outputs, and may be cross-drilled to connect one or more outlets together to increase the quantity available for each cycle.

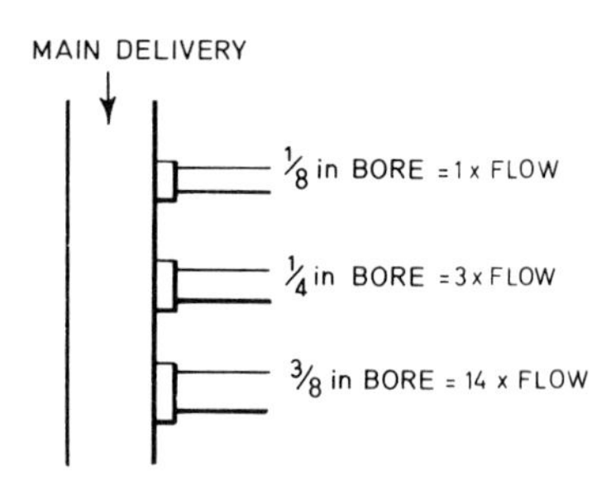

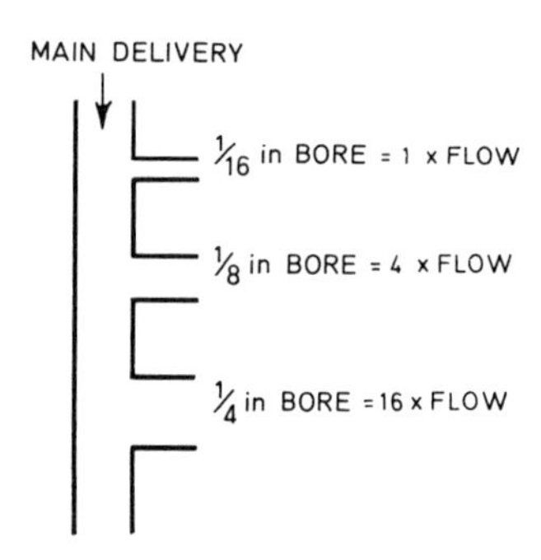

Figure 18.13 Typical flow ratios

Orifice plates may be used at the entry to the bearing or gear system. The actual flow rates will vary with viscosity unless knife-edge orifices are used, in which case the viscosity variation is negligible.

With larger flow rates it may be adequate, with a controlled pressure and oil temperature, simply to alter the bore of the pipe through which the supply is taken. The actual flow rates will vary with viscosity, and pipework configuration, i.e. increased number of fittings and directional changes.

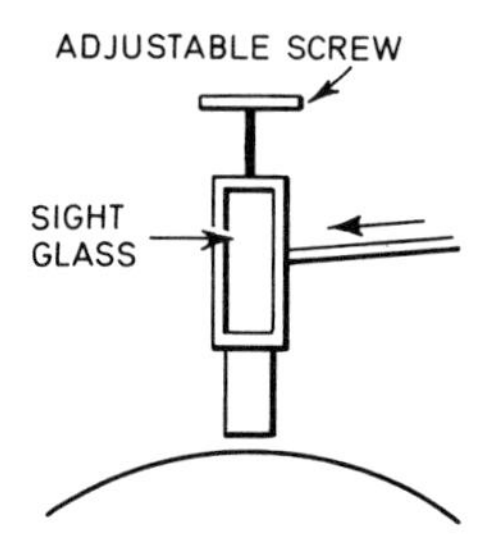

Figure 18.14

Combined needle and sight flow indicators used for adjusting small quantities of lubricant giving only a visual indication of the flow of lubricant into the top of a bearing.

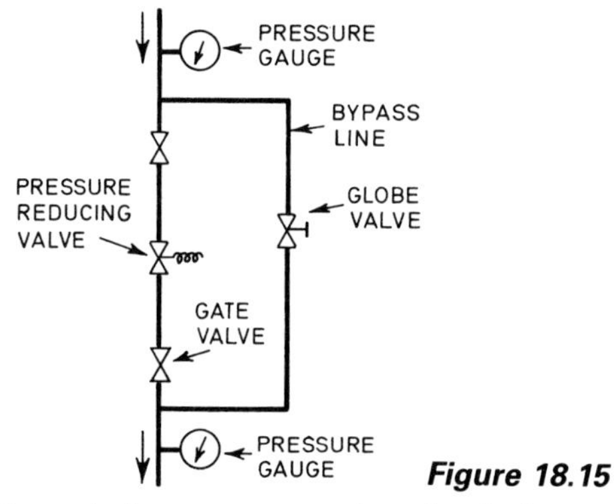

Figure 18.15

The layout of a typical pressure control station is shown above.

TOTAL-LOSS SYSTEMS

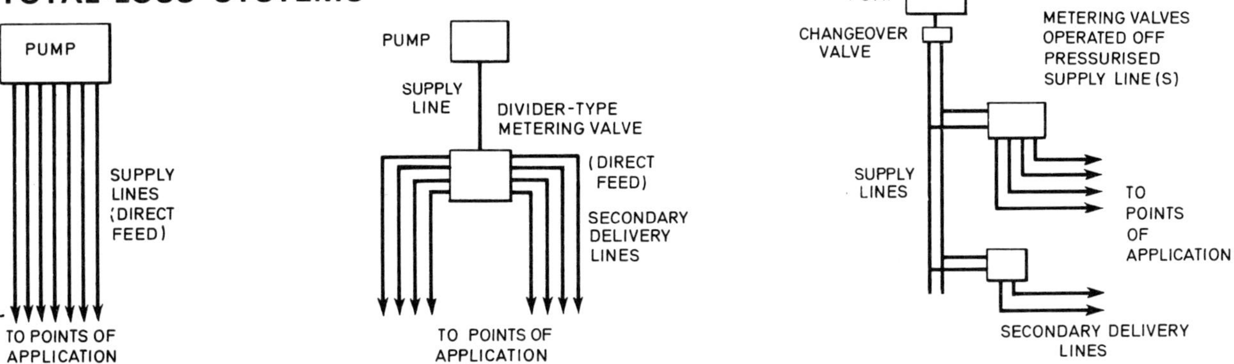

Figure 19.1 Schematic diagrams of typical total-loss systems – lubricant is discharged to points of application and not recovered

Commissioning procedure

1 Check pumping unit.
2 Fill and bleed system. *Note:* it is not normally considered practicable to flush a total-loss system.
3 Check and set operating pressures.
4 Test-run and adjust.

No special equipment is required to carry out the above procedure but spare pressure gauges should be available for checking system pressures.

Pumping unit

PRIME MOVER

For systems other than those manually operated, check for correct operation of prime mover, as follows.

(*a*) Mechanically operated pump—check mechanical linkage or cam.
(*b*) Air or hydraulic pump:
  (*i*) check air or hydraulic circuit,
  (*ii*) ascertain that correct operating pressure is available.
(*c*) Motor-operated pump:
  (*i*) check for correct current characteristics,
  (*ii*) check electrical connections,
  (*iii*) check electrical circuits.

PUMP

(*a*) If pump is unidirectional, check for correct direction of rotation.
(*b*) If a gearbox is incorporated, check and fill with correct grade of lubricant.

CONTROLS

Check for correct operation of control circuits if incorporated in the system, i.e. timeclock.

RESERVOIR

(*a*) Check that the lubricant supplied for filling the reservoir is the correct type and grade specified for the application concerned.
(*b*) If the design of the reservoir permits, it should be filled by means of a transfer pump through a bottom fill connection via a sealed circuit.
(*c*) In the case of grease, it is often an advantage first to introduce a small quantity of oil to assist initial priming.

Filling of system

SUPPLY LINES

These are filled direct from the pumping unit or by the transfer pump, after first blowing the lines through with compressed air.

In the case of direct-feed systems, leave connections to the bearings open and pump lubricant through until clean air-free lubricant is expelled.

In the case of systems incorporating metering valves, leave end-plugs or connections to these valves and any other 'dead-end' points in the system open until lubricant is purged through.

With two-line systems, fill each line independently, one being completely filled before switching to the second line via the changeover valve incorporated in this type of system.

SECONDARY LINES (Systems incorporating metering or dividing valves)

Once the main line(s) is/are filled, secure all open ends and after prefilling the secondary lines connect the metering valves to the bearings.

System-operating pressures

PUMP PRESSURE

This is normally determined by the pressure losses in the system plus back pressure in the bearings.

Systems are designed on this basis within the limits of the pressure capability of the pump.

Check that the pump develops sufficient pressure to overcome bearing back pressure either directly or through the metering valves.

In the case of two-line type systems, with metering valves operating 'off' pressurised supply line(s), pressures should be checked and set to ensure positive operation of all the metering valves.

Running tests and adjustments

SYSTEM OPERATION

Operate system until lubricant is seen to be discharging at all bearings. If systems incorporate metering valves, each valve should be individually inspected for correct operation.

ADJUSTMENT

In the case of direct-feed systems, adjust as necessary the discharge(s) from the pump and, in the case of systems operating from a pressure line, adjust the discharge from the metering valves.

RELIEF OR BYPASS VALVE

Check that relief or bypass valve holds at normal system-operating pressure and that it will open at the specified relief pressure.

CONTROLS

Where adjustable electrical controls are incorporated, e.g. timeclock, these should be set as specified.

ALARM

Electrical or mechanical alarms should be tested by simulating system faults and checking that the appropriate alarm functions. Set alarms as specified.

Fault finding

Action recommended in the event of trouble is best determined by reference to a simple fault finding chart as illustrated in Table 19.2.

CIRCULATION SYSTEMS

Commissioning procedure

1 Flush system. *Note:* circulation systems must be thoroughly flushed through to remove foreign solids.
2 Check main items of equipment.
3 Test-run and adjust.

No special equipment is required to carry out the above but spare pressure gauges for checking system pressures, etc., and flexible hoses for bypassing items of equipment, should be available.

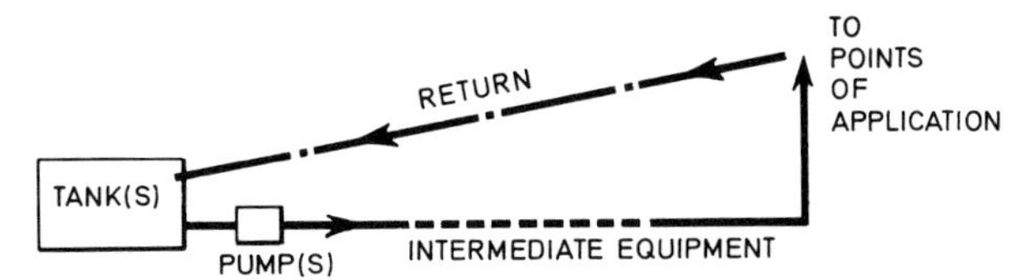

***Figure 19.2 Schematic diagram of typical oil-circulation system**. Oil is discharged to points of application, returned and re-circulated*

Flushing

1 Use the same type of oil as for the final fill or flushing oil as recommended by the lubricant supplier.
2 Before commencing flushing, bypass or isolate bearings or equipment which could be damaged by loosened abrasive matter.
3 Heat oil to 60–70°C and continue to circulate until the minimum specified design pressure drop across the filter is achieved over an eight-hour period.
4 During flushing, tap pipes and flanges and alternate oil on an eight-hour heating and cooling cycle.
5 After flushing drain oil, clean reservoir, filters, etc.
6 Re-connect bearings and equipment previously isolated and refill system with running charge of oil.

Main items of equipment

RESERVOIR

(*a*) Check reservoir is at least two-thirds full.
(*b*) Check oil is the type and grade specified.
(*c*) Where heating is incorporated, set temperature-regulating instruments as specified and bring heating into operation at least four hours prior to commencement of commissioning.

ISOLATING AND CONTROL VALVES

(*a*) Where fitted, the following valves must initially be left open: main suction; pump(s) isolation; filter isolation; cooler isolation; pressure-regulator bypass.
(*b*) Where fitted, the following valves must initially be closed: low suction; filter bypass; cooler bypass; pressure-regulator isolation; pressure-vessel isolation.
(*c*) For initial test of items of equipment, isolate as required.

MOTOR-DRIVEN PUMP(S)

(*a*) Where fitted, check coupling alignment.
(*b*) Check for correct current characteristics.
(*c*) Check electrical circuits.
(*d*) Check for correct direction of rotation.

PUMP RELIEF VALVE

Note setting of pump relief valve, then release spring to its fullest extent, run pump motor in short bursts and check system for leaks.

Reset relief valve to original position.

CENTRIFUGE

Where a centrifuge is incoporated in the system, this is normally commissioned by the manufacturer's engineer, but it should be checked that it is set for 'clarification' or 'purification' as specified.

FILTER

(a) Basket and cartridge type—check for cleanliness.
(b) Edge type (manually operated)—rotate several times to check operation.
(c) Edge type (motorised)—check rotation and verify correct operation.
(d) Where differential pressure gauges or switches are fitted, simulate blocked filter condition and set accordingly.

PRESSURE VESSEL

(a) Check to ensure safety relief valve functions correctly.
(b) Make sure there are no leaks in air piping.

PRESSURE-REGULATING VALVE

(a) Diaphragm-operated type—with pump motor switched on, set pressure-regulating valve by opening isolation valves and diaphragm control valve and slowly closing bypass valve.
Adjust initially to system-pressure requirements as specified.
(b) Spring-pattern type—set valve initially to system-pressure requirements as specified.

COOLER

Check water supply is available as specified.

Running tests and adjustments

(1) Run pump(s) check output at points of application, and finally adjust pressure-regulating valve to suit operating requirements.
(2) Where fitted, set pressure and flow switches as specified in conjunction with operating requirements.
(3) Items incorporating an alarm failure warning should be tested separately by simulating the appropriate alarm condition.

Fault finding

Action in the event of trouble is best determined by reference to a simple fault finding chart illustrated in Table 19.1

FAULT FINDING

Table 19.1 Fault finding – circulation systems

Condition	*Cause*	*Action*
(1) System will not start	Incorrect electrical supply to pump motor	Check electrics
(2) System will not build up required operating pressure	(a) Leaking pipework	Find break or leak and repair
	(b) Regulating bypass valve open	Check and close
	(c) Pressure-regulating valve wrongly adjusted	Check and re-set to increase pressure
	(d) Loss of pump prime	Check suction pipework for leaks
	(e) Blocked system filter	Inspect and clean or replace
	(f) Low level of oil in reservoir	Top up
	(g) Pump relief valve bypassing	Check and repair or replace as necessary
	(h) Faulty pump	Check and repair or replace as necessary
(3) System builds up abnormal operating pressure	(a) Pressure-regulating isolation and diaphragm-control valves closed	Check and open
	(b) Pressure-regulating valve wrongly adjusted	Check and re-set to decrease pressure
(4) Oil fails to reach points of application		
(i) via orifices or sprays	Blockage or restriction	Clean out
(ii) via flow control valves	(a) Control valve wrongly adjusted	Check and re-set
	(b) Blockage or restriction in delivery line	Clean out or replace as necessary
(iii) via metering valves	(a) Metering valve contaminated or faulty	Clean out or replace
	(b) Blockage or restriction in delivery line	Clean out or replace
(iv) via pressurised sight feed indicators	Air supply failure	Check and restore correct air supply

Table 19.2 Fault finding – total-loss systems

Condition	Cause	Action
1 System will not start (automatically operated systems)	(*a*) Incorrect supply to pump prime mover	Check electrical, pneumatic or hydraulic supply
	(*b*) System controls not functioning or functioning incorrectly	Check controls and test each function separately
2 System will not build up required operating pressure, or normal pumping time is prolonged	(*a*) Faulty pump	Isolate pump, check performance and repair or replace as necessary
	(*b*) Loss of pump prime	(1) Bleed air from pump (2) Check for blockage in reservoir line strainer
	(*c*) Relief valve bypassing	If with relief valve isolated pump builds pressure, repair or replace relief valve
	(*d*) Broken or leaking supply line(s)	Find break or leak and repair
	(*e*) Air in supply line(s)	Bleed air from line(s) as for filling
(systems incorporating metering valves)	(*f*) Bypass leakage in metering valve(s)	Trace by isolating metering valves systematically and repair or replace faulty valve(s)
3 System builds up operating pressure at pump but will not complete lubrication cycle	(*a*) Changeover control device faulty	Isolate from system, check and repair or replace
(systems incorporating electrical and/or pressure control devices in their operation)	(*b*) Incorrect piping to control device	Check and rectify as necessary
	(*c*) Incorrect wiring to control device	Check and rectify as necessary
4 System builds correct operating pressures but metering valve(s) fails/fail to operate (systems incorporating metering valves operating 'off' pressurised supply line/s)	If with delivery line(s) disconnected, valve(s) then operates/operate:	
	(*a*) Delivery piping too small	Change to larger piping
	(*b*) Blockage or restriction in delivery line	Clean out or replace
	(*c*) Blockage in bearing	Clean out bearing entry and/or bearing
	If with delivery line(s) disconnected, valve(s) still fails/fail to operate	Clean out or replace
	(*d*) Metering valve contaminated or faulty	
5 System builds excessive pressure	(*a*) Blockage or restriction in main supply line	Trace and clean out or replace faulty section of pipework
	(*b*) Supply pipe too small	(1) Change to larger piping (2) Lubricant too heavy Seek agreement to use lighter lubricant providing it is suitable for application
(direct-feed systems)	(*c*) Delivery line piping too small	Change to larger piping
,,	(*d*) Blockage or restriction in delivery line	Clean out or replace
,,	(*e*) Blockage in bearing	Clean out bearing entry and/or bearing
(direct-feed systems incorporating divider-type metering valves)	(*f*) Metering valve(s) contaminated or faulty [see 4(*d*) above]	Clean out or replace

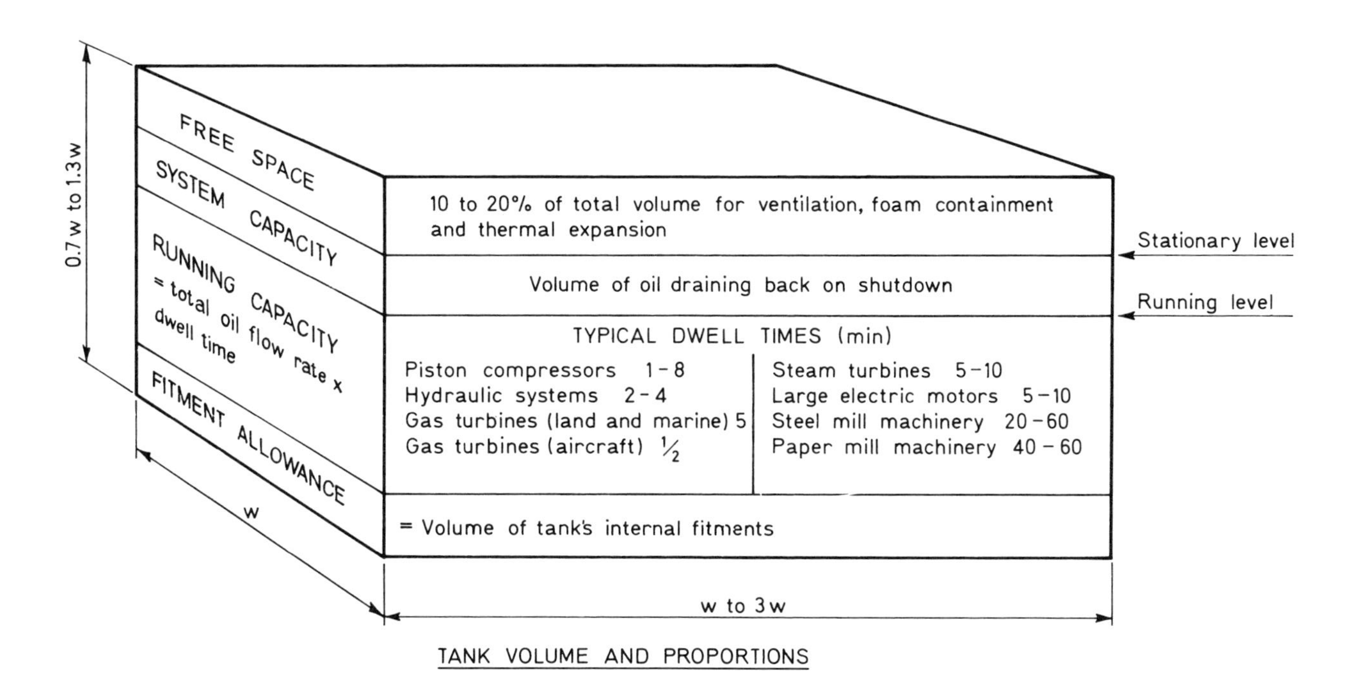

TANK VOLUME AND PROPORTIONS

Table 20.1 Tank materials

Stainless steel or anodised aluminium alloy	*Mild steel plate*
Material cost relatively high, but expensive preparation and surface protective treatment is not needed. Maintenance costs relatively low. Thinner gauge stainless steel may be used	Most widely used material for tanks; low material cost, but surface requires cleaning and treatment against corrosion, e.g. by shot or sand blasting and (*a*) lanolin-based rust preventative (*b*) oil resistant paint (*c*) coating with plastic (epoxy resin) (*d*) aluminium spraying

Table 20.2 Tank components

Component	*Design features*	*Diagram*
RETURN LINE	*Location:* at or just above running level *Size and slope:* chosen to run less than half full to let foam drain, and with least velocity to avoid turbulence *Alternative:* allow full bore return below running level *Refinement:* perforated tray prevents aeration due to plunging	PERFORATED TRAY RUNNING OIL LEVEL
SUCTION LINE	*Location:* remote from return line *Inlet depth:* shallow inlet draws clean oil; deep inlet prolongs delivery in emergency. Compromise usually two-thirds down from running level *Refinement:* filter or strainer; floating suction to avoid depth compromise (illustrated); anti-vortex baffle to avoid drawing in air	PIVOT STOP FLOAT
BAFFLES AND WEIRS	*Purpose:* see diagram. Also provide structural stiffening, inhibit sloshing in mobile systems, prevent direct flow between return and suction *Design:* Settling needs long, slow, uniform flow; baffles increase flow velocity. Avoid causing constrictions or 'dead pockets'. Arrange baffles to separate off cleanest layer. Consider drainage and venting needs	① WEIR TO LOCALISE TURBULENCE ② BAFFLE TO TRAP FOAM AND FLOATING CONTAMINANTS ③ BAFFLE TO TRAP SINKING CONTAMINANTS

Table 20.2 Tank components *(continued)*

VENTILATORS	*Purpose:* to allow volume changes in oil; to remove volatile acidic breakdown products and water vapour *Number:* normally one per 5 m^2 (50 ft^2) of tank top *Filtration:* treated paper or felt filters, to be inspected regularly, desirable. Forced ventilation, by blower or exhauster, helps remove excess water vapour provided air humidity is low
DRAINAGE POINTS AND ACCESS	*Drainage:* Located at lowest point of tank. Is helped by bottom slope of 1:10 to 1:30. Gravity or syphon or suction pump depending on space available below tank. Baffle over drain helps drainage of bottom layer first *Access:* Size and position to allow removal of fittings for repair and to let all parts be easily cleaned. Manholes with ladders needed in large tanks
LEVEL INDICATORS	*Dipstick:* suitable only for small tanks *Sight-glass:* simple and reliable unless heavy or dirty oil obscures glass. Needs protection or 'unbreakable' glass. Ball-check stops draining if glass breaks *Float and pressure gauges:* various proprietary gauges can give local or remote reading or audible warning of high or low levels BALL CHECK
DE-AERATION SCREEN	*Purpose:* removal of entrained air in relatively clean systems, to prevent recirculation *Position:* typically as shown. Should be completely immersed as surface foam can penetrate finest screen *Material:* wire gauze of finest mesh that will not clog too quickly with solid contaminants. Typically 100-mesh WIRE GAUZE
HEATERS	*Purpose:* aid cold-start circulation, promote settling by reducing viscosity, assist water removal *Design features:* Prevent debris covering elements. Provide cut-out in case of oil loss. Take care that convection currents do not hinder settling. Consider economics of tank lagging
COOLERS	*Purpose:* reducing temperature under high ambient temperature conditions. Usually better fitted as separate unit
STIFFENERS	*Purpose:* prevention of excessive bulging of sides and reduction of stresses at edges in large tanks *Location:* external stiffening preferred for easy internal cleaning and avoidance of dirt traps, but baffles may eliminate need for separate stiffening
INSTRUMENTS, ETC.	*Thermometer or temperature gauge:* desirable in tanks fitted with heaters *Magnetic drain-plug or strainers:* often fitted both for removing ferrous particles and to collect them for fault analysis *Sampling points:* May be required at different levels for analysis

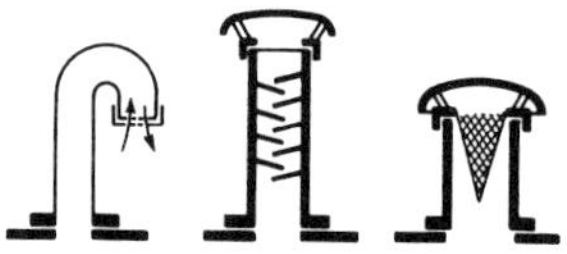

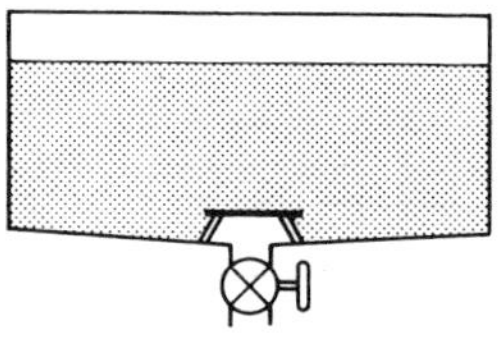

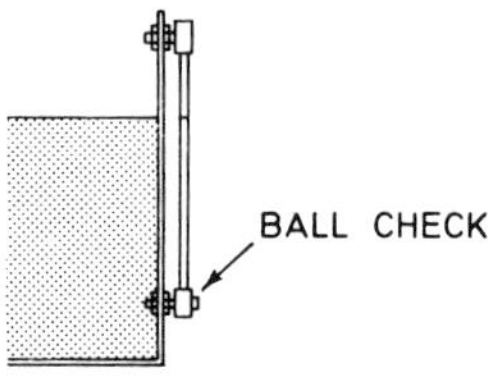

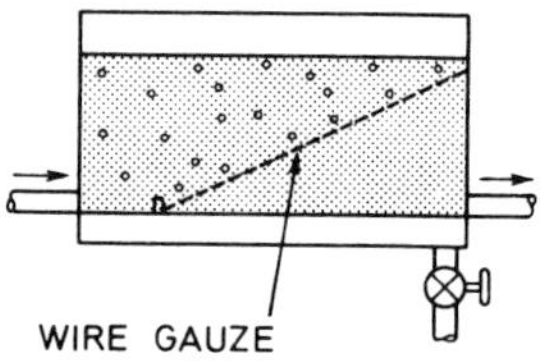

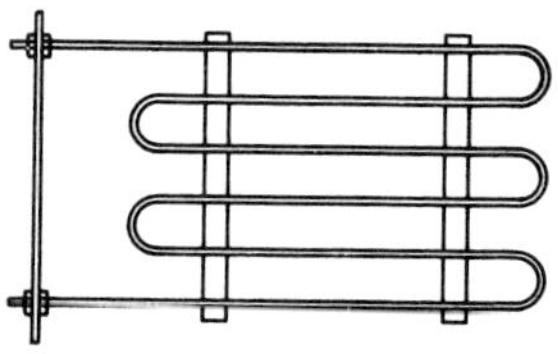

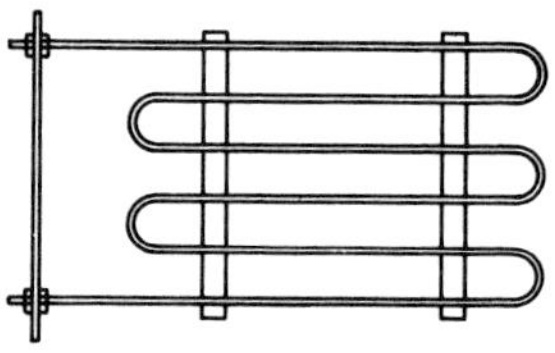

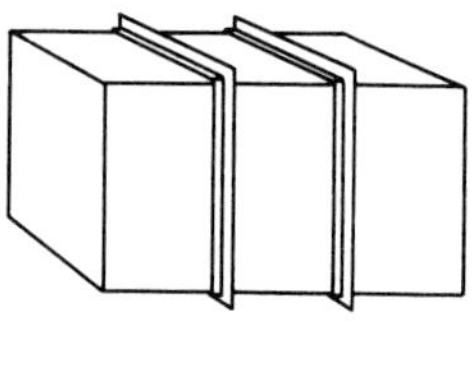

Table 21.1 System factors affecting choice of pump type

Factor	*Way in which pump choice is affected*	*Remarks*
Rate of flow	Total pump capacity = maximum equipment requirement + known future increase in demand (if applicable) + 10 to 25% excess capacity to cater for unexpected changes in system demand after long service through wear in bearings, seals and pump Determines pump size and contributes to determining the driving power	Actual selection may exceed this because of standard pumps available. Capacity of over 200% may result for small systems
Viscosity	Lowest viscosity (highest expected operating temperature) is contributing factor in determining pump size Highest viscosity (lowest expected operating temperature) is contributing factor in determining driving power	May influence decision whether reservoir heating is necessary
Suction conditions	May govern selection of pump type and/or its positioning in system Losses in inlet pipe and fittings with highest expected operating oil viscosity + static suction lift (if applicable) not to exceed pump suction capability	See Figure 21.1 below for determining positive suction head, or total suction lift
Delivery pressure	Total pressure at pump = pressure at point of application + static head + losses in delivery pipe, fittings, filter, cooler, etc. with maximum equipment oil requirement at normal viscosity Determines physical robustness of pump and contributes in deciding driving power	See Figure 21.1 below for determining delivery head
Relief valve pressure rating (positive displacement pumps)	Relief valve sized to pass total flow at pressure 25% above 'set-pressure'. Set pressure = pumping pressure + 70 kN/m^2 (10 p.s.i.) for operating range 0–700 kN/m^2 (0–100 lbf/in^2) or, +10% for operating pressure above 700 kN/m^2 (100 lbf/in^2) Determine actual pressure at which full flow passes through selected valve	
Driving power	Maximum absorbed power is determined when considering: (1) total flow (2) pressure with total flow through relief valve (3) highest expected operating oil viscosity Driver size can then be selected	

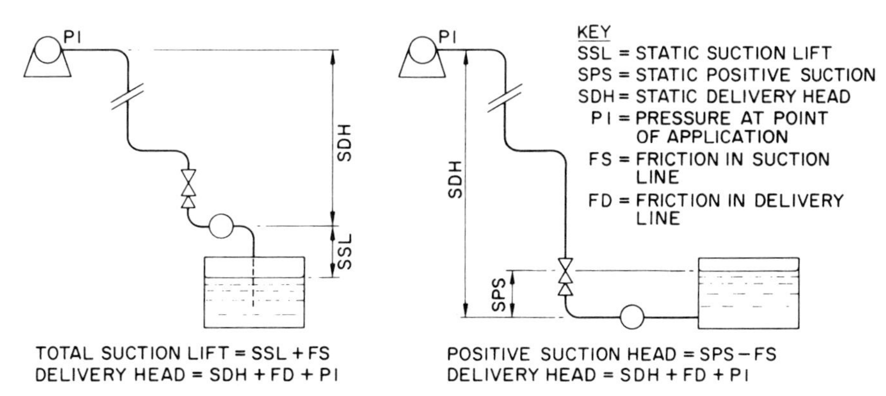

Figure 21.1 Definition of pump heads

Table 21.2 Comparison of the various types of pump

Gear pump

Spur gear relatively cheap, compact, simple in design. Where quieter operation is necessary helical or double helical pattern may be used. Both types capable of handling dirty oil. Available to deliver up to about 0.02 m^3/s (300 g.p.m.).

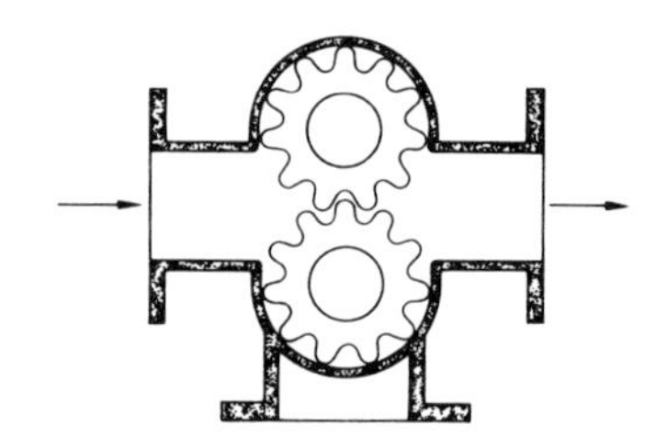

Lobe pump

Can handle oils of very viscous nature at reduced speeds.

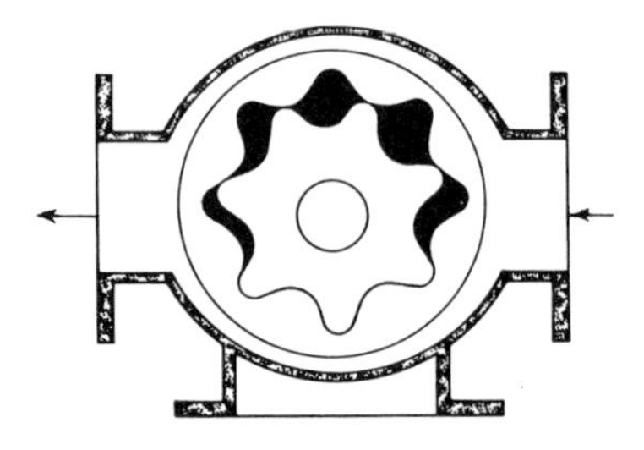

Screw pump

Quiet running, pulseless flow, capable of high suction lift, ideal for pumping low viscosity oils, can operate continuously at high speeds over very long periods, low power consumption. Adaptable to turbine drive. Available to deliver up to and above 0.075 m^3/s (1000 g.p.m.).

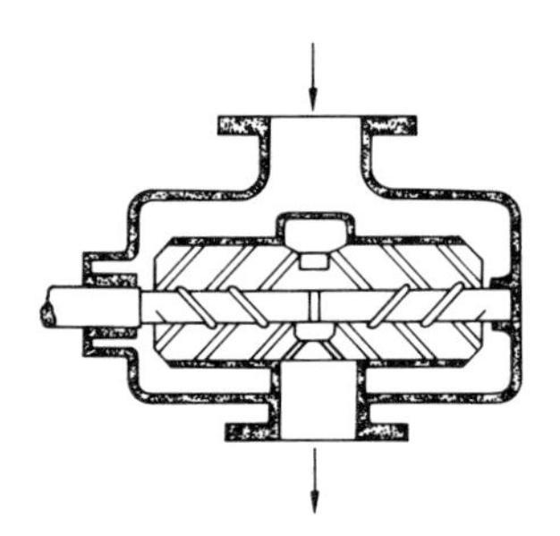

Vane pump

Compact, simple in design, high delivery pressure capability, usually limited to systems which also perform high pressure hydraulic duties.

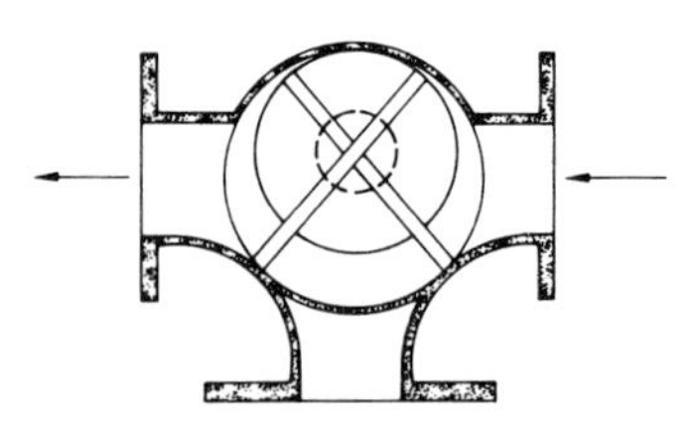

Centrifugal pump

High rate of delivery at moderate pressure, can operate with greatly restricted output, but protection against overheating necessary with no-flow condition. Will handle dirty oil.

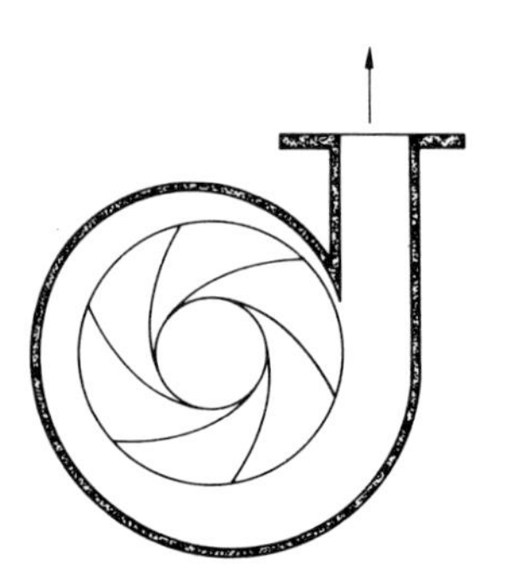

Table 21.3 Pump performance factors affecting choice of pump type

Positive displacement	*Centrifugal*
Rate of delivery at given speed substantially unaffected by changes in delivery pressure	Rate of delivery affected by change in delivery pressure. Therefore flow demand and temperature which influence pressure must be accurately controlled
Rate of delivery varies nearly directly with speed	Because wide variation in output results from change of pressure, pump is well suited for installations requiring large flows and subject to occasional transient surge conditions, e.g. turbine hydraulic controls
Delivery pressure may be increased within material strength limitation of the pump, by increasing drive power	
Very high delivery pressures can be produced by pumps designed to reduce internal leakages	

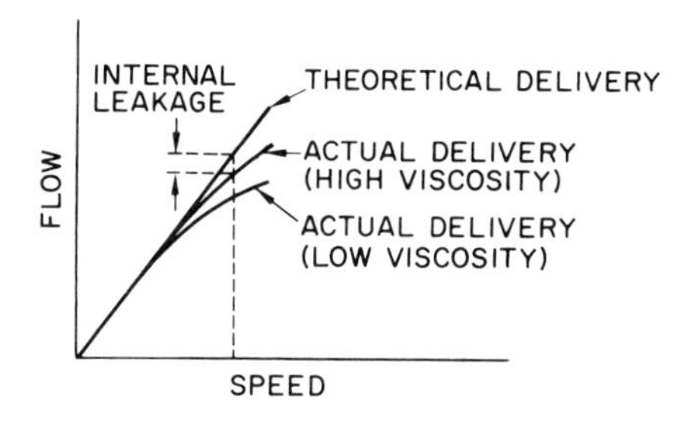

Figure 21.2 Delivery against speed and viscosity for a positive displacement pump

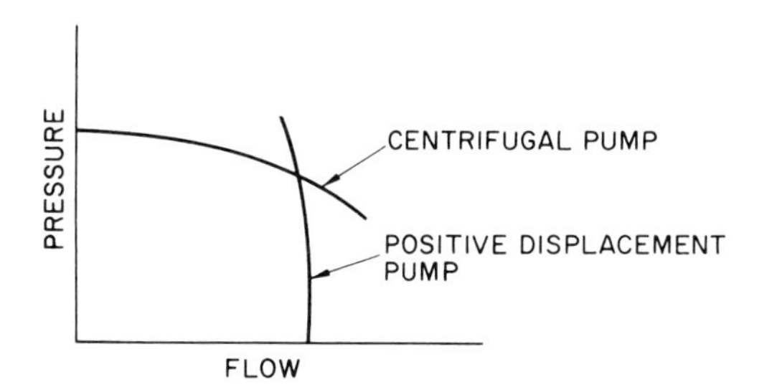

Figure 21.3 Pressure against delivery for positive displacement and centrifugal pumps

Table 21.4 Selection by suction characteristics

Pump type	*Maximum suction lift, m* (bar chart, scale 2, 4, 6, 8, 10)	*Self-priming*
Gear	bar to approx. 7.4	Yes if wet (Gear, Lobe rotor, Screw)
Lobe rotor	bar to approx. 7.4	Depends on speed and viscosity (Gear, Lobe rotor, Screw)
Screw	bar to approx. 7.4	
Vane	bar to 2	Yes
Radial piston	bar to approx. 1.5	Depends on viscosity
Centrifugal	bar to 8	No

Table 21.5 Selection by head or pressure

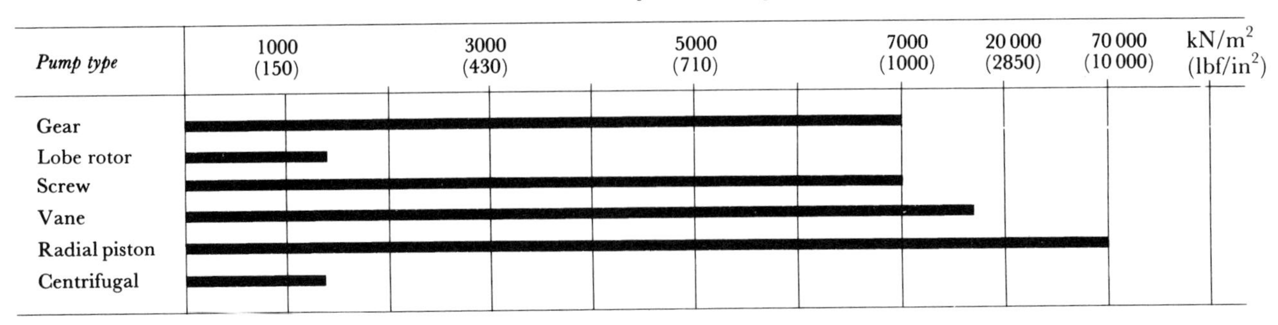

Table 21.6 Selection by capacity

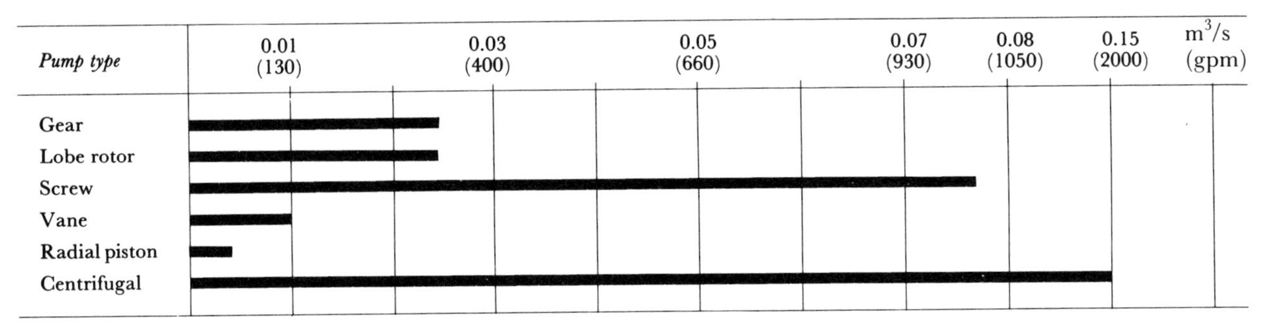

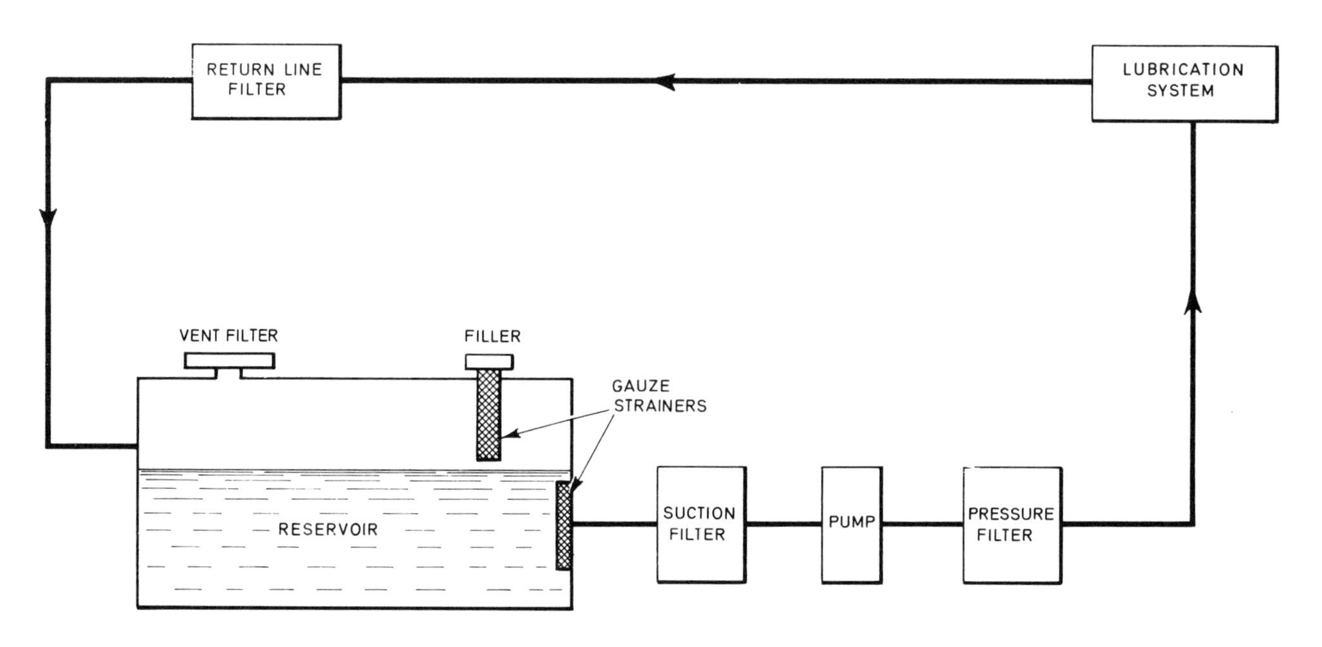

Figure 22.1 Typical circuit showing positions of various filters

Table 22.1 Location and purpose of filter in circuit

Location	*Degree of filtration*	*Type*	*Purpose*
Oil reservoir vent	Coarse	Wire wool Paper Oil bath	Removal of airborne contaminant
Oil reservoir filler	Coarse	Gauze	Prevention of ingress of coarse solids
Suction side of pump	Medium	Paper Gauze	Protection of pump
Delivery side of pump	Fine	Sintered metal Felt Paper	Protection of bearings/system
Return line to reservoir	Medium	Gauze Paper	Prevention of ingress of wear products to reservoir
Separate from system	Very fine	Centrifuge	Bulk cleaning of whole volume of lubricant

Table 22.2 Range of particle sizes which can be removed by various filtration methods

Filtration method	*Examples*	*Range of minimum particle size trapped* micrometres (μm)
Solid fabrications	Scalloped washers, wire-wound tubes	5–200
Rigid porous media	Ceramics and stoneware Sintered metal	1–100 3–100
Metal sheets	Perforated Woven wire	100–1000 5–200
Porous plastics	Plastic pads, sheets, etc. Membranes	3–100 0.005–5
Woven fabrics	Cloths of natural and synthetic fibres	10–200
Cartridges	Yarn-wound spools, graded fibres	2–100
Non-woven sheets	Felts, lap, etc. Paper—cellulose —glass Sheets and mats	10–200 5–200 2–100 0.5–5
Forces	Gravity settling, cyclones, centrifuges	Sub-micrometre

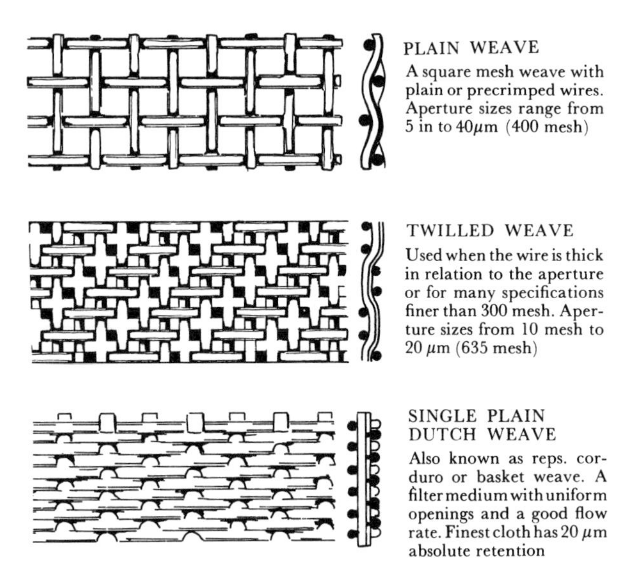

Figure 22.2 Various forms of woven wire mesh

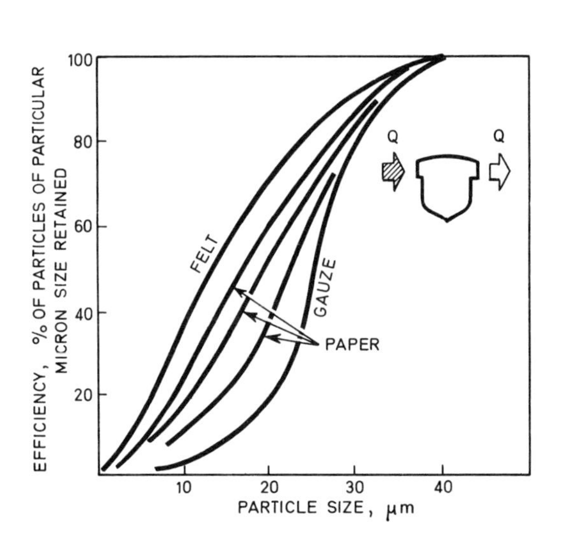

Figure 22.3 Typical filter efficiency curves

PRESSURE FILTERS

Pressure filter specification and use

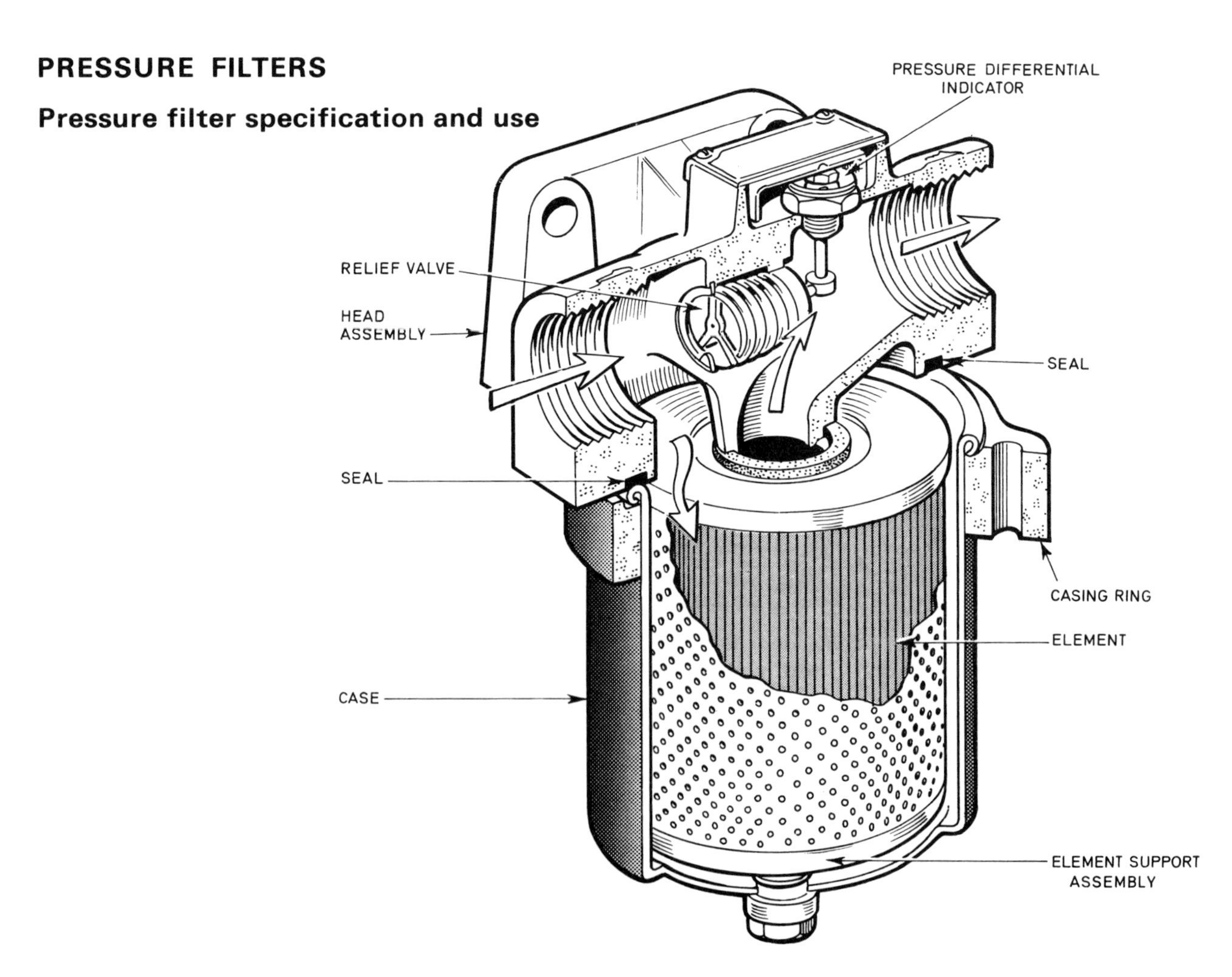

Figure 22.4 Typical full-flow pressure filter with integral bypass and pressure differential indicator

In specifying the requirements of a filter in a particular application the following points must be taken into account:

1 Maximum acceptable particle size downstream of the filter.
2 Allowable pressure drop across the filter.
3 Range of flow rates.
4 Range of operating temperatures.
5 Viscosity range of the fluid to be filtered.
6 Maximum working pressure.
7 Compatibility of the fluid, element and filter materials.

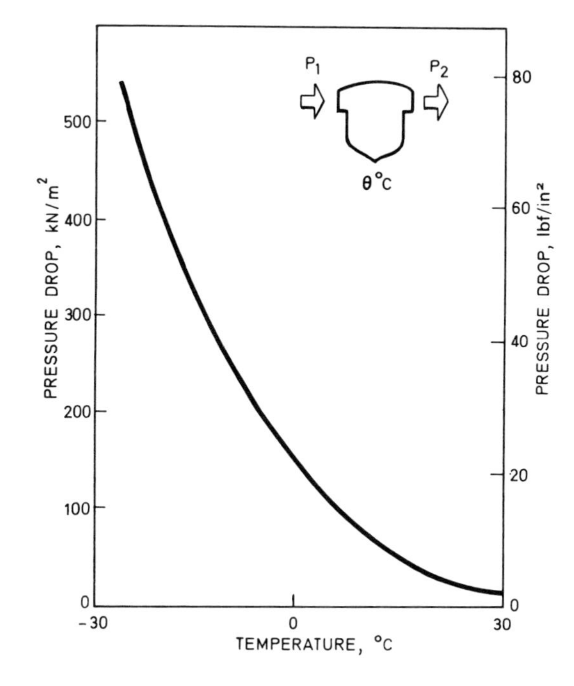

Figure 22.5 Curve showing effect of temperature on pressure drop when filtering lubricating oil

In-line filtration

In many systems, the lubricating oil flows under pressure around a closed circuit, being drawn from and returned to a reservoir. The same oil will then pass through the system continuously for long periods and effective filtration by one of two approaches is possible, i.e. full-flow filtration and bypass filtration.

Full-flow filtration

A full-flow filter will handle the total flow in the circuit and is situated downstream of the pump. All of the lubricant is filtered during each circuit.

ADVANTAGE OF FULL FLOW

All particles down to specified level are removed.

Figure 22.6 Simplified circuit of full-flow filter

Bypass filtration

In bypass filtration only a proportion of the oil passes through the filter, the rest being bypassed unfiltered. In theory, all of the oil will eventually be filtered but the prevention of the passage of particles from reservoir to bearings, via the bypass, cannot be guaranteed.

ADVANTAGES OF BYPASS

Small filter may be used. System not starved of oil under cold (high viscosity) conditions. Lower pressure drop for given level of particle retention. Filter cannot cut off lubricant supply when completely choked.

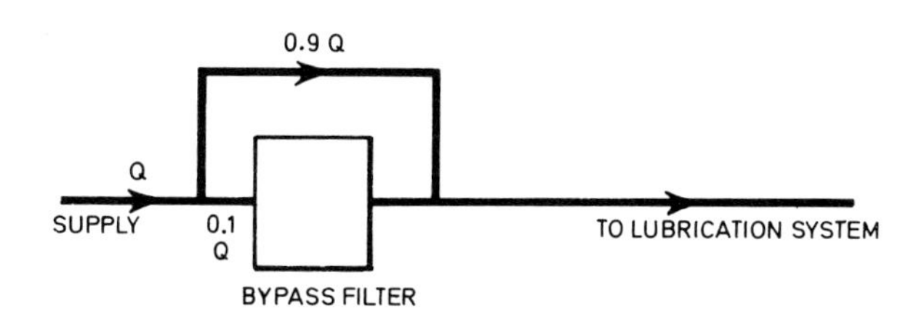

Figure 22.7 Simplified circuit of bypass filter

CENTRIFUGAL SEPARATION

Throughput specification

Selection of a centrifugal separator of appropriate throughput will depend on the type of oil and the system employed. A typical unit of nominal 3000 l/h (660 gal/h) should be used at the following throughput levels:

	l/h	gal/h
Lubricating oil, straight:		
Bypass system, maximum	3000	660
Bypass system, optimum	1200	265
Batch system recommended	1900	420
Lubricating oil, detergent:		
Bypass system, maximum	1800	395
Bypass system, maximum	750	165
Batch system, recommended	1150	255

Operating throughputs of other units may be scaled in proportion.

Recommended separating temperatures

Straight mineral oils, 75°C (165°F).
Detergent-type oils, 80°C (175°F).

Fresh-water washing

Water washing of oil in a centrifuge is sometimes advantageous, the following criteria to be used to determine the hot fresh-water requirement:

	Quantity	*Requirement*
Straight mineral oil	3–5% of oil flow	About 5°C (9°F) above oil temperature
Detergent-type oil*	Max. 1% of oil flow	About 5°C (9°F) above oil temperature

* Only on oil company recommendation.

Lubricating oil heaters and coolers are available in many different forms. The most common type uses steam or water for heating or cooling the oil, and consists of a stack of tubes fitted inside a tubular shell. This section gives guidance on the selection of units of this type.

LUBRICATING OIL HEATERS

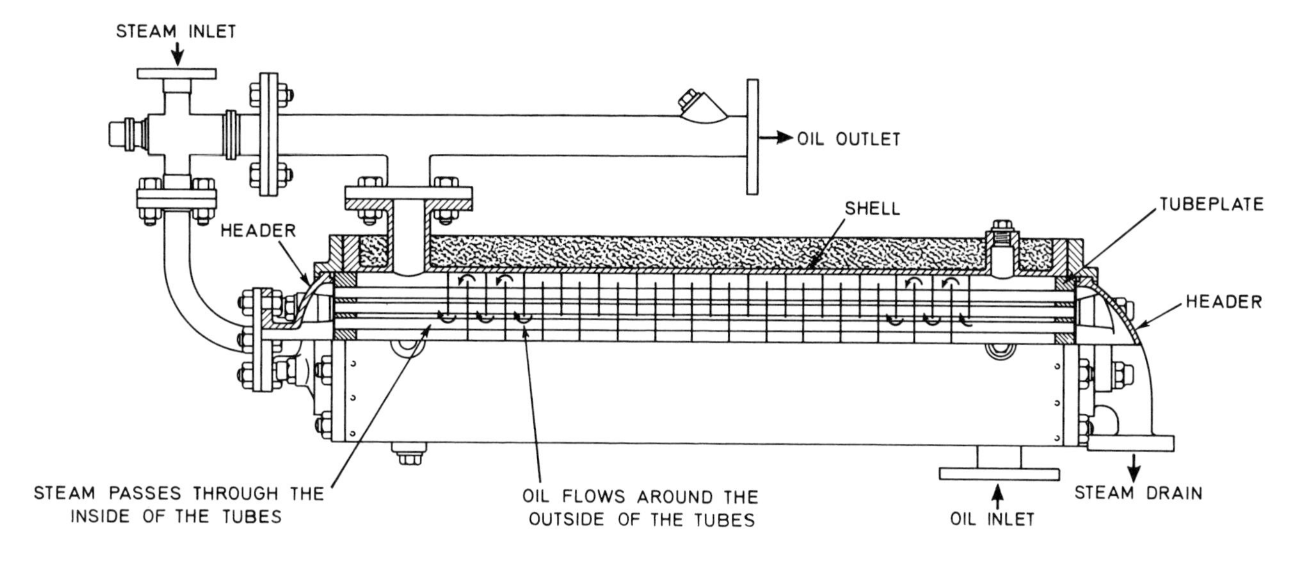

Figure 23.1 Cross section through a typical oil heater

The required size of the heater and the materials of construction are influenced by factors such as:

Lubricating oil circulation rate.
Lubricating oil pressure and grade or viscosity
Maximum allowable pressure drop across the heater.
Inlet lubricating oil temperature to heater.
Outlet lubricating oil temperature from heater.
Heating medium, steam or hot water.
Inlet pressure of the steam or hot water to the heater.
Inlet steam or hot water temperature.

Table 23.1 Guidance on materials of construction

Component	Suitable material	Remarks
Shell	Cast iron Mild steel fabrication	In contact with oil
Tubes	Mild steel	In contact with steam and hot water, but corrosion is not usually a problem with treated boiler water
Tube plates	Mild steel	
Headers	Gunmetal Cast iron Mild steel fabrication	

Guidance on size of heat transfer surface required

The graph shows how the required heat transfer surface area varies with the heat flow rate and the oil velocity, for a typical industrial steam heated lubricating oil heater, and is based on:

Heating medium	Dry saturated steam at 700 kN/m^2 (100 p.s.i.)
Oil velocity	Not exceeding 1 m/s
Oil viscosity	SAE 30
Oil inlet temperature	20°C
Oil outlet temperature	70°C

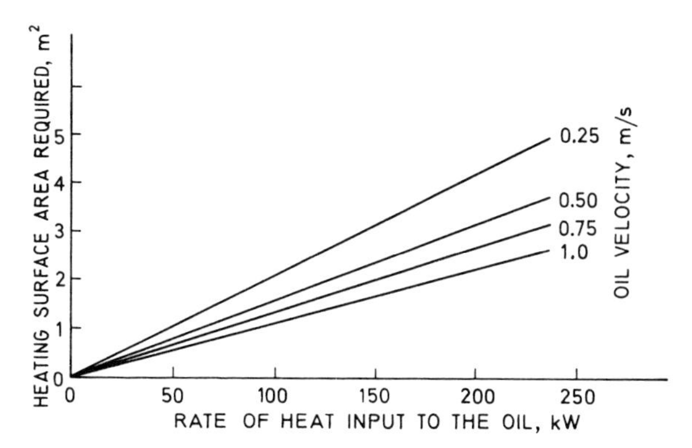

Figure 23.2 Guide to the heating surface area for a desired rate of heat input to oil flowing at various velocities

LUBRICATING OIL COOLERS

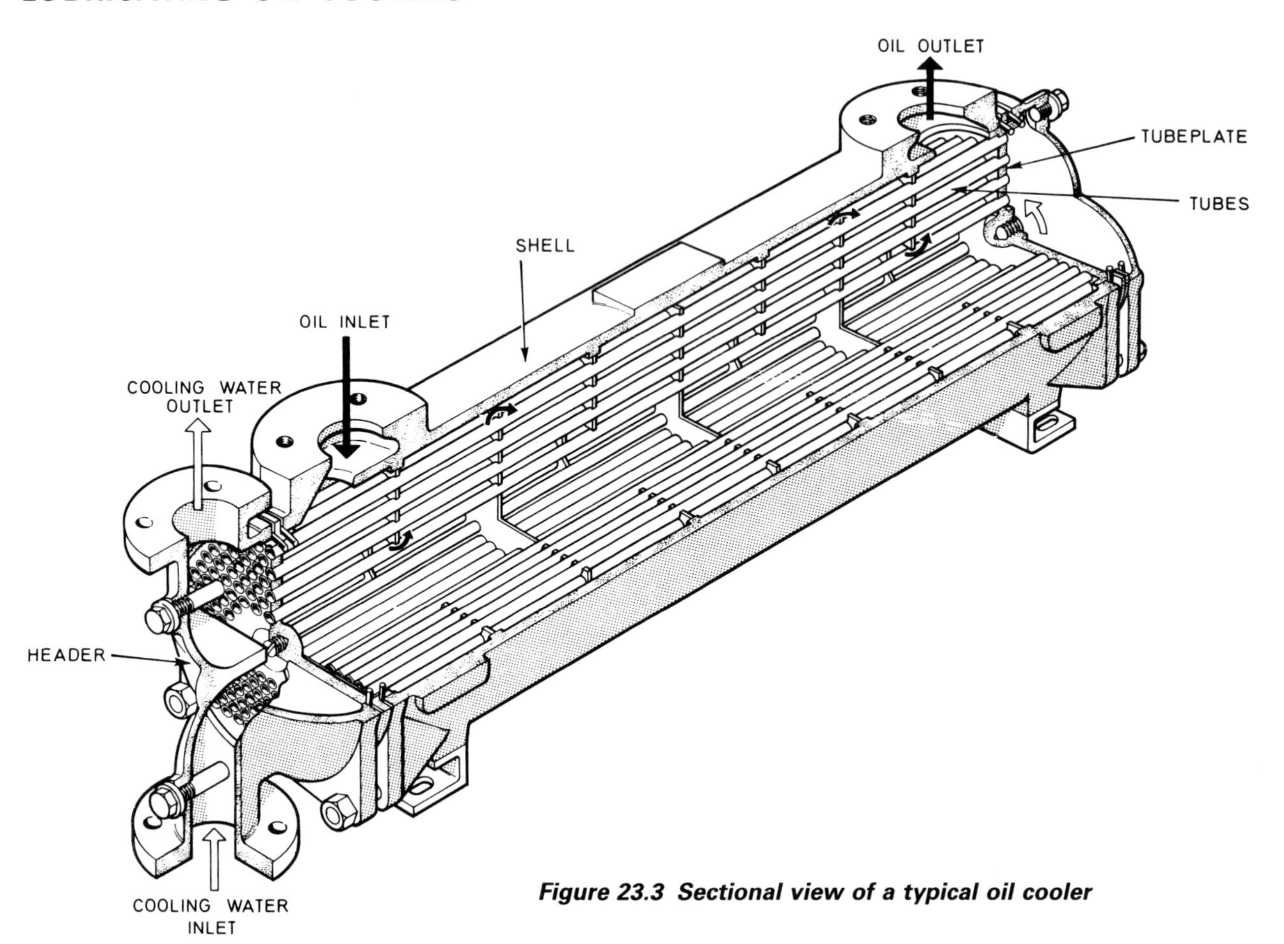

Figure 23.3 Sectional view of a typical oil cooler

The required size of cooler and the materials of construction are influenced by factors such as:

- Lubricating oil circulation rate.
- Lubricating oil pressure and grade or viscosity.
- Maximum allowable pressure drop across the cooler.
- Inlet lubricating oil temperature to cooler.
- Outlet lubricating oil temperature from cooler.
- Cooling medium (sea water, river water, town water, etc.)
- Cooling medium pressure.
- Cooling medium inlet temperature to cooler.
- Cooling medium circulation rate available.

Table 23.2 Guidance on materials of construction

Component	*Suitable material*	*Remarks*
Shell	Cast iron Aluminium Gunmetal Mild steel fabrication	In contact with oil
Headers	Cast iron Gunmetal Bronze	In contact with cooling water but are of thick section so corrosion is less important
Tubes	Copper based alloys Titanium	The risk of corrosion by the cooling water is a major factor in material selection. Guidance is given in the next table
Tube plates	Copper based alloys Titanium	

Table 23.3 Choice of tube materials for use with various types of cooling water

Material	*Cooling water*		
	Aerated non saline waters	*Aerated saline waters*	*Polluted waters*
Copper	S	NR	NR
70/30 Brass	S	NR	NR
Admiralty brass	R	NR	NR
Aluminium brass	R	R	NR
90/10 Copper nickel	X	X	S
68/30/1/1 Copper nickel	X	X	R
Titanium	X	X	R
	River, canal, town's main water, deionised and distilled water	Estuarine waters seawater	Polluted river canal, harbour and estuarine waters, often non aerated

NR = Not recommended, S = Satisfactory, R = Recommended, X = Satisfactory, but more expensive

Guidance on the size of cooling surface area required

The graph shows how the cooling area required varies with the heat dissipation required, and the cooling water temperature for typical lubricating oil system conditions of:

Oil velocity	0.7 m/s
Oil viscosity	SAE 30
Water velocity	1 m/s
Oil inlet temperature	70°C
Oil outlet temperature	60°C

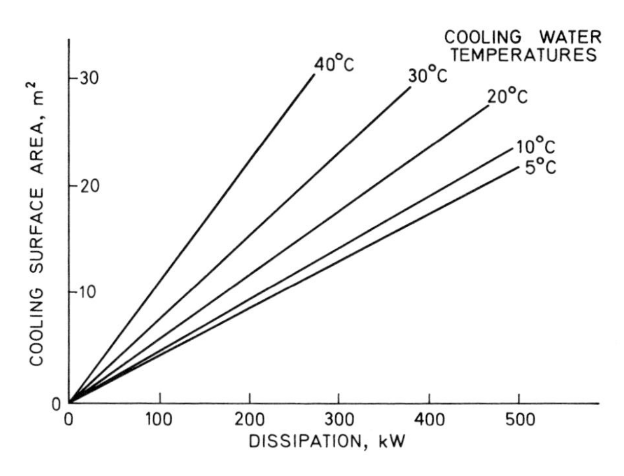

Figure 23.4 Guide to the cooling surface area required for a desired dissipation rate at various cooling water temperatures

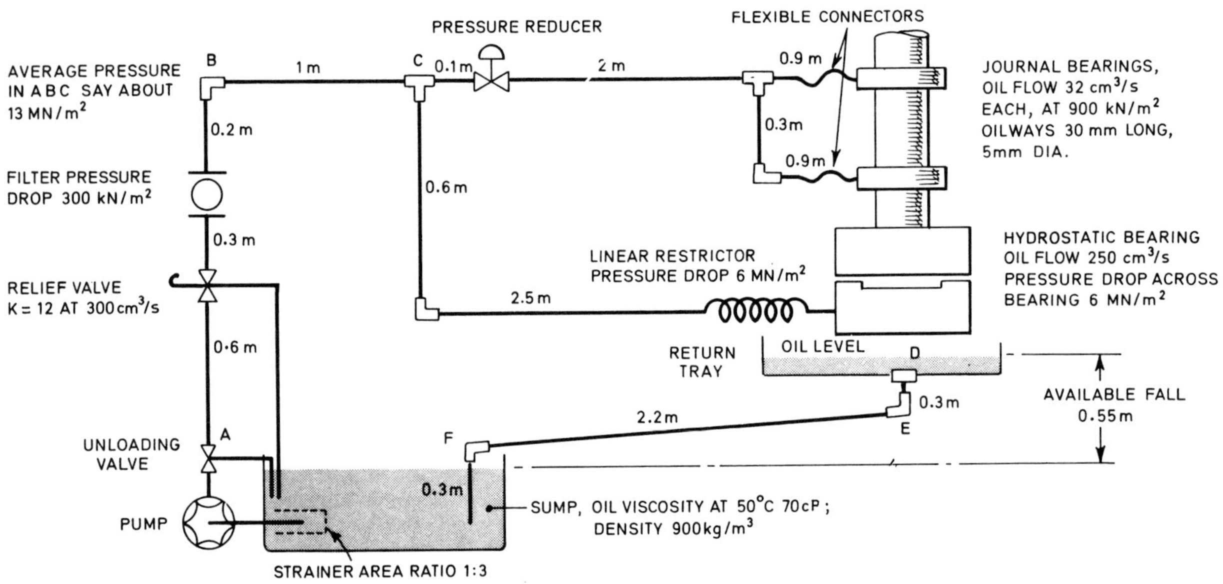

Figure 24.1 Typical lubrication system

Table 24.1 Selection of pipe materials

Material	*Range*		*Approx. relative price*		*Applications*	*Notes*
	Bore dia. in	*Pressure** lbf/in²	*Material only*	*Installed*		
Mild steel	No limit	No limit	2	3	Not recommended. Occasionally used for mains in large high-pressure systems where cost important	Thoroughly clean and de-scale by acid pickling. Seal ends after pickling until installation
Stainless steel	No limit	No limit	12	6	Large permanent systems	Best material for high flow rates. Can be untidy on small systems
Half-hard copper and brass	0.07–2	5000–1000	4	4	Universal. Especially favoured for low flow rates	Brass resists corrosion better than copper. Neat runs and joints possible
Flexible reinforced high-pressure hose	0.125–0.75	5000–2500	9	4	As final coupling to bearing to assist maintenance. Connection to moving part or where subject to heavy vibration	Nylon or rubber base obtainable
Hard nylon	0.125–1.125	700–350	2	2	As cheap form of flexible coupling. Large centralised low-flow lubrication systems for cheapness	Deteriorates in acid atmosphere. Heavily pigmented variety should be used in strong light
PVC reinforced	0.125–0.75	150	2	1	Low-pressure systems. Gravity returns	Readily flexible. Pigmented variety best in sunlight
PVC unreinforced	0.125–2	20–10	1	1	Gravity returns	Pigmented variety best in sunlight. Can be untidy. Relatively easily damaged

* The pressures quoted are approximate working pressures corresponding to the minimum and maximum bores, respectively, and are intended as a guide to selection only. Use manufacturers' values for design purposes.

Table 24.2 Selection of control valves

Valve type	*Operation*	*Common use*	*Notes*
Bypass (unloading)	Bleeds to drain when primary pressure exceeds given value	Control main circuit pressure	Designed to operate in open position without oil heating. Often included in pump unit, otherwise requires return to sump 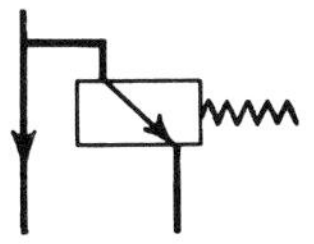
Relief (safety)	As above	Safety device to protect pipework fittings and pump	If used to control circuit pressure, oil heating may result. Do not undersize or forget return to sump 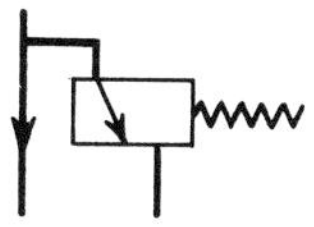
Non- return (check)	Prevents reverse flow	Prevent back-flow through a bypass	Sometimes incorporated in bypass or relief valve 
Pressure-reducing (regulating)	Gives fixed pressure drop or fixed reduced pressure	Cater for different bearing needs from one supply pressure	Requires drain to sump 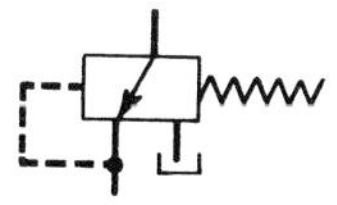
Flow-regulating and -dividing	Controls or divides flow rate irrespective of supply pressure	Pass fraction of flow to open (non-pressure) system or to filter	Approximate control only. Orifices are cheaper alternative, as effective for many applications 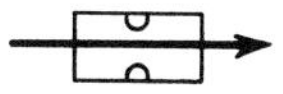
Sequence	Admits oil to secondary circuit only after primary reaches given pressure	Protection of a particular bearing circuit in complex system	Rather uncommon. May be used in hydraulic power/bearing circuit 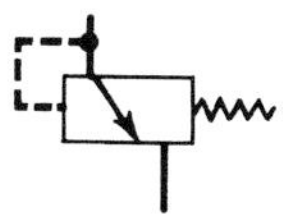
Directional	Switches flow on receipt of signal	Activate various sections of circuit as required	Remote or manual operation. For simple systems, stop valves may be better 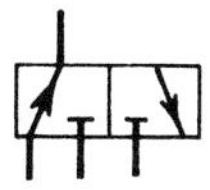
Volume metering	Meter quantity of lubricant to bearing on receipt of pressure pulse	Intermittent lubrication system only, placed at lubrication points	

Table 24.3 Pipe sizes and pressure-drop calculations

Item	*Equation,* figure*	*Example*	*Notes*
Determination of pipe bore			
(a) Draw flow diagram		See Figure 24.1	Essential for all but the simplest cases
(b) Size lines assuming velocities: (i) 3 m/s (10 ft/s) for delivery (ii) 1 m/s (3 ft/s) for pump suction (iii) 0.3 m/s (1 ft/s) for return (iv) 20 m/s (60 ft/s) for linear restrictors	$d = \sqrt{\frac{4Q}{\pi v}}$ or use Figure 24.2	For ABC in Figure 24.1: $Q = 314$ cm^3/s From Figure 24.2, $d \simeq 12$ mm or 0.5 in	Adjust sizes, if necessary, after calculation of pressure drops. Use reducer, if necessary, in pump suction. If air entrainment likely in return line, size to run half full
Pressure drop—delivery and pump suction lines			Keep suction lines short, free of fittings
(a) Determine correct viscosity for working conditions of temperature and pressure	$\eta = X \times \eta_0$ for pressure, Figure 24.3	From Figure 24.3 at 13 MN/m^2: $X = 1.4$, $\eta = 1.4 \times 70 = 98$ cP	Use supplier's value for viscosity at working temperature (usually 30–50°C above ambient)
(b) Evaluate pipe viscous losses	$P_p = 32\frac{\eta v l}{d^2}$ or use Figure 24.4	From Figure 24.4 for ABC: $P_p = 65$ kN/m^2/m $\times 2.1 = 137$ kN/m^2	Check to ensure flow is laminar (Reynolds number < 2000). Find Re from Fig. 24.4
(c) Determine losses in valves, fittings and strainers	$P_c = k \times \frac{1}{2}\rho v^2$, find K from Table 24.4 Manufacturer's data	For the 90° bend at B: $P_c = 2 \times \frac{1}{2} \times 900 \times 9 = 8.1$ kN/m^2 For the relief valve: $P_c = 12 \times \frac{1}{2} \times 900 \times 9 = 48.6$ kN/m^2 For the filter: $P_c = 300$ kN/m^2	Use manufacturers' figures for K or P_c, where available. Where the flow changes direction, e.g. in a bend, losses increase at low Reynolds number (2× at Re = 200, 4× at Re = 100 and 8× at Re = 50)
(d) Determine total loss	$P = \sum P_p + \sum P_c$	For ABC: $P = 137 + 48.6 + 300 + 8.1 \simeq 494$ kN/m^2	The total loss is simply the arithmetical sum of the individual losses
Pressure drop—return lines			
(a) Calculate pressure losses as for delivery lines	As for 2(a) to (d)	For DEF: $P_p = 2.8 \times 0.8 = 2.24$ kN/m^2 K (2 bends, entrance and exit) — $2+2+1+1 = 6$ $P_c = 6 \times \frac{1}{2} \times 900 \times 0.3^2 = 0.25$ kN/m^2 $P = 2.49$ kN/m^2	Keep lines as direct as possible, with minimum fittings. Self-draining lines are best (1½° minimum slope)
(b) Express P as hydrostatic head of oil	$h = \frac{P}{\rho g}$	$h = \frac{2490}{900 \times 9.81} = 0.28$ m	h must be less than vertical drop available, i.e. <0.55 m in the example of Figure 24.1
(c) Check to ensure positive pressure after each fitting		For the entrance D: $K = 1$, $P_c = 40$ N/m^2, $h = 45$ mm Thus, return tray must be at least 45 mm deep	Increase bore size if this condition cannot be met. The entrance to the return pipe is often a source of difficulty
Pressure drop—coiled pipes			
Linear restrictors for hydrostatic bearing circuits are often coiled capillary tubing. Coiling increases the pressure drop and must be allowed for when calculating the length of the restrictor	$P_p = \frac{32\eta v l}{d^2}$ or use Fig. 4 $P_t = P_p \times Z$, Z from Fig. 6	For G in Figure 24.1: $Q = 250$ cm^3/s; $d = 4$ mm, say, when $v = 18$ m/s (Figure 24.2) $P_p = 3.5$ MN/m^2/m (Figure 24.4), giving $l = 1.71$ m for 6 MN/m^2 Now Re = 650 (Figure 24.5) and if D (dia. of coil) = 36 mm, $Z = 2$ Thus required $l = 1.71/2 = 0.85$ m	Fine bore tubes make better restrictors than orifices for large pressure drops Base calculations on actual bore sizes not nominal bores If l impractically long or short, increase or reduce v, respectively, and repeat calculation

* When using the equations with Imperial units, these must be self-consistent. For example, Q in in^3/s; v in in/s; d, h, e in in; η in reyns (1 reyn = 69 000 poise); ρ in slugs/in^3 (1 slug/in^3 = 32.2 lb/in^3); p in lbf/in^2; g = 386 in/s^2

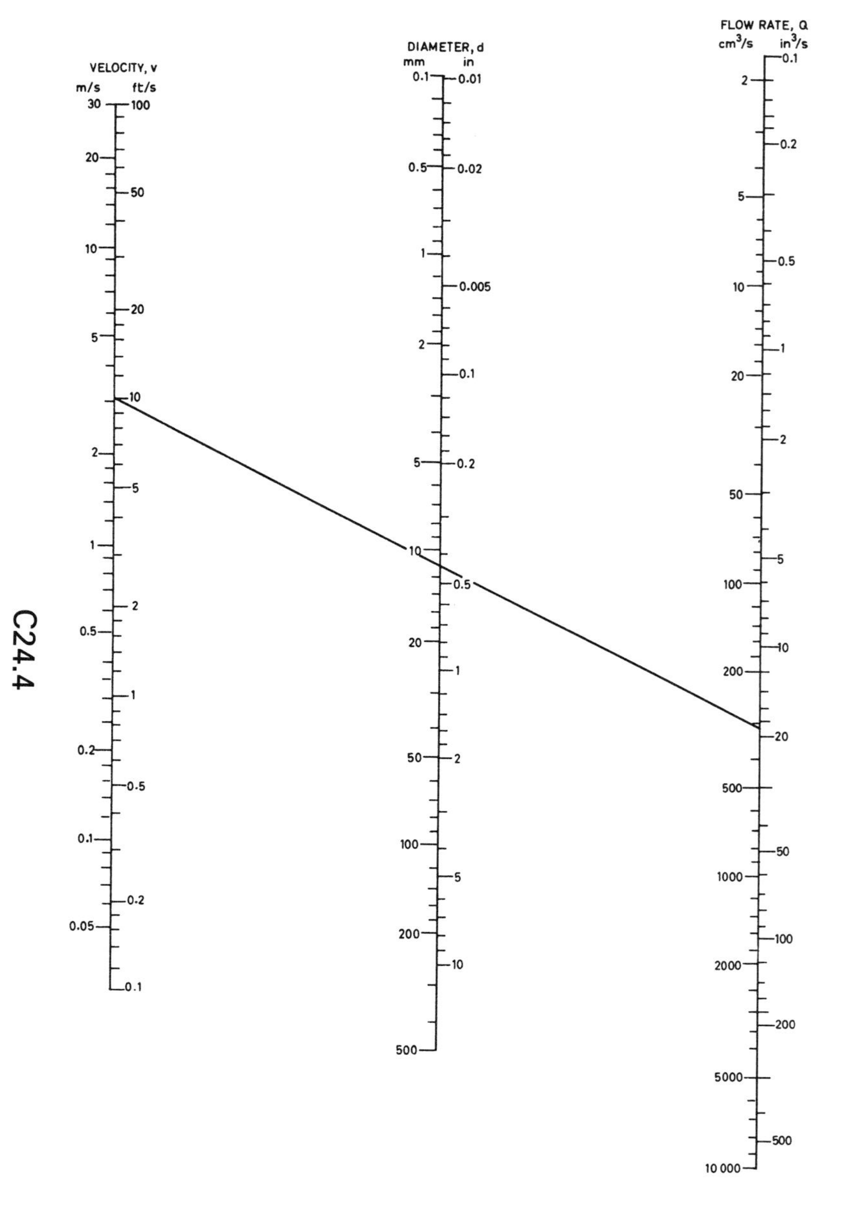

Figure 24.2 Nomogram for determination of pipe bore

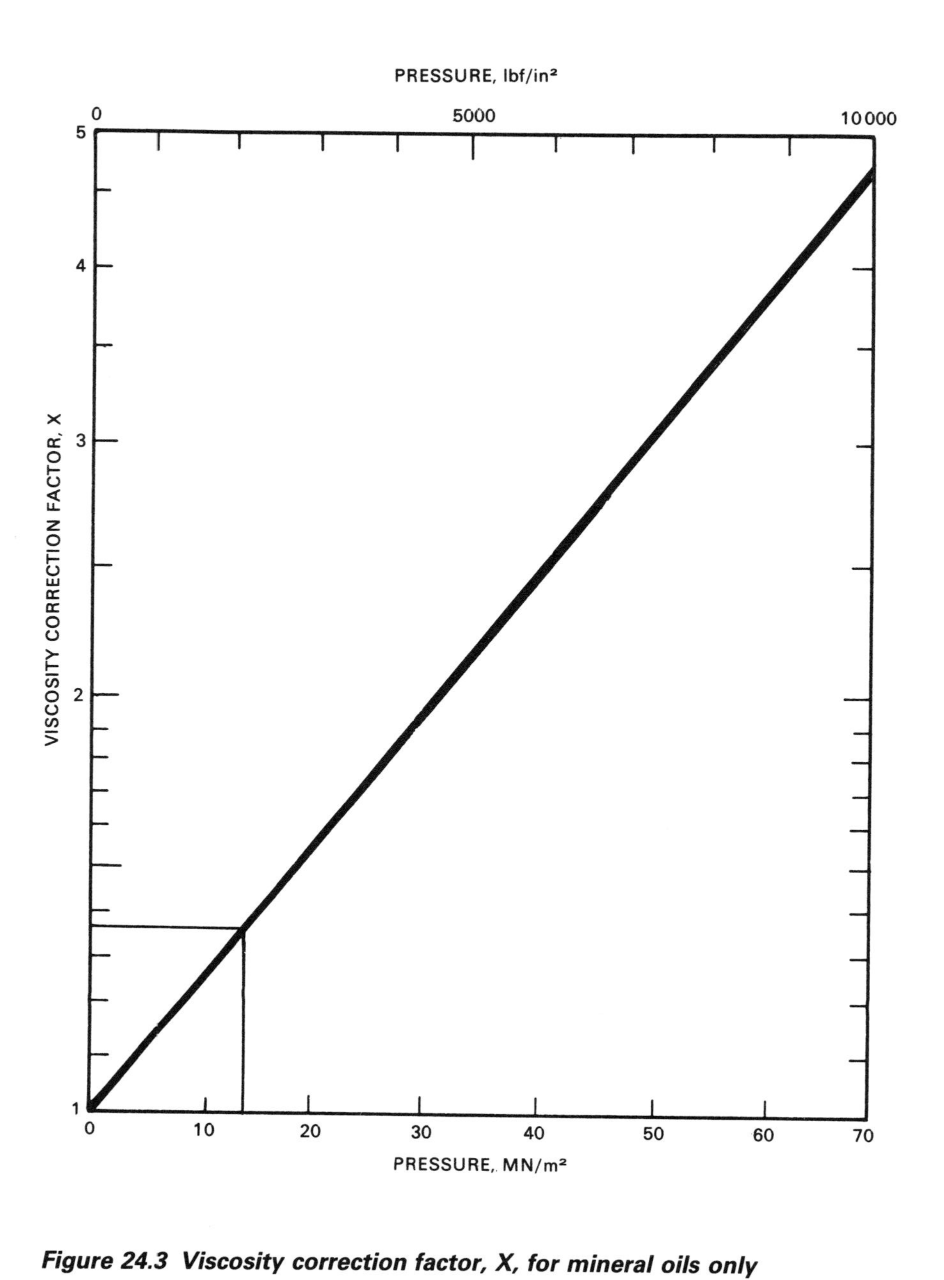

Figure 24.3 Viscosity correction factor, X, for mineral oils only

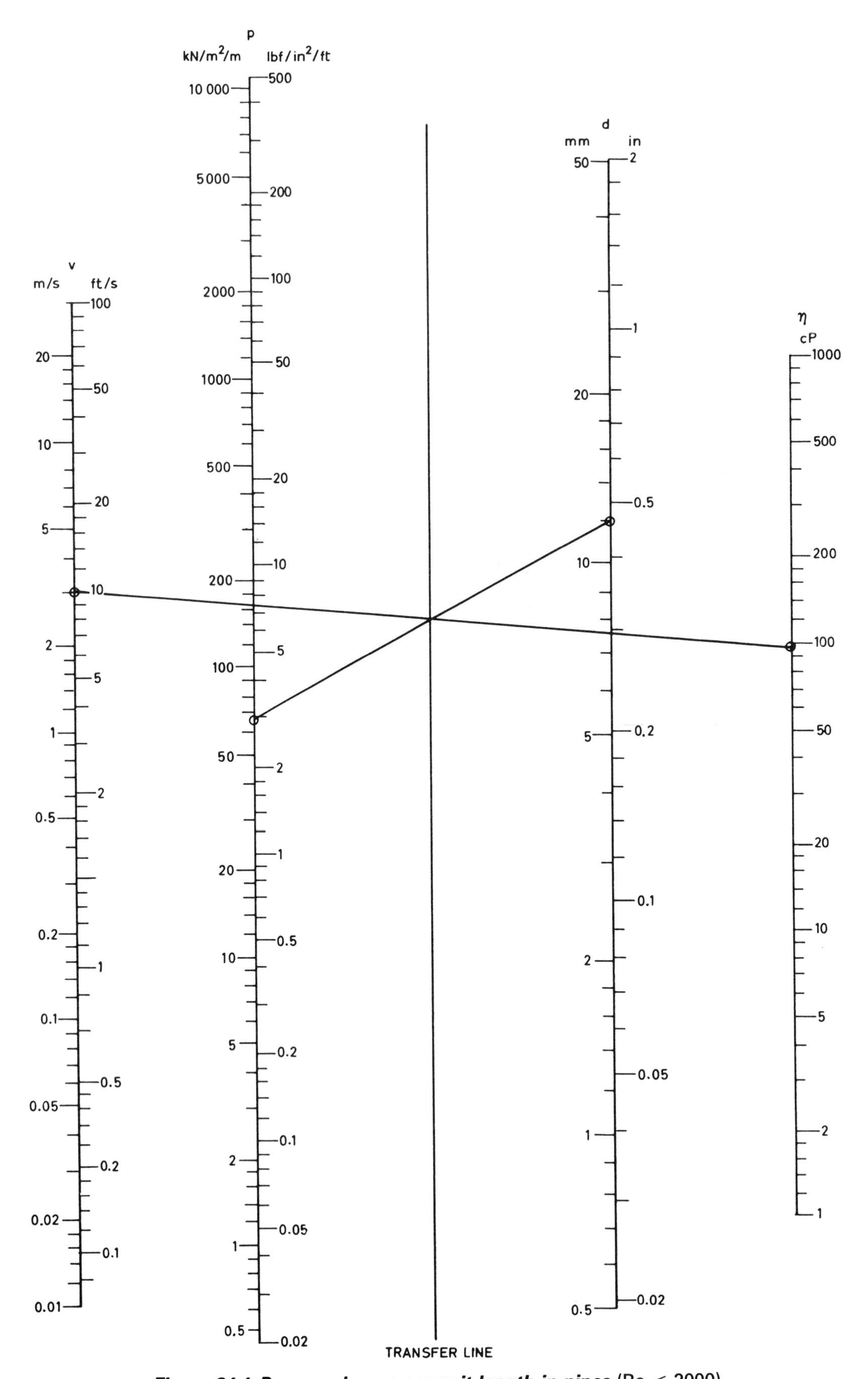

Figure 24.4 Pressure losses per unit length in pipes (Re < 2000)

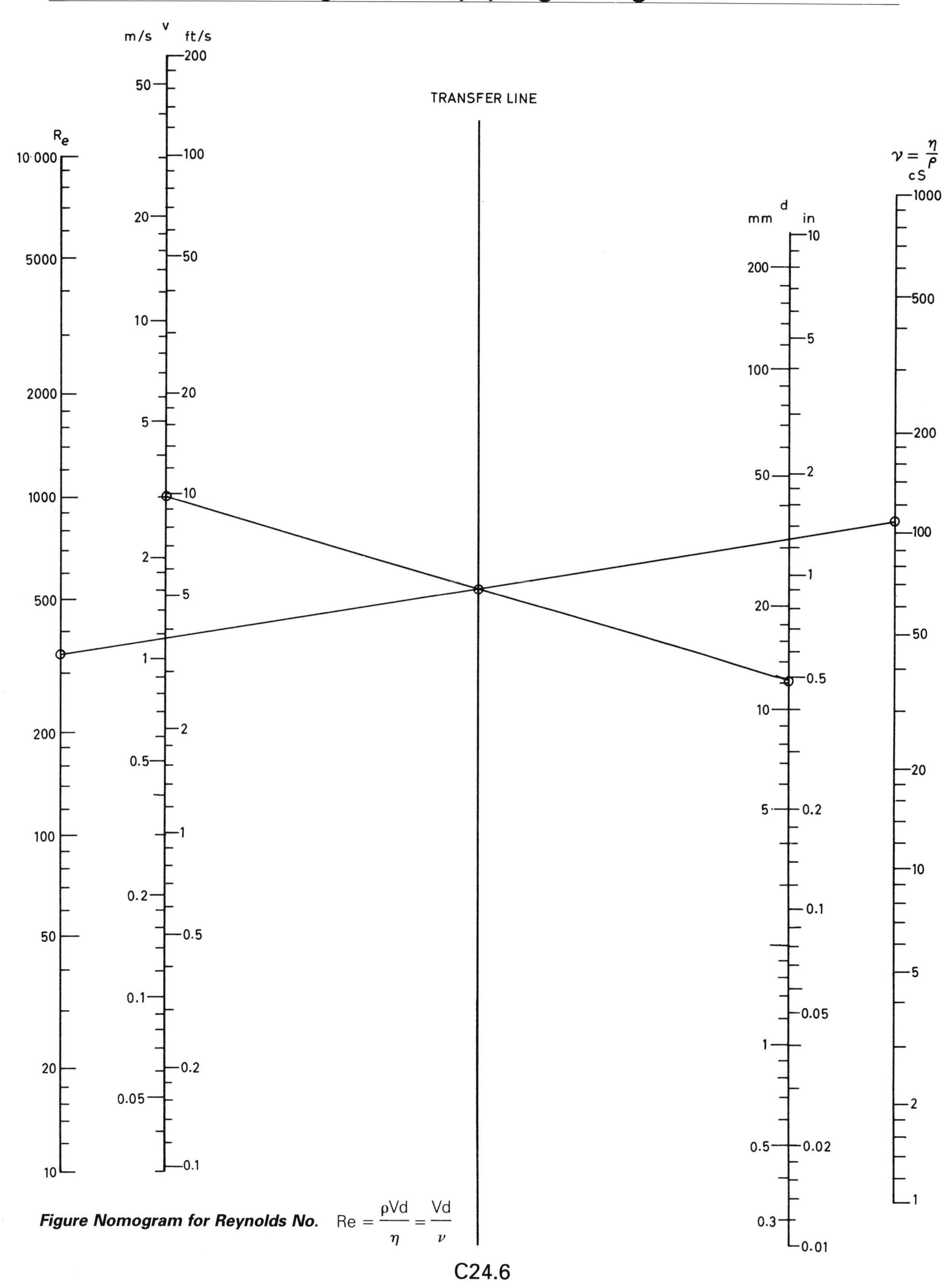

Figure Nomogram for Reynolds No. $Re = \frac{\rho V d}{\eta} = \frac{Vd}{\nu}$

Table 24.4 Loss coefficients

(a) FITTINGS

	Enlargement	*Contraction*	*Restrictions*			*Bends*
			A→a	→d, l, l/d < 5	a	
$K =$	1	0·5	$K \simeq (A/a)^2$ where A is the area of the pipe and a the open area available			2

	Entrances			*Exit*	*T-junction*	
$K =$	0	0.5	1	1	3.0	0.5

(b) VALVES

	Stop			*Non-return*		*Spool*
	Gate	*Plug*	*Needle*	*Flap*	*Ball*	
$K =$	2	20	60	10	50–100	5–100

Use manufacturers' figures, where available. Approximate loss coefficient in doubtful cases may be obtained from the formula

$$K = \left(\frac{\text{Cross-sectional area of approach pipe}}{\text{Minimum cross-sectional area of valve}}\right)^2$$

(c) STRAINERS

Approximate loss coefficient $K = \left(\frac{\text{Cross-sectional area of pipe}}{\text{Open area of strainer}}\right)^2$

(d) FILTERS

Not susceptible to calculation—use manufacturers' data. If a relief bypass valve is included the actual pressure drop may be two or three times the nominal setting.

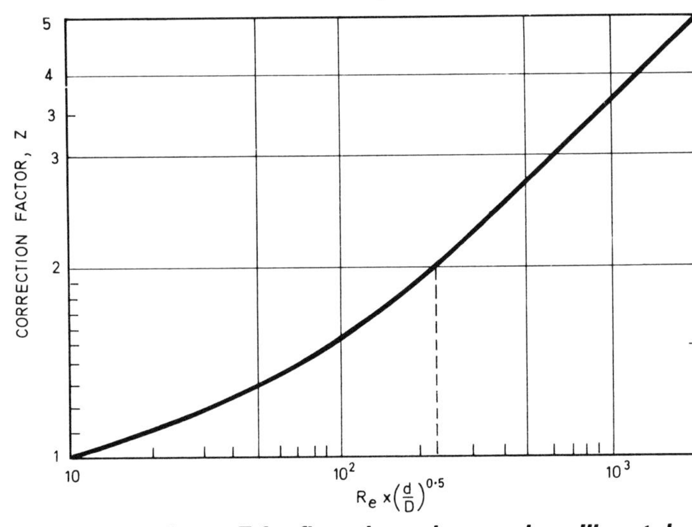

Figure 24.6 Correction factor Z for flow through curved capillary tubes of bore diameter d and coil diameter D

Satisfactory operation of a centralised recirculatory lubrication system requires adequate control and instrumentation to ensure continuous delivery of the correct volume of clean oil at the design pressure and temperature.

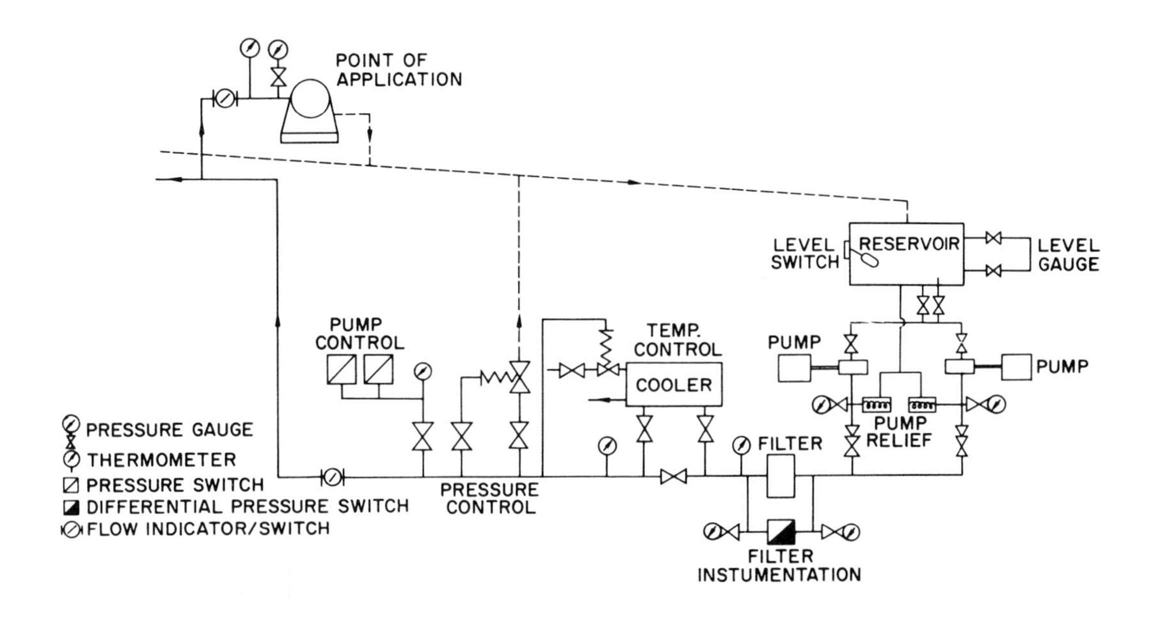

Figure 25.1 A basic lubrication system complete with warning and protection devices

Table 25.1 The function of each major system component and the device required to provide the information or control necessary to maintain that function

System component	*Function of component*	*Information or control required to maintain function*	*Device required*	*Comment on application*
Reservoir	Maintenance of required volume of oil	Level indication	Level gauge	Visual indication of contents
		Level control	Level switch	(1) High level warning of overfilling or ingress of water (2) Low level warning of system leak or normal consumption
Pump(s)	Delivery of the required quantity of oil at prescribed pressure	Pressure indication	Pressure gauge	(1) Situated at pump discharge, is necessary when adjusting pump relief valve (2) Situated at a point downstream of system equipment, is necessary when adjusting pressure control valve
		Pressure control	(1) Pressure switch(es)	To switch in standby pump on falling pressure. Give warning of falling pressure
			(2) Pressure control valve or other method	To spill-off surplus oil and regulate pressure variations due to temperature fluctuations and changes in system demand
		Flow indication and control	Flow switch	Indicates satisfactory flow is established
		Pump protection	Spring relief valve	Protection from overpressure due to system malfunction

Table 25.1 The function of each major system component and the device required to provide the information or control necessary to maintain that function (continued)

System component	*Function of component*	*Information or control required to maintain function*	*Device required*	*Comment on application*
Filter(s)	Maintenance of cleanliness of oil	Condition of filter	(1) Pressure gauges	Visual indication of pressure drop across filter
			(2) Differential pressure switch	To signal when pressure drop has increased to a pre-determined value and filter requires cleaning
Cooler(s)	Maintenance of prescribed temperature of oil supply	Temperature indication	Thermometer	Visual indication of temperature of oil into and out of cooler. Is necessary when adjusting temperature control valve
		Temperature control	Temperature control valve	Regulates temperature of oil leaving cooler
Point of application to lubricated parts	Dispense correct quantity of oil at prescribed pressure and temperature	Oil pressure, oil temperature, oil flow	Pressure gauge Thermometer Flow indicator	Local visual indication or fitted with contacts to give malfunction signal

Table 25.2 Some protective devices available with guidance on their selection and installation

Device	*Some types*	*Comments on selection*	*Installation*
Level gauge	(1) Dip-stick	Cheap, simple to make	Through reservoir top
	(2) Glass tube	Direct reading, simple in design, requires protecting with metal tube or cage	On reservoir side complete with shut-off cocks
	(3) Dial. Float actuated	Direct reading. For accuracy of calibration reservoir dimensions, shape and any internal obstructions, also specific gravity of oil must be considered	On reservoir side or top
	(4) Dial. Hydrostatic, pneumatic operated	Remote reading. Comments as for (3). Absolutely essential capillary joints are positively sealed against leak	Through reservoir top or side with dial remote panel mounted
Level switch	(1) Float actuated	Usually magnetic operation, glandless, therefore leakage from reservoir into switch housing not possible. Some types are level adjustable	Through reservoir top or side, or, on reservoir side in float chamber complete with isolating valves
	(2) Sensing probe	Techniques generally used; capacitance, conductivity or resistance. Complete lubricant characteristics must be provided to supplier for selection purposes. Not adjustable after fitting	Through reservoir top or side
Pressure gauge	(1) Bourdon tube	Available for front flange, back flange or stem mounting. Any normal pressure range	Local stem mounted or remote panel mounted, complete with shut-off cock
	(2) Diaphragm actuated	More robust construction, more positive indication than Bourdon tube type, mainly stem mounting. Withstand sudden pressure surges, overload pressures, overheating. Any normal pressure range	As for (1)
Pressure switch	(1) Bellows actuated (2) Piston actuated (3) Bourdon tube	Select switch to satisfy pressure and differential range requirements, also oil temperature. (1) and (3) suitable for oil and air. (2) limited to oil applications; will withstand higher pressures	Local stem mounted or remote panel mounted

***Table 25.2 Some protective devices available with guidance on their selection and installation** (continued)*

Device	*Some types*	*Comments on selection*	*Installation*
Pressure control valve	(1) Standard high-lift spring-loaded relief	Simple design, cheap. Is viscosity sensitive but normally satisfactory for control of small simple systems	Piped-in branch from main delivery after filter. Discharge back to reservoir
	(2) Direct operated diaphragm	Used on larger systems, this valve will maintain control within acceptable limits provided viscosity remains reasonably stable	Fit as for (1). Pressure signal transmitted to diaphragm from tapping downstream of filter and cooler
	(3) Pneumatically controlled diaphragm	Employs same type of valve as in (2) but the diaphragm is actuated by low pressure air via a control instrument, providing more positive control. Used in place of a direct operated valve, where large diaphragm loads might occur	Fit as (1). Air control remote panel mounted. Pressure signal transmitted as (2)
Header tank		Control by use of static-head and provide emergency oil supply on failure of pumps	Erect at height equivalent to system pressure requirement
Flow switch	(1) Vane actuated	Simple design, robust construction. Can be obtained to give visual as well as electrical indication. Vane angle should be between 30° to 60° for maximum sensitivity	Pipe in line
	(2) Orifice with differential pressure switch	Used when very precise control is required. Orifice size and positioning of pressure tappings is critical	Pipe in line
Spring relief valve	(1) Standard safety valve	Simple design, either chamfered or flat seat	Integral with pump arranged for internal relief, or, fitted remote for external relief
	(2) High lift	Skirted flat seat, designed to give small difference between opening pressure and full flow pressure. Use where pressure accumulation must be minimal	
Differential pressure switch	(1) Opposed bellows actuated	Comments as for pressure switches	Local or remote panel mounted complete with shut-off cocks
	(2) Opposed piston actuated	Comments as for pressure switches	Pressure tappings close to filter inlet and outlet
Thermometer	(1) Bi-metallic	Dial calibration, almost equal spacing	Local vertical or co-axially mounted in separable pocket
	(2) Vapour-pressure	Dial calibration, logarithmic-scale. Changes of ambient have no effect on reading Allowance must be made if difference in height between bulb and dial exceeds 2 m. Maximum capillary length about 10 m	Bulb in separable pocket. Dial either local or remote panel mounted
	(3) Mercury-in-steel	Dial calibration, equally spaced. Ambient changes have minimal effect on reading No error of importance results from difference of height between bulb and dial. Maximum capillary length about 50 m	Install as for (2)
Temperature control valve	(1) Direct operating with integral bi-metallic element	Reverse acting. Control is not instantaneous, but is generally acceptable for many systems. Restricts flexibility of pipe routing. Care necessary in adjusting when control is operating	Valve in cooling water supply, operating element in the cooler oil outlet
	(2) Vapour pressure actuated, plunger operated	Reverse acting, reliable, protected against over-temperature. Can adjust when control is operating	Valve in cooling water supply. Remote sensing bulb in separable pocket in cooler oil outlet
	(3) Pneumatically controlled diaphragm	Employs valve similar to pressure control, reverse acting Diaphragm is actuated by low pressure air, providing more sensitive control	Valve as (2). Air controller remote panel mounted. Remote sensing bulb in separable pocket in cooler oil outlet

1 GENERAL REQUIREMENTS

Condition required	*Requirements not met by new assemblies because of:*	*Condition produced by running-in*
Macro-conformity	Latitude in fits, tolerances, assembly, alignment Mechanical deflections of shafts, gear teeth, etc. Thermal and mechanical distortion of bearing housings, cylinder liners, etc. Differential expansion of bearing materials	Increase in apparent area of contact and the relief of contact stresses Reduction in friction and temperature
Micro-conformity	Initial composite surface roughness exceeds oil film thickness obtainable with smooth surfaces	Decrease in roughness. Increase in effective bearing area (Figure 26.1) coherence of oil film and, ideally, full film lubrication
Cleanliness	Casting sand, machining swarf, wear products of running-in, etc.	Safe expulsion of contaminants to filters
Protective layers	Scuff- and wear-resistant low-friction surfaces are not naturally present on freshly machined surfaces	Safe formation of work-hardened layers, low-friction oxide films, oil-additive reaction films, substrate extrusion films, e.g. graphite

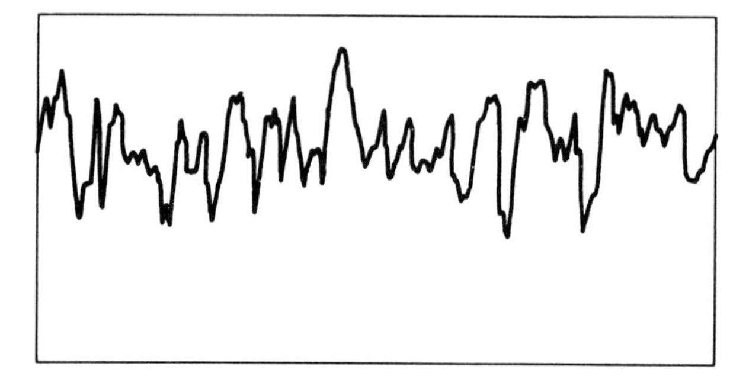

Un-run cylinder liner

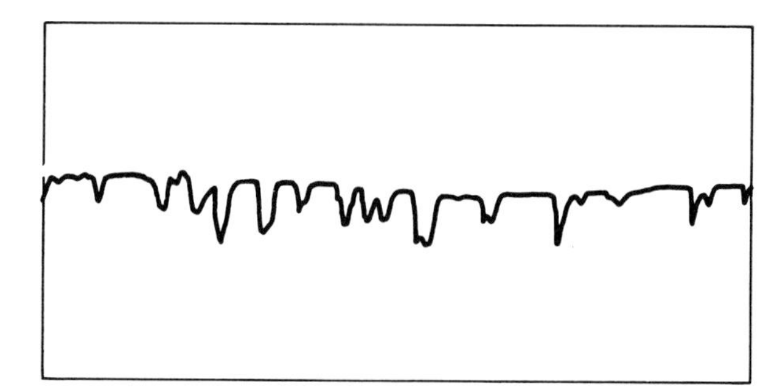

Run-in cylinder liner

Figure 26.1 Profilometer traces *(vertical magnification 5 times the horizontal)*

Running-in to achieve micro-conformity can be monitored by surface finish measurement and analysis before and after the running-in process. Surface finish criteria such as R_a (CLA) and bearing area curves are likely to be the best. The comparison of these parameters with subsequent reliability data can guide manufacturers on any improvements needed in surface finish and in running-in procedures. No generally applicable rule of thumb can be given.

2 RELATIVE REQUIREMENTS

The running-in requirement of assembled machinery is that of its most critical part. The list below rates the ease of running-in of common tribological contacts.

Most critical ↑	Piston rings and liners	Highly critical especially in IC engines—see subsection 3
	Gears	Watch hypoids—see subsection 4
	Cams and tappets	Commonly catered for by needs of piston rings, otherwise see subsection 4
	Rubbing plain bearings Porous plain bearings Fluid film plain bearings	See subsection 5
↓ *Least critical*	Hydrodynamic thrust bearings Rolling element bearings Externally pressurised bearings Gas bearings	Assuming correct design, finish and assembly no running-in is required. For certain high-precision rolling element (e.g. gyro) bearings, and for operation in reactive gas environment, consult makers

3 RUNNING-IN OF INTERNAL COMBUSTION ENGINES

The most effective running-in schedule for new and rebuilt engines depends to a large extent on the individual design of engine and materials used. It is therefore important to follow the maker's recommendations. In the absence of a specific schedule the following practice is recommended.

Running-in on dynamometer

Stage	*Duration* h (see Note 2)	% *Max. b.m.e.p.*	% *Rated max. speed*	*Remarks*
1	Until temperatures stabilise	0	30	
2	0.5	25	30	
3	0.5	50	50	
4	1.0	75	75	
5	0.1	100	100	Final power check
6	10	75	75	In service: limited max. duty period

Notes:

1 Where possible Stage 1 should be preceded by 0.5 h motoring at 30% rated max. speed.

2 Durations should be increased by (cumulative) factors according to engine design as follows:

Factor	*Increase durations by:*
Bores exceeding 150 mm (6 in) dia.	2 times
Bores exceeding 600mm (25 in) dia.	4 times
Max. load exceeding 10 bar (150 lb/in^2) b.m.e.p.	2 times
Max. load exceeding 15 bar (200 lb/in^2) b.m.e.p	4 times
Max. mean piston speed exceeding 20 m/s (3000 ft/min)	2 times

3 Change oil, tighten cylinder head, re-set valve clearances, etc., after Stage 6.

4 It may be possible to shorten these running-in times by progressive analysis of the amount and shape of wear particles in the lubricating oil. Ferrography, described later in this section, can be used for this.

Running-in in a road vehicle

Stage	*Duration* h	*Duty*
1	Until temperatures stabilise	Vehicle stationary. 30% max. rated speed
2	2	Main road (avoiding hills, cities, motorways). No payload. Max. speed 50% rated max. Avoid 'slogging'. Set fast idle
3	10	Main road. With payload. Max. speed 75% rated max. Avoid 'slogging'. Set fast idle. Change oil, tighten cylinder head, reset valve clearances, etc., after this stage

Note: Durations should be modified according to engine design, as in dynamometer schedule—see previous Note 2.

Monitoring running-in

The following observations provide a guide as to the completeness of the running-in process:

Observe	*Expect*
Coolant temperature	Possible overheating
Oil temperature	Gradual decrease and stabilisation
Blow-by rate	Target: 20–40 l/kW(0.5–1 ft^3/h per b.h.p.)
Crankcase pressure	Gradual decrease and stabilisation
Exhaust smoke	Gradually clears
Specific fuel consumption Maximum power	Approach rated values
Oil consumption	Target: 0.5–1% of fuel consumption
Compression pressure	Gradual increase and stabilisation

In research and development the following additional observations provide valuable guidance:

Observe	*By*
Surface roughness	Profilometer traces on liner replicas (Figure 26.1)
Surface appearance	Photographs of liner replicas (Figure 26.2)
Wear debris generation	Radioactive tracers, iron and chromium analysis of oil samples
Engine friction	Willan's line,* Morse test,† hot motoring

* *Willan's line*—Applicable to diesel engines. At constant speed, plot fuel consumption *v.* b.m.e.p. Extrapolation to zero consumption gives friction m.e.p.

† *Morse test*—Applicable to multi-cylinder gasoline engines. At constant speed, observe reduction in torque when cutting each cylinder in turn. Assume reduction is the indicated m.e.p. of idle cylinder. Calculate friction m.e.p. as difference between average indicated m.e.p.'s and b.m.e.p.

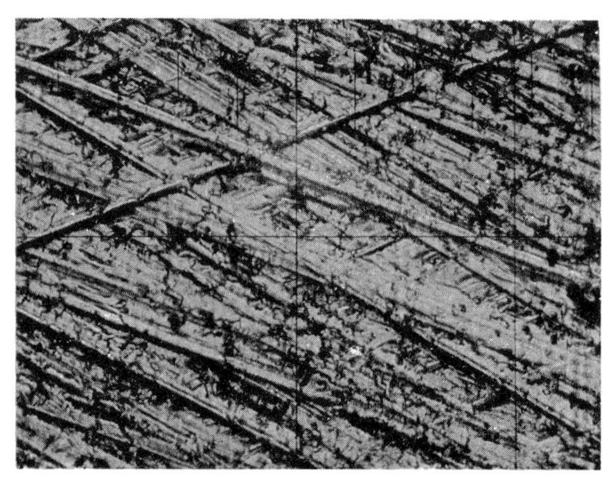

Figure 26.2(a) Un-run cylinder liner × 140

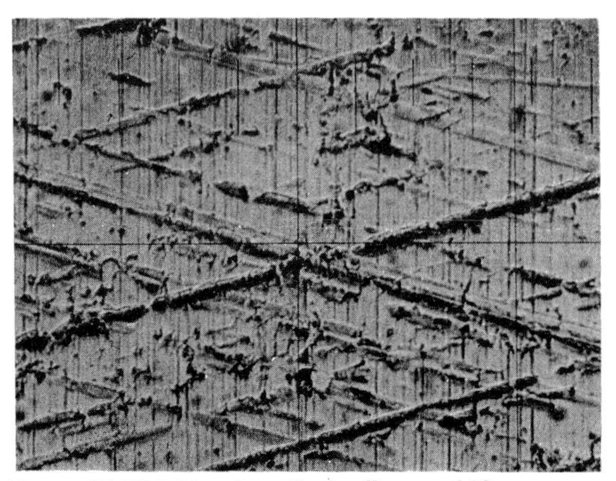

Figure 26.2(b) Run-in cylinder liner × 140

Running-in accelerators

Running-in accelerators should only be used in consultation with the engine maker. Improper use can cause serious damage.

Accelerator	*Precautions*
Abrasive added to intake air	Ensure even distribution and slow feed rate Danger of general abrasive wear Change oil and filters afterwards
Running-in additive in fuel	Consult oil supplier and engine maker to establish dosage and procedure Change oil and filters afterwards
Running-in oil	Use as recommended. Ensure change to service oil afterwards
Abrasive and other ring coatings	No special precautions
Taper-faced rings	Ensure that they are assembled with lower edge in contact with bore

Ferrography

Ferrography is a technique of passing a diluted sample of the lubricating oil over a magnet to extract ferrous particles. It has found useful application in running-in studies aimed at shortening running-in of production engines and so making possible large cost savings. The principle is to examine suitably diluted samples of engine oil to obtain, during the process of running-in, a measure of the content of large (L) and small (S) particles. Over a large number of dynamometer tests on new production engines a trend of 'Wear Severity Index' ($I_s = L^2 - S^2$) with time may be discerned which allows comparison to be made between the effectiveness of running-in schedules.

4 RUNNING-IN OF GEARS

Procedures

It is not feasible to lay down any generally applicable running-in procedure. The following guiding principles should be applied in particular cases:

Avoid	*Otherwise*
High speed—high load combinations	High temperatures develop—scuffing
Low speed—high load combinations	Poor elastohydrodynamic conditions and possible plastic flow
High speed, shock loading	Scuffing—particularly met on over-run of hypoids

Observing progress of running-in

Observe	*Remarks*
Oil temperature	See Fig. 3
Wear debris analysis Depletion of oil additives	Ensure sample is homogeneous and representative of bulk oil
Surface roughness (of replicas) Additive deposition by autoradiography Gear efficiency increase	Research and development techniques

Materials and lubricants

See also Sections A23, 24, 25. Running-in has been found to be influenced by materials and lubricants broadly as follows:

Factor	*Effect*
Excessive hardness (>400 VPN) Carburising Nitriding Anti-wear additives in oil	Can delay running-in
EP additives in oil Thermo-chemical treatments, e.g. phosphating Increased oil viscosity	Can prevent scuffing during running-in

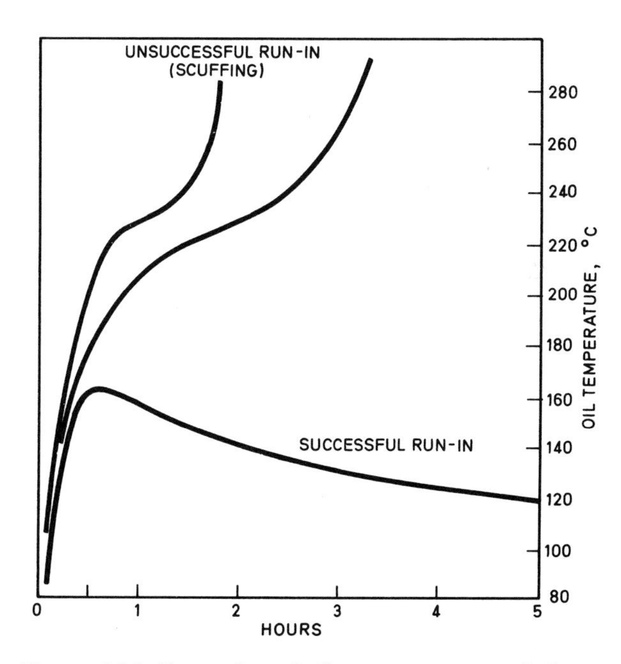

Figure 26.3 Examples of oil temperature variation during early life of hypoid axles

5 RUNNING-IN OF PLAIN BEARINGS

Special running-in requirements

Type of bearing	*Requirement*
Rubbing (non-metallic)	Running-in desirable for successful transfer of protective layer, e.g. PTFE to countersurface
Porous	Running-in desirable to achieve conformity without overheating impregnant oil
Fluid film	Usually none unless marginally lubricated—mainly a precaution against minor assembly errors. For piston rings, see subsection 3

Procedure

Principle	*Restrict speed and load to limit temperature rise*
Practice* (in order of preference)	1 Run at reduced speed and load 2 Run at reduced speed, normal load 3 Run at reduced load, normal speed 4 Run at normal load and speed for short bursts, cooling between
Observe	Bearing temperature (Figure 26.4) When curves repeat, run-in at those conditions is complete

* In all cases run with increased oil flow if possible

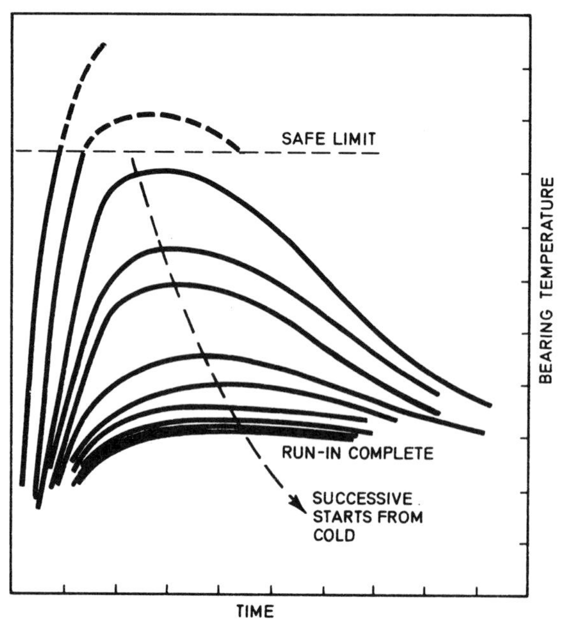

Figure 26.4 Typical effects of running-in on warm-up of plain journal bearings

6 RUNNING-IN OF SEALS

Rubbing seals, both moulded and compression, undergo a bedding-in process. No general recommendations can be given but the following table summarises experience:

Typical bedding-in duration	2 h
Typical friction reduction	50%
Reduced speed operation during bed-in	Desirable but not essential
Pre-lubrication during fitting	Essential
Renewal of seals	If possible, never refit a bedded-in seal*
Compression seals	Allow excess leakage to start with, tighten as bedding-in proceeds

* This is because seals often harden in service, making re-accommodation difficult; flexing during removal may cause cracking

THE NEED FOR LUBRICANT CHANGES

Reason for changing lubricant	*Cause of the problem*	*Effect of the problem*	*Methods of extending lubricant life*
DEGRADATION	*Approximate life of refined mineral oils*	Alteration of the physical characteristics of the lubricant Formation of sludges that can block oil-ways	Where possible bulk oil temperature should be kept below 60°C and unnecessary aeration should be avoided
CONTAMINATION	Wear debris, external solids and water entering the system Access of working fluids to the lubricant	Solids and immiscible liquids interfere with lubrication and cause wear Soluble contaminants, such as hydrocarbons and organic solvents, alter the physical properties of the oil	Contamination can be reduced or eliminated by use of filters, centrifuges and regular draining of separated material from the bottom of the lubricant reservoir

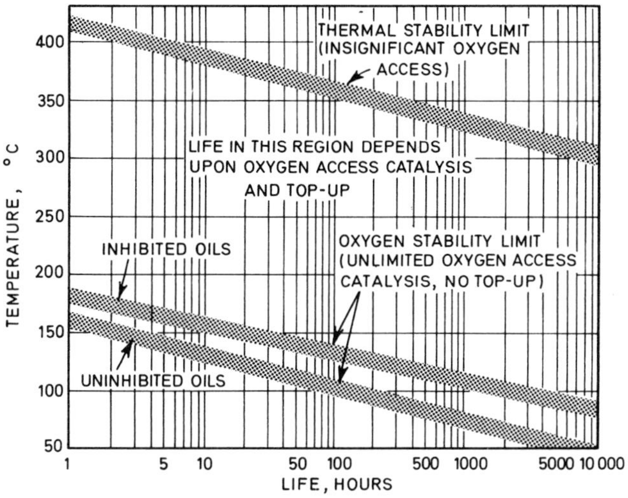

CHANGE PERIODS

Systems containing less than 250 litre (50 gal)

Analytical testing is not justified and change periods are best based on experience. The following examples in the opposite column are typical of industrial practice:

System	*Operating conditions*	*Change period* years
Oil bath	All	1
Well-sealed systems (e.g. gearboxes, hydraulic systems, compressor crankcases)	40–50°C 60°C	2–3 1
Grease-packed ball and cylindrical roller bearings[1]	Up to $dn^{(2)} = 150\,000$ $dn = 200\,000$	2 1

Notes: (1) Horizontally mounted, medium series bearings lubricated with a rolling bearing quality grease operating in normal ambient temperature conditions.
(2) d = bore in mm, n = rev/min.

Systems containing more than 250 litre (50 gal)

Regular testing should be carried out to determine when the lubricant is approaching the end of its useful service life. A combination of visual examination and laboratory testing is recommended.

Visual examination	*Laboratory tests*
1 Carry out at moderately short intervals, say weekly	1 Longer term tests, say every 6–12 months
2 Can be carried out on the plant with simple equipment and little experience	2 If suitable facilities are not available, oil supplier may be able to help
3 Allows immediate action to be taken and stimulates interest on the plant	3 Takes some time before results are available, but gives quantitative values for assessment of lubricant condition
4 Samples should be kept for comparison with the next set	4 Records of laboratory tests enable more subtle changes in the oil to be followed and allow prediction of further useful life

The results obtained are only representative of the sample. This should preferably be taken when the system is running, and a clean container must be used. Guidance on interpreting the results is given in the following tables.

VISUAL EXAMINATION OF USED LUBRICATING OIL

1 Take sample of circulating oil in clean glass bottle (50–100 ml).
2 If dirty or opaque, stand for 1 h, preferably at 60°C (an office radiator provides a convenient source of heat).

Appearance of sample		*Reason*	*Action to be taken*	
When taken	*After* 1 h		*System without filter*[(1)] *or centrifuge*	*System with filter*[(1)] *or centrifuge*
Clear	—	—	None	None
Opaque[(2)]	Clear	Foaming	Cause of foaming to be sought[(3)]	Cause of foaming to be sought[(3)]
	Clear oil with separated water layer	Unstable emulsion[(4)]	Run off water (and sludge) from drain[(5)]	Check centrifuge[(6)]
	No change	Stable emulsion	Submit sample for analysis[(7)]	Check centrifuge;[(6)] if centrifuge fails to clear change oil
Dirty	Solids separated[(8)]	Contamination	Submit sample for analysis[(7)]	Check filter or centrifuge
Black (acrid smell)	No change	Oil oxidised	Submit sample for analysis[(7)]	Submit sample for analysis[(7)]

Notes: (1) The term filter is restricted to units able to remove particles less than 50 μm; coarse strainers, which are frequently fitted in oil pump suctions to protect the pump, do not remove all particles liable to damage bearings, etc.
(2) Both foams (mixtures of air and oil) and emulsions (mixtures of water and oil) render the oil opaque.
(3) Foaming is usually mechanical in origin, being caused by excessive churning, impingement of high-pressure return oil on the reservoir surface, etc. Foam can be stabilised by the presence of minor amounts of certain contaminants, e.g. solvents, corrosion preventives, grease. If no mechanical reason can be found for excessive foam generation, it is necessary to change the oil.
(4) Steps should be taken to remove the water as soon as possible. Not only is water liable to cause lubrication failure, but it will also cause rusting; the presence of finely divided rust tends to stabilise emulsions.
(5) Failure of water to separate from oil in service may be the result of inadequate lubricant capacity or the oil pump suction being too close to the lowest part of the reservoir. More commonly it results from re-entrainment of separated water from the bottom of the sump when, by neglect, it has been allowed to build up in the system.
(6) The usual reason for a centrifuge failing to remove water is that the temperature is too low. The oil should be heated to 80°C before centrifuging.
(7) It is not always possible to decide visually whether the oil is satisfactory or not. In doubtful cases it is necessary to have laboratory analysis (see next table).
(8) In a dark oil, solids can be seen by inverting the bottle and examining the bottom.

LABORATORY TESTS FOR USED MINERAL LUBRICATING OILS

Property	*Test method*[1]	*Type of oil*	*Action level*
Viscosity	IP71/ASTM D445	All	15% increase from new oil value (cSt at 40°C)[2]
Acidity (neutralisation value)	IP1	Straight oil	2–3 mg KOH/g[3]
	IP139/ASTM D947 or IP177/ASTM D664	Turbine oil	1 mg KOH/g[4]
	IP139/ASTM D947 or IP177/ASTM D664	hd hydraulic oil, ep gear oil	Discuss with oil supplier[5]
Ash (check for contamination by solids)	IP4	Straight oil, Turbine oil	> 0.2%[6]
	ASTM D893 or IP4	hd hydraulic oil, ep gear oil	> 0.2%[7]
Water level	ASTM D1744	All	200 ppm[8]
Additive level (checks for additive loss)		Turbine oil, hd hydraulic oil, ep gear oil, Engine oil	Discuss with oil supplier[9]

Notes: (1) *I.P. Standards for Petroleum and Its Products*, Institute of Petroleum and also ASTM Standards.

(2) Change oil if viscosity has increased by 15% as a result of degradation. (Where increase or decrease is caused by contamination with miscible liquid or topping up with the wrong grade of oil, much greater changes can be tolerated and limits have to be chosen to suit particular cases.)

(3) Acidity develops as a result of oxidation; the oxidation reaction is auto-catalytic so that once it has started it proceeds at an increasing rate. The acids formed are not corrosive to materials of construction, but measurement of acidity gives a useful guide to the condition of the oil. The exact value at which the oil should be changed is not critical, but experience shows that when a value of 2–3 mg KOH/g is reached the useful life of the oil is limited.

(4) In oils containing oxidation inhibitors there is a long induction period before acidity develops; once this occurs it develops rapidly and the figure given provides a useful indication of the end of the service life of the oil.

(5) Additives in the oil interfere with the determination of acidity.

(6) This is an arbitrary value and particular cases may require individual judgement.

(7) Oil-soluble ash-forming additives render the simple ash test inapplicable in these cases.

(8) This figure applies to systems with critical components, e.g. rolling bearings, hydraulic control valves, gears. In less critical systems, e.g. steam-engine crankcases, much higher levels can be tolerated.

(9) In circulation systems with a large inventory of oil it may be worth checking the depletion of additives in additive-containing oils. No general tests for these are available, but oil suppliers should be able to help in providing test methods for the particular additives they use.

NOTES ON GOOD MAINTENANCE PRACTICE

Attention to detail will give improved performance of oils in lubrication systems. The following points should be noted:

1. Oil systems should be checked weekly and topped up as necessary. Systems should not be over-filled as this may lead to overheating through excessive churning.
2. Oil levels in splash-lubricated gearboxes may be different when the machine is running from when it is stationary. For continuously running machines the correct running level should be marked to avoid the risk of over- or under-filling.
3. Degradation is a function of temperature. Where possible the bulk oil temperature in systems should not exceed 60°C. The outside of small enclosed systems should be kept clean to promote maximum convection cooling.
4. Care must be exercised to prevent the ingress of dirt during topping up.

C28 Biological deterioration of lubricants

The ability of micro-organisms to use petroleum products as nourishment is relatively common. When they do so in very large numbers a microbiological problem may arise in the use of the petroleum product. Oil emulsions are particularly prone to infection, as water is essential for growth, but problems also arise in straight oils.

CHARACTERISTICS OF MICROBIAL PROBLEMS

1 They are most severe between 20°C and 40°C.
2 They get worse.
3 They are 'infectious' and can spread from one system to another.
4 Malodours and discolorations occur, particularly after a stagnation period.
5 Degradation of additives by the organisms may result in changes in viscosity, lubricity, stability and corrosiveness.
6 Masses of organisms agglomerate as 'slimes' and 'scums'.
7 Water is an essential requirement.

Factors affecting level of infection of emulsions

The severity of a problem is related to the numbers and types of organisms present. Most of the factors in the following table also influence straight oil infections.

Increase infection level	*Decrease infection level*
1 High initial 'inoculum' from dregs of previous charge and equipment slimes	1 Temperatures consistently outside optimum range for growth
2 Oil formulation containing complete diet for organisms	2 Local temperatures in sterilising range (above 70°C), for example in a purifier
3 Growth stimulation from added sandwiches, urine, etc.	3 High pH (above 9.5)
4 Leaks or carry-over into the system from other infected systems	4 Organisms removed as slimes in centrifuges and filters by adhering to metal swarf and machined parts
5 Poor quality water used in preparing charge	5 Anti-microbial inhibitors present in oil formulation
6 Over diluted emulsion in use for prolonged period	6 Variety of oil formulations used in successive charges

Characteristics of principal infecting organisms (generalised scheme)

Organism	*pH Relationship*	*Products of growth*	*Type of growth*
Aerobic bacteria (use oxygen)	Prefer neutral to alkaline pH	Completely oxidised products (CO_2 and H_2O) and some acids. Occasionally generate ammonia	Separate rods, forming slime when agglomerated. Size usually below 5 μm in length
Anerobic bacteria (grow in absence of oxygen)	Prefer neutral to alkaline pH	Incompletely oxidised and reduced products including CH_4, H_2, and H_2S	As above. Often adhere to steel surfaces, particularly swarf
Yeasts	Prefer acid pH	Oxidised and incompletely oxidised products. pH falls	Usually separate cells, 5–10 μm, often follow bacterial infections or occur when bacteria have been inhibited. Sometimes filamentous
Fungi (moulds)	Prefer acid pH but some flourish at alkaline pH in synthetic metal working fluids	Incompletely oxidised products—organic acids accumulate	Filaments of cells forming visible mats of growth. Spores may resemble yeasts. Both yeasts and moulds grow more slowly than bacteria

Comparison of microbial infections in oil emulsions and straight oils

Oil emulsions	*Straight oils*
Organisms live in the continous (water) phase	Organisms live in a separate discrete water phase or a dispersed water phase
Alkaline environment, usually changing *slowly* to neutral or even slightly acidic	Water often becoming rapidly acidic
Micro-organisms predominantly bacteria except in synthetic formulations	Bacteria, fungi and yeasts commonly occur
Oil-in-water emulsion becomes unstable due to degradation of emulsifiers. Effect is less pronounced in bio-stable formulations	Water-in-oil emulsions become stabilised due to surface activity of micro-organisms
Anti-microbials (biocides) often included in formulation and give some initial protection	Anti-microbials rarely included, but can be added in use
Temperature (just above ambient) favours growth	Temperature often so high that systems are self-sterilising. Heated purifiers also pasteurise the oil
Anti-microbial treatment usually involves water-soluble biocides	Biocides must have some oil solubility

ECONOMICS OF INFECTION

The total cost of a problem is rarely concerned with the cost of the petroleum product infected but is made up from some of the following components:

1 Direct cost of replacing spoiled oil or emulsion.
2 Loss of production time during change and consequential down-time in associated operations.
3 Direct labour and power costs of change.
4 Disposal costs of spoiled oil or emulsion.
5 Deterioration of product performance particularly:
   (*a*) surface finish and corrosion of product in machining;
   (*b*) staining and rust spotting in steel rolling;
   (*c*) 'pick-up in aluminium rolling.
6 Cost of excessive slime accumulation, e.g. overloading centrifuges, 'blinding' grinding wheels, blocking filters.
7 Wear and corrosion of production machinery, blocked pipe-lines, valve and pump failure.
8 Staff problems due to smell and possibly health.

ANTI-MICROBIAL MEASURES

These may involve:

1 Cleaning and sterilising machine tools, pipework, etc., between charges.
2 Addition of anti-microbial chemicals to new charges of oil or emulsion.
3 Changes in procedures, such as:
   (*a*) use of clean or even de-ionised water;
   (*b*) continuous aeration or circulation to avoid malodours;
   (*c*) prevention of cross-infection;
   (*d*) re-siting tanks, pipes and ducts, eliminating dead legs;
   (*e*) frequent draining of free water from straight oils;
   (*f*) change to less vulnerable formulations
4 Continuous laboratory or on-site evaluation of infection levels.

Physical methods of controlling infection (heat, u.v., hard irradiation) are feasible but chemical methods are more generally practised for metal working fluids. Heat is sometimes preferred for straight oils. There is no chemical 'cure-all', but for any requirement the following important factors will affect choice of biocide.

1 Whether water or oil solubility is required—or both.
2 Speed of action required. Quick for a 'clean-up', slow for preventing re-infection.
3 pH of system—this will affect the activity of the biocide and, conversely, the biocide may affect the pH of the system (most biocides are alkaline).
4 Identity of infecting organisms.
5 Ease of addition—powders are more difficult to measure and disperse than liquids.
6 Affect of biocide on engineering process; e.g. reaction with oil formulation, corrosive to metals present.
7 Toxicity of biocide to personnel—most care needed where contact and inhalation can occur—least potential hazard in closed systems, e.g. hydraulic oils.
8 Overall costs over a period.
9 Environmental impact on disposal.

Most major chemical suppliers can offer one or more industrial biocides and some may offer an advisory service.

C29 Lubricant hazards; fire, explosion and health

FIRES AND EXPLOSIONS

Mineral lubricating oils, though not highly flammable materials, can be made to burn in air and in certain circumstances can give rise to serious fires and explosions. The risk depends on the spontaneous ignition conditions for mixtures of oil vapour (or mist). Figure 29.1 shows the ignition limits at atmospheric pressure and Figure 29.2 the ignition limits for the most flammable mixture as a function of pressure. Curves showing the equilibrium oil vapour mixtures for typical oils are also included in the illustrations to show the values likely to be experienced.

Explosions can occur in enclosed lubricated mechanisms in which a flammable oil vapour–air mixture can be formed, e.g. crankcases of diesel engines, steam engines and reciprocating compressors, and large gearboxes.

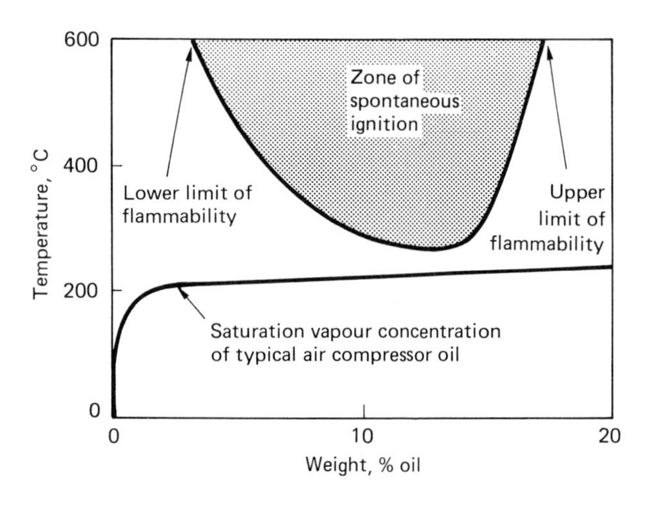

Figure 29.1 Spontaneous ignition limits for mineral-oil vapour (mist) air mixtures at atmospheric pressure

Figure 29.2 Spontaneous ignition limits for 12% mineral-oil vapour (mist) air mixture as a function of pressure

Crankcase (gearcase) explosions

Cause	*Cure*		*Comments*
	Action	*Method*	
Oil mist is formed by oil coming into contact with a hot spot, the vapour condensing to form mist as it is swept away from the hot spot by windage. Figure 29.1 shows that the equilibrium vapour mixture will not ignite spontaneously, and that for temperatures above about 230°C an over-rich (non-flammable) mixture is formed. Explosions occur if there is insufficient oil at the hot spot to produce an over-rich mixture and rising temperature brings the mixture at the hot spot into a spontaneous ignition range, or if the mixture at the hot spot is diluted by removing a cover and allowing access of air*	Prevention	(*a*) Inert gas blanketing	Nitrogen or carbon dioxide added to reduce oxygen content in vapour space to 10%
		(*b*) Inert gas injection	Development of oil mist detected and nitrogen or carbon dioxide injected at set level
	Protection	(*a*) Explosion containment	Casing must be strong enough to withstand 800 kN/m^2. Doors and vent covers must be adequately secured
		(*b*) Explosion relief	Relief valve with relief area at least 35 mm^2/l crankcase volume or preferably 70 mm^2/l. Flame trap must be fitted

* If blue oil smoke is seen emerging from a machine, do not remove doors or covers until sufficient time has elapsed for any hot spot to have cooled down.

Note: Explosion relief is considered to be the best practical solution. Suitable relief valves incorporating flame traps are available commercially.

Air compressor fires and explosions

Figure 29.2 shows that spontaneous ignition can occur in equilibrium mixtures of air and oil at about 260°C at 200 kN/m^2, falling to 240°C at 800 kN/m^2.

Compressor type	*Cause of fire*	*Cure*		*Comment*
Reciprocating	Exothermal oxidation of oil degradation deposits in the delivery lines of oil-lubricated compressors, raising the temperature to the spontaneous ignition limit	Delivery temperature control	(*a*) 140°C max.	No deposit formation
		(use lowest viscosity oil compatible with lubrication requirements and particularly high volatility oils resistant to deposit formation)*	(*b*) 160°C max.	Routine removal of deposits to prevent build-up greater than 3 mm thick
Oil-cooled rotary	Exothermal oxidation of the thin oil film on the oil reclaimer pad, raising the temperature to the spontaneous ignition limit	Use of low volatility, oxidation-resistant lubricant†		Reduce oil loss and build-up in reclaimer

*Oil to DIN 51506 VD-L
†Oil to DIN 51506 VC-L

Note: The creation of a shock wave on ignition may result in detonative explosions in oil-wetted delivery lines.

Lagging fires

Oil-wetted lagging can ignite even though the lagging temperature is below the minimum spontaneous ignition temperature given in Figure 29.1.

Cause	*Cures*	*Comments*
The wicking action of the lagging produces a thin film of oil that oxidises exothermally raising the temperature to the spontaneous ignition region	Use an impermeable material (e.g. foam glass). Where this is not practicable flanges should be left unlagged, provided the resulting heat loss is acceptable	Because of the poor access of air to the interior of the lagging it is easy for an over-rich oil–air mixture to be formed that only ignites if the hot lagging is stripped off. This can be a hazard with lagging that is glowing on the outside; stripping it off only exposes more oil and can give rise to a more serious fire
	Use a fire-resistant phosphate–ester fluid (See Hydraulic Oil Fires section) instead of a mineral oil	For example, phosphate–esters are used in the hydraulic control systems of high-temperature steam turbines

Hydraulic oil fires

Hydraulic systems present a fire hazard because a leak of high-pressure oil will produce a finely atomised spray that is liable to ignite if it impinges on a hot surface where the necessary conditions for ignition shown in Figure 29.1 can be realised.

Protection against fires can be obtained by the use of fire-resistant hydraulic fluids. These are of two general types; water-containing fluids that prevent ignition by forming a steam blanket at the hot spot, and synthetic lubricants that are less flammable than mineral oils and, in normal circumstances, do not support combustion when the heat source is removed.

The following table shows the general characteristics of the principal types of fire-resistant hydraulic fluids and points out some restrictions associated with their use. This information should indicate the most suitable fluid for a particular application. Detailed design points are not covered and full discussion with the fluid manufacturer is recommended before a fire-resistant fluid is adopted.

Fire-resistant hydraulic fluids

	Water-containing fluids			*Synthetic fluids*	
	Soluble-oil emulsions (2% oil)	*Water-in-oil emulsions* (40% water)	*Water–glycol blends* (45% water)	*Phosphate esters*	*Phosphate–ester chlorinated hydrocarbon blends*
Maximum system temperature, °C	65	65	65	100	100
Restrictions on materials used in normal oil systems:					
(*i*) internal paints	None	None	Special paints required	Special paints required	Special paints required
(*ii*) rubber seals	None	None	Normally no problem	Special seals required	Special seals required
(*iii*) materials of construction	None	None	Avoid magnesium zinc and cadmium plating	Avoid aluminium rubbing contacts	Avoid aluminium rubbing contacts
Lubrication:					
(*i*) rolling bearings—apply factor to load for design calculations	Not suitable	2.0	2.5	1.2	1.2
(*ii*) gear pumps	Not recommended	Limit pressure to 3.5 MN/m^2 (500 lbf/in^2)	Limit pressure to 3.5 MN/m^2 (500 lbf/in^2)	Satisfactory	Satisfactory
Maintenance	—	Water content must be maintained*	Water content must be maintained	Should be kept dry	Should be kept dry
Cost relative to mineral oil	—	1.5–2	4–5	5–7	7–9

* Some separation of water droplets may occur on standing. The emulsion can, however, be readily restored by agitation. Care must be taken to avoid contamination by water–glycol or phosphate–ester fluids as these will cause permanent breakdown of the emulsion.

HEALTH HAZARDS

The major risk is from prolonged skin contact; this is predominantly a problem in machine shops where the risk of continued exposure to cutting oils and lubricants is greatest. These should be available in all workshops.

The risks to health are, however, small if such reasonable hygiene precautions are taken as outlined below.

Hazard	*Comment*	*Recommended practice*
Toxicity	Mineral oils are not toxic, though certain additives which can be used in them may be	(*i*) Avoid ingestion (*ii*) Wash hands before eating
Dermatitis	Prolonged skin contact with neat or soluble cutting oils is liable to cause dermatitis, though individual susceptibility varies considerably	(*i*) Use solvent-refined oils (*ii*) Use barrier creams on the hands and forearms and wear protective clothing where there is a risk of wetting by oil
Acne	Mainly caused by neat cutting and grinding oils	(*iii*) Treat and cover skin-abrasions
Cancer	Some mineral oil constituents may cause cancer after prolonged exposure of the skin. Certain types of refining, of which solvent refining is the best known, lessen the risk by reducing the carcinogens in the oil	(*iv*) Wash thoroughly with soap and hot water after each spell of work to free skin from oil (*v*) Do not put oil-wetted rags in trouser pockets (*vi*) Do not wear oil-soaked clothing. Work- and underclothes should be regularly laundered (*vii*) Do not use solvents for degreasing hands and other contaminated parts

Note: The water–glycol phosphate ester and phosphate ester–chlorinated hydrocarbon fire-resistant hydraulic fluids are more toxic than mineral oils, but should not be a hazard if sensible handling precautions are taken. If the synthetic fluids contact very hot surfaces, copious fumes may be evolved. These fumes are toxic and unpleasant and should not be inhaled.

To achieve efficient planning and scheduling of lubrication a great deal of time and effort can be saved by following a constructive routine. Three basic steps are required:

(*a*) A detailed and accurate survey of the plant to be lubricated including a consistent description of the various items, with the lubricant grade currently used or recommended, and the method of application and frequency

(*b*) A study of the information collected to attempt to rationalise the lubricant grades and methods of application

(*c*) Planning of a methodical system to apply lubrication

THE PLANT SURVEY

Plant identification

A clearly identifiable plant reference number should be fixed to the machinery. The number can incorporate a code of age, value and other facts which can later facilitate information retrieval.

A procedure to deal with newly commissioned or existing plant and a typical reference document is illustrated below, Figure 30.1.

Table 30.1 A convenient standardised code to describe the method of lubricant application

Main method	*Code*	*Application*
Representing:		
Oil can	← OC–C	Cup
	–H	Hole
	–B	Bottle (gravity) (wick feed) (drip feed) (syphon)
	–S	Surface: slideways hand oiled
	–A	Air mist
	–W	Well
Oil gun	← OG–N	Nipple
	–M	Multivalve
Oil filled	← OF–B	Bath. Sump. Ring oiler
	–L	Mechanical lubricator
	–S	Circulating system
	–H	Hydraulic
Grease gun	← GG–N	Nipple
	–M	Multivalve
	–S	Spring feed—grease cups
Grease hand	← GH–H	Hand applied
	–C	Cups—stauffer
Grease filled	← GF–S	System

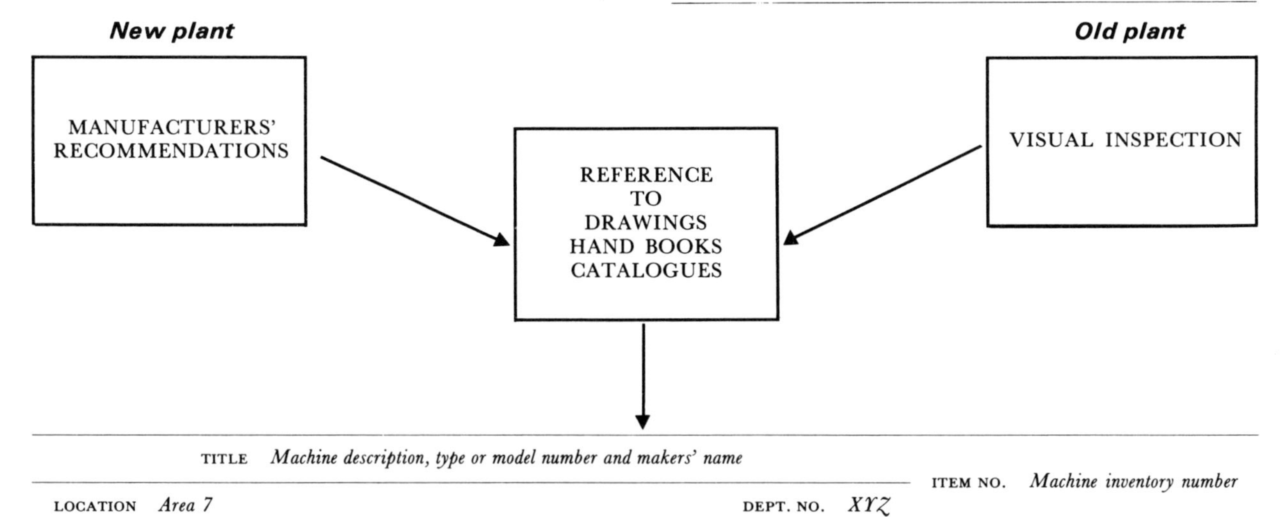

PLANNED LUBRICATION MAINTENANCE INVENTORY SHEET

NO.	ITEM OF PLANT	METHOD TABLE 1	NO. OF POINTS	GRADE	QUANTITY	FREQUENCY
	(Details of part to be lubricated)					
1234	*Electric Motor Bearing*	*GGN*	*2*	*Grease No. 3*	*2 shots*	*A*
3256	*Spindle Bearings*	*OGN*	*2*	*Oil No. 927*	*2 shots*	*D*
3221	*Gear Box*	*OFB*	*1*	*Oil No. 41*	*10 gallon sump*	*W*

Figure 30.1

Method of application

In the case of new plant the proposed methods of lubrication should be subjected to careful scrutiny bearing in mind subsequent maintenance requirements. Manufacturers are sometimes preoccupied with capital costs when selling their equipment and so designed-out maintenance should be negotiated early on when the tribological conditions are studied. In this context it is possible to economise on the application costs of lubrication and problems of contamination and fire hazards can be forestalled. A standardised code for describing the method of application is given in Table 30.1. Confusion can arise unless a discipline is maintained both on surveying and scheduling.

Number of application points

The number of application points must be carefully noted.

(*a*) By adequate description—group together numbers of identical points wherever possible when individual point description serves no purpose. This simplifies the subsequent planning of daily work schedules.

(*b*) Highlighting of critical points by symbol or code identification as necessary.

Factors for lubricant selection

For the purpose of assessing the grade of lubricant, the following table suggests the engineering details required to determine the most suitable lubricant.

Table 30.2 Some factors affecting lubricant selection

Element	*Type*	*Size*	*Material*	*Operating temperature*	*Operating conditions*	*Velocity*	*Remarks*
Bearings	Plain, needle roller, ball	Shaft diameter				rev/min	
Chain drives	Links. Number and pitch	PCD of all wheels and distance between centres				Chain speed ft/min	
Cocks and valves	Plug, ball etc.		Fluid being controlled				Depends on properties of the fluid
Compressors	BHP, manufacturer's name			Gas temperature	Max gas pressure	rev/min	
Couplings	Universal or constant velocity					rev/min	
Cylinders		Bore, Stroke	Cylinder, piston, rings	Combustion and exhaust gas temperature	Combustion and exhaust gas pressure	Crank speed rev/min	
Gears	Spur, worm, helical, hyperbolic	BHP, distance between centres			Radiated heat and heat generated	rev/min	Method of lubricant application
Glands and seals	Stuffing box		Fluid being sealed				Depends on design
Hydraulic systems	BHP Pump type (gear, piston vane)		Hydraulic fluid materials 'O' rings and cups etc.				Lubricant type adjusted to loss rate
Linkages				Environmental heat conditions		Relative link speeds ft/s, angular vel. rad/s	
Ropes	Steel hawser	Diameter			Frequency of use and pollution etc.		
Slideways and guides						Surface relative speed ft/min	

LUBRICANT RATIONALISATION

Recommended grade of lubricant

Manufacturers recommended grades may have to be acceded to during the guarantee period for critical applications. However, a compromise must be reached in order to ensure the maintenance of an optimum list of grades which is essential to the economic sorting, handling and application of lubricants. In arriving at this rationalised list of grades, speeds, tolerances, wear of moving parts and seals create conditions where the viscosity and quality of lubricant required may vary. For a balanced and economic rationalisation, all tribological factors have therefore to be assessed. Where a special lubricant has to be retained, if economically viable it may form a compromise solution that will satisfy future development projects, particularly where demands are likely to be more critical than for existing equipment. Generally speaking, in most industries 98% of the bulk of lubrication can be met by six grades of oil and three greases.

A considerable range of lubricant grades exists largely blended to meet specific demands of manufacturers. Table 30.3 illustrates a typical selection. There are viscosity ranges, indices, inherent characteristics and additive improvers to be considered. Generally speaking, the more complex the grade, the more expensive, but often the more comprehensive its application. Advice is readily available from oil companies.

Quantity and frequency

In the main the quantity of lubricant applied is subjected to so many variable conditions that any general scale of recommendations would be misleading. 'Little and often' has an in-built safeguard for most applications (particularly new plant), but as this can be uneconomic in manpower, and certain items can be over-lubricated planning should be flexible to optimise on frequency and work loads. Utilisation of the machinery must also be allowed for.

Knowledge of the capacity and quantity required will naturally help when assessing the optimum frequency of application and a rough guide is given in Table 30.4.

Table 30.3 Range of lubricant grades commonly available showing factors to be taken into account for economic rationalisation

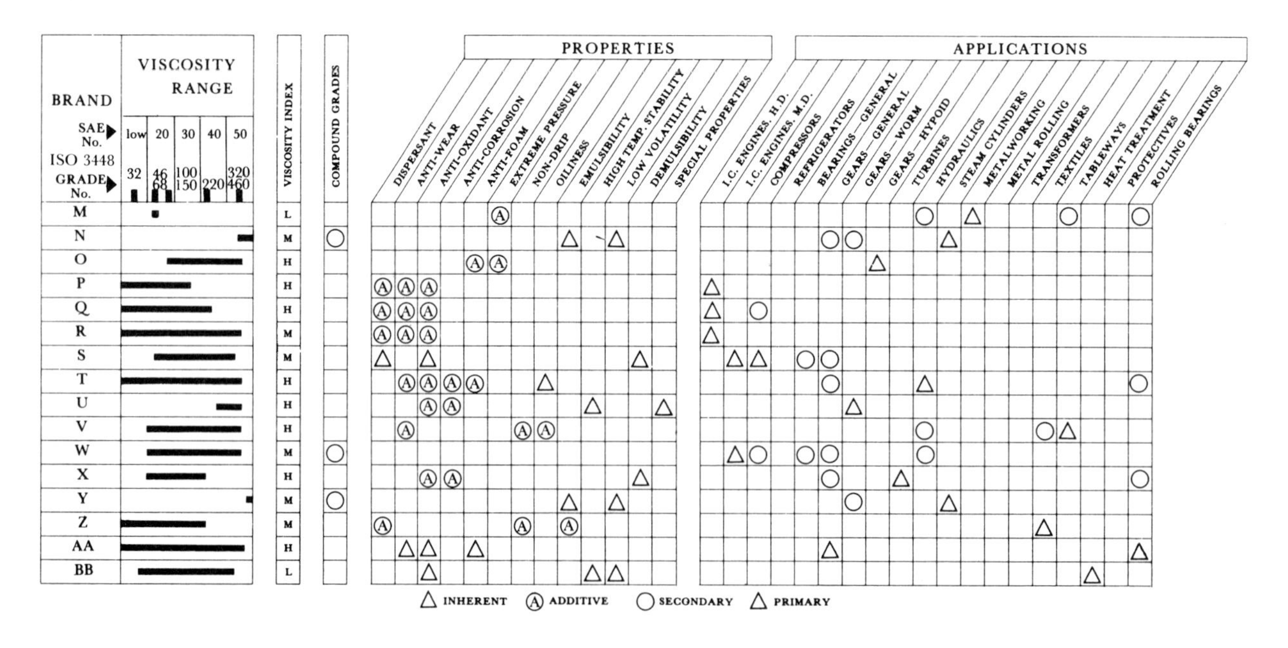

PROPERTIES

BRAND	VISCOSITY RANGE (SAE No.: low, 20, 30, 40, 50; ISO 3448 GRADE No.: 32, 46, 68, 100, 150, 220, 320, 460)	VISCOSITY INDEX	COMPOUND GRADES	DISPERSANT	ANTI-WEAR	ANTI-OXIDANT	ANTI-CORROSION	ANTI-FOAM	EXTREME PRESSURE	NON-DRIP	OILINESS	EMULSIBILITY	HIGH TEMP. STABILITY	LOW VOLATILITY	DEMULSIBILITY	SPECIAL PROPERTIES
M		L							Ⓐ							
N		M	○									Δ		Δ		
O		H						Ⓐ	Ⓐ							
P		H		Ⓐ	Ⓐ	Ⓐ										
Q		H		Ⓐ	Ⓐ	Ⓐ										
R		M		Ⓐ	Ⓐ	Ⓐ										
S		M		Δ		Δ									Δ	
T		H			Ⓐ	Ⓐ	Ⓐ	Ⓐ			Δ					
U		H				Ⓐ	Ⓐ						Δ			Δ
V		H			Ⓐ					Ⓐ	Ⓐ					
W		M	○													
X		H				Ⓐ	Ⓐ								Δ	
Y		M	○									Δ		Δ		
Z		M		Ⓐ						Ⓐ		Ⓐ				
AA		H			Δ	Δ		Δ								
BB		L				Δ							Δ	Δ		

APPLICATIONS

BRAND	I.C. ENGINES, H.D.	I.C. ENGINES, M.D.	COMPRESSORS	REFRIGERATORS	BEARINGS—GENERAL	GEARS—GENERAL	GEARS—WORM	GEARS—HYPOID	TURBINES	HYDRAULICS	STEAM CYLINDERS	METALWORKING	METAL ROLLING	TRANSFORMERS	TEXTILES	TABLEWAYS	HEAT TREATMENT	PROTECTIVES	ROLLING BEARINGS
M										○		Δ				○			○
N						○	○				Δ								
O								Δ											
P	Δ																		
Q	Δ		○																
R	Δ																		
S		Δ	Δ		○	○													
T						○				Δ									○
U							Δ												
V										○					○	Δ			
W		Δ	○		○	○				○									
X						○			Δ										○
Y							○				Δ								
Z															Δ				
AA						Δ													Δ
BB																	Δ		

Δ INHERENT Ⓐ ADDITIVE ○ SECONDARY Δ PRIMARY

Table 30.4 Some factors affecting lubrication frequency
(This is a general guide only – affected by local conditions and environment)

Frequency / *Element*	*Shift*	*Daily*	*Weekly*	*Fortnightly*	*Monthly*	*Bi-annually*	*Biennially*	*Remarks*
High-speed bearings	Oil top-up	Oil top-up	Oil top-up					High frequency where dia.* × rpm is greater than 6000. Where dia. × rpm is less than 6000 lower frequencies may be used unless extreme dirt conditions or temperatures above 60°C prevail
Low speed bearings			Apply grease with gun until slight resistance is felt		Grease if very slow speed or limited movement		Clean and repack, or at major overhaul	Weekly and fortnightly lubrication for relatively high speeds (dia.* × rpm) greater than 3000 but less than 6000
Chain drives		Clean off and renew lubricant						Use high frequency with very dirty conditions
Cocks and valves			When used frequently		When used less than ten times per day			Where usage less than once per day lubricate just prior to use
Compressors cylinders and sumps		Top-up if required	Oil change after 250h if sealing poor		Oil change after 500 h if sealing good			Change more frequently if very adverse conditions prevail
Couplings				Grease or top-up depending on sealing				Grease nipples 2–3 shots until resistance felt
Gears—open		Clean off lubricant and apply new			Check and top-up if necessary			Depending upon environment
Gears—enclosed					Check and top-up if necessary	Change oil		Depending upon operation conditions, temperature etc.
Glands and seals		For soft packed glands						Especially those handling fluids which re-act with the lubricant give one or two shots of grease
Hydraulic systems		Top-up if required				Change oil depending on operating conditions, temperatures etc.		Hydraulic fluid may be changed more frequently if the colour indicates contamination from dissolving seals etc.
Ropes			Clean and apply new lubricants					Previous experience will determine variations depending upon dirt and usage
Slideways, guides and linkages	Apply lubricants				Guides, lifts and hoists only			More or less frequently depending on conditions of dirt, swarf and usage

* Dia. in inches.

PLANNING A METHODICAL LUBRICATION SYSTEM

When planning a methodical system for plant lubrication, the following techniques for sorting out the work to be done may be helpful:

(1) Divide the work in terms of the frequency of lubricant application.
(2) Divide the work by method of application and lubricant grade.
(3) Consider the optimum route for the lubrication personnel, accounting for walking distances where necessary.

By tabulating the data in this way, simple work loads can be established.

The following diagram outlines a possible plan.

Control

To assist in the control of the system, the following items have been found to be useful:

(1) Central plant records ranging from pages in a loose leaf book to comprehensive filing systems, depending on the size of the plant.
(2) A steel-bound book for keeping records in the plant itself.
(3) Wall charts to show progress through the lubrication year.
(4) Route cards showing weekly and monthly work for each day.
(5) Record cards attached to the machines.

Additional advantages

The personnel carrying out the lubrication should report back machine defects, and the planned lubrication system can be used to initiate repair work.

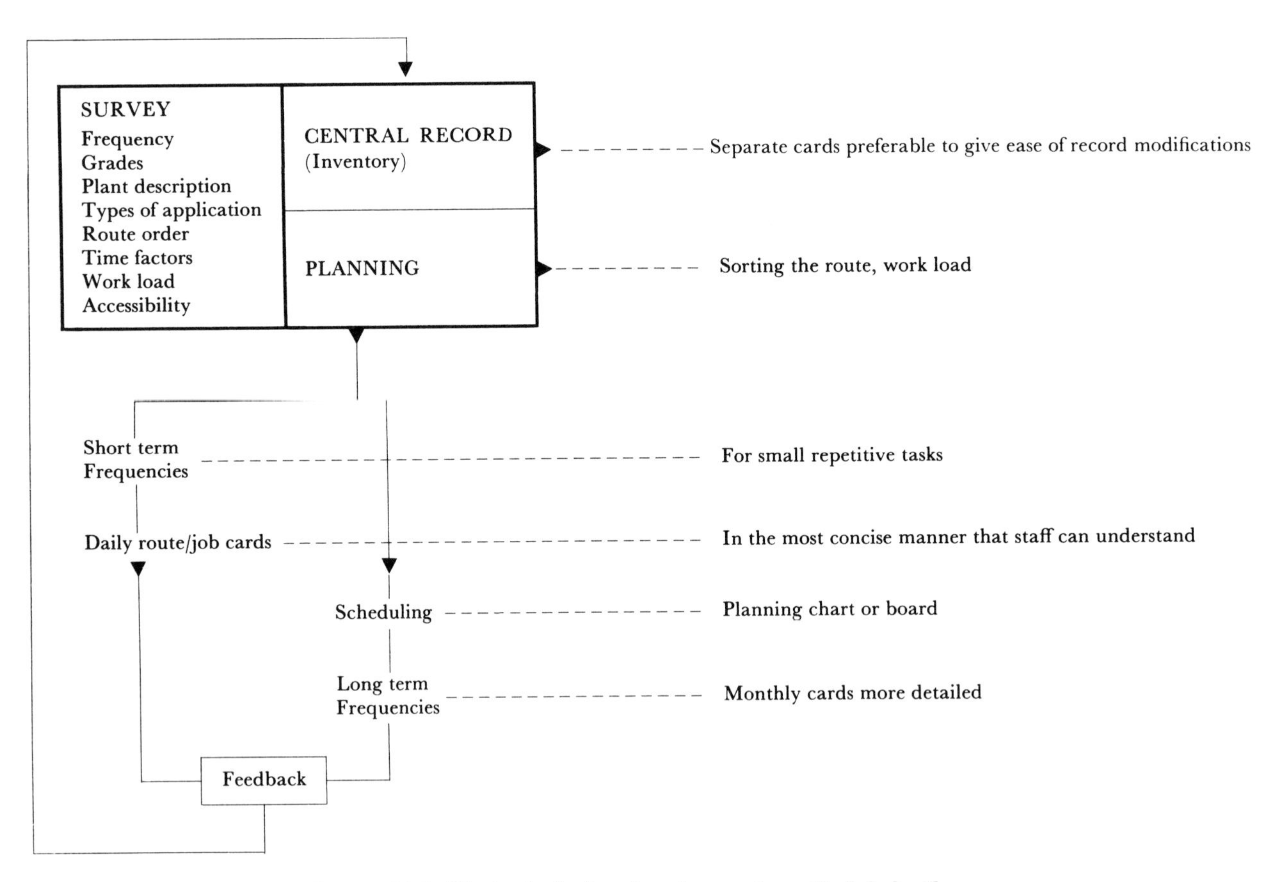

Figure 30.2 Method of planning for systematic lubrication

COMPUTERISED MANAGEMENT SYSTEMS

All the above activities can be controlled by a computer based system dedicated to asset and maintenance management activities. Successful implementation will be enhanced if the key aspects and items above are already identified or it is intended to augment an existing working manual system. Feedback mechanisms must be ensured and previous maintenance histories input to achieve efficient utilisation of the computerised system.

PRESSURE

Level of pressure	*Effect*
Pressures in tens of MPa (thousands of psi)	Oil lubricants increase considerably in viscosity and density; they may even go solid
Pressures in MPa (hundreds of psi)	Gases in contact with lubricants dissolve in them causing property changes and foaming on decompression

Effect of pressure on lubricants

Type of lubricant	*Factor*	*Effect*	*Tribological significance*
Gas	Pressure	Increased density	Aerodynamic gas-lubricated bearings
Oil	Pressure	Increased density (volume change) (Figure 31.3) Increased viscosity (Figures 31.1 and 31.2) and raised pour point	Very high pressure hydraulic systems: elastohydrodynamic lubrication
	Gas environment	Solubility of gas is increased with consequent fall in viscosity	Compressors where lubricant is in contact with gas (e.g. reciprocating piston-ring compressors, sliding vane rotary compressors)
Solid	Pressure	None	—

Figure 31.1 Effect of pressure on viscosity of HVI paraffinic oils

Oil	*Viscosity*, cS		
	100°F	210°F	*VI*
A	33.1	5.32	102
B	55.3	7.17	96
C	165.3	15.12	99

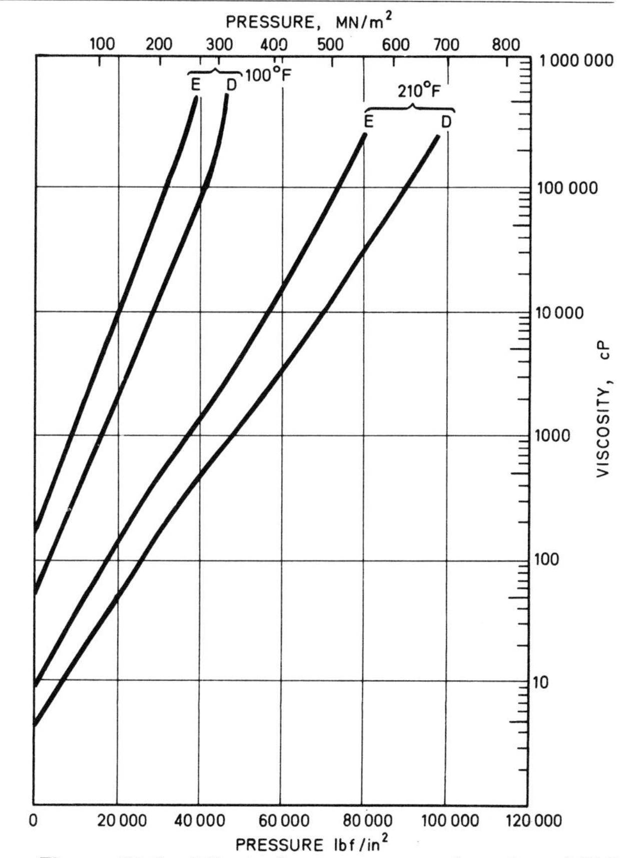

Figure 31.2 Effect of pressure on viscosity of LVI naphthenic oils

Oil	*Viscosity*, cS		
	100°F	210°F	*VI*
D	55.1	5.87	23
E	143.1	9.48	8

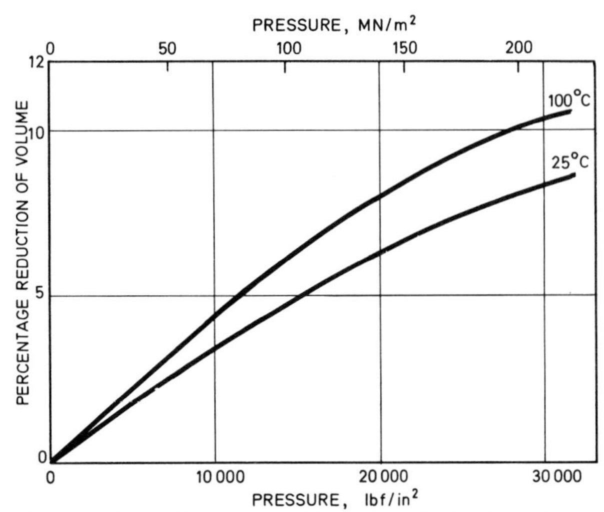

Figure 31.3 Compressibility of typical mineral oils

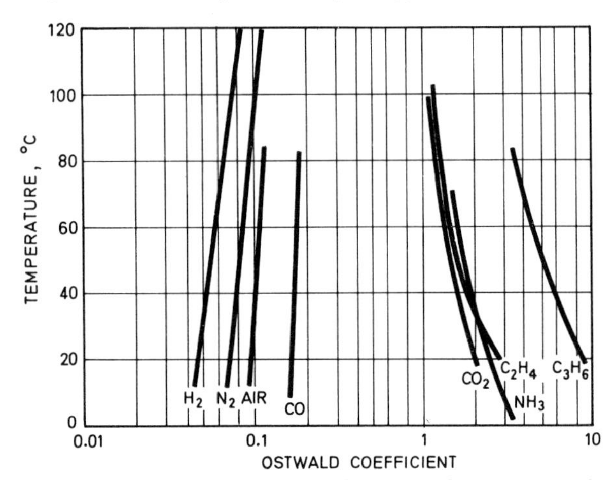

Figure 31.4 Ostwald coefficients for gases in mineral lubricating oils

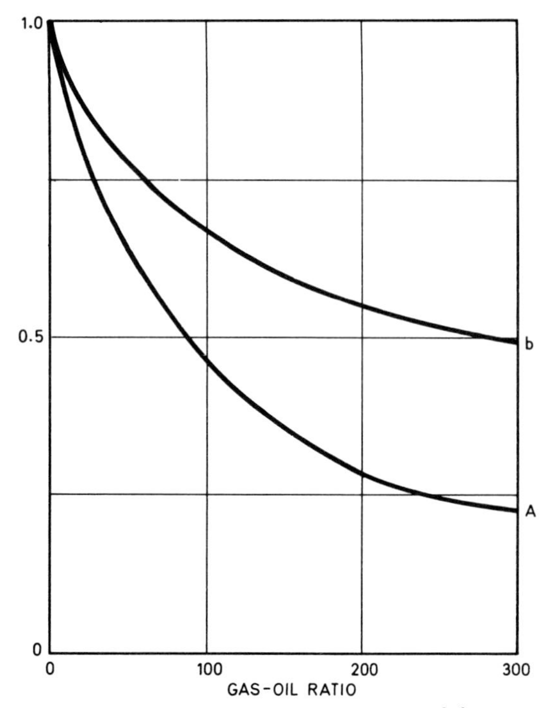

Figure 31.5 Constants for eqn (2)

Effect of dissolved gases on the viscosity of mineral oils

An estimate of the viscosity of oils saturated with gas can be obtained as follows:

(*i*) Determine Ostwald coefficient for gas in mineral oil from Figure 31.4.

(*ii*) Calculate gas : oil ratio from:

$$\text{Gas:oil ratio} = \text{Ostwald coefficient} \times p \cdot \frac{293}{\theta + 273} \quad \ldots \quad (1)$$

where p is the mean gas pressure (bar), and θ the mean temperature (°C).

(*iii*) Obtain viscosity of oil saturated with gas(es) from:

$$v_s = A v_0^b \quad \ldots \quad (2)$$

where v_0 is the viscosity of oil at normal atmospheric pressure (CSt); and A, b are constants obtained from Figure 31.5.

VACUUM

Level of pressure	*Effect*
Moderate vacuum to 10^{-4} torr	Liquid lubricants tend to evaporate
High vacuum below 10^{-5} torr	Surface films are lost, and metals in contact can seize

Lubricant loss by evaporation

Effect	*Method of control*
As the pressure is reduced and the vapour pressure of a liquid lubricant is approached, its rate of evaporation increases	1 Use a lubricant with a very low vapour pressure.* The rate of loss can then be very low 2 If the pressure is below 10^{-6} torr lubricant evaporation can be reduced by a simple labyrinth seal because the mean free path of its vapour molecules will be greater than the labyrinth paths
At very low pressure the lubricant may evaporate too quickly to be usable	3 If the space around the lubricated component can be sealed, the local pressure will stabilise at the lubricant vapour pressure

Sealing Method	*Vacuum obtainable*
Single lip seal†	10 torr
Double or triple lip seals†	10^{-3} torr

* For example, the materials in Table 31.1.
† Seals should be lubricated with a smear of vacuum grease.

Table 31.1 Lubricants and coatings which have been used in high vacuum

Material	*Vacuum*	*Temperature, °C*	*Speed*	*Load*	*Remarks*
Versilube F50	10^{-6}	160	High	Medium	Ball bearing and gear
Versilube G–300 Grease	10^{-6}	160	High	Medium	Ball bearing
Apiezon 'T'	10^{-6}	160	High	Medium	Ball bearing
Barium film	10^{-6}	200	High	Light	Ball bearings
PbO, PbI_2 or other halides in graphite	10^{-6}	750	High	Medium	Brushes
Everlube 811 (phenolic bonded MoS_2)	10^{-6}–10^{-7}	40–85	High	Medium	(Remove initial wear debris)
BR2S Grease	10^{-7}	Ambient	Medium	Medium	Ball bearing
MoS_2–graphite–silicate	10^{-7}	Max. 200	Low	Medium	Ball bearings, pin-on-disc, gears
15% tin in nickel	10^{-7}	Ambient	Medium	Medium	Slider
Apiezon 'L'	10^{-7}–10^{-8}	<50	Low	High radial	100 mm ball bearings, no atmospheric contamination
Cu–PTFE–WSe_2	10^{-7}–10^{-8}	−50 to +150	Medium	Medium	Rolling/sliding
Gold plating, silver plating	10^{-7}–10^{-8}	<50	Low	Low	Gears
AeroShell Grease 15	10^{-8}	<120	8000 rev/min	Low	Instrument bearings (sensitive to mis-alignment)
MoS_2–burnished 24 ct Au film	10^{-8}	Ambient	Low	Medium	Ball bearings
In situ MoS_2	10^{-9}	Ambient	Medium	Medium	Slider
Lead film	10^{-9}	−20 to +80	Medium	Low	Ball bearing, gear
Silver–copper–MoS_9	10^{-9}	Ambient	Medium	Low	Brushes
MoS_2 or Cu in polyimide	10^{-10}	Ambient	Medium	Medium	Slider

Loss of surface films in high vacuum

Surface contaminant films of soaps, oils and water, etc., and surface layers of oxides, etc., enable components to rub together without seizure under normal atmospheric conditions. Increasing vacuum causes the films to be lost, and reduces the rate at which oxide layers reform after rubbing. The chance of seizure is therefore increased.

Seizure can be minimised by using pairs of metals which are not mutually soluble, and Table 31.2 shows some compatible common metals under high vacuum conditions, but detailed design advice should usually be obtained.

Vacuum level	*Effect on surfaces*
0.5 bar (pressure)	Minimum pressure for there to be sufficient water vapour in average room air to enable graphite to work successfully
6 torr	Minimum pressure for graphite to work successfully in pure water vapour
10^{-1} torr	Water lost from surface
10^{-5} torr	Most soaps and oils lost from surface
10^{-5} torr	Oxide films become difficult to replace after rubbing
10^{-8} torr	Oxide films no longer replaced after rubbing
Below 10^{-10} torr	Oxide film may be lost without rubbing

Table 31.2 Some compatible metal pairs for vacuum use

Material	*Satisfactory partner*
(*a*) Stainless steel (martensitic)*	Polyimide + 20% Cu fibre
(*b*) Stainless steel (austenitic)*†	Rhenium; cobalt (below 300°C), cobalt + 25% molybdenum (up to 700°C)
Tool steel*†	Tool steel, 700 VPN; nickel alloy–MoS_2 composite
Mild steel*† Soft irons	Silver; lead
Cast irons	Assuming grey irons, graphite on its own is not recommended for vacuum work and this may apply to structures containing free graphite
Copper‡	Molybdenum; chromium; tungsten
Tin	Iron; nickel; cobalt; chromium
Lead	Chromium; cobalt; nickel; iron; copper; zinc; aluminium
Tungsten	Silver; copper
Molybdenum	Copper
Aluminium	Indium; lead; cadmium
Cadmium	Aluminium; iron; nickel
Nickel§	Tin; silver; lead
Chromium	Copper; lead; tin; silver
Gold	Rhenium; lead
Silver	Plain carbon steel; chromium; cobalt; nickel

* High sulphur contents of these will help reduction of wear under vacuum.

† Generally, for steels, PTFE-based composites are advised. PTFE is the film former + lamellar solids such as MoS_2, WS_2, $CdCl_2$, CdI_2, $CdBr_2$ or selenides but not carbons, graphites or BN. A binder phase of Ag or Cu can also be included. At very high vacuum all plastics will out-gas.

‡ Unlike normal atmospheric conditions, copper and copper alloys give high wear and friction against ferrous materials in a vacuum.

§ If sintered, dispersed oxide in the nickel will be beneficial.

HIGH TEMPERATURE

Temperature limitations of liquid lubricants

The chief properties of liquid lubricants which impose temperature limits are, in usual order of importance, (1) oxidation stability; (2) viscosity; (3) thermal stability; (4) volatility; (5) flammability.

Oxidation is the most common cause of lubricant failure. Figure 32.1 gives typical upper temperature limits when oxygen supply is unrestricted.

Compared with mineral oils most synthetic lubricants, though more expensive, have higher oxidation limits, lower volatility and less dependence of viscosity on temperature (i.e. higher viscosity index).

For greases (oil plus thickener) the usable temperature range of the thickener should also be considered (Figure 32.2).

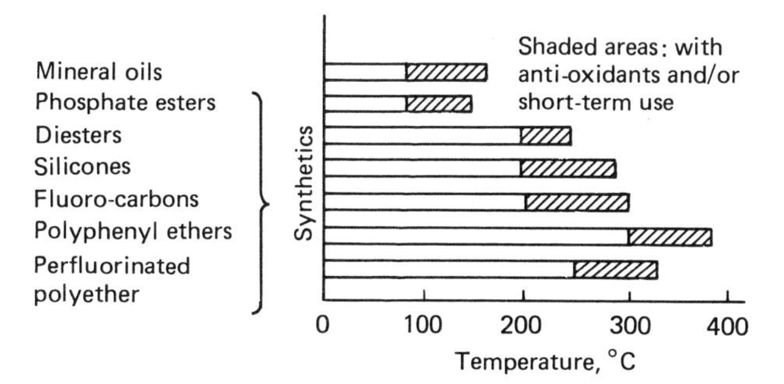

Figure 31.1

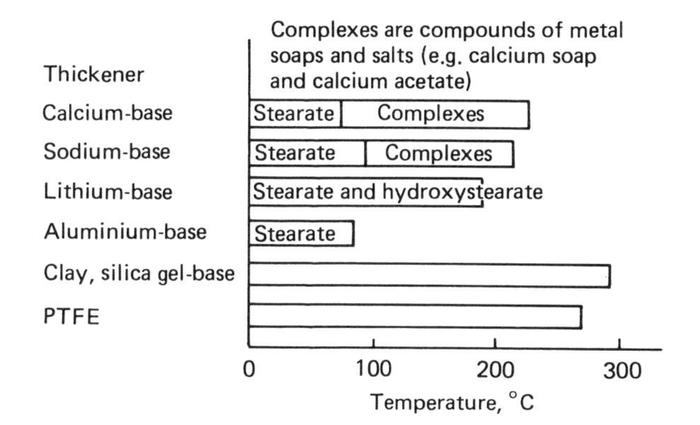

Figure 31.2

Temperature limitations of solid lubricants

All solid lubricants are intended to protect surfaces from wear or to control friction when oil lubrication is either not feasible or undesirable (e.g. because of excessive contact pressure, temperature or cleanliness requirements).

There are two main groups of solid lubricant, as given in Table 32.1.

Table 32.1

		Example	*Practical temperature limit, °C**	*Common usages*
1	Boundary lubricants and 'extreme pressure' additives (surface active)	Metal soap (e.g. stearate)	150	Metal cutting, drawing and shaping. Highly-loaded gears
		Chloride (as $Fe\,Cl_3$)	300	
		Sulphide (as FeS)	750	
		Phthalocyanine (with Cu and Fe)	550	Anti-seizure
2	Lamellar solids and/or low shear strength solids	Graphite	600	General, metal working, anti-seizure and anti-scuffing
		Molybdenum disulphide	350	
		Tungsten disulphide	500	
		Lead monoxide†	650	
		Calcium fluoride	1000	
		Vermiculite	900	Anti-seizure
		PTFE	250	Low friction as bonded film or reinforced composite

* The limit refers to use in air or other oxidising atmospheres.
† Bonded with silica to retard oxidation.

Dry wear

When oil, grease or solid lubrication is not possible, some metallic wear may be inevitable but oxide films can be beneficial. These may be formed either by high ambient temperature or by high 'hot spot' temperature at asperities, the latter being caused by high speed or load.

Examples of ambient temperature effects are given in Figures 32.3 and 32.4, and examples of asperity temperature effects are given in Figures 32.5 and 32.6.

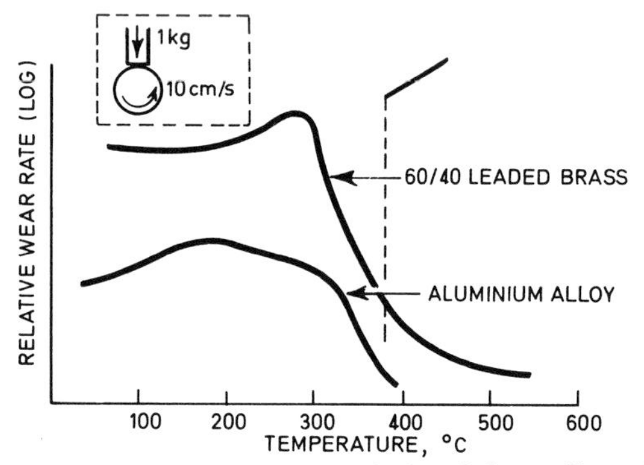

Figure 32.3 Wear of brass and aluminium alloy pins on tool steel cylinder, demonstrating oxide protection (negative slope region). *Oxide on aluminium alloy breaks down at about 400°C, giving severe wear*

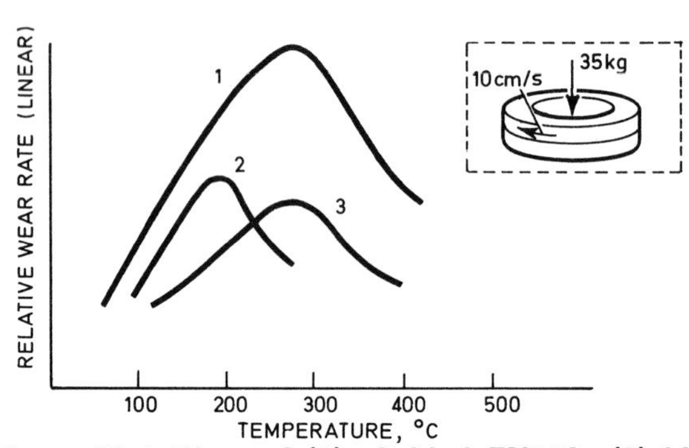

Figure 32.4 Wear of (1) nitrided EN41A; (2) high carbon tool steel; (3) tungsten tool steel. *Oxides: αFe_2O_3 below maxima; αFe_3O_4-type above maxima*

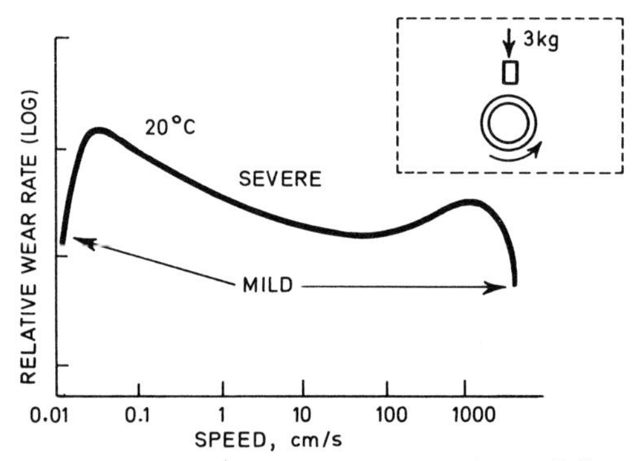

Figure 32.5 Wear of brass pin on tool steel-ring. *At low speed wear is mild because time is available for oxidation. At high speed wear is again mild because of hot-spot temperatures inducing oxidation*

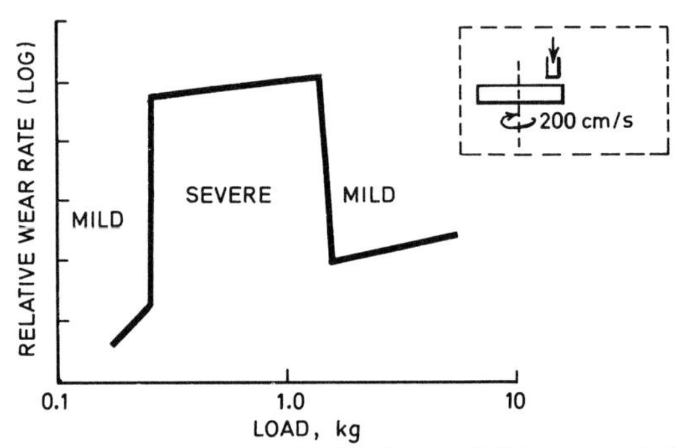

Figure 32.6 Transition behaviour of 3% Cr steel. *Mild wear region characterised by oxide debris: severe wear region characterised by metallic debris*

Bearing materials for high temperature use

When wear resistance, rather than low friction, is important, the required properties (see Table 32.2) of bearing materials depend upon the type of bearing.

Table 32.2

Rolling contact	*Sliding contact*	*Gas bearings*
High hot hardness (> 600 VPN). Dimensional stability, resistance to: oxidation, phase change, residual stress and creep. Thermal shock resistance	Moderate hot hardness. Good thermal conductivity and shock resistance. Resistance to oxidation and scaling	Extreme dimensional stability. Low thermal expansion and porosity. High elastic modulus. Capable of fine surface finish

Hot hardness, particularly in rolling contact bearings, is of high importance and Figure 32.7 shows maximum hardness for various classes of material.

Some practical bearing materials for use in oxidising atmospheres are shown in Table 32.3.

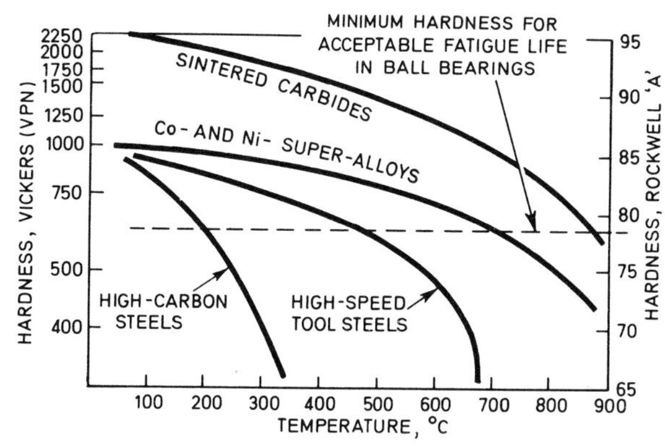

Figure 32.7

Table 32.3

Rolling contact	*Sliding contact*	*Approximate temp. limit,* °C
High-speed tool steel (Mo and W types)	PH stainless steel Nitrided steel	500
	Hastelloy (Ni super-alloy)	750
Stellite (Co super alloy) Titanium carbide	Stellite	850
Dense α-alumina Zirconia	Alumina–Cr–W cermet Silicon carbide Silicon nitride	1000 or above

LOW TEMPERATURE

General

'Low temperature' may conveniently be subdivided into the three classes shown in Table 32.4. In Class 1, oils are usable depending upon the minimum temperature at which they will flow, or the 'pour point'. Some typical values are given in Table 32.5. Classes 2 and 3 of Table 32.4 embrace most industrially important gases (or cryogenic fluids) with the properties shown in Table 32.6.

Because of their very low viscosity (compare to 7×10^{-2} Ns/m^2 for SAE 30 oil at 35°C) these fluids are impractical as 'lubricants' for hydrodynamic journal bearings. (Very high speed bearings are theoretically possible but the required dimensional stability and conductivity are severe restrictions.)

Table 32.5

Type of lubricant	*Pour point,* °C
Mineral oil	−57 (min.)
Diester	−60
Phosphate ester	−57
Silicate ester	−65
Di-siloxane	−70
Silicone	−70
Polyphenyl-ether	−7
Perfluorinated polyether	− 75 to −90

Table 32.4

Class	*Temperature range*	*Usages involving tribology*
1	0°C to −80°C (190 K)	Domestic refrigeration, transport over snow and ice
2	−80°C to −196°C (77 K)	Liquefaction and handling of industrial gases, rocket propulsion (turbo-pumps, seals)
3	−196°C to −273°C (0 K)	Space exploration, liquid hydrogen and helium systems

Table 32.6

Fluid	*Boiling point* K	*Viscosity at boiling point* Ns/m^2
Oxygen	90.2	1.9×10^{-4}
Nitrogen	77.4	1.6×10^{-4}
Argon	87.3	—
Methane	111.7	—
Hydrogen	20.4	1.3×10^{-5}
Helium	4.2	0.5×10^{-5}

Unlubricated metals

In non-oxidising fluids, despite low temperature, metals show adhesive wear (galling, etc.) but in oxygen the wear is often less severe because oxide films may be formed. Where there is condensation on shafts, seals or ball bearings (dry lubricated) a corrosion-resistant hard steel (e.g. 440 C) is preferable.

Plain bearing materials

As bushes and thrust bearings, filled PTFE/metal and filled graphite/metal combinations are often used – see Table 32.8.

Safety

Aspects of safety are summarised in Table 32.7.

Table 32.7

Hazard	*Precaution*
Heat generated at bearings or seals may cause local boiling of liquid or ignition	Design adequate venting system. For fuel liquids (e.g. methane, hydrogen) and oxygen particularly; ensure total compatibility of bearing materials under extreme conditions
Fine wear debris or grease residues	Thorough check on ignition aspects and/or extreme cleanliness in installation, particularly for liquid oxygen

Table 32.8 Some successful plain bearing materials for cryogenic fluids

Bush or face	*Suitable journal or counterface*	*Remarks*
Bronze/graphite-filled PTFE PTFE/lead-filled bronze (steel backed)	18/8 Stainless steel Martensitic steel (e.g. S.80) Chromium plate	Suitable for all fluids
Copper/lead-filled graphite	Chromium plate	Soft stainless steel is scored. Combination is best in liquid oxygen
Phenolic-impregnated carbon	Carbon	—
Pure PTFE	Duralumin or bronze	Thermal conductivity of counterface important

Ball bearings and seals for cryogenic temperatures

Table 32.9 Recommended tribological practice at cryogenic temperatures

Component	*Recommended material*
High speed ball bearings (> 10 000 RPM)	The raceway coating should include MoS_2 or PTFE, and the cage should be woven glass fibre reinforced PTFE
Low speed ball bearings	Either the cage should be PTFE filled with MoS_2 and chopped glass fibre, or a film of magnetron-sputtered MoS_2 (or ion-plated lead) should be present on the raceways and balls
Reciprocating seals	Use a seal manufactured from PTFE filled with chopped glass fibre or from a PTFE and bronze composite
Rotary seals	Use a carbon-graphite face loaded against a tungsten carbide or hard chromium plated face

GENERAL NOTES

Ambient temperatures and humidities can vary widely over short geographic distances especially in mountain or coastal areas. The maps in this section can only indicate whether temperature or humidity is likely to be a problem in any area.

AVERAGE SUMMER TEMPERATURES

The map relates to the average of temperatures throughout the day and night in the warmest month of the year.

EXTREME TEMPERATURES

The map shows the highest and lowest recorded air temperatures. The average highest and lowest each year are typically about 5°C less extreme than shown. Much higher temperatures can be attained by equipment standing in the sunshine.

HUMIDITY

Relative humidity is very variable between seasons and at different times of day, with pronounced local variations, particularly in coastal and mountain regions. RH values below 20% and above 90% are to be expected in almost any part of the world. In particular, early morning humidities of 80 to 100% are common in most coastal and low lying areas. The map in Figure 33.3 shows areas in which exceptionally high or low relative humidities are maintained throughout the day for long periods, defined as follows:

VERY HUMID: Mean daily humidity in most humid month of the year exceeds 90% RH.

HUMID: Mean daily humidity in most humid month of the year exceeds 85% RH.

DRY: Mean daily humidity in driest month of the year below 40% RH.

VERY DRY: Mean daily humidity in driest month of the year below 20% RH.

Note that the data apply to open-air conditions. Heating reduces the humidity, so that even in the 'very humid' areas of Canada and North Europe the humidity inside a heated building may well be 'very dry'.

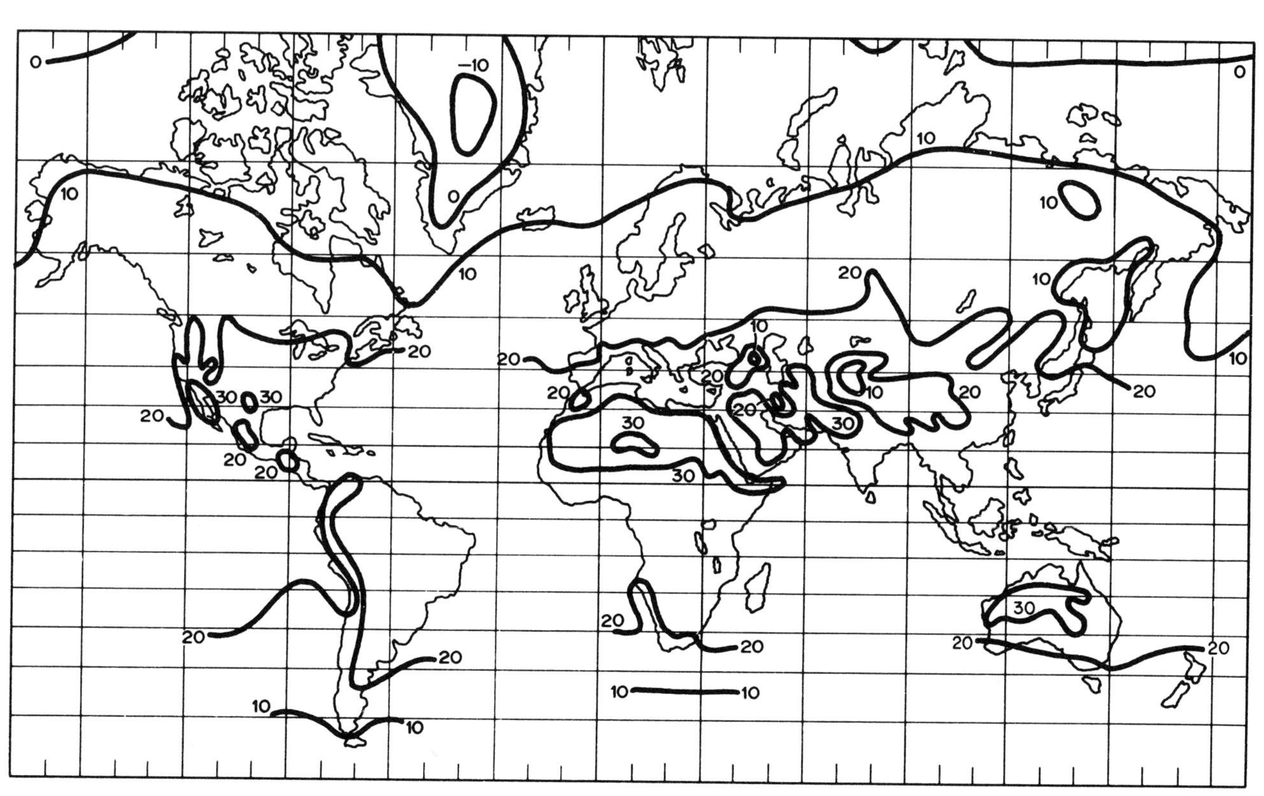

Figure 33.1 Average summer temperature, °C

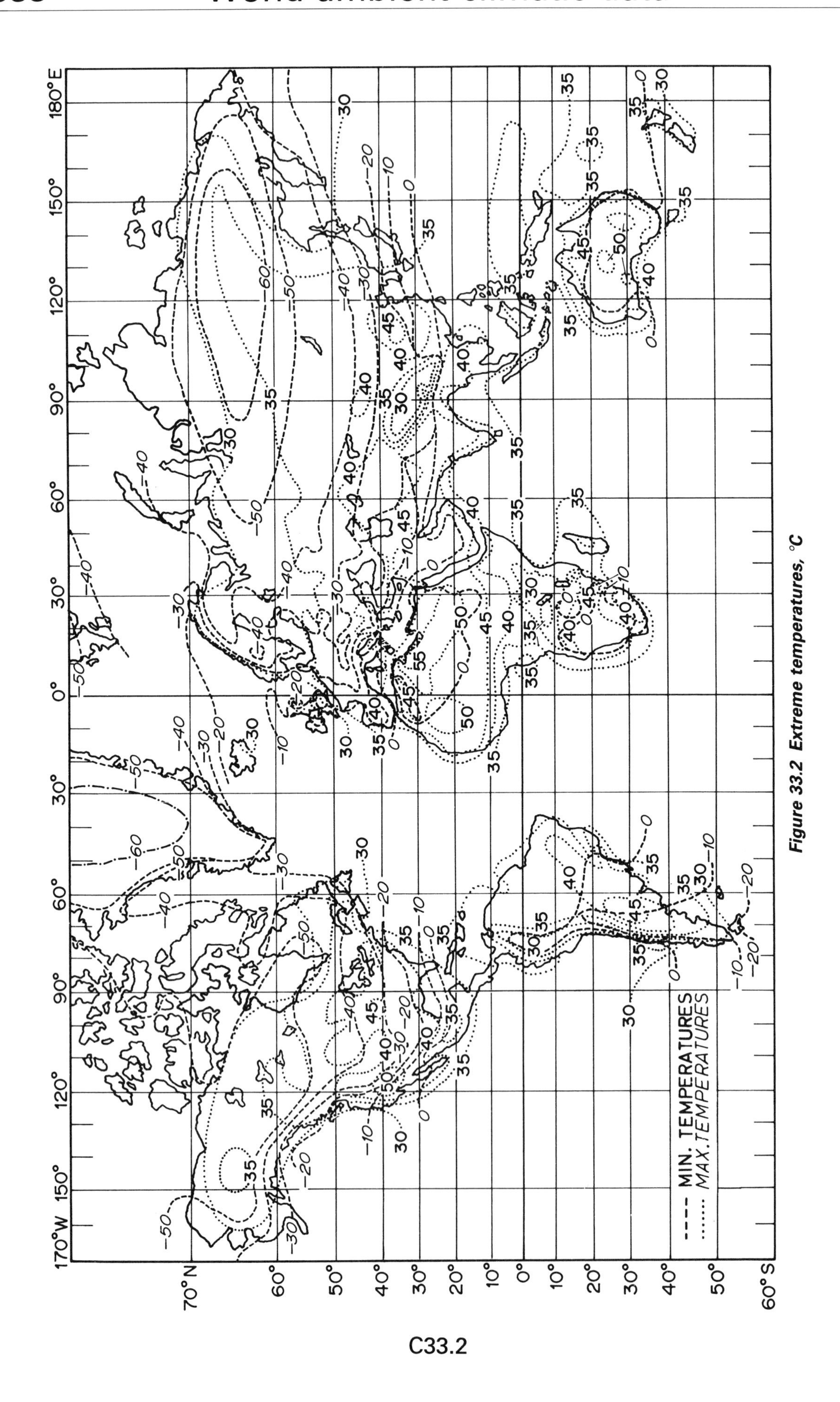

Figure 33.2 Extreme temperatures, °C

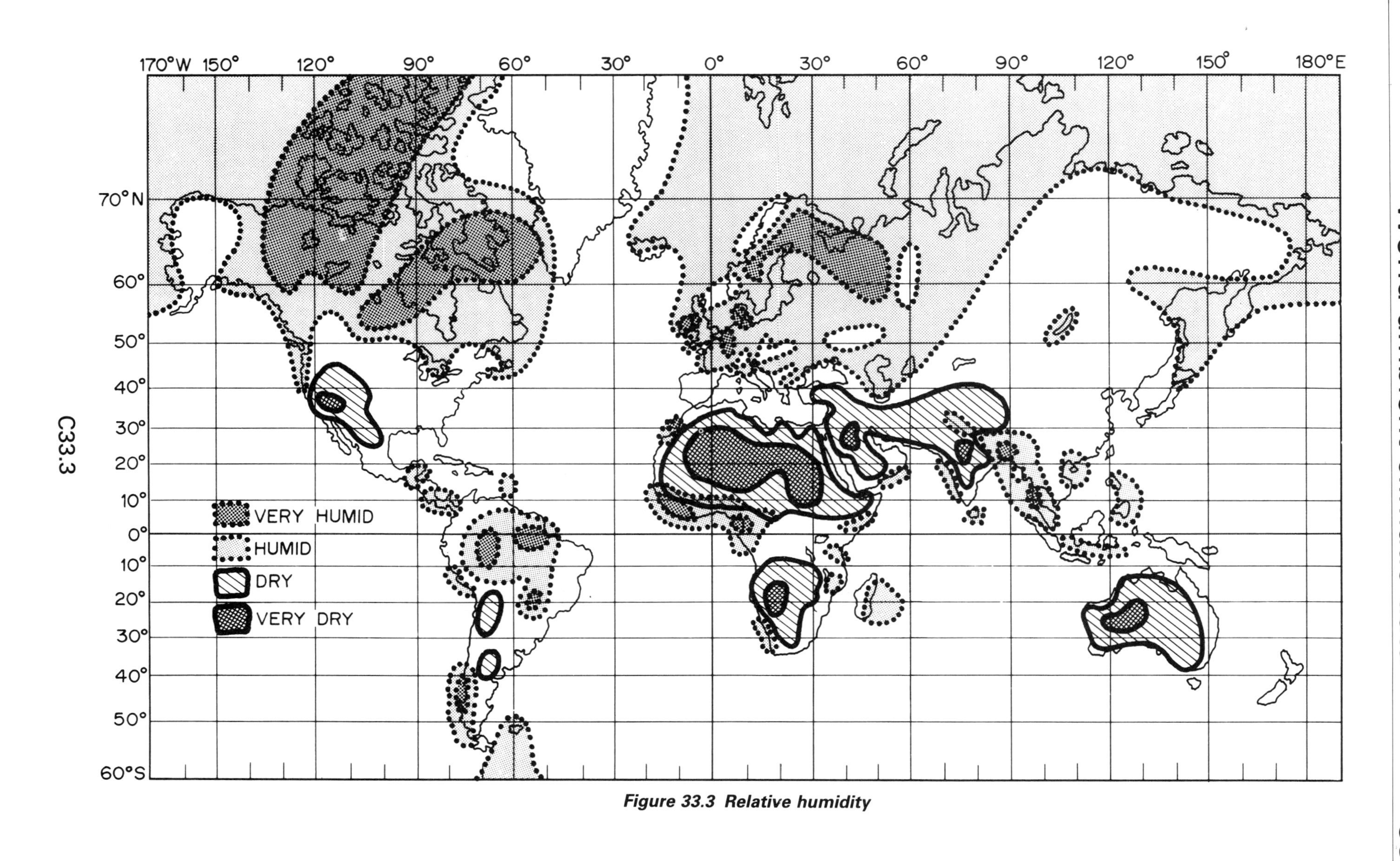

Figure 33.3 Relative humidity

C34 Industrial plant environmental data

Table 34.1 Effects of atmospheric conditions

Atmosphere	*Effect*	*Precautionary measures*
High temperature, ambient and radiation from hot sources	Lowers lubricant viscosity. Plastics and lubricants deteriorate. Expansion problems	Use cooling system. Choose high temperature materials. Allow adequate clearances in design
High temperature, radiation from hot sources	Unequal expansion in addition to above	Keep source/object distance large. Use heat reflectors or insulation. Surfaces preferably reflective, smooth and of light colour. Use large characteristic volume
High humidity	Etching, corrosion and electrical breakdown	Use protective coatings or materials that do not rust or corrode. Keep temperature changes to a minimum
Corrosive gases and vapours	Corrosion	Use inert materials or protective coatings
Dust	Accelerates wear. Movement seizure	Use adequate seals or filter air

TEMPERATURE

The main problems in industry arise with radiation from hot processes. Typical examples of heat sources are as follows:

Table 34.2 Temperatures of some industrial processes

Sources of heat	*Temperature* °C	*Sources of heat*	*Temperature* °C
Drying steel wire	150	Calorising (baking in aluminium powder)	930
Drying lacquer	150	Heating sheet bars	930
Japanning	80–230	Normalising sheet steel	930
Core baking for iron castings	150–230	Carburising	950
Blueing	150–230	Pack-heating sheet steel	950
Carbolic and creosote oil distillation	200–300	Coke oven	950–1050
Hot-dip tinning	260	Short cycle annealing, malleable iron	980
Tempering in oil	260	Cyaniding	980
Tempering high-speed steel	330	Glasing porcelain	1000
Hydrocarbon synthesis	300–400	Gas turbines	1000–1200
Tissue paper drying	400	Vitreous enamelling (castings)	1010
Annealing aluminium	400–500	Bar and pack heating, stainless steel	1040
Cracking petroleum	400	Normalising stainless steel	930–1090
Heating aluminium for rolling	450	Rolling stainless steel	950–1230
Petroleum processing	500–600	Forging titanium alloys	870–1060
Nitriding steel	510	Annealing manganese-steel castings	1040
Annealing brass	540	Heating tool steel for rolling	1040
Annealing glass	620	Heating sheet steel for pressing	1050
Annealing copper	620–700	Heating spring steel for rolling	1090
Annealing german silver	650	Glost-firing porcelain	1120
Enamelling wet process	650	Firebrick kiln	1150–1400
Strain relieving	650–700	Hardening high-speed steel	1200
Annealing cold-rolled strip	680–760	Bisque-firing porcelain	1230
Porcelain decorating	760	Heating steel blooms and billets for rolling, also rivet heating	1250
Heating brass for rolling	790		
Annealing nickel or monel wire or sheets	800	Heating steel for drop forging or die pressing	1300
Annealing high-carbon steel	820	Calcining limestone	1370
Heat-treating medium-carbon steel	840	Welding steel tubes from preformed skelp	1400
Patenting wire	840	Burning firebrick	1320–1480
Box-annealing sheet steel	840	Burning portland cement	1430
Vitreous-enamelling sheet steel	870	Glass melting	1300–1500
Annealing malleable iron, long cycle	870	Basic refractories	1550–1750
Heating copper for rolling	870	Steel melting	1680
Forging commercially pure titanium	870	Melting chromium steel	1790
Annealing steel castings	900	High alumina brick manufacture	1900–2000
Building brick kiln	900–1050	Spray steel making	2100
Normalising steel pipes	900		

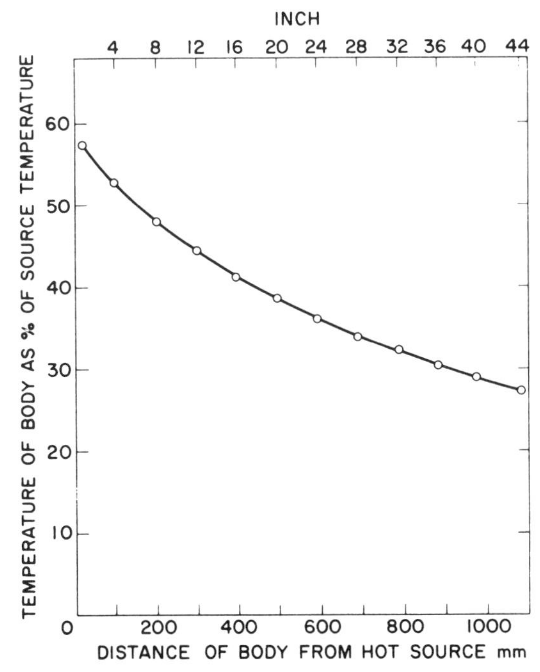

Figure 34.1 Applicable to furnace walls from 150 to 300°C

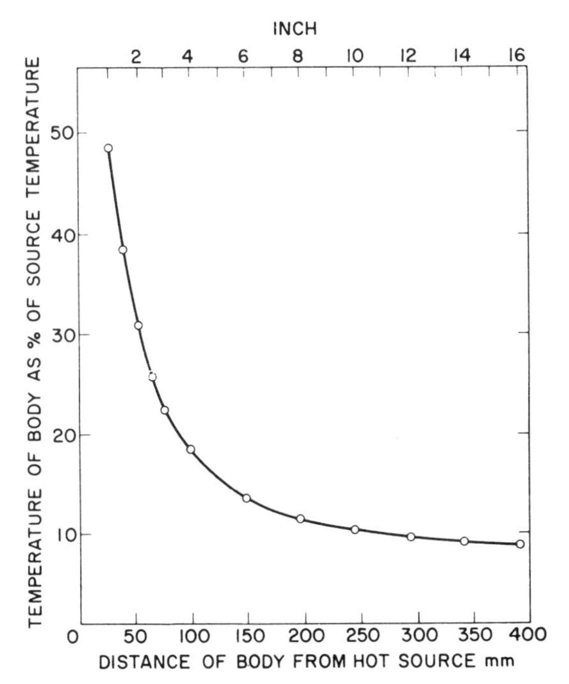

Figure 34.2 Applicable to sources from 300 to 1400°C

These graphs are based on laboratory and field measurements where a blackened metallic body was used with convective cooling. Figure 34.2 is for a source area of 20 in^2. Increasing the source area will reduce the slope of the graph towards that of Figure 34.1 which approximates to an infinite plane source. The multiplicity of variables associated with radiative heat transfer precludes a simple accurate calculation of the temperature any body will reach when placed near any source of heat. However, the graphs will indicate if temperature is likely to be a problem. The heat generated by the body itself must, of course, not be overlooked.

HUMIDITY

Relative humidities above 45% often lead to condensation problems.

Table 34.3 Typical values of relative humidity and dry bulb temperatures for working areas found in industry

Industry	*Relative humidity %*			*Dry bulb temperature °C*		
	Minimum	Maximum	Average	Minimum	Maximum	Average
Abrasive manufacture	—	—	50	—	—	24
Bakery	40	85	70	4	41	24
Brewery	55	75	70	0	10	7
Confectionery	13	63	45	4	66	24
Ceramics	35	90	60	16	66	27
Coke ovens	30	70	55	13	46	27
Distilling	35	60	55	0	24	21
Electrical products	15	70	50	20	24	23
Engineering	35	50	40	18	38	24
Ordnance	—	—	40	21	88	24
Pharmaceuticals	15	50	35	21	32	26
Plastics	25	65	45	4	27	24
Printing	45	55	50	21	27	24
Rubber goods	25	50	40	16	32	27
Textiles	50	85	60	21	29	27
Tobacco	60	85	70	24	32	27
Steel	35	90	55	7	35	21

CORROSIVE ATMOSPHERES

Table 34.4 Industries and processes with which corrosive atmospheres are often associated

Anodising and bright dipping aluminium
Batteries and electrolytic cell production
Chemical manufacture
Electroplating
Explosive manufacture
Fertiliser manufacture
Pharmaceutical manufacture
Plastics manufacture
Sheet steel pickling and phosphating
Smelting processes
Textile and paper bleaching

DUST

Table 34.5 Industries in which dust problems may be excessive

Abrasive manufacture
Asbestos manufacture and usage
Brick manufacture including refractories
Cement manufacture
Foundries of all kinds
Gas and coke production
Glass fibre manufacture and usage
Metal smelting
Millers of grain
Potteries and earthenware manufacture
Quarries, mining and stone working
Roofing felt manufacturers
Steel manufacture
Users of mineral fillers in mastic and rubber industries

Table 34.6 Particle sizes of common materials as a guide to the specification of seals and air filters

Particles	*Size* μm
Carbon black	0.01–2
Cement dust	5–120
Coal dust	10–5000
Flour*	5–110
Foundry dust	1–1000
Ground limestone	30–800
Metallurgical dust	0.5–100
Metallurgical fume	0.01–2
Milk powder	1–10
Oil smoke	0.03–1
Pigments	0.07–7
Pulverised coal	10–400
Pulverised coal fly ash	1–50
Sand blasting dust	0.5–2
Stoker fly ash	10–800

Note: * Flour dust can be a particular problem when it drifts into small gaps, becomes damp and then sets into a hard solid mass.

This section is restricted to chemical effects on metals.
Chemical effects can arise whenever metals are in contact with chemicals, either alone or as a contaminant in a lubricant. Wear in the presence of a corrosive liquid can lead to accelerated damage due to corrosive wear. This problem is complex, and specialist advice should be taken.
Corrosion or corrosive wear are often caused by condensation water in an otherwise clean system.

Table 35.1 Some specific contamination situations

	Problem	Typical occurrence	Solution
1	Contamination by reactive gases	Internal combustion engines	Sophisticated lubricants
2	Dilution by petroleum gases	Gas compressors, valves, flowmeters	Use of higher viscosity or non petroleum oil.
3	Inert or reducing gases	Gas compressors, valves, flowmeters	Use of self-lubricating materials

Table 35.2 Typical corrosion resistant materials

Material	Nominal composition	Reference number in next table
Corrosion resisting steel[1]	0.15 C max. 12 Cr 1 Ni max.	1
Corrosion resisting steel[1]	0.08 C max. 19 Cr 9 Ni	2
Corrosion resisting steel[1]	0.1 C max. 18 Cr 10 Ni 2 Mo	3
Corrosion resisting steel[1]	0.07 C max. 22 Cr 29 Ni 3 Mo	4
Corrosion resisting steel[1]	0.35 C max. 27 Cr 3 Ni max.	5
Grey cast irons[2]		6
High silicon irons	14 Si min.	7
Austenitic cast irons	Total Ni Cr Cu 22 min.	8
Bronzes		9
Nickel, chrome/molybdenum alloys	Fe 20 max. Some Cu, W, Si Mn	10
Monel metal		11
Nickel		12
Lead		13
Non metallic materials		14

Notes
The resistance of these materials to corrosion will be substantially affected by the velocity of corrosive fluids in contact with them. High fluid velocities increase corrosion rates.
1. Stainless steel materials are prone to seizure if allowed to rub together. The risk may be reduced by maintaining the highest practical hardness of both surfaces, or by MoS_2 lubrication of the interface.
2. With grey cast iron precautions must be taken against corrosion in storage or when temporarily out of action.

SELECTION OF CORROSION RESISTANT MATERIALS FOR CONTACT WITH VARIOUS LIQUIDS

Liquid	Condition	Materials suitable under the majority of conditions
Acetaldehyde		6
Acetate Solvents		2 3 4 6 9 10
Acetone		6
Acid, Acetic and Anhydride	Cold	2 3 4 7 10
Acid, Acetic	Boiling	3 4 7 10
Acid, Arsenic		2 3 4 7 10
Acid, Benzoic		2 3 4 10
Acid, Boric	Aqueous sol.	2 3 4 7 9 10
Acid, Butyric	Conc.	2 3 4 10
Acid, Carbolic	Conc. (m.p.106°F)	2 3 4 6 10
Acid, Carbolic	Aqueous sol.	2 3 4 10
Acid, Carbonic	Aqueous sol.	9
Acid, Chromic	Aqueous sol.	2 3 4 7 10
Acid, Citric	Aqueous sol.	2 3 4 7 9 10
Acids, Fatty (Oleic, Palmitic, Stearic, etc.)		2 3 4 9 10
Acid, Formic		3 4 10
Acid, Fruit		2 3 4 9 10 11
Acid, Hydrochloric	Dil. cold	4 7 10 11 12 14
Acid, Hydrochloric	Dil. hot or conc.	7 10 14
Acid, Hydrocyanic		2 3 4 6 10
Acid, Hydroflouric	Anhydrous, with hydro carbon	11
Acid, Hydroflouric	Aqueous sol.	9 11

Liquid	Condition	Materials suitable under the majority of conditions
Acid, Hydrofluosilicic		9 11
Acid, Lactic		2 3 4 7 9 10
Acid, Mixed		2 3 4 6 7 10
Acid, Naphthenic		1 2 3 4 6 10
Acid, Nitric	Conc. boiling	4 5 7
Acid, Nitric	Dilute	1 2 3 4 5 7
Acid, Oxalic	Cold	2 3 4 7 10
Acid, Oxalic	Hot	4 7 10
Acid, Ortho-Phosphoric		3 4 10
Acid, Picric		2 3 4 7 10
Acid, Pyrogallic		2 3 4 10
Acid, Pyroligneous		2 3 4 9 10
Acid, Sulphuric	Cold	4 7 10
Acid, Sulphuric	Hot	7 10
Acid, Sulphuric (Oleum)	Fuming	4 10
Acid, Sulphurous		2 3 4 9 10 13
Acid, Tannic		2 3 4 9 10 11
Acid, Tartaric	Aqueous sol.	2 3 4 9 10 11
Alcohols		9
Alum	(See Aluminum Sulphate and Potash Alum)	
Aluminum Sulphate	Aqueous sol.	4 7 10 11 13
Ammonia, Aqua		6
Ammonium Bicarbonate	Aqueous sol.	6
Ammonium Chloride	Aqueous sol.	3 4 7 10 11
Ammonium Nitrate	Aqueous sol.	2 3 4 6 10 11
Ammonium Phosphate	Aqueous sol.	2 3 4 6 10 11
Ammonium Sulphate	Aqueous sol.	2 3 4 6 10
Ammonium Sulphate	With H_2SO_4	3 4 7 9 10 13
Aniline		6
Aniline Hydrochloride	Aqueous sol.	7 10 14
Asphalt	Hot	1 6
Barium Chloride or Nitrate	Aqueous sol.	2 3 4 6 10
Beer or Beer Wort		2 9
Beet Juice or Pulp		2 9
Benzene (Benzol)		6
Blood		9
Brine, Calcium Chloride	$pH>8$	6
Brine, Calcium Chloride	$pH<8$	4 8 9 10 11
Brine, Calcium and Magnesium Chlorides	Aqueous sol.	4 8 9 10 11
Brine, Calcium and Sodium Chloride	Aqueous sol.	4 8 9 10 11
Brine, Sodium Chloride	Under 3% Salt, cold	6 8 9
Brine, Sodium Chloride	Over 3% Salt, Cold	2 3 4 8 9 10 11
Brine, Sodium Chloride	Over 3% Salt, hot	3 4 7 10 11
Brine, Sea Water		6 9
Butane		6
Calcium Bisulphite	Paper mill	3 4 10 13
Calcium Chlorate	Aqueous sol.	4 7 10 14
Calcium Hypochlorite		4 6 7 10
Cane Juice		8 9
Carbon Bisulphide		6

Liquid	Condition	Materials suitable under the majority of conditions
Carbon Tetrachloride	Anhydrous	6
Carbon Tetrachloride	Plus water	2 9
Caustic Potash	(see Potassium Hydroxide)	
Caustic Soda	(see Sodium Hydroxide)	
Cellulose Acetate		3 4 10
Chloride Water	(Depending on conc.)	3 4 7 10 13 14
Chlorobenzene		2 9
Chloroform		2 3 4 9 10 11
Chrome Alum	Aqueous sol.	4 7 10
Condensate	(see Water Distilled)	
Copper Ammonium Acetate	Aqueous sol.	2 3 4 6 10
Copper Chloride (Cupric)	Aqueous sol.	7 10 14
Copper Nitrite		2 3 4 10
Copper Sulphate, Blue Vitriol	Aqueous sol.	2 3 4 7 10 13
Creosote		6
Cresol, Meta		4 6 10
Cyanogen	In water	6
Diphenyl		6
Enamel		6
Ethanol	(see Alcohols)	
Ethylene Chloride (Dichloride)	Cold	2 3 4 9 10 11
Ferric Chloride	Aqueous sol.	7 10 14
Ferric Sulphate	Aqueous sol.	2 3 4 7 10
Ferrous Chloride	Cold. aqueous	7 10 14
Ferrous Sulphate (Green Copperas)	Aqueous sol.	3 4 7 10 11 13
Formaldehyde		2 3 4 9 10
Fruit Juices		2 3 4 9 10 11
Furfural		2 3 4 6 9 10
Glaubers Salt	(see Sodium Sulphate)	
Glucose		9
Glue	Hot	6
Glue Sizing		9
Glycerol (Glycerin)		6 9
Heptane		6
Hydrogen Peroxide	Aqueous sol.	2 3 4 10
Hydrogen Sulphide	Aqueous sol.	2 3 4 10
Hydrosulphite of Soda	(see Sodium Hydrosulphite)	
Hyposulphite of Soda	(see Sodium Thiosulphate)	
Kaolin Slip	Suspension in water	6
Kaolin Slip	Suspension in acid	4 7 10
Lard	Hot	6
Lead Acetate (Sugar of Lead)	Aqueous sol.	3 4 10 11
Lead	Molten	6
Lime Water (Milk of Lime)		6
Liquors—Pulp Mill:		3 4 6 8 10 11
Sulphite		3 4 10 13
Lithium Chloride	Aqueous sol.	6
Magnesium Chloride	Aqueous sol.	4 7 10 14
Magnesium Sulphate (Epsom Salts)	Aqueous sol.	2 3 4 6 10

Liquid	*Condition*	*Materials suitable under the majority of conditions*
Manganese Chloride	Aqueous sol.	2 3 4 7 9 10
Manganese Sulphate	Aqueous sol.	2 3 4 6 9 10
Mash		2 9
Mercuric Chloride	Aqueous sol.	7 10
Mercuric Sulphate	In H_2SO_4	4 7 10 14
Mercurous Sulphate	In H_2SO_4	4 7 10 14
Methyl Chloride		6
Methylene Chloride		2 6
Milk		2
Molasses		9
Mustard		2 3 4 7 9 10
Nicotine Sulphate		4 7 10 11
Nitre	(see Potassium Nitrate)	
Nitre Cake	(see Sodium Bisulphate)	
Nitro Ethane		6
Nitro Methane		6
Oil, Coal Tar		2 3 4 6 10
Oil, Coconut		2 3 4 6 9 10 11
Oil, Crude	Cold	6
Oil, Crude	Hot	6
Oil, Linseed		2 3 4 6 9 10 11
Oil, Mineral (including kerosene, fuel, naptha, lubricating, paraffin)		6
Oil, vegetable (including olive, palm, soya, turpentine)		6
Oil, quenching		6
Oil, rapeseed		2 3 4 9 10 11
Perhydrol	(see Hydrogen Peroxide)	
Petrol (Petroleum Ether)		6
Phenol	(see Acid, Carbolic)	
Photographic Developers		2 3 4 10 14
Potash	Plant liquor	2 3 4 8 9 10 11
Potash Alum	Aqueous sol.	3 4 7 8 9 10 11
Potassium Bichromate	Aqueous sol.	6
Potassium Carbonate	Aqueous sol.	6
Potassium Chlorate	Aqueous sol.	2 3 4 7 10
Potassium Chloride	Aqueous sol.	2 3 4 9 10 11
Potassium Cyanide	Aqueous sol.	6
Potassium Hydroxide	Aqueous sol.	1 2 3 4 6 8 10 11 12
Potassium Nitrate	Aqueous sol.	1 2 3 4 6 10
Potassium Sulphate	Aqueous sol.	2 3 4 9 10
Propane		6
Pyridine		6
Pyridine Sulphate		4 7 13
Resin (Colophony)	Paper Mill	6
Salt Cake	Aqueous sol.	2 3 4 7 9 10
Sea Water	(see Brines)	
Sewage		6 9
Shellac		9
Silver Nitrate	Aqueous sol.	2 3 4 7 10
Soap Liquor		6
Soda Ash	Cold	6
Soda Ash	Hot	2 3 4 8 10 11
Sodium Bicarbonate	Aqueous sol.	2 3 4 6 8 10
Sodium Bisulphate	Aqueous sol.	4 7 10 13
Sodium Carbonate	(see Soda Ash)	
Sodium Chlorate	Aqueous sol.	2 3 4 7 10

Liquid	*Conditions*	*Materials suitable under the majority of conditions*
Sodium Chloride	(see Brines)	
Sodium Cyanide	Aqueous sol.	6
Sodium Hydroxide	Aqueous sol.	1 2 3 4 6 8 10 11 12
Sodium Hydrosulphite	Aqueous sol.	2 3 4 10 13
Sodium Hypochlorite		4 7 10 14
Sodium Hyposulphite	(see Sodium Thiosulphate)	
Sodium Meta Silicate		6
Sodium Nitrate	Aqueous sol.	1 2 3 4 6 10
Sodium Phosphate:		
Monobasic, Dibasic	Aqueous sol.	2 3 4 9 10
Tribasic	Aqueous sol.	6
Meta	Aqueous sol.	2 3 4 9 10
Hexameta	Aqueous sol.	2 3 4 10
Sodium Plumbite	Aqueous sol.	6
Sodium Sulphate	Aqueous sol.	2 3 4 9 10
Sodium Sulphide	Aqueous sol.	2 3 4 6 10 13
Sodium Sulphite	Aqueous sol.	2 3 4 9 10 13
Sodium Thiosulphate	Aqueous sol.	2 3 4 10 14
Stannic Stannous Chloride	Aqueous sol.	7 10 13 14
Starch		9
Strontium Nitrate	Aqueous sol.	2 6
Sugar	Aqueous sol.	2 3 4 8 9 10
Sulphite Liquors	(see Liquors, Pulp Mill)	
Sulphur	In water	6 8 9
Sulphur	Molten	6
Sulphur Chloride	Cold	6 13
Syrup	(see Sugar)	
Tallow	Hot	6
Tanning Liquors		2 3 4 7 9 10 11
Tar	Hot	6
Tar and Ammonia	In water	6
Tetraethyl Lead		6
Toluene (Toluol)		6
Trichloroethylene		2 6 9
Varnish		2 6 9 11
Vegetable Juices		2 3 4 9 10 11
Vinegar		2 3 4 7 9 10
Vitriol, Blue	(see Copper Sulphate)	
Vitriol, Green	(see Ferrous Sulphate)	
Vitriol, Oil of	(see Acid, Sulphuric)	
Water, Boiler Feed High Makeup	Not evaporated	6
Water, Boiler Feed Low Makeup	Evaporated	1 2 11
Water, Distilled	High purity	2 9
	Condensate	9
Water, Fresh		6
Wine		2 9
Wood Pulp (Stock)		6 9
Wood Vinegar	(see Acid Pyroligneous)	
Xylol (Xylene		2 3 4 6 10
Yeast		2 9
Zinc Chloride	Aqueous sol.	3 4 7 10
Zinc Sulphate	Aqueous sol.	3 4 9 10

(*Data in table courtesy:* American Hydraulic Institute)

Most materials deteriorate with time; therefore, the art of storage is always to have materials available when required, and to ensure a stock turnover so that materials are used up before any significant loss of working life has occurred.

CAUSES OF DETERIORATION AND THEIR PREVENTION

Cause	*Components affected*	*Effect*	*Prevention*
ATMOSPHERIC			
Oxygen	Lubricants	Forms gums, resins and acidic products with viscosity increase	Use lubricant containing anti-oxidation additive. Keep in sealed containers
	Engines and components	When moisture present causes corrosion, particularly to ferrous components	Coat with lubricant or temporary protective. Wrap in airtight packages additionally using vapour phase inhibitors. In sealed units include desiccants
	Cables and wires	Corrosion in the presence of water	Coat with lubricant or temporary protective
	Seals	Promotes slow cracking of natural rubber and some similar materials. Negligible normally at ambient temperatures	Use a different polymer. Do not store in a hot place
Pollutants (e.g. sulphur dioxide, hydrogen sulphide)	Engines and components Cables and wires Brakes and clutches	Rapid corrosion of most metals	Store in sealed containers. Coat metals with temporary protective or lubricant. Filter air supply to remove pollutants
Dust and Dirt	Lubricants	Increased rate of wear between bearing surfaces	Keep covered, or in containers
	Engines and components Cables and wires Brakes and clutches	Increased rate of wear between bearing surfaces. Promotes corrosion in the presence of moisture	Keep covered, or in containers
TEMPERATURE			
Heat	Lubricants	Increases rate of deterioration as under 'Oxygen'. Will increase oil separation from greases	Keep store temperature no higher than 20°C
	Seals	Increases deterioration rate as above	Keep store temperature no higher than 20°C
Cold	Lubricants	In water-containing materials (e.g. cutting oils and certain fire-resisting hydraulic 'oils') water could separate out	Keep temperature above freezing point
HUMIDITY	Seals	Could become brittle	
	Engines and components Cables and wires	Promotes corrosion. More severe when ferrous and non-ferrous metals present. See 'Oxygen'	Coat metal parts with lubricant or temporary protective
	Brakes and clutches Belts and ropes Seals	Promotes fungus/bacterial growth	Store in the dry
LIGHT	Lubricants	Promotes formation of gums, resins and acidity	Store in metal or opaque containers
FUNGUS/BACTERIA	Lubricants	Growth occurs at water/oil interface	Keep water out of containers. In certain cases biocide and fungicides can be added
	Brakes and clutches Belts and ropes Seals	Surface covered and attacked by mould growth	Store in the dry. Treat with biocide and fungicide
VIBRATION	Engines and components	Ball bearings and to a lesser extent roller bearings suffer false Brinelling	Do not store where there is vibration. Resilient mountings can reduce effects of vibration

PACKAGING

Unseasoned timber, certain plastics and glues (e.g. phenol formaldehyde), and certain papers (containing chlorine and sulphur compounds) give off vapours corrosive to ferrous and many non-ferrous metals. Therefore it is important to choose the right type of material for packaging.

Vapour phase inhibitors (VPIs) have been designed to prevent atmospheric corrosion of steel. They are volatile and therefore the component should be in a container, ideally sealed. VPIs should not be used with lead or cadmium, and can cause tarnishing of copper and its alloys.

Frequency of inspection

Ideal general storage conditions would be a temperature of 15°C, relative humidity of 40%, an atmosphere free from pollutants, dirt and dust, no vibration, no direct sunlight, and components in sealed containers or packages.

The following recommendations are based on components stored under reasonably good conditions. If conditions differ widely from the ideal, and facing table indicates that this could affect components, more frequent inspection should be carried out.

Component	*How stored*	*Inspect*
Lubricants	Bulk tanks	Each year and when refilled
	Sealed containers	Check annually for damage to containers. Limited tests for serviceability of contents after three years
Engines and components	General storage	Annually. Hand turn engines where possible
	Packaged or sealed containers	Two to four years
Cables and wires Brakes and clutches Belts and seals	General storage	Visual inspection annually
Ropes	General storage	Turn annually and in addition,
rot proofed		every four years (Test to destruction)
untreated		every two years (Test to destruction)
synthetic fibre		every four years (Test to destruction)

RELUBRICATION AND REPROTECTION

Relubrication and reprotection should be carried out under as good conditions as possible, otherwise contaminants (e.g. dust, dirt and pollutants) can attack the component and cause more damage than if relubrication and/or reprotection had not been attempted.

Components	*How stored*	*Inspection and treatment*
In general:		
Engines and components	Coated with lubricant or temporary protective	Recoat annually. Lubricate when brought into use
	Packaged or sealed containers	Replace VPI, desiccants, or lubricant, every two to four years
Special items:		
Grease-packed ball and roller bearing	After prolonged or adverse storage, or if oil 'bleeding' has occurred	Before use clean out old grease with solvent,* remove surplus solvent, and replace with new grease
Oil-impregnated porous metal bearings	After prolonged or adverse storage	Before use soak in warm oil of the same type as originally impregnated
Small mechanisms in their own cases, e.g. watches, servos		Relubricate every five years
Small mechanisms and components, e.g. gas bearings, watch components		These will require specialised cleaning and lubrication before being brought into use
Cables and wires	Coated with lubricant	Replace every two to three years

* For example, paraffin, trichlorethylene. *Note:* traces of chlorinated solvents such as trichlorethylene, particularly in the presence of moisture, can cause corrosion of most metals. Therefore, after cleaning with chlorinated solvents all traces should be removed, ideally by blowing with warm dry air.

THE SIGNIFICANCE OF FAILURE

Failure is only one of three ways in which engineering devices may reach the end of their useful life.

The way in which the end of useful life is reached		*Typical devices which can end their life this way*
Failure	Slow	Seals leaking progressively
	Sudden	Electric lamp bulbs
Obsolescence		Gas lamps Steam locomotives
Completion		Bullets and bombs Packaging

In the design process an attempt is usually made to ensure that failure does not occur before a specified life has been reached, or before a life limit has been reached by obsolescence or completion. The occurrence of a failure, without loss of life, is not so much a disaster, as the ultimate result of a design compromise between perfection and economics.

When a limit to operation without failure is accepted, the choice of this limit depends on the availability required from the device.

Availability is the average percentage of the time that a device is available to give satisfactory performance during its required operating period. The availability of a device depends on its reliability and maintainability.

Reliability is the average time that devices of a particular design will operate without failure.

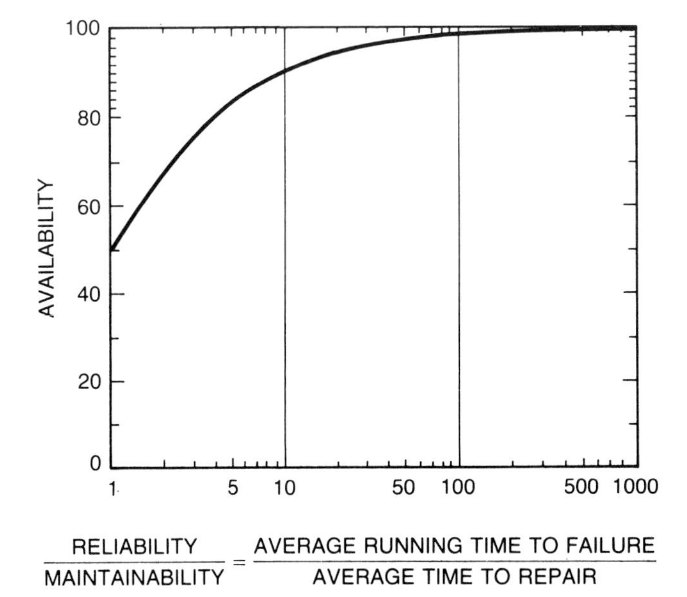

Figure 1.1 The relationship between availability, reliability and maintainability. *High availabilities can only be obtained by long lives or short repair times or both*

Maintainability is measured by the average time that devices of a particular design take to repair after a failure.

The availability required, is largely determined by the application and the capital cost.

FAILURE ANALYSIS

The techniques to be applied to the analysis of the failures of tribological components depend on whether the failures are isolated events or repetitive incidents. Both require detailed examination to determine the primary cause, but, in the case of repeated failures, establishing the temporal pattern of failure can be a powerful additional tool.

Investigating failures

When investigating failures it is worth remembering the following points:

(a) Most failures have several causes which combine together to give the observed result. A single cause failure is a very rare occurrence.

(b) In large machines tribological problems often arise because deflections increase with size, while in general oil film thicknesses do not.

(c) Temperature has a very major effect on the performance of tribological components both directly, and indirectly due to differential expansions and thermal distortions. It is therefore important to check:

Temperatures
Steady temperature gradients
Temperature transients

Causes of failure

To determine the most probable causes of failure of components, which exist either in small numbers, or involve mass produced items the following procedure may be helpful:

1. Examine the failed specimens using the following sections of this Handbook as guidance, in order to determine the probable mode of failure.
2. Collect data on the actual operating conditions and double check the information wherever possible.
3. Study the design, and where possible analyse its probable performance in terms of the operating conditions to see whether it is likely that it could fail by the mode which has been observed.
4. If this suggests that the component should have operated satisfactorily, examine the various operating conditions to see how much each needs to be changed to produce the observed failure. Investigate each operating condition in turn to see whether there are any factors previously neglected which could produce sufficient change to cause the failure.

Repetitive failures

Two statistics are commonly used:-

1. MTBF (mean time between failures)

$$= \frac{L_1 + L_2 + ... + L_n}{n}$$

where L_1, L_2, etc., are the times to failure and n the number of failures

2. L_{10} Life, is the running time at which the number of failures from a sample population of components reaches 10%. (Other values can also be used, e.g. L_1 Life, viz the time to 1% failures, where extreme reliability is required.)

MTBF is of value in quantifying failure rates, particularly of machines involving more than one failing component. It is of most use in maintenance planning, costing and in assessing the effect of remedial measures.

L_{10} Life is a more rigorous statistic that can only be applied to a statistically homogeneous population, i.e. nominally identical items subject to nominally identical operating conditions.

Failure patterns

Repetitive failures can be divided by time to failure according to the familiar 'bath-tub' curve, comprising the three regions: early-life failures (infantile mortality), 'mid-life' (random) failures and 'wear-out'.

Early-life failures are normally caused by built-in defects, installation errors, incorrect materials, etc.

Mid-life failures are caused by random effects external to the component, e.g. operating changes, (overload) lightning strikes, etc.

Wear-out can be the result of mechanical wear, fatigue, corrosion, etc.

The ability to identify which of these effects is dominant in the failure pattern can provide an insight into the mechanism of failure.

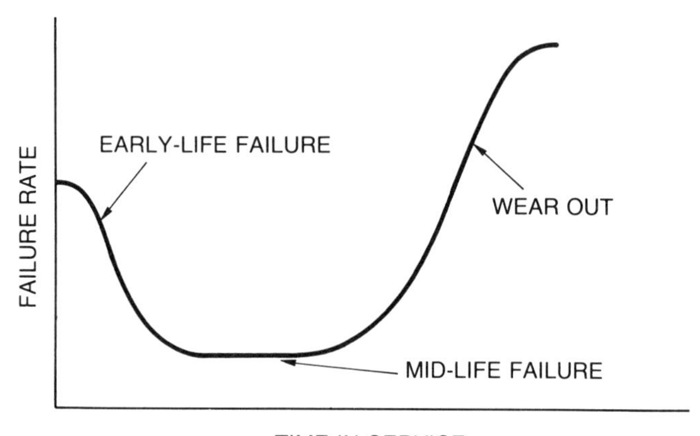

Figure 1.2 The failure rate with time of a group of similar components

As a guide to the general cause of failure it can be useful to plot failure rate against life to see whether the relationship is falling or rising.

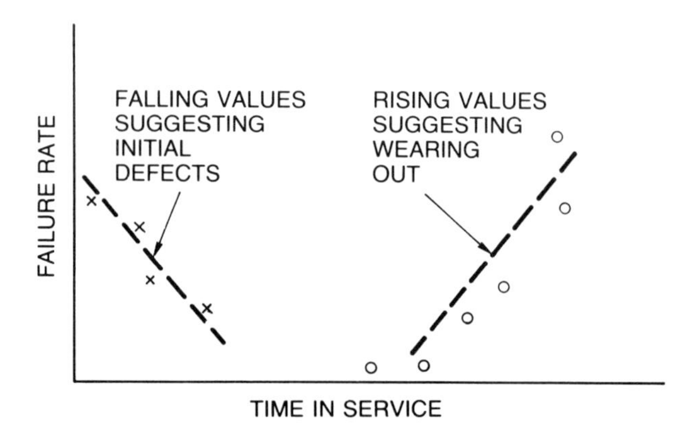

Figure 1.3 The failure rate with time used as an investigative method

Weibull analysis

Weibull analysis is a more precise technique. Its power is such that it can provide useful guidance with as few as five repeat failures. The following form of the Weibull probability equation is useful in component failure analysis:

$$F(t) = 1 - \exp[\alpha(t - \gamma)^{\beta}]$$

where $F(t)$ is the cumulative percentage failure, t the time to failure of individual items and the three constants are the scale parameter (α), the Weibull Index (β) and the location parameter (γ).

For components that do not have a shelf life, i.e. there is no deterioration before the component goes into service, $\gamma = 0$ and the expression simplifies to:

$$F(t) = 1 - \exp[\alpha t^{\beta}].$$

The value of the Weibull Index depends on the temporal pattern of failure, viz:

early-life failures	$\beta = 0.5$
random failures	$\beta = 1$
wear out	$\beta = 3.4$

Weibull analysis can be carried out simply and quickly as follows:

1. Obtain the values of $F(t)$ for the sample size from Table 1.1
2. Plot the observed times to failure against the appropriate value of $F(t)$ on Weibull probability paper (Figure 1.5).
3. Draw best fit straight line through points.
4. Drop normal from 'Estimation Point' to the best fit straight line.
5. Read off β value from intersection on scale.

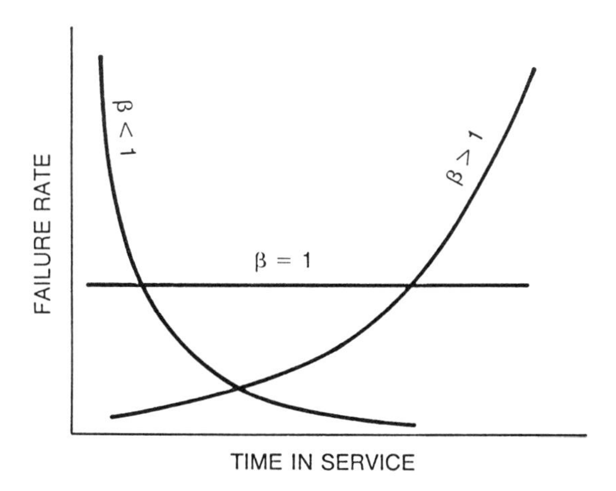

Figure 1.4 The relationship between the value of β and the shape of the failure rate curve

For $n > 20$ – Calculate approximate values of $F(t)$ from

$$\frac{100(i - 0.3)}{n + 0.4}$$

where: i is the ith measurement in a sample of n arranged in increasing order.

Sample size																				
5	12.9	31.3	50.0	68.8	87.0															
6	10.9	26.4	42.1	57.8	73.5	89.0														
7	9.4	22.8	36.4	50.0	63.5	77.1	90.5													
8	8.3	20.1	32.0	44.0	55.9	67.9	79.8	91.7												
9	7.4	17.9	28.6	39.3	50.0	60.6	71.3	82.0	92.5											
10	6.6	16.2	25.8	35.5	45.1	54.8	64.4	74.1	83.7	93.3										
11	6.1	14.7	23.5	32.3	41.1	50.0	58.8	67.6	76.4	85.2	93.8									
12	5.6	13.5	21.6	29.7	37.8	45.9	54.0	62.1	70.2	78.3	86.4	94.3								
13	5.1	12.5	20.0	27.5	35.0	42.5	50.0	57.4	64.9	72.4	79.9	87.4	94.8							
14	4.8	11.7	18.6	25.6	32.5	39.5	46.5	53.4	60.4	67.4	74.3	81.3	88.2	95.1						
15	4.5	10.9	17.4	23.9	30.4	36.9	43.4	50.0	56.5	63.0	69.5	76.0	82.5	89.0	95.4					
16	4.2	10.2	16.3	22.4	28.5	34.7	40.8	46.9	53.0	59.1	65.2	71.4	77.5	83.6	89.7	95.7				
17	3.9	9.6	15.4	21.1	26.9	32.7	38.4	44.2	50.0	55.7	61.5	67.2	73.0	78.8	84.5	90.3	96.0			
18	3.7	9.1	14.5	20.0	25.4	30.9	36.3	41.8	47.2	52.7	58.1	63.6	69.0	74.5	79.9	85.4	90.8	96.2		
19	3.5	8.6	13.8	18.9	24.1	29.3	34.4	39.6	44.8	50.0	55.1	60.3	65.5	70.6	75.8	81.0	86.1	91.3	96.4	
20	3.4	8.2	13.1	18.0	22.9	27.8	32.7	37.7	42.6	47.5	52.4	57.3	62.2	67.2	72.1	77.0	81.9	86.8	91.7	96.5

Table 1.1 Values of the cumulative percent failure F(t) for the individual failures in a range of sample sizes

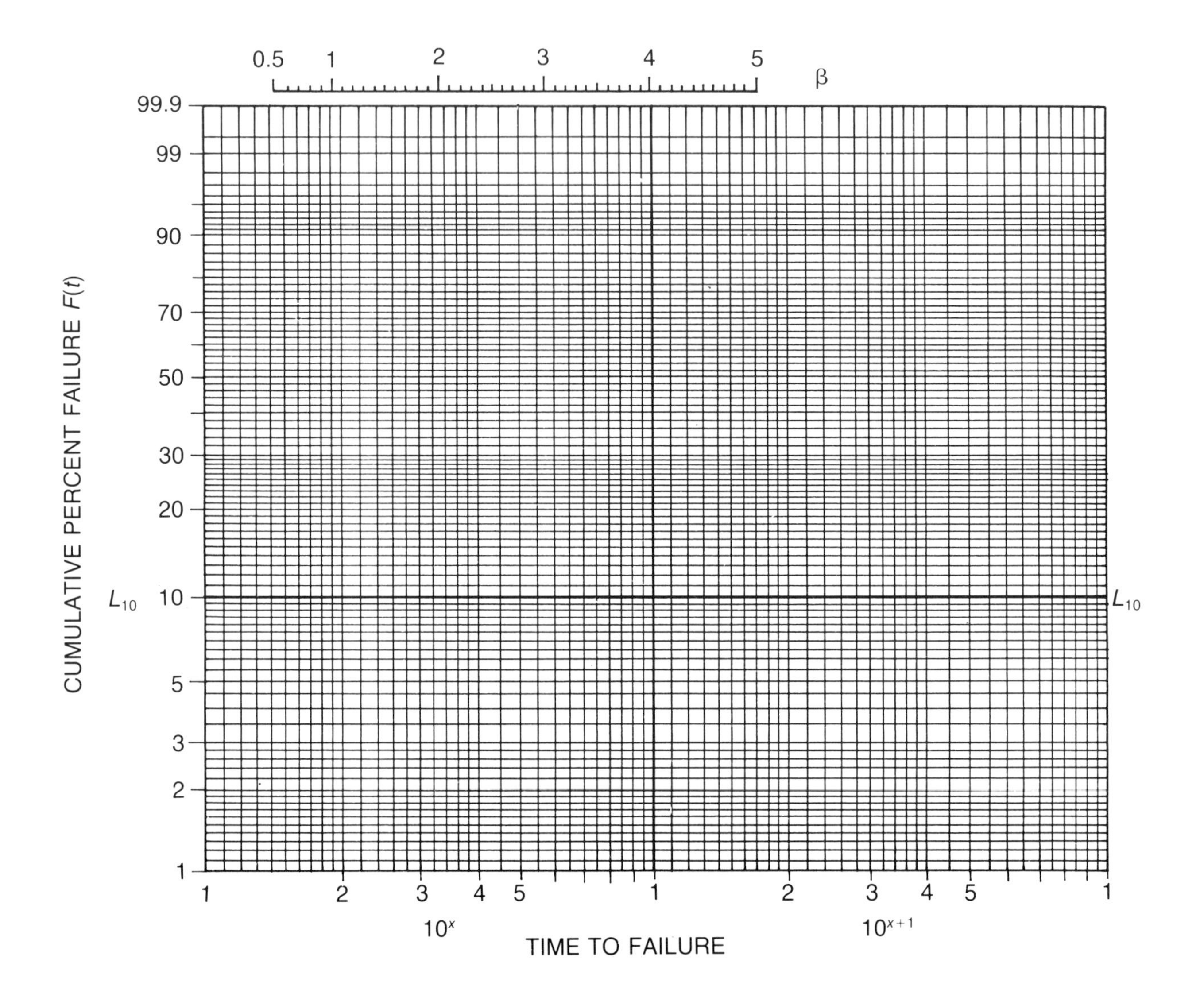

Figure 1.5 Weibull probability graph paper

Figure 1.6 gives an example of 9 failures of spherical roller bearings in an extruder gearbox. The β value of 2.7 suggests wear-out (fatigue) failure. This was confirmed by examination of the failed components. The L_{10} Life corresponds to a 10% cumulative failure. L_{10} Life for rolling bearings operating at constant speed is given by:

$$L_{10}\text{ Life (hours)} = \frac{10^6}{n}\frac{C^x}{P}$$

Where $n =$ speed (rev/min), $C =$ bearing capacity, $P =$ equivalent radial load, $x = 3$ for ball bearings, 10/3 for roller bearings.

Determination of the L_{10} Life from the Weibull analysis allows an estimate to be made of the actual load. This can be used to verify the design value. In this particular example, the exceptionally low value of L_{10} Life (2500 hours) identified excessive load as the cause of failure.

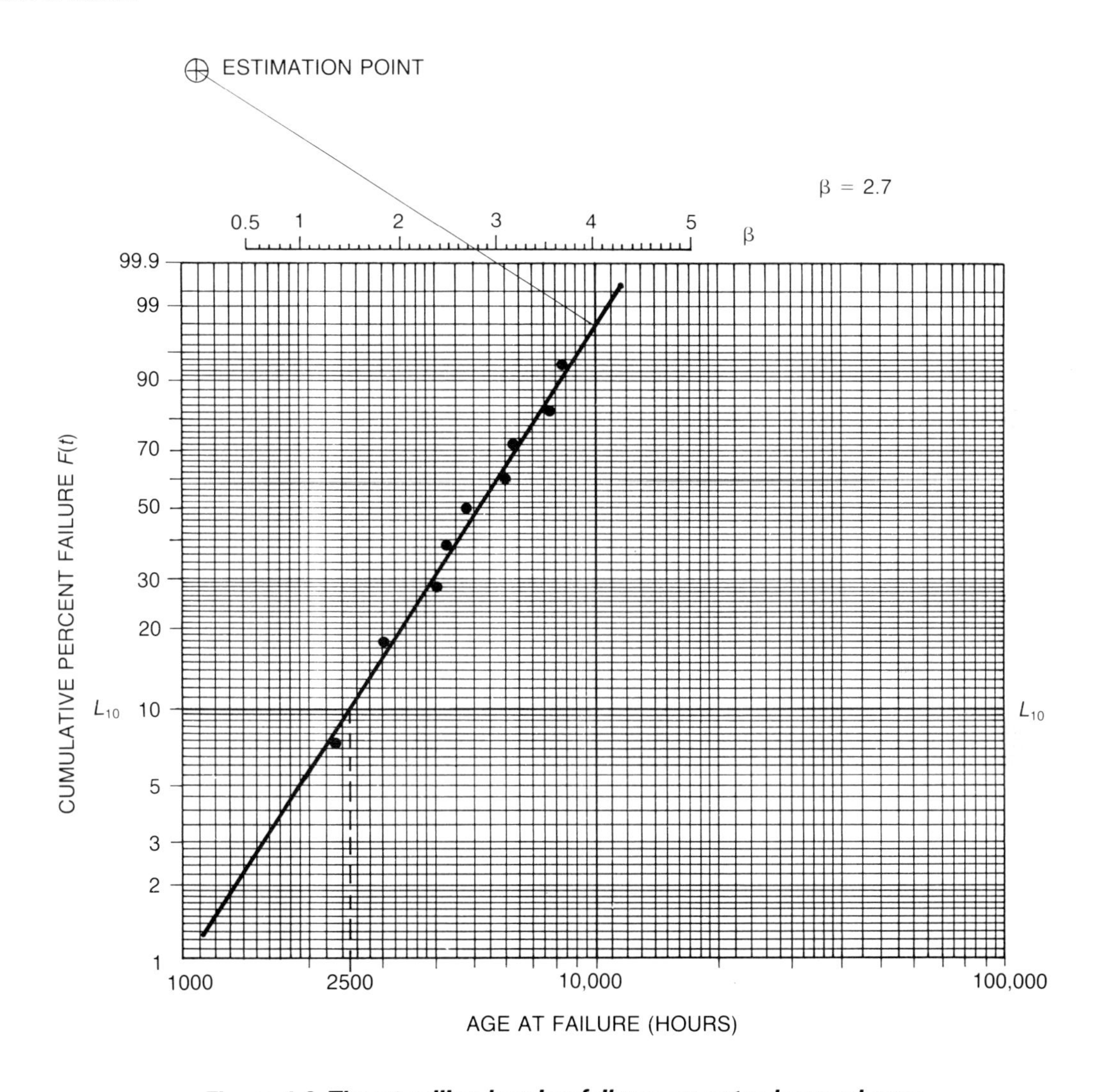

Figure 1.6 Thrust rolling bearing failures on extruder gearboxes

Figure 1.7 gives an example for 17 plain thrust bearing failures on three centrifugal air compressors. The β value of 0.7 suggests a combination of early-life and random failures. Detailed examination of the failures showed that they were caused in part by assembly errors, in part of machine surges.

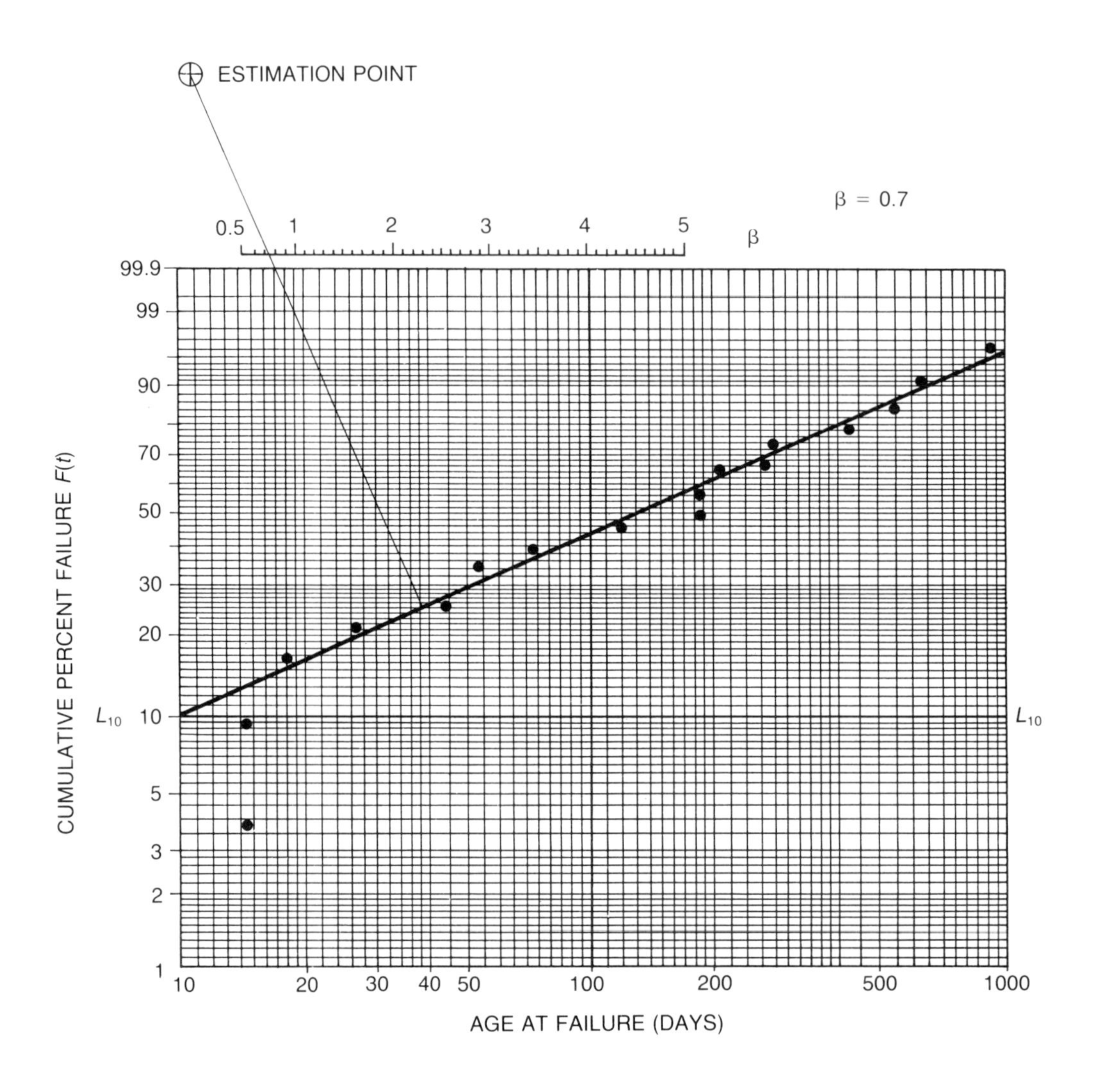

Figure 1.7 Plain thrust bearing failures on centrifugal air compressors

Foreign matter

Characteristics

Fine score marks or scratches in direction of motion, often with embedded particles and haloes.

Causes

Dirt particles in lubricant exceeding the minimum oil film thickness.

Foreign matter

Characteristics

Severe scoring and erosion of bearing surface in the line of motion, or along lines of local oil flow.

Causes

Contamination of lubricant by excessive amounts of dirt particularly non-metallic particles which can roll between the surfaces.

Wiping

Characteristics

Surface melting and flow of bearing material, especially when of low-melting point, e.g. whitemetals, overlays.

Causes

Inadequate clearance, overheating, insufficient oil supply, excessive load, or operation with a non-cylindrical journal.

Fatigue

Characteristics

Cracking, often in mosaic pattern, and loss of areas of lining.

Causes

Excessive dynamic loading or overheating causing reduction of fatigue strength; overspeeding causing imposition of excessive centrifugal loading.

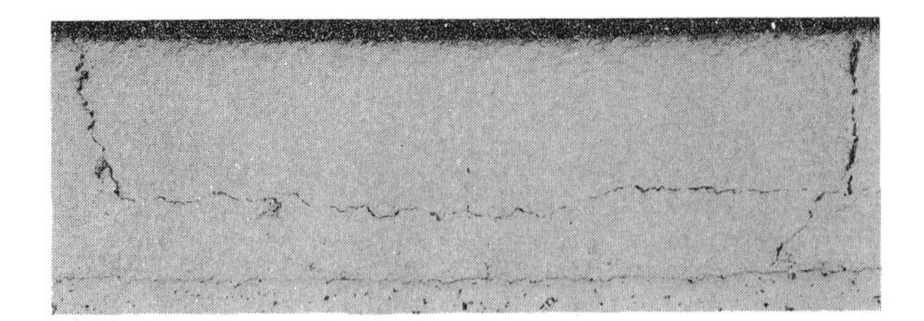

Fatigue

Characteristics

Loss of areas of lining by propagation of cracks initially at right angles to the bearing surface, and then progressing parallel to the surface, leading to isolation of pieces of the bearing material.

Causes

Excessive dynamic loading which exceeds the fatigue strength at the operating temperature.

Fretting

Characteristics

Welding, or pick-up of metal from the housing on the back of bearing. Can also occur on the joint faces. Production and oxidation of fine wear debris, which in severe cases can give red staining.

Causes

Inadequate interference fit; flimsy housing design; permitting small sliding movements between surfaces under operating loads.

Excessive interference

Characteristics

Distortion of bearing bore causing overheating and fatigue at the bearing joint faces.

Causes

Excessive interference fit or stagger at joint faces during assembly.

Misalignment

Characteristics

Uneven wear of bearing surface, or fatigue in diagonally opposed areas in top and bottom halves.

Causes

Misalignment of bearing housings on assembly, or journal deflection under load.

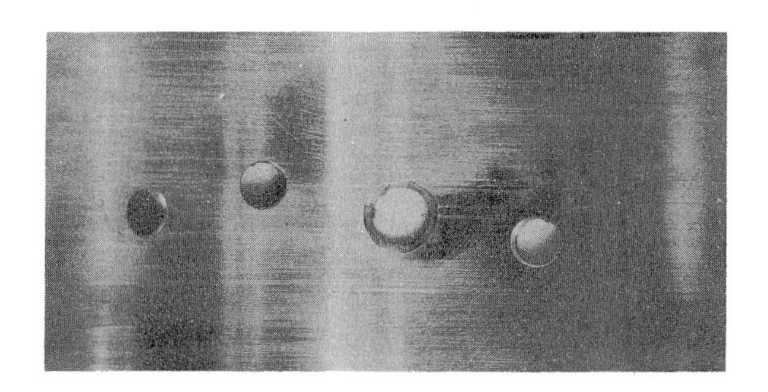

Dirty assembly

Characteristics

Localised overheating of the bearing surface and fatigue in extreme cases, sometimes in nominally lightly loaded areas.

Causes

Entrapment of large particles of dirt (e.g. swarf), between bearing and housing, causing distortion of the shell, impairment of heat transfer and reduction of clearance (see next column).

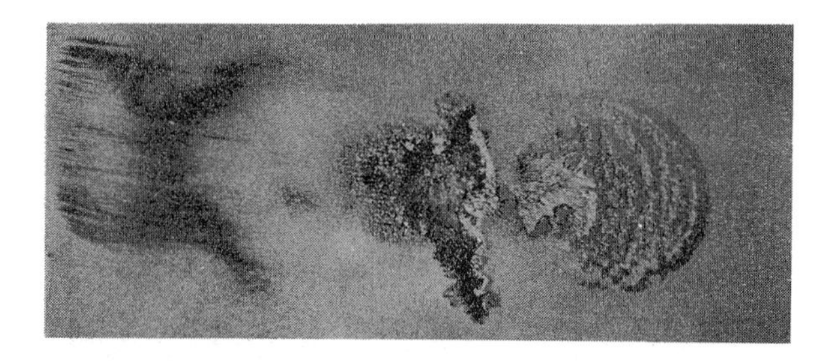

Cavitation erosion

Characteristics

Removal of bearing material, especially soft overlays or whitemetal in regions near joint faces or grooves, leaving a roughened bright surface.

Causes

Changes of pressure in oil film associated with interrupted flow.

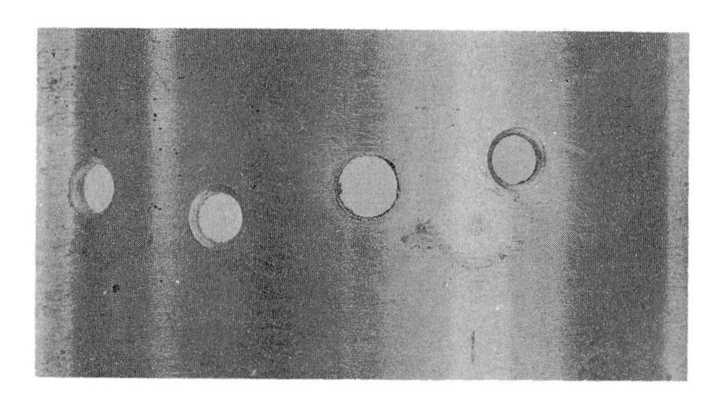

Dirty assembly

Characteristics

Local areas of poor bedding on the back of the bearing shell, often around a 'hard' spot.

Causes

Entrapment of dirt particles between bearing and housing. Bore of bearing is shown in previous column illustrating local overheating due to distortion of shell, causing reduction of clearance and impaired heat transfer.

Discharge cavitation erosion

Characteristics

Formation of pitting or grooving of the bearing material in a V-formation pointing in the direction of rotation.

Causes

Rapid advance and retreat of journal in clearance during cycle. It is usually associated with the operation of a centrally grooved bearing at an excessive operating clearance.

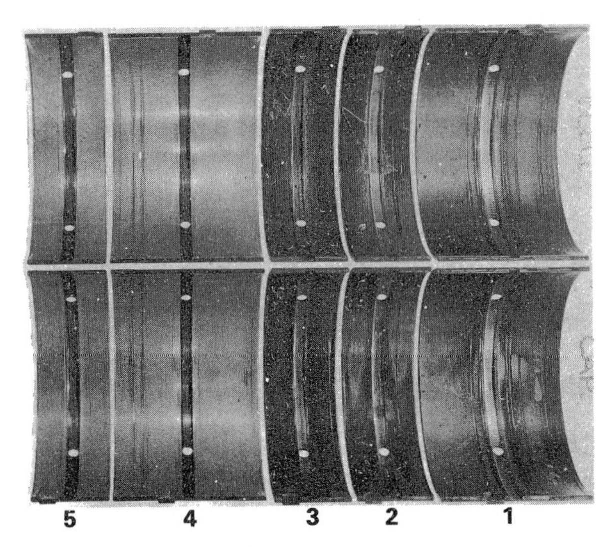

Cavitation erosion

Characteristics

Attack of bearing material in isolated areas, in random pattern, sometimes associated with grooves.

Causes

Impact fatigue caused by collapse of vapour bubbles in oil film due to rapid pressure changes. Softer overlay (Nos 1, 2 and 3 bearings) attacked. Harder aluminium –20% tin (Nos 4 and 5 bearings) not attacked under these particular conditions.

Corrosion

Characteristics

Removal of lead phase from unplated copper–lead or lead–bronze, usually leading on to fatigue of the weakened material.

Causes

Formation of organic acids by oxidation of lubricating oil in service. Consult oil suppliers; investigate possible coolant leakage into oil.

Tin dioxide corrosion

Characteristics

Formation of hard black deposit on surface of white-metal lining, especially in marine turbine bearings. Tin attacked, no tin-antimony and copper-tin constituents.

Causes

Electrolyte (sea water) in oil.

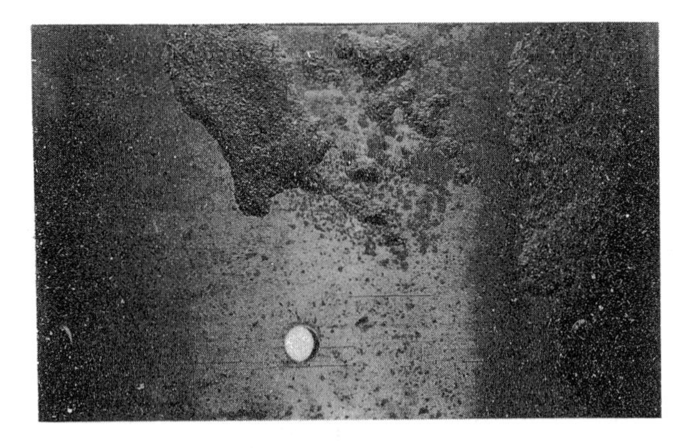

'Sulphur' corrosion

Characteristics

Deep pitting and attack or copper-base alloys, especially phosphor–bronze, in high temperature zones such as small-end bushes. Black coloration due to the formation of copper sulphide.

Causes

Attack by sulphur-compounds from oil additives or fuel combustion products.

'Wire wool' damage

Characteristics

Formation of hard black scab on whitemetal bearing surface, and severe machining away of journal in way of scab, as shown on the right.

Causes

It is usually initiated by a large dirt particle embedded in the whitemetal, in contact with journal, especially chromium steel.

'Wire wool' damage

Characteristics

Severe catastrophic machining of journal by 'black scab' formed in whitemetal lining of bearing. The machining 'debris' looks like wire wool.

Causes

Self-propagation of scab, expecially with 'susceptible' journals steels, e.g. some chromium steels.

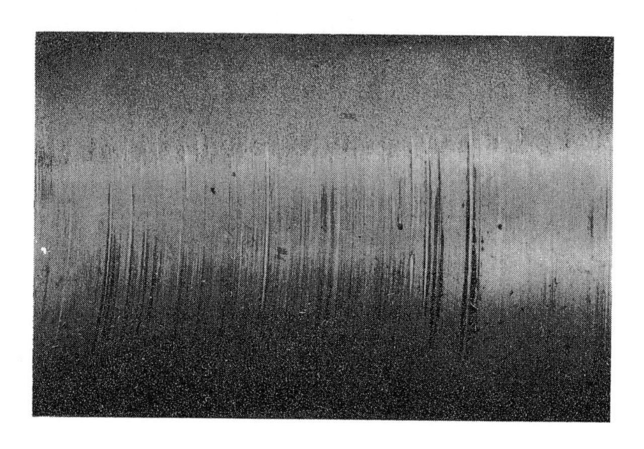

Electrical discharge

Characteristics

Pitting of bearing surface and of journal; may cause rapid failure in extreme cases

Causes

Electrical currents from rotor to stator through oil film, often caused by faulty earthing.

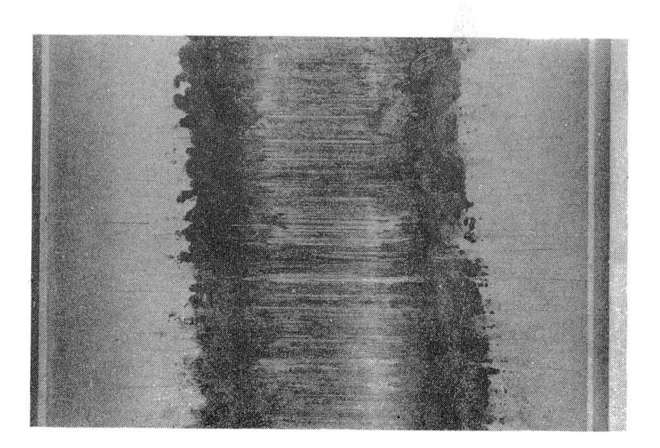

Fretting due to external vibration

Characteristics

Pitting and pick-up on bearing surface.

Causes

vibration transmitted from extgernal sources, causing damage while journal is stationary.

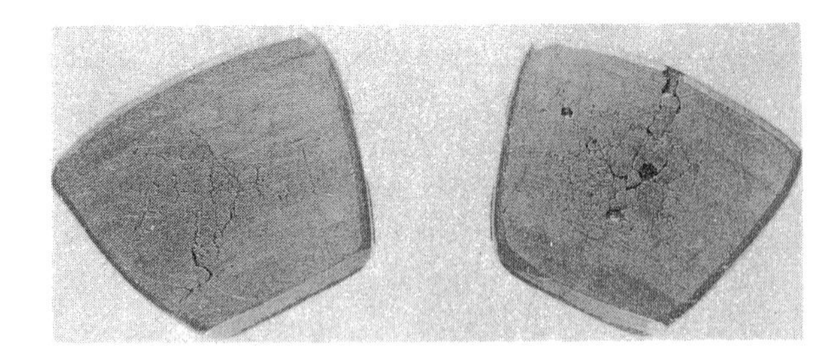

Overheating

Characteristics

Extrusion and cracking, especially of whitemetal-lined bearings.

Causes

Operation at escessibe temperatures.

Thermal cycling

Characteristics

Surface rumpling and grain-boundary cracking of tin-base whitemetal bearings.

Causes

Thermal cycling in service, causing plastic deformation, associated with the non-uniform thermal expansion of tin crystals.

Faulty assembly

Characteristics

Localised fatigue or wiping in nominally lightly loaded areas.

Causes

Stagger at joint faces during assembly, due to excessive bolt clearances, or incorrect bolt disposition (bolts too far out).

Faulty assembly

Characteristics

Overheating and pick-up at the sides of the bearings.

Causes

Incorrect grinding of journal radii, causing fouling at fillets.

Incorrect journal grinding

Characteristics

Severe wiping and tearing-up of bearing surface.

Causes

Too coarse a surface finish, or in the case of SG iron shafts, the final grinding of journal in wrong direction relative to rotation in bearing.

Inadequate oil film thickness

Characteristics

Fatigue cracking in proximity of a groove.

Causes

Incorrect groove design, e.g. positioning a groove in the loaded area of the bearing.

Inadequate lubrication

Characteristics

Seizure of bearing.

Causes

Inadequate pump capacity or oil gallery or oilway dimensions. Blockage or cessation of oil supply.

Bad bonding

Characteristics

Loss of lining, sometimes in large areas, even in lightly loaded regions, and showing full exposure of the backing material.

Causes

Poor tinning of shells; incorrect metallurgical control of lining technique.

All photographs courtesy of Glacier Metal Co. Ltd

FATIGUE FLAKE

Characteristics

Flaking with conchoidal or ripple pattern extending evenly across the loaded part of the race.

Causes

Fatigue due to repeated stressing of the metal. This is not a fault condition but it is the form by which a rolling element bearing should eventually fail. The multitude of small dents are caused by the debris and are a secondary effect.

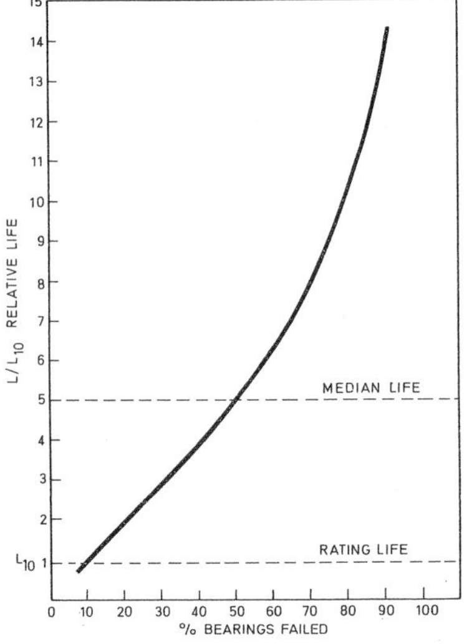

EARLY FATIGUE FLAKE

Characteristics

A normal fatigue flake but occurring in a comparatively short time. Appearance as for fatigue flake.

Causes

Wide life-expectancy of rolling bearings. The graph shows approximate distribution for all types. Unless repeated, there is no fault. If repeated, load is probably higher than estimated; check thermal expansion and centrifugal loads.

ATMOSPHERIC CORROSION

Characteristics

Numerous irregular pits, reddish brown to dark brown in colour. Pits have rough irregular bottoms.

Causes

Exposure to moist conditions, use of a grease giving inadequate protection against water corrosion.

ROLLER STAINING

Characteristics

Dark patches on rolling surfaces and end faces of rollers in bearings with yellow metal cages. The patches usually conform in shape to the cage bars.

Causes

Bi-metallic corrosion in storage. May be due to poor storage conditions or insufficient cleaning during manufacture. Special packings are available for severe conditions. Staining, as shown, can be removed by the manufacturer, to whom the bearing should be returned.

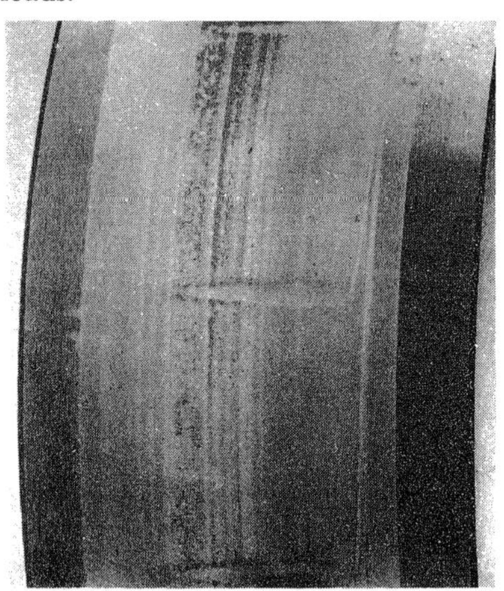

BRUISING (OR TRUE BRINELLING)

Characteristics

Dents or grooves in the bearing track conforming to the shape of the rolling elements. Grinding marks not obliterated and the metal at the edges of the dents has been slightly raised.

Causes

The rolling elements have been brought into violent contact with the race; in this case during assembly using impact.

FALSE BRINELLING

Characteristics

Depressions in the tracks which may vary from shallow marks to deep cavities. Close inspection reveals that the depressions have a roughened surface texture and that the grinding marks have been removed. There is usually no tendency for the metal at the groove edges to have been displaced.

Causes

Vibration while the bearing is stationary or a small oscillating movement while under load.

FRACTURED FLANGE

Characteristics

Pieces broken from the inner race guiding flange. General damage to cage and shields.

Causes

Bad fitting. The bearing was pressed into housing by applying load to the inner race causing cracking of the flange. During running the cracks extended and the flange collapsed. A bearing must never be fitted so that the fitting load is transmitted via the rolling elements.

OUTER RACE FRETTING

Characteristics

A patchy discoloration of the outer surface and the presence of reddish brown debris ('cocoa'). The race is not softened but cracks may extend inwards from the fretted zone.

Causes

Insufficient interference between race and housing. Particularly noticeable with heavily loaded bearings having thin outer races.

INNER RACE FRETTING

Characteristics

Heavy fretting of the shaft often with gross scalloping; presence of brown debris ('cocoa'). Inner race may show some fretting marks.

Causes

Too little interference, often slight clearance, between the inner race and the shaft combined with heavy axial clamping. Axial clamping alone will not prevent a heavily loaded inner race precessing slowly on the shaft.

INNER RACE SPINNING

Characteristics

Softening and scoring of the inner race and the shaft, overheating leading to carbonisation of lubricant in severe cases, may lead to complete seizure.

Causes

Inner race fitted with too little interference on shaft and with light axial clamping.

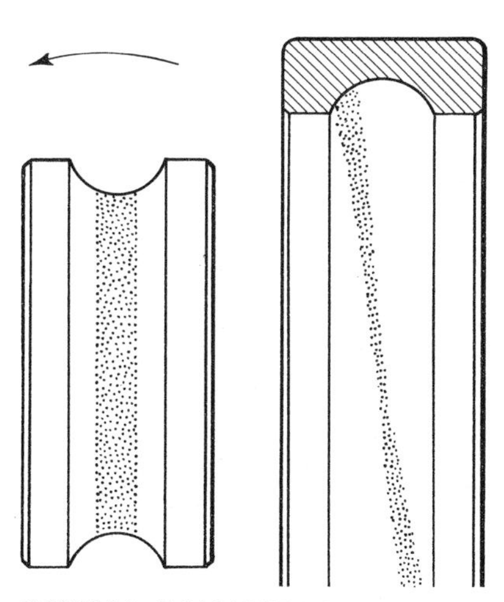

SKEW RUNNING MARKS

Characteristics

The running marks on the stationary race are not parallel to the faces of the race. In the figure the outer race is stationary.

Causes

Misalignment. The bearing has not failed but may do so if allowed to continue to run out of line.

UNEVEN FATIGUE

Characteristics

Normal fatigue flaking but limited to, or much more severe on, one side of the running track.

Causes

Misalignment.

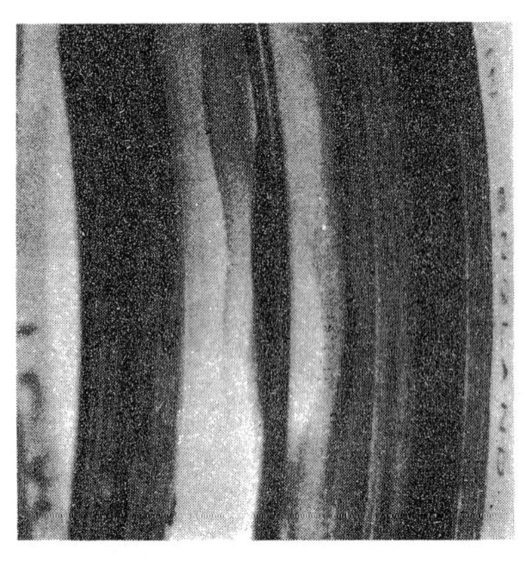

UNEVEN WEAR MARKS

Characteristics

The running or wear marks have an uneven width and may have a wavy outline instead of being a uniform dark band.

Causes

Ball skidding due to a variable rotating load or local distortion of the races.

ROLLER END COLLAPSE

Characteristics

Flaking near the roller-end radius at one end only. Microscopic examination reveals roundish smooth-bottomed pits.

Causes

Electrical damage with some misalignment. If the pits are absent then the probable cause is roller end bruising which can usually be detected on the undamaged shoulder. Although misalignment accentuates this type of damage it has rarely been proved to be the sole cause.

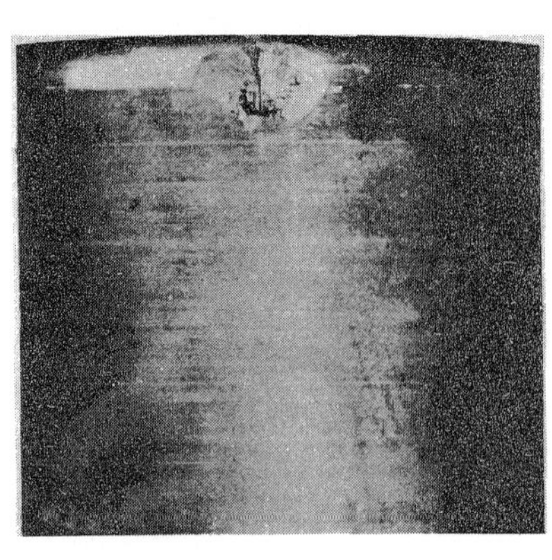

ROLLER END CHIPPING

Characteristics

A collapse of the material near the corner radii of the roller. In this instance chipping occurred simultaneously at opposite ends of the roller. A well-defined sub-surface crack can be seen.

Causes

Subcutaneous inclusions running the length of the roller. This type of failure is more usually found in the larger sizes of bearing.

Chipping at one end only may be caused by bruising during manufacture, or by electrical currents, and accentuated by misalignment.

ROLLER PEELING

Characteristics

Patches of the surface of the rollers are removed to a depth of about 0.0005 in.

Causes

This condition usually follows from an initial mild surface damage such as light electrical pitting; this could be confirmed by microscopic examination. It has also been observed on rollers which were slightly corroded before use.

If the cause is removed this damage does not usually develop into total failure.

ROLLER BREAKAGE

Characteristics

One roller breaks into large fragments which may hold together. Cage pocket damaged.

Causes

Random fatigue. May be due to faults or inclusions in the roller material. Replacement bearing usually performs satisfactorily.

MAGNETIC DAMAGE

Characteristics

Softening of the rotating track and rolling elements leading to premature fatigue flaking.

Causes

Bearing has been rotating in a magnetic field (in this case, 230 kilolines (230×10^{-5} Wb), 300 rev/min, 860 h).

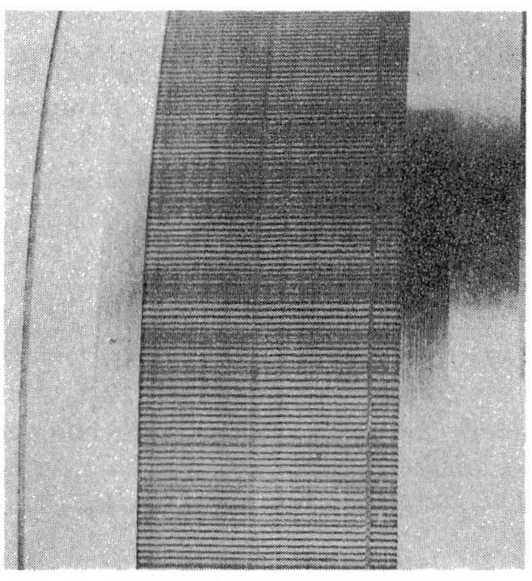

LADDER MARKING OR WASHBOARD EROSION

Characteristics

A regular pattern of dark and light bands which may have developed into definite grooves. Microscopic examination shows numerous small, almost round, pits.

Causes

An electric current has passed across the bearing; a.c. or d.c. currents will cause this effect which may be found on either race or on the rolling elements.

GREASE FAILURE

Characteristics

Cage pockets and rims worn. Remaining grease dry and hard; bearing shows signs of overheating.

Causes

Use of unsuitable grease. Common type of failure where temperatures are too high for the grease in use.

MOLTEN CAGE

Characteristics

Cage melted down to the rivets, inner race shows temper colours.

Causes

Lubrication failure on a high-speed bearing. In this case an oil failure at 26 000 rev/min. In a slower bearing the damage would not have been so localised.

OVERHEATING

Characteristics

All parts of the bearing are blackened or show temper colours. Lubricant either absent or charred. Loss of hardness on all parts.

Causes

Gross overheating. Mild overheating may only show up as a loss of hardness.

SMEARING

Characteristics

Scuff marks, discoloration and metal transfer on non-rolling surfaces. Usually some loss of hardness and evidence of detrioration of lubricant. Often found on the ends of rollers and the corresponding guide face on the flanges.

Causes

Heavy loads and/or poor lubrication.

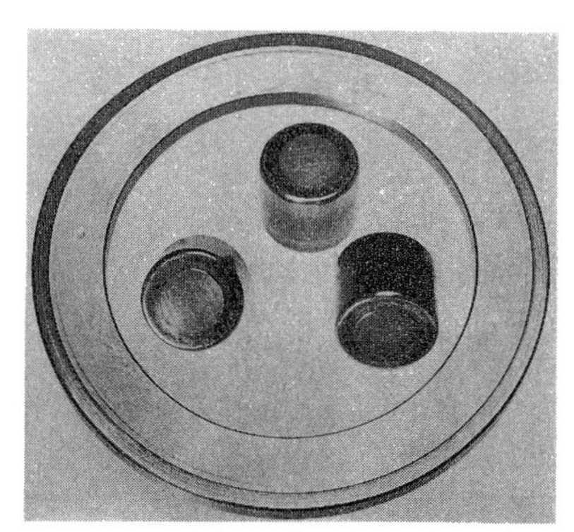

ABRASIVE WEAR

Characteristics

Dulling of the working surfaces and the removal of metal without loss of hardness.

Causes

Abrasive particles in the lubricant, usually non-metallic.

Gear failures

Gear failures rarely occur. A gear pair has not failed until it can no longer be run. This condition is reached when (*a*) one or more teeth have broken away, preventing transmission of motion between the pair or (*b*) teeth are so badly damaged that vibration and noise are unacceptable when the gears are run.

By no means all tooth damage leads to failure and immediately it is observed, damaged teeth should be examined to determine whether the gears can safely continue in service.

SURFACE FATIGUE

This includes case exfoliation in skin-hardened gears and pitting which is the commonest form of damage, especially with unhardened gears. Pitting, of which four types are distinguished, is indicated by the development of relatively smooth-bottomed cavities generally on or below the pitch line. In isolation they are generally conchoidal in appearance but an accumulation may disguise this.

Case exfoliation

Case exfoliation on a spiral bevel pinion

Characteristics

Appreciable areas of the skin on surface hardened teeth flake away from the parent metal in heavily loaded gears. Carburised and hardened, nitrided and induction hardened materials are affected.

Causes

Case exfoliation often indicates a hardened skin that is too thin to support the tooth load. Cracks sometimes originate on the plane of maximum Hertzian shear stress and subsequently break out to the surface, but more often a surface crack initiates the damage. Another possible reason for case exfoliation is the high residual stress resulting from too severe a hardness gradient between case and core. Exfoliation may be prevented by providing adequate case depth and tempering the gear material after hardening.

Initial or arrested pitting

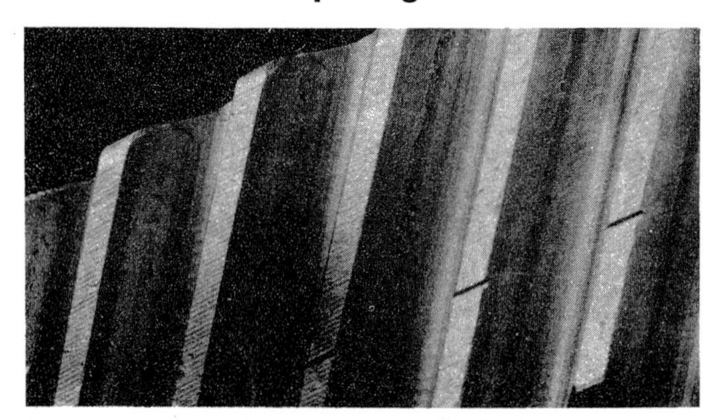

Initial or arrested pitting on a single helical gear

Characteristics

Initial pitting usually occurs on gears that are not skin hardened. It may be randomly distributed over the whole tooth flank, but more often is found around the pitch line or in the dedendum. Single pits rarely exceed 2 mm across and pitting appears in the early running life of a gear.

Causes

Discrete irregularities in profile or surface asperities are subjected to repeated overstress as the line of contact sweeps across a tooth to produce small surface cracks and clefts. In the dedendum area the oil under the high pressure of the contact can enter these defects and extend them little by little, eventually reaching the surface again so that a pit is formed and a small piece of metal is dislodged. Removal of areas of overstress in this way spreads the load on the teeth to a level where further crack or cleft formation no longer occurs and pitting ceases.

Progressive or potentially destructive pitting

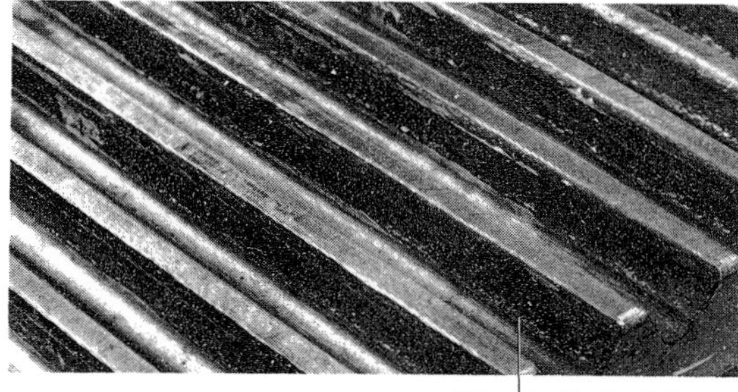

Progressive pitting on single helical gear teeth

Characteristics

Pits continue to form with continued running, especially in the dedendum area. Observation on marked teeth will indicate the rate of progress which may be intermittent. A rapid increase, particularly in the root area, may cause complete failure by increasing the stress there to the point where large pieces of teeth break away.

Causes

Essentially the gear material is generally overstressed, often by repeated shock loads. With destructive pitting the propagating cracks branch at about the plane of maximum Hertzian shear stress; one follows the normal initial pitting process but the other penetrates deeper into the metal.

Remedial action is to remove the cause of the overload by correcting alignment or using resilient couplings to remove the effect of shock loads. The life of a gear based on surface fatigue is greatly influenced by surface stress. Thus, if the load is carried on only half the face width the life will only be a small fraction of the normal value. In slow and medium speed gears it may be possible to ameliorate conditions by using a more viscous oil, but this is generally ineffective with high speed gears.

In skin-hardened gears pits of very large area resembling case exfoliation may be formed by excessive surface friction due to the use of an oil lacking sufficient viscosity.

Dedendum attrition

Dedendum attrition on a large single helical gear

Characteristics

The dedendum is covered by a large number of small pits and has a matt appearance. Both gears are equally affected and with continued running the dedenda are worn away and a step is formed at the pitch line to a depth of perhaps 0.5 mm. The metal may be detached as pit particles or as thin flakes. The wear may cease at this stage but may run in cycles, the dedenda becoming smooth before pitting restarts. If attrition is permitted to continue vibration and noise may become intolerable. Pitting may not necessarily be present in the addendum.

Causes

The cause of this type of deterioration is not fully understood but appears to be associated with vibration in the gear unit. Damage may be mitigated by the use of a more viscous oil.

Micro-pitting

Characteristics

Found predominantly on the dedendum but also to a considerable extent on the addendum of skin-hardened gears. To the naked eye affected areas have a dull grey, matt or 'frosted' appearance but under the microscope they are seen to be covered by a myriad of tiny pits ranging in size from about 0.03 to 0.08 mm and about 0.01 mm deep.

Depending on the position of the affected areas, micro-pitting may be corrective, especially with helical gears.

Causes

Overloading of very thin, brittle and super-hard surface layers, as in nitrided surfaces, or where a white-etching layer has formed, by normal and tangential loads. Coarse surface finishes and low oil viscosity can be predisposing factors. In some cases it may be accelerated by unsuitable load-carrying additives in the oil.

SMOOTH CHEMICAL WEAR

Can arise where gears using extreme pressure oil run under sustained heavy loads, at high temperatures.

Smooth chemical wear

Hypoid pinion showing smooth chemical wear

Characteristics

The working surfaces of the teeth, especially of the pinion, are worn and have a burnished appearance.

Causes

Very high surface temperatures cause the scuff resistant surface produced by chemical reaction with the steel to be removed and replaced very rapidly. The remedies are to reduce the operating temperatures, to reduce tooth friction by using a more viscous oil and to use a less active load-carrying additive.

SCUFFING

Scuffing occurs at peripheral speeds above about 3 m/s and is the result of either the complete absence of a lubricant film or its disruption by overheating. Damage may range from a lightly etched appearance (slight scuffing) to severe welding and tearing of engaging teeth (heavy scuffing). Scuffing can lead to complete destruction if not arrested.

Light scuffing

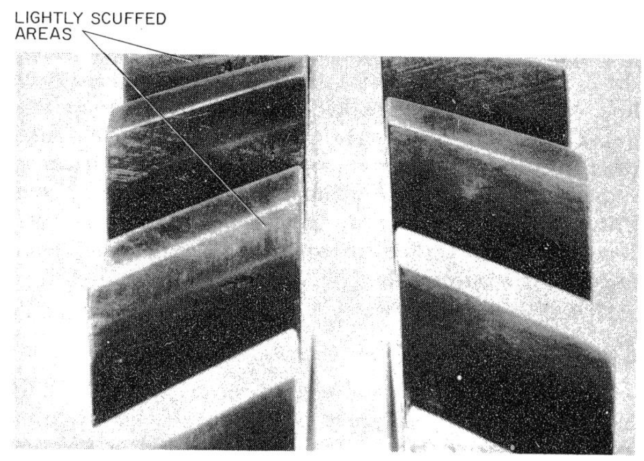

Light scuffing

Characteristics

Tooth surfaces affected appear dull and slightly rough in comparison with unaffected areas. Low magnification of a scuffed zone reveals small welded areas subsequently torn apart in the direction of sliding, usually at the tip and root of the engaging teeth where sliding speed is a maximum.

Causes

Disruption of the lubricant film occurs when the gear tooth surfaces reach a critical temperature associated with a particular oil and direct contact between the sliding surfaces permits discrete welding to take place. Low viscosity plain oils are more liable to permit scuffing than oils of higher viscosity. Extreme pressure oils almost always prevent it.

Heavy scuffing

Heavy scuffing on a case hardened hypoid wheel

Characteristics

Tooth surfaces are severely roughened and torn as the result of unchecked adhesive wear.

Causes

This is the result of maintaining the conditions that produced light scuffing. The temperature of the contacting surfaces rises so far above the critical temperature for the lubricant that continual welding and tearing of the gear material persists.

Spur, helical and bevel gears, may show so much displacement of the metal that a groove is formed along the pitch line of the driving gear and a corresponding ridge on that of the driven gear. It may be due to the complete absence of lubricant, even if only temporarily. Otherwise, the use of a more viscous oil, or one with extreme pressure properties is called for.

GENERAL COMMENTS ON GEAR TOOTH DAMAGE

Contact marking is the acceptance criterion for all toothed gearing, and periodic examination of this feature until the running pattern has been established, is the most satisfactory method of determining service performance. It is therefore advisable to look at the tooth surfaces on a gear pair soon after it has been run under normal working conditions. If any surface damage is found it is essential that the probable cause is recognised quickly and remedial action taken if necessary, before serious damage has resulted. Finding the principal cause may be more difficult when more than one form of damage is present, but it is usually possible to consider each characteristic separately.

The most prolific sources of trouble are faulty lubrication and misalignment. Both can be corrected if present, but unless scuffing has occurred, further periodic observation of any damaged tooth surfaces should be made before taking action which may not be immediately necessary.

ABRASIVE WEAR

During normal operation, engaging gear teeth are separated from one another by a lubricant film, commonly about 0.5μm thick. Where both gears are unhardened and abrasive particles dimensionally larger than the film thickness contaminate the lubricant, especially if it is a grease, both sets of tooth surfaces are affected (three-body abrasion). Where one gear has very hard tooth surfaces and surface roughness greater than the film thickness, two-body abrasive wear occurs and the softer gear only becomes worn. For example, a rough case-hardened steel worm mating with a bronze worm wheel, or a rough steel pinion engaging a plastic wheel.

Foreign matter in the lubricant

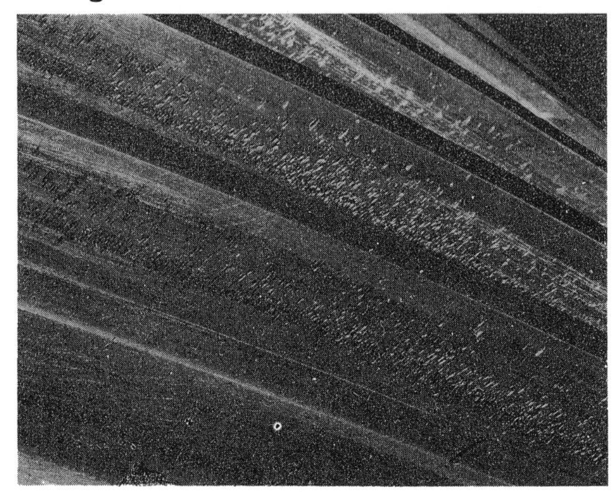

Effect of foreign matter in lubricant

Characteristics

Grooves are cut in the tooth flanks in the direction of sliding and their size corresponds to the size of the contaminant present. Displaced material piles up along the sides of a groove or is removed as a fine cutting. Usually scratches are short and do not extend to the tooth tips.

Causes

The usual causes of three-body abrasion are gritty materials falling into an open gear unit or, in an enclosed unit, inadequate cleaning of the gear case and oil supply pipes of such materials as casting sand, loose scale, shot-blast grit, etc.

Attrition caused by fine foreign matter in oil

Spur gear virtually destroyed by foreign matter in the oil

Characteristics

These are essentially similar to lapping. Very fine foreign matter suspended in a lubricant can pass through the gear mesh with little effect when normal film lubrication prevails. Unfavourable conditions permit abrasive wear; tooth surfaces appear dull and scratched in the direction of sliding. If unchecked, destruction of tooth profiles results from the lapping.

Causes

The size of the foreign matter permits bridging through the oil film. Most frequently, the origin of the abrasive material is environmental. Both gears and bearings suffer and systems should be cleaned, flushed, refilled with clean oil and protected from further contamination as soon as possible after discovery.

TOOTH BREAKAGE

If a whole tooth breaks away the gear has failed but in some instances a corner of a tooth may be broken and the gear can continue to run. The cause of a fracture should influence an assessment of the future performance of a gear.

Brittle fracture resulting from high shock load

Brittle fracture on spiral bevel wheel teeth

Characteristics

More than one tooth may be affected. With hard steels the entire fracture surface appears to be granular denoting a brittle fracture. With more ductile materials the surface has a fibrous and torn appearance.

Causes

A sudden and severe shock load has been applied to one or other member of a gear pair which has greatly exceeded the impact characteristics of the material. A brittle fracture may also indicate too low an Izod value in the gear material, though this is a very rare occurrence. A brittle fracture in bronze gears indicates the additional effect of overheating.

Tooth end and tip loading

Tooth end and tip loading

Characteristics

Spiral bevel and hypoid gears are particularly liable to heel end tooth breakage and other types of skin hardened gears may have the tooth tips breaking away. Fractured surfaces often exhibit rapid fatigue characteristics.

Causes

The immediate cause is excessive local loading. This may be produced by very high transmitted torque, incorrect meshing or insufficient tip relief.

Impact or excessive loading causing fatigue fracture

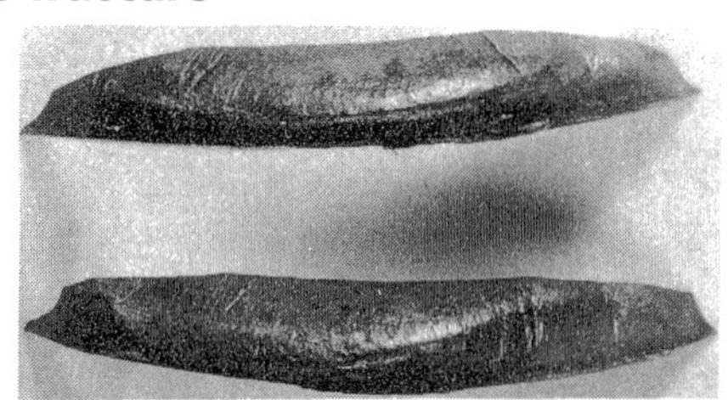

Slow fatigue on a through-hardened helical wheel

Characteristics

Often exhibits cracks in the roots on the loaded side of a number of teeth. If teeth have broken out the fracture surfaces show two phases; a very fine-grained, silky, conchoidal zone starting from the loaded side followed, where the final failure has suddenly occurred, by a coarse-grained brittle fracture.

Causes

The loading has been so intense as to exceed the tensile bending stress limit resulting in root cracking. Often stress-raisers in the roots such as blowholes, bruises, deep machining marks or non-metallic inclusions, etc. are involved. If the excessive loading continues the teeth will break away by slow fatigue and final sudden fracture.

Fatigue failure resulting from progressive pitting

Fatigue failure from progressive pitting

Characteristics

Broken tooth surfaces exhibit slow fatigue markings, with the origin of the break at pits in the dedendum of the affected gear.

Causes

Progressive pitting indicates that the gears are being run with a surface stress intensity above the fatigue limit. Cracks originating at the surface continue to penetrate into the material.

PLASTIC DEFORMATION

Plastic deformation occurs on gear teeth due to the surface layers yielding under heavy loads through an intact oil film. It is unlikely to occur with hardnesses above HV 350.

Severe plastic flow in steel gears

Severe plastic flow in helical gears

Characteristics

A flash or knife-edge is formed on the tips of the driving teeth often with a hollow at the pitch cylinder and a corresponding swelling on the driven teeth. The ends of the teeth can also develop a flash and the flanks are normally highly burnished.

Causes

The main causes are heavy steady or repeated shock loading which raises the surface stress above the elastic limit of the material, the surface layers being displaced while in the plastic state, especially in the direction of sliding. Since a work-hardened skin tends to develop, the phenomenon is not necessarily detrimental, especially in helical gears, unless the tooth profiles are severely damaged. A more viscous oil is often advantageous, particularly with shock-loading, but the best remedy is to reduce the transmitted load, possibly by correcting the alignment.

CASE CRACKING

With correctly manufactured case hardened gears case cracking is a rare occurrence. It may appear as the result of severe shock or excessive overload leading to tooth breakage or as a condition peculiar to worm gears.

Heat/load cracking on worms

RADIAL CRACKS IN POLISHED CONTACT ZONE

Heat/load cracking on a worm wheel

Characteristics

On extremely heavily loaded worms the highly polished contact zone may carry a series of radial cracks. Spacing of the cracks is widest where the contact band is wide and they are correspondingly closer spaced as the band narrows. Edges rarely rise above the general level of the surface.

Causes

The cracks are thought to be the result of high local temperatures induced by the load. Case hardened worms made from high core strength material (En39 steel) resist this type of cracking.

FAILURES OF PLASTIC GEARS

Gears made from plastic materials are meshed with either another plastic gear or more often, with a cast iron or steel gear; non ferrous metals are seldom used. When applicable, failures generally resemble those described for metal gears.

Severe plastic flow, scoring and tooth fracture indicate excessive loading, possibly associated with inadequate lubrication. Tempering colours on steel members are the sign of unsatisfactory heat dispersal by the lubricant.

Wear on the metallic member of a plastic/metal gear pair usually suggests the presence of abrasive material embedded in the plastic gear teeth. This condition may derive from a dusty atmosphere or from foreign matter carried in the lubricant.

When the plastic member exhibits wear the cause is commonly attributable to a defective engaging surface on the metallic gear teeth. Surface texture should preferably not be rougher than 16μin (0.4μm) cla.

PISTON PROBLEMS

Piston problems usually arise from three main causes and these are:
1. Unsatisfactory rubbing conditions between the piston and the cylinder.
2. Excessive operating temperature, usually caused by inadequate cooling or possibly by poor combustion conditions.
3. Inadequate strength or stiffness of the piston or associated components at the loads which are being applied in operation.

Skirt scratching and scoring

Characteristics

The piston skirt shows axial scoring marks predominantly on the thrust side. In severe cases there may be local areas showing incipient seizure.

Causes

Abrasive particles entering the space between the piston and cylinder. This can be due to operation in a dusty environment with poor air filtration. Similar damage can arise if piston ring scuffing has occurred since this can generate hard particulate debris. More rarely the problem can arise from an excessively rough cylinder surface finish.

Skirt scratching

Piston skirt seizure

Characteristics

Severe scuffing damage, particularly on the piston skirt but often extending to the crown and ring lands. The damage is often worse on the thrust side.

Causes

Operation with an inadequate clearance between the piston and cylinder. This can be associated with inadequate cooling or a poor piston profile. Similar damage could also arise if there was an inadequate rate of lubricant feed up the bore from crankshaft bearing splash.

Skirt seizure

Piston crown and ring land damage

Characteristics

The crown may show cracking and the crown land and lands between the rings may show major distortion, often with the ring ends digging in to the lands.

Causes

Major overheating caused by poor cooling and in diesel engines defective injectors and combustion. The problem may arise from inadequate cylinder coolant flow or from the failure of piston cooling arising from blocked oil cooling jets.

Misaligned pistons

Characteristics

The bedding on the skirt is not purely axial but shows diagonal bedding.

Causes

Crankshaft deflections or connecting rod bending. Misalignment of rod or gudgeon pin bores.

Diagonal skirt bedding

Cracking inside the piston

Characteristics

Cracks near the gudgeon pin bosses and behind the ring grooves.

Causes

Inadequate gudgeon pin stiffness can cause cracking in adjacent parts of the piston, or parts of the piston cross section may be of inadequate area.

RING PROBLEMS

The most common problem with piston rings is scuffing of their running surfaces. Slight local scuffing is not uncommon in the first 20 to 50 hours of running from new when the rings are bedding in to an appropriate operating profile. However the condition of the ring surfaces should progressively improve and scuffing damage should not spread all round the rings.

Scuffing of cast iron rings

Characteristics

Local zones around the ring surface where there are axial dragging marks and associated surface roughening. Detailed examination often shows thin surface layers of material with a hardness exceeding 1000 Hv and composed of non-etching fine grained martensite (white layer).

Causes

Can arise from an unsuitable initial finish on the cylinder surface. It can also arise if the rings tend to bed at the top of their running surface due to unsuitable profiling or from thermal distortion of the piston.

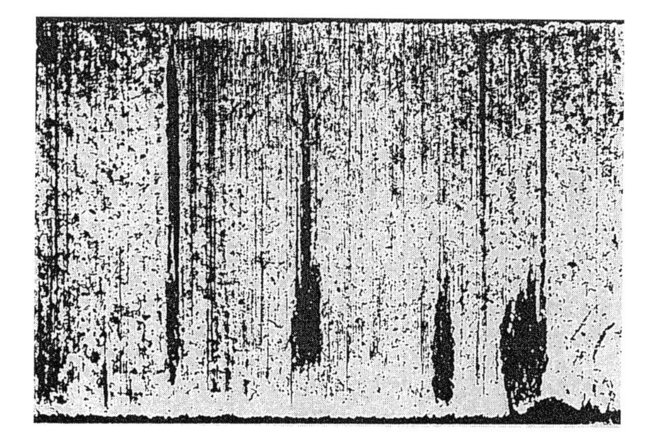

Scuffed cast iron rings

Scuffing of chromium plated piston rings

Characteristics

The presence of dark bands running across the width of the ring surface usually associated with transverse circumferential cracks. In severe cases portions of the chromium plating may be dragged from the surface.

Causes

Unsuitable cylinder surface finish or poor profiling of the piston rings. Chromium plated top rings need to have a barelled profile as installed to avoid hard bedding at the edges.

In some cases the problem can also arise from poor quality plating in which the plated surface is excessively rough or globular and can give local sharp areas on the ring edges after machining.

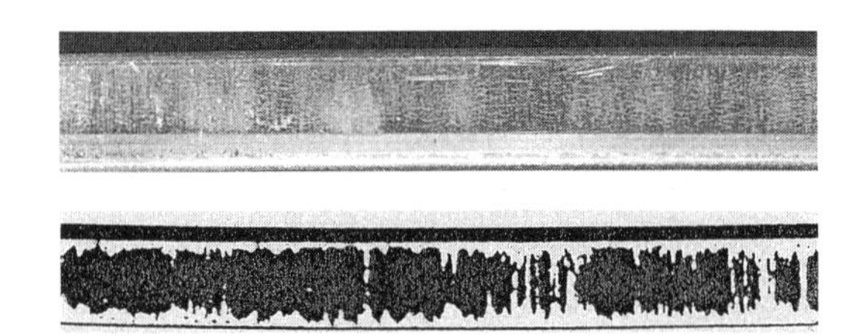

Scuffed chromium plated rings

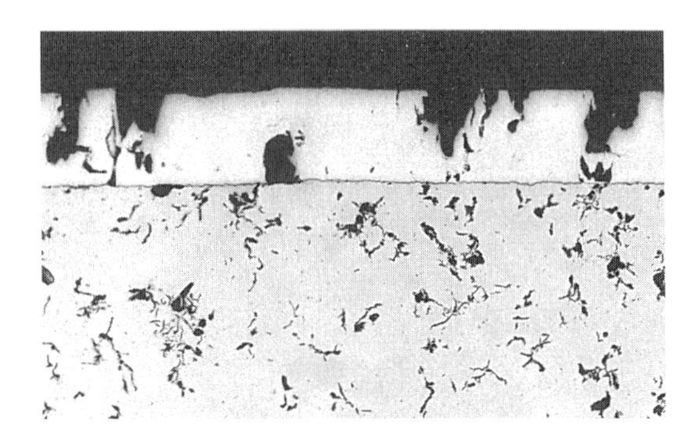

Severely damaged chromium plate

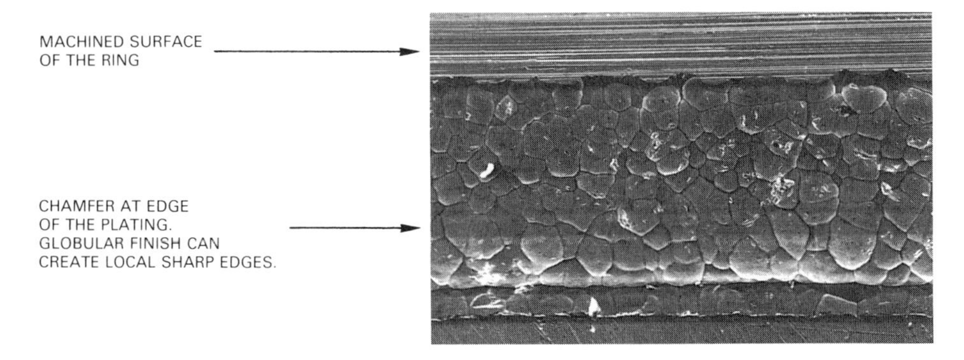

The edge of a piston ring

Rings sticking in their grooves

Characteristics

The rings are found to be fixed in their grooves or very sluggish in motion. There may be excessive blow by or oil consumption.

Causes

The ring groove temperatures are too high due to conditions of operation or poor cooling. The use of a lubricating oil of inadequate quality can also aggravate the problem.

A stuck piston ring

CYLINDER PROBLEMS

Problems with cylinders tend to be of three types:

1. Running in problems such as bore polishing or in some cases scuffing.
2. Rates of wear in service which are high and give reduced life.
3. Other problems such as bore distortion arising from the engine design or cavitation erosion damage of the water side of a cylinder liner, which can penetrate through to the bore.

Bore polishing

Characteristics

Local areas of the bore surface become polished and oil consumption and blow by tend to increase because the piston rings do not then bed evenly around the bore. The polished areas can be very hard thin, wear-resistant 'white' layers.

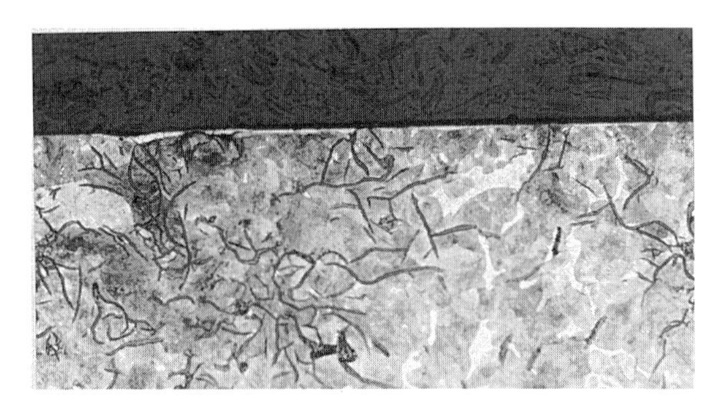

Bore polishing

Causes

The build up of hard carbon deposits on the top land of the piston can rub away local areas of the bore surface and remove the controlled surface roughness required to bed in the piston rings.

If there is noticeable bore distortion from structural deflections or thermal effects, the resulting high spots will be preferentially smoothed by the piston rings.

The chemical nature of the lubricating oil can be a significant factor in both the hard carbon build-up and in the polishing action.

High wear of cast iron cylinders

Characteristics

Cylinder liners wear in normal service due to the action of fine abrasive particles drawn in by the intake air. The greatest wear occurs near to the TDC position of the top ring.

Corrosion of a cast iron bore surface can however release hard flake-like particles of iron carbide from the pearlite in the iron. These give a greatly increased rate of abrasive wear.

Causes

Inadequate air filtration when engines are operated in dusty environments.

Engines operating at too low a coolant temperature, i.e. below about 80°C, since this allows the internal condensation of water vapour from the combustion process, and the formation of corrosion pits in the cylinder surface.

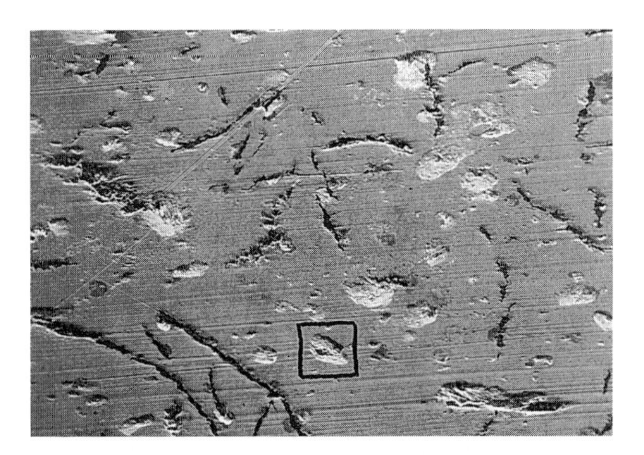

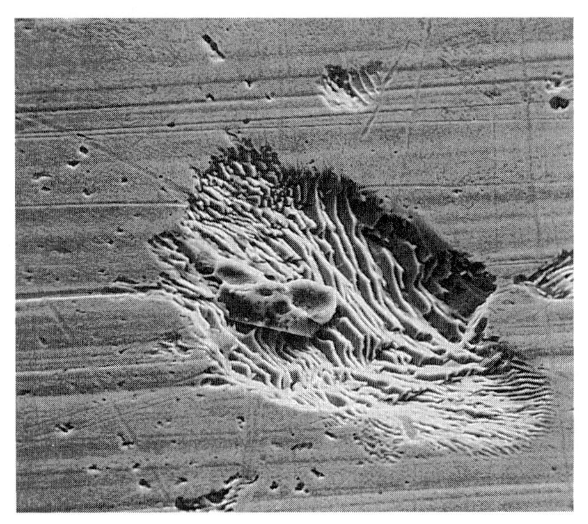

Corrosion of a cast iron bore

High wear of chromium plated cylinders

Characteristics

An increasing rate of wear with operating time associated with the loss of the surface profiling which provides a dispersed lubricant supply. The surface becomes smooth initially and then scuffs because of the unsatisfactory surface profile. This then results in a major increase in wear rate.

Causes

High rates of abrasive particle ingestion from the environment can cause this problem. A more likely cause may be inadequate quality of chromium plating and its finishing process aimed at providing surface porosity. Some finishing processes can leave relatively loose particles of chromium in the surface which become loose in service and accelerate the wear process.

Abrasive turn round marks at TDC

Bore scuffing

Characteristics

Occurs in conjunction with piston ring scuffing. The surface of the cylinder shows areas where the metal has been dragged in an axial direction with associated surface roughening.

Causes

The same as for piston ring scuffing but in addition the problem can be accentuated if the metalurgical structure of the cylinder surface is unsatisfactory.

In the case of cast iron the material must be pearlitic and should contain dispersed hard constituents derived from phosphorous, chromium or vanadium constituents. The surface finish must also be of the correct roughness to give satisfactory bedding in of the piston rings.

In the case of chromium plated cylinder liners it is essential that the surface has an undulating or grooved profile to provide dispersed lubricant feeding to the surface.

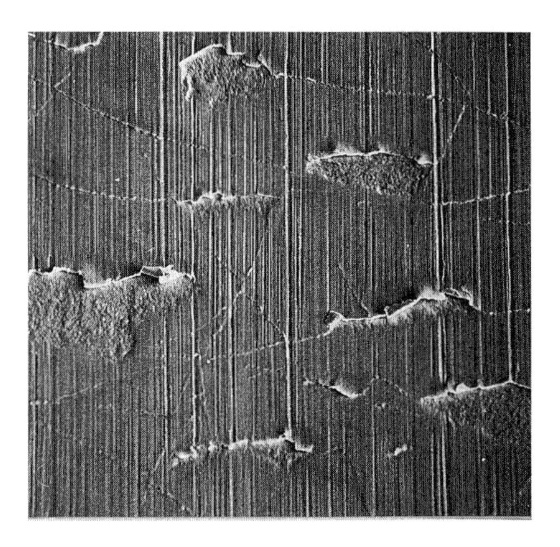

A chromium plated liner which has scuffed after losing its surface profiling by wear

Cavitation erosion of cylinder liners

Characteristics

If separate cylinder liners are used with coolant in contact with their outside surface, areas of cavitation attack can occur on the outside. The material removal by cavitation continues and eventually the liner is perforated and allows the coolant to enter the inside of the engine.

Causes

Vibration of the cylinder liner under the influence of piston impact forces is the main cause of this problem but it is accentuated by crevice corrosion effects if the outside of the liner has dead areas away from the coolant flow.

ROTARY MECHANICAL SEALS

Table 6.1 Common failure mechanisms of mechanical seals

Special conditions	Likely symptoms: High leakage	High friction	High wear	Seizure	Failure mechanism	Remedy
OPERATING CONDITIONS						
Speed of sliding high		X	X	X	Excessive frictional heating, film vaporises	Provide cooling
"	X				Thermal stress cracking of the face (Figure 6.1)	Use material with higher conductivity or higher tensile strength
"	X	X	X	X	Thermal distortion of seal (Figure 6.2)	Provide cooling
Speed of sliding low		X	X	X	Poor hydrodynamic lubrication, solid contact	Use face with good boundary lubrication capacity
Appreciable vibration present	X		X	X	Face separation unstable	Try to reduce vibration, avoid bellows seals, fit damper
Low pressure differential	(X)				Fluid pumped by seal against pressure	Try reversing seal to redirect flow
High pressure differential		X	X	X	Hydrodynamic film overloaded	Modify area ratio of seal to reduce load
"	X	X	X	X	Seal or housing distorting (Figure 6.2)	Stiffen seal and/or housing
Sterilisation or cleaning cycle used	X				High temperature or solvents incompatible with seal materials, especially rubbers	Use compatible materials
Exposure to sunlight, ozone, radiation	X				Seal materials (rubber) fail	Protect seal from exposure, consider other materials
FLUID						
Viscosity of fluid high		X	X	X	Excessive frictional heating, film vaporises	Provide cooling
	X	X	X	X	Excessive frictional heating, seal distorts	Provide cooling
Viscosity of fluid low		X	X	X	Poor hydrodynamic lubrication, solid contact	Use faces with good boundary lubrication capacity
Lubricity of fluid poor	X	X	X	X	Surfaces seize or 'pick-up'	Use faces with good boundary lubrication capacity
Abrasives in fluid	X		X		Solids in interface film	Circulate clean fluid round seal
Crystallisable fluid	X		X		Crystals form at seal face	Raise temperature or flush fluid outside seal
Polymerisable fluid	X		X		Solids form at seal face	Raise temperature or flush fluid outside seal
Ionic fluid, e.g. salt solutions	X	X	X	X	Corrosion damages seal faces	Select resistant materials
Non-Newtonian fluid, e.g. suspension, colloids, etc.	(X)				Fluid behaves unpredictably, leakage may be reversed	Try reversing seal to redirect flow in acceptable direction
DESIGN						
Auxiliary cooling, flushing, etc.	X	X	X	X	Stoppage in auxiliary circuit	Overhaul auxiliaries
Double seals	X	X	X	X	Pressure build-up between seals if there is no provision for pressure control	Provide pressure control
Housing flexes due to pressure or temperature changes	X	X	X	X	Seal faces out of alignment, non-uniform wear	Stiffen housing and/or mount seal flexibly
Seal face flatness poor	X				Excessive seal gap (Figure 6.3)	Lap faces flatter
Seal faces rough	X	X	X	X	Asperities make solid contact	Lap or grind faces
Bellows type seal	X				Floating seal member vibrates	Fit damping device to bellows

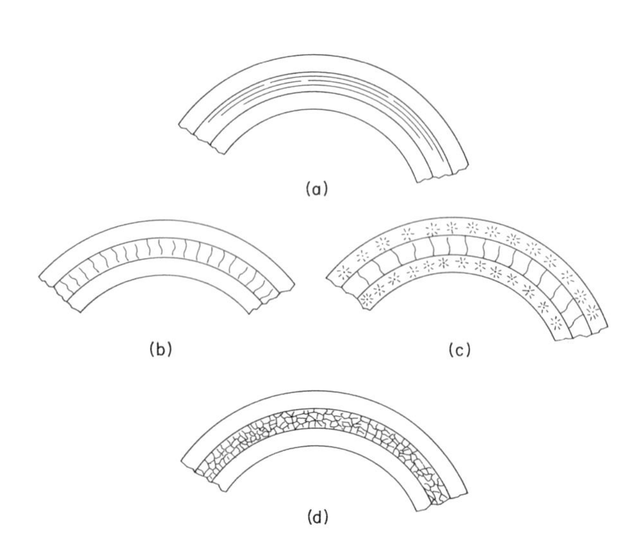

Figure 6.1 Mechanical seal faces after use: *(a) normal appearance, some circumferential scoring; (b) parallel radial cracks; (c) radial cracks with blisters; (d) surface crazing; (b) – (d) are due to overheating, particularly characteristic of ceramic seal faces*

Figure 6.2 Tungsten carbide mechanical seal face showing symmetrical surface polishing characteristic of mild hydraulic or thermal distortion

Figure 6.3 Tungsten carbide mechanical seal face showing localised polishing due to lack of flatness; this seal leaked badly. *The inset illustrates a typical non-flat seal face viewed in sodium light using an optical flat to give contour lines at 11 micro-inch increments of height*

RUBBER SEALS OF ALL TYPES

Table 6.2 Common failure mechanisms of rubber seals

Symptoms	*Cause*	*Remedy*
Rubber brittle, possibly cracked, seal leaks	Rubber ageing. Exposure to ozone/sunlight. Overheated due to high fluid temperature or high speed (Figure 6.4)	Renew seal, consider change of rubber compound; consider improving seal environmental or operating conditions
Rubber softened, possibly swollen	Rubber incompatible with sealed fluid (Figure 6.5)	Change rubber compound or fluid
Seal motion irregular, jerky, vibration, or audible squeal	'Stick-slip' (Figure 6.6)	Higher or lower speed may avoid problem; change fluid temperature; change rubber compound
Seal friction very high on starting	'Stiction', i.e. static friction is time dependent and much higher than kinetic friction (Figure 6.7)	Probably inevitable in some degree, time effect slowed by softer rubber and/or more viscous fluid
Seal permanently deformed	'Permanent set', a characteristic of rubbers, some more than others (Figure 6.4)	Change rubber compound

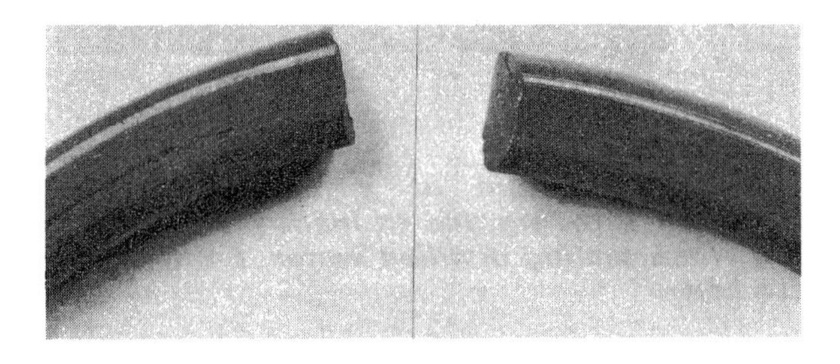

Figure 6.4 Rubber O-ring failure due to overheating. *The brittle fracture is due to hardening of the originally soft rubber and the flattened appearance is a typical example of compression set*

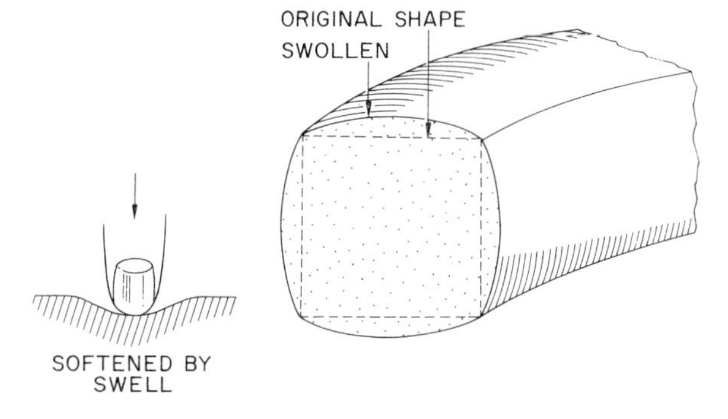

Figure 6.5 Rubber-fluid incompatibility

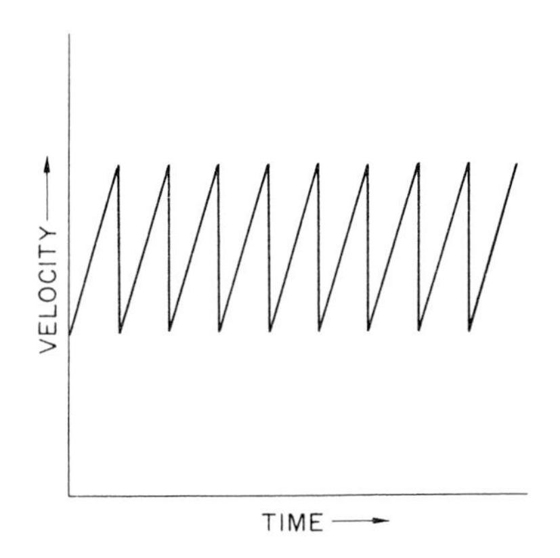

Figure 6.6 'Stick-slip'

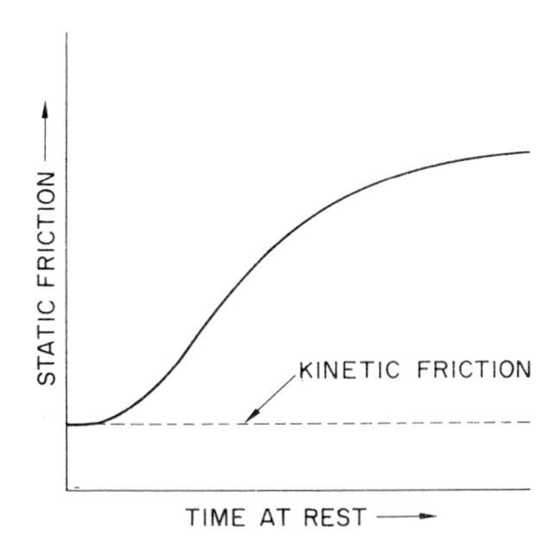

Figure 6.7 'Stiction'

O-Rings, Rectangular Rubber Rings, etc.

Table 6.3 Common failure mechanisms

Symptoms	Cause	Remedy
Fine circumferential cut set back slightly from sliding contact zone. Rubber 'nibbled' on one side. Ring completely ejected from its groove	Extrusion damage (Figure 6.8)	Reduce back clearance, check concentricity of parts; fit back-up ring; use reinforced seal; use harder rubber
Wear, not restricted to sliding contact zone. Partial or total fracture	Ring rolling or twisting in groove	Replace O-ring by rectangular section ring or a lobed type ring

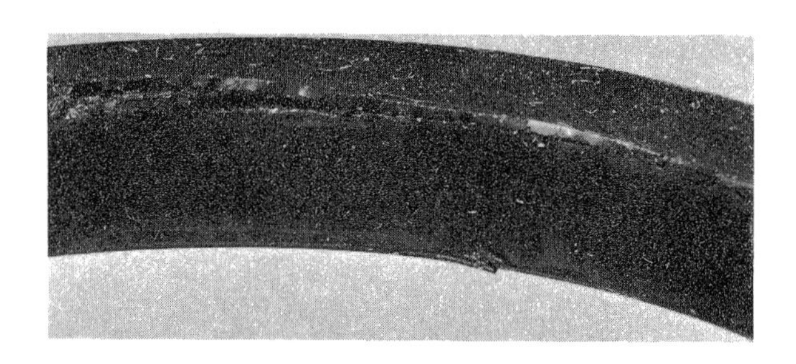

Figure 6.8 A rectangular-section rubber seal ring showing extrusion damage. *Where damage is less severe (r.h.s.) only a knife-cut is visible, but material has been nibbled away where extrusion was severe. The circumferential variation of the damage indicates eccentricity of the sealed components*

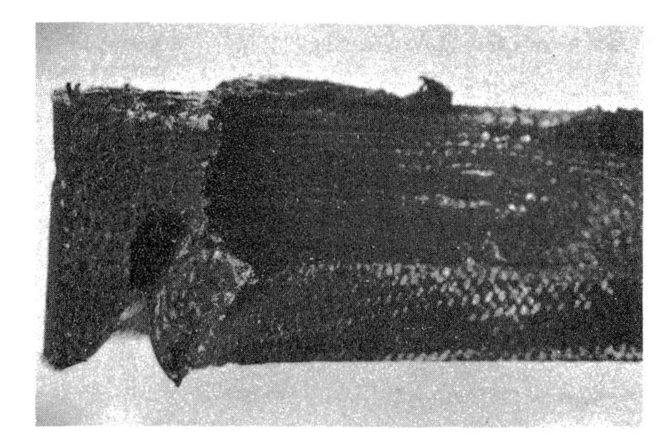

Figure 6.9 Wear failure of a rubberised-fabric square-back U-ring due to inadequate lubrication when sealing distilled water. Friction was also bad

Reciprocating Seals

Table 6.4 Common failure mechanisms

Symptoms	Cause	Remedy
Excessive wear and/or high friction	Poor lubrication. (Figure 6.9) Seal overloaded	If multiple seals are in use replace with single seal. Heavy duty seal may cure overloading. With aqueous fluids leather may be better than rubber
Non-uniform wear circumferentially	Dirt ingress Deposits on rod Side load	Fit wiper or scraper " Check bearings

Rotary Lip Seals

Table 6.5 Common failure mechanisms

Symptoms	Cause	Remedy
Excessive leakage	Damaged lip. Machining has left a spiral lead on shaft Unsuitable shaft surface	Check for damage and cause, e.g. contact with splines or other rough surface during assembly or careless handling/storage. Use mandrel for assembly. Eliminate shaft lead Finish the surface to R_A 0.1–0.5μm
Lip cracked in places	Excessive speed. Poor lubrication. Hot environment	Consider alternative rubber compounds. Improve lubrication. Reduce environmental temperature

PACKED GLANDS

Table 6.6 Common failure mechanisms of packed glands

Symptoms	*Cause*	*Remedy*
Packing extruded into clearance between shaft and housing or gland follower (Figure 6.10)	Designed clearance excessive or parts worn by abrasives or shaft bearings inadequate	Reduce clearances, check bearings
Leakage along outside of gland follower (Figure 6.11)	Packing improperly fitted or housing bore condition bad	Repack with care after checking bore condition
Used packing scored on outside surface, possibly leakage along outside of gland follower	Packing rotating with shaft due to being undersized	Check dimensions of housing and packing
Packing rings near gland follower very compressed (Figure 6.12) or rings embedded into each other	Packing incorrectly sized	Repack with accurately sized packing rings
Bore of used packing charred or blackened possibly shaft material adhering to packing	Lubrication failure	Change packing to one with more suitable lubricants or fit lantern ring with lubricant feed
Shaft badly worn along its length (Figure 6.13)	Lubrication failure or abrasive solids present	Change packing to one with more suitable lubricants or fit lantern ring with lubricant feed. Flush abrasives

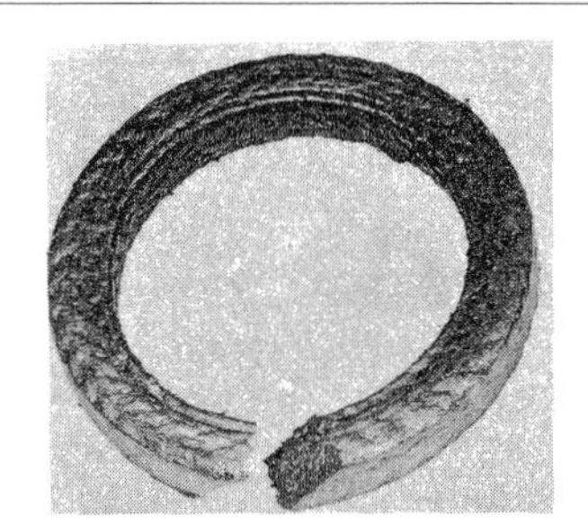

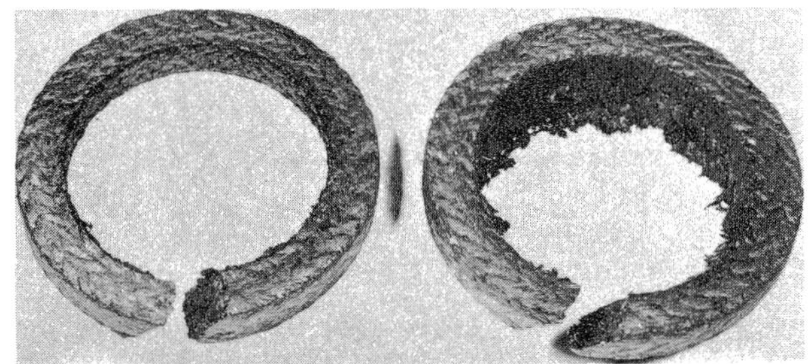

Figure 6.10 Soft packing rings after use. Left: *normal appearance*; top: *scored ring due to rotation of the ring in its housing;* right: *extruded ring due to excessive clearance between housing and shaft*

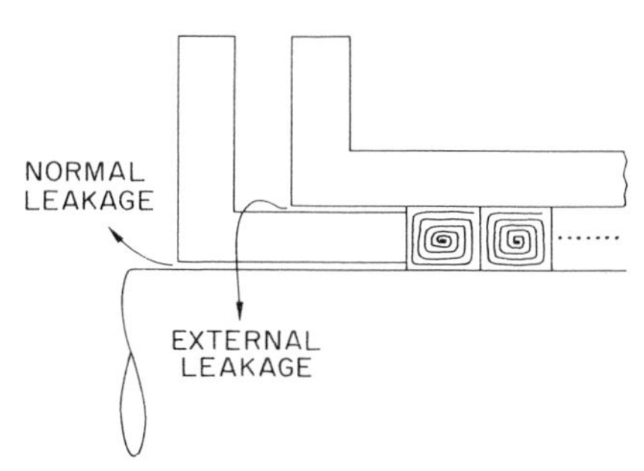

Figure 6.11 Packed gland showing abnormal leakage outside the gland follower

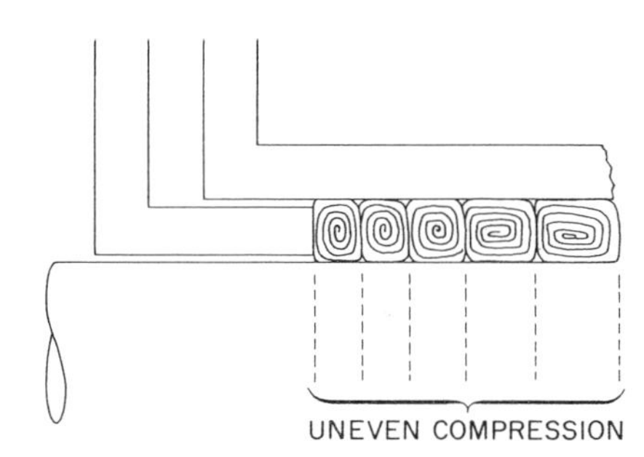

Figure 6.12 Packed gland showing uneven compression due to incorrect installation

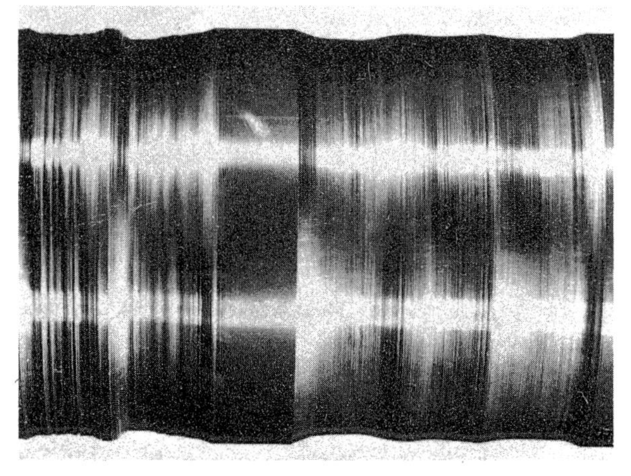

Figure 6.13 Severe wear of a bronze shaft caused by a soft packing with inadequate boundary lubricant

Wire rope failures

A wire rope is said to have failed when the condition of either the wire strands, core or termination has deteriorated to an unacceptable extent. Each application has to be considered individually in terms of the degree of degradation allowable; certain applications may allow for a greater degree of deterioration than others.

Complete wire rope failures rarely occur. The more common modes of failure/deterioration are described below.

DETERIORATION

Mechanical damage

Characteristics

Damage to exposed wires or complete strands, often associated with gross plastic deformation of the steel material. Damage may be localised or distributed along the length of the rope.

Inspection by visual means only.

Causes

There are many potential causes of mechanical damage, such as:

- rubbing against a static structure whilst under load
- impact or collision by a heavy object
- misuse or bad handling practices

External wear

Characteristics

Flattened areas formed on outer wires. Wear may be distributed over the entire surface or concentrated in narrow axial zones. Severe loss of worn wires under direct tension. Choice of rope construction can be significant in increasing wear resistance (e.g. Lang's lay ropes are usually superior to ordinary lay ropes).

Assess condition visually and also by measuring the reduction in rope diameter.

Causes

Abrasive wear between rope and pulleys, or between successive rope layers in multi-coiled applications, particularly in dirty or contaminated conditions (e.g. mining). Small oscillations, as a result of vibration, can cause localised wear at pulley positions.

Regular rope lubrication (dressings) can help to reduce this type of wear.

External fatigue

Characteristics

Transverse fractures of individual wires which may subsequently become worn. Fatigue failures of individual wires occur at the position of maximum rope diameter ('crown' fractures).

Condition is assessed by counting the number of broken wires over a given length of rope (e.g. one lay length, 10 diameters, 1 metre).

Causes

Fatigue failures of wires is caused by cyclic stresses induced by bending, often superimposed on the direct stress under tension. Tight bend radii on pulleys increases the stresses and hence the risk of fatigue. Localised Hertzian stresses resulting from ropes operating in oversize or undersize grooves can also promote premature fatigue failures.

Internal damage

Characteristics

Wear of internal wires generates debris which when oxidised may give the rope a rusty (or 'rouged') appearance, particularly noticeable in the valleys between strands.

Actual internal condition can only be inspected directly by unwinding the rope using clamps while under no load.

As well as a visual assessment of condition, a reduction in rope diameter can give an indication of rope deterioration.

Causes

Movement between strands within the rope due to bending or varying tension causes wear to the strand cross-over points (nicks). Failure at these positions due to fatigue or direct stress leads to fracture of individual wires. Gradual loss of lubricant in fibre core ropes accelerates this type of damage.

Regular application of rope dressings minimises the risk of this type of damage.

Corrosion

Characteristics

Degradation of steel wires evenly distributed over all exposed surfaces. Ropes constructed with galvanised wires can be used where there is a risk of severe corrosion.

Causes

Chemical attack of steel surface by corrosive environment e.g. seawater.

Regular application of rope dressings can be beneficial in protecting exposed surfaces.

Deterioration at rope terminations

Characteristics

Failure of wires in the region adjacent to the fitting. Under severe loading conditions, the fitting may also sustain damage.

Causes

Damage to the termination fitting or to the rope adjacent to the fitting can be caused by localised stresses resulting from sideways loads on the rope.

Overloading or shock loads can result in damage in the region of the termination.

Poor assembly techniques (e.g. incorrect mounting of termination fitting) can give rise to premature deterioration at the rope termination.

All photographs courtesy of Bridon Ropes Ltd., Doncaster

INSPECTION

To ensure safety and reliability of equipment using wire ropes, the condition of the ropes needs to be regularly assessed. High standards of maintenance generally result in increased rope lives, particularly where corrosion or fatigue are the main causes of deterioration.

The frequency of inspections may be determined by either the manufacturer's recommendations, or based on experience of the rate of rope deterioration for the equipment and the results from previous inspections. In situations where the usage is variable, this may be taken into consideration also.

Inspection of rope condition should address the following items:

- mechanical damage or rope distortions
- external wear
- internal wear and core condition
- broken wires (external and internal)
- corrosion
- rope terminations
- degree of lubrication
- equality of rope tension in multiple-rope installations
- condition of pulleys and sheaves

During inspection, particular attention should be paid to the following areas:

- point of attachment to the structure or drums
- the portions of the rope at the entry and exit positions on pulleys and sheaves
- lengths of rope subject to reverse or multiple bends

In order to inspect the internal condition of wire ropes, special tools may be required.

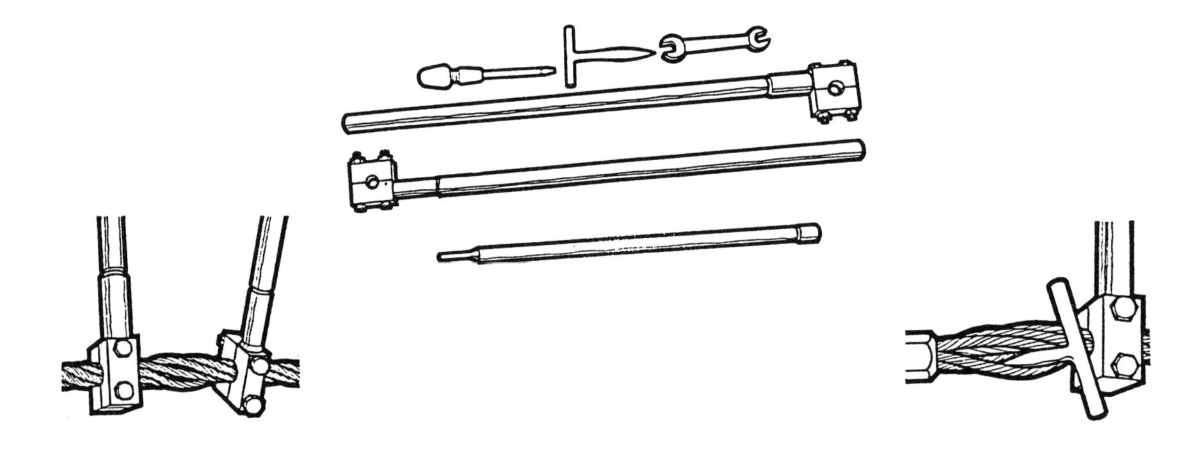

Figure 7.1 Special tools for internal examination of wire rope

MAINTENANCE

Maintenance of wire ropes is largely confined to the application of rope dressings, general cleaning, and the removal of occasional broken wires.

Wire rope dressings are usually based on mineral oils, and may contain anti-wear additives, corrosion inhibiting agents or tackiness additives. Solvents may be used as part of the overall formulation in order to improve the penetrability of the dressing into the core of the rope. Advice from rope manufacturers should be sought in order to ensure that selected dressings are compatible with the lubricant used during manufacture.

The frequency of rope lubrication depends on the rate of rope deterioration identified by regular inspection. Dressings should be applied at regular intervals and certainly before there are signs of corrosion or dryness.

Dressings can be applied by brushing, spraying, dripfeed, or by automatic applicators. For best results, the dressing should be applied at a position where the rope strands are opened up such as when the rope passes over a pulley.

When necessary and practicable ropes can be cleaned using a wire brush in order to remove any particles such as dirt, sand or grit.

Occasional broken wires should be removed by using a pair of pliers to bend the wire end backwards and forwards until it breaks at the strand cross-over point.

REPLACEMENT CRITERIA

Although the assessment of rope condition is mainly qualitative, it is possible to quantify particular modes of deterioration and apply a criterion for replacement. In particular the following parameters can be quantified:

- the number of wire breaks over a given length
- the change in rope diameter

Guidance for the acceptable density of broken wires in six and eight strand ropes is given below.

Table 7.1 Criterion for replacement based on the maximum number of distributed broken wires in six and eight strand ropes operating with metal sheaves

Total number of wires in outer strands (including filler wires)	*Example of rope construction*	*Number of visible broken wires necessitating discard in a wire rope operating with metal sheaves when measured over a length of 10× nominal rope diameter*	
		Factor of safety <5	*Factor of safety >5*
Less than 50	6 × 7 (6/1)	2	4
51–120	6 × 19 (9/9/1)	3–5	6–10
121–160	6 × 19F (12/6+6F/1)	6–7	12–14
161–220	8 × 19F (12/6+6F/1)	8–10	16–20
221–260	6 × 37 (18/12/6/1)	11–12	22–24

Rope manufacturers should be consulted regarding other types of rope construction.

Guidance for the allowable change in rope diameter is given below.

Table 7.2 Criterion for replacement based on the change in diameter of a wire rope

Rope construction	*Replacement criteria related to rope diameter**
Six and eight strand ropes	Replacement necessary when the rope diameter is reduced to 90% of the nominal diameter at any position.
Multi-strand ropes	A more detailed rope examination is necessary when the rope diameter is either reduced to 97%, or has increased to 105%, of the nominal diameter.

*The diameter of the rope is measured across the tips of the strands (i.e. the maximum rope diameter).

Some of the more common brake and clutch troubles are pictorially presented in subsequent sections; although these faults can affect performance and shorten the life of the components, only in exceptional circumstances do they result in complete failure.

BRAKING TROUBLES

Metal surface

Heat Spotting

Characteristics

Small isolated discoloured regions on the friction surface. Often cracks are formed in these regions owing to structural changes in the metal, and may penetrate into the component.

Causes

Friction material not sufficiently conformable to the metal member; or latter is distorted so that contact occurs only at small heavily loaded areas.

Crazing

Characteristics

Randomly orientated cracks on the rubbing surface of a mating component, with main cracks approximately perpendicular to the direction of rubbing. These can cause severe lining wear.

Causes

Overheating and repeated stress-cycling from compression to tension of the metal component as it is continually heated and cooled.

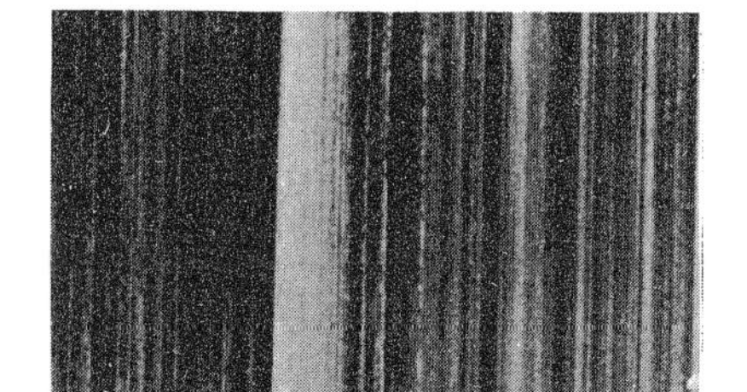

Scoring

Characteristics

Scratches on the rubbing path in the line of movement.

Causes

Metal too soft for the friction material; abrasive debris embedded in the lining material.

Friction material surface

Heat Spotting

Characteristics

Heavy gouging caused by hard proud spots on drum resulting in high localised work rates giving rise to rapid lining wear.

Causes

Material rubbing against a heat-spotted metal member.

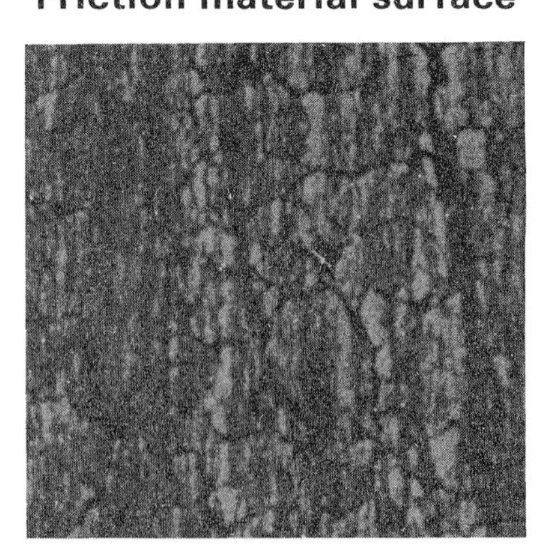

Crazing

Characteristics

Randomly orientated cracks on the friction material, resulting in a high rate of wear.

Causes

Overheating of the braking surface from overloading or by the brakes dragging.

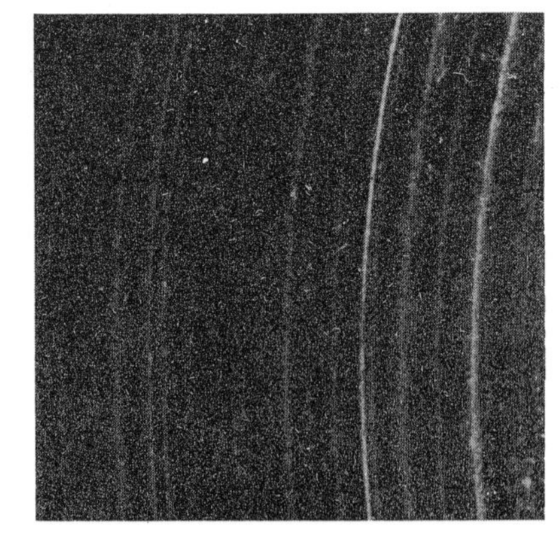

Scoring

Characteristics

Grooves formed on the friction material in the line of movement, resulting in a reduction of life.

Causes

As for metal surface or using new friction material against metal member which needs regrinding.

Fade

Characteristics

Material degrades at the friction surface, resulting in a decrease in μ and a loss in performance, which may recover.

Causes

Overheating caused by excessive braking, or by brakes dragging.

Metal Pick-up

Characteristics

Metal plucked from the mating member and embedded in the lining.

Causes

Unsuitable combination of materials.

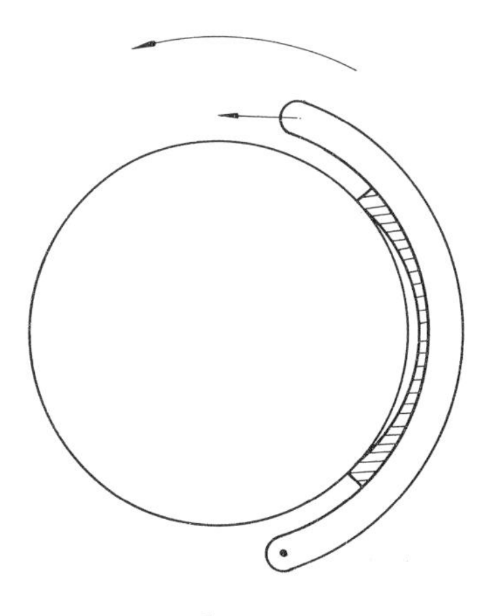

Grab

Characteristics

Linings contacting at ends only ('heel and toe' contact) giving high servo effect and erratic performance. The brake is often noisy.

Causes

Incorrect radiusing of lining.

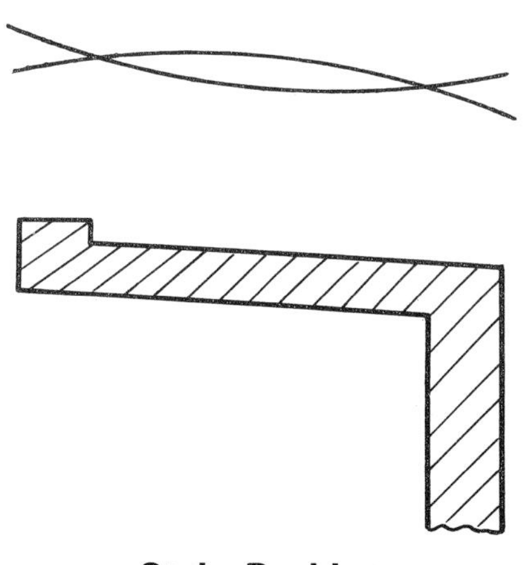

Strip Braking

Characteristics

Braking over a small strip of the rubbing path giving localised heating and preferential wear at these areas.

Causes

Distortion of the brake path making it concave or convex to the lining, or by a drum bell mouthing.

Neglect

Characteristics

Material completely worn off the shoe giving a reduced performance and producing severe scoring or damage to the mating component, and is very dangerous.

Causes

Failure to provide any maintenance.

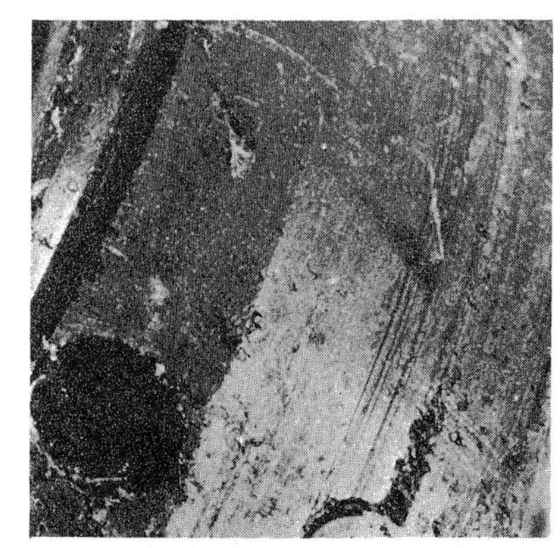

Misalignment

Characteristics

Excessive grooving and wear at preferential areas of the lining surface, often resulting in damage to the metal member.

Causes

Slovenly workmanship in not fitting the lining correctly to the shoe platform, or fitting a twisted shoe or band.

CLUTCH TROUBLES

As with brakes, heat spotting, crazing and scoring can occur with clutches; other clutch troubles are shown below.

*Dishing

Characteristics

Clutch plates distorted into a conical shape. The plates then continually drag when the clutch is disengaged, and overheating occurs resulting in thermal damage and failure. More likely in multi-disc clutches.

Causes

Lack of conformability. The temperature of the outer region of the plate is higher than the inner region. On cooling the outside diameter shrinks and the inner area is forced outwards in an axial direction causing dishing.

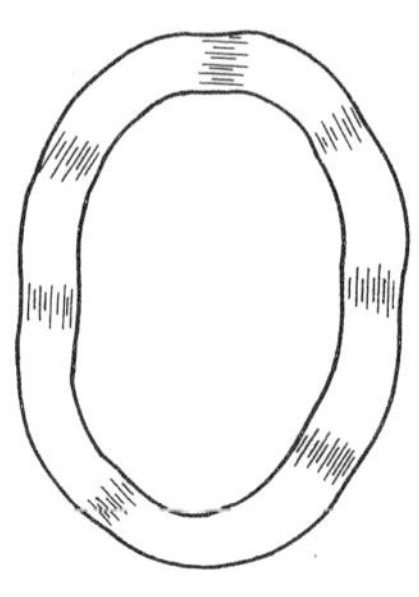

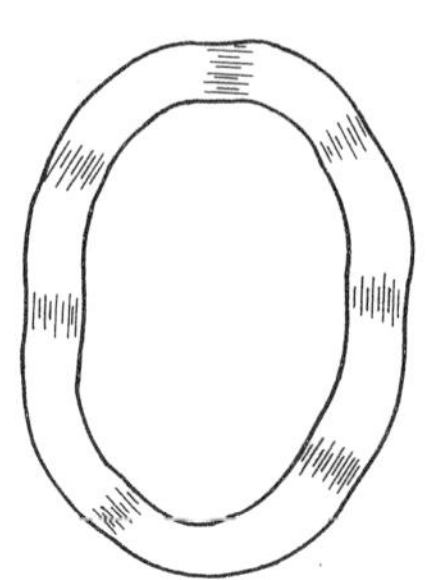

*Waviness or Buckling

Characteristics

Clutch plates become buckled into a wavy pattern. Preferential heating then occurs giving rise to thermal damage and failure. More likely in multi-disc clutches.

Causes

Lack of conformability. The inner area is hotter than the outer area and on cooling the inner diameter contracts and compressive stresses occur in the outer area giving rise to buckling.

*Band Crushing

Characteristics

Loss of friction material at the ends of a band in a band clutch. Usually results in grooving and excessive wear of the opposing member.

Causes

Crushing and excessive wear of the friction material owing to the high loads developed at the ends of a band of a positive servo band clutch.

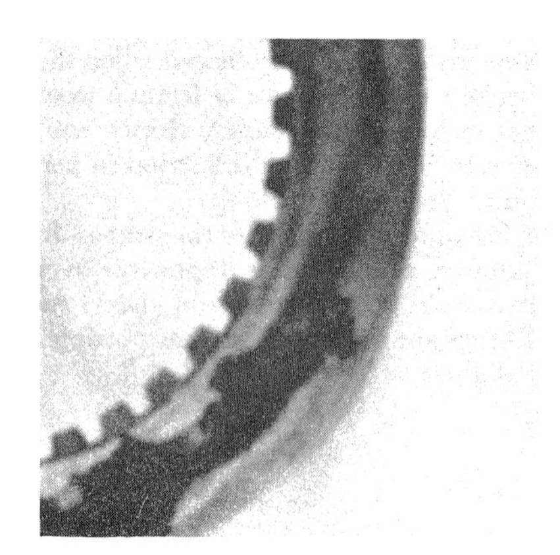

*Bond Failure

Characteristics

Material parting at the bond to the core plate causing loss of performance and damage to components.

Causes

Poor bonding or overheating, the high temperatures affecting bonding agent.

Material Transfer

Characteristics

Friction material adhering to opposing plate, often giving rise to excessive wear.

Causes

Overheating and unsuitable friction material.

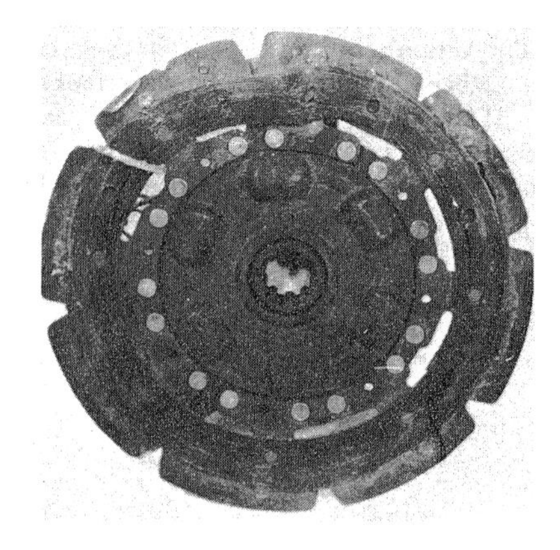

Burst Failure

Characteristics

Material splitting and removed from the spinner plate.

Causes

High stresses on a facing when continually working at high rates of energy dissipation, and high speeds.

*These refer to oil immersed applications.

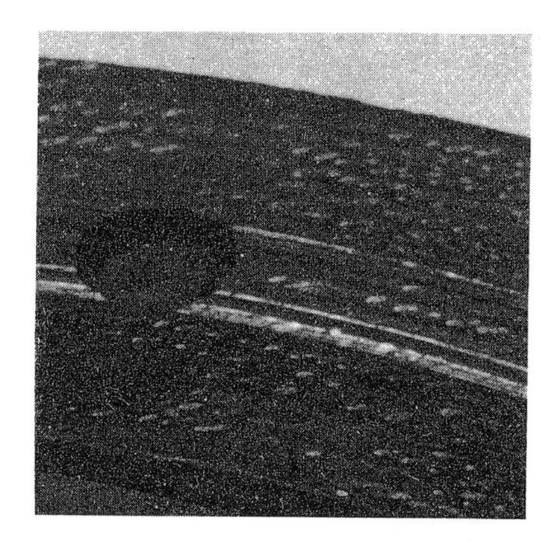

Grooving

Characteristics

Grooving of the facing material on the line of movement.

Causes

Material transfer to opposing plate.

Reduced Performance

Characteristics

Decrease in coefficient of friction giving a permanent loss in performance in a dry clutch.

Causes

Excess oil or grease on friction material or on the opposing surface.

Distortion

Characteristics

Facings out of flatness after high operating temperatures giving rise to erratic clutch engagement.

Causes

Unsuitable friction material.

GENERAL NOTES

The action required to prevent these failures recurring is usually obvious when the causes, as listed in this section, are known.

Other difficulties can be experienced unless the correct choice of friction material is made for the operating conditions.

If the lining fitted has too low a coefficient of friction the friction device will suffer loss of effectiveness. Oil and grease deposited on dry linings and facings can have an even more marked reduction in performance by a factor of up to 3. If the μ is too high or if a badly matched set of linings are fitted, the brake may grab or squeal.

The torque developed by the brake is also influenced by the way the linings are bedded so that linings should be initially ground to the radius of the drum to ensure contact is made as far as possible over their complete length.

If after fitting, the brake is noisy the lining should be checked for correct seating and the rivets checked for tightness. All bolts should be tightened and checks made that the alignment is correct, that all shoes have been correctly adjusted and the linings are as fully bedded as possible. Similarly, a clutch can behave erratically or judder if the mechanism is not correctly aligned.

BASIC MECHANISMS

Fretting occurs where two contacting surfaces, often nominally at rest, undergo minute oscillatory tangential relative motion, which is known as 'slip'. It may manifest itself by debris oozing from the contact, particularly if the contact is lubricated with oil.

Colour of debris: red on iron and steel, black on aluminium and its alloys.

On inspection the fretted surfaces show shallow pits filled and surrounded with debris. Where the debris can escape from the contact, loss of fit may eventually result. If the debris is trapped, seizure can occur which is serious where the contact has to move occasionally, e.g. a machine governor.

The movement may be caused by vibration, or very often it results from one of the contacting members undergoing cyclic stressing. In this case fatigue cracks may be observed in the fretted area. Fatigue cracks generated by fretting start at an oblique angle to the surface. When they pass out of the influence of the fretting they usually continue to propagate straight across the component. This means that where the component breaks, there is a small tongue of metal on one of the fracture surfaces corresponding to the growth of the initial part of the crack.

Fretting can reduce the fatigue strength by 70–80%. It reaches a maximum at an amplitude of slip of about 8 μm. At higher amplitudes of slip the reduction is less as the amount of material abraded away increases.

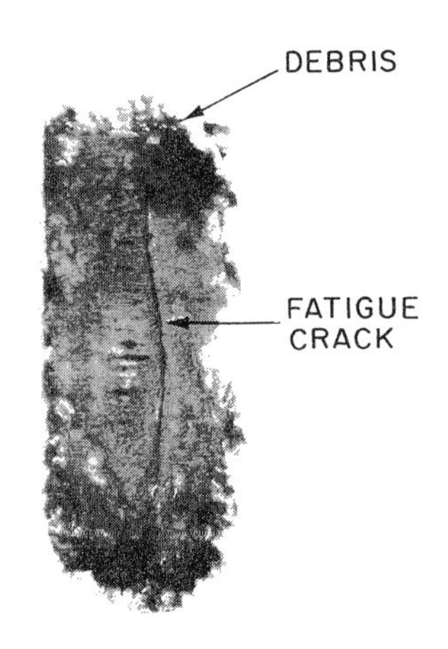

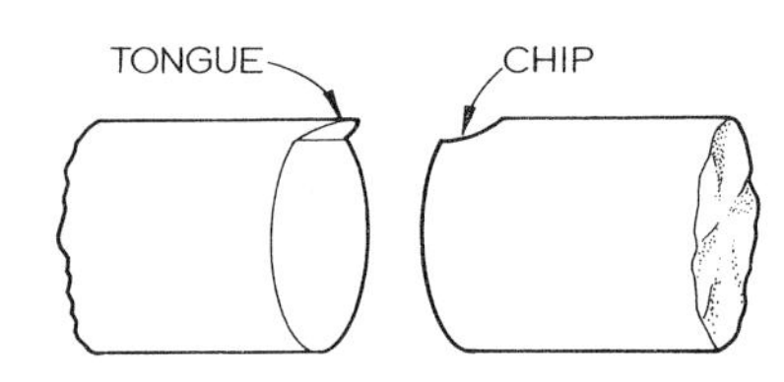

Figure 9.1 A typical fatigue fracture initiated by fretting

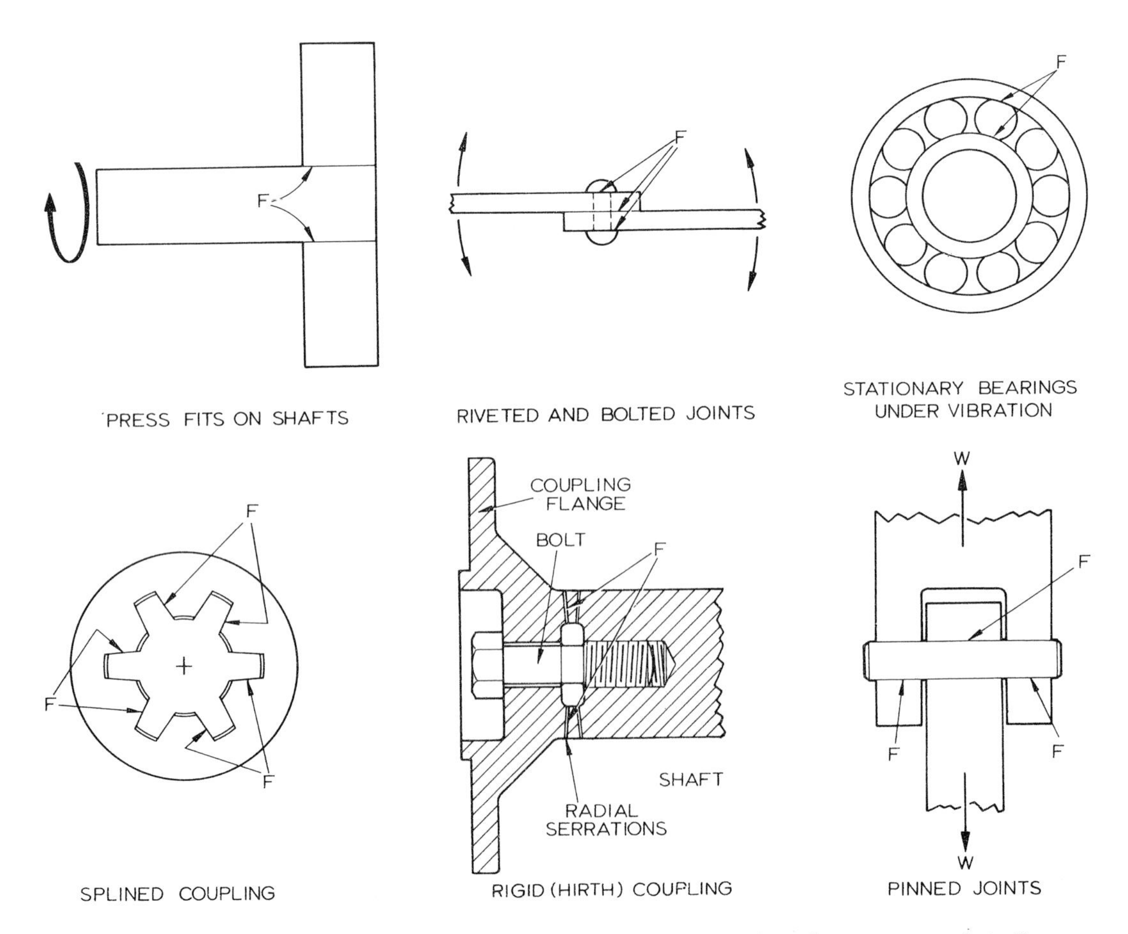

Figure 9.2 Typical situations in which fretting occurs. Fretting sites are at points F.

Detailed mechanisms

Rupture of oxide films results in formation of local welds which are subjected to high strain fatigue. This results in the growth of fatigue cracks oblique to the surface. If they run together a loose particle is formed. One of the fatigue cracks may continue to propagate and lead to failure. Oxidation of the metallic particles forms hard oxide debris, i.e. Fe_2O_3 on steel, Al_2O_3 on aluminium. Spreading of this oxide debris causes further damage by abrasion. If the debris is compacted on the surfaces the damage rate becomes low.

Where the slip is forced, fretting wear damage increases roughly linearly with normal load, amplitude of slip, and number of cycles. Damage rate on mild steel—approx. 0.1 mg per 10^6 cycles, per MN/m^2 normal load, per μm amplitude of slip. Increasing the pressure can, in some instances, reduce or prevent slip and hence reduce fretting damage.

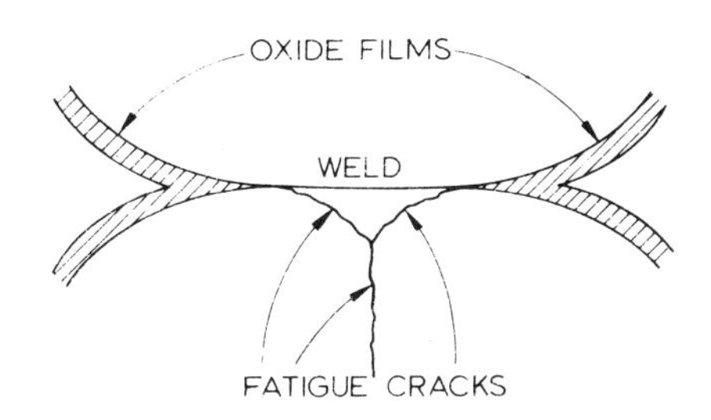

Figure 9.3 Oxide film rupture and the development of fatigue cracks

PREVENTION

Design

(*a*) elimination of stress concentrations which cause slip
(*b*) separating surfaces where fretting is occurring
(*c*) increasing pressure by reducing area of contact

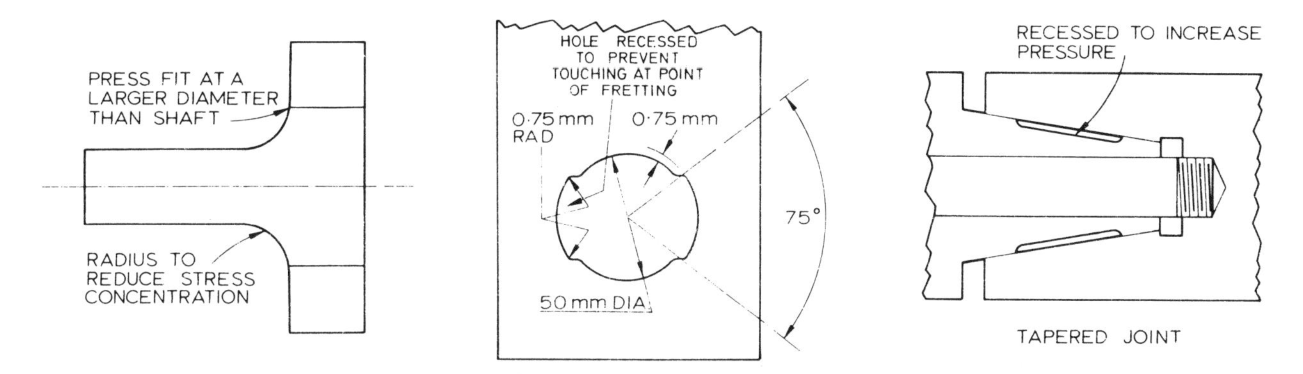

Figure 9.4 Design changes to reduce the risk of fretting

Lubrication

Where the contact can be continuously fed with oil, the lubricant prevents access of oxygen which is advantageous in reducing the damage. Oxygen diffusion decreases as the viscosity increases. Therefore as high a viscosity as is compatible with adequate feeding is desirable. The flow of lubricant also carries away any debris which may be formed. In other situations greases must be used. Shear-susceptible greases with a worked penetration of 320 are recommended. E.P. additives and MoS_2 appear to have little further beneficial effect, but anti-oxidants may be of value. Baked-on MoS_2 films are initially effective but gradually wear away.

Non-metallic coatings

Phosphate and sulphidised coatings on steel and anodised coatings on aluminium prevent metal-to-metal contact. Their performance may be improved by impregnating them with lubricants, particularly oil-in-water emulsions.

Metallic coatings

Electrodeposited coatings of soft metals, e.g. Cu, Ag, Sn or In or sprayed coatings of Al allow the relative movement to be taken up within the coating. Chromium plating is generally not recommended.

Non-metallic inserts

Inserts of rubber, or PTFE can sometimes be used to separate the surfaces and take up the relative movement.

Choice of metal combinations

Unlike metals in contact are recommended—preferably a soft metal with low work hardenability and low recrystallisation temperature (such as Cu) in contact with a hard surface, e.g. carburised steel.

D10 Maintenance methods

The purpose of maintenance is to preserve plant and machinery in a condition in which it can operate, and can do so safely and economically.

Table 10.1 Situations requiring maintenance action

Situation	*Basis of decision*	*Decision mechanism*
Loss of operating efficiency	Recognition of the problem associated with a study of the costs	The economic loss arising from the reduced efficiency is greater than the effective cost of the repair.
Loss of function	Safety risk	This must be avoided by prior maintenance action. There must therefore be a method of predicting the approach of a failure condition.
	Downtime cost Repair cost	The maintenance method that is selected needs to be chosen to keep total costs to a minimum, while maximising the profit and market opportunities. The ongoing cost of maintenance versus total replacement also needs to be considered.

It is the components of plant and machinery which fail individually, and can lead to the loss of function of the whole unit or system. Maintenance activity needs therefore to be concentrated on those components that are critical.

Table 10.2 Factors in the selection of critical components

Important factor	*Typical examples*	*Guide to selection*
Likelihood of failure	Components subject to: Occasional overload Fatigue loading Wear Corrosion or other environmental effects	Can be identified by a review of the design of the system or by an analysis of relevant operating experience. Components which are more likely to fail should be selected for maintenance attention.
Effect of the component failure on the system	Components which, when they fail, cause a failure of the whole system, possibly with some consequential damage	These components need to be selected for special attention. They must be selected if there are safety implications.
	Components which, when they fail, may allow other components to be overloaded, and then cause failure of the system	An analysis of the effects of the failure needs to be carried out to detect any safety implications.
	Components which, when they fail, only produce a reduction in the performance of the system	The likely economic effects of the failure need to be analysed to decide whether these components merit special attention.
The time required to replace a failed component or to rectify the effect of its failure	Components that can be replaced quickly such as bulbs, fuses and some printed circuit boards, or components whose function can be taken over immediately by a standby system	If there are no safety implications, no particular maintenance action is required other than component replacement or repair after failure
	Components which take a long time to replace or repair	Their condition needs to be monitored and their maintenance planned in advance

Table 10.3 Maintenance methods which can be used

Maintenance method	*Advantages*	*Disadvantages*
Allow the equipment to break down and then repair it	Can be the cheapest solution if the repair is easy and there is no safety risk or possibility of consequential damage	The cost of consequential damage can be high and there might be safety risks. The plant may be of a type which cannot be allowed to stop suddenly because of product solidification or deterioration, etc. The failure may occur at an inconvenient time or, if the plant is mobile, at an inconvenient place. The necessary staff and replacement components may not then be available.
Preventive maintenance carried out at regular intervals	Reduces the risk of failure in service Enables the work and availability of special tools and spare parts to be planned well in advance Is particularly appropriate for components which need changing because of capacity absorption such as filter elements	Some failures will continue to occur in service because components do not fail at regular intervals as shown in Figure 10.2. The failures can be unexpected and inconvenient. During the overhaul many components in good condition will be stripped and inspected unnecessarily. Mistakes can be made on re-assembly so that the plant can end up in a worse condition after the overhaul. The overhaul process can take a considerable time resulting in a major loss of profitable production.
Opportunistic maintenance carried out when the plant or equipment happens to become available	The maintenance can be carried out with no loss of effective operating time.	Staff and spare parts have to be kept available which may cause increased costs. Is only practical in combination with condition based maintenance or some preventive maintenance, because otherwise any work which needs to be done is not known about. Breakdowns are still likely to occur in service.
Condition-based maintenance	Enables the plant to be left in service until a failure is about to occur. It can then be withdrawn from service in a planned manner and repaired at minimum cost. Consequential damage from a failure can be avoided.	Can only detect failures which show a progression to failure, which enables the incipient failure to be detected. Components with a sudden failure mode cannot provide the required advanced detection. Failures of components due to an unexpected random overload in service cannot be avoided, even by this method.
Maintenance/replacement based on expected component life	Is the only safe method usable for components which have a sudden failure mode	Requires monitoring of the operating conditions of the component so that its expected life can be estimated. If all failures are to be avoided many components will need to be changed when individually they have a useful life still remaining, e.g. in Figure 10.2 the working life has to be kept below the range in which failures may occur.

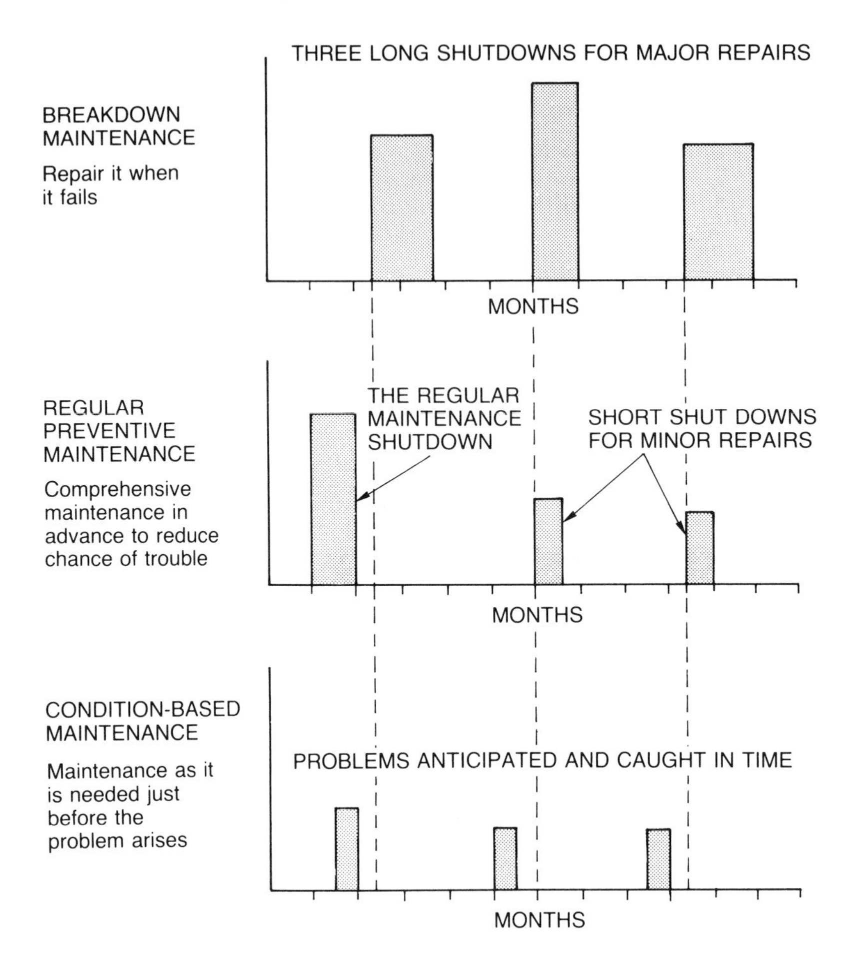

Figure 10.1 The results of maintaining the same industrial plant in different ways
The height of the bars indicates the amount of maintenance effort required

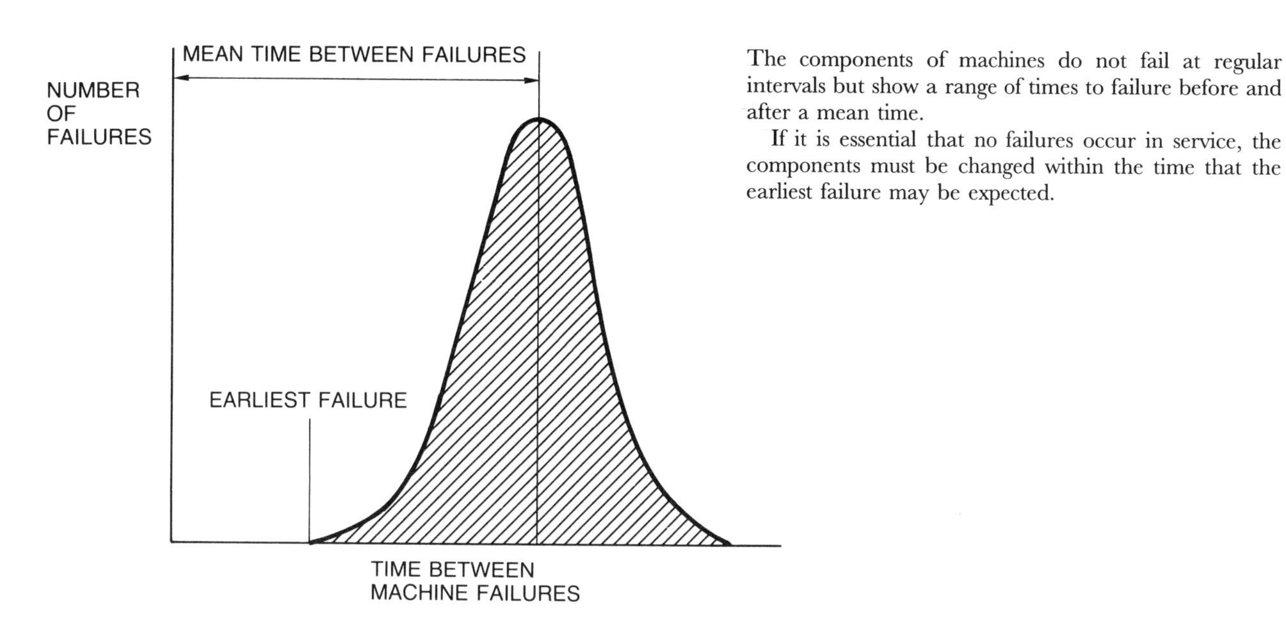

The components of machines do not fail at regular intervals but show a range of times to failure before and after a mean time.

If it is essential that no failures occur in service, the components must be changed within the time that the earliest failure may be expected.

Figure 10.2 The distribution of the time to failure for a typical component

Table 10.4 Feedback from maintenance to equipment design

Opportunity	*Remarks*
The maintenance activity provides an opportunity to record and analyse the problems and failures which occur during the service life of equipment. In particular, records need to be kept of the operating time to failure, the failure mode and any particular operating conditions which may have contributed to the failure.	This is a unique source of information which is of very high value to the process of design and development of improved equipment. It is a profit centre of the maintenance activity.
An intelligent review of maintenance operations can enable repair times to be determined for various components. The nature and cause of accessibility problems can be determined together with the definition of the design principles which need to be used to improve it.	This is useful for maintenance planning but it is particularly important as a method of getting new equipment that is designed to make maintenance more simple and of reduced cost.
The maintenance activity can be reviewed to identify opportunities in which modular design can be used to reduce downtime caused by in service faults. It can also often simplify the process of equipment manufacture.	If sub-systems can be designed to be self-contained and readily removable, this can enable the sub-system to be exchanged when a fault occurs. The actual problem can then be corrected off-line at a comprehensive test facility and the sub-system returned for further use as an exchange unit.

Table 10.5 Maintenance management

Technique	*Description*	*Remarks*
A separate maintenance department	Called in by the operating departments when something goes wrong	Inherently operates on a breakdown basis. Has little chance to contribute to a major uplift in plant performance. An outdated technique.
Terotechnology	Viewing plant and equipment in terms of their total life cost and thus placing more emphasis on cost effective maintainability as distinct from first cost alone	A level of management philosophy which prepares the way for efficient maintenance systems
Computerised maintenance information	Systems covering: Plant inventory and data Operating history Spares usage and storage On-going condition data Repair work records Failure analysis and reliability data Cost records	Provides the basis for an effective management system that is able to concentrate on problem areas, plan ahead, reduce costs and contribute to the specification of improved future plant
Total productive maintenance	A technique developed in Japan in which the production operators carry out the first line of problem detection and maintenance. There is intentionally no clear interface between the production and maintenance departments.	An effective way of involving all the people in a company and achieving recognition of the importance of process plant availability. It can reduce in-service failures by 25%.
Reliability centred maintenance	A technique developed in the airlines. A system which looks at component functions, failure modes and effects in order to concentrate on key issues. Safety in operation is paramount followed closely by the control of downtime, quality and customer service.	An effective means of concentrating maintenance effort cost effectively, with an ability to identify problems and any need for design improvements
Mobile equipment maintenance	The basic objective is to ensure that the equipment does not break down away from its base. Also, when visiting its base, any essential work that is needed should be carried out.	It is particularly important to monitor the condition of such plant and its expected component lives. It is sometimes necessary to carry out some maintenance work before it becomes essential in order to match the time of access to the equipment, and the overhaul department's workload.

Condition monitoring is a technique used to monitor the condition of equipment in order to give an advanced warning of failure. It is an essential component of condition-based maintenance in which equipment is maintained on the basis of its condition.

MONITORING METHODS

The basic principle of condition monitoring is to select a physical measurement which indicates that deterioration is occurring, and then to take readings at regular intervals. Any upward trend can then be detected and taken as an indication that a problem exists. This is illustrated in Figure 11.1 which shows a typical trend curve and the way in which this provides an alert that an incipient failure is approaching. It also gives a lead time in which to plan and implement a repair.

Since failures occur to individual components, the monitoring measurements need to focus on the particular failure modes of the critical components.

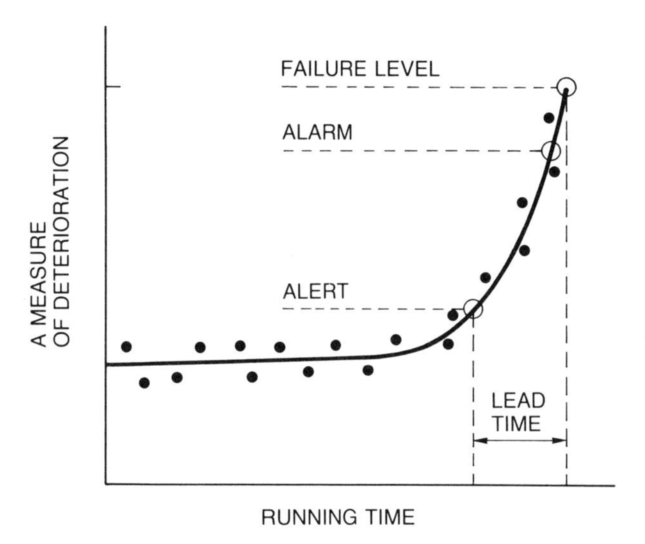

Figure 11.1 The principle of condition monitoring measurements which give an indication of the deterioration of the equipment

Table 11.1 Monitoring methods and the components for which they are suitable

Method	*Principle*	*Application examples*
Wear debris monitoring	The collection and analysis of wear debris derived from component surfaces, and carried away in the lubricating oil	Components such as bearings or other rubbing parts which wear, or suffer from surface pitting due to fatigue
Vibration monitoring	The detection of faults in moving components, from the change in the dynamic forces which they generate, and which affect vibration levels at externally accessible points	Rotating components such as gears and high speed rotors in turbines and pumps
Performance monitoring	Checking that the machine components and the complete machine system are performing their intended functions	The temperature of a bearing indicates whether it is operating with low friction. The pressure and flow rate of a pump indicate whether its internal components are in good condition.

The monitoring measurements give an indication of the existence of a problem as shown in Figure 11.1. More detailed analysis can indicate the nature of the problem so that rectification action can be planned. Other sections of this handbook give more details about these methods of monitoring.

In wear debris monitoring, the amount of the debris and its rate of generation indicate when there is a problem. The material and shape of the debris particles can indicate the source and the failure mechanism.

The overall level of a vibration measurement can indicate the existence of a problem. The form and frequency of the vibration signal can indicate where the problem is occurring and what it is likely to be.

Introducing condition monitoring

If an organisation has been operating with breakdown maintenance or regular planned maintenance, a change over to condition-based maintenance can result in major improvements in plant availability and in reduced costs. There are, however, up front costs for organisation and training and for the purchase of appropriate instrumentation. There are operational circumstances which can favour or retard the potential for the introduction of condition-based maintenance.

Table 11.2 Factors which can assist the introduction of condition-based maintenance

Factor	*Mechanism of action*
Where a safety risk is particularly likely to arise from the breakdown of machinery	Typical examples are plant handling dangerous materials, and machines for the transport of people.
Where accurate advanced planning of maintenance is essential	Typical examples are equipment situated in a remote place which is visited only occasionally for maintenance, and mobile equipment which makes only occasional visits to its base.
Where plant or equipment is of recent design, and may have some residual development problems	Condition monitoring enables faults to be detected early while damage is still slight, thus providing useful evidence to guide design improvements. It also improves the negotiating position with the plant manufacturer.
Where relatively insensitive operators use expensive equipment whose breakdown may result in serious damage	Condition monitoring enables a fault to be detected in sufficient time for an instruction to be issued for the withdrawal of the equipment before expensive damage is done.
Where the manufacturer can offer a condition monitoring service to several users of his equipment	The cost to each user can be reduced in this way, and the manufacturer gets a useful feed-back to guide his product design and development.
Where instruments or other equipment required for condition monitoring can be used, or is already being used, for another purpose	Other applications of the instruments or equipment may be process control or some servicing activity such as rotor balancing.

Table 11.3 Factors which can retard the introduction of condition-based maintenance

Factor	*Mechanism of action*
Where an industry is operating at a low level of activity, or operates seasonally, so that plant and machinery is often idle	If the plant is only operating part of the time, there is generally plenty of opportunity for inspection and maintenance during idle periods.
Where there is too small a number of similar machines or components being monitored by one engineer or group of engineers to enable sufficient experience to be built up for the effective interpretation of readings and for correct decisions on their significance	To gain experience in a reasonable time, the minimum number of machines tends to vary between 4 and 10 depending on the type of machine or component. The problem may be overcome by pooling monitoring services with other companies, or by involving machine manufacturers or external monitoring services.
Where skilled operators have close physical contact with their machines, and can use their own senses for subjective monitoring	Machine tools and ships can be examples of this situation, but any trends towards the use of less skilled operators or supervisory engineers, favours the application of condition monitoring.

Table 11.4 A procedure for setting up a plant condition monitoring activity

Activity	*Remarks*
1. Check that the plant is large enough to justify having its own internal system.	If the total plant value is less than £2M it may be worth sub-contracting the activity.
2. Consider the cost of setting up.	For most plant, a setting up cost of 1% of the plant value can be justified. If there is a major safety risk, up to 5% of the plant value may be appropriate.
3. Select the machines in the plant that should be monitored.	The important machines for monitoring will tend to be those which: (a) Are in continuous operation. (b) Are involved in single stream processes. (c) Have minimum parallel or stand-by capacity. (d) Have the minimum product storage capacity on either side of them. (e) Handle dangerous or toxic materials. (f) Operate to particularly high pressures or speeds.
4. Select the components of the critical machines on which the monitoring needs to be focussed.	The important components will be those where: (a) A failure is possible. (b) The consequences of the failure are serious in terms of safety or machine operation. (c) If a failure is allowed to occur the time required for a repair is likely to be long.
5. Choose the monitoring method or methods to be used.	List the possible techniques for each critical component and try to settle for two or at the most three techniques for use on the plant.

Table 11.5 Problems which can arise

Problem	*Solution*
Regular measurements need to be taken, often for months or years before a critical situation arises. The operators can therefore get bored.	The management need to keep the staff motivated by stressing the importance of their work. The use of portable electronic data collectors partially automates the collection process, provides a convenient interface with a computer for data analysis, and can also monitor the tour of duty of the operators.
One of the measurements indicates that an alert situation has arisen and a decision has to be made on whether to shut down the plant and incur high costs from loss of use, or whether it is a false alarm.	To avoid this situation install at least two physically different systems for monitoring really critical components. e.g. measure bearing temperature and vibration. In any event always recheck deviant readings and re-examine past trends.
The operators take a long time to acquire the necessary experience in detection and diagnosis, and can create false alarms.	Start taking the measurements while still operating a planned regular maintenance procedure. Take many measurements just prior to shut down and then check the components to see whether the diagnosis was correct.

Table 11.6 The benefits that can arise from the use of condition monitoring

Benefit	*Mechanism*
1. Increased plant availability resulting in greater output from the capital invested. 2. Reduced maintenance costs.	Machine running time can be increased by maximising the time between overhauls. Overhaul time can be reduced because the nature of the problem is known, and the spares and men can be ready. Consequential damage can be reduced or eliminated.
3. Improved operator and passenger safety.	The lead time given by condition monitoring enables machines to be stopped before they reach a critical condition, especially if instant shut-down is not permitted.
4. More efficient plant operation, and more consistent quality, obtained by matching the rate of output to the plant condition.	The operating load and speed on some machines can be varied to obtain a better compromise between output, and operating life to the next overhaul.
5. More effective negotiations with plant manufacturers or repairers, backed up by systematic measurements of plant condition.	Measurements of plant when new, at the end of the guarantee period, and after overhaul, give useful comparative values.
6. Better customer relations following from the avoidance of inconvenient breakdowns which would otherwise have occurred.	The lead time given by condition monitoring enables such breakdowns to be avoided.
7. The opportunity to specify and design better plant in the future.	The recorded experience of the operation of the present machinery is used for this purpose.

Table 12.1 Maximum contact temperatures for typical tribological components

Component	Maximum temperature	Reason for limitation
White metal bearing	200°C at 1.5 MN/m^2 to 130°C at 7 MN/m^2	Failure by incipient melting at low loading (1.5 MN/m^2); by plastic deformation at high loading (7 MN/m^2)
Rolling bearing	125°C	Normal tempering temperature (special bearings are available for higher temperature operation)
Steel gear	150–250°C	Scuffing; the temperature at which scuffing occurs is a function of both the lubricant and the steel and cannot be defined more closely

The temperatures in Table 12.1 are indicative of design limits. In practice it may be difficult to measure the contact temperature. Table 12.2 indicates practical methods of measuring temperatures and the limits that can be accepted.

Table 12.2 Temperature as an indication of component failure

Component	Method of temperature measurement	Comments	Action limits(1)(4)
White metal bearing	Thermocouple in contact with back of white metal in thrust pad or at load line in journal bearings(5)	Extremely sensitive, giving immediate response to changes in load. Failure is indicated by rapid temperature rise	Alarm at rise of 10°C above normal running temperature. Trip at rise of 20°C
	Thermometer/thermocouple in oil bleed from bearing (viz. through hole drilled in bearing land)	Reasonably sensitive, may be preferable for journal bearings where there is difficulty in fitting a thermocouple into the back of the bearing in the loaded area	Alarm at rise of 10°C above normal running temperature. Trip at rise of 20°C
	Thermometer in bearing pocket or in drain oil	Relatively insensitive as majority of heat is carried away in oil that passes through bearing contact and this is rapidly cooled by excess oil that is fed to bearing. Can be useful in commissioning or checking replacements	Normal design 60°C Acceptable limit 80°C
Rolling bearing	Thermocouple or thermometer in contact with outer race (inner race rotating)	Two failure mechanisms cause temperature rise(2)	
		(*a*) breakdown of lubrication	Slow rise of temperature from steady value is indicative of deterioration of lubrication: Alarm at 10°C rise. Acceptable limit 100°C(3)
		(*b*) loss of internal clearance	Failure occurs so rapidly that there is insufficient time for warning of failure to be obtained from temperature indication of outer race
	Thermometer in oil		Acceptable limit 100°C
Gears	Thermometer in oil		Acceptable limit 80°C above ambient
Metallic packing	Contact thermometer on rod		Acceptable limit 80°C

(1) Temperature rise above normal value is more useful as an indication of trouble than the absolute value. The more the running value is below the acceptable limit the greater the margin of safety.
(2) Failure by fatigue or wear of raceways does not give temperature rise. They may be detected by an increase in noise level.
(3) Temperature in grease-packed bearing will rise to peak value until grease clears into housing and then fall to normal running value. Peak value may be 10–20°C above normal and attainment of equilibrium may take up to six hours. With bearing with grease relief valve a similar cycle will occur on each re-lubrication.
(4) Running-in. Higher than normal temperatures may occur during the initial running. Equilibrium temperatures can be expected after about twenty-four hours. The acceptable limits given should not be exceeded; if the limit is reached the machine should be stopped and and allowed to cool before proceeding with the run-in.
(5) Care must be taken to avoid deforming the bearing surface as this will result in a falsely high reading.

PRINCIPLES

Vibration analysis uses vibration measurements taken at an accessible position on a machine, and analyses these measurements in order to infer the condition of moving components inside the machine.

Table 13.1 The generation and transmission of vibration

The signal	*Mechanism*	*Examples*
Generation of the signal	The mass centres of moving parts move during machine operation, generally in a cyclic manner. This gives rise to cyclic force variations.	Unbalanced shafts. Bent shafts and resonant shafts Rolling elements in rolling bearings moving unevenly. Gear tooth meshing cycles. Loose components. Cyclic forces generated by fluid interactions.
Transmission of the signal	From the moving components via their supporting bearing components to the machine casing	Ideally, there should be a relatively rigid connecting path between the area where the vibration is transmitted internally to the machine casing, and the points on the outside of the machine where the measurement is taken.
Transmission problems	If the moving parts are very light and the machine casing is very heavy and rigid, the signal measured externally may be too small for accurate analysis and diagnosis.	High speeds rotors in high pressure machines with rigid barrel casings can have this problem. A solution is to take a direct measurement of the cyclic movement of the shaft, relative to the casing at its supporting bearings.

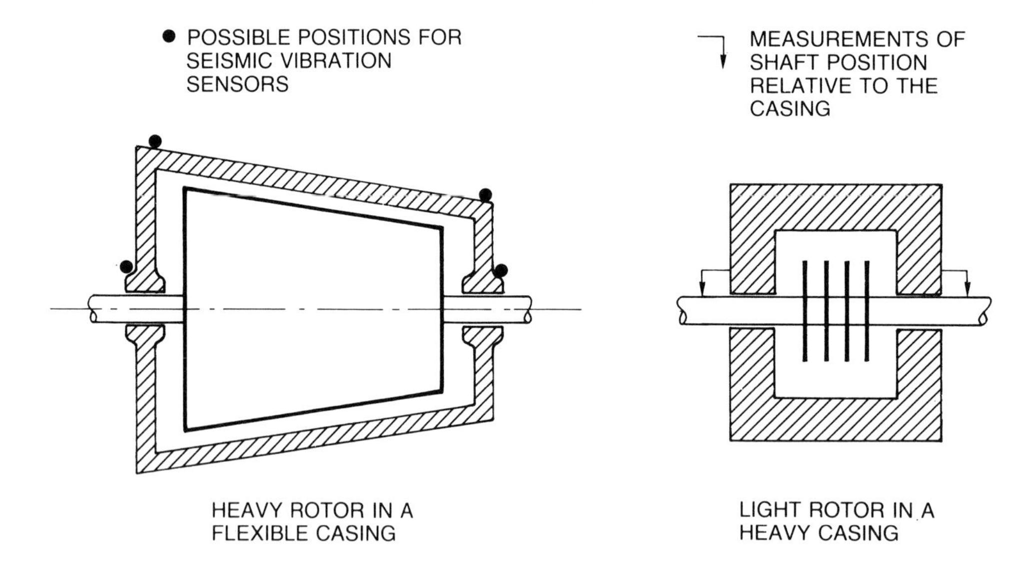

Figure 13.1 Vibration measurements on machines

Table 13.2 Categories of vibration measurement

Measurement	*The principle behind the technique*	*Applications*
Overall level of vibration (see subsequent section)	The general level of vibration over a wide frequency band. It determines the degree to which the machine may be running roughly. It is a means of quantifying the technique of feeling a machine by hand.	All kinds of rotating machines but with particular application to higher speed machines Not usually applicable to reciprocating machines
Spectral analysis of vibration (see subsequent section on vibration frequency monitoring)	The vibration signal is analysed to determine any frequencies where there is a substantial component of the vibration level. It is equivalent to scanning the frequency bands on a radio receiver to see if any station is transmitting.	From the value of the frequencies where there is a signal peak, the likely source of the vibration can be determined. Such a frequency might be the rotational speed of a particular shaft, or the tooth meshing frequency of a particular pair of gear wheels.
Discrete frequency monitoring	A method of monitoring a particular machine component by measuring the vibration level generated at the particular frequency which that component would be expected to generate	If a particular shaft in a machine is to be examined for any problems, the monitoring would be tuned to its rotational speed.
Shock pulse monitoring	Using a vibration probe, with a natural resonant frequency that is excited by the shocks generated in rolling element bearings, when they operate with fatigue pits in the surfaces of their races	The monitoring of rolling element bearings with a simple hand held instrument
Kurtosis measurement	This is a technique that looks at the 'spikyness' of a vibration signal, i.e. the number of sharp peaks as distinct from a smoother sinusoidal profile.	The monitoring of fatigue development in rolling bearings with a simple portable instrument, that is widely applicable to all types and sizes of bearing
Signal averaging (see subsequent section)	The accumulation over a few seconds of the parts of a cyclic vibration signal, which contain a particular frequency. Parts of the signal at other frequencies are averaged out. By matching the particular frequency to, for example, the rotational speed of a particular machine component, the resulting diagram will show the characteristics of that component.	The monitoring of a gear by signal averaging, relative to its rotational speed, will show the cyclic action of each tooth. A tooth with a major crack could be detected by its increased flexibility.
Cepstrum analysis	If two vibration frequencies are superimposed in one signal, sideband frequencies are generated on either side of the higher frequency peak, with a spacing related to that of the lower frequency involved. Cepstrum analysis looks at these sidebands in order to understand the underlying frequency patterns and their relative effects.	Interactions between the rotational frequency of bladed rotors and the blade passing frequency Also between gear tooth meshing frequencies and gear rotational speeds

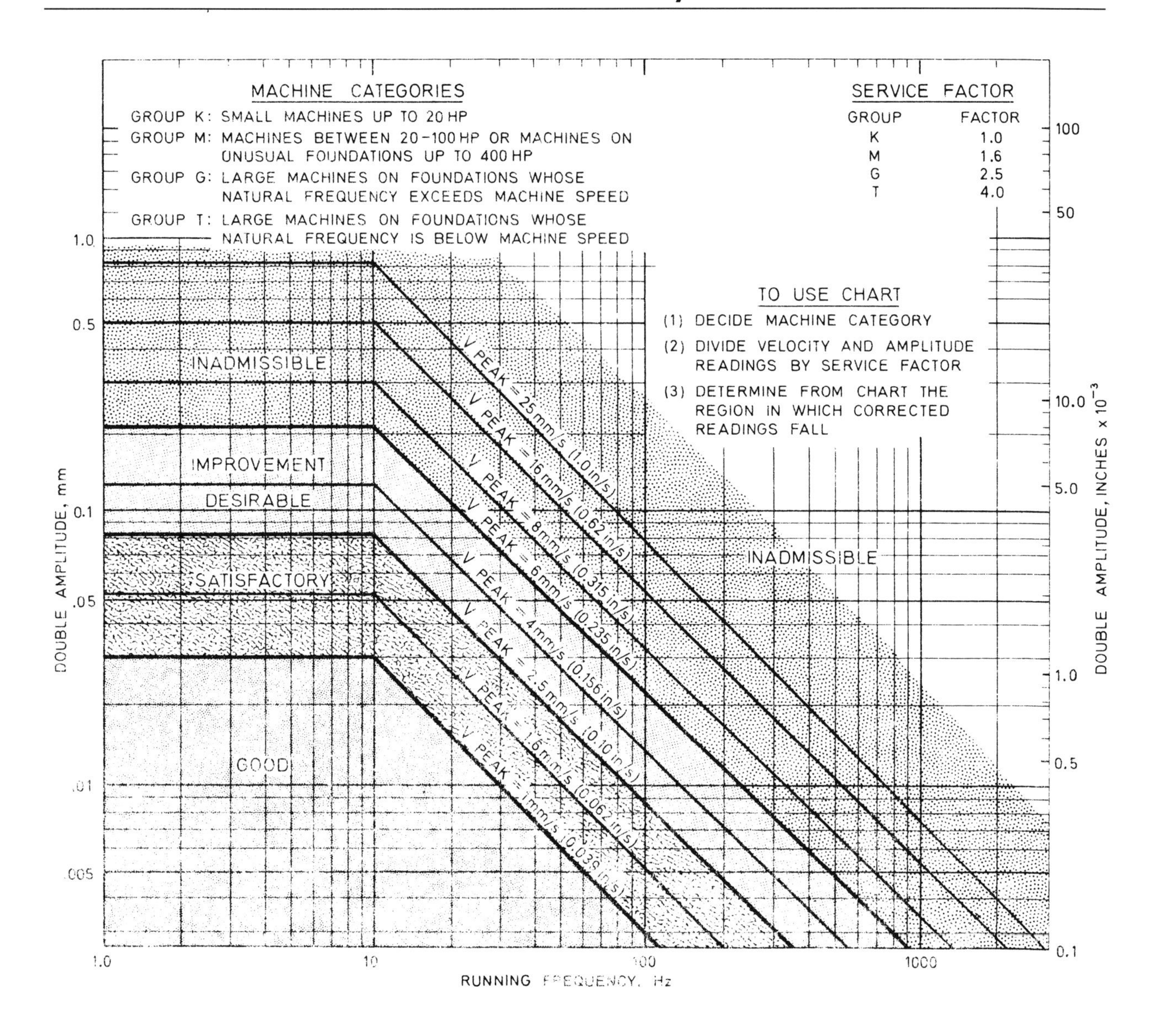

Figure 13.2 Guidance on the levels of overall vibration of machines

OVERALL LEVEL MONITORING

This is the simplest method for the vibration monitoring of complete machines. It uses the cheapest and most compact equipment. It has the disadvantage however that it is relatively insensitive, compared with other methods, which focus more closely on to the individual components of a machine.

The overall vibration level can be presented as a peak to peak amplitude of vibration, as a peak velocity or as a peak acceleration. Over the speed range of common machines from 10 Hz to 1000 Hz vibration velocity is probably the most appropriate measure of vibration level. The vibration velocity combines displacement and frequency and is thus likely to relate to fatigue stresses.

The normal procedure is to measure the vertical, horizontal and axial vibration of a bearing housing or machine casing and take the largest value as being the most significant.

As in all condition monitoring methods, it is the trend in successive readings that is particularly significant. Figure 13.2, however, gives general guidance on acceptable overall vibration levels allowing for the size of a machine and the flexibility of its mounting arrangements.

For machine with light rotors in heavy casings, where it is more usual to make a direct measurement of shaft vibration displacement relative to the bearing housing, the maximum generally acceptable displacement is indicated in the following table.

Table 13.3 Allowable vibrational displacements of shafts

Ratio *Vibration displacement / Diametral clearance*	*Speed rev/min*
0.5	300
0.25	3000
0.1	12000

VIBRATION FREQUENCY MONITORING

The various components of a machine generate vibration at characteristic frequencies. If a vibration signal is analysed in terms of its frequency content, this can give guidance on its source, and therefore on the cause of any related problem. This spectral analysis is a useful technique for problem diagnosis and is often applied, when the overall level of vibration of a machine exceeds normal values.

In spectral analysis the vibration signal is converted into a graphical plot of signal strength against frequency as shown in Figure 13.3, in this case for a single reduction gearbox.

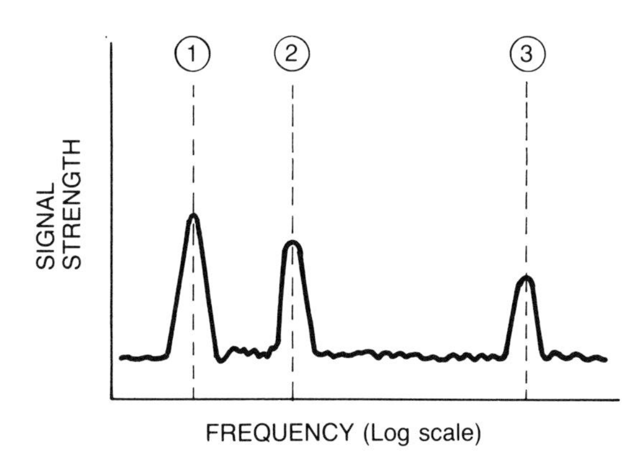

Figure 13.3 The spectral analysis of the vibration signal from a single reduction gearbox.

In Figure 13.3 there are three particular frequencies which contribute to most of the vibration signal and, as shown in Figure 13.4, they will usually correspond to the shaft speeds and gear tooth meshing frequencies.

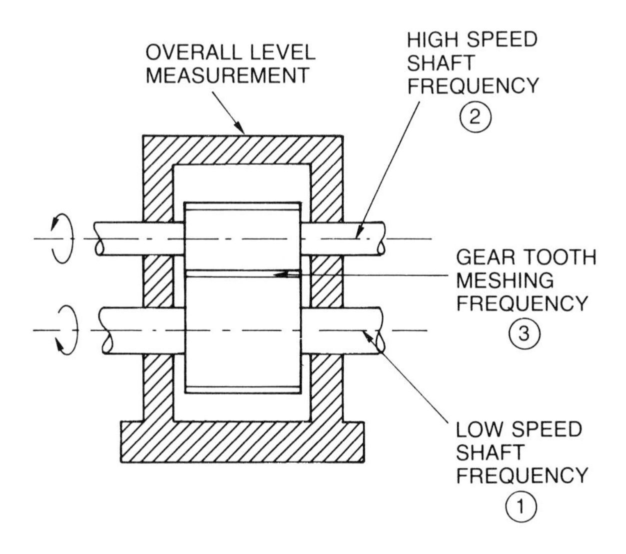

Figure 13.4 An example of the sources of discrete frequencies observable in a spectral analysis

Discrete frequency monitoring

If it is required to monitor a particular critical component the measuring system can be turned to signals at its characteristic frequency in order to achieve the maximum sensitivity. This discrete frequency monitoring is particularly appropriate for use with portable data collectors, particularly if these can be preset to measure the critical frequencies at each measuring point. The recorded values can then be fed into a base computer for conversion into trends of the readings with the running time of the machine.

Table 13.4 Typical discrete frequencies corresponding to various components and problems

Component/problem	*Frequency*	*Characteristics*
Unbalance in rotating parts	Shaft speed	Tends to increase with speed and when passing through a resonance such as a critical speed
Bent shaft	Shaft speed	Usually mainly axial vibration
Shaft misalignment	Shaft speed or 2 × shaft speed	Often associated with high levels of axial vibration
Shaft rubs	Shaft speed and 2 × shaft speed	Can excite higher resonant frequencies. May vary in level between runs.
Oil film whirl	0.45 to 0.5 × shaft speed	Only on machines with lubricated sleeve bearings
Gear tooth problems	Tooth meshing frequency	Generally also associated with noise
Reciprocating components	Running speed and 2 × running speed.	Inherent in reciprocating machinery
Rolling element bearing fatigue damage	Shock pulses at high frequency	Caused by the rolling elements hitting the fatigue pits
Cavitation in fluid machines	High frequency similar to shock pulses	Can be mistaken for rolling element bearing problems

SIGNAL AVERAGING

If a rotating component carries a number of similar peripheral sub-units, such as the teeth on a gear wheel or the blades on a rotor which interact with a fluid, then signal averaging can be used as an additional monitoring method.

A probe is used to measure the vibrations being generated and the output from this is fed to a signal averaging circuit, which extracts the components of the signal which have a frequency base corresponding to the rotational speed of the rotating component which is to be monitored. This makes it possible to build up a diagram which shows how the vibration forces vary during one rotation of the component. Some typical diagrams of this kind are shown in Figure 13.5 which indicates the contribution to the vibration signal that is made by each tooth on a gear. An outline of the technique for doing this is shown in Figure 13.6.

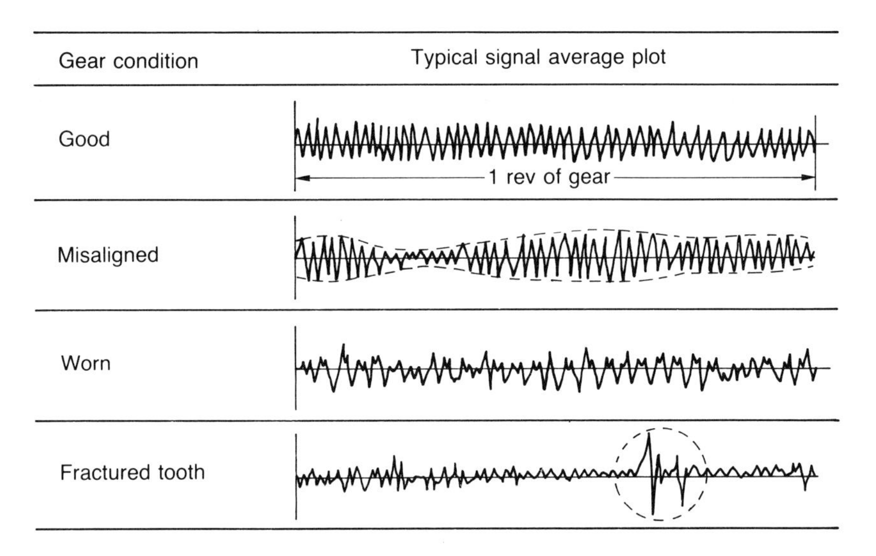

Figure 13.5 Signal average plots used to monitor a gear and showing the contribution from each tooth

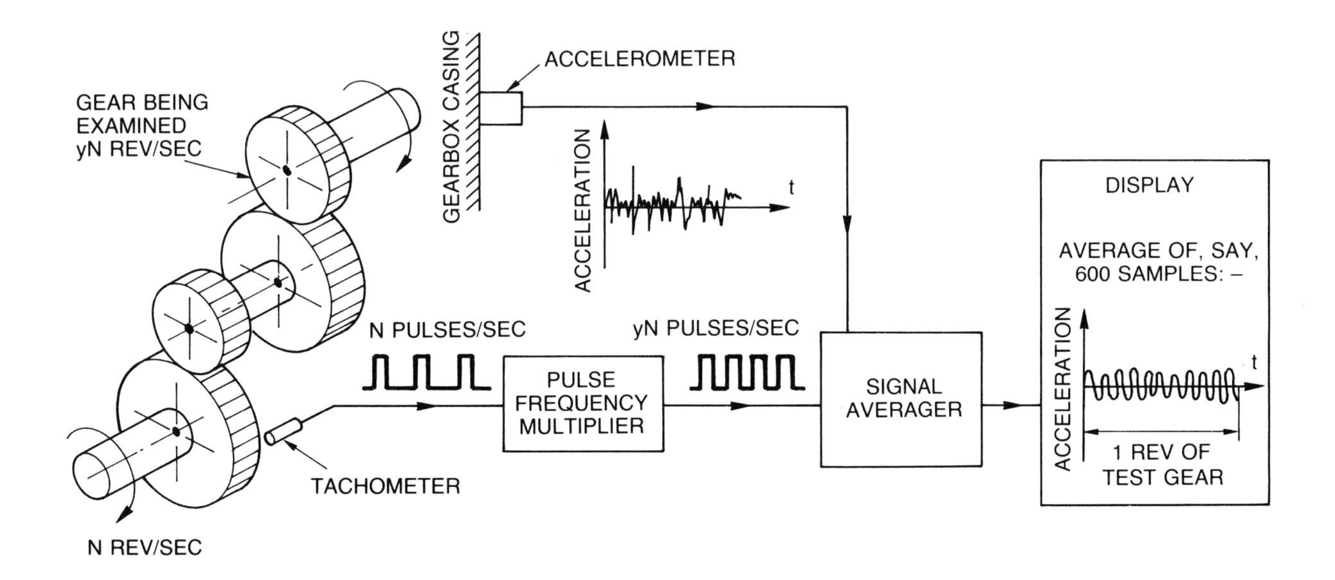

Figure 13.6 A typical layout of a signal averaging system for monitoring a particular gear in a transmission system

In wear debris analysis machine lubricants are monitored for the presence of particles derived from the deterioration of machine components. The lubricant itself may also be analysed, to indicate its own condition and that of the machine.

WEAR DEBRIS ANALYSIS

Table 14.1 Wear debris monitoring methods

Type	*Mechanism of operation*
IN LINE	
Monitoring the main flow of oil through the machine	Magnetic plugs or systems which draw ferro-magnetic particles from the oil flow for inspection, or on to an inductive sensor, that produces a signal indicating the mass of material captured
	Inductive systems using measuring coils to assess the amount of ferrous material in circulation
	Measurement of pressure drop across the main full flow filter
ON LINE	
Monitoring a by-passed portion of the main oil flow	Optical measurement of turbidity as an indicator of particle concentration
	Pressure drop across filters of various pore sizes to indicate particle size distribution
	Discoloration of a filter strip after the passage of a fixed sample volume
	Resistance change between the grid wires of a filter to indicate the presence of metallic particles
OFF LINE	
Extracting a representative sample from the oil volume and analysing it remotely from the machine	Spectrometric analysis of the elemental content of the wear debris in order to determine its source
	Magnetic gradient separation of wear particles from a sample to determine their relative size, as a measure of problem severity
	Microscopic examination of the shape and size of the particles to determine the wear mechanisms involved
	Inductive sensor to give a direct numerical measurement of the level of ferrous debris in a sample of oil
	Optical particle counting on a diluted oil sample

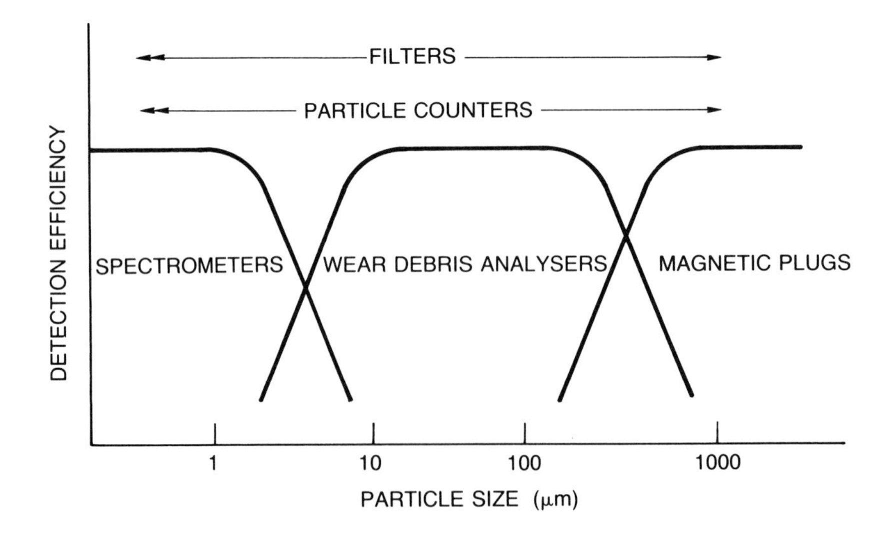

Figure 14.1 The relative efficiency of various wear debris monitoring methods

Table 14.2 Off-line wear debris analysis techniques

Technique	*Description*	*Application*
Atomic absorption spectroscopy	Oil sample is burnt in a flame and light beams of a wave length characteristic of each element are passed through the flame. The amount of light absorbed is a measure of the amount of the element that is present in the oil sample	Detects most common engineering metals. Detects particles smaller than 10μm only. Accurate at low concentrations of less than 5 ppm.
Atomic emission spectroscopy	Oil sample is burnt in an electric arc and the spectral colours in the arc are analysed for intensity by a bank of photomultipliers. Gives a direct reading of the content of many elements in the oil.	Detects most common engineering metals. Detects wear particles smaller than 10μm only. Accuracy is poor below 5 ppm.
Ferrography	A diluted oil sample is flowed across a glass slide above a powerful magnet. The particles deposit out on the slide with a distribution related to their size.	The size distribution can indicate the severity of the wear. The particle shapes indicate the wear mechanism.
Rotary particle depositor	A diluted oil sample is placed on a glass cover above rotating magnets. Wear particles are deposited as a set of concentric rings.	As for ferrography with the added advantage that it can be linked directly to a particle quantifier
X-ray fluorescence	When the sample is exposed to a radioactive beam, X-rays characteristic of the material content are emitted.	Detects most engineering metals. Accurate to only ±5 ppm.
Inductively coupled plasma emission spectroscopy	The oil sample is sprayed into an argon plasma torch. The spectral colours of the emitted light and its intensity are then measured to indicate the amount of various elements that are present.	Detects most engineering metals. Accurate down to parts per billion.

Table 14.3 Problems with wear debris analysis

Problem	*Solution*
Poor quality or variable samples	Ensure that sample bottles are clean and properly labelled. Use sampling valves or suction syringes. Hydraulic fluid sampling methods defined in ISO 3722.
Unrecorded oil change or a large oil addition	Often indicated by a sudden drop in contaminant levels. The importance of recording oil top-ups needs to be emphasised to the operators.
Addition of the wrong oil to the machine detected by an increase of elements commonly used in oil additives	Take a second sample and discuss the problem with the machine operator in order to avoid a recurrence

Table 14.4 Sources of materials found in wear debris analysis

Material	*Likely worn component or other source*	*Material*	*Likely worn component or other sources*
Aluminium	Light alloy pistons Aluminium tin crankshaft bearings Components rubbing on aluminium casings	Lead Magnesium	Plain bearings Wear of plastic components with talc fillers Seawater intrusion
Antimony	White metal plain bearings	Nickel	Valve seats Alloy steels
Boron	Coolant leaks Can be present as an oil additive	Potassium	Coolant leaks
Chromium	Piston rings or cylinder liners Valve seats	Silicon Silver	Mineral dust intrusion Silver-plated bearing surfaces Fretting of silver soldered joints
Cobalt	Valve seats Hard coatings		
Copper	Copper-lead or bronze bearings Rolling element bearing cages	Sodium Tin	Coolant leakage Seawater intrusion Plain bearings
Indium	Crankshaft bearings	Vanadium	Intrusion of heavy fuel oil
Iron	Gears Shafts Cast iron cylinder bores	Zinc	A common oil additive

Table 14.5 Quick tests for metallic debris from filters

Metal	*Test method*	*Method*	*Test method*
Steel and nickel	Can be attracted by a permanent magnet	Silver	Dissolves to form a white fog in nitric acid
Tin	Fuses with tin solder on a soldering iron	Copper and Bronze	Dissolves to produce a blue/green cloud in nitric acid
Aluminium	Dissolves rapidly in sodium or potassium hydroxide solution to form a white cloud	Chromium	Dissolves to produce a green cloud in hydrochloric acid

Physical characteristrics of wear debris

Rubbing wear

The normal particles of benign wear of sliding surfaces. Rubbing wear particles are platelets from the shear mixed layer which exhibits super-ductility. Opposing surfaces are roughly of the same hardness. Generally the maximum size of normal rubbing wear is 15μm.

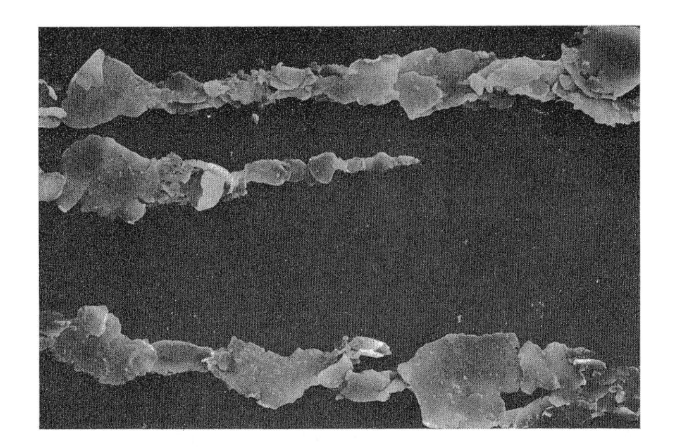

Break-in wear particles are typical of components having a ground or machined surface finish. During the break-in period the ridges on the wear surface are flattened and elongated platelets become detached from the surface often 50μm long.

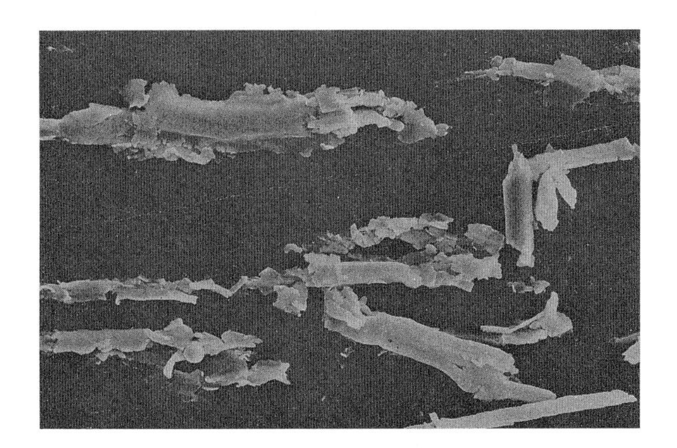

Cutting wear

Wear particles which have been generated as a result of one surface penetrating another. The effect is to generate particles much as a lathe tool creates machining swarf. Abrasive particles which have become embedded in a soft surface, penetrate the opposing surface generating cutting wear particles. Alternatively a hard sharp edge or a hard component may penentrate the softer surface. Particles may range in size from 2–5μm wide and 25 to 100μm long.

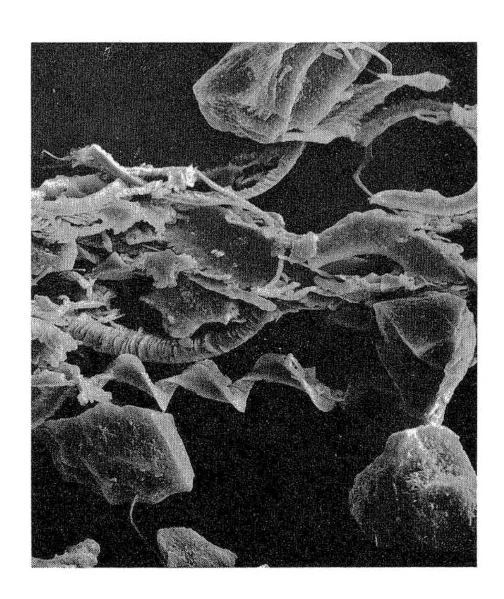

Rolling fatigue wear

Fatigue spall particles are released from the stressed surface as a pit is formed. Particles have a maximum size of 100μm during the initial microspalling process. These flat platelets have a major dimension to thickness ratio greater than 10:1.

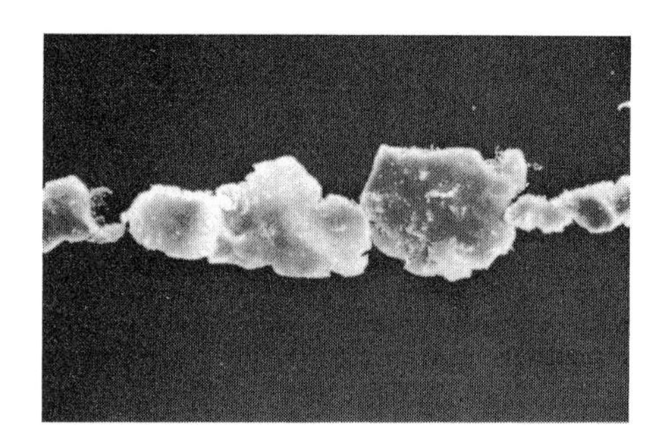

Spherical particles associated with rolling bearing fatigue are generated in the bearing fatigue cracks. The spheres are usually less than 3μm in diameter.

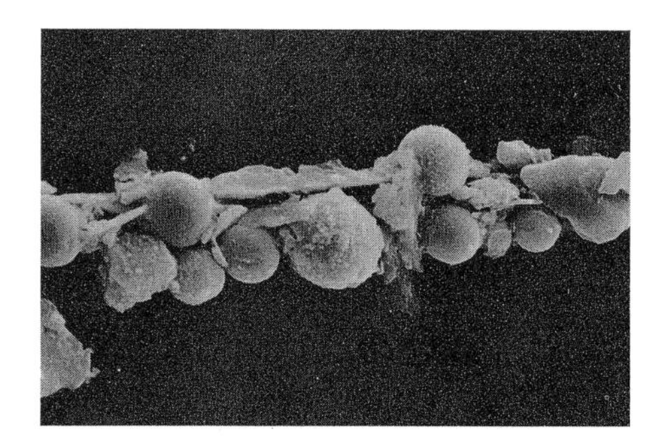

Laminar particles are very thin free metal particles between 20–50μm major dimension with a thickness ratio approximately 30:1. Laminar particles may be formed by their passage through the rolling contact region.

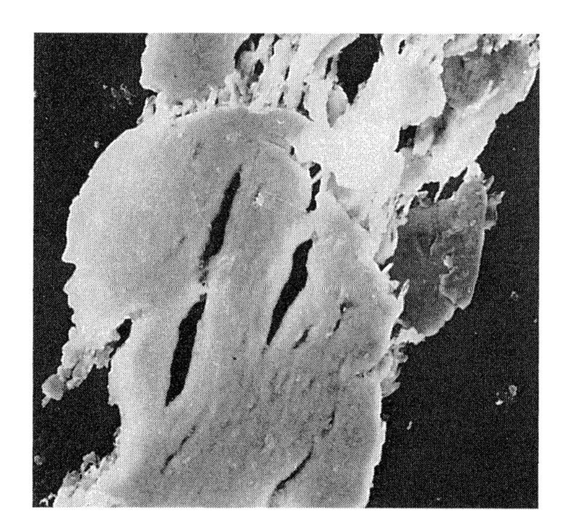

Combined rolling and sliding (gear systems)

There is a large variation in both sliding and rolling velocities at the wear contacts; there are corresponding variations in the characteristics of the particles generated. Fatigue particles from the gear pitch line have similar characteristics to rolling bearing fatigue particles. The particles may have a major dimension to thickness ratio between 4:1 and 10:1. The chunkier particles result from tensile stresses on the gear surface causing fatigue cracks to propagate deeper into the gear tooth prior to pitting. A high ratio of large (20μm) particles to small (2μm) particles is usually evident.

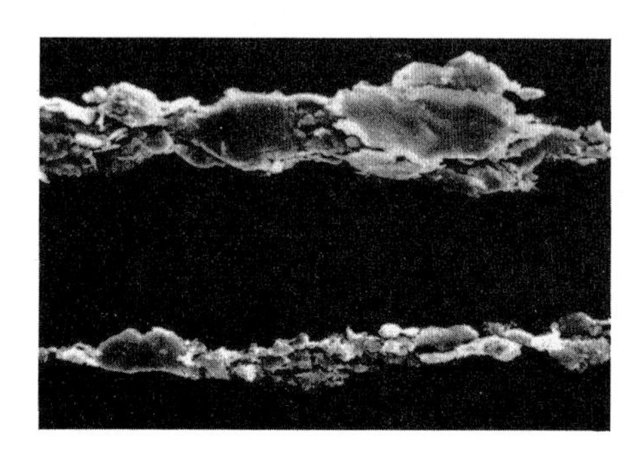

Severe sliding wear

Severe sliding wear particles range in size from 20μm and larger. Some of these particles have surface striations as a result of sliding. They frequently have straight edges and their major dimension to thickness ratio is approximately 10:1.

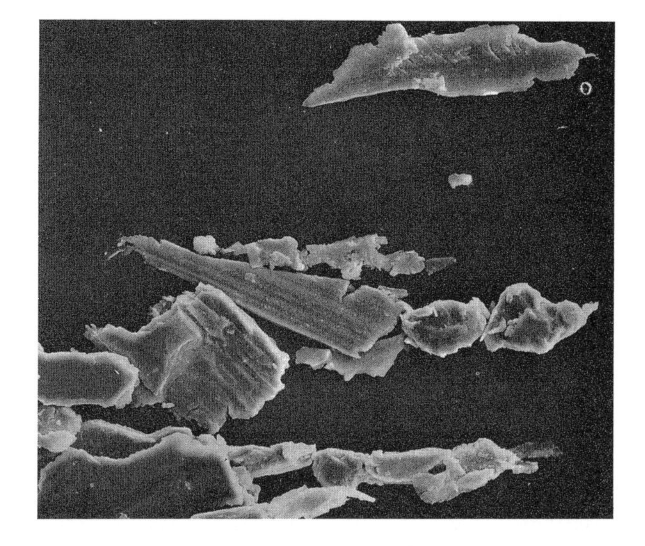

Crystalline material

Crystals appear bright and changing the direction of polarisation or rotating the stage causes the light intensity to vary. Sand appears optically active under polarised light.

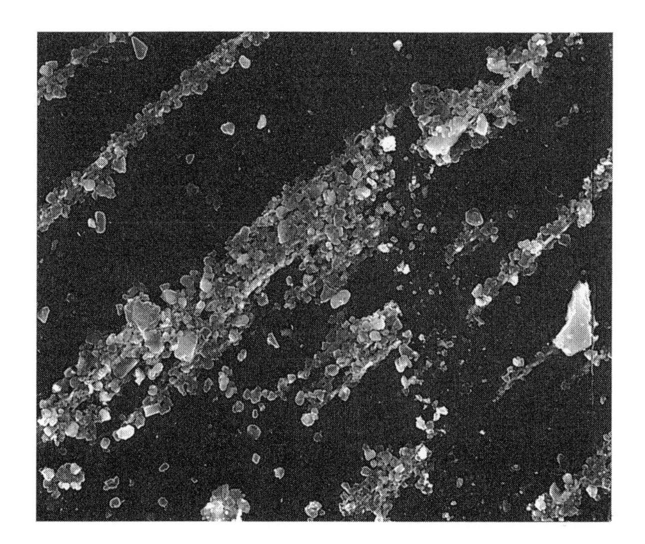

Weak magnetic materials

The size and position of the particles after magnetic separation on a slide indicates their magnetic susceptibility. Ferromagnetic particles (Fe, Co, Ni) larger than 15μm are always deposited at the entry or inner ring zone of the slide. Particles of low susceptibility such as aluminium, bronze, lead, etc, show little tendency to form strings and are deposited over the whole of the slide.

Polymers

Extruded plastics such as nylon fibres appear very bright when viewed under polarised light.

Examples of problems detected by wear debris analysis

Crankshaft bearings from a diesel engine

Rapid wear of the bearings occurred in a heavy duty cycle transport operation. The copper, lead and tin levels relate to a combination of wear of the bearing material and its overlay plating.

Sample no.	*1*	*2*	*3*	*4*
Iron ppm	35	40	42	40
Copper ppm	15	25	35	45
Lead ppm	20	28	32	46
Tin ppm	4	4	10	15

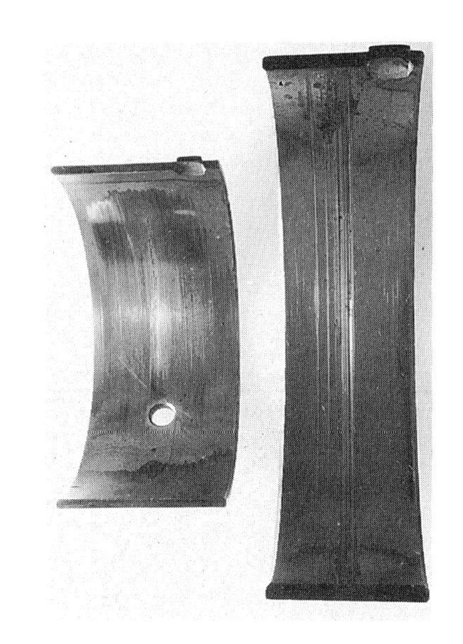

Grease lubricated screwdown bearing

The ratio of chromium to nickel, corresponding broadly to that in the material composition, indicated severe damage to the large conical thrust bearing.

Sample no.	*1*	*2*	*3*	*4*
Iron ppm	150	240	1280	1540
Chromium ppm	1	2	11	31
Nickel ppm	2	4	23	67

Differential damage in an Intercity bus

Excessive iron and the combination of chromium and nickel resulted from the disintegration of a nose cone bearing

Sample no.	*1*	*2*	*3*	*4*
Iron ppm	273	383	249	71000
Chromium ppm	2	3	2	21
Nickel ppm	0	1	0	5

Large journal bearing in a gas turbine pumping installation

The lead based white metal wore continuously.

Sample no.	*1*	*2*	*3*	*4*
Iron ppm	0	2	2	2
Copper ppm	4	5	11	15
Lead ppm	10	24	59	82

Piston rings from an excavator diesel engine

Bore polishing resulted in rapid wear of the piston rings. The operating lands of the oil control rings were worn away.

Sample no.	*1*	*2*	*3*	*4*
Iron ppm	5	5	9	10
Chromium ppm	15	25	40	120

Engine cylinder head cracked

The presence of sodium originates from the use of a corrosion inhibitor in the cooling water. A crack was detected in the cylinder head allowing coolant to enter the lubricant system.

Sample no.	*1*	*2*	*3*	*4*
Iron ppm	60	84	104	203
Chromium ppm	7	12	12	53
Nickel ppm	7	10	12	27
Sodium ppm	19	160	330	209

LUBRICANT ANALYSIS

Table 14.6 Off-line lubricant analysis techniques

Technique	*Description*	*Interpretation*
Viscosity measurement	Higher viscosity than a new sample Lower viscosity than a new sample	Oxidation of the oil and/or heavy particulate contamination Fuel dilution in the case of engine oils
Total acid number, TAN ASTM D974 D664 IP 139 177	Indicates the level of organic acidity in the oil	A measure of oil oxidation level For hydraulic oils the value should not exceed twice the level for the new oil
Strong acid number, SAN ASTM D974 D664 IP 139 177	Indicates the level of strong inorganic acidity in the oil	In engine oils indicates the presence of sulphur-based acids from fuel combustion. Synthetic hydraulic oils can show high values if they deteriorate
Total base number, TBN ASTM D664 D2896 IP 177 276	Indicates the reserves of alkalinity present in the oil	Running engines on higher sulphur fuels creates acids which are neutralised by the oil, as long as it continues to be alkaline
Infra-red spectroanalysis	Measures molecular compounds in the oil such as water, glycol, refrigerants, blow by gases, liquid fuels, etc. Also additive content.	A very versatile monitoring method for lubricant condition

Table 14.7 Analysis techniques for the oil from various types of machine

** essential * useful	*Diesel engine*	*Gasoline engine*	*Gears*	*Hydraulic systems*	*Air compressor*	*Refrigeration compressor*	*Gas turbine*	*Steam turbine*	*Trans-formers*	*Heat transfer*
Spectro-chemical analysis	**	**	**	**	**	**	**	**		
Infra-red analysis	**	**	*	*	*	**	*	*	*	*
Wear debris quantifier	**	**	**	*	*	*	*	*		
Viscosity at 40°C	**	**	**	**	**	**	**	**		*
Viscosity at 100°C	*	*								
Total base number	**	**								
Total acid number			*	**	*	*	**	*	*	**
Water %	**	**	*	**	*			**	**	
Total solids	*	*							*	*
Fuel dilution	**	**								
Particle counting				**			*			

A useful condition monitoring technique is to check the performance of components, and of complete machines and plant, to check that they are performing their intended function correctly.

COMPONENT PERFORMANCE

The technique for selecting a method of monitoring is to decide what function a component is required to perform and then to consider the various ways in which that function can be measured.

Table 15.1 Methods of monitoring the performance of fixed components for fault detection

Component	*Function*	*Monitoring method*
Casings and frameworks	Rigid support and transfer of loads to foundations	Crack detection by: Visual inspection Dye penetrants Ultrasonic tests Eddy current probes Magnetic flux Radiography Tests for deflection under a known applied load. Visual checks for material loss by corrosion
Cold pressure vessels	The containment of fluid under pressure	Detection of external surface cracks by: Visual inspection Dye penetrants Eddy current probes Magnetic flux Detection of internal cracks by: Ultrasonic tests Radiography Boroscopes (when out of service) Detection of loss of wall thickness by corrosion: Ultrasonic tests Electrochemical probes Sacrificial coupons Small sentinel holes (where permissible) Detection of strain growth by acoustic emission
Boilers and thermal reactors	The heating of fluids and containment of pressure	Detection of surface cracks and leakage by visual inspection Detection of hidden cracks by ultrasonic tests Hammer testing of the shell Chemical check of feed water and boiler water samples, to indicate likely corrosion or deposit build up
Nozzle blades	The profiled flow of fluids and transmission of forces	Checking for profile changes and integrity by boroscopes (when out of service)
Ducts	Guiding the flow of air or other gases	Detection of leaks by gas sniffer detectors If gas is hot and carrying fine solids, partial blockages can be detected by: Infrared thermography Thermographic paints
Pipes	Guiding the flow of fluids under pressure	Checking for reduction in wall thickness by: Ultrasonics Corrosion coupons Electrochemical probes Sentinel holes

Table 15.2 Methods of monitoring the performance of moving components for fault detection

Component	*Function*	*Monitoring method*
Journal bearings	Locating rotating shafts radially with the minimum friction	Checking low friction operation by temperature measurement
Thrust bearings	Locating rotating shafts axially with the minimum friction	Checking low friction operation by temperature measurement. Checking location by shaft position probes
Shafts and couplings	Smooth rotation while transmitting torques	Stroboscopes for visual inspection while rotating
Seals	Allowing rotating shafts to enter pressurised fluid containments with minimum leakage	Checking for leakage: Visually via deposits Gas sniffer detectors Liquid leakage pools
Pistons and cylinders	The interchange of fluid pressure and axial force with minimum leakage	Listening for leakage at ultrasonic frequencies
I.C. engine combustion	The provision of regular power pulses to drive the crankshaft	Toothed wheel fitted to the free end of the crankshaft to check for even rotation and pulses
Belt drives	The smooth transmission of power by differential tension	Stroboscopes for visual inspection while rotating. Proximity probes to detect excess slackness
Springs	Allowing controlled deflection with increasing loads	Deflection measurement at known loads. Crack detection by visual inspection and dye penetrants

Table 15.3 Methods of monitoring the performance of machines and systems for fault detection

Machine/System	*Function*	*Monitoring method*
Hydraulic power systems	Converting fluid pressure and flow into mechanical power	Measuring the relationship between pressure and flow in the system
Pumps	Converting mechanical power into fluid pressure and flow	Measuring the relationship between delivery pressure and flow, to detect any deterioration in the internal components
Textile machines	The production of fabric from threads	Checking the fabric for any patterns which indicate inconsistencies in machine operation
Coal or ore crushing mills	Reducing the particle size of materials	Checking changes in the size distribution of the output material
Motor vehicle	Consuming fuel to provide transportation	Measuring the distance covered per unit of fuel consumption
Heat exchangers	Producing temperature changes between two fluid flow streams	Monitoring the relationship between flow rate and temperature change
Thermodynamic and chemical process systems	Using pressure, temperature and volume/flow changes to interchange energy or change materials	The measurement and comparison against datums of pressure, temperature and flow relationships in the system
Systems with automatic control	The control of system variables to obtain a required output	The monitoring of the control actions taken by the system in normal operation to determine extreme values or combinations which indicate a system fault. Giving the system a control exercise designed to detect likely faults

BALL AND ROLLER BEARINGS

If there is evidence of pitting on the balls, rollers or races, suspect fatigue, corrosion or the passage of electrical current. Investigate the cause and renew the bearing.

If there is observable wear or scuffing on the balls, rollers or races, or on the cage or other rubbing surfaces, suspect inadequate lubrication, an unacceptable load or misalignment. Investigate the cause and renew the bearing.

ALL OTHER COMPONENTS

Wear weakens components and causes loss of efficiency. Wear in a bearing may also cause unexpected loads to be thrown on other members such as seals or other bearings due to misalignment. No general rules are possible because conditions vary so widely. If in doubt about strength or efficiency, consult the manufacturer. If in doubt about misalignment or loss of accuracy, experience of the particular application is the only sure guide.

Bearings as such are considered in more detail below.

JOURNAL BEARINGS, THRUST BEARINGS, CAMS, SLIDERS, etc.

Debris

If wear debris is likely to remain in the clearance spaces and cause jamming, the volume of material worn away in intervals between cleaning should be limited to $\frac{1}{5}$ of the available volume in the clearance spaces.

Surface treatments

Wear must not completely remove hardened or other wear resistant layers.

Note that some bearing materials work by allowing lubricant to bleed from the bulk to the surface. No wear is normally detectable up to the moment of failure. In these cases follow the manufacturer's maintenance recommendations strictly.

Some typical figures for other treatments are:

Treatment	*Allowable wear depth*
Good quality carburising	2.5 mm (0.1 in)
Gas nitriding	0.25 mm (0.01 in)
Salt bath nitriding	None
Cyanide hardening (shallow)	0.025 mm (0.001 in)
Cyanide hardening (deep)	0.25 mm (0.01 in)
Graphite or MoS_2 films	None
White metal, etc.	up to 50% of original thickness

Surface condition

Roughening (apart from light scoring in the direction of motion) usually indicates inadequate lubrication, overloading or poor surfaces. Investigate the cause and renew the bearing.

Pitting usually indicates fatigue, corrosion, cavitation or the passage of electrical current. Investigate the cause. If a straight line can be drawn (by eye) across the bearing area such that 10% or more of the metal is missing due to pits, then renew the components.

Scoring usually indicates abrasives either in the lubricant or in the general surroundings.

Journal bearings with smoothly-worn surfaces

The allowable increase in clearance depends very much on the application, type of loading, machine flexibilities, importance of noise, etc., but as a general guide, an increase of clearance which more than doubles the original value may be taken as a limit.

Wear is gnerally acceptable up to these limits, subject to the preceding paragraphs and provided that more than 50% of the original thickness of the bearing material remains at all points.

Thrust bearings, cams, sliders, etc. with smoothly-worn surfaces

Wear is generally acceptable, subject to the preceding paragraphs, provided that no surface features (for example jacking orifices, oil grooves or load generating profiles) are significantly altered in size, and provided that more than 50% of the original thickness of the bearing material remains at all points.

CHAINS AND SPROCKETS

For effects of wear on efficiency consult the manufacturers. Some components may be case-hardened in which case data on surface treatments will apply.

CABLES AND WIRES

For effects of wear on efficiency consult the manufacturer. Unless there is previous experience to the contrary any visible wear on cables, wires or pulleys should be investigated further.

METAL WORKING AND CUTTING TOOLS

Life is normally set by loss of form which leads to unacceptable accuracy or efficiency and poor surface finish on the workpiece.

The surfaces of most components which have been worn, corroded or mis-machined can be built up by depositing new material on the surface. The new material may be applied by many different processes which include weld deposition, thermal spraying and electroplating.

The choice of a suitable process depends on the base material and the final surface properties required. It will be influenced by the size and shape of the component, the degree of surface preparation and final finishing required, and by the availability of the appropriate equipment, materials and skills. The following tables give guidance on the selection of suitable methods of repair.

Table 17.1 Common requirements for all processes

Requirement	*Characteristic*
Re-surfacing should be cheaper than replacement with a new part	Larger components often tend to be the most suitable and usually can be repaired on site. Also, the design may be unique or changes may have rendered the part obsolete, making replacement impossible. If the worn area was not surfaced originally the replacement surface may give better performance than that obtained initially. Repair, and re-surfacing are conservationally desirable processes, saving raw materials and energy and often proving environmentally friendly.
Appropriate surface condition prior to coating	Whatever process is considered applicable the surfaces for treatment should be in sound metallurgical condition, i.e. free from scale, corrosion products and mechanically damaged or work-hardened metal. The part may have the remains of a previous coating and unless it is possible to apply the new coating on top of this the old coating must be removed, generally by machining or grinding, sometimes by grit-blasting, anodic dissolution or heat treatment. Surfaces which have been nitrided, carburised, etc., may possibly be modified by heat treatment, otherwise mechanical removal will be necessary.
Preparatory machining of the surface prior to coating	If the component has to be restored to its original dimensions it will need to be undercut to allow for the coating thickness. In many cases this undercutting should be no more than 250 μm/500 μm deeper than the maximum wear which can be tolerated on the new surface. With badly damaged parts it will be necessary to undercut to the extent of the damage and this may limit the choice of repair processes which could be used. For precision components which require finishing by machining it is important to preserve reference surfaces to ensure that the finished component is dimensionally correct.

Table 17.2 Factors affecting the choice of process

Factor	*Effect*
SIZE AND SHAPE Size and weight of component Thickness of deposit Weight to be applied Location of surfaces to be protected Accessibility	For small parts hand processes suitable. Spraying is a line-of-sight process limiting coating in bores, but some processes will operate in quite small diameters. Electroplating requires suitable anodes. Generally no limitation with weld deposition other than access of rod or torch in bores, however, some processes do not permit positional welding.
DESIGN OF ASSEMBLY Has brazing, welding, interference fits, riveting been used in the assembly? Are there temperature sensitive areas nearby? Have coaxial tolerances to be maintained?	Some processes put very little heat into an assembly so brazing filler metals would not be melted. Welds need careful cleaning and smoothing. Consideration must be given to the effect of heat on fitted parts. Suitable reference points are essential if tolerances are critical.
DISTORTION PROBLEMS What distortion is permissible? Is the part distorted as received? Will preparation cause distortion? Will surfacing cause distortion? Can distortion occur on cooling?	Eliminate distortion in part for repair. Remove cold-work stresses. Bond coats may replace grit-blasting. Some processes do not require surface roughening. Some processes provide more severe temperature gradients giving greater distortion. Allowance must be made for thermal expansion. Slow cooling must be available. In some cases pre-setting before coating may be used. Straightening or matching after coating may also be possible.
PROPERTIES OF BASE METAL Can chemical composition be determined? Have there been previous surface treatments/coating? What is surface condition? What is the surface hardness? What metallographic structure and base metal properties are ultimately required?	The depth of penetration and degree of surface melting can eliminate surface treatments, but previous removal may be necessary. If too hard for grit-blasting softening may be possible or a bond coat used. Suitable post-surfacing heat treatment may be needed.
DEPOSIT REQUIREMENTS Dimensional tolerances acceptable Surface roughness allowable/desirable Need for machining Further assembly to be carried out Additional surface treatments required, e.g. plating, sealing, painting, polishing Possible treatment of adjacent areas such as carburising Thermal treatments needed to restore base metal properties	Arc welding not suitable for thin coatings. To retain dimensions, a low or uniform heat input essential. If machining is needed a smooth deposit saves time and cost. Further assembly demands a dense deposit as do many surface treatments. Painting may benefit from some surface roughness, lubrication from porosity, some surfaces need to be smooth, low friction, others need frictional grip. These are some of the many factors needing consideration in selecting the process.
EFFECT ON BASE METAL PROPERTIES Consideration must be given to the effect of the process on base metal strength. Some sprayed coatings can reduce base metal fatigue strength as can hard, low ductility fused deposits in some environments. The same applies to some electro-deposited coats. Such coatings on high UTS steels can produce hydrogen embrittlement. Applied coatings do not usually contribute to base metal strength and allowance must be made for this, particularly where wear damage is machined out, prior to reclamation of the part. Welding will produce a heat affected zone which must be assessed. Fusing sprayed deposits can cause metallurgical changes in the base metal and the possibility of restoring previous properties must be considered. Unless coatings can be used as deposited, finish machining must be employed. Some coatings, e.g. flame sprayed or electro-deposited coatings must be finish ground, many dense adherent coatings may be machined. If porosity undesirable the coating must be sealed or densified.	
OTHER CONSIDERATIONS The number of components to be treated at one time or the frequency of repairs required, influence the expenditure permissible on surfacing equipment, handling devices, manipulators and other equipment. The re-surfacing of equipment in situ such as quarry equipment, steelmill plant, paper making machinery, military equipment and civil engineering plant may limit or dictate choice of process.	

Table 17.3 The surfacing processes that are available

Type	*Symbol*	*Process*	*Coating materials and their form*	*Energy source and other materials*	*Surface preparation and coating procedure*	*Coating thickness*	*Treatment after coating*
Gas Welding	GW	Gas welding	Rods, wires, tungsten carbide containing rods	Oxygen, acetylene	Flux application may be necessary if substrate contains elements forming refractory oxides, e.g. Cr	Up to 3 mm	Slow cooling Machining Grinding Flux removal
	PW	Powder welding	Self-fluxing alloy powder	Oxygen, acetylene. Possibly flux	Flux application to high Cr steels. Pre-heating in some cases	Up to 5 mm	Slow cooling Machining Grinding Filing softer deposits
Arc Welding AC and/or DC	MMA	Arc welding Manual (stick) welding	Bare rods, pre-alloyed wire electrodes or with alloying in the coating or tungsten carbide containing tubes	AC and/or DC Flux	May need pre-setting and/ or pre-heating. If wear excessive, preliminary build-up with low alloy steel economic	Thick coatings possible	Possibly slow cooling. If suitable heating treatment. If required straighten and/or machine. Generally flux/slag removal
	TIG	Tungsten inert gas	Bare rods or wire	Argon. Tungsten electrodes	May need pre-setting and/ or pre-heating. May use manipulator	2 mm upwards	Possibly slow cooling/heat treatment. Straightening, stress relieving, machining
	MIG	Metal inert gas, gas shielded arc	Rods, wires or tubes	Carbon dioxide	As TIG	3 mm upwards	As TIG
	FCA	Flux-cored arc	Tubes, cored wires		As TIG	3 mm upwards	As TIG, flux removal may be necessary
	SA	Submerged arc	Wires, strips	Granulated flux	As TIG. Will need use of an automatic manipulator	3 mm upwards	As TIG. Flux removal needed
	PTA	Plasma transferred arc	Metal or alloy powder	Argon, hydrogen and/or nitrogen, helium	As TIG. Use of manipulator/robot general	2–5 mm	As TIG
Plating	EP	Electroplating	Ingots, plates, wires	Electric DC. Controlled chemical solutions. Cr, Ni possibly Cu anodes	Generally smooth turning or grinding. Protect areas where plating not desired. If steel has tensile strength exceeding 1000 N/mm^2 or fatigue strength is important shot peen	Wide range	Heat treat to reduce coating stresses Machine, grind

Table 17.3 *(continued)*

Type	*Symbol*	*Process*	*Coating materials and their form*	*Energy source and other materials*	*Surface preparation and coating procedure*	*Coating thickness*	*Treatment after coating*
Thermal spraying/Metallising	MP	Powder spraying	Metal or ceramic powder	Oxygen, acetylene sometimes hydrogen. Compressed air	Grit-blasting with or without grooving, rough threading or application of bond coat. Mask area not to be blasted or sprayed. Warm to prevent condensation from flame. Heat to reduce contraction stresses or improve adhesion	2 mm upwards	May machine, impregnate, seal, use as sprayed or densify mechanically or by HIPping
	MW	Wire spraying	Metal wire, ceramic rods, powder/plastic wire	Oxygen, acetylene sometimes hydrogen. Compressed air	As MP	75 μm upwards	As MP
	MA	Arc spraying	Solid or tubular wires	Electric power. Compressed air	Generally as MP. Al bronze an additional bond coat. Adjustment of spray conditions may provide coarse adherent deposit as initial bond coat. Mechanical handling often used. Warming/heating not used	2 mm upwards	Use as sprayed, machine, seal
	HVOF	High velocity oxy/flame spraying	Metal powder, sometimes ceramic powder	Oxygen, composite gas, acetylene, hydrogen. Carrier gas Ar, He or N_2	As MP	Wide range	As MP
	PF	Plasma flame spraying	Metal powder, perhaps wire or rod	Electric power. Argon, hydrogen nitrogen helium	As MP	Up to 2 mm	As MP
Sprayed and fused	SW	Sprayed and fused coatings	Self-fluxing alloy powders, occasionally bonded into wire	Oxygen, acetylene, (hydrogen), compressed air. Pre-heating torches/burners	Use only chilled iron or nickel alloy grit which must be clean and free from fines. Do not use bond coats. If needed pre-set/pre-heat. Mask areas not to be blasted/sprayed. Often lathe mounted and thermal expansion must be accommodated	0.4–2.0 mm	Fuse. Slow cool. Possibly heat treat. Use as fused or machine

Table 17.4 Characteristics of surfacing processes

Process	*Advantages*	*Disadvantages*
Gas welding	Small areas built up easily with thick deposits, grooves and recesses can be filled accurately. Thin, smooth coatings can be deposited. Minimum melting of the parent metal is possible with low dilution of the surfacing alloy. This is advantageous if using highly alloyed consumables or if a thin coating only desired. The process is under close control by the operator. Equipment is inexpensive and requires only fuel gases and possibly flux. A wide range of consumables available. There is minimum solution of carbon/carbide granules from tubular rods. Self-fluxing alloys can be deposited without flux, except on highly allowed substrates, simplifying post-welding cleaning.	Process slow and not suitable for surfacing large areas. Build up of heat may overheat component and lead to distortion. Careful control of flame adjustment. As it is a manual process results are dependent on operator skill, fitness and degree of fatigue. Good technique is essential to ensure sound bond with the interface, especially if fusion with the substrate is not involved. There is a lack of NDT methods to check adhesion between coating and base metal.
Powder welding	Requires less skill than gas welding. Equipment not expensive but the self-fluxing powders are. The consumables have a wide temperature range between liquidus and solidus and between these temperatures the pasty consistency of the deposit enables thin, sharp edges and worn corners to be built up. Once an initial, thin coating has been built up high deposition rates of subsequent coatings are easily maintained. The ability to put down a thin first coating makes it possible to coat small areas on large components. Smooth or contoured deposits are possible requiring little finish machining. Soft deposits may be finished by hand filing. The process leaves one hand free to manipulate the work.	As for gas welding. Additionally there is a limited range of suitable consumables. It is essential that the interface temperature reaches 1000°C to ensure that the coating is bonded to and not just cast on the base metal and that there is enough heat and time for the self-fluxing action of the coating alloy to clean the surface of interfering oxides.
Manual metal arc welding	Equipment cost low, requiring very little maintenance. It is adaptable to small or large complex parts, can be used with limited access and positional welding possible, i.e. vertical. A wide range of consumables available. Deposition rates up to 5 $kghr^{-1}$. Ideal for one off and small series work and is useful when only small quantities of hardfacing alloys are required.	A skilled operator is needed for high quality deposits and slag removal is necessary. Dilution tends to be high. Granular carbides in tubular electrodes are usually melted.
TIG welding	The process can be closely controlled by the operator using hand-held torch and hand-held filler rod but can be mechanised for special applications. Small areas can be surfaced e.g. small pores in hard surfacing deposits. A deposit thickness of 2 mm upwards is achievable and a deposition rate up to 2 $kghr^{-1}$. High quality deposits can be made.	The process is slow and unsuitable for surfacing large areas. A limited range of consumables is available. The equipment is expensive and not suitable for site work, needing a workshop in which the shielding gas can be protected from disturbing draughts.
MIG welding	A continuous process which is used semi-automatically with a hand-held gun or is wholly mechanised by traversing the gun and/or the workpiece. Slag removal is not needed. It provides a positional surfacing capability and guns are available for internal bore work. High deposition rates 3–8 $kghr^{-1}$ with deposit thickness 3 mm upwards.	Equipment relatively expensive requiring regular maintenance. Use of shielding gas makes the process marginally less transportable than MMA and the gas must be selected to suit the surfacing alloy. Alloys for hard surfacing are not generally available in wire form so consumables restricted to mild steel (for build-up), stainless steels, aluminium bronze or tin bronze. High levels of UV radiation produced especially using high peak current pulse welding.
Flux-cored arc welding	Similar in principle to MIG surfacing but uses a tubular electrode containing a flux which decomposes to provide a shield to protect the molten pool. A separate shielding gas supply is not required. It is a continuous process used semi-automatically with a hand-held gun or is wholly mechanised. A wide range of consumables is available. High deposition rates up to 8 $kghr^{-1}$ for CO_2 shielded process and 11 $kghr^{-1}$ higher for self-shielded process.	With dilution of 15–30%, depending on technique, the process is not suitable for non-ferrous surfacing alloys. Equipment is fairly expensive and needs regular maintenance. Deposit quality may be lower than with MIG surfacing. Restrictions on use and transportation similar to the MIG process.

Table 17.4 *(continued)*

Process	*Advantages*	*Disadvantages*
Submerged arc	It is a fully automatic process providing high deposition rates 10 $kghr^{-1}$ upwards on suitable workpieces. Deposit thickness 3 mm and over. A wide range of consumables available giving high quality deposits of excellent appearance requiring minimum finishing. Slag removal easy.	Intended primarily for workshop use with a fixed installation the equipment is very expensive and needs regular maintenance. Applications are limited. Generally, to large cylindrical or flat components; there is limited access to internal surfaces or larger bores.
PTA surfacing	A mechanised process giving close control of surface profile with minimum finishing required. Penetration and dilution low. Deposit thickness in range 2–5 mm. Deposition rate higher than with TIG process: 3.5 $kghr^{-1}$. Torches available for surfacing small bores down to about 25 mm diameter up to 400 mm deep. Very useful for surfacing internal valve seats. Balancing the power input between the arc plasma and the transferred arc enables deposits to be made on a wide range of components varying greatly in thickness and base metal composition.	Equipment expensive and not readily portable. Process costs high requiring a very skilled operator.
Electroplating	As operating temperatures never exceed 100°C work should not distort or suffer undesirable metallurgical changes. Coatings dense and adherent to subtrate, molecular bonding may be as strong as 1000 N/mm^2. Plating conditions may be adjusted to modify hardness, internal stress and metallurgical characteristics of the deposits. No technical limits to thickness of deposits but most applications require thin coatings. Areas not requiring build-up can be masked. Brush plating can be used on localised areas, the equipment is portable and can be taken to the work. Deposition rates can exceed those of vat plating and may reach 200–400 $mmhr^{-1}$.	The thickness of deposit is proportional to current density and plating time, seldom exceeding 75 μmhr^{-1}. As current density over workpiece surface is seldom uniform coatings tend to be thicker at edges and corners and thinner in recesses and the centre of large flat areas. Although application of coatings is not confined to line-of-sight the ability to plate round corners may be limited, however anode design and location may assist. The size of the vat limits dimensions of the work. Brush plating is labour intensive and requires considerable skill. The electrolytes are expensive but small volumes generally only required.
Powder spraying	Low cost equipment which can be used by semi-skilled operators. The most useful flame spray process for high alloy and self-fluxing surfacing materials continuously fed to the pistol. Coatings can be provided of materials which cannot be produced as rods or wires. Spray rates relatively high. Can provide coatings of uniform, controlled thickness, ideal for large cylindrical components but also possible on small and/or irregular parts. Low heat input to the base material minimises distortion and adverse metallurgical changes. Machining if needed can be minimal.	Most suitable for workshop use with adequate dust extraction available. Requires rough surface preparation. A line-of-sight process with limited access to bores and limitations on deposit thickness. Provides only a mechanical bond to the substrate. Deposit porous and may need sealing or densification. Needs compressed air as well as combustion gases. Not suitable for ceramic spraying.
Wire spraying	Generally as with powder spraying but much more suitable for site work. Wire reels can be at considerable distance from the spraying torch making it possible to work inside large constructions. Spray rates are fast and a wide range of surfacing materials can be sprayed. The spraying can be mechanised. Large areas easily sprayed given adequate design of work station. Can give useful porosity, thin coatings (75–150 μm) easily applied. For corrosion resistance sealed Zn, Al, or ZnAl coatings generally better than multi-layer paint systems. Ceramic spraying possible using rods.	As with powder spraying but does not require controlled mesh size of the consumables.

Table 17.4 *(continued)*

Process	*Advantages*	*Disadvantages*
Arc spraying	Faster than gas spraying, providing extremely good bond and denser coatings. Either solid or tubular wires can be used giving a wide range of coating alloys. Using two different wires, composite or 'pseudo-alloy' coatings are possible with compositions/structures unobtainable by other means. Can be used for thick build-ups on cylindrical parts. Surface preparation not as critical as for flame spraying.	Equipment more expensive than for flame spraying. Weight of gun and great spraying rate makes mechanical handling desirable. Limited to consumables available in wire form. Density and adhesion lower than the plasma spraying or HVOF process.
High velocity oxyflame spraying	Produces dense, high bond strength coatings and thick, low stress coatings independent of coating hardness. Tight spray pattern allows accurate placement of the deposit. Low heat input to the component. Torches available for spraying internal surfaces down to 100 mm diameter. Some equipments can spray ceramics. Adhesion is high eliminating need for a bond coat.	More expensive than flame spraying the process needs careful control and monitoring. Not really suitable for manual spraying, manipulators should be used. Consumables used are in powder form.
Plasma spraying	The high temperature enables almost all materials to be sprayed giving high density coatings strongly bonded to the substrate, with low heat input to it. This can cause problems with differential contraction between coating and base metal. High density, high adhesion, low oxide coatings with properties reproducable to close limits are possible by spraying in a vacuum chamber back-filled with argon. Heat input to the component is then much greater.	Capital cost higher than gas arc spraying. A spray booth is desirable. Vacuum spraying requires pumping equipment with a suitable manipulator. The size of the chamber limits work dimensions. Process time slow. High heat input to the workpiece may cause distortion or metallurgical changes in the substrate.
Spray, fused coatings	The coating is dense, non-porous and metallurgically bonded to the substrate. Spraying thickness and uniformity easily monitored and controlled. Although restricted to Ni- and Co-base self-fluxing alloys a large range of these available with a wide spread of hardness and other properties. Often used 'as fused' the smooth accurate surface facilitates machining with minimum wastage of the surface alloy. Tungsten carbide particles often incorporated in the coating to enhance wear resistance. Torches may be held manually or mechanically manipulated. Coatings, generally 0.4–2 mm thick can be put on components varying greatly in size and shape. Fusing can be carried out manually, in furnaces (generally vacuum), by induction heating or using lasers.	A lot of heat is introduced into the part. The interface must reach about 1000°C to ensure metallurgical bonding at the interface. This can cause distortion and metallurgical changes in the base material. Cooling from above 1000°C can provide problems due to the differential thermal contaction of coating and substrate metal. Coatings on transformable steels can provide cracking problems.

Table 17.5 General guidance on the choice of process

Process	Cost of equipment	Ease of operation	Operator skill	Applicability of the process to:											Heat input on surfacing			Suitable surfacing alloys
				Large areas	Small areas	Thick deposits	Thin deposits	Thick sections	Thin sections	Massive work	Small parts	Edge build-up	Site work	Machined components	Before	During	After	
Gas welding	Low	Easy	Medium	3	1	1	2	3	1	3	1	2	1	2	Fair	High diffuse	May be desirable	Many. Excellent for WC containing rods.
Powder welding	Low	Easy	Low	3	1	2	1	2	1	2	1	1	1	1	Low	Fairly high	None	Restricted to self-fluxing alloys.
Manual metal arc welding	Fairly low	Fairly easy	Fairly high	2	2	1	3	1	3	1	3	3	1	3	Fair	Very high, steep gradients	May be desirable	Wide range available.
TIG welding	Medium high	Moderately easy	Fairly high	3	1	2	2	1	2	2	2	2	2	2	Low	Very high, fairly confined	May be desirable	Wide range available.
MIG welding	Medium	Moderately easy	Medium	2	3	2	3	1	3	2	3	3	2	3	Low	Very high, steep gradients	May be desirable	Restricted to alloys available as wires.
Flux-cored arc welding	Medium	Moderately easy	Medium	1	3	2	3	1	3	1	4	3	2	3	Low	Very high, steep gradients	May be desirable	Needs cored wires.
Submerged arc welding	Very high	Moderately hard	Fairly high	1	4	1	4	1	4	1	4	4	4	4	Low	Very high, diffuse	Some	Needs suitable coating materials or flux additions.
PTA surfacing	High	Hard	High	4	1	3	2	3	1	3	2	3	4	1	Low	High, moderate gradient	May be desirable	Many alloys possible.
Electroplating	High	Fairly easy	Medium	1	1	2	1	1	1	2	1	3	4	1	Low	Low	Low	Generally Cr, Ni and Cu for resurfacing.
Powder spraying	Fairly low	Easy	Low	2	2	4	1	1	1	2	2	4	1	1	Very low	Low	None	Limited, other than self-fluxing alloys.
Wire spraying	Fairly low	Easy	Low	1	2	4	1	1	1	2	2	4	1	1	Very low	Low	None	Considerable range including bond coats. Some ceramic rods.
Arc spraying	High	Easy	Fairly low	1	3	2	2	1	1	1	2	4	1	1	Very low	Fairly low	None	Limited range of wires.
HVOF spraying	Fairly high	Moderately easy	Medium	2	2	2	1	1	1	2	2	3	2	1	Very low	Fairly low	None	Considerable range of metallic powders. Some equipments spray plastics.
Plasma spraying	High	Fairly hard	High	3	1	3	1	2	1	2	1	3	4	1	Low	Fairly low	None	Many alloys and ceramics, usually powder
Sprayed and fused coatings	Fairly low	Fairly easy	Medium	2	2	3	1	3	1	3	2	2	3	1	Low	Fairly low	Fairly high, uniform	Needs self-fluxing alloys.

1, Very applicable; 2, Not very applicable; 3, Not really suitable; 4, Unsuitable.

Table 17.6 Available coating materials

Group	Sub-group	Alloy system	Important properties	Process suitability														Typical applications
				GW	PW	MMA	TIG	MIG	FCA	SA	PTA	EP	MP	MW	MA	HVOF PF	SW	
STEEL (up to 1.7% C)	Pearlitic	Low carbon	Crack resistant, low cost, good base for hard-surfacing	1		X		X	X					1	1			Build up to restore dimensions. Track links, rollers, idlers
	Martensitic	Low, medium or high carbon. Up to 9% alloying elements	Abrasion resistance increases with carbon content, resistance to impact decreases. Economical	1		X		X	X	1				1	X			Bulldozer blades, excavator teeth, bucket lips, impellers, conveyor screws, tractor sprockets, steel mill wobblers, etc
	High speed	Complex alloy	Heat treatable to high hardness	X		1												Working and conveying equipment
	Semi-austenitic	Manganese chromium	Tough crack resistant. Air and work hardenable	X		1				1								Mining equipment, especially softer rocks
	Austenitic	Manganese	Work hardening			X	1	1	1									Rock crushing equipment
		Alloyed manganese	Work hardening, less susceptible to thermal embrittlement. Useful build-up			X		1	1									Build up normal manganese steel prior to application of other hard-surfacing alloys
		Chromium nickel	Stainless, tough, high temperature and corrosion resistant (low carbon)			1	X	X	X	X				1	X			Furnace parts, chemical plant
IRON (above 1.7% C)	High chromium	Martensitic	Show improved hot hardness and increased abrasion resistance	1		X		1	1	1								Steelworks equipment, scraper blades, bucket teeth
		Multiple alloy	Hardenable. Can anneal for machining and re-harden. Good hot hardness	1		X	1	1	1	1								Mining equipment, dredger parts
		Austenitic	Wide plastic range, can be hot shaped, brittle. Oxidation resistant	1		X		1	1	1								Low stress abrasion and metal-to-metal wear. Agricultural equipment
	Martensitic alloy	Chromium tungsten Chromium molybdenum Nickel chromium	Very good abrasion resistance, very high compressive strength so can resist light impact. Can be heat-treated. Considerable variation in properties between gas and arcweld deposits	X		X	1	1	1	1								Cutting tools, shear blades, rolls for cold rolling
	Austenitic alloy	Chromium molybdenum Nickel chromium	Lower compressive strength and abrasion resistance. Less susceptible to cracking. Will work harden	X		X	1	1	1	1	1							Mixing and steelworks equipment, agricultural implements

Table 17.6 *(continued)*

Group	Sub-group	Alloy system	Important properties	Process suitability															Typical applications
				GW	PW	MMA	TIG	MIG	FCA	SA	PTA	EP	MP	MW	MA	HVOF	PF	SW	
CARBIDE	Iron base	Tungsten carbides in steel matrix	Resistant to severe abrasive wear. Care required in selection and application	X		X	1	1	1	1									Rock drill bits. Earth handling and digging equipment. Extruder screw augers
	Cobalt base	Tungsten carbides in cobalt alloy matrix	Matrix gives improved high temperature properties and corrosion resistance	X	X	1	1	1	1		1							X	Oil refinery components, etc
	Nickel base	Tungsten carbides in nickel alloy matrix	Matrix gives improved corrosion resistance		X						X							X	Screws, pump sleeves etc. in corrosive environments
	Copper base	Tungsten carbides in copper alloy matrix	Often larger carbide particles to give cutting and sizing properties	X															Oil field equipment
NICKEL BASE	Nickel		High corrosion resistance	1		1	1					X		1	1				Chemical plant. Bond costs for ceramics
	Nickel Chromium Boron	With Fe, Si and C; W or Mo may be added	Self-fluxing alloys available in wide range of hardnesses. Abrasion, corrosion, oxidation resistant. Can be applied as thin, dense, impervious layers. Metallurgical bond to substrate	X	X	1	1				X		1	1		1	1	X	Glass mould equipment, engineering components, chemical and petrochemical industries
	Nickel Chromium	With possibly C, Mo or W to improve hardness and hot strength. Fe modifies thermal expansion improves creep resistance	Relatively soft and ductile. Good hot gas corrosion resistance. Very good corrosion resistance			X	1		1		X								High temperature engineering applications. Chemical industry. I.C. engine valves
	Nickel Iron molydenum	60% Ni, 20% Mo, 20% Fe or 65% Ni, 30% Mo, 5% Fe	Resistant to HCl, also sulphuric, formic and acetic acids	1		1	1	X	X		1								Chemical plant
	Nickel Copper	Monel	Corrosion resistant	1		1	1	1	X					1	1				Chemical plant
COBALT	Cobalt Chromium Tungsten	About 30% Cr, increasing W and C increases hardness	Superlative high temperature properties. Wear, oxidation and corrosion resistant	X		X	X	X	1		X						1		Oilwell equipment, steelworks, chemical engineering plant, textile machinery
	Self-fluxing	With B, Si, Ni	Modified to provide self-fluxing properties. Good abrasion, corrosion resistance	1	X		1				1		X		1	1	1	X	Chemical and petrochemical industries. Extrusion screws

Table 17.6 *(continued)*

Group	Sub-group	Alloy system	Important properties	Process suitability															Typical applications
				GW	PW	MMA	TIG	MIG	FCA	SA	PTA	EP	MP	MW	MA	HVOF	PF	SW	
COPPER BASE	Copper		Electrical conductivity	1		1						X		X	1				Electrical equipment, paper-working machinery
	Bronze	Aluminium manganese, tobin, phosphor, commercial	Resistance to frictional wear and some chemical corrosion. Al-bronze excellent bond coat arc metallising	1		1								X	X				Bearing shells, shafts, slides, valves, propellers, etc
	Brass		Bearing properties, decorative finishes	1										X					Water tight seals. Electric discharge machining electrodes
CHROMIUM			Wear, corrosion resistant									X							Engineering components
MOLYBDENUM			High adhesion to base metal											X	1				Bond coats. Engineering parts reclamation
ALUMINIUM			Corrosion and heat resistance										1	1	X				Steel structures. Furnace parts
ZINC			Corrosion resistance											X	1				Steel structures, gas cylinders, tanks, etc. HF shielding
ZINC/ ALUMINIUM ALLOYS			Corrosion resistance											X	X				Steel structures
LEAD BASE	Lead		Resistance to chemical attack	1								1		X	1				Resistance to sulphuric acid. Radiation shielding
	Solder	Often 60/40% Pb/Sn	Joining	1										X					Tinning surfaces for subsequent joining
TIN BASE	Tin		High corrosion resistance	1								X		X	1				Electrical contacts. Food industry plant
	Babbitt	Sn, Sb, Cu	Bearing alloys	1										X	1				Bearing shells
OXIDES		Principally of Al, Zn, Cr, Ti and mixtures	High temperature oxidation resistance. Wear resistance											X		1	X		Pump sleeves, acrospace parts
REFRACTORY METALS		W, Ti, Ta, Cr	Good, high temperature properties. Often develop stable, protective oxide films											1	1	1	X		Electrical contacts. High density areas, corrosion protection
CARBIDES/ BORIDES		Cr. W. B. possibly + Co or Ni	Very high wear resistance	1		X	1							1		X	X		Thin cutting edges. Wear resistant areas
COMPLEX CERAMICS		Silicides, titanates, zirconates, etc	Wear, oxidation, and erosion resistance											1		X	X		Thermal barriers, coating equipment handling molten metal and glass

X frequently used, 1 sometimes used.

Table 17.7 Factors affecting choice of coating material

Factor	*Effect*
PROPERTIES REQUIRED OF DEPOSIT Function of surface. Nature of adjacent surface or rubbing materials. What is the service temperature? What is the working environment (corrosive, oxidising, abrasive, etc.) What coefficient of friction required?	Is porosity desirable or to be avoided? Sprayed coatings are porous; fused or welded coatings are impervious. Porosity can be advantageous if lubrication required. What compressive stresses are involved? A porous deposit can be deformed and detached from the substrate. What degree of adhesion to the substrate is necessary? In many applications corrosion protection is satisfactory with mechanical adhesion – in some cases a metallurgical, bond will be required. How important is macro-hardness, metallographic structure – this often affects abrasion resistance. Is there danger of electro-chemical corrosion? If abrasion resistance is required, is it high or low stress? Many different alloys are available and needed to meet the many wear conditions which may be encountered. Are similar applications known?
PROPERTIES NEEDED IN ALLOY Physical Chemical Metallurgical Mechanical	Thermal expansion, or contraction, compared with the base metal, affects distortion. Chemical composition relates to corrosion, erosion and oxidisation resistance. The metallurgical structure is very important, influencing properties such as abrasion resistance and frictional properties. Mechanical properties can be critical. It is necessary to assess the relative importance of properties needed such as resistance to abrasion, corrosion, oxidation, erosion, seizure, impact and to what extent machinability, ductility, thermal conductivity or resistance, electrical conductivity or resistance may be required.
PROCESS CONSIDERATIONS Surfacing processes available. Surface preparation methods available and feasible; auxiliary services which can be used. Location and size of work. Thickness of deposit required.	Some consumables can be produced only as wires – or powders – or cast rods. Choice could be limited by the equipment in use and operator experience. Lack of specific surface preparation methods could limit choice of process and this, in turn, prohibits use of certain materials. Similar limitations could arise from lack of necessary pre-heating or post-heat treatment plant. Can resurfacing be carried out without dismantling the assembly? Can work be carried out on site? Are only small areas on a large part to be repaired?
ECONOMICS How much cost will the repair bear?	Often much more important than a direct cost comparison – cost of the repair compared to the purchase of a new part – are many other features, e.g. saving in subsequent maintenance labour and material costs, value of the lost production which is avoided, lower scrap rate during subsequent processing, time saved in associated production units, improved quality of component and product, increased production rates, reduction in consumption of raw materials.

Table 17.8 Methods of machining electroplated coatings

Deposited metal	*Machining with cutting tools*	*Grinding*
Chromium	Chromium is too hard	Grinding or related procedures are the only suitable process. Soft or medium wheels should be used at the highest speed consistent with the limits of safety. Coolant must be continuous and copious. Light cuts only—preferably not exceeding 0.0075 mm should be taken. Heavy cuts can cause cracking or splintering of the deposit
Nickel	High-speed steel is, in general, the most satisfactory material for cutting nickel. Tipped tools are not recommended. The shape of the tools should be similar to those used for steel but with somewhat increased rake and clearance. Nickel easily work hardens therefore tools must be kept sharp and well supported to ensure that the cut is continuous	To avoid glazing, use an open textured wheel with a peripheral speed of 25–32 m/s. Coolant as for grinding steel

Table 17.9 Bearing materials compatible with electroplated coatings

Material	*Excellent*	*Good*	*Avoid*
Chromium	Copper–lead, lead–bronze, white metal, Fine grain cast iron	Rubber or plastic, water-lubricated. Soft or medium hard steel with good lubrication and low speed. Brass, gun metal	Hard steel, phosphor bronze,* light alloys*
Nickel		Bronze, brass, gun metal, white metal	Ferrous metals, phosphor bronze

*May be satisfactory in some conditions.

When using deposited metals in sliding or rotating contact with other metals, adequate lubrication must be assured at all times.

Table 17.10 Examples of successful repairs

Application	*Coating material*	*Process used*
Cast iron glass container moulds	Ni base self fluxing alloy in rod form Ni base alloy	Gas welding Powder welding
Steel plungers on molten glass pumps	Ni base alloy	Metal spraying
Excavator teeth	Bulk welding all over with chromium carbide alloy	Bulk welding
Excavator tooth tips	Hard alloy	Flux cored welding
Metering pump pistons	Aluminium oxide titanium oxide composite	Plasma flame spraying
Fire pump shafts	Stainless steel	Plasma wire spraying
Printing press rolls	Stainless steel on a nickel alumnide bond coat	Arc spraying
Large hydraulic press ram	Martensitic steel on a bond coat and sealed with a vinyl sealer	Wire spraying
Shafts of pumps handling slurries	Ni and Co base alloys with tungsten carbide	Plasma spraying
Screw conveyors	Ni base alloy rods	Gas welding
Small valves less than 12 cm diameter	Co–Cr–Ni–W alloys	Gas welding
Medium size valves	Co–Cr–Ni–W alloys	Powder welding
Very large valves	Co–Cr–Ni–W alloys	Submerged arc welding
Paper mill drums	Stainless steel wire	Arc spraying
Ammonia compressor pistons	Whitemetal	Wire spraying
Hydraulic pump plungers	Ni base alloy	Sprayed and fused
Guides on steel mills	Ni base alloy	Gas welding

ABRASIVE WEAR

Abrasive wear is the loss of material from a surface that results from the motion of a hard material across this surface.

There are several types of abrasive wear. Since the properties required of a wear-resistant material will depend on the type of wear the material has to withstand, a brief mention of these types of wear may be useful.

There are three main types of wear generally considered: gouging abrasion (impact), Figure 18.1; high-stress abrasion (crushing), Figure 18.2; and low-stress abrasion (sliding), Figure 18.3. This classification is made more on the basis of operating stresses than on the actual abrading action.

Gouging abrasion

This is wear that occurs when coarse material tears off sizeable particles from wearing surfaces. This normally involves high imposed stresses and is most often encountered when handling large lumps.

High-stress abrasion

This is encountered when two working surfaces rub together to crush granular abrasive materials. Gross loads may be low, while localised stresses are high. Moderate metal toughness is required; medium abrasion resistance is attainable.

Rubber now competes with metals as rod and ball mill linings with some success. Main advantages claimed are longer lifer at a given cost, with no reduction in throughput, lower noise level, reduced driving power consumption, less load on mill bearings and more uniform wear on rods.

Low-stress abrasion

This occurs mainly where an abrasive material slides freely over a surface, such as in chutes, bunkers, hoppers, skip cars, or in erosive conditions. Toughness requirements are low, and the attainable abrasion resistance is high.

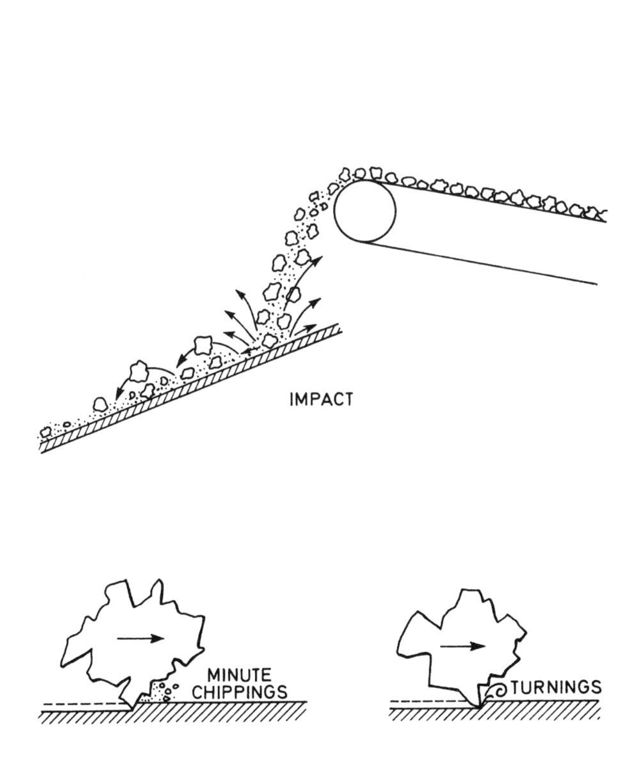

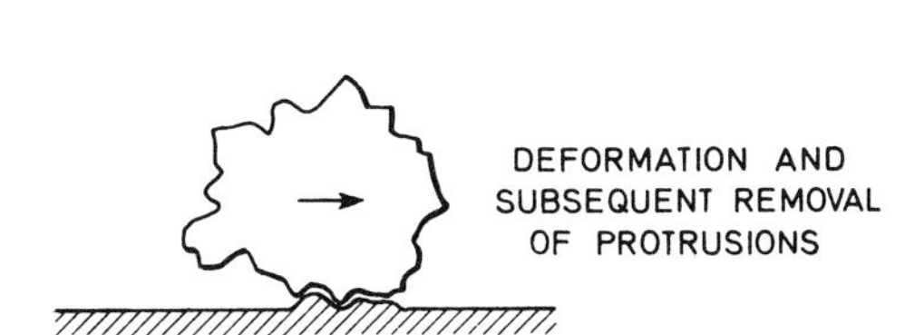

Figure 18.1 Types of gouging abrasion

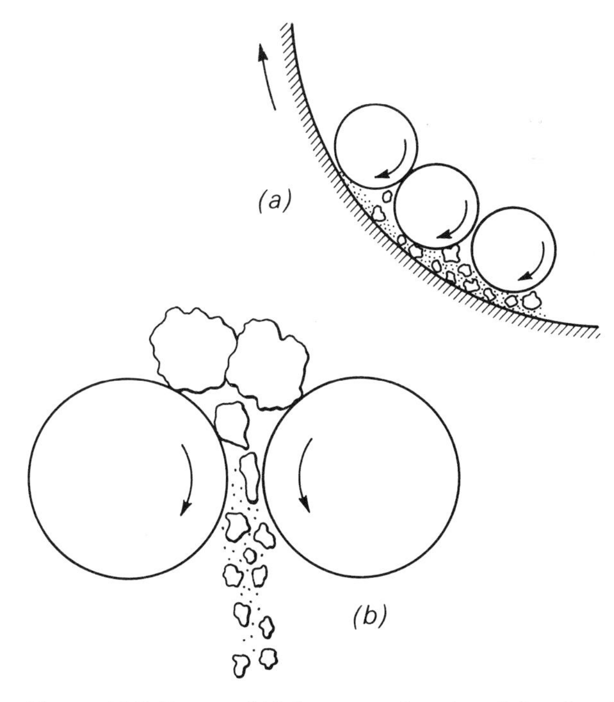

Figure 18.2 Types of high-stress abrasion: (a) rod and ball mills; (b) roll crushing

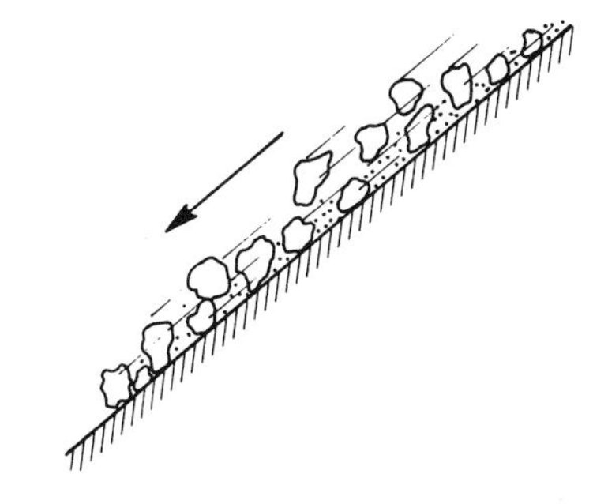

Figure 18.3 Low-stress abrasion

MATERIAL SELECTION

Very generally speaking the property required of a wear-resistant material is the right combination of hardness and toughness. Since these are often conflicting requirements, the selection of the best material will always be a compromise. Apart from the two properties mentioned above, there are few *general* properties. Usually the right material for a given wear-resistant application can only be selected after taking into consideration other factors that determine the rate of wear. Of these the most important are:

Ambient temperature, or temperature of material in contact with the wear surface.
Size distribution of particles flowing over the wear surface.
Abrasiveness of these particles.
Type of wear to which wear surface is subjected (i.e. gouging, sliding, impact, etc.).
Velocity of flow of material in contact with wear surface.
Moisture content or level of corrosive conditions.
General conditions (e.g. design of equipment, head-room available, accessibility, acceptable periods of non-availability of equipment).

Tables 18.1 and 18.2 give some general guidance on material selection and methods of attaching replaceable components.

Table 18.3 gives examples of actual wear rates of various materials when handling abrasive materials.

The subsequent tables give more detailed information on the various wear resistant materials.

Table 18.1 Suggested materials for various operating conditions

Operating conditions	*Properties required*	*Material*
High stress, impact	Great toughness; work-hardening properties	Austenitic manganese steel, rubber of adequate thickness
Low stress, sliding	1, Great hardness; 2, toughness less important; 3, quick replacement	Hardened and/or heat-treated metals, hardfacing, ceramics
	1, Cheapness of basic material; 2, replacing time less important	Ceramics, quarry tiles, concretes
	1, Maximum wear resistance; cost is immaterial	Tungsten carbide
Gouging wear	High toughness	Usually metals, i.e. irons and steels, hardfacing
Wet and corrosive conditions	Corrosion resistance	Stainless metals, ceramics, rubbers, plastics
Low stress; contact of fine particles; low abrasiveness	Low coefficient of friction	Polyurethane, PTFE, smooth metal surfaces
High temperature	Resistance to cracking, spalling, thermal shocks; general resistance to elevated temperatures	Chromium-containing alloys of iron and steel; some ceramics
Minimum periods of shut-down of plant	Ease of replacement	Any material that can be bolted in position and/or does not require curing
Curved, non-uniform irregular surface and shapes	Any one or a combination of the above properties	Hardfacing weld metal; most trowellable materials
Arduous and hot conditions		Hardfacing weld metal

Table 18.2 Methods of attachment of replaceable wear-resistant components

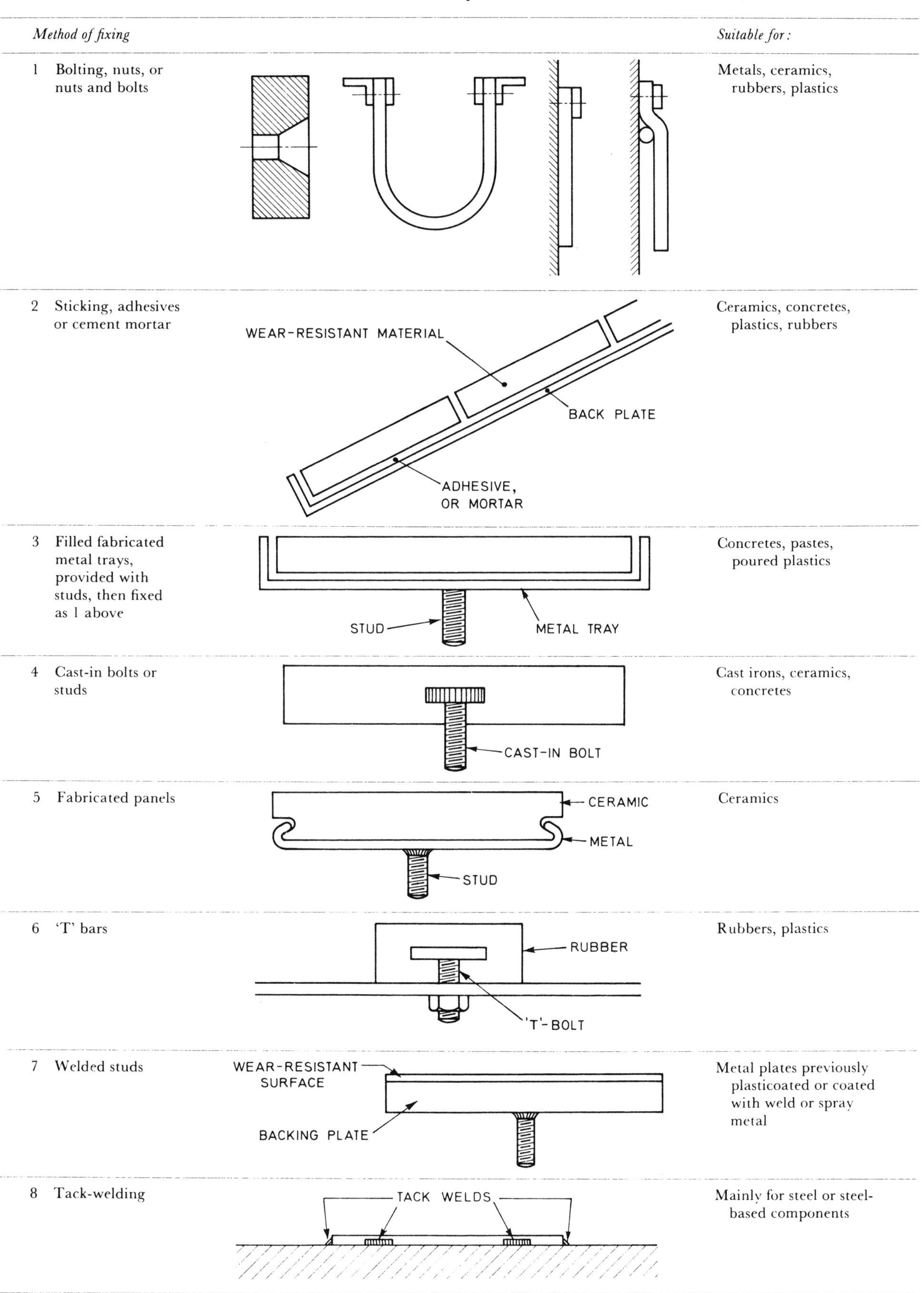

Method of fixing	*Suitable for:*
1 Bolting, nuts, or nuts and bolts	Metals, ceramics, rubbers, plastics
2 Sticking, adhesives or cement mortar	Ceramics, concretes, plastics, rubbers
3 Filled fabricated metal trays, provided with studs, then fixed as 1 above	Concretes, pastes, poured plastics
4 Cast-in bolts or studs	Cast irons, ceramics, concretes
5 Fabricated panels	Ceramics
6 'T' bars	Rubbers, plastics
7 Welded studs	Metal plates previously plasticoated or coated with weld or spray metal
8 Tack-welding	Mainly for steel or steel-based components

Table 18.3 Typical performance of some wear-resistant materials as a guide to selection

Type	*Some typical materials*	*Sliding wear-rate** *by coke*	*by sinter*	*Temperature limitations*	*Ease and convenience of replacement*	*General comments*
Cast irons	Ni-hard type martensitic white iron High chrome martensitic white irons Spheroidal graphite-based cast iron High phosphorus pig iron Low alloy cast iron	0.11 0.12 0.22 0.32 –	0.06 0.11 0.09 0.91 0.44		Yes	The most versatile of the materials which, now, by varying alloying elements, method of manufacture and application are able to give a wide range of properties. Their main advantage is the obtainable combination of strength, i.e. toughness and hardness, which accommodates a certain amount of abuse. Other products are sintered metal and metal coatings
Cast steels	$3\frac{1}{4}$ Cr–Mo cast steel 13 Mn austenitic cast steel $1\frac{1}{2}$ Cr–Mo cast steel	0.17 0.22 0.43	– – –			
Rolled steels	Armour plate Work-hardened Mn steel Low alloy steel plate, quenched and tempered EN8 steel	0.12 0.13 0.31 0.43	– – 0.30–0.84 0.63			
Hard facings	High chrome hardfacing welds, various	0.09–0.16	0.05–0.14	No	Could be difficult if applied *in situ*	
Ceramics	Fusion-cast alumina-zirconia-silica Slagceram Fusion-cast basalt Acid-resisting ceramic tile Plate glass Quarry floor tiles	0.05 0.15 0.17 0.19 0.81 2.2–3.4	0.11–0.14 0.33 0.53 1.27 – –		Yes, if bolted. Not so convenient if fixed by adhesive or cement mortar, as long curing times may be unacceptable	Great range of hardness. Most suitable for low-stress abrasion by low-density materials, and powders. Disadvantage: brittleness
Concretes	Aluminous cement concrete. Quartz-granite aggregate-based concrete	0.42 0.87	4.0–4.4 6.5		Could be messy. Might be difficult under dirty conditions	Advantages: cheapness, castability. Disadvantage: long curing or drying-out times
Rubbers	Rubbers, various	2.1–3.2	–		Bonded and bolted. Stuck with adhesive, could be difficult under dirty conditions	Main advantage is resilience and low density, with a corresponding loss in bulk hardness. The most useful materials where full advantage at the design stage can be taken of their resilience and anti-sticking properties
Rubber-like plastics	Polyurethanes, various	2.3–5.4	2.3			
Other plastics	High-density polyethylene Polytetrafluoroethylene (PTFE)	6.4 8.2	– –	Yes	In sheet form it is difficult to stick	Low coefficient of friction, good antisticking properties. Best for low-stress abrasion by fine particles
Resin-bonded compounds	Resin-bonded clacined bauxite	2.3	–		Trowelled; could be messy. Difficult in dirty and inaccessible situations	These materials are only as strong as their bonding matrix and therefore find more application where low-stress wear by powders or small particles (grain, rice) takes place

*Wear rate is expressed in in^3 of material worn away per 1000 tons of the given bulk materials per ft^2 of area in contact with the abrading material. The results were obtained from field trials in a chute feeding a conveyor belt.

This data is provided as examples of the relative wear rates of the various materials when handling abrasive bulk materials.

The following tables give more detailed information on the materials listed in Table 18.3 with examples of some typical applications in which they have been used successfully.

When selecting the materials for other applications, it is important to identify the wear mechanism involved as this is a major factor in the choice of an optimum material. Further guidance on this is given in Table 18.1.

Table 18.4 Cast irons

Type	*Nominal composition*	*Hardness Brinell*	*Characteristics*	*Typical application*
Grey irons BS 1452 ASTM A48	Various	150–300	Graphite gives lubrication	Brake blocks and drums, pumps
Spheroidal graphite	Meehanite WSH2	Up to 650	Heat-treatable. Can be lined with glass, rubber, enamels	Many engineering parts, crusher cones, gears, wear plates
High phosphorus	3.5%C 2.0%P	Up to 650	Brittle, can be reinforced with steel mesh.	Sliding wear
Low alloy cast iron	3%C 2%Cr 1%Ni	250–700		Sliding wear, grate bars, cement handling plant, heat-treatment
NiCr Martenstic irons Typical examples: NiHard, BF 954	2.8–3.5%C, 1.5–10% Cr 3–6% Ni	470–650		High abrasion Ore handling, sand and gravel
Ni Hard 4	7–9% Cr, 5–6.5% Ni, 4–7% Mn			More toughness. Resists fracture and corrosion
CrMoNi Martensitic irons Typical examples: Paraboloy	14–22% Cr, 1.5% Ni, 3.0% Mo	500–850		Ball and rod mills, wear plates for fans, chutes, etc.
High Chromium irons Typical examples: BF 253 HC 250	22–28% Cr	425–800	Cast as austenite Heat-treated to martensite	Crushing and grinding Plant Ball and Rod mills Shot blast equipment, pumps

Table 18.5 Cast steels

Type	*Nominal composition*	*Hardness*	*Characteristics*	*Typical applications*
Carbon steel BS 3100 Grade A		Up to 250		Use as backing for coatings
Low alloy steels BS 3100 Grade B	Additions of Ni, Cr, Mo up to 5%	370–550		For engineering 'lubricated' wear conditions
Austenitic BS 3100 BW10	11% Mn min.	200 soft Up to 600 when work-hardened		For heavy impact wear, Jaw and Cone crushers, Hammer mills
High alloy steels BS 3100 Grade C	30% Cr 65% Ni + Mo, Nb etc.	500		Special alloys for wear at high temperature and corrosive media
Tool steels Many individual specifications	17% Cr, 4% Ni, 9% Mo, 22% W, 10% Co	Up to 1000		Very special applications, usually as brazed-on plates.

Table 18.6 Rolled steels

Type	*Nominal composition*	*Hardness*	*Characteristics*	*Typical application*
Carbon steels BS 1449 Part 1 Typical examples: BS 1449 Grade 40 (En8 plate) Abrazo 60	.06/1.0% C, 1.7% Mn	160–260	Higher carbon for low/ medium wear	For use as backing for hard coatings
Low alloy steels Many commercial specifications Typical examples: ARQ Grades, Tenbor 25 30, Wp 300 and 500, Creusabro Grades, Abro 321 and 500, OXAR 320 and 450, Red Diamond 20 & 21, Compass B555	Up to 3.5% Cr, 4% Ni, 1% Mo	250–500	Quenched and tempered. Are weldable with care	Use for hopper liners, chutes, etc.
Austenitic manganese steel Typical examples: Cyclops 11/14 Mn, Red Diamond 14	11/16% Mn	200 in soft condition 600 skin hardened by rolling		
High alloy and stainless steel BS 1449 Part 2	Up to 10% Mn, 26% Cr, 22% Ni, + Nb, Ti	Up to 600	Heat and corrosion resistant	Stainless steels

Table 18.7 Wear resistant coatings for steel

	Method	*Technique*	*Materials*	*Characteristics and applications*
Weld applied surfaces	Gas welding	Manual	Rods of wide composition. Mainly alloys of Ni, Cr, Co, W etc.	For severe wear, on small areas. Thickness up to 3 mm
	Arc welding	Manual	Coated tubular electrodes. Specifications as above	Wear, corrosion and impact resistant. Up to 6 mm thickness.
		Semi- or full automatic	Solid or flux-cored wire, or by bulk-weld Tapco process. Wide range of materials	As above, but use for high production heavy overlays 10–15 mm.
	Fused paste	Paste spread onto surface, then fused with oxy-fuel flame or carbon-arc.	Chromium boride in paste mix	Excellent wear resistance. Thin (1 mm) coat. Useful for thin fabrications: fans, chutes, pump impellors, screw conveyors, agricultural implements
Flame spray processes	Oxy-fuel	Consumable in form of powder, wire, cord or rod, fed through oxy-fuel gun. Deposit may be 'as-sprayed' or afterwards fused to give greater adhesion	Materials very varied, formulated for service duty	Wear, corrosion heat, galling, and impact resistance. Thicknesses vary depending on material. Best on cylindrical parts or plates
	Arc spray	Wire consumable fed through electric arc with air jet to propel molten metal	Only those which can be drawn into wire	High deposition rate, avoid dew-point problems, therefore suited to larger components
Plasma spray processes	Non-transferred	Similar to flame spray, but plasma generated by arc discharge in gun	Materials as for flame-spray, but refractory metals, ceramics and cermets in addition, due to very high temperatures developed	High density coatings. Very wide choice of materials. Application for high temperature resistance and chemical inertness
	Transferred	As above but part of plasma passes through the deposit causing fusion	Mainly metal alloys	High adhesion Low dilution Extremely good for valve seats
	Detonation gun	Patent process of Union Carbide Corp. Powder in special gun, propelled explosively at work	Mainly hard carbides and oxides	Very high density. Requires special facilities
	High velocity oxy-fuel	Development of flame gun, gives deposit of comparable quality to plasma spray	Similar to plasma spray	More economic than plasma
Others	Hard chromium plating	Electrode position	Hard chrome Up to 950 HV (70 Rc)	Wear, corrosion and sticking resistant
	Electroless nickel	Chemical immersion	Nickel phosphide 850 HV after heat treatment	Similar to hard chrome
	Putty or paint	Applied with spatula or brush	Epoxy or polyester resins, or self-curing plastics, filled with wear resistant materials	Wear and chemical resistant

Table 18.8 Some typical wear resistant hardfacing rods and electrodes

Material type	*Name*	*Typical application*
Low alloy steels	Vodex 6013, Fortrex 7018, Saffire Range. Tenosoudo 50, Tenosoudo 75, Eutectic 2010	Build-up, and alternate layering in laminated surfaces
Low alloy steels	Brinal Dymal range. Deloro Multipass range. EASB Chromtrode and Hardmat. Metrode Met-Hard 250, 350, 450. Eutectic N6200, N6256. Murex Hardex 350, 450, 650, Bostrand S3Mo. Filarc 350, Filarc PZ6152/ PZ6352. Suodometal Soudokay 242-0, 252-0, 258-0, Tenosoudo 105, Soudodur 400/600, Abrasoduril. Welding Alloys WAF50 range Welding Rods Hardrod 250, 350, 650	Punches, dies, gear teeth, railway points
Martensitic chromium steels	Brinal Chromal 3, ESAB Wearod, Metrode Met-Hard 650. Murex Muraloy S13Cr. Filarc PZ6162. Oelikon Citochrom 11/13. Soudometal Soudokay 420, Welding Rods Serno 420FM. Welding Alloys WAF420	Metal to metal wear at up to 600C. High C types for shear blades, hot work dies and punches, etc.
High speed steels	Brinal Dyma H. ESAB OK Harmet HS. Metrode Methard 750TS. Murex-Hardex 800, Oerlikon Fontargen 715. Soudometal Duroterm 8, 12, 20, Soudostel 1, 12, 21. Soudodur MR	Hot work dies, punches, shear blades, ingot tongs
Austenitic stainless steels	Murex Nicrex E316, Hardex MnP, Duroid 11, Bostrand 309. Metrode Met-Max 20.9.3, Met-Max 307, Met-Max 29.9 Soudometal Soudocrom D	Ductile buttering layer for High Mn steels on to carbon steel base. Furnace parts, chemical plant
Austenitic manganese steels	Brinal Mangal 2. Murex Hardex MnNi Metrode Workhard 13 Mn, Workhard 17 MnMo, Workhard 12MnCrMo. Soudometal Soudomanganese, Filarc PZ6358	Hammer and cone crushers, railway points and crossings
Austenitic chromium manganese steels	Metrode Workhard 11Cr9Mn, Workhard 14Cr14Mn. Soudometal Comet MC, Comet 624S	As above but can be deposited on to carbon steels. More abrasion resistant than Mn steels
Austenitic irons	Soudometal Abrasodur 44. Deloro Stellite Delcrome 11	Buttering layer on chrome irons, crushing equipment, pump casings and impellors
Martensitic irons	Murex Hardex 800. Soudometal Abrasodur 16. Eutectic Eutectdur N700	For adhesive wear, forming tools, scrapers, cutting tools
High chromium austenitic irons	Murex Cobalarc 1A, Soudometal Abrasodur 35, 38. Oerlikon Hardfacing 100, Wear Resistance WRC. Deloro Stellite Delcrome 91	Shovel teeth, screen plate, grizzly bars, bucket tips
High chromium martensitic irons	Metrode Met-hard 850, Deloro Stellite Delcrome 90	Ball mill liners, scrapers, screens, impellors
High complex irons	Brinal Niobal. Metrode Met-hard 950, Met-Hard 1050. Soudometal Abrasodur 40, 43, 45, 46	Hot wear applications, sinter breakers and screens
Nickel alloys	Metrode 14.75Nb, Soudonel BS, Incoloy 600. Metrode 14.75MnNb, Soudonel C, Incoloy 800. Metrode HAS C, Comet 95, 97, Hastelloy types	Valve seats, pump shafts, chemical plant
Cobalt alloys	EutecTrode 90, EutecRod 91	Involving hot hardness requirement: Valve seats, hot shear blades
Copper alloys	Saffire Al Bronze 90/10, Citobronze, Soudobronze	Bearings, slideways, shafts, propellers
Tungsten carbide	Cobalarc 4, Diadur range	Extreme abrasion: fan impellors, scrapers

Table 18.9 Wear resistant non-metallic materials

	Type	Nominal composition	Hardness	Characteristics	Typical application
Ceramic materials	**Extruded ceramic** Indusco Vesuvius	High density Alumina	9 Moh	Process limits size to 100 × 300 × 50 mm. Low stress wear, also at high temperature	
	Ceramic plates Hexagon-shaped, cast Indusco				Suitable for lining curved surfaces
	Sintered Alumina Alumina 1542	96%Al_2O_3 2,4%SiO_2	9 Moh	Low stress wear also at high temperatures	
	Isoden 90	90%Al_2O_3			
	Isoden 95	95%Al_2O_3			
	Fusion-cast Alumina Zac 1681	50%Al_2O_3 32.5%ZrO_2 16%SiO_2		Can be produced in thick blocks to any shape. Low stress and medium impact, also at high temperatures	
	Concrete Alag Ciment Fondu	Mainly calcium silico-aluminates 40%Al_2O_3		Low cost wear-resistant material. High heat and chemical resistance	Floors, coke wharves, slurry conveyors, chutes
	Cast Basalt Heat-treated	Remelted natural basalt	7–8 Moh	Low stress abrasion. Brittle	Floors, coke chutes, bunker, pipe linings, usually 50 mm thick minimum. Therefore needs strong support
	Plate Glass		Glass	Very brittle	Best suited for fine powders, grain, rice etc.
Rubber and plastics	**Rubber** Trellex Skega	Various grades	Various	Resilient, flexible	Particularly suitable for round particles, water borne flow of materials
	Linatex	95% Natural rubber	Fairly soft		
	Ceramic ballsheet Hoverdale	Rubber filled with ceramic balls		Enhanced wear resistance	
	Plastics Duthane Flexane Tivarthane (Polyhi-Solidor) Scandurathane (Scandura) Supron (Slater)	Polyurethane based, rubber-like materials		Low stress abrasion applications	Floors, chute liners screens for fine materials
	Duplex PTFE	Polytetra-fluorethylene		Low coefficient of friction	For fine powders light, small particles
	Resins Belzona Devcon Greenbank AD1 Thortex Systems Nordbak	Resin-based materials with various wear-resistant aggregates		Can be trowelled. Specially suitable for curved and awkward surfaces but not for lumpy materials	Floors, walls, chutes, vessels. In-situ repairs

Repair of plain bearings

In general, the repair of bearings by relining is confined to the low melting-point whitemetals, as the high pouring temperatures necessary with the copper or aluminium based alloys may cause damage or distortion of the bearing housing or insert liner. However, certain specialist bearing manufacturers claim that relining with high melting-point copper base alloys, such as lead bronze, is practicable, and these claims merit investigation in appropriate cases.

For the relining and repair of whitemetal-lined bearings three methods are available:

(1) Static or hand pouring.
(2) Centrifugal lining.
(3) Local repair by patching or spraying.

Table 19.1 Guidance on choice of lining method

Type of bearing	*Relining method*	*Field of application*
Direct lined housings	Static pouring or centrifugal lining	Massive housings. To achieve dynamic balance during rotation, parts of irregular shape are often 'paired' for the lining operation, e.g. two cap half marine type big-end bearings lined together, ditto the rod halves
INSERT LINERS		
'Solid inserts'	Not applicable	New machined castings or pressings required
Lined inserts		
Thick walled	Static pouring or centrifugal lining	Method adopted depends on size and thickness of liner, and upon quantities required and facilities available
Medium walled		
Thin walled	Not recommended	Relining not recommended owing to risk of distortion and loss of peripheral length of backing. If relining essential (e.g. shortage of supplies) special lining jigs and protective measures essential

(1) PREPARATION FOR RELINING

(*a*) Degrease surface with trichlorethylene or similar solvent degreaser. If size permits, degrease in solvent tank, otherwise swab contaminated surfaces thoroughly.

(*b*) Melt off old whitemetal with blowpipe, or by immersion in melting-off pot containing old whitemetal from previous bearings, if size permits.

(*c*) Burn out oil with blowpipe if surface heavily contaminated even after above treatment.

(*d*) File or grind any portions of bearing surface which remain contaminated or highly polished by movement of broken whitemetal.

(*e*) Protect parts which are not to be lined by coating with whitewash or washable distemper, and drying. Plug bolt holes, water jacket apertures, etc., with asbestos cement or similar filler, and dry.

(2) TINNING

Use pure tin for tinning steel and cast iron surfaces; use 50% tin, 50% lead solder for tinning bronze, gunmetal or brass surfaces.

Flux surfaces to be tinned by swabbing with 'killed spirit' (saturated solution of zinc in concentrated commercial hydrochloric acid, with addition of about 5% free acid), or suitable proprietary flux.

Tinning cast iron presents particular difficulty due to the presence of graphite and, in the case of used bearings, absorption of oil. It may be necessary to burn off the oil, scratch brush, and flux repeatedly, to tin satisfactorily. Modern methods of manufacture embodying molten salt bath treatment to eliminate surface graphite enable good tinning to be achieved, and such bearings may be retinned several times without difficulty.

Tin bath

(*i*) Where size of bearing permits, a bath of pure tin held at a temperature of 280°–300°C or of solder at 270°–300°C should be used.

(*ii*) Flux and skim surface of tinning metal and immerse bearing only long enough to attain temperature of bath. Prolonged immersion will impair bond strength of lining and cause contamination of bath, especially with copper base alloy housings or shells.

(*iii*) Flux and skim surface of bath to remove dross, etc., before removing bearing.

(*iv*) Examine tinned bearing surface. Wire brush any areas which have not tinned completely, reflux and re-immerse.

Stick tinning

(*i*) If bearing is too large, or tin bath is not available, the bearing or shell should be heated by blowpipes or over a gas flame as uniformly as possible.

(*ii*) A stick of pure tin, or of 50/50 solder is dipped in flux and applied to the surface to be lined. The tin or solder should melt readily, but excessively high shell or bearing temperatures should be avoided, as this will cause oxidation and discoloration of the tinned surfaces, and impairment of bond.

(*iii*) If any areas have not tinned completely, reheat locally, rub areas with sal-ammoniac (ammonium chloride) powder, reflux with killed spirit, and retin.

(3) LINING METHODS

(a) Static lining

(*i*) *Direct lined bearings*

The lining set-up depends upon the type of bearing. Massive housings may have to be relined *in situ*, after preheating and tinning as described in sections (1) and (2). In some cases the actual journal is used as the mandrel (see Figures 19.1 and 19.2).

Journal or mandrel should be given a coating of graphite to prevent adhesion of the whitemetal, and should be preheated before assembly.

Sealing is effected by asbestos cement or similar sealing compounds.

(*ii*) *Lined shells*

The size and thickness of shell will determine the type of lining fixture used. A typical fixture, comprising face plate and mandrel, with clamps to hold shell, is shown in Figure 19.5 while Figure 19.6 shows the pouring operation.

Figure 19.1 Location of mandrel in end face of direct lined housing

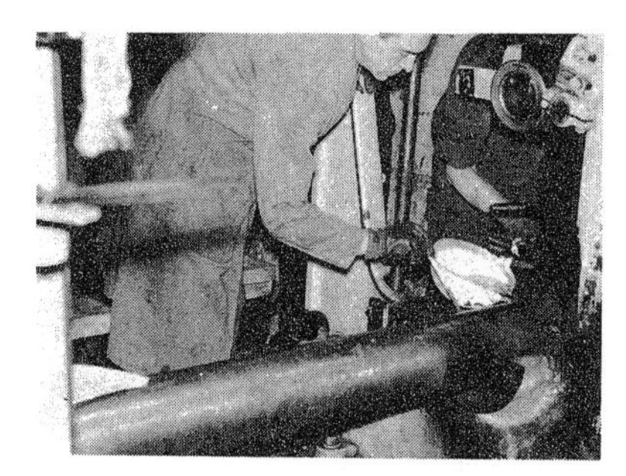

Figure 19.3 Direct lined housing. Pouring of white-metal

Figure 19.2 Outside register plate, and inside plate machined to form radius

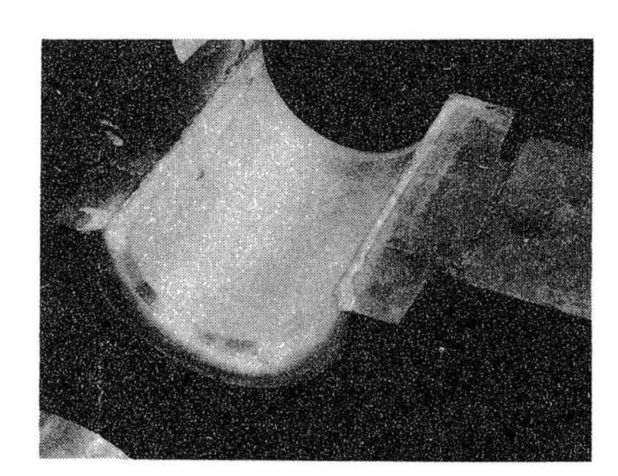

Figure 19.4 Direct lined housing, as lined

(b) Centrifugal lining

This method is to be preferred if size and shape of bearing are suitable, and if economic quantities require relining.

(*i*) *Centrifugal lining equipment*

For small bearings a lathe bed may be adapted if suitable speed control is provided. For larger bearings, or if production quantities merit, special machines with variable speed control and cooling facilities, are built by specialists in the manufacture or repair of bearings.

(*ii*) *Speed and temperature control*

Rotational speed and pouring temperature must be related to bearing bore diameter, to minimise segregation and eliminate shrinkage porosity.

Rotational speed must be determined by experiment on the actual equipment used. It should be sufficient to prevent 'raining' (i.e. dropping) of the molten metal during rotation, but not excessive, as this increases segregation. Pouring temperatures are dealt with in a subsequent section.

(*iii*) *Cooling facilities*

Water or air–water sprays must be provided to effect directional cooling from the outside as soon as pouring is complete.

Figure 19.5 Lining fixture for relining of shell type bearing

Figure 19.6 Pouring operation in relining of shell type bearing

(*iv*) *Control of volume of metal poured*

This is related to size of bearing, and may vary from a few grams for small bearings to many kilograms for large bearings.

The quantity of metal poured should be such that the bore will clean up satisfactorily, without leaving dross or surface porosity after final machining.

Excessively thick metal wastes fuel for melting, and increases segregation.

(*v*) *Advantages*

Excellent bonding of whitemetal to shell or housing.
Freedom from porosity and dross.
Economy in quantity of metal poured
Directional cooling.
Control of metal structure.

(*vi*) *Precautions*

High degree of metallurgical control of pouring temperatures and shell temperatures required.

Close control of rotational speed essential to minimise segregation.

Measurement or control of quantity of metal poured necessary.

Control of timing and method of cooling important.

Figure 19.7 Purpose-build centrifugal lining machine for large bearings

Figure 19.8 Assembling a stem tube bush 680 mm bore by 2150 mm long into a centrifugal lining machine

(4) POURING TEMPERATURES

(a) Objective

In general the minimum pouring temperature should be not less than about 80°C above the liquidus temperature of the whitemetal, i.e. that temperature at which the whitemetal becomes completely molten, but small and thin 'as cast' linings may require higher pouring temperatures than thick linings in massive direct lined housings or large and thick bearing shells.

The objective is to pour at the minimum temperature consistent with adequate 'feeding' of the lining, in order to minimise shrinkage porosity and segregation during the long freezing range characteristic of many whitemetals. Table 19.2 gives the freezing range (liquidus and solidus temperatures) and recommended minimum pouring temperatures of a selection of typical tin-base and lead-base whitemetals. However, the recommendations of manufacturers of proprietary brands of whitemetal should be followed.

Table 19.2 Whitemetals, solidification range and pouring temperatures

Specification	*Nominal composition %*					*Solidus temp °C*	*Liquidus temp °C*	*Min. pouring temp °C*
	Antimony	*Copper*	*Other*	*Tin*	*Lead*			
ISO 4381	12	6	0	Remainder	2	183	400	480
Tin base	7	3	0	Remainder	0.4	233	360	440
Alloys	7	3	1.0 Cd	Remainder	0.4	233	360	440
ISO 4381	14	0.7	1.0 As	1.3	Remainder	240	350	450
Lead base	15	0.7	0.7 Cd 0.6 As	10	Remainder	240	380	480
Alloys	14	1.1	0.5 Cd 0.6 As	9	Remainder	240	400	480
	10	0.7	0.25 As	6	Remainder	240	380	480

(b) Pouring

The whitemetal heated to the recommended pouring temperature in the whitemetal bath, should be thoroughly mixed by stirring, without undue agitation. The surface should be fluxed and cleared of dross immediately before ladling or tapping. Pouring should be carried out as soon as possible after assembly of the preheated shell and jig.

(c) Puddling

In the case of large statically lined bearings or housings, puddling of the molten metal with an iron rod to assist the escape of entrapped air, and to prevent the formation of contraction cavities, may be necessary. Puddling must be carried out with great care, to avoid disturbance of the structure of the freezing whitemetal. Freezing should commence at the bottom and proceed gradually upwards, and the progress of solidification may be felt by the puddler. When freezing has nearly reached the top of the assembly, fresh molten metal should be added to compensate for thermal contraction during solidification, and any leakage which may have occurred from the assembly.

(d) Cooling

Careful cooling from the back and bottom of the shell or housing, by means of air–water spray or the application of damp cloths, promotes directional solidification, minimises shrinkage porosity, and improves adhesion.

(5) BOND TESTING

The quality of the bond between lining and shell or housing is of paramount importance in bearing performance. Non destructive methods of bond testing include:

(a) Ringing test

This is particularly applicable to insert or shell bearings. The shell is struck by a small hammer and should give a clear ringing sound if the adhesion of the lining is good. A 'cracked' note indicates poor bonding.

(b) Oil test

The bearing is immersed in oil, and on removal is wiped clean. The lining is then pressed by hand on to the shell or housing adjacent to the joint faces or split of the bearing. If oil exudes from the bond line, the bonding is imperfect.

(c) Ultrasonic test

This requires specialised equipment. A probe is held against the lined surface of the bearing, and the echo pattern resulting from ultrasonic vibration of the probe is observed on a cathode ray tube. If the bond is satisfactory the echo occurs from the back of the shell or housing, and its position is noted on the C.R.T. If the bond is imperfect, i.e. discontinuous, the echo occurs at the interface between lining and backing, and the different position on the C.R.T. is clearly observable. This is a very searching method on linings of appropriate thickness, and will detect small local areas of poor bonding. However, training of the operator in the use of the equipment, and advice regarding suitable bearing sizes and lining thicknesses, must be obtained from the equipment manufacturers.

This method of test which is applicable to steel backed bearings is described in ISO 4386-1 (BS 7585 Pt 1). It is not very suitable for cast iron backed bearings because the cast iron dissipates the signal rather than reflecting it. For this material it is better to use a gamma ray source calibrated by the use of step wedges.

(d) Galvanometer method

An electric current is passed through the lining by probes pressed against the lining bore, and the resistance between intermediate probes is measured on an ohm-meter. Discontinuities at the bond line cause a change of resistance. Again, specialised equipment and operator training and advice are required, but the method is searching and rapid within the scope laid down by the equipment manufacturers.

(6) LOCAL REPAIR BY PATCHING OR SPRAYING

In the case of large bearings, localised repair of small areas of whitemetal, which have cracked or broken out, may be carried out by patching using stick whitemetal and a blowpipe, or by spraying whitemetal into the cavity and remelting with a blowpipe. In both cases great care must be taken to avoid disruption of the bond in the vicinity of the affected area, while ensuring that fusion of the deposited metal to the adjacent lining is achieved.

The surface to be repaired should be fluxed as described in section (2) prior to deposition of the patching metal. Entrapment of flux must be avoided.

The whitemetal used for patching should, if possible, be of the same composition as the original lining.

Patching of areas situated in the positions of peak loadings of heavy duty bearings, such as main propulsion diesel engine big-end bearings, is not recommended. For such cases complete relining by one of the methods described previously is to be preferred.

THE PRINCIPLE OF REPLACEMENT BEARING SHELLS

Replacement bearing shells, usually steel-backed, and lined with whitemetal (tin or lead-base), copper lead, lead bronze, or aluminium alloy, are precision components, finish machined on the backs and joint faces to close tolerances such that they may be fitted directly into appropriate housings machined to specified dimensions.

The bores of the shells may also be finish machined, in which case they are called 'prefinished bearings' ready for assembly with shafts or journals of specified dimensions to provide the appropriate running clearance for the given application.

In cases where it is desired to bore *in situ*, to compensate for misalignment or housing distortion, the shells may be provided with a boring allowance and are then known as 'reboreable' liners or shells.

The advantages of replacement bearing shells may be summarised as follows:

(1) Elimination of hand fitting during assembly with consequent labour saving, and greater precision of bearing contour.
(2) Close control of interference fit and running clearance.
(3) Easy replacement.
(4) Elimination of necessity for provision of relining and machining facilities.
(5) Spares may be carried, with saving of bulk and weight.
(6) Lower ultimate cost than that of direct lined housings or rods.

Special Note

'Prefinished' bearing shells must not be rebored *in situ* unless specifically stated in the maker's catalogue, as many modern bearings have very thin linings to enhance load carrying capacity, or may be of the overlay plated type. In the first case reboring could result in complete removal of the lining, while reboring of overlay-plated bearings would remove the overlay and change the characteristics of the bearing.

Linings are attached to their shoes by riveting or bonding, or by using metal-backed segments which can be bolted or locked on to the shoes. Riveting is normally used on clutch facings and is still widely used on car drum brake linings and on some industrial disc brake pads. Bonding is used on automotive disc brake pads, on lined drum shoes in passenger car sizes and also on light industrial equipment.

For larger assemblies it is more economical to use bolted-on or locked-on segments and these are widely used on heavy industrial equipment. Some guidance on the selection of the most appropriate method, and of the precautions to be taken during relining, are given in the following tables.

Table 20.1 Ways of attaching friction material

	Riveting	*Bonding*	*Bolted-on-segments*	*Locked-on-segments*
	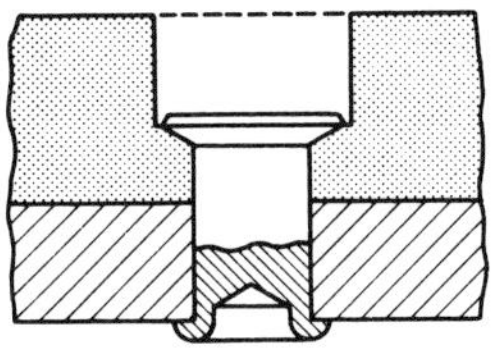	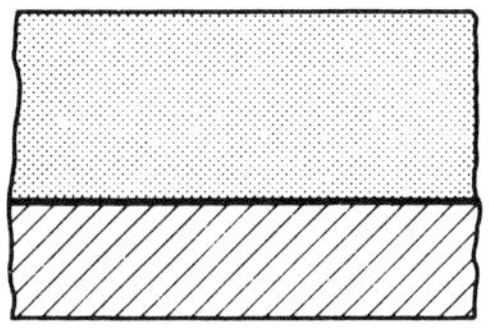	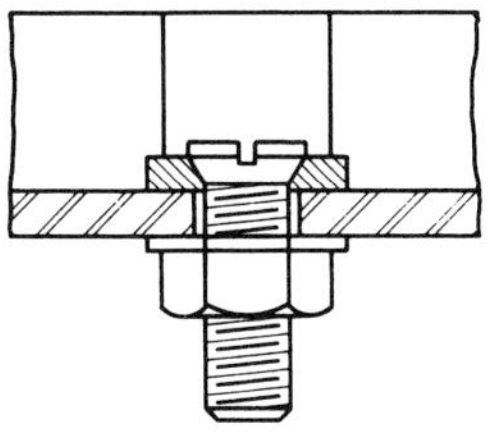	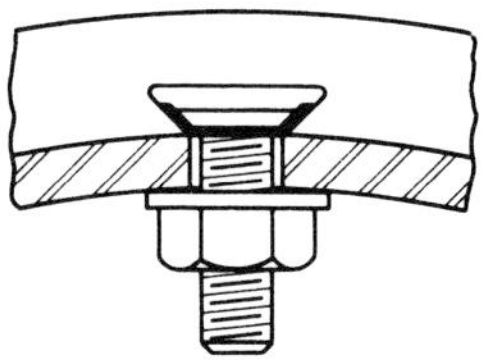
Shoe relining	De-riveting old linings and riveting on new linings can be done on site. General guidance is given in BS 3575 (1981) SAE J660	Shoes must be returned to factory. Cannot be done on site	Can be relined on site without dismantling brake assembly. Bolts have to be removed	Can be relined on site without dismantling brake assembly by slackening off bolts
Plate clutch relining	As above	As above	Not applicable	Not applicable
Use of replacement shoes already lined	Quick. Old shoes returned in part exchange	Quick. Old shoes returned in part exchange	Quick	Quick
Use of replacement clutch plates already lined	As above	As above	Not applicable	Not applicable
Friction surface	Reduced by rivet holes	Complete unbroken surface, giving full friction lining area	As for bonding apart from bolt slots in side of lining	As for bonding
Life	Amount of wear governed by depth of rivet head from working surface. If linings are worn to less than 0.8 mm (0.031 in) above rivets they should be replaced	Not affected by rivets or rivet holes. Can be worn right down. If linings are worn to less than 1.6 mm (0.062 in) above shoe they should be replaced	Governed by thickness of tie plates. Advantage over rivets, or countersunk screws. If linings are worn to less than 0.8 mm (0.031 in) above tie plates they should be replaced	Governed by depth of keeper plates. Comparable with use of rivets or countersunk screws. If linings are worn to less than 0.8 mm (0.031 in) above countersunk screws they should be replaced

Table 20.1 (continued)

	Riveting	*Bonding*	*Bolted-on segments*	*Locked-on segments*
Spares	Good. Linings drilled or undrilled can be supplied ex-stock together with rivets. Small space required for stocks	Bulky. Complete shoe or plate with lining attached required. Where large metal shoes or plates are involved there is a high cost outlay and extra storage space	Good. Only linings with tie plates bonded into them required	Good. Only linings required suitably grooved
Limitations	Less suitable for low-speed, high-torque applications	Less suitable for high ambient temperatures, corrosive atmospheres, or where bonding to alloy with copper content of over 0.4%	12 mm (½ in) thick or over, up to 610 mm (24 in) long	12 mm (½ in) thick or over, up to 610 mm (24 in) long
Suitable applications	General automotive and industrial	Large production runs. Attachment of thin linings. Attachment to shoes where other methods are not practicable	For attachment to shoes which are not readily dismantled such as large winding engines and excavating machinery brakes, also high torque applications	As for bolted-on segments

Table 20.2 Practical techniques and precautions during relining

	Riveting	*Bonding*	*Bolted-on segments*	*Locked-on segments*
Removing lining or facings	Best to strip old linings and facings by drilling out rivets, taking care to avoid damage to the rivet holes and shoe platform	Best done as a factory job	Slacken off brake adjustment. Slacken off nuts on main side. Remove nuts and bolts from outer side and slide linings across the slots in side of the lining	Slacken off brake adjustment. Slacken off bolts sufficiently to allow linings to slide along the keeper plates. If necessary tap the linings with a wooden drift to assist removal
Replacing lining or facings	Clean shoes and spinner plate, replace if distorted or damaged. If new linings or facings are drilled clamp to new shoe or pressure plate, insert rivets, clench lightly. Insert all rivets before securely fastening. If undrilled, clamp to shoe or spinner plate, locate in correct position and drill holes using the drilled metal part as a template. Counterbore on opposite side. Use same procedure as for drilled linings or drilled facings to complete the riveting. During relining particular attention should be given to rivet and hole size and also to the clench length	See above	Replace by the reverse procedure using the slots to locate the linings. Tighten up all bolts and readjust the brake	Replace by the reverse procedure. Afterwards re-adjust the brake

Table 20.2 (continued)

	Riveting	*Bonding*	*Bolted-on segments*	*Locked-on segments*
Precautions	Use brass or brass-coated steel rivets to avoid corrosion problems. Copper rivets may be used for passenger car linings and also in light industrial applications. There must be good support for the rivet and the correct punch must be used. Allow one third the thickness of lining material under rivet head	Best done as a factory job	Best to use high-tensile steel socket-head type of bolt. Can be applied to all linings over 12 mm ($\frac{1}{2}$ in) thick	Avoid over-tightening the bolts so as not to distort the heads or crack the linings. Generally linings must be 12 mm ($\frac{1}{2}$ in) thick or over for rigid material and at least 25 mm (1 in) thick for flexible material
	Care must be taken during relining to avoid the lining becoming soaked or contaminated by oil or grease, as it will be necessary to replace the lining, or segment of lining, for if not its performance will be reduced throughout its life			

Table 20.3 Methods of working the lining and finishing the mating surfaces

Material	*Cutting*	*Drilling*	*Surface finishing*	*Handling*
Woven materials	Hacksaw or bandsaw. Grinding is not recommended as it gives a scuffed surface and possible fire hazard	HSS tools are suitable. A burnishing tool is necessary to remove ragged edges	HSS tools are suitable for turning or boring of facings	Can be more easily bent to radius by heating to around 60°C When machining and handling asbestos based materials, work must be carried out within the relevant asbestos dust regulations.
Moulded materials	Hacksaw, bandsaw or abrasive wheels	Tungsten carbide (WC) tipped tools are usually needed	Tungsten carbide (WC) tipped tools are usually needed for turning or boring of facings	Care must be taken to give adequate support during machining because of their brittle nature
Mating members	Fine/medium ground to 0.63–1.52 μm (25–60 μin) cla surface finish is best. Avoid chatter marks, keep drum ovality to within 0.127 mm (0.005 in) and discs parallel to within 0.076 mm (0.0003 in). Surface should be cleaned up if rust, heat damage or deep scoring is evident but shallow scoring can be tolerated. If possible the job should be done *in situ* or with discs or drums mounted on hubs or mandrels. The total amount removed by griding from the disc thickness or the drum bore diameter should not exceed: 1.27 mm (0.05 in) on passenger cars 2.54 mm (0.1 in) on commercial vehicles. If these values are exceeded a replacement part should be fitted. When components have been ground, a thicker lining should be fitted to compensate for the loss of metal. With manual clutches the metal face can be skimmed by amounts up to 0.25 mm (0.01 in). For guidance on the reconditioning of vehicle disc and drum brakes, reference should be made to the vehicle manufacturer's handbook.			

Industrial flooring materials

Factors to consider in the selection of a suitable flooring material

Factor	*Remarks*
Resistance to abrasion	This is usually the most important property of a flooring material because in many cases it determines the effective life of the surface. Very hard materials which resist abrasion may, however, have low impact resistance
Resistance to impact	In heavy engineering workshops this is often the determining factor in the choice of flooring material
Resistance to chemicals and solvents	In certain industrial environments where particular chemicals are likely to be spilled on the floor, the floor surfacing must not be attacked or dissolved
Resistance to indentation	Any permanent indentation by shoe heels or temporarily positioned equipment is unsightly, makes cleaning difficult and, in severe cases, can cause accidents.
Slipperiness	Slipperiness depends not only on the floor surfacing material, but on its environment. Cleanliness is important. Any adjacent floors which are wax polished, can result in wax layers being transferred to an otherwise non slip surface by foot traffic. Adjacent floors with different degrees of slipperiness can cause accidents to unaccustomed users
Other safety aspects	Potholing, cracking and lifting can occur in badly laid floors. In high fire risk areas, floors which do not generate static electricity are required. Non absorbent floors are normally necessary in sterile areas
Ease of cleaning	This is a key factor in total flooring cost, and in maintaining the required properties of the floor
Comfort	In light engineering workshops, laboratories and offices, comfort can usually be taken more into account without sacrificing the performance of the floor from other aspects
Initial laying	The standard of workmanship and the familiarity of trained operators with the laying process can have a major effect on floor performance. Faults in foundations, or in a sub floor can result in faults in the surface
Subsequent repair	This must be considered when selecting a floor material, particularly in applications where damage is inevitable. Small units like tiles can usually be repaired quickly. Asphalt floors can be used as soon as they are cool, but need space for heating equipment and specialised labour. Cement and concrete can be repaired by local labour, but production time is lost while waiting for hardening and drying
Cost	The initial relative cost of different materials should be compared with their probable life. British Standards and Codes of Practice describe non proprietary materials. Manufacturers of proprietary materials often have independent test data available

Comparative properties of some common floor finishes

Type of finish	Wear resistance: Abrasion	Impact	Resistance to indentation	Slipperiness	Ease of cleaning	Resistance to: Acid	Alkalis	Sulphates
Portland Cement concrete	G—P	G—P	VG	G—F	F	VP	G	VP
Portland cement precast	G	G—F	VG	G—F	F	VP	G	P
High alumina cement concrete	G—P	G—P	VG	G—F	F	VP	P	G
Granolithic concrete	VG—F	G	VG	G—F	G	VP	P	P
Mastic asphalt	G—F	G—F	F—P	G—F	G—F	VG—F	G—F	VG
Cement bitumen	G—F	G—F	F—P	G—F	G—F	P	G	P
Pitch mastic	G—F	G—F	F—P	G—F	G—F	G—F	G—F	VG
Steel or cast iron tile	VG	VG	VG	F	G—F	VP	F	F—P
Steel anchor plates in PCC	VG	VG[1]	VG	G—F	F	VP	G	VP
Steel grid in PCC	VG	G[2]	VG	G[2]	F—P	VP	G	VP
Steel grid in mastic asphalt	G	G—F	G	G	F—P	P	G	G
Rubber sheet	G	VG	VG	G—F[3]	G	F—P	F—P	VG
Linoleum sheet or tile	G—F	F	F	G	G	P	P	G
PVC sheet and tile	G	G—F	G	G	G	G	G	G
Magnesium oxy chloride	G—F	F	F	F	G	P	F	F
Terrazzo	G	G—F	VG	G[4]	G	P	G—F	G—F
Thermo plastic tile	F	F	F	G	G[5]	F	F	G
Timber softwood board	F	G	F	G	G	F—P	F—P	F—P
Timber softwood block	F	G	F	G	G	F—P	F—P	F—P
Timber hardwood strip	G	G	G	G	G	F	F	F
Timber hardwood block	G	G	G	G	G	F	F	F
Wood chip board and block	F	F	VG	G	G	F	F	F
Clay tile and bricks	G	G—F	VG	G—F[4]	G	G—F	G—F	VG—G
Cement PVA emulsion	G—F	G—F	F	G	G—F	P	G	F
Cement rubber latex	G—F	G—F	F	G	G—F	P	G	F
Composition blocks	G	G	VG	F	F	F	F	F
Concrete tiles	G	G—F	VG	G—F	F	VP	G	P
Cork carpet	G	G	P	VG	F	G	G	G
Cork tiles	G	G	P	VG	F	G	G	G
Granite slab	VG	VG	VG	G	G	G	G	VG
Sand stone slab	G—F	G—F	VG	G	G—F	G—P	G—P	G—F
Sawdust cement	F	F	F	G	F	P	P	VP

VG = Very good G = Good F = Fair P = Poor VP = Very poor

[1] Particularly suitable for heavy engineering workshops
[2] The grid size should be chosen to expose sufficient concrete for non slip purposes, but in small enough areas to reduce damage by impact
[3] Rubber can be slippery when wet, particularly with rubber soled shoes
[4] Clay tiles and terrazzo become slippery when polished or oiled
[5] Thermoplastic tiles require a special type of polish

REAL AREA OF CONTACT

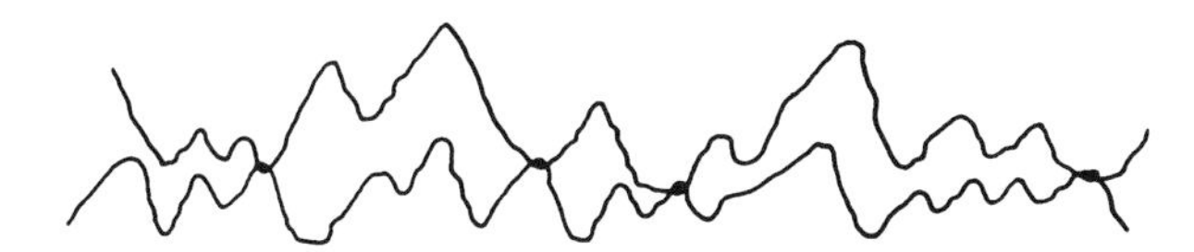

Contact between flat surfaces at light loads occurs at asperity tips only — the scale of the surface roughness does not matter.

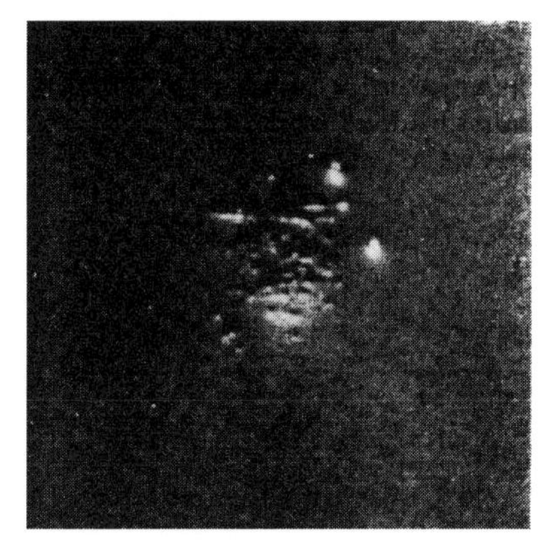

0·2 N
12·7mm (0·5 in)

Steel ball loaded against aluminised glass block, viewed through the block* × *200

In concentrated contacts, as between a ball and its race, contact still occurs at discrete points.

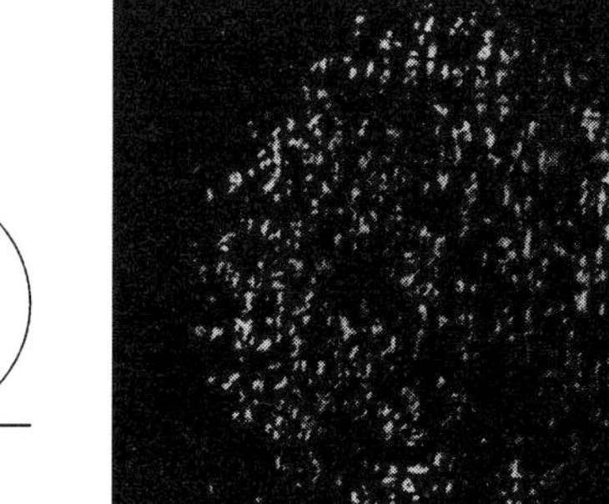

Metal transfer from a copper-coated steel ball loaded against steel plate* ×*38 *(courtesy* K. L. Johnson)

At practical loads, contact is still non-uniform.

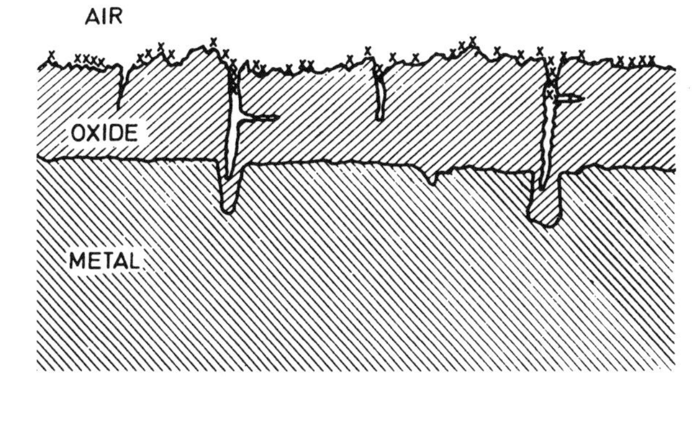

The surface of a metal consists of a thin, often transparent oxide film (0.01–0.1 μm thick), containing cracks and pores. Molecules of water, oxygen and grease are weakly attached to the oxide. Below the oxide may be a layer of mixed-up oxide and metal, often extremely hard, (perhaps 0.1 μm thick), and below this the metal will be work-hardened to a depth of 1–10 μm.

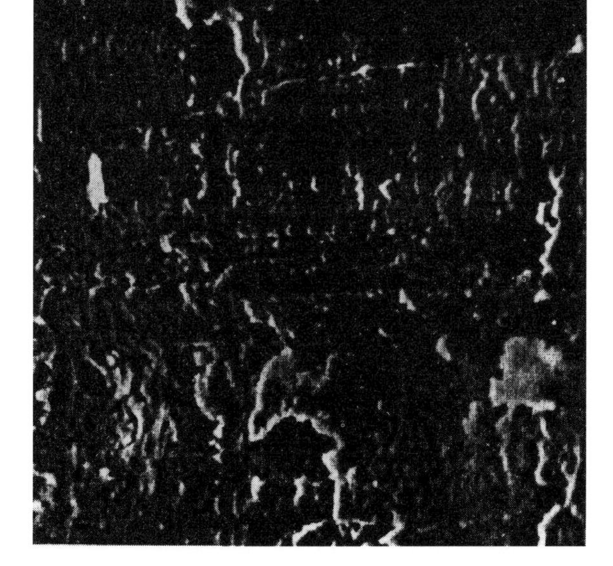

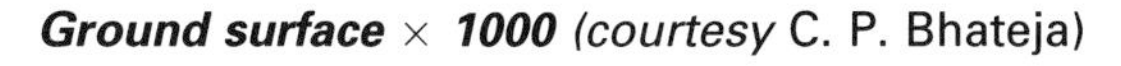

Ground surface* × *1000 *(courtesy* C. P. Bhateja)

All surfaces, even those which feel smooth and give good reflections, are rough (see Fig. above). The earth's surface provides a good model, since most slopes are only a few degrees, though local features can be very much steeper. But metal surfaces often have overhangs and tenuously attached flakes; indeed chemical measurements show that the real surface area of an abraded metal can be three or more times the apparent area.

Even in a Brinell indentation, contact occurs at the asperity tips, and the asperities persist, though rather deformed

Note that the magnification of the vertical scale is 50 times the horizontal scale and consequently the sharpness of the peaks and valleys is exaggerated.

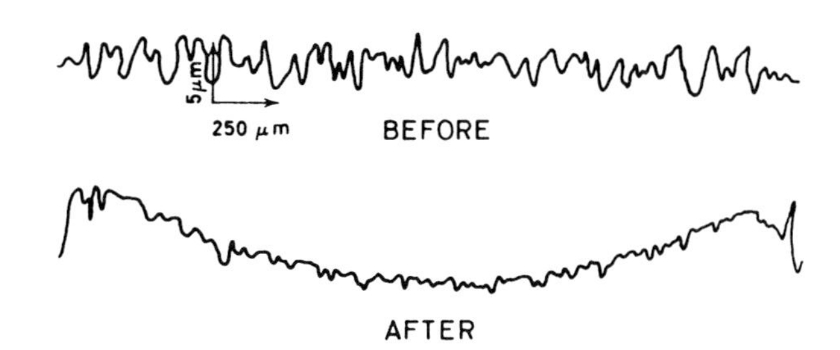

Talysurf trace of a bead-blastered surface before and after indentation

Estimate of contact area

The real area of contact depends on the load and not on the apparent area of contact. A useful estimate (in mm^2) is given by:

$$\frac{W}{10H}$$

where W is the load in Newtons and H the hardness (Brinell or Vickers) of the softer member.

ELASTIC AND PLASTIC CONTACT

Elastic contact between a ball and a plane, or between two balls, is described by the Hertz equations:

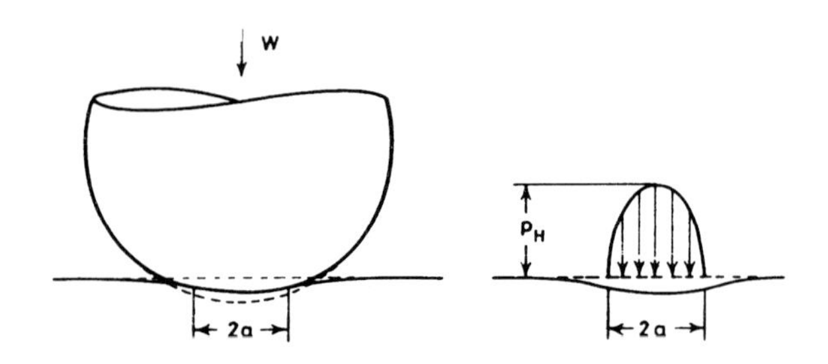

Contact geometry. Contact pressure

contact radius $\quad a = \frac{3}{4}\left(\frac{WR}{E}\right)^{\frac{1}{3}}$

maximum (or Hertz) pressure

$$p_H \simeq 0.6\left(\frac{WE^2}{R^2}\right)^{\frac{1}{3}}$$

where $\quad \frac{1}{E} = \frac{1-\nu_1^2}{E_1} + \frac{1-\nu_2^2}{E_2}$

E_1 and E_2 being Young's moduli and ν_1 and ν_2 Poisson's ratios for the two bodies.

$$\frac{1}{R} = \frac{1}{R_1} + \frac{1}{R_2}$$

R_1 and R_2 the radii of curvature of the two bodies. For a ball in a cup, the cup radius of curvature is taken negative.

Even though the real areas of contact are discrete points within the Hertz area and the load is actually transmitted through the real areas with much higher pressures, these equations give the overall dimensions and overall pressure correctly, except for rough surfaces at light loads when the contacts are dispersed over a larger area.

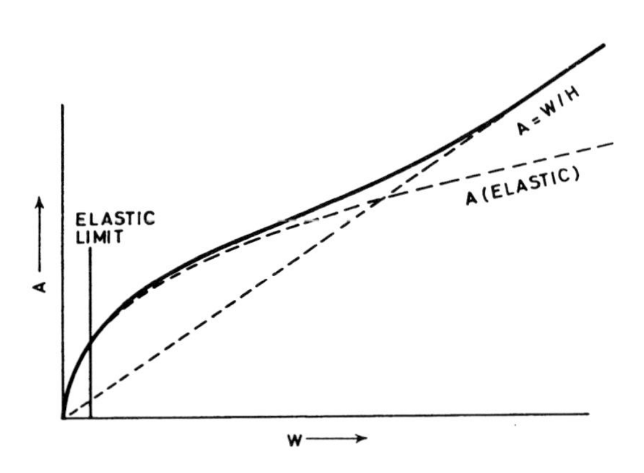

At heavy loads plastic flow takes place, beginning when p_H reaches 1.8 × (yield stress in tension of softer body). The area then increases more rapidly with load, approaching the value (load/hardness), and on unloading leaves a permanent impression ('Brinelling'). But even in this range, the real area of contact is only $\frac{1}{3}$ to $\frac{1}{2}$ of the area of the impression.

Asperity deformation

Individual asperity contacts probably behave like the ball and plane just described, so that each contact deforms elastically when it carries a small load and plastically when it carries a high one.

But as the total load increases the surfaces move closer together and the number of asperity contacts increases. The new small contacts which form balance the growth of the existing contacts, and the average contact size is unchanged. The number of contacts is roughly proportional to the load, and the fraction of the load carried by elastic contacts will not change — even though the original elastic contacts have become plastic. Contact between flat surfaces is therefore elastic or plastic depending on the surface geometry and material properties, but does not change from elastic at low loads to plastic at high ones.

Plasticity index

The plasticity index is $\frac{E}{H}\left(\frac{\sigma}{\beta}\right)^{\frac{1}{2}}$ where E is Young's modulus, H is the hardness of softer component, σ is the standard deviation of asperity heights (approximately equal to the cla (see Section E2) for new surfaces but less for worn ones) and β the mean asperity radius of curvature. Values below 1 indicate mainly elastic contacts; values above 3 indicate mainly plastic contacts. Very few manufactured surfaces come below 10 — ball and roller bearings being an important exception with an index around 1.

Running-in produces smoother surfaces (σ decreases and β increases) and contact then becomes elastic, though this is partly due to better *conformity* between the surfaces as well as lower roughness. But non-geometric effects like *toughening* and *surface oxidation* are also involved in running in.

CONTACT SIZES

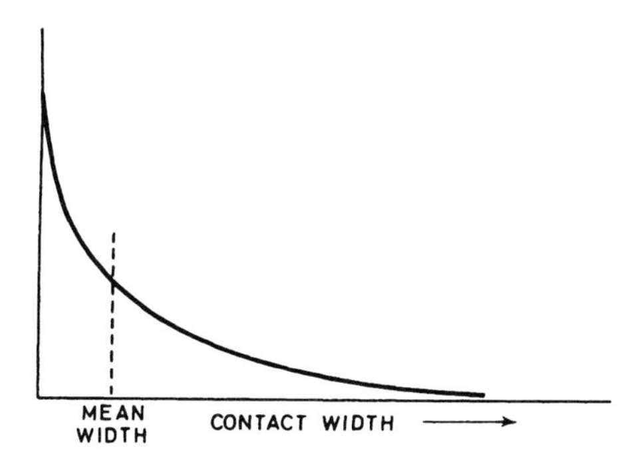

Individual contacts have a large range of sizes with a few large ones and very many small ones. But the *mean* contact width does not vary greatly with the pressure, provided this is less than one tenth of the hardness, nor even between surfaces finished in different ways.

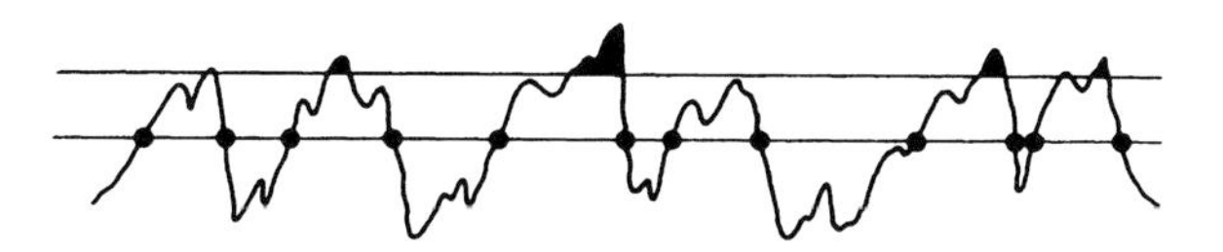

Zero-crossings and contact widths

A good approximation is to count the number of times n the surface profile crosses the mean line per unit length: the mean contact width is about $0.2/n$. The contacts in the figure follow this rule. For many surfaces, especially surface ground ones, n is about 100/mm, giving a mean contact width of 2 μm. This value can generally be used except at high pressures when the contacts get bigger and closer together (as in the photographs of contact areas under heavily loaded balls).

In practice surface waviness and misalignment can often give high apparent pressures locally when the nominal pressures are low, and the real contact areas tend to be grouped instead of occurring randomly.

INTRODUCTION

Manufacturing processes tend to leave on the surface of the workpiece characteristic patterns of hills and valleys known as the texture. The texture produced by stock removal processes is deemed to have components of roughness and waviness. These may be superimposed on further deviations from the intended geometrical form, for example, those of flatness, roundness, cylindricity, etc.

Functional considerations generally involve not only the topographic features of the surface, each having its own effect, but also such factors as the properties especially of the outer layers of the workpiece material, the operating conditions, and often the charcteristics of a second surface with which contact is made.

While the properties of the outer layers may not differ from those of the material in bulk, significant changes can result from the high temperatures and stresses often associated particularly with the cutting and abrasive processes.

Optimised surface specification thus becomes a highly complex matter that often calls for experiment and research, and may sometimes involve details of the process of manufacture.

Surface profiles

The hills and valleys, although very small in size, can be visualised in the same way as can those on the surface of the earth. They have height, shape and spacing from one peak to the next. They can be portrayed in various ways.

An ordinary microscope will give useful information about their direction (the lay) and their spacing, but little or none

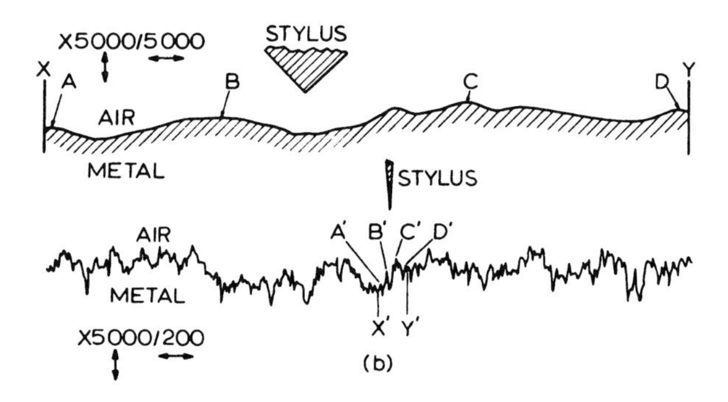

Fig. 2.1. Effect of horizontal compression

about their height. The scanning electron microscope can give vivid monoscopic or stereoscopic information about important details of topographic structure, but is generally limited to small specimens. Optical interference methods are used to show contours and cross-sections of fine surfaces. The stylus method, which has a wide range of application, uses a sharply pointed diamond stylus to trace the profile of a cross-section of the surface.

The peak-to-valley heights of the roughness component of the texture may range from around 0.05 μm for fine lapped, through 1 μm to 10 μm for ground, and up to 50 μm for rough machined surfaces, with peak spacings along the surface ranging from 0.5 μm to 5 mm. The height of the associated waviness component, resulting for example from machine vibration, should be less than that of the roughness when good machines in good order are used, but the peak spacing is generally much greater.

Because of the need for portraying on a profile graph a sufficient length of surface to form a representative sample, and the small height of the texture compared with its spacing, it is generally necessary to use far greater vertical than horizontal magnification. The effect of this on the appearance of the graph, especially on the slopes of the flanks, is shown in Fig. 2.1. The horizontal compression must always be remembered.

The principle of the stylus method is basically the same as that of the telescopic level and staff used by the terrestrial surveyor, and sketched in Fig. 2.2(a). In Fig. 2.2(b), the stylus T is equivalent to the staff and the smooth datum surface P is equivalent to the axis of the telescope. The vertical displacements of the stylus are usually determined by some form of electric transducer and amplifying system.

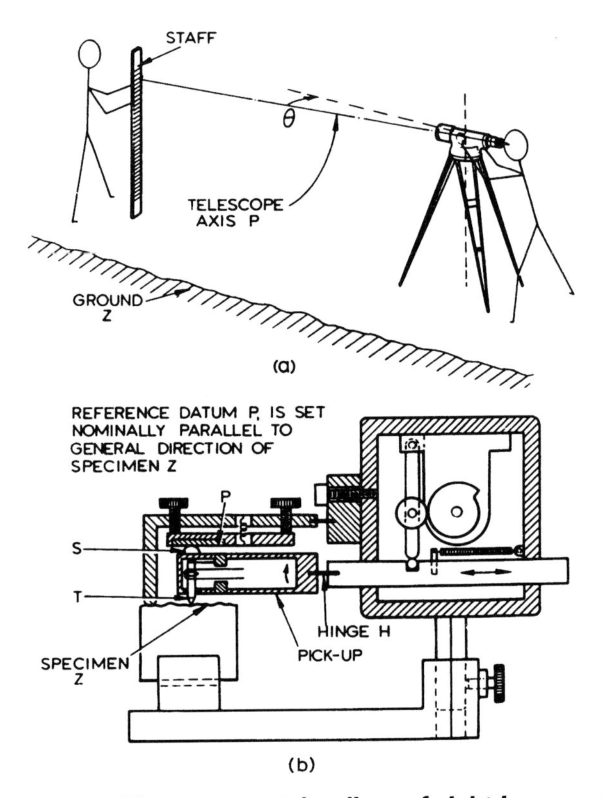

Fig. 2.2. The surveyor takes lines of sight in many directions to plot contours. The engineer plots one or more continuous, but generally unrelated cross-sections (a) Telescope axis usually set tangential to mean sea level by use of bubble in telescope (b) Skid S slides along the reference surface. Stylus T is carried on flexure links or a hinge

For convenience, the datum surface P of Fig. 2.2(b) is often replaced by another form of datum provided by the surface of the workpiece itself (Fig. 2.3), over which slides a rounded skid. This form of datum may be quicker to use but it is only approximate, as the unwanted vertical excursions of the skid combine in various indeterminate ways with the wanted excursions of the stylus, and in practice the combination can be accepted only when the asperities are deep enough and close enough together for the excursions of the skid of given radius to be small compared with those of the stylus.

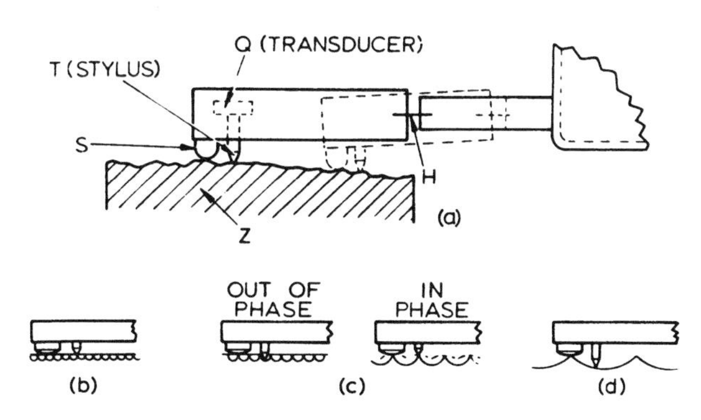

Fig. 2.3. (a) Skid S slides over crests of specimen (b) acceptable approximation to independent datum (c) significant skid error, negative or positive according to whether the skid and stylus move in or out of phase (d) acceptable mechanical filtration of wavelengths which are long compared with separation of skid and stylus

The diamond stylus may have the form of a 90° cone or 4-sided pyramid, with standardised equivalent tip radii and operative forces of 2 μm (0.7mN) or 10 μm (16mN). While these tips are sharp enough for the general run of engineering surfaces, values down to 0.1 μm (10 μN) can be used in specialised instruments to give better resolution of the finest textures (e.g. those of gauge blocks).

Some typical profiles of roughness textures, horizontally compressed in the usual way are shown in Fig. 2.4.

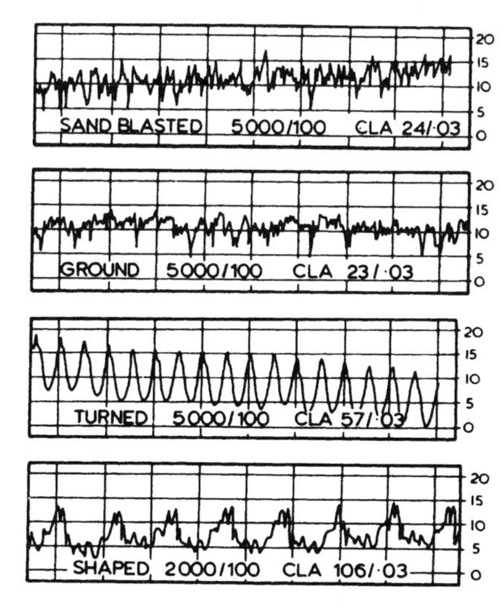

Fig. 2.4. Typical profiles. Magnification is shown thus: vertical/horizontal

NUMERICAL ASSESSMENT

Significance and preparation

For purposes of communication, especially on drawings, it is necessary to describe surface texture numerically, and many approaches to this have been considered. A numerical evaluation of some aspect of the texture is often referred to as a 'parameter'. The height, spacing, slope, crest curvatures of the asperities, and various distributions and correlation factors of the roughness and waviness can all be significant and contribute to the sum total of information that may be required; but no single parameter dependent on a single variable can completely describe the surface, because surfaces having quite different profiles can be numerically equal with respect to one such parameter while being unequal with respect to others. Economic considerations dictate that the number admitted to workshop use should be minimised.

The profile found by the pick-up may exhibit (*a*) tilt relative to the instrument datum, (*b*) general curvature, (*c*) long wavelengths classed as 'waviness' and (*d*) the shorter wavelengths classed as roughness, shown collectively by the profile in Fig. 2.5(c). The first two of these being irrelevant to the third and fourth, some preparation of the profile is required before useful measurement can begin. Preparation involves recognition and isolation of the irregularities to be measured, and the establishment of a suitable reference line from which to measure those selected.

Recognition of the different kinds of texture and of their boundaries may involve some degree of judgement and experience. Broadly, roughness is deemed to include all those irregularities normally produced by the process, these being identified primarily on the basis of their peak spacing. Cutting processes generally leave feed marks of which the spacing can be recognised immediately. Abrasive textures are more difficult because of their random nature, but experience has shown that the peak spacings of the finer ones can generally be assumed to be less than 0.8 mm, while those of the coarser ones, which may be greater, tend to become reasonably visible.

Isolation is effected on a wavelength basis by some form of filtering process which has the effect of ironing out (i.e. attenuating) the longer wavelengths that do not form part of the roughness texture.

The reference line used for the assessment of parameters is generally not the instrument datum of Figs 2.1 and 2.2, but a reference line derived from the profile itself, this line taking the form of a mean line passing through those irregularities of the profile that have been isolated by the filtering process.

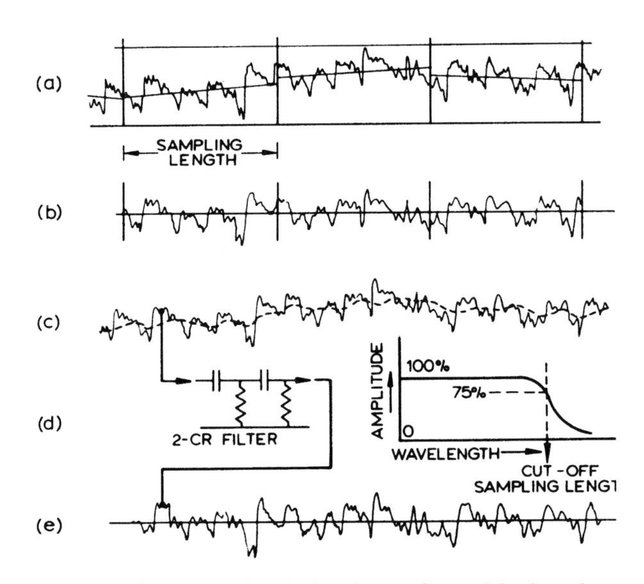

Fig. 2.5. Comparative behaviour of graphical and electrical methods of filtering. Residual differences between (b) and (e) are referred to as Method Divergence

The profile can be filtered graphically by restricting the measurement to a succession of very short sampling lengths, as shown in Fig. 2.5(a), through each of which is drawn a straight mean line parallel to its general direction. The length of each sample must be not less than the dominant spacing of the texture to be measured. If the individual samples are redrawn with their mean lines in line, the filtering effect becomes immediately apparent (Fig. 2.5(b)).

In the case of meter instruments, the alternating electric current representing the whole profile (Fig. 2.5(c)) is passed through an electric wave filter which transmits the shorter but attenuates the longer wavelengths. The standardised 2-CR filter network, its rate of attenuation and the position of the long wavelength cut-off accepted by convention (which is known as the *meter* cut-off) are shown in Fig. 2.5(d). When the meter cut-off is made equal to the graphical sampling length, the two methods are found to give, on average, equal numerical assessments.

The electric wave filter determines automatically a mean line which weaves its way through the input profile according to the way in which the filter reacts with the rates of change of the profile. This wave filter mean line is shown dotted in Fig. 2.5(c), where the profile is a repetition of Fig. 2.5(a). Relative to the output from the filter, the mean line becomes a straight line representing zero current (Fig. 2.5(e), cf. Fig. 2.5(b)). It is from this line that the meter operates and about which the filtered profile would be displayed on a recorder having a sufficient frequency response. Sampling lengths and filter characteristics to suit the whole range of textures are standardised in British, US, ISO and other Standards the usual values being 0.25mm, 0.8mm and 2.5mm.

Several graphical samples are usually taken consecutively to provide a good statistical basis. Meter instruments do the equivalent of this automatically; but care must be taken not to confuse the total length of traverse with the much shorter meter cut-off, for it is the latter that decides the greatest spacing on the surface to which the meter reading refers.

The 2-CR filter shown in Fig. 2.5(d) lends itself to simple instrumentation, but can distort the residual waveform that is transmitted for measurement. The amount of distortion is generally not sufficient to affect seriously the numerical assessments that are made.

An important point is that the signal fed to a recorder is generally *not* filtered, so that the graph can show as true a cross-section as the stylus, transducer and datum permit.

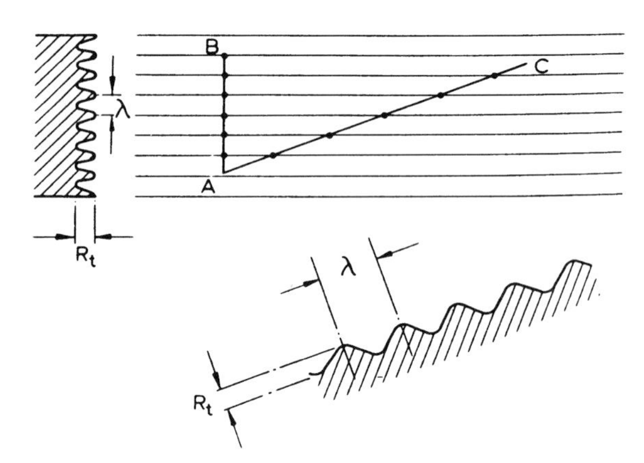

Fig. 2.6. Oblique traverse increases λ but not R_t.

It is generally best to measure in the direction in which a maximum of information can be collected from the shortest possible traverse. In Fig. 2.6, representing a surface finished by a cutting process, that would be in the direction *AB* lying across the lay. In the oblique direction *AC* the peaks are farther apart, the slopes are less and the radii of curvature of the peaks are greater, but the *height* is the same. Thus, if a meter reading of the height is taken transversely with a just sufficient meter cut-off, and then obliquely without corresponding increase in the meter cut-off, a low reading is likely to result, even though the total traverse is increased. This would also apply to ground surfaces. In the case of textures having a random lay (e.g. shot blast or lapped with a criss-cross motion) the spacing, and hence the minimum meter cut-off, may be much the same in all directions.

Parameters

Except for the fairly periodic textures sometimes produced by cutting processes, surface textures tend to vary randomly in height and spacing. The problem of describing the different kinds of texture offeres full scope for the devices of the statistician, from the simplest forms of averaging to the complexities of correlation functions.

Height

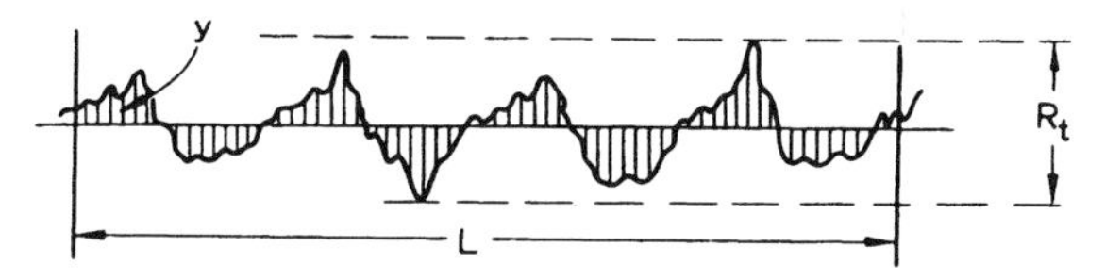

Fig. 2.7 $cla = \frac{1}{L}\int_0^L |y|\,dL$, $rms = \left(\frac{1}{L}\int_0^L y^2\,dL\right)^{\frac{1}{2}}$

The most commonly used and easily measured height parameter is the average departure from the mean line of the filtered profile. It is known in the USA as the AA value. In Great Britain it was known as the cla value, but it is now known as the R_a value to line up with ISO terminology.

The rms value has been used in approximate form, obtained by multiplying the AA value by 1.11 which is the conversion factor for sine waves.

The height from the highest peak to deepest valley found anywhere along a selected part of the profile, known as R_t, is used especially in Europe, but it is unsatisfactory because the single extremes are too greatly dependent on the chance position selected.

An averaged overall height is more representative than R_t. The 'ten point height R_z' is obtained by averaging the five highest peaks and five deepest valleys in the total traverse. In a German variant, known as R_{tm}, the average of the R_t values in five consecutive samples of equal length is taken.

Although the ratio of one height measure to another varies with the shape of the profile, some degree of conversion is generally possible as shown below:

Surfaces	rms/cla	10pt/cla	Peak-to-valley/cla
Turned*	1.1 to 1.15	4 to 5	4 to 5
Ground	1.18 to 1.30	5 to 7	7 to 14
Lapped	1.3 to 1.5	—	7 to 14
Statistically random	1.25	—	8.0

*With clean cut from round-nose tool.

Spacing

The spacing of the more significant peaks along the surface, may be functionally important.

Peaks per unit length have been counted, a peak being rated as such only if the adjacent valley exceeds a given depth. Bearing intercepts per unit length at a given level have also been counted.

Bearing area

A concept frequently encountered is that of 'bearing area', shown in abstract principle in Fig. 2.8, the level of the intersecting line being expressed either as a depth below the highest peak (an uncertain reference point) or as an offset from the mean line.

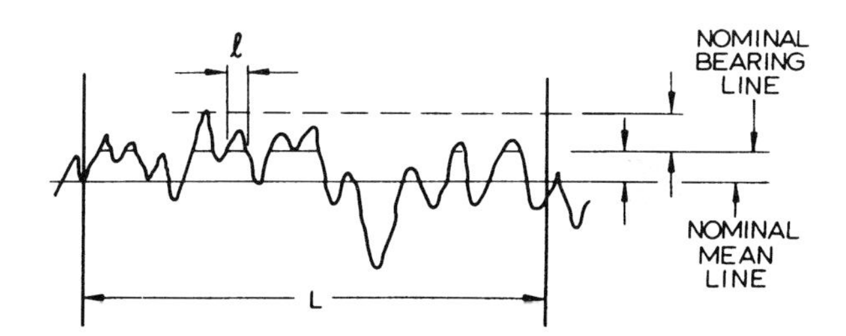

Fig. 2.8 Nominal bearing area of sample

$$\% = \frac{\sum l}{L} \times 100$$

It must be remembered that this measure is confined to a small sample of the surface and does not represent the overall bearing area taking waviness and errors of form into account, nor does it allow for elastic deformation of the peaks under load. These and other considerations limit its value.

Crest curvatures and height distribution

The average radius of curvature of the crests, and their height distribution, are referred to in connection with the plasticity index in Section E1. While meter indications are not generally available, these quantities can readily be computed from digitised profile records.

STRAIGHTNESS, FLATNESS, ROUNDNESS, CYLINDRICITY AND ALIGNMENT

Although these aspects are generally considered separately from surface texture, they can be highly significant to the functioning of surfaces, and must therefore receive at least some mention in the present section.

These aspects are generally measured with a blunt stylus that traces only the crests of the roughness, and does not appreciably enter the narrower scratch marks.

Straightness and flatness

Straightness and flatness are measured with the same basic type of apparatus as for texture in Fig. 2.1, but with the instrument datum made much longer. Horizontal magnifications are generally lower and compression ratios often higher than for texture measurement.

The normal engineering way of describing the deviations is in terms of the separation of two parallel lines or planes between which all deviations are contained. This is a maximum peak-to-valley measure that would not distinguish between the two surfaces in Fig. 2.9. A distinction can be made by measuring over the whole and also over a fraction of the length. This resembles the sampling length procedure used for surface texture.

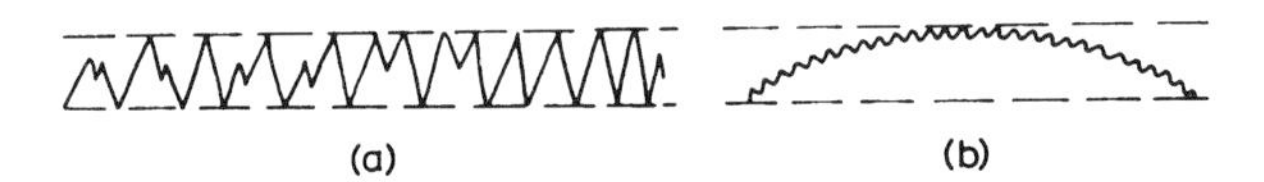

Fig. 2.9. (a) Rough but straight (b) Smooth but curved

Roundness

Roundness is generally measured by rotation of the pick-up or workpiece round a precisely generated axis (Fig. 2.10). Variations in the radius of the workpiece are plotted on a polar chart on which can be superposed a least-squares reference circle from which the radial deviations are determined. They are expressed in terms of the separation of two circles, drawn from a specified centre, that just contain the undulations. Four centres are possible, the two standardised being (*a*) the centre giving the minimum separation and (*b*) the least-squares centre (Fig. 2.11). The measure is again a maximum peak-to-valley value. For control of vibration in rolling bearings, the amplitude is sometimes assessed in three or four sharply defined wavebands.

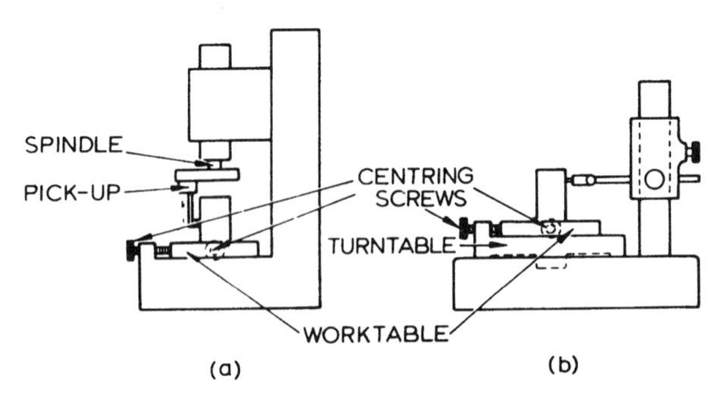

Fig. 2.10. Principle of roundness instruments (a) rotating pick-up (b) rotating workpiece

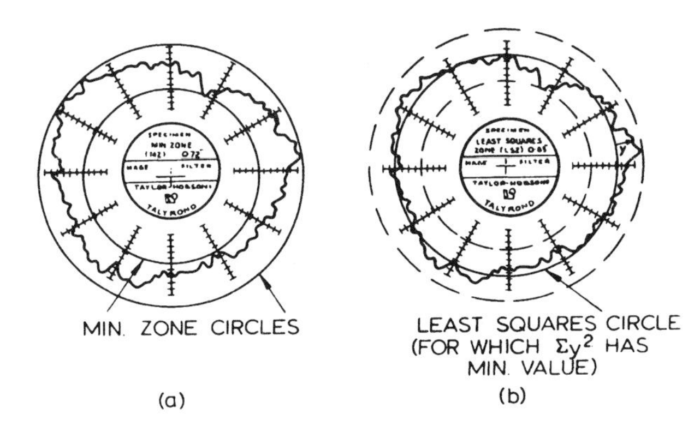

Fig. 2.11. Polar graphs showing methods of assessing radial variations

Since the angular magnification is always unity, the effective compression ratio is generally very much greater than for texture measurement, and the resulting forms of distortion must be fully understood.

Cylindricity

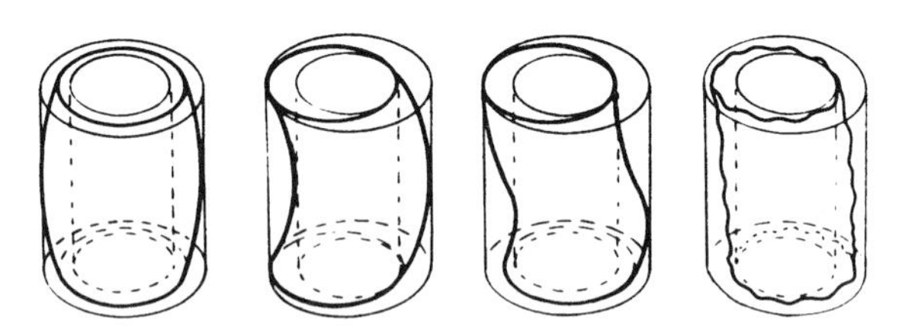

Fig. 2.12. Errors in cylindricity

The expression of cylindricity (Fig. 2.12) requires suitable instrumentation and display. The magnitude of the error is generally conceived in the same way as for straightness and roundness, but in this case as lying between two co-axial cylinders (or cones).

USE AND INTERPRETATION OF SURFACE MEASUREMENTS

The industrial requirement is (1) To investigate the characteristics of surfaces and identify the types that are functionally acceptable. (2) To specify the dominant requirements on drawings. (3) To control manufacture.

Ideally, the most economical surface or surfaces for a given application should be determined through the medium of experimental models closely allied to the intended method of manufacture. Surface texture tends to change as the surfaces are run in, and initially smooth surfaces are not necessarily best. The peaks can get smoothed down even after the first pass, but the run-in surface may retain traces of the original valleys (which may assist the distribution of lubricant) throughout the life of the machine. A recitation of the machining data used for the most successful model, in conjunction with the simplest topographic data (e.g. R_a) may provide a serviceable basis for manufacture, reflecting not only the topographic, but also the probable physical characteristics of the surface. If it is subsequently found expedient to change the process of manufacture, it may be advisable to reconsider both the functional consequences and the R_a value.

When the cost of fully experimental evolution cannot be accepted, it may suffice to rely on general experience, in which case the problem may arise of knowing what surface texture values would be descriptive of surfaces already familiar by sight and touch. Standard roughness comparison specimens can then be helpful. Electro-formed reproductions are available which show progressive grades of roughness of the usual machining processes, each grade having twice the R_a value of the previous one. This is about the smallest increment that can readily be detected by touch, and is often the smallest that matters functionally.

Sets of Roughness Comparison Specimens generally give a fair idea of the range of R_a values that can be achieved by each process, though it is as well to remember that as the fine end is approached, the cost of component production may rise rapidly, and have to be offset against other benefits that may accrue, for example in assembly or performance. The best surface of which any process is capable will involve many factors such as the stiffness of the workpiece, the material being worked, the condition of the machine and tool or wheel, the uniformity of the preceding process, the amount of stock to be removed, the time allowed for appropriately gentle cutting or grinding, and the care that is given to every detail of the process. The last two factors may determine the economic limit.

Eventually it may be possible to select optimum characteristics from tables, as is done for dimensional tolerances, but so many factors are involved that this seems a long way off. The evolution of a composite index, like the plasticity index in Section E1, may lead to a solution.

The simple R_a parameter must be interpreted with full awareness of what it can and cannot tell about surface texture. It is best regarded as a practical index for comparing the heights of similar profiles on a linear basis—twice the index, twice the height. To this extent it has proved generally serviceable for process control. Its ability to compare dissimilar profiles is more limited. It corresponds with the sensory impression of roughness only for similar textures and over a limited range of heights and peak spacings, as examination of sets of roughness comparison specimens will show. It cannot provide a direct measure of the functional quality of a surface, because this aspect depends on many factors and often involves those of a second surface.

Statements of parameter values

A statement of height will have little meaning if it is not accompanied by a statement of the maximum peak spacing (generally expressed by the meter cut-off) to which it refers. The standard British way is to recite the height value and the parameter, followed by the meter cut-off in brackets. Thus, in metric units using micrometres for height and millimetres for the cut-off, an example would be 0.2 μm R_a (2.5). A standard cut-off value often found suitable for the finer surfaces is 0.8 mm, and if this value was used or is to be used, the standards allow it to be assumed and direct statement omitted; but this does not mean that its significance can be ignored.

For fully co-ordinated control, the cut-off indicated on a drawing should be taken not only as the value to be used for inspection but also as the maximum significant spacing (e.g. traverse feed) that may result from production.

A further point is that the R_a value given on a drawing is often taken not as a target figure but as an upper limit, anything smoother being acceptable unless a lower limit is also given, so that manufacture must aim at something less if half the product is not to be rejected. On the other hand the R_a value marked on a roughness comparison specimen is the nominal value of the specimen. This difference in usage must be allowed for when choosing the required texture from a set of specimens, and indicating the choice on a drawing.

INTRODUCTION

The hardness of the surface of components is an important property affecting their tribological performance. For components with non conformal contacts such as rolling bearings and gears, the hardness, and the corresponding compressive strength, of the surface material must be above a critical value. For components with conformal contacts such as plain bearings, the two sides of the contact require a hardness difference typically with a hardness ratio of 3:1 and ideally with 5:1. The component with the surface, which extends outside the close contact area, needs to be the hardest of the two, in order to avoid any incipient indentation at the edge of the contact. Shafts and thrust collars must therefore generally be harder than their associated support bearings.

HARDNESS MEASUREMENT

The hardness of component surfaces is measured by indenting the surface with a small indenter made from a harder material.

The hardness can then be inferred from the width or area of the indentation or from its depth.

The Brinell hardness test generally uses a steel ball 10 mm diameter which is pressed into the surface under a load of 30 kN. In the Vickers hardness test, a pyramid shaped indenter is pressed into the surface, usually under a load of 500 N. In both cases the hardness is then inferred from a comparison of the load and the dimensions of the indentation.

The Rockwell test infers the hardness from the depth of penetration and thus enables a direct reading of hardness to be obtained from the instrument. Hard materials are measured on the Rockwell C scale using a diamond rounded tip cone indenter and a load of 1.5 kN. Softer metals are measured on the Rockwell B scale using a steel ball of about 1.5 mm diameter and a load of 1 kN.

Table 3.1 gives a comparison of the various scales of hardness measurement, for the convenience of conversion from one scale to another. The values are reasonable for most metals but conversion errors can occur if the material is prone to work hardening.

Table 3.1 Approximate comparison of scales of hardness

Brinell	*Vickers*	*Rockwell B*	*Rockwell C*
120	120	67	—
150	150	88	—
200	200	92	—
240	250	100	22
300	320	—	32
350	370	—	38
400	420	—	43
450	475	—	47
500	530	—	51
600	680	—	59
650	750	—	62

SURFACE INTERACTION AND THE CAUSE OF FRICTION

Most surfaces are rough on an atomic scale and when placed in contact touch only at the tips of their asperities. The real area of contact will generally be much smaller than the apparent (see Section E1). At these regions of real contact, if the surfaces are clean, the atoms on one surface will attract those on the other and produce strong adhesion. With metals this may be referred to as cold welding and is particularly marked in mechanisms operating in high vacuum, e.g. outer space. When sliding occurs these adhesions have to be overcome, that is, the junctions have to be sheared. The force to shear the junctions is the primary cause of the friction between clean surfaces. If in addition one surface is harder than the other the roughnesses on it will plough out grooves in the softer and this constitutes a second cause of friction. In general we may write: force of friction = force to shear junctions + force to plough the asperities on one surface through the other, or

$$F = F_{adh} + F_{ploughing} \qquad \ldots \quad (1)$$

For clean surfaces the adhesion easily dominates. For lubricated surfaces the ploughing term may be an appreciable fraction of F.

With metals the junctions have a shear strength comparable to the bulk shear strength of the softer of the two metals in sliding contact. With similar metals, the junctions, due to heavy work-hardening during shear, are stronger than both and large fragments may be torn out of both surfaces. For this reason it is generally bad practice to slide similar metals together. If they must be similar it is desirable to choose hard materials which do not workharden further during sliding and which have limited ductility: or to use materials of inhomogeneous structure such as cast iron.

With non-metals the adhesion is less easy to describe—a metallurgist would not welcome the concept of cold welding of wood to steel—but it again arises from interatomic forces and can be strong. Again the shear strength of these adhesions is comparable with the bulk strength of the weaker of the pair.

THE LAWS OF DRY FRICTION

The adhesion component of friction can be estimated as follows. If the real area of contact is A and the specific shear strength of the junction is s, we have

$$F = As \qquad \ldots \quad (2)$$

For a given sliding pair s is approximately constant. Since A is nearly proportional to the load for a very wide range of surface conditions (see Section E1), and does not depend markedly on the overall size of the bodies, the frictional force too is proportional to the load and independent of size of the bodies. These are the two basic laws of dry friction.

With metals where the contact regions flow plastically

$$A = \frac{W}{p} \qquad \ldots \quad (3)$$

where p is the yield pressure or hardness of the softer metal. Then the coefficient of friction μ is given by

$$\mu = \frac{F}{W} = \frac{As}{W} = \frac{s}{p} \qquad (4)$$

For ductile solids p is of order 5s so that this gives a value $\mu = 0.2$. Most metals in air have a value nearer $\mu = 1$. There are two reasons for this. First, work-hardening during shear may increases s in the surface layers without appreciably affecting p. Secondly, under the influence of the combined normal and tangential stresses junction growth may occur producing an area of contact much larger than that given by eqn (3). This can be very marked with clean ductile metals and can lead to enormous coefficients of friction or even to gross seizure.

With non-metals the laws of deformation may give a different dependence of A on W than that given by eqn (3). In addition s itself may depend on the contact pressure. With polymers and rubbers both these factors usually lead to a coefficient of friction which decreases with increasing load.

EFFECT OF LUBRICANTS AND SURFACE FILMS

The most effective way of reducing friction and wear is to prevent contact between the surfaces. This is achieved, ideally, if the surfaces can run under hydrodynamic or elasto-hydrodynamic conditions with the equilibrium film thickness greater than the height of the surface asperities. In this case the chemical nature of the lubricant is immaterial, only its viscous properties are relevant. If, however, contact occurs through the liquid film it is desirable to add to the lubricant, additives of specific chemical properties. For example, polar molecules can adsorb on the surfaces to give a thin protective film which will prevent the surfaces themselves from coming into contact. If the rubbing surfaces are metals, fatty acids which can react to form metal soaps are particularly effective. Again, if the lubricant contains reactive chlorine or sulphur compounds these will react to give protective metal chlorides and sulphides.

The action of these surface films is, in principle, easy to explain—they provide a surface layer which is very much easier to shear than the underlying solids. For example, if hard metal surfaces are covered with a very thin film of a softer metal, the area of contact A will be determined by the hardness of the underlying metal whereas shearing will occur in the weaker metal layer. Thus a thin film of indium on tool steel can reduce the friction by a factor of 10. Such films are gradually worn away. There are two ways of replenishing them. One is by incorporating the soft metal as a separate phase in the harder matrix. The soft metal is extruded by the rubbing process itself. This is essentially the way in which copper-lead alloys function.

The other is to fall back on a chemically-formed surface film which can be replenished as it is worn away. This involves the use of 'oiliness' or extreme pressure additives dissolved in the lubricating oil. Here it is important to avoid too reactive an additive since it may lead to excessive corrosion.

Another approach to the reduction of friction and wear is to coat the rubbing surfaces with lamellar materials such as graphite or molybdenum disulphide. These substances are strong in compression and weak in shear. However, because the cleavage planes are low-energy surfaces one of the major difficulties is to achieve strong adhesion to the underlying solids.

Yet another way of reducing friction, and especially wear, is to coat the solids with layers of very hard materials such as those that are produced by plasma or flame spraying. There is evidence that because of their extreme hardness the surfaces are scarcely deformed during sliding so that even adsorbed films of oxygen or water vapour can provide some lubricating action. In addition, because of their limited ductility they will tend to be wear-resistant. On the other hand, if ductility is completely absent the films may crack and fragment during sliding and the consequent behaviour will be catastrophic.

Broadly speaking, the coefficient of friction for metals in air, when dry, is of order $\mu = 1$ for the softer metals and $\mu \simeq 0.4$ for harder metals such as steel. For surfaces operating with a good boundary lubricant such as a fatty acid, μ is of order 0.05 to 0.1, and the amount of wear may be 1000 to 10 000 times less than for unlubricated surfaces. However, this is only achieved when the adsorbed film is in a condensed form. As the temperature is increased these films are melted or desorbed and for most surface-active materials the films cease to be effective as friction- and wear-reducing agents when the surface temperature reaches 150 to 200°C. At higher temperatures E.P. additives containing sulphur or chlorine or phosphorus must be used. The friction is of order $\mu = 0.1$ to 0.2. Under very severe running conditions oxygen can often provide a useful protective film and in many cases this is provided by air dissolved in the lubricant itself.

FRICTION AND LUBRICATION OF POLYMERS

Because polymers are viscoelastic materials their deformation characteristics and hence their frictional properties are load, speed and temperature dependent. However, a representative value for most thermoplastics at room temperature is $\mu \simeq 0.4$ and there is often lumpy transfer. In this class of materials PTFE (Teflon) and high-density Polythene are exceptional. Once some orientation has been produced at the sliding interface the friction may be extremely small ($\mu \simeq 0.06$) and the transfer consists of extremely thin drawn-out highly-oriented polymer fibres. If the molecule is modified to incorporate bulky side groups (e.g. Teflon-HFP copolymer, or low-density Polythene) the low friction and light wear properties are lost. The special behaviour of the unmodified materials is associated with a smooth molecular profile.

Cross-linked polymers generally have a lower friction than thermoplastics, but they tend to be brittle and to fragment during sliding.

Because polymers are viscoelastic materials, deformation losses (i.e. the ploughing term) may be appreciable. At engineering speeds this may lead to excessive heating *beneath* the surface causing subsurface failure and spalling. By contrast, frictional heating with metals may produce softening of the surface layers or structural changes or even the formation of surface compounds such as carbides, oxides or nitrides.

Polymers may be lubricated with boundary films as with metals, though the reduction in friction is not in general as marked. With some polymers such as Polythene it is possible to incorporate surface active materials in the polymer itself (e.g. stearamide or oleamide); these can diffuse to the surface and provide a lubricating film which is replenished by further diffusion as the film is worn away.

Rubber is an extreme case. The friction is usually high and is generally higher the softer the rubber. This is mainly because a soft rubber gives a larger area of true contact. Values of μ greater than 1 are often observed, e.g. with tyres on road surfaces. At low speeds grease greatly reduces the friction; at higher speeds water provides effective hydrodynamic lubrication since the contact pressures are so low. This is highly desirable in bearings but disastrous in tyre road interactions. In the latter case a rough road surface and a high-loss rubber give a larger contribution to the deformation component of friction. In addition, suitable tread patterns can facilitate the extrusion of water from between the tyre and the road surface. Under exceptional conditions tyre and road may be completely separated by a continuous water film (aquaplaning) and the grip of the tyre on the road is then negligible.

For any given pair of surfaces the friction is roughly proportional to the load so that the coefficient of friction, μ, is a constant. To a first approximation the friction is also independent of the area of the bodies. However, the coefficient of friction must not be regarded as a fundamental 'design' property of the materials in contact, since friction depends on many variables such as: the macroscopic shape of the surfaces; surface roughness; films, if any, present on the surfaces; speed of sliding. For this reason the friction coefficients for assorted pairs of materials given below should be regarded only as representative values and care should be taken in applying them to specific cases.

UNLUBRICATED SURFACES

Metal surfaces cleaned *in vacuo* adhere where they touch; consequently, any attempt to slide them over one another results in a greater frictional force, and in some cases complete seizure. Where the latter occurs, *S* has been inserted in the appropriate column. The smallest trace of gas or vapour produces a considerable reduction in the friction coefficient μ_s, as shown in the first table.

Table 5.1 Friction of metals

Conditions	*Metals*											
	Ag	Al	Cd	Cu	Cr	Fe	In	Mg	Mo	Ni	Pb	Pt
μ_s on itself in air	1.4	1.3	0.5	1.3	0.4	1.0	2	0.5	0.9	0.7	1.5	1.3
μ_s on itself *in vacuo*	*S*	*S*	—	*S*	1.5	1.5	—	0.8	1.1	2.4	—	4
μ_s metal on steel (0.13% C, 3.42% Ni normalised) in air	0.5	0.5	0.4	0.8	0.5	—	2	—	0.5	0.5	1.2	—

Most of the data given is based on experiments between a curved slider and a flat surface at loads of the order of 1–10 N and sliding speeds of a few millimetres per second. The reason for this choice of variables is as follows. A curved slider on a flat localises the interaction and enables the deformation and damage to be studied in greater detail. Further, the stress situation is more definite than for flat on flat.

Moderate loads and low sliding speeds have been used to avoid the complication of frictional heating. This makes it easier to interpret frictional behaviour in terms of the properties of the materials. At higher sliding speeds, such as are common in engineering practice, the interfacial temperature can soften or even melt the surface layers. This may profoundly affect the friction.

The coefficient of kinetic friction, μ_k, is generally less than the static friction. With most metals it decreases as the speed increases and if the system is susceptible to vibration this may generate intermittent or 'stick-slip' motion. Stick-slip motion can be reduced by introducing sufficient damping or by reducing the elastic compliance. With softer metals and polymers the kinetic friction may increase with increasing speed up to some rather low critical velocity; at higher velocities the friction then decreases. Below this critical speed such sliding combinations will not cause natural oscillations in the system, although some type of intermittent motion may still be generated.

At high sliding speeds the kinetic friction of practically all materials decreases with increasing speed. If surface melting is produced this can lead to low coefficients of friction. For example, with ice at 0.1 ms^{-1}, μ_k is less than 0.05; with clean iron or steel surfaces at 500 ms^{-1}, $\mu_k \simeq 0.1$, whereas the static value is of order $\mu_s = 1$.

Table 5.2 The friction of alloys on steel (0.13% C, 3.4% Ni)

Alloys	μ_s
Copper-lead	0.2
White metal (tin base)	0.8
White metal (lead base)	0.5
Phosphor bronze	0.3
Brass (Cu 70, Zn 30)	0.5
Constantan	0.4
Steel (0.13% C, 3.42% Ni)	0.8
Cast iron	0.4

Table 5.3 The friction of ferrous alloys on themselves

Alloys	μ_s
Ball race steel	0.7
Austenitic steel	1.0
Cast iron	0.4
Chromium plate	0.7
Tool steel	0.4
$3\frac{1}{2}$% silicon iron	0.2

Table 5.4 The friction of very hard solids

Surface	*μ in air*
Bonded WC (Co binder) on:	0.2
itself	
copper	0.4
cadmium	0.8–1.0
iron	0.4–0.8
cobalt	0.3

Surface (on itself)	*μ in air*	*μ in vacuo*
Unbounded WC	0.15	0.4
Unbonded TiC	0.15	1.0
Unbonded B_4C	0.1	0.9
Diamond	0.1	0.9
Sapphire	0.2	0.9

Table 5.5 Friction of steel on polymers: room temperature, low sliding speeds

Material	*Condition*	*μ*
Nylon	Dry	0.4
Nylon	Wet	0.15
Perspex (Plexiglass)	Dry	0.5
PVC	Dry	0.5
Polypropylene	Dry	0.3
Polystyrene	Dry	0.5
Polyethylene (no plasticiser)	Dry or wet	0.4
Polyethylene (plasticiser)	Dry or wet	0.1
Polyethylene (high density)	Dry or wet	0.08
KelF	Dry	0.3
Teflon-FEP Copolymer	Dry	0.2
Lignum vitae	Natural state	0.1
PTFE	Dry or wet	0.05
PTFE (high loads)	Dry or wet	0.08
PTFE (high speeds)	Dry or wet	0.3
Filled PTFE (15% glass fibre)	Dry	0.12
Filled PTFE (15% graphite)	Dry	0.09
Filled PTFE (60% bronze)	Dry	0.09
Rubber (polyurethane)	Dry	1.6
Rubber (isoprene)	Dry	3 to 10
Rubber (isoprene)	Wet (water-alcohol solution)	2 to 4

LUBRICATED SURFACES

Ideal hydrodynamic lubrication exists when the moving surfaces are separated by a relatively thick film of lubricant and the friction depends on the viscosity of this layer. At higher pressures elastohydrodynamic lubrication may occur. If the lubricant film is thinner than the height of the surface asperities, the film will be penetrated and contact will occur at surfaces which, at most, are covered with a very thin residual lubricant film. These are the conditions of boundary lubrication; as for unlubricated surfaces the friction force is roughly proportional to load and independent of area.

Table 5.6 Lubrication of steel by various mineral oils

Lubricant	μ
Light machine	0.16
Thick gear	0.12
Solvent refined	0.15
Heavy motor	0.2
BP paraffin	0.18
Extreme pressure	0.10
Graphited oil	0.13
Oleic acid	0.08
Trichloroethylene	0.3
Alcohol	0.4
Benzene	0.5
Glycerine	0.2

When alcohols, paraffins and fatty acids are used as lubricants on steel surfaces the friction coefficient decreases as the length of the carbon chain increases, until a limiting value of something less than 0.1 is reached.

Table 5.7 Lubrication of various metals on steel by mineral oil

Surface	μ
Axle steel	0.16
Cast iron	0.21
Gunmetal	0.21
Bronze	0.16
Pure lead	0.5
Lead-base white metal	0.1
Pure tin	0.6
Tin-base white metal	0.11
Brass	0.19

Table 5.8 Extreme pressure (EP) lubrication

Protective film	μ_s	*Temperature up to which film is effective*
PTFE (Fluon, Teflon)	0.05	~ 320°C
Graphite	0.07–0.13	~ 600°C
Molybdenum disulphide	0.07–0.1	~ 400°C

Other protective films, such as sulphide and chlorides, may be formed by a chemical reaction between the surface and a suitable additive in the lubricating oil. In general, the chlorides give a lower friction than the sulphides. In the presence of moisture, however, they may hydrolyse and the resulting HCl may cause excessive corrosion

DEFINITION OF VISCOSITY

Viscosity is a measure of the internal friction of a fluid. It is the most important physical property of a fluid in the context of lubrication. The viscosity of a lubricant varies with temperature and pressure and, in some cases, with the rate at which it is sheared.

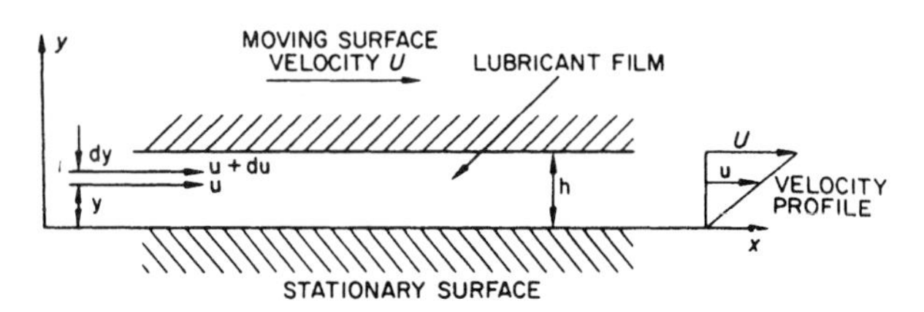

Fig. 6.1. Lubricant film between parallel plates

Dynamic viscosity

Dynamic viscosity is the lubricant property involved in tribological calculations. It provides a relationship between the shear stress and the rate of shear which may be expressed as:

Shear stress
= Coefficient of Dynamic Viscosity × Rate of Shear

or
$$\tau = \eta \frac{\partial u}{\partial y} = \eta D,$$

where τ = shear stress,
η = dynamic viscosity,
$\frac{\partial u}{\partial y} = D$ = rate of shear.

For the parallel-plate situation illustrated in Fig. 6.1.

$$\frac{\partial u}{\partial y} = \frac{U}{h}$$

and
$$\tau = \eta \frac{U}{h}$$

If τ is expressed in N/m^2 and $\frac{\partial u}{\partial y}$ in s^{-1}
then η is expressed in Ns/m^2, i.e. viscosity in SI units.

The unit of dynamic viscosity in the metric system is the poise $\left(\frac{g}{cm\ s}\right)$:

$$1 \frac{Ns}{m^2} = 10 \text{ poise}.$$

Kinematic viscosity

Kinematic viscosity is defined as $\nu = \frac{\eta}{\rho}$ where ρ is the density of the liquid.

If ρ is expressed in kg/m^3, then ν is expressed in m^2/s, i.e. in SI units.

The unit of kinematic viscosity in the metric system is the stoke $\left(\frac{cm^2}{s}\right)$.

$$1 \frac{m^2}{s} = 10^4 \text{ stokes}.$$

Table 1 gives the factors for converting from SI to other units.

Table 1 Viscosity conversion factors

Dynamic viscosity $\left(\text{SI unit is } \frac{Ns}{m^2}\right)$	
$\frac{Ns}{m^2} \times 10$	$= \text{poise}\left(\frac{g}{cm.s}\right)$
poise × 0.1	$= \frac{N\,s}{m^2}$
Kinetic viscosity $\left(\text{SI unit is } \frac{m^2}{s}\right)$	
$\frac{m^2}{s} \times 10^4$	$= \text{stokes}\left(\frac{cm^2}{s}\right)$
stokes × 0.1^{-4}	$= \frac{m^2}{s}$

ANALYTICAL REPRESENTATION OF VISCOSITY

The viscosities of most liquids decrease with increasing temperature and increase with increasing pressure. In most lubricants, e.g. mineral oils and most synthetic oils, these changes are large. Effects of temperature and pressure on the viscosities of typical lubricants are shown in Figs 6.2 and 6.3. Numerous expressions are available which describe these effects mathematically with varying degrees of accuracy. In general, the more tractable the mathematical expression the less accurate is the description. The simplest expression is:

$$\eta = \eta_o \exp(\gamma p - \beta t)$$

where $\eta_o =$ viscosity at some reference temperature and pressure, $p =$ pressure, $t =$ temperature, and γ and β are constants determined from measured viscosity data. A more accurate representation is obtained from the expression:

$$\eta = \eta_o \exp\left[\frac{A + Bp}{t + t_o}\right]$$

where A and B are constants.

Numerical methods can be employed to give a greater degree of accuracy.

A useful expression for the variation of density with temperature used in the calculation of kinematic viscosities is:

$$\rho_t = \rho_s - a(t - t_s) + b(t - t_s)^2$$

where ρ_s is the density at temperature t_s, and a and b are constants.

The change in density with pressure may be estimated from the equation:

$$\frac{V_o P}{V_o - V} = K_o + mp$$

where V_o is the initial volume, V is the volume at pressure p, and K_o and m are constants.

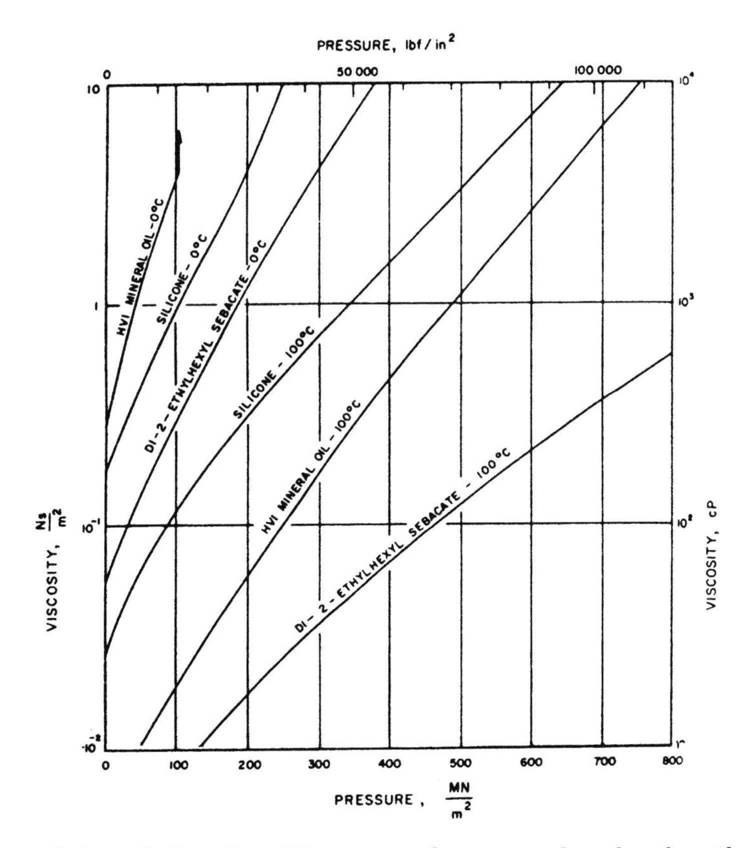

Fig. 6.2. The variation of viscosity with pressure for some mineral and synthetic oils.

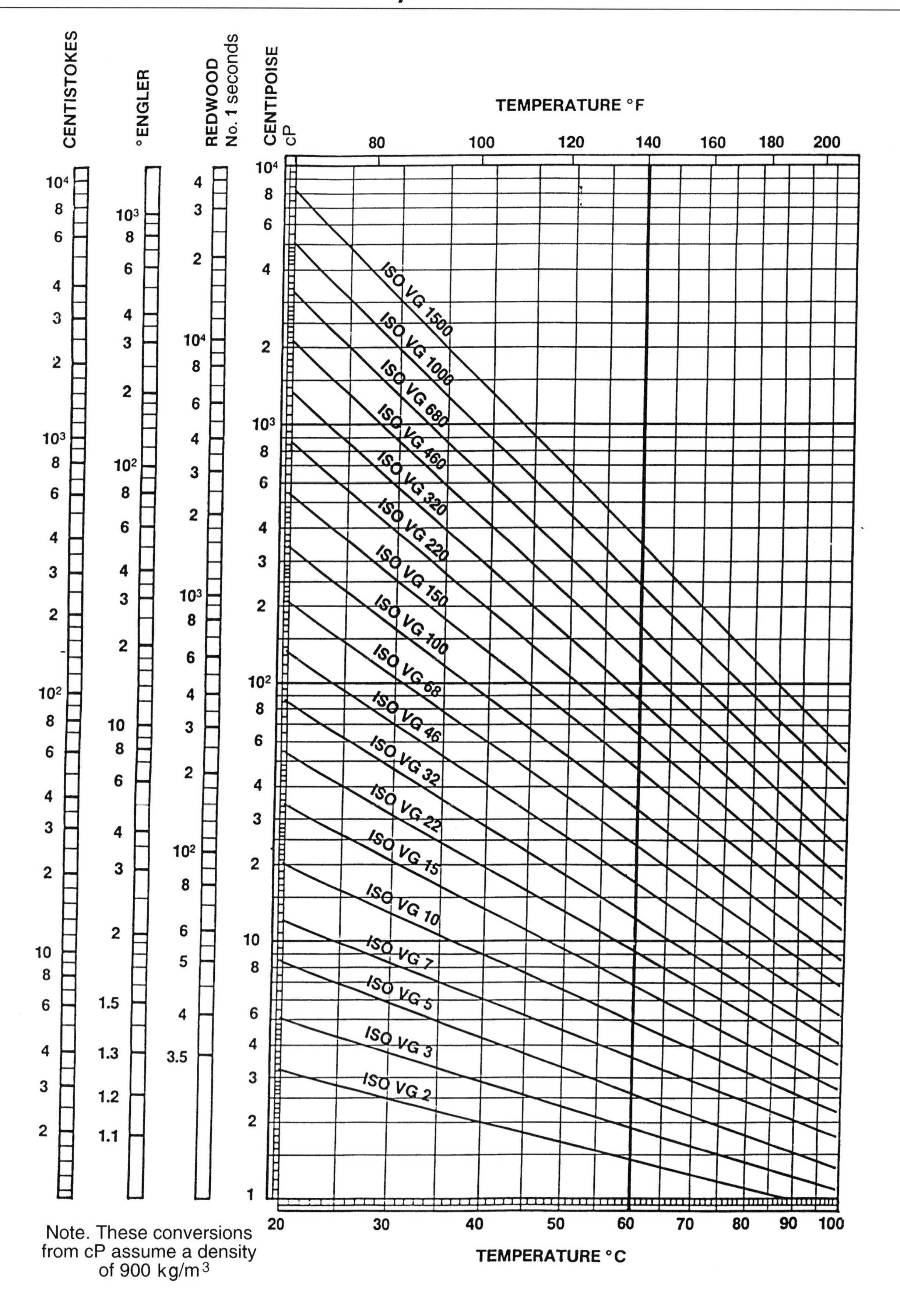

Fig. 6.3. The viscosity of lubricating oils to ISO 3448 at atmospheric pressure

VISCOSITY OF NON-NEWTONIAN LUBRICANTS

If the viscosity of a fluid is independent of its rate of shear, the fluid is said to be Newtonian. Mineral lubricating oils and synthetic oils of low molecular weight are Newtonian under almost all practical working conditions.

Polymeric liquids of high molecular weight (e.g. silicones, molten plastics, etc.) and liquids containing such polymers may exhibit non Newtonian behaviour at relatively low rates of shear. This behaviour is shown diagrammatically in Fig. 6.4. Liquids that behave in this way may often be described approximately in the non linear region by a power–law relationship of the kind:

$$\tau = (\phi s)^n$$

where ϕ and n are constants. For a Newtonian liquid $n = 1$ and $\phi \equiv \eta$, and typically for a silicone, n ?? 0.95.

Greases are non-Newtonian in the above sense but, in addition, they exhibit a yield stress the magnitude of which depends on their constitution. The stress/strain rate characteristics for a typical grease is also indicated in Fig 6.4. This characteristic may be represented approximately in the non linear region by an expression of the form,

$$\tau = \tau_1 + (\phi s)^n$$

where τ_1 is the yield stress and ϕ and n are constants.

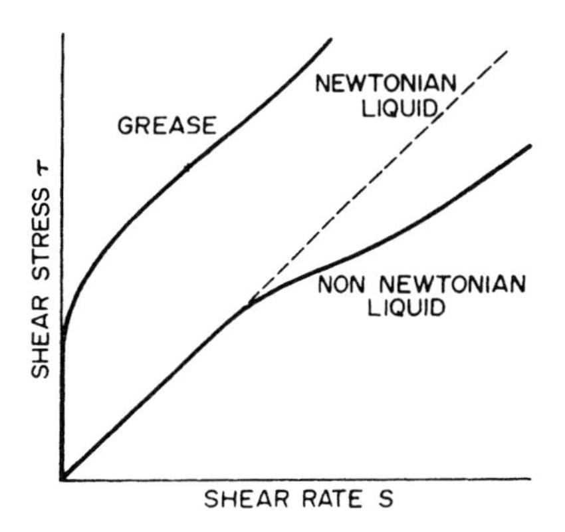

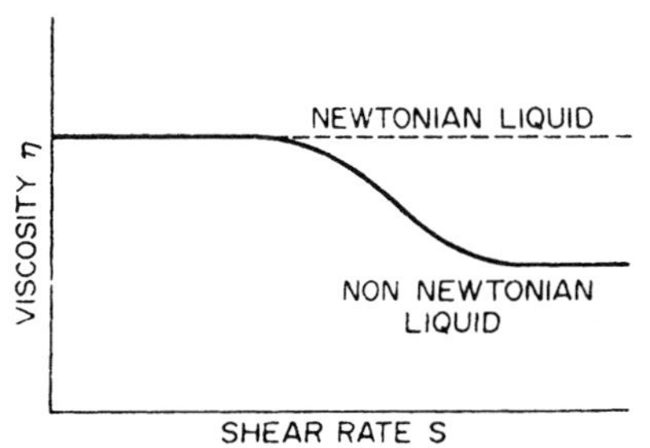

Fig. 6.4. Shear stress/viscosity/shear rate characteristics of non-Newtonian liquids

MEASUREMENT OF VISCOSITY

Viscosity is now almost universally measured by standard methods that use a suspended-level capillary viscometer. Several types of viscometer are available and typical examples are shown in Fig. 6.5. Such instruments measure the kinematic viscosity of the liquid. If the dynamic viscosity is required the density must also be measured, both kinematic viscosity and density measurements being made at the same temperature.

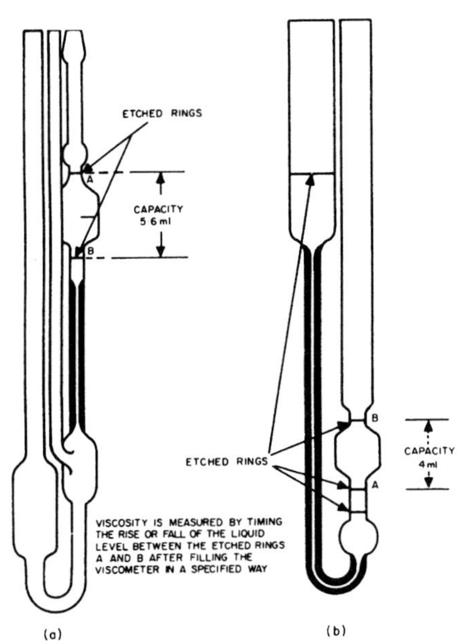

Fig. 6.5. Typical glass suspended-level viscometers

If the viscosity/rate-of-shear characteristics of a liquid are required a variable-shear-rate instrument must be used. The cone-and-plate viscometer is the one most frequently employed in practice. The viscosity of the liquid contained in the gap between the cone and the plate is obtained by measuring the torque required to rotate the cone at a given speed. The geometry, illustrated in Fig. 6.6, ensures that the liquid sample is exposed to a uniform shear rate given:

$$D = \frac{U}{h} = \frac{r\omega}{r\alpha} = \frac{\omega}{\alpha}$$

where r = cone radius, ω = angular velocity and α = angle of gap.

From the torque, M, on the rotating cone the viscosity is then calculated from the expression:

$$\eta = \frac{3M\alpha}{2\pi r^3 \omega}$$

This instrument is thus an absolute viscometer measuring dynamic viscosity directly.

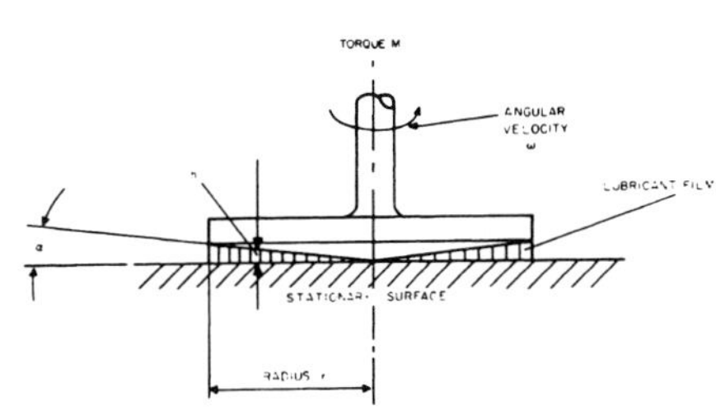

Fig. 6.6. Cone-and-plate viscometer

Methods of fluid film formation

MODES OF LUBRICATION

Before considering the methods by which fluid films can be formed between bearing surfaces it is necessary to distinguish between the three major modes of lubrication; boundary, mixed and fluid film. The friction and wear characteristics of bearings often indicate the mode of lubrication and a useful further guide is given by the ratio,

$$\lambda = \frac{\text{thickness of lubricating film or protective layer}}{\Sigma \text{ surface roughnesses } (R_a)}$$

This ratio not only indicates the mode of lubrication, but it has a direct bearing upon the effective life of lubricated machine components.

Boundary lubrication

The friction and wear characteristics of the lubricated contact are determined by the properties of the surface layers, often of molecular proportions, and the underlying solids. The viscosity of the bulk lubricant has little effect upon the performance of boundary lubricated contacts and the frictional behaviour broadly follows the well-known laws for unlubricated surfaces.

This mode of lubrication is encountered in door hinges and many machine tool slideways.

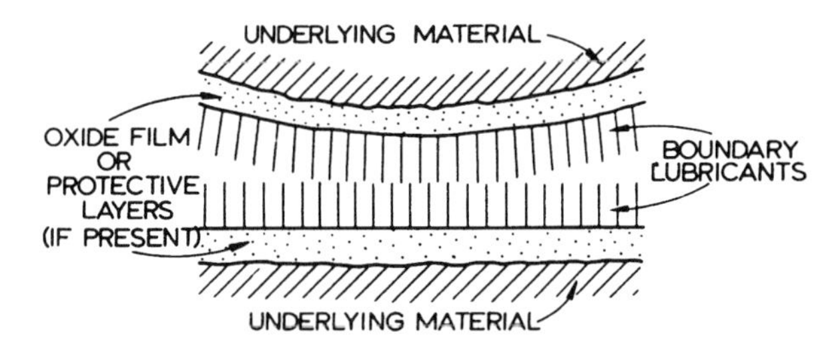

The thickness of the protective layers formed by physical and chemical reaction between the solids and the surrounding bulk lubricant, additives or atmosphere is usually small compared with the roughness of the solid surfaces. The length of fatty acid molecules, which are frequently used as boundary lubricants, and the thickness of protective oxide films is often as small as 2 nm (10^{-7} in). Hence, for boundary lubrications,

$$\lambda \leqslant 1$$

Mixed lubrication

Although it is usual to classify the mode of lubrication in many machine elements as either 'boundary' or 'fluid film', it

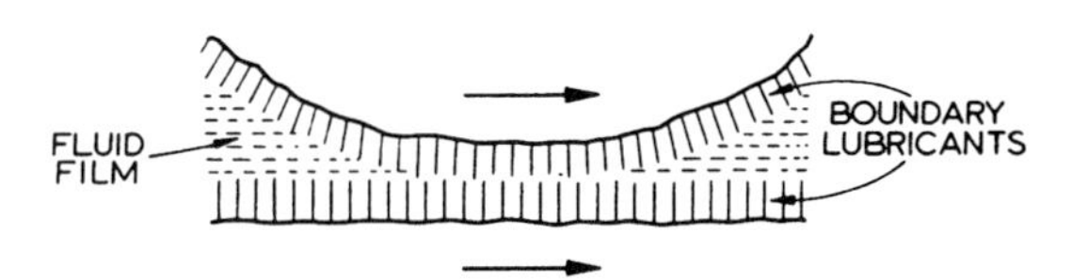

is not generally known that a very large proportion of them operate with a mixture of both mechanisms at the same instant. There may be regions of close approach where surface interactions and boundary lubrication contributes to the overall friction and wear characteristics in addition to a substantial fluid film lubrication action from most of the contact. In addition it is recognised that local hydrodynamic effects between surface irregularities can contribute to the total load carrying action; a mechanism known as asperity lubrication.

In mixed lubrication it is necessary to consider both the physical properties of the bulk lubricant and the chemical interactions between the bulk lubricant or additives and the adjacent solids.

This mode of lubrication is encountered in many gears, ball and roller bearings, seals and even some conventional plain bearings. It is now recognised that it is difficult to eliminate 'fluid film' action from boundary lubrication experiments and 'boundary' effects occur in 'fluid film' investigations more often than is generally acknowledged. This indicates the growing importance of a recognition of the regime of 'mixed' lubrication.

If mixed lubrication is to be avoided there must be no possibility of asperity interaction and hence the total fluid and boundary film thickness must exceed the sum of the surface R_a values by a factor which varies from about two to five depending upon the method of surface manufacture. There is a possibility of mixed lubrication whenever,

$$\lambda \leqslant 5$$

Fluid film lubrication

The best way to minimise wear and surface damage in rolling or sliding contacts in machines is to separate the solids by a film of lubricant. The lubricant can be a liquid or a gas and the load supporting film can be created by the motion of the solids (self-acting or hydrodynamic) or by a source of pressure outside the bearing (externally pressurised or hydrostatic).

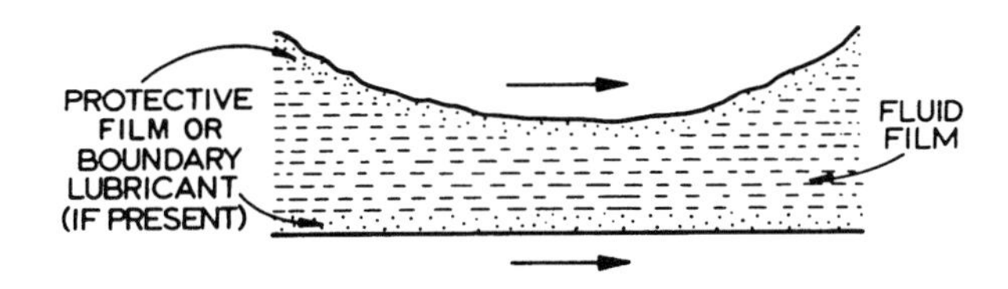

The main feature of this mode of lubrication is that the bearing solids are separated by a fluid film which is considerably thicker than the dimensions of the surface irregularities or protective surface films formed by boundary lubricants or chemical reaction. The films are normally many thousands of times thicker than the size of single molecules and this allows them to behave like the bulk lubricant applied to the bearing. They can nearly always be analysed according to the laws of slow viscous flow and the frictional resistance arises solely from the viscous shearing of the fluid. Viscosity of the bulk lubricant is the most important physical property of fluid film lubricants, but density is also important in gas bearings and some very highly stressed liquid lubricated contacts. It can usually be assumed that liquid lubricants adequately wet solid bearing surfaces and the main role of surface tension in fluid film lubrication is found in the mixedphase flow regions where cavitation has occurred in bearings, its effect on foaming and on viscous lifting in devices like ring oiled bearings.

Since a requirement for successful fluid lubrication is the absence of asperity interaction the film thickness is usually at least two to five times greater than the sum of the surface R_a values. It is more difficult to give an upper limit to the film thickness in relation to roughness, since there is no physical boundary to the definition presented earlier. However, thick films do not normally carry much load and in engineering situations it is rare to find a fluid lubricating film thicker than one hundred times the sum of the surface roughnesses. Fluid film lubrication can reliably be expected when $\lambda \geqslant 5$ and hence, in engineering situations,

$$5 \leqslant \lambda \leqslant 100$$

A special form of fluid film lubrication in which the development of effective films is encouraged by local elastic deformation of the bearing solids is known as 'elastohydrodynamic lubrication'.

It is now recognised that this is the principal mode of lubrication in many gears, ball and roller bearings, cams and some soft rubber seals. In counterformal contacts where the local pressures are high an additional important feature is the effect of high pressure upon the viscosity of many liquid lubricants. In all cases the elastic deformation creates a near parallel lubricating film in the central region of the conjunction with a restriction giving the minimum film thickness near the outlet or sides of the effective load bearing region.

Many nominal elastohydrodynamic contacts operate with film thicknesses which are small by conventional fluid-film lubrication standards and there is often an associated 'mixed' lubrication action. If local elastic distortion of the bearing surfaces occurs in the load carrying region the range of effective fluid film lubrication is often extended well into the region normally associated with 'mixed' or even 'boundary' lubrication. A typical range of film thickness ratios for elastohydrodynamic conditions is given by,

$$1 \leqslant \lambda \leqslant 10$$

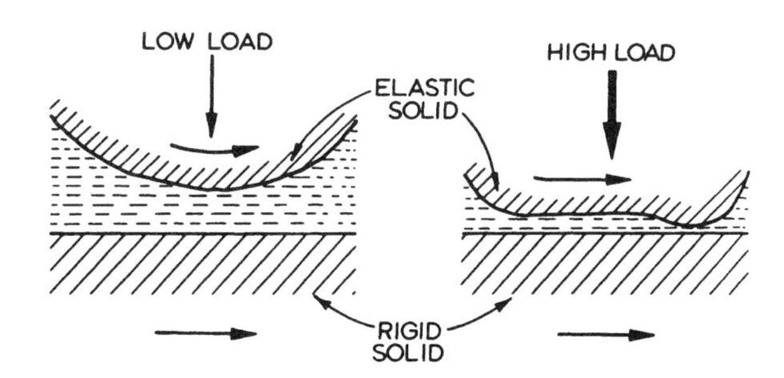

FORMATION OF FLUID FILMS

Flow mechanisms

If the bearing surfaces are completely separated by a lubricant the film usually behaves like a normal (Newtonian) viscous fluid. There is no slip between the solids and the lubricant adjacent to the bearing surfaces. Body (magnetic and gravity) forces are normally negligible compared with the forces arising from viscous shearing and since acceleration of the lubricant gives rise to negligible inertia forces, fluid film lubrication can be treated as a case of viscous flow in which pressure and viscous forces are everywhere in balance. With these assumptions the development of load bearing fluid films can readily be understood if it is recalled that there are only two mechanisms for creating the flow of viscous lubricant; Couette (surface motion) and Poiseuille (pressure gradients).

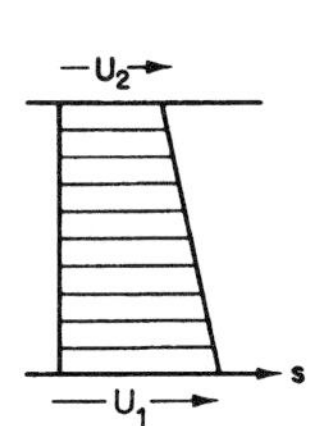

CFouette flow

Due to surface motion. Velocity distribution linear.

Volume rate of flow per unit width

$$q_s = \tfrac{1}{2}(U_1 + U_2)h$$

Viscous shear stress on solids

$$\pm\frac{\eta}{h}(U_1 - U_2)$$

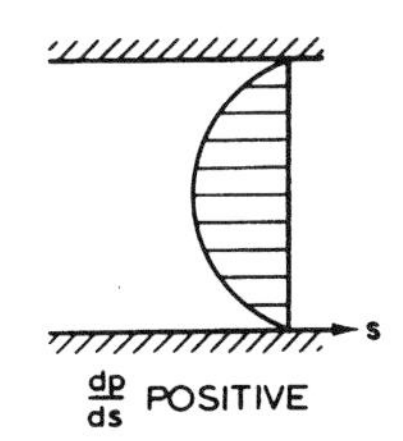

Poiseuille flow

Due to pressure gradients. Velocity distribution parabolic.

Volume rate of flow per unit width

$$q_x = -\frac{h^3}{12\eta}\left(\frac{dp}{ds}\right)$$

Viscous shear stress on solids

$$-\frac{h}{2}\frac{dp}{ds}$$

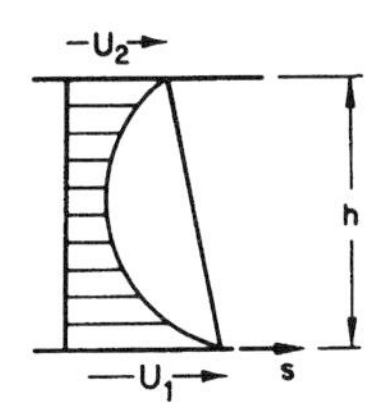

Combined flow

Due to both surface motion and pressure gradients.

Velocity distribution parabolic superimposed upon linear profile.

Volume rate of flow per unit width

$$q_s = \tfrac{1}{2}(U_1 + U_2)h - \frac{h^3}{12\eta}\left(\frac{dp}{ds}\right)$$

Viscous shear stress on solids

$$\pm\frac{\eta}{h}(U_1 - U_2) - \frac{h}{2}\left(\frac{dp}{ds}\right)$$

Pressure generation

The most important pressure generating mechanisms in self-acting fluid film bearings are the 'physical wedge' and 'squeeze film.

Physical wedge

Consider the general form of bearing shown in the diagram in which the upper pad is stationary and the lower surface

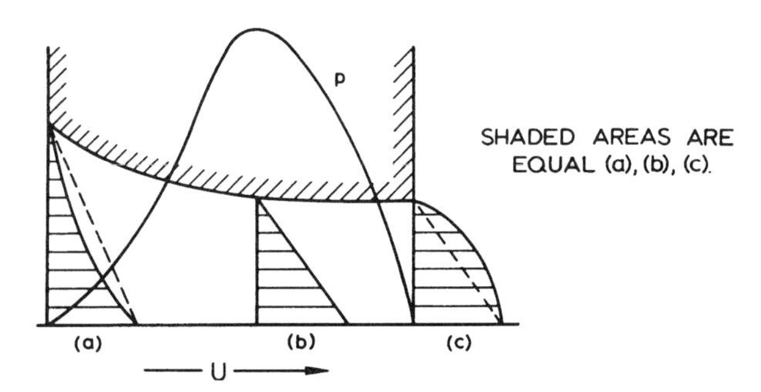

moves with velocity U. It will be assumed that all the flow takes place in the plane of the paper. Since the film thickness decreases in the direction of sliding the Couette action (linear velocity distribution) carries a diminishing volume of lubricant past each of the sections (a), (b) and (c). This action alone would violate the requirement for continuity of flow and hence the second form of flow mechanism must be introduced in such a way that the Poiseuille flow in the direction of sliding increases from (a) to (b) to (c). The only way in which this can be achieved whilst retaining ambient pressure at inlet and outlet is for a pressure distribution of the general form shown in the diagram to be generated. The pressure gradient at inlet holds the fluid back whilst that at outlet helps to push it out to maintain a constant volume rate of flow past each section.

Osborne Reynolds first demonstrated the need for decreasing film thickness in the direction of sliding in successful steadily loaded self-acting bearings in 1886, and this vital film geometry is referred to as a physical wedge. The optimum wedge geometry gives a long thin channel of small taper. The ratio of pad length to minimum film thickness is usually in the range of 10^3 to 10^4, whilst the ratio of maximum to minimum film thickness is about 2 for optimum thrust bearings and between 2 and 10 for journal bearings.

In journal bearings the applied load forces the journal into an eccentric position within the bearing, thus providing a converging film as shown in the diagram. Liquid lubricants generally rupture in the divergent clearance space beyond the point of minimum film thickness, giving rise to cavitation and the associated creation and collection of gas and vapour bubbles in this region.

For journal bearings the load carrying capacity W can be expressed in the form

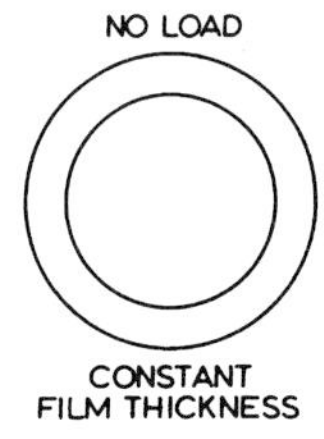

$$W = \eta U b \left(\frac{d}{c}\right)^2 \times f\left(\varepsilon, \frac{b}{d}\right)$$

where $\eta =$ dynamic viscosity, $U =$ shaft surface speed, $b =$ axial length of the bearing, $d =$ shaft diameter, $c =$ radial clearance, $\varepsilon =$ eccentricity ratio.

Thrust bearings have to be designed to create a converging film shape and this explains why plain parallel thrust rings and the simple form of bearings used prior to the work of Reynolds have limited load carrying capacity.

A common form of thrust bearing is the pivoted pad shown in the central diagram (b), but fixed inclination pads like that shown at (a) are sometimes used. The offset pivot, which is placed at the centre of pressure about 60% of the pad length from inlet, is used for uni-directional motion, whilst the central pivot is used for reversible machinery. An important feature of

THRUST BEARING PADS

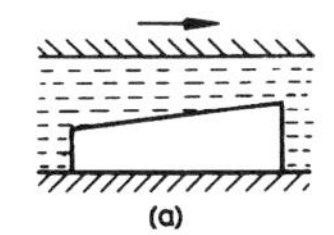

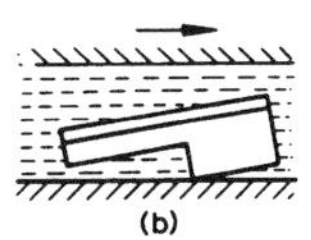

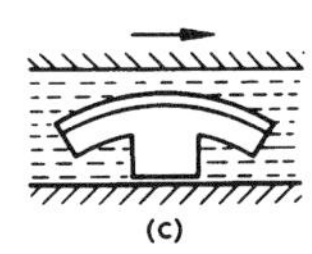

the offset pivoted pad is that it tilts with varying loads to provide an approximately constant ratio of inlet/outlet film thickness. This action provides optimum performance over a wide range of operating conditions.

The centrally pivoted pad (c) relies upon thermal and elastic distortion to present a suitable film geometry (usually convergent-divergent) since the centre of pressure normally lies downstream of the front of the supporting shoulder and hence tilting is inhibited.

It can be shown that the load carrying capacity of a thrust bearing pad W is given by

$$W = \frac{\eta U b l^2}{h_o^2} f\left(\frac{h_i}{h_o}, \frac{b}{l}\right)$$

where $\eta =$ absolute viscosity, $U =$ surface speed of thrust plate, $b =$ breadth of pad perpendicular to the direction of sliding, $l =$ length of pad in the direction of sliding, $h_o =$ minimum film thickness, $h_i =$ inlet film thickness.

It is found that $h_o \propto \left(\frac{U}{W}\right)^m$ with $\frac{1}{2} \leqslant m \leqslant 1$ for a wide variety of thrust bearing pad profiles.

Squeeze films

This effect arises when fluid is trapped between approaching surfaces. When the films are as thin as normal lubricating films, the action provides a valuable cushioning effect when excessive loads are applied for short intervals of time.

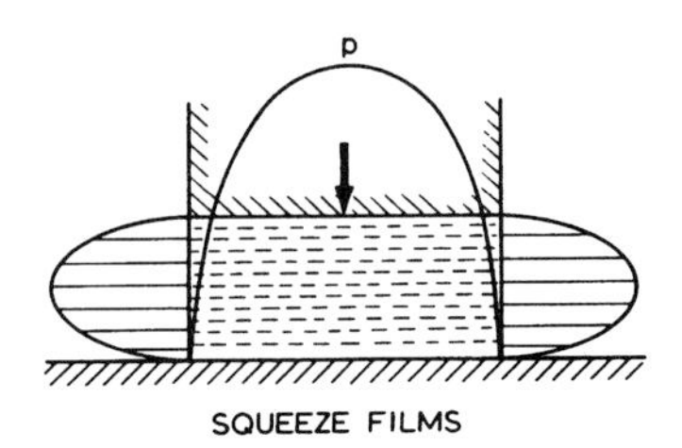

SQUEEZE FILMS

Positive pressures are generated as the surfaces approach each other but the low pressures created by separation often lead to rupture of liquid films and the occurrence of cavitation.

Squeeze film action is relied upon in the design of bearings for reciprocating machinery and for the protection of plain bearings subjected to impact loading.

The squeeze film time t required for the film thickness between parallel flat plates to reduce from h_1 to h_2 under a constant load W is given by

$$t = \frac{C\eta}{W}\left(\frac{1}{h_2^2} - \frac{1}{h_1^2}\right)$$

where η is the absolute viscosity and C is a function of the plate geometry having dimensions of $(\text{length})^4$.

Other pressure generating mechanisms which are rarely significant are:

Stretch mechanism

Encountered when the surface velocities change in magnitude within the bearing. The effect might occur if the bearing surfaces were elastic and the extent to which they were stretched varied as they traversed the bearing. For positive pressures to be generated the surface velocity has to decrease in the direction of sliding. The action is not encountered in conventional bearings.

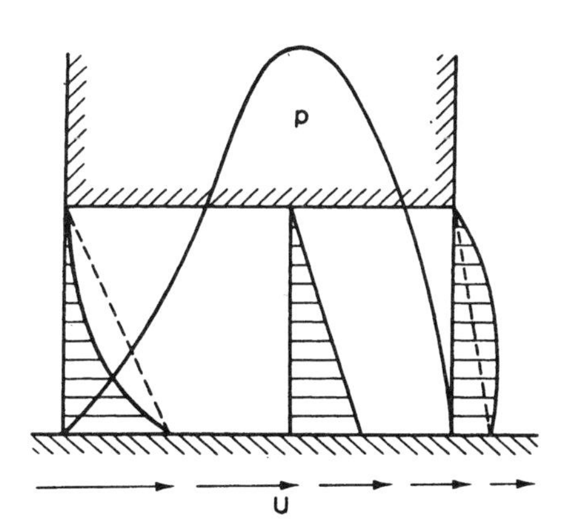

Density wedge

If the density of the lubricant changes in the direction of sliding the Couette mass flow at entry will be different to the efflux due to the same action. For continuity of mass flow this discrepancy must be eliminated by the generation of a balancing Poiseuille flow.

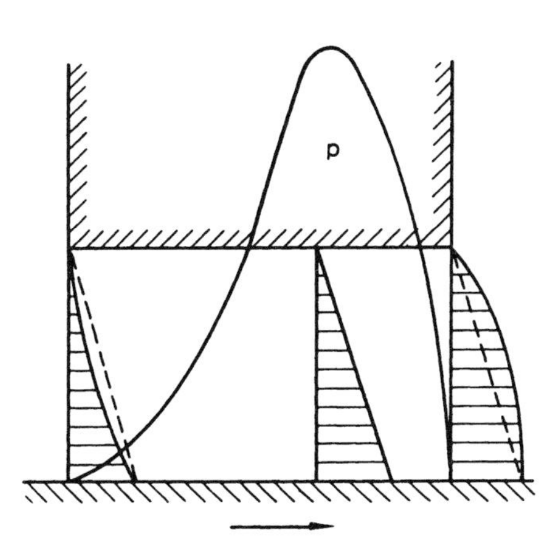

For this action to be effective the density must decrease in the direction of sliding. This effect could arise from a rise in temperature of the lubricant as it passes through the bearing and although normally insignificant compared with the physical wedge the action does afford some load carrying potential to parallel surface bearings.

Viscosity wedge

This action arises from a variation of viscosity associated with a variation in temperature across the thickness of the lubricating film. The less viscous lubricant shears more readily under constant stress and Couette action alone would give an imbalance of flow rates at inlet and outlet in the case illustrated.

The effect is rarely encountered and since the temperature rise along the moving surface in a thrust bearing is usually small compared with the rise along the stationary surface a 'negative' load carrying effect might occur. There is, however, some evidence that the action occurs in films between contra-rotating discs.

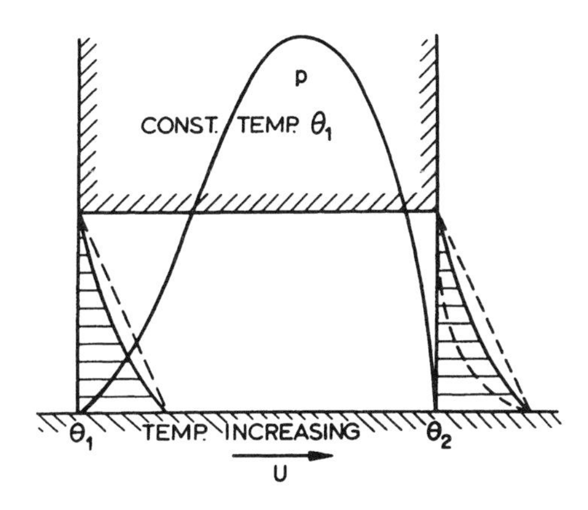

Local expansion

If the lubricant between stationary plates is heated by an external source of heat it will expand and the excess volume has to be expelled from the edges by Poiseuille action. This effect gives rise to load carrying pressures but it is a transient phenomenon of no significance in bearing performance.

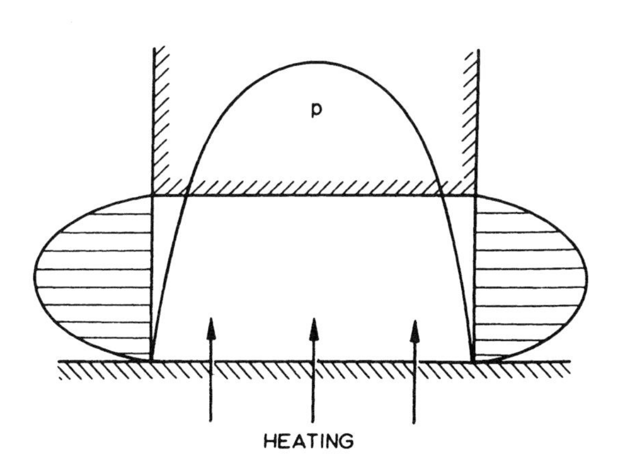

Wear can be defined as the progressive loss of substance resulting from mechanical interaction between two contacting surfaces. In general these surfaces will be in relative motion, either sliding or rolling, and under load. Wear occurs because of the local mechanical failure of highly stressed interfacial zones and the failure mode will often be influenced by environmental factors. Surface deterioration can lead to the production of wear particles by a series of events characterised by adhesion and particle transfer mechanisms or by a process of direct particle production akin to machining or, in certain cases, a surface fatigue form of failure. These three mechanisms are referred to as adhesive, abrasive and fatigue wear and are the three most important.

In all three cases stress transfer is principally via a solid–solid interface, but fluids can also impose or transfer high stresses when their impact velocity is high. Fluid erosion and cavitation are typical examples of fluid wear mechanisms. Chemical wear has been omitted from the list because environmental factors, such as chemical reaction, influence almost every aspect of tribology and it is difficult to place this subject in a special isolated category. Chemical reaction does not itself constitute a wear mechanism; it must always be accompanied by some mechanical action to remove the chemical products that have been formed. However, chemical effects rarely act in such a simple manner; usually they interact with an influence a wear process, sometimes beneficially and sometimes adversely.

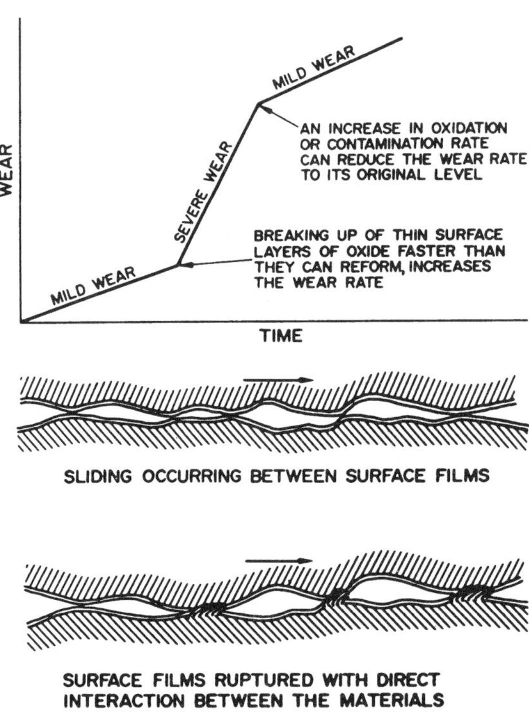

A simplified picture of adhesive wear

ADHESIVE WEAR

The terms cohesion and adhesion refer to the ability of atomic structures to hold themselves together and form surface bonds with other atoms or surfaces with which they in intimate contact. Two clean surfaces of similar crystal structure will adhere strongly to one another simply by placing them in contact. No normal stress is theoretically required to ensure a complete bond. In practice a number of factors interfere with this state of affairs, particularly surface contamination, and measurable adhesion is only shown when the surfaces are loaded and translated with respect to each other causing the surface films to break up. Plastic deformation frequently occurs at the contacting areas because of the high loading of these regions, and this greatly assists with the disruption of oxide films.

Since the frictional force required to shear the bonded regions is proportional to their total area, and this area is proportional to the load under plastic contact conditions (also with multiple elastic contacts), a direct relationship exists between these two forces; the ratio being termed the coefficient of friction. However, it is important to realise that the coefficient of friction is not a fundamental property of a pair of materials, since strong frictional forces can be experienced without a normal load so long as the surfaces are clean and have an intrinsic adhesive capability.

Any factor which changes the area of intimate contact of two surfaces will influence the frictional force and the simple picture of plastic contact outlined above is only an approximation to the real behaviour of surfaces. Plasticity theory predicts that when a tangential traction is applied to a system already in a state of plastic contact, the junction area will grow as the two surfaces are slid against each other. The surfaces rarely weld completely because of the remarkable controlling influence exerted by the interfacial contaminating layers. Even a small degree of contamination can reduce the shear strength of the interface sufficiently to discourage continuous growth of the bonded area. Coefficients of friction therefore tend to remain finite. Control of the growth of contact regions can also be encouraged by using heterogeneous rather than homogeneous bearing surfaces, whilst the provision of a suitable finish can assist matters greatly. The direction of the finishing marks should be across the line of motion so that frequent interruptions occur.

The actual establishment of a bond, or cold weld as it is sometimes called, is only the first stage of a wear mechanism and does not lead directly to the loss of any material from the system. The bonded region may be strengthened by work hardening and shear may occur within the body of one of the bearing components, thus allowing a fragment of material to be transferred from one surface to another. Recent observations indicate that the bond plane may rotate as well as grow when a tangential traction is applied, the axis of rotation being such that the two surfaces appear to interlock and the deformation bulges formed on each surface act like prow waves to each other. If the result of a bond fracture is material transfer, then no wear occurs until some secondary mechanism encourages this particle to break away. Often transferred material resides on a surface and may even back transfer to the original surface. Quite frequently groups of particles are formed and they break away as a single entity. Numerous explanations have been put forward to explain this final stage of the wear process, but the stability of a group of particles will be affected by the environment. One view is that break-away occurs when the elastic energy just exceeds the surface energy; the latter being greatly reduced by environmental reaction.

It is useful to look upon the adhesive wear system as being in a state of dynamic equilibrium with its environment. Continuous sliding and the exposure of fresh surfaces cannot go on indefinitely and the situation is usually stabilised by the healing reaction of the air or other active components of the surrounding fluid. The balance between the rupturing and healing processes can be upset by changing the operating parameters, and surfaces may abruptly change from a low to a high wearing stage. Increasing the speed of sliding, for instance, reduces the time available for healing reactions to occur, but it also encourages higher surface temperatures which may accelerate chemical reactions or desorb weakly bound adsorbants. The particular course which any system will take will thus depend greatly upon the nature of the materials employed.

Many wear processes start off as adhesive mechanisms, but the fact that the wear process leads to the generation of debris inevitably means that there is always a possibility that it may change to one of abrasion. In most cases, wear debris becomes, or is formed, as oxide, and such products are invariably hard and hence abrasive. A typical situation where this can arise is when two contacting surfaces are subjected to very small oscillatory slip movements. This action is referred to as fretting and the small slip excursion allows the debris to build up rapidly between the surfaces. This debris is often in a highly oxidised condition. The actual rate of wear tends to slow down because the debris acts as a buffer between the two surfaces. Subsequent wear may occur by abrasion or by fatigue.

ABRASION

Wear caused by hard protrusions or particles is very similar to that which occurs during grinding and can be likened to a cutting or machining operation, though a very inefficient one by comparison. Most abrasive grits present negative rake angles to the rubbed material and the cutting operation is generally accompanied by a large amount of material deformation and displacement which does not directly lead to loose debris or chips. The cutting efficiency varies considerably from one grit to another and on average only a small amount, 15–20% of the groove volume is actually removed during a single passage.

During abrasion a metal undergoes extensive work hardening and for this reason initial hardness is not a particularly important factor so long as the hardness of the abrasive grit is always substantially greater than that of the metal surface. Under this special condition there is a relatively simple relationship between wear resistance and hardness. For instance, pure metals show an almost linear relationship between wear resistance and hardness in the annealed state.

When the hardness of a metal surface approaches that of the abrasive grains, blunting of the latter occurs and the wear resistance of the metal rises. The form of the relationship between wear resistance and the relative hardnesses of the metal and abrasive is of considerable technical importance. As an abrasive grain begins to blunt so the mode of wear changes from one of chip formation, perhaps aided by a plastic fatigue mechanism, to one which must be largely an adhesive-fatigue process. The change is quite rapid and is usually fully accomplished over a range of $H_{metal}/H_{abrasive}$ of 0.8 to 1.3, where H_{metal} is the actual surface hardness.

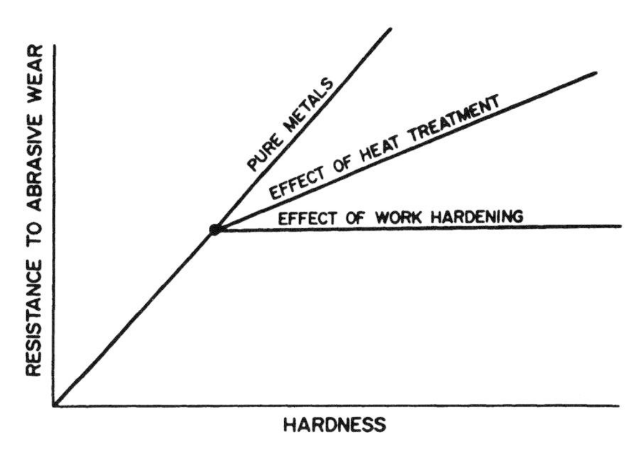

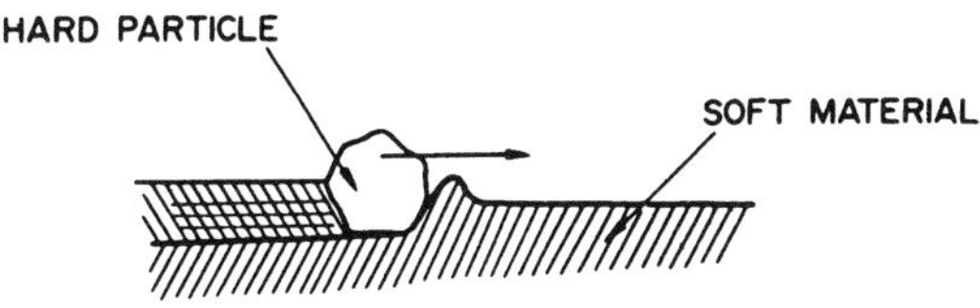

Some aspects of abrasive wear

Heterogeneous materials composed of phases with a considerable difference in hardness from a common and important class of wear resistant materials. When the abrasive is finely divided, the presence of relatively coarse, hard, material in these alloys increases the wear resistance considerably, but, when the abrasive size increases and becomes comparable with the scale of the heterogeneity of the structure, such alloys can prove disappointing. The reason seems to be that the coarse abrasive grits are able to gouge out the hard wear resistant material from the structure.

Brittle non metallic solids behave in a somewhat different way to the ductile metals. In general, the abrasion is marked by extensive fracture along the tracks and the wear rates can exceed those shown by metals of equivalent hardness by a factor of ten. With very fine abrasive material, brittle solids can exhibit a ductile form of abrasion. As in the case of metals, the effective wear resistance of a brittle material is a function of the relative hardnesses of the solid and the abrasive, but whereas with metals the effect is negligible until H_m/H_a reaches 0.8, with non metallic brittle solids blunting appears to take place at much lower values of this hardness ratio, indeed there seems to be no threshold. The wear resistance climbs slowly over a very wide range of $H_{brittle\ solid}/H_{abrasive}$.

Abrasion is usually caused either by particles which are embedded or attached to some opposing surface, or by particles which are free to slide and roll between two surfaces. The latter arrangement causing far less wear than the former. However, the abrasive grits may also be conveyed by a fluid stream and the impact of the abrasive laden fluid will give rise to erosive wear of any interposed surface. The magnitude and type of wear experienced now depends very much upon the impinging angle of the particles and the level of ductility, brittleness or elasticity of the surface. Many erosive wear mechanisms are similar to those encountered under sliding conditions, although they are modified by the ability of the particles to rebound and the fact that the energy available is limited to that of the kinetic energy given to them by the fluid stream. Rotation of the particles can also occur, but this is a feature of any loose abrasive action.

CONTACT FATIGUE

Although fatigue mechanisms can operate under sliding wear conditions, they tend to occupy a much more prominent position in rolling contact where the stresses are high and slip is small. Such contacts are also capable of effective elastohydrodynamic lubrication so that metal to metal contact and hence adhesive interaction is reduced or absent altogether. Ball and roller bearings, as well as gears and cams, are examples where a fatigue mechanism of wear is commonly observed and gives rise to pitting or spalling of the surfaces.

The mechanisms of rolling contact fatigue can be understood in terms of the elastic stress fields established within the surface material of the rolling elements. Elastic stress analysis indicates that the most probable critical stress in contact fatigue is the maximum cyclic orthogonal shear stress rather than the unidirectional shear stress which occurs at somewhat greater depths. The Hertzian stress distribution is adversely affected by numerous factors, including such features as impurity inclusions, surface flaws, general misalignment problems and other geometrical discontinuities, as well as the elastohydrodynamic pressure profile of the lubricant and the tangential traction.

Although the maximum cyclic stress occurs below the immediate surface, the presence of surface flaws may mean that surface crack nucleation will become competitive with those of sub-surface origin and hence a very wide range of surface spalls can arise. Furthermore, it is important to remember that if the surfaces are subjected to considerable tangential traction forces then the positions of the shear stress maxima slowly move towards the surface. This last condition is likely to arise under inferior conditions of lubrication, as when the elastohydrodynamic film thickness is unable to prevent asperity contact between the rolling elements. Pure sub-surface fatigue indicates good lubrication and smooth surfaces, or potent stress raising inclusions beneath the surface.

As in other aspects of fatigue, the environment can determine not only the stress required for surface crack nucleation, but more significantly the rate of crack propagation once a crack has reached the surface. The presence of even small amounts of water in a lubricant can have very serious consequences if suitable lubricant additives are not incorporated. It has also been suggested that a lubricant can accelerate crack propagation by the purely physical effect of becoming trapped and developing high fluid pressures in the wedge formed by the opening and closing crack.

FLUID AND CAVITATION EROSION

Both these wear mechanisms arise from essentially the same cause, namely the impact of fluids at high velocities. In the case of fluid erosion, the damage is caused by small drops of liquid, whilst in the case of cavitation, the impact arises from the collapse of vapour or gas bubbles formed in contact with a rapidly moving or vibrating surface.

Fluid erosion frequently occurs in steam turbines and fast flying aircraft through the impact of water droplets. The duration of impact is generally extremely small so that very sharp intense compression pulses are transferred to the surface material. This can generate ring cracks in the case of such brittle materials as perspex, or form plastic depressions in a surface. As the liquid flows away from the deformation zone, it can cause strong shear deformation in the peripheral areas. Repeated deformation of this nature gives rise to a fatigue form of damage and pitting or roughening of the surfaces soon becomes apparent.

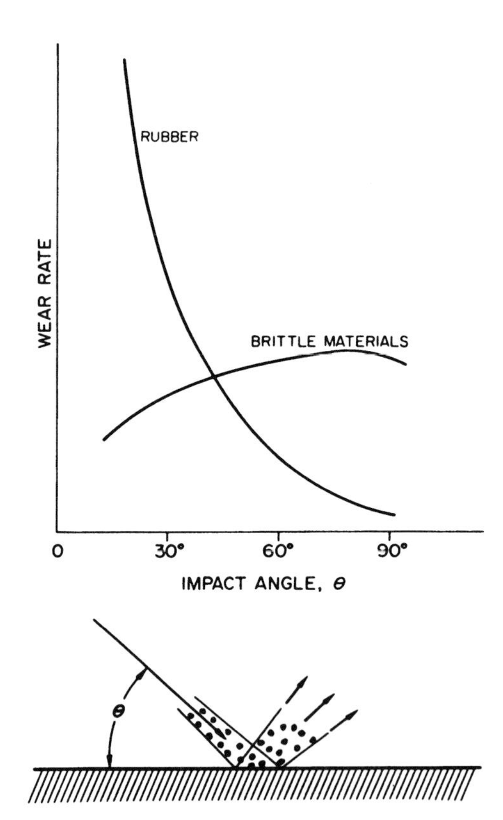

Wear by fluids containing abrasive particles

With cavitation erosion, damage is caused by fluid cavities becoming unstable and collapsing in regions of high pressure. The cavities may be vaporous, or gaseous if the liquid contains a lot of gas. The damage caused by the latter will be less than the former. The physical instability of the bubbles is determined by the difference in pressure across the bubble interface so that factors such as surface tension and fluid vapour pressure become important. The surface energy of the bubble is a measure of the damage which is likely to occur, but other factors such as viscosity play a role. Surface tension depressants have been used successfully in the case of cavitation attack on Diesel engine cylinder liners. Liquid density and bulk modulus, as well as corrosion, may be significant in cavitation, but since many of these factors are interrelated it is difficult to assess their individual significance.

Attempts to correlate damage with material properties has lead to the examination of the ultimate resilience characteristic of a material. This is essentially the energy that can be dissipated by a material before any appreciable deformation or cracking occurs and is measured by $\frac{1}{2}$ (tensile strength)^2/ elastic modulus. Good correlation has been shown with many materials. The physical damage to metals is of a pitting nature and obviously has a fatigue origin.

Heat is dissipated from a bearing assembly by:

(a) Heat transfer from the bearing housing, H_h

(b) Heat transfer along and from the shaft, H_s

(c) Heat transfer to a lubricant/coolant flowing through the assembly, H_l

HEAT TRANSFER FROM THE BEARING HOUSING H_h

$$H_h = hA_s f(\theta_b - \theta_a)$$

where h is the total heat transfer coefficient (see Fig. 9.1), A_s is the housing surface area and can be estimated from the surface area of an annular disc of similar overall size (see examples in Fig. 9.2), f is a factor which depends on the housing internal and external thermal resistances and can be estimated from the dimensions of the equivalent annular disc (see Fig. 9.3), θ_a is the temperature of the surroundings and θ_b is the bearing temperature.

Notes: 1 θ_b is not normally known; values of θ_b must be assumed initially to obtain a curve relating H_h to θ_b; this curve, together with others relating H_s, H_l and the heat generation in the bearing to θ_b, can then be used to determine the actual value of θ_b (see paragraph entitled Bearing Operating Temperature).

2 For housings which are an integral part of a machine or structure some experience and judgement may be required to decide appropriate effective boundaries of the housing and hence the size of the equivalent disc. Joints between the housing and the rest of the structure can only be taken as boundaries of the housing if they lie across the direction of heat flow and have low conductivity. Factors which reduce joint conductivity are low interface pressure (e.g. bolted joints), rough surface finish and low temperature difference across the joint.

3 When working with British units, the units which must be used for lengths and areas in all the formulae of this section of the Handbook are feet and square feet respectively.

HEAT TRANSFER ALONG AND FROM THE SHAFT, H_s

The shaft may either supply heat to the bearing assembly or remove heat from it, depending on the temperature and location on the shaft of any other heat sources or sinks (such as the rotor of an electric motor or the impeller of an air circulating fan). In the absence of heat sources, the shaft will remove heat from the bearing. In estimating the heat transfer to or from the assembly the parts of the shaft extending on each side of the bearing must be considered separately and their contributions added.

Most practical situations are covered by one or other of the two cases following.

Case 1 Heat source or sink of known temperature on the shaft

This case includes any situation where the shaft temperature is known or can be assumed at some known distance from the bearing.

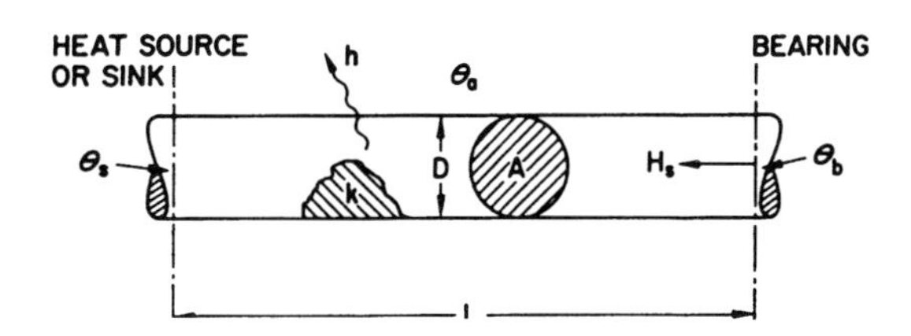

Heat flow from the bearing is given by:

$$H_s = \frac{kA}{L}\{C_1(\theta_b - \theta_a) + D_1(\theta_a - \theta_s)\}$$

where k = thermal conductivity of shaft material (Table 9.1)
A = shaft cross-sectional area
θ_a = temperature of surrounding air
θ_s = temperature of heat source or sink
θ_b = bearing temperature (see Note 1 opposite)
C_1 and D_1 are factors (Fig. 9.4) depending on mL,
where $m = (4h/kD)^{\frac{1}{2}}$ for solid shafts, diameter D,
and h = total heat transfer coefficient for shaft (Fig. 9.1).

Note that H_s may be negative, denoting heat flow into the bearing, if θ_s is sufficiently high.

Case 2 Shaft end free or insulated

This case includes situations in which either the shaft is thermally insulated at some section, or the heat flow along the shaft can be assumed negligible at a specified section, or the shaft end is free.

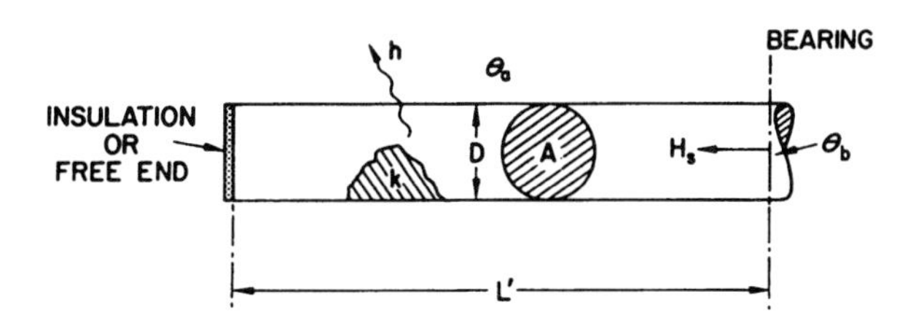

Heat flow from the bearing is given by:

$$H_s = \frac{kA}{L}\{C_2(\theta_b - \theta_a)\}$$

where the terms have the same meaning as in Case 1 except for:

$L = L' + D/4$ if the shaft end is free,
or $L = L'$ if the shaft end is insulated,
and C_2 is another dimensionless factor depending on mL see Fig. 9.4)

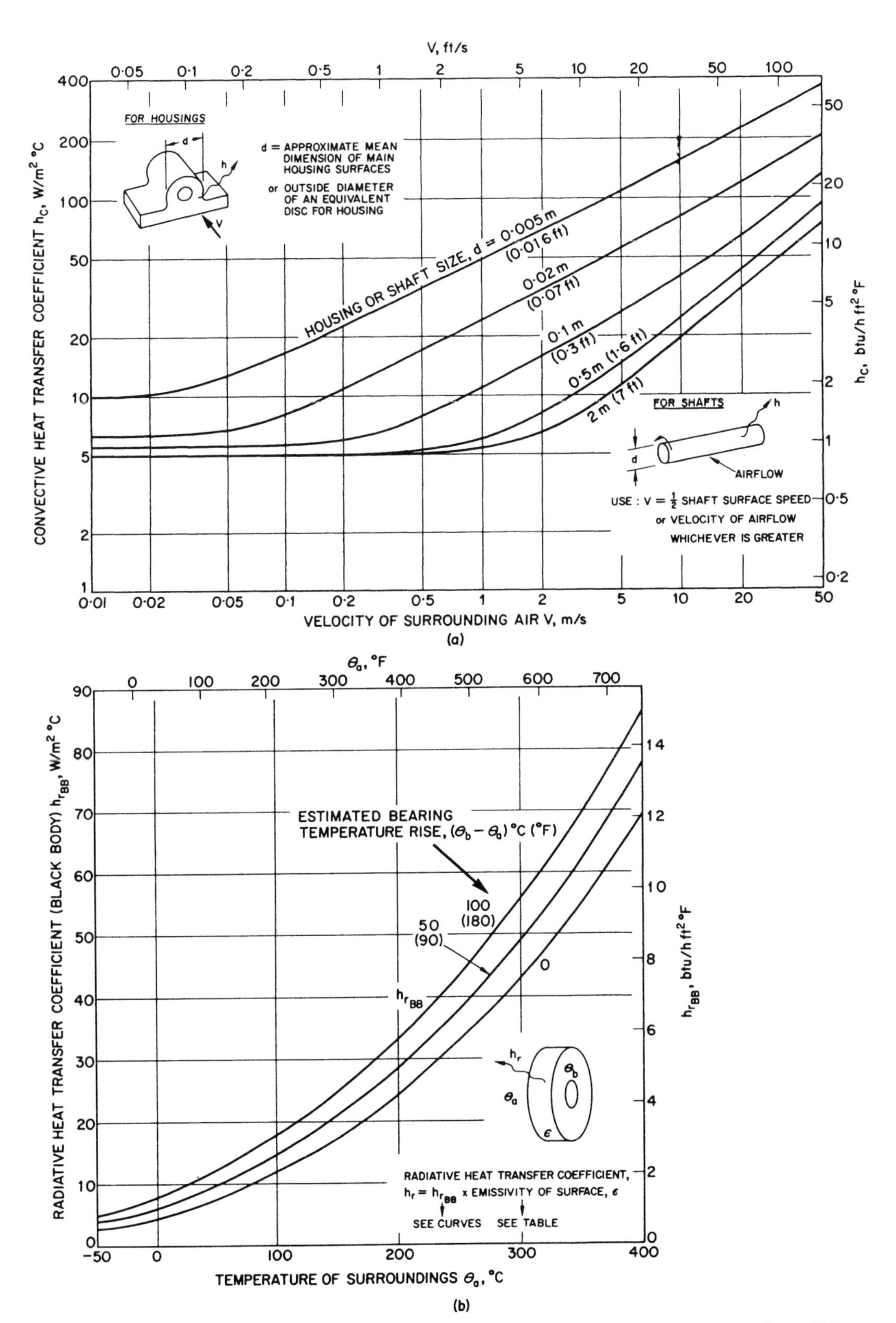

Fig. 9.1. Total heat transfer coefficient $h = h_c + h_r$; h_c is obtained from (a) and h_r from (b)

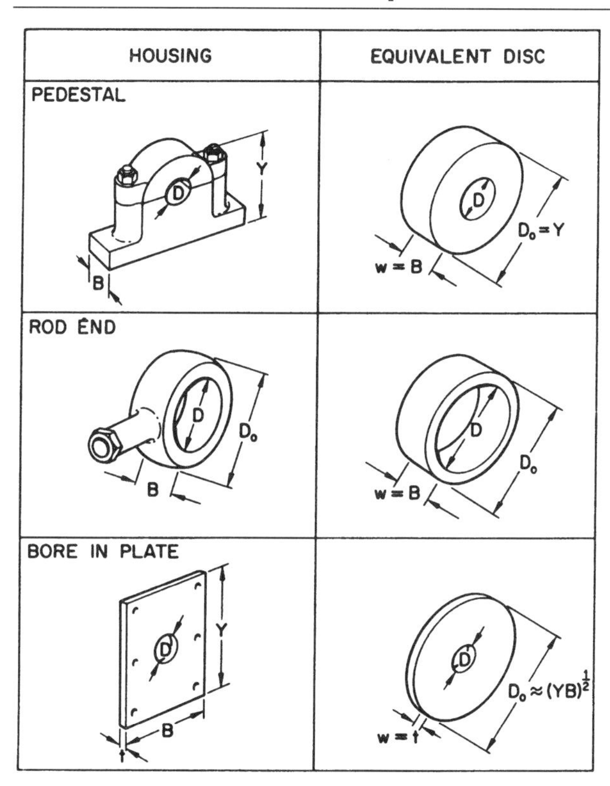

Fig. 9.2. Equivalent discs for three simple housing types

Housing surface area, $A_s \simeq \frac{\pi}{2}(D_o^2 - D^2) + \pi D_o w$

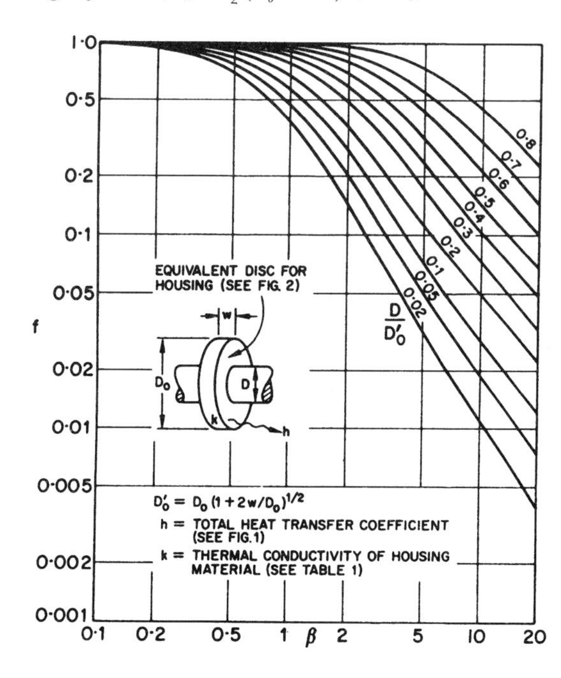

$$\beta = \begin{cases} \left(\dfrac{h}{2kw}\right)^{\frac{1}{2}} D_o' & \textit{for solid bearings} \\ 2\left(\dfrac{h}{2kw}\right)^{\frac{1}{2}} D_o' & \textit{for housings with substantial internal air spaces} \end{cases}$$

Fig. 9.3 f factors for housings

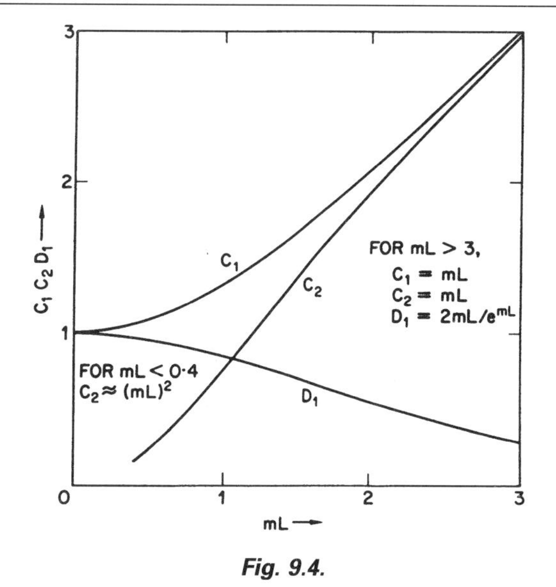

Fig. 9.4.

Table 9.1 Thermal conductivities of a selection of metallic materials

Metal	*Thermal conductivity*, k, *at* 20°C (68° F)	
	W/m°C	Btu/ft h°F
Aluminium (pure)	229	132
Aluminium Alloys*	110–180	63–104
Wrought Iron	59	34
Cast Iron	52	30
Steel (low carbon)	56	32
(high carbon)	43	25
Stainless Steel (18/8)	16	9
Alloy Steels*	10–60	6–35
Copper (pure)	386	223
Bronze (75 Cu, 25 Sn)	26	15
Brass (70 Cu, 30 Zn)	111	64

Note: alloying and impurities can have a large effect on k and a value for the particular material should be obtained if possible.

Table 9.2 Emissivities of a selection of surfaces

Surface	*Emissivity*, ε
Iron: sheet	0.5
wrought	0.7
red-rusted	0.8
Metals (smooth, oxidised):	
aluminium	0.2
galvanised iron	0.3
copper	0.7
nickel	0.9
Metals: emery rubbed	0.2
Paint: dull black	0.9
Carbon lampblack: thick	1.0
Varnished white enamel	0.9
Varnished aluminium	0.4
Oil: 0.02 mm thick	0.2
0.1 mm thick	0.6
Ceramics	0.7–1.0

HEAT TRANSFER TO A LUBRICANT/COOLANT, H_l

$$H_l = Q\rho c(\theta_o - \theta_i)$$

where Q = volume flow rate

ρ = density, c = specific heat — of lubricant/coolant

θ_i = inlet or supply temperature

θ_o = outlet temperature

Provided that flow rate is not excessive, θ_o will normally be nearly equal to the bearing temperature, θ_b (see Note 1 above).

EXPERIMENTAL METHOD FOR DETERMINING THE HEAT DISSIPATION FROM BEARING ASSEMBLIES

An experimental method can be used where calculation methods are unreliable, and normally consists of substituting an electrical heating source for the frictional heat generation at the bearing surfaces. For good accuracy the installation and operating conditions used in the test must approximate those intended for service as closely as possible and direct heat transfer from the heater to the surroundings must be avoided. The steady bearing temperatures (measured near the bearing surface), corresponding to a series of recorded electrical power inputs, should be noted, and can be used to plot a curve of heat dissipation against bearing temperature rise above ambient, which is the required heat dissipation characteristic for the assembly.

BEARING OPERATING TEMPERATURE

The normal purpose of determining the heat dissipation characteristics of bearing assemblies is to estimate the bearing operating temperture for which the procedure is as follows:

1. Decide the range within which the bearing temperature, θ_b, is expected to lie.
2. For several values of θ_b covering this range, determine the total heat dissipation from the bearing assembly H_d, either by the experimental method (see above) or by adding togehter H_h, H_s and H_l, as calculated by the methods given earlier. Plot H_d against θ_b.
3. For the type and design of bearing under consideration use appropriate sections of this Handbook to estimate the friction coefficient, and hence frictional heat generation, H_f, in the bearing. Estimate any other heat inputs to the assembly, for example, due to friction at bearing seals, and add these to H_f to obtain the total heat generation, H_g. If H_g varies with θ_b calculate H_g at several values of θ_b. Plot H_g against θ_b on the same axes as used for H_d above.
4. The bearing operating temperature is that corresponding to the intersection of the curves for H_d and H_g.

Methods for reducing the operating temperature of bearings

These include:

1. Positioning the housing for unrestricted access of the surrounding air to as much as possible of its surface.
2. Provision of substantial heat flow paths, unimpeded by joints, between the bearing surface and housing exterior.
3. Use of a material for the housing having a high thermal conductivity.
4. Fins on the housing surface.
5. Forced air cooling—often conveniently achieved by means of a fan on the shaft.
6. Internal cooling—for example by water coil or passages.
7. Fitting thermal insulation in any joints between the housing and hot structure or components.

TO FIND THE DEFLECTION AND SLOPE OF A STEPPED SHAFT

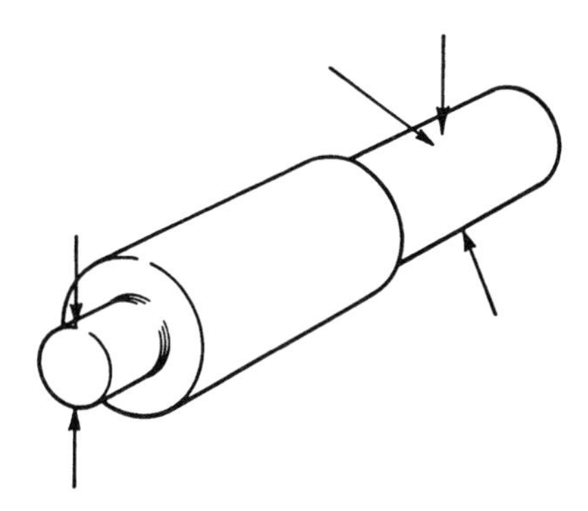

(**1**) Find all the loading and bearing forces acting on the shaft.

(**2**) Set up a z axis along the axis of the shaft, and right-angle axes x and y perpendicular to this axis. Choose the orientation of x, and y to coincide with one direction in which the shaft deflection is wanted.

(**3**) Resolve all the forces into components along each of the axies.

(**4**) Consider the shaft split into sections, each section containing one step of the shaft (i.e. each section consists of a piece of shaft of constant diameter). Number the left-hand section 1, the next section 2, etc.

(**5**) Consider only the component forces in the y direction.

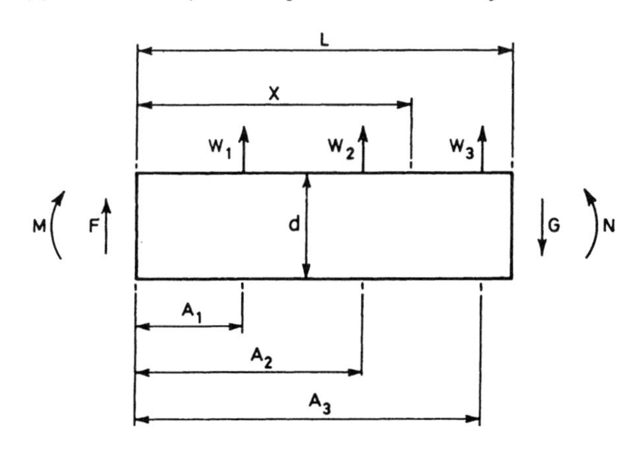

(**6**) Each section will be as shown above. W_1, W_2, W_3, etc. are the components of forces in one direction, and F, G, M and N are the internal shear forces and bending moments maintaining the section in equilibrium. For equilibrium:

$$G = F + W_1 + W_2 + W_3, \text{ etc.}$$

$$N = (G \times L) + M - (W_1 \times A_1) - (W_2 \times A_2) - (W_3 \times A_3), \text{ etc.}$$

(**7**) Consider each section in turn beginning at section 1. $M = 0$ and $F = 0$ for this section. Hence, calculate G and N using the above formulae. For section 2, F and M are the same as G and N of section 1. Hence, calculate G and N for section 2. Similarly calculate G and N for all the sections, in each case F and M are equal to the G and N of the preceding section. For the last section G and N should be zero. If this is not so, check the calculations for an error.

(**8**) The deflection DE and slope SL at any position within a section are

$$\text{DE} = \frac{1}{E \times I}\left(\frac{F \times X^3}{6} + \frac{M \times X^2}{2}\right) + C \times X + D + Y_1 + Y_2 + Y_3, \text{ etc.}$$

$$\text{SL} = \frac{1}{E \times I}\left(\frac{F \times X^2}{2} + M \times X\right) + C + S_1 + S_2 + S_3, \text{ etc.}$$

where I is the second moment of area about a diameter

$$= \frac{\pi \times d^4}{64} \text{ (for a solid shaft)}$$

and

$$Y_1 = \frac{W_1}{6 \times E \times I}(X - A_1)^3, \quad S_1 = \frac{W_1}{2 \times E \times I}(X - A_1)^2,$$

$$Y_2 = \frac{W_2}{6 \times E \times I}(X - A_2)^3, \quad S_2 = \frac{W_2}{2 \times E \times I}(X - A_2)^2, \text{ etc.}$$

for each load W_1, W_2, etc.

If in calculating Ys and Ss the terms $(X - A)$ become negative, then the corresponding Ys and Ss are zero.

C is the slope at the beginning of the section and D is the deflection at the beginning of the section.

(**9**) Again consider each section in turn. At this stage the slope and deflection at the beginning of section 1 are not known, so let the slope be J and the deflection K. Calculate the slope and deflection at each loading position, or other positions of interest, and at the end of the section (i.e. put $X = A_1$, then $X = A_2$, etc., and finally $X = L$). Each value will be calculated in terms of numbers and J and K.

The slope and deflection at the beginning of section 2 (i.e. C and D) are the same as at the end of section 1 as the two sections are joined together. But the values of the end of section 1 have just been calculated and thus C and D for section 2 are known (in terms of numbers and J and K). Calculate the slope and deflection at each loading position and at the end of the section.

Similarly for all other sections C and D are equal to the previously calculated slope and deflection at the end of the preceding section. Hence all values can be calculated.

(**10**) At the two datum points, which are usually the two bearings, the deflection is zero. In stage (9) these deflections have been calculated in terms of numbers and J and K. Thus, by making these equal to zero, two simultaneous equations for J and K are created. These should then be solved. The resulting numerical values of J and K can be substituted in all the expressions for slope and deflection, and hence find their numerical value.

(**11**) Repeat the whole process from paragraph (6) for component forces in the x direction.

(**12**) The slopes and deflections at any one point and in any direction can be found by compounding the values in the x and y directions. The greatest slope or deflection is given by the square root of the sum of the squared values in the x and y direction.

Example

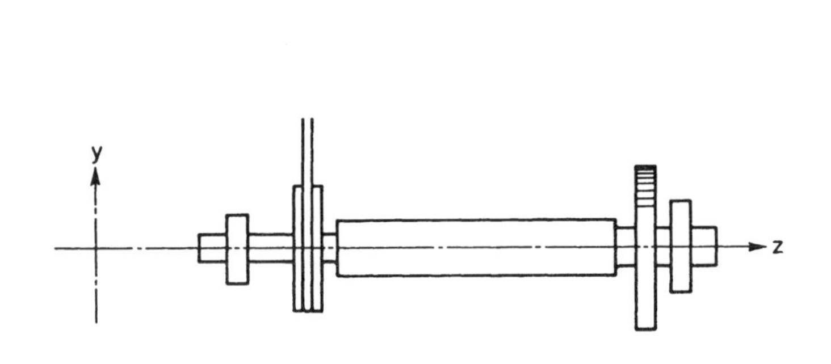

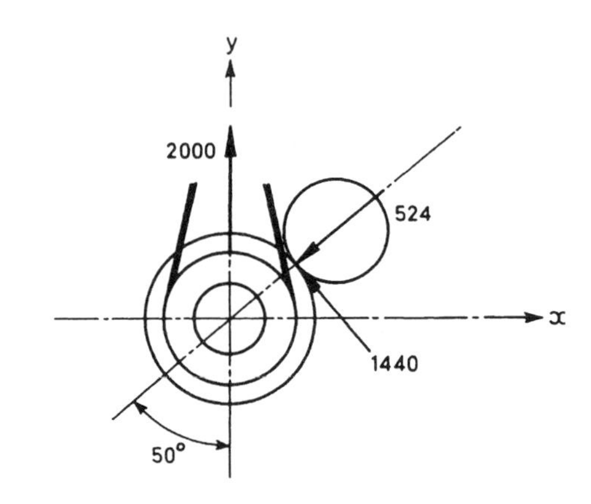

Resolve forces:

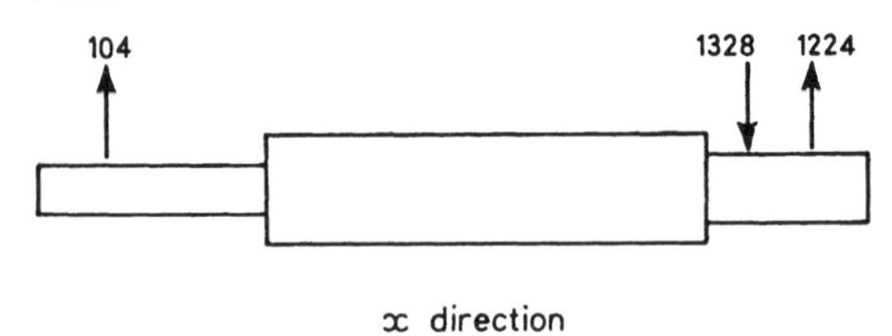

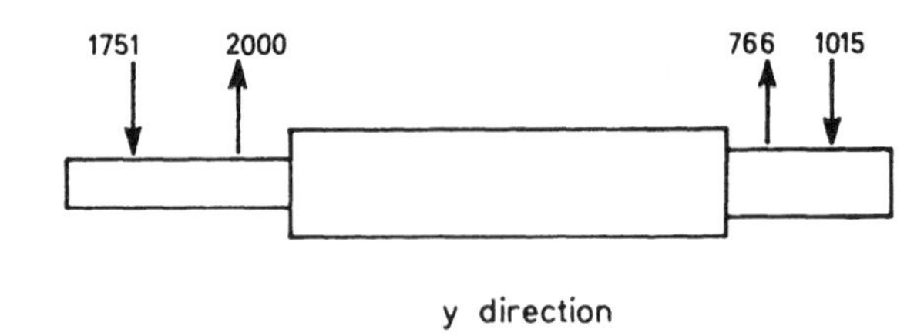

Consider the y direction

Split the shaft into sections:

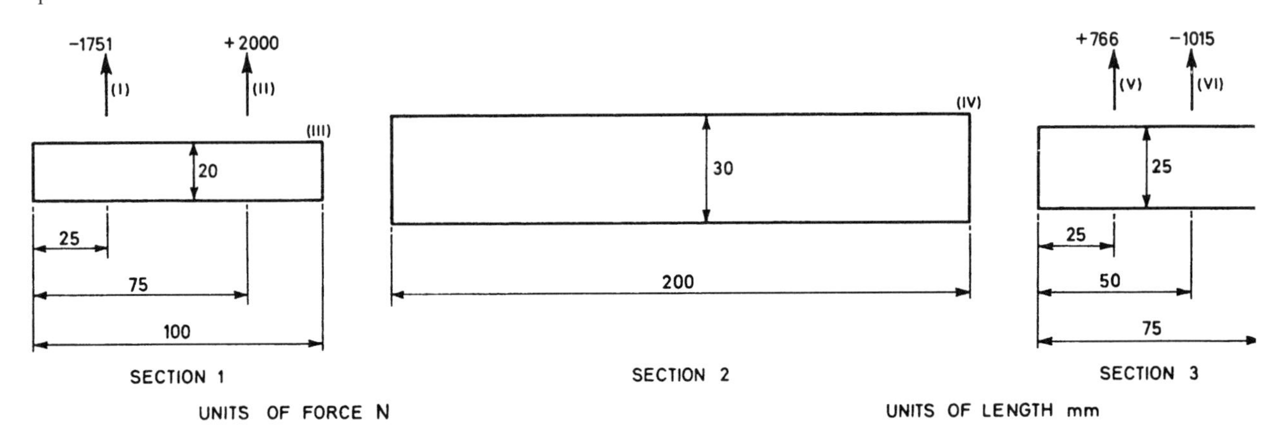

Calculate the values of F, G, M and N

For section 1;

$F = 0,\ M = 0$

thus $G = 0 - 1751 + 2000$ i.e. $G = 249$

and $N = 249 \times 100 + 0 + 1751 \times 25 - 2000 \times 75$

i.e. $N = -81\,300$

For section 2;

$F = 249,\ M = -81\,300,$ thus $G = 249$ and $N = -31\,500$

For section 3;

$F = 249,\ M = -31\,500,$ thus $G = 0$ and $N = 0$

Calculate the slopes and deflections

In section 1,

$C = J,\ D = K$

$1/(E \times I) = (6.15 \times 10^{-10},\ F = 0,\ M = 0$

(assuming $E = 200$ GN/m^2)

$W_1 = -1751,\ A_1 = 25,\ W_2 = 2000$ and $A_2 = 75$

At position (*i*)

$X = 25$

thus $\text{DE} = 6.15 \times 10^{-10} \times (0 + 0) + 25J + K + 0 + 0$

i.e. $\text{DE} = 25J + K$

and $SL = 6.15 \times 10^{-10} \times (0 + 0) + J + 0 + 0$

i.e. $\text{SL} = J$

At position (*ii*)

$X = 75$

thus $\text{DE} = 75J + K - 0.00224$ and $\text{SL} = J - 0.001\,35$

At position (*iii*)

$X = 100$

thus $\text{DE} = 100J + K - 0.0727$ and $\text{SL} = J - 0.002\,65$

In section 2,

$C = J - 0.002\,65,\ D = 100J + K - 0.0727,$

$1/(E \times I) = 1.21 \times 10^{-10}$

$F = 249,\ M = -81\,300$

At position (*iv*)

$X = 200$

thus $\text{DE} = 300J + K - 0.759$ and $\text{SL} = J - 0.004\,02$

In section 3,

$C = J - 0.004\,02,\ D = 300J + K - 0.759,$

$1/(E \times I) = 2.65 \times 10^{-10},\ F = 249,\ M = -31\,500,$

$W_1 = 766,\ A_1 = 25,\ W_2 = -1015$ and $A_2 = 50$

At position (*v*)

$X = 25$

thus $\text{DE} = 325J + K - 0.852$ and $\text{SL} = J - 0.004\,21$

At position (*vi*)

$X = 50$

thus $\text{DE} = 350J + K - 0.969$ and $\text{SL} = J - 0.004\,35$

Find J and K

The bearings are at positions (*i*) and (*vi*). The deflection datum is zero at these points, thus:

at position (*i*)

$\text{DE} = 25J + K = 0$

and at position (*vi*)

$\text{DE} = 350J + K - 0.969 = 0$

thus $J = 0.002\,98$ and $K = -0.0746$

Find the slopes and deflections

Substituting the value of J and K in the various equations for deflection and slope yield the following table

Position	*Deflection*	*Slope*
(*i*)	0	0.003
(*ii*)	0.13	0.0016
(*iii*)	0.15	0.000 33
(*iv*)	0.061	−0.001 0
(*v*)	0.033	−0.001 2
(*vi*)	0	−0.001 3

Final results

Similarly the slopes and deflections for the x direction can be found and from this:

The maximum slope at bearing (*i*) is 0.003 mm/mm

The maximum slope at bearing (*vi*) is 0.0014 mm/mm

The pulley (*ii*) is deflected 0.13 mm

And the gear (*v*) is deflected 0.004 mm.

E11 Shape tolerances of typical components

It is not possible to quote definite values for the tolerances of geometry maintainable by manufacturing process: values for any process can vary, not only from workshop to workshop, but within a workshop. The type of shop, the rate and quantity of production, the expected quality of work, the type of labour, the sequence of operations, the equipment available (and its condition) are among factors which must be taken into account. It is very difficult to apply values to these influencing factors, but as general guidance it might be expected to halve or double the maintainable tolerances by either better or worse practice. There are obviously no hard and fast rules.

The working values tabulated below are in good general agreement with modern practice. Achievable values can be better: they can be worse.

EXPLANATION OF TOLERANCES

Feature	*Symbol**	*The tolerance zone is limited by:*	*Diagrammatically:*
Flatness	▱	Two parallel planes distance t apart	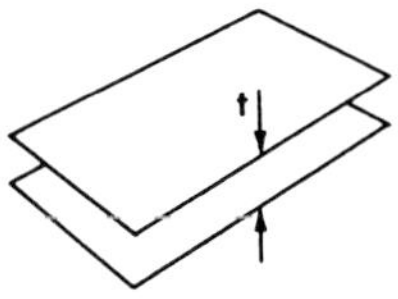
Straightness	—	A cylinder of diameter t	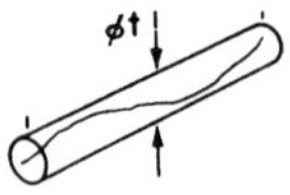
Parallelism	//	Two parallel straight lines/planes distance t apart and parallel to the datum line/plane	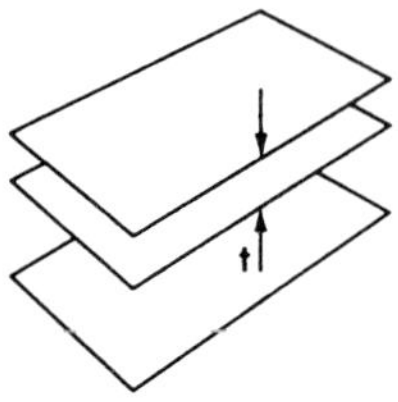
Perpendicularity	⊥	Two parallel lines/planes distance t apart and perpendicular to the datum line/plane	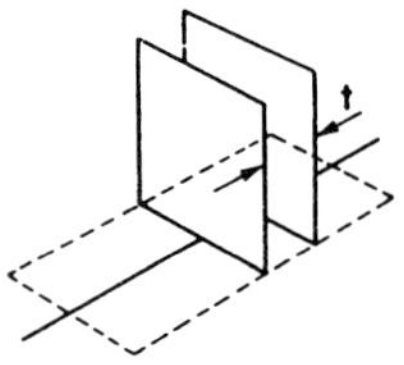
Concentricity	◎	A circle of diameter t the centre of which coincides with the datum point	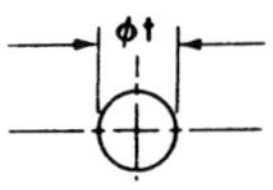
Circularity, roundness	○	Two concentric circles distance t apart	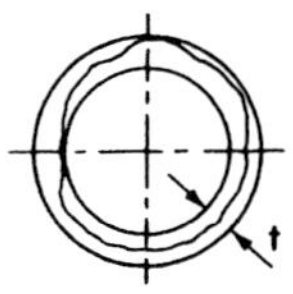

TYPICAL TOLERANCE VALUE/SIZE RELATIONSHIPS

	Geometric tolerance feature					
	Flatness of surface	Parallelism of cylinders or taper of cones on diameter (cylindricity, or conicity)	Straightness of cylinders or cones	Parallelism and/or squareness of flat surfaces	Parallelism or squareness of cylinders and flats	Roundness
Manufacturing process	*Orders of tolerance t in mm/mm, or inches/inch length in surface or cylinder*					
Turn, bore	0.00005	0.0001	0.0001	0.0001	0.0001	0.00004
	50×10^{-6}	100×10^{-6}	100×10^{-6}	100×10^{-6}	100×10^{-6}	40×10^{-6}
Fine turn, fine bore	0.00003	0.00004	0.00004	0.00005	0.00005	0.00003
	30×10^{-6}	40×10^{-6}	40×10^{-6}	50×10^{-6}	50×10^{-6}	30×10^{-6}
Cylindrical grind	0.00003	0.00005	0.00005	0.00005	0.00005	0.00002
	30×10^{-6}	50×10^{-6}	50×10^{-6}	50×10^{-6}	50×10^{-6}	20×10^{-6}
Fine cylindrical grind	0.00002	0.00002	0.00002	0.00003	0.00002	0.00001
	20×10^{-6}	20×10^{-6}	20×10^{-6}	30×10^{-6}	20×10^{-6}	10×10^{-6}

Achievable tolerance values

These can be obtained from the above table by multiplying the tolerance t in mm/mm or inches/inch by the size of the feature, bearing in mind that there will be a reasonable *minimum value*. These minimum values usually correspond to the values obtained by applying the above rules using a feature size of 25 mm or 1 inch except for *roundness*, where 50 mm or 2 inches gives more satisfactory values.

Examples

1 The minimum maintainable roundness tolerance for the diametral size of a turned bore would be

$$0.00004 \times 50 = 0.0020\,\text{mm}$$

2 The minimum straightness tolerance for a cylindrically-ground bore would be

$$0.00005 \times 25 = 0.0013\,\text{mm}$$

3 The tolerance on diametral parallelism appropriate to a cylindrically-ground parallel bore 200 mm long would be

$$\begin{aligned}&\text{proportional value of } t \times \text{length}\\&= 0.00005 \times 200\\&= 0.010\,\text{mm}\end{aligned}$$

E12 SI units and conversion factors

The International System of Units (SI—Système International d'Unites) is used as a common system throughout this handbook. The International System of Units is based on the following seven basic units.

Physical quantity	*SI unit*	*Symbol for the unit*
length	metre	m
mass	kilogram	kg
time	second	s
electric current	ampere	A
thermodynamic temperature	kelvin	K
luminous intensity	candela	cd
amount of substance	mole	mol

The remaining mechanical engineering units are derived from these, and the most important derived unit is the unit of force. This is called the newton, and is the force required to accelerate a mass of 1 kilogram at 1 metre/second^2. The acceleration due to gravity does not come into the basic unit system, and any engineering formulae in SI units no longer need *g* correction factors. The whole system of units is consistent, so that it is no longer necessary to have conversion factors between, for example, the various forms of energy such as mechanical, electrical, potential, kinetic or heat energy. These are all measured in joules in the SI system.

Other SI units frequently used in mechanical engineering have the names and symbols given in the following table.

Physical quantity	*SI unit*	*Symbol for the unit*
force	newton	N = kg m/s^2
work, energy or quantity of heat	joule	J = N m
power	watt	W = J/s
velocity	metre/second	m/s
angular velocity	radian/second	rad/s
acceleration	metre/second^2	m/s^2
density	kilogramme/metre^3	kg/m^3
absolute or dynamic viscosity	†newton second/metre^2	N s/m^2
kinematic viscosity	†metre^2/second	m^2/s
volumetric flow rate	metre^3/second	m^3/s
pressure	newton/metre^2	n/m^2
torque	newton metre	N m

† The centipoise and centistokes are also acceptable as units with the SI system.

In many cases the basic SI unit for a physical quantity will be found to be an unsatisfactory size and multiples of the units are therefore used as follows:

10^{12}	tera	T	10^{-3}	milli	m
10^{9}	giga	G	10^{-6}	micro	μ
10^{6}	mega	M	10^{-9}	nano	n
10^{3}	kilo	k	10^{-12}	pico	p

A typical example is the watt, which for mechanical engineering is too small as a unit of power. For most purposes the kilowatt (kW i.e. 10^3 W) is used, while for really large powers the megawatt (MW i.e. 10^6 W) is more convenient.

It should be noted that the prefix symbol denoting a multiple of the basic SI unit is placed immediately to the left of the basic unit symbol without any intervening space or mark. The multiple unit is treated as a single entity, e.g. mm^2 means (mm)^2 i.e. $(10^{-3} \times \text{m})^2$ or $10^{-6} \times \text{m}^2$ and *not* m(m)^2 i.e. *not* $10^3 \times \text{m}^2$.

The following table of conversion factors is arranged in a form that provides a simple means for converting a quantity of SI units into a quantity of the previous British units.

This table also makes it possible to get a feel for the size of the SI units, e.g. that one newton is just less than a quarter of a pound (about the weight of an apple).

Physical quantity	*Symbol for SI unit*	*Conversion factor**	*Symbol for familiar British units*
acceleration	m/s^2	3.28	ft/s^2
angular acceleration	rad/s^2	57.3	deg/s^2
angular velocity	rad/s	57.3	deg/s
area	m^2	10.8	ft^2
coefficient of heat transfer	W/m^2 K	0.176	Btu/h ft^2°F
coefficient of linear expansion	1/K	0.556	1/°F
density	kg/m^3	6.24×10^{-2}	lb/ft^3
dynamic viscosity	N s/m^2	10^3	cP
energy, work	J	0.737	ft lbf
		2.78×10^{-7}	kW h
force	N	0.225	lbf
heat capacity	J/K	5.27×10^{-4}	Btu/°F or CHU/°C
heat flow rate	W	3.41	Btu/h
heat flux	W/m^2	0.317	Btu/h ft^2
heat quantity	J	9.48×10^{-4}	Btu
		5.27×10^{-4}	CHU
kinematic viscosity	m^2/s	10^6	cSt
		10.8	ft^2/s
length	m	3.28	ft
mass	kg	2.20	lb
moment, torque	N m	0.738	lbf ft
moment of inertia	kg m^2	23.7	lb ft^2
power	W	1.34×10^{-3}	hp
pressure	N/m^2	1.45×10^{-4}	lbf/in^2
second moment of area	m^4	2.40×10^6	in^4
specific heat capacity	J/kg K	2.39×10^{-4}	Btu/lb°F
specific heat/unit volume	J/m^3 K	1.49×10^{-5}	Btu/ft^3°F
stress	N/m^2	1.45×10^{-4}	lbf/in^2
surface tension	N/m	6.85×10^{-2}	lbf/ft
thermal conductivity	W/m K	0.578	Btu/h ft°F
velocity	m/s	3.28	ft/s
volume	m^3	35.3	ft^3
volumetric flow rate	m^3/s	1.32×10^4	Imp. gall/min

* Multiply the number of SI units by this conversion factor to obtain the number of familiar British units.

e.g. 100 metres = $100 \times 3.28 = 328$ ft

The conversion factors in this table are correct to only three significant figures.